Introduction to Biology Practical Skills

Second Edition

PEARSON

We work with leading authors to develop the strongest educational materials bringing cutting-edge thinking and best learning practice to a global market.

Under a range of well-known imprints, including Financial Times/Prentice Hall, Addison Wesley and Longman, we craft high quality print and electronic publications which help readers to understand and apply their content, whether studying or at work.

Pearson Custom Publishing enables our customers to access a wide and expanding range of market-leading content from world-renowned authors and develop their own tailor-made book. You choose the content that meets your needs and Pearson Custom Publishing produces a high-quality printed book.

To find out more about custom publishing, visit www.pearsoncustom.co.uk

PEARSON CUSTOM PUBLISHING

Introduction to Biology Practical Skills

Second Edition

Compiled from:

Practical Skills in Chemistry
by John R. Dean, Alan M. Jones, David Holmes,
Rob Reed, Jonathan Weyers, and Allan Jones

Practical Skills in Biology
Fourth Edition
by Allan Jones, Rob Reed and Jonathan Weyers

Practical Skills in Biomolecular Sciences
Third Edition
by Rob Reed, David Holmes, Jonathan Weyers
and Allan Jones

ALWAYS LEARNING — PEARSON

Harlow, England • London • New York • Boston • San Francisco • Toronto • Sydney • Auckland • Singapore • Hong Kong
Tokyo • Seoul • Taipei • New Delhi • Cape Town • Sao Paulo • Mexico City • Madrid • Amsterdam • Munich • Paris • Milan

Pearson Education Limited
Edinburgh Gate
Harlow
Essex CM20 2JE

And associated companies throughout the world

Visit us on the World Wide Web at:
www.pearsoned.co.uk

This Custom Book Edition © Pearson Education Limited 2012

Compiled from:

Practical Skills in Chemistry
by John R. Dean, Alan M. Jones, David Holmes, Rob Reed,
Jonathan Weyers and Allan Jones
ISBN 978 0 13 028002 2
Copyright © Pearson Education Limited 2002

Practical Skills in Biology
Fourth Edition
by Allan Jones, Rob Reed and Jonathan Weyers
ISBN 978 0 13 175509 3
Copyright © Pearson Education Limited 1994, 1998, 2003, 2007

Practical Skills in Biomolecular Sciences
Third Edition
by Rob Reed, David Holmes, Jonathan Weyers and Allan Jones
ISBN 978 0 13 239115 3
Copyright © Addison Wesley Longman Limited 1998
Copyright © Pearson Education Limited 2003, 2007

All rights reserved. No part of this publication may be reproduced, stored in a retrieval
system, or transmitted in any form or by any means, electronic, mechanical, photocopying,
recording or o+therwise, without either the prior written permission of the publisher
or a licence permitting restricted copying in the United Kingdom issued by the Licensing
Agency Ltd, Saffron House, 6–10 Kirby Street, London EC1N 8TS.

ISBN 978 1078016 024 5

Printed and bound in Great Britain by Antony Rowe.

Contents

Study and Examination Skills

The Importance of Transferable Skills — 3
Chapter 1 in *Practical Skills in Biomolecular Sciences* Third Edition
Rob Reed, David Holmes, Jonathan Weyers and Allan Jones

Managing Your Time — 8
Chapter 2 in *Practical Skills in Biomolecular Sciences* Third Edition
Rob Reed, David Holmes, Jonathan Weyers and Allan Jones

Working with Others — 12
Chapter 3 in *Practical Skills in Biomolecular Sciences* Third Edition
Rob Reed, David Holmes, Jonathan Weyers and Allan Jones

Taking Notes from Lectures and Texts — 16
Chapter 4 in *Practical Skills in Biomolecular Sciences* Third Edition
Rob Reed, David Holmes, Jonathan Weyers and Allan Jones

Learning and Revising — 20
Chapter 5 in *Practical Skills in Biomolecular Sciences* Third Edition
Rob Reed, David Holmes, Jonathan Weyers and Allan Jones

Curriculum Options, Assessments and Exams — 29
Chapter 6 in *Practical Skills in Biomolecular Sciences* Third Edition
Rob Reed, David Holmes, Jonathan Weyers and Allan Jones

Preparing your Curriculum Vitae — 38
Chapter 7 in *Practical Skills in Biomolecular Sciences* Third Edition
Rob Reed, David Holmes, Jonathan Weyers and Allan Jones

Communicating Information

Organising a Poster Display — 45
Chapter 14 in *Practical Skills in Biomolecular Sciences* Third Edition
Rob Reed, David Holmes, Jonathan Weyers and Allan Jones

Giving a Spoken Presentation — 50
Chapter 15 in *Practical Skills in Biomolecular Sciences* Third Edition
Rob Reed, David Holmes, Jonathan Weyers and Allan Jones

General Aspects of Scientific Writing — 56
Chapter 16 in *Practical Skills in Biomolecular Sciences* Third Edition
Rob Reed, David Holmes, Jonathan Weyers and Allan Jones

Writing Essays — 63
Chapter 17 in *Practical Skills in Biomolecular Sciences* Third Edition
Rob Reed, David Holmes, Jonathan Weyers and Allan Jones

Reporting Practical and Project Work — 66
Chapter 18 in *Practical Skills in Biomolecular Sciences* Third Edition
Rob Reed, David Holmes, Jonathan Weyers and Allan Jones

Writing Literature Surveys and Reviews — 71
Chapter 19 in *Practical Skills in Biomolecular Sciences* Third Edition
Rob Reed, David Holmes, Jonathan Weyers and Allan Jones

Fundamental Laboratory Techniques

Your Approach to Practical Work — 77
Chapter 20 in *Practical Skills in Biomolecular Sciences* Third Edition
Rob Reed, David Holmes, Jonathan Weyers and Allan Jones

Health and Safety — 81
Chapter 21 in *Practical Skills in Biomolecular Sciences* Third Edition
Rob Reed, David Holmes, Jonathan Weyers and Allan Jones

Working With Liquids — 84
Chapter 22 in *Practical Skills in Biomolecular Sciences* Third Edition
Rob Reed, David Holmes, Jonathan Weyers and Allan Jones

Basic Laboratory Procedures — 90
Chapter 23 in *Practical Skills in Biomolecular Sciences* Third Edition
Rob Reed, David Holmes, Jonathan Weyers and Allan Jones

Principles of Solution Chemistry — 100
Chapter 24 in *Practical Skills in Biomolecular Sciences* Third Edition
Rob Reed, David Holmes, Jonathan Weyers and Allan Jones

pH and Buffer Solutions — 108
Chapter 25 in *Practical Skills in Biomolecular Sciences* Third Edition
Rob Reed, David Holmes, Jonathan Weyers and Allan Jones

The Investigative Approach

The Principles of Measurement — 117
Chapter 25 in *Practical Skills in Biology* Fourth Edition
Allan Jones, Rob Reed and Jonathan Weyers

Making Observations — 121
Chapter 27 in *Practical Skills in Biology* Fourth Edition
Allan Jones, Rob Reed and Jonathan Weyers

Making Notes of Practical Work — 125
Chapter 32 in *Practical Skills in Biology* Fourth Edition
Allan Jones, Rob Reed and Jonathan Weyers

SI Units and Their Use — 130
Chapter 29 in *Practical Skills in Biomolecular Sciences* Third Edition
Rob Reed, David Holmes, Jonathan Weyers and Allan Jones

Scientific Method and Design of Experiments — 135
Chapter 30 in *Practical Skills in Biomolecular Sciences* Third Edition
Rob Reed, David Holmes, Jonathan Weyers and Allan Jones

Project Work — 143
Chapter 31 in *Practical Skills in Biomolecular Sciences* Third Edition
Rob Reed, David Holmes, Jonathan Weyers and Allan Jones

Drawing and Diagrams — 147
Chapter 28 in *Practical Skills in Biology* Fourth Edition
Allan Jones, Rob Reed and Jonathan Weyers

Basic Fieldwork Procedures — 154
Chapter 29 in *Practical Skills in Biology* Fourth Edition
Allan Jones, Rob Reed and Jonathan Weyers

Samples and Sampling — 158
Chapter 30 in *Practical Skills in Biology* Fourth Edition
Allan Jones, Rob Reed and Jonathan Weyers

Obtaining and Identifying Specimens

Collecting Animals and Plants — 165
Chapter 34 in *Practical Skills in Biology* Fourth Edition
Allan Jones, Rob Reed and Jonathan Weyers

Fixing and Preserving Animals and Plants — 168
Chapter 35 in *Practical Skills in Biology* Fourth Edition
Allan Jones, Rob Reed and Jonathan Weyers

Naming and Classifying Organisms — 173
Chapter 37 in *Practical Skills in Biology* Fourth Edition
Allan Jones, Rob Reed and Jonathan Weyers

Identifying Plants and Animals 178
Chapter 38 in *Practical Skills in Biology* Fourth Edition
Allan Jones, Rob Reed and Jonathan Weyers

Identifying Microbes 185
Chapter 39 in *Practical Skills in Biology* Fourth Edition
Allan Jones, Rob Reed and Jonathan Weyers

The Purpose and Practice of Dissection 191
Chapter 40 in *Practical Skills in Biology* Fourth Edition
Allan Jones, Rob Reed and Jonathan Weyers

Preparing Specimens for Light Microscopy 196
Chapter 42 in *Practical Skills in Biology* Fourth Edition
Allan Jones, Rob Reed and Jonathan Weyers

Handling Cells and Tissues

Sterile Technique and Microbial Culture 205
Chapter 32 in *Practical Skills in Biomolecular Sciences* Third Edition
Rob Reed, David Holmes, Jonathan Weyers and Allan Jones

Isolating, Identifying and Naming Microbes 212
Chapter 33 in *Practical Skills in Biomolecular Sciences* Third Edition
Rob Reed, David Holmes, Jonathan Weyers and Allan Jones

Working with Animal and Plant Tissues and Cells 225
Chapter 34 in *Practical Skills in Biomolecular Sciences* Third Edition
Rob Reed, David Holmes, Jonathan Weyers and Allan Jones

Culture Systems and Growth Measurement 234
Chapter 35 in *Practical Skills in Biomolecular Sciences* Third Edition
Rob Reed, David Holmes, Jonathan Weyers and Allan Jones

Homogenisation and Fractionation of
Cells and Tissues 245
Chapter 36 in *Practical Skills in Biomolecular Sciences* Third Edition
Rob Reed, David Holmes, Jonathan Weyers and Allan Jones

Analytical Techniques

Calibration and its Application to 253
Quantitative Analysis
Chapter 37 in *Practical Skills in Biomolecular Sciences* Third Edition
Rob Reed, David Holmes, Jonathan Weyers and Allan Jones

Immunological Methods 259
Chapter 38 in *Practical Skills in Biomolecular Sciences* Third Edition
Rob Reed, David Holmes, Jonathan Weyers and Allan Jones

Radioactive Isotopes and their uses 269
Chapter 39 in *Practical Skills in Biomolecular Sciences* Third Edition
Rob Reed, David Holmes, Jonathan Weyers and Allan Jones

Light Measurement
Chapter 40 in *Practical Skills in Biomolecular Sciences* Third Edition
Rob Reed, David Holmes, Jonathan Weyers and Allan Jones
278

Basic Spectroscopy
Chapter 41 in *Practical Skills in Biomolecular Sciences* Third Edition
Rob Reed, David Holmes, Jonathan Weyers and Allan Jones
282

Advanced Spectroscopy and Spectrometry
Chapter 42 in *Practical Skills in Biomolecular Sciences* Third Edition
Rob Reed, David Holmes, Jonathan Weyers and Allan Jones
291

Centrifugation
Chapter 43 in *Practical Skills in Biomolecular Sciences* Third Edition
Rob Reed, David Holmes, Jonathan Weyers and Allan Jones
299

Chromatography – Separation Methods
Chapter 44 in *Practical Skills in Biomolecular Sciences* Third Edition
Rob Reed, David Holmes, Jonathan Weyers and Allan Jones
305

Chromatography – Detection and Analysis
Chapter 45 in *Practical Skills in Biomolecular Sciences* Third Edition
Rob Reed, David Holmes, Jonathan Weyers and Allan Jones
316

Principles and Practice of Electrophoresis
Chapter 46 in *Practical Skills in Biomolecular Sciences* Third Edition
Rob Reed, David Holmes, Jonathan Weyers and Allan Jones
322

Advanced Electrophoretic Techniques
Chapter 47 in *Practical Skills in Biomolecular Sciences* Third Edition
Rob Reed, David Holmes, Jonathan Weyers and Allan Jones
332

Electroanalytical Techniques
Chapter 48 in *Practical Skills in Biomolecular Sciences* Third Edition
Rob Reed, David Holmes, Jonathan Weyers and Allan Jones
338

Assaying Biomolecules and Studying Metabolism

Analysis of Biomolecules: Fundamental Principles
Chapter 49 in *Practical Skills in Biomolecular Sciences* Third Edition
Rob Reed, David Holmes, Jonathan Weyers and Allan Jones
351

Assaying Amino Acids, Peptides and Proteins
Chapter 50 in *Practical Skills in Biomolecular Sciences* Third Edition
Rob Reed, David Holmes, Jonathan Weyers and Allan Jones
354

Assaying Lipids
Chapter 51 in *Practical Skills in Biomolecular Sciences* Third Edition
Rob Reed, David Holmes, Jonathan Weyers and Allan Jones
359

Assaying Carbohydrates
Chapter 52 in *Practical Skills in Biomolecular Sciences* Third Edition
Rob Reed, David Holmes, Jonathan Weyers and Allan Jones
365

Assaying Nucleic Acids and Nucleotides — 370
Chapter 53 in *Practical Skills in Biomolecular Sciences* Third Edition
Rob Reed, David Holmes, Jonathan Weyers and Allan Jones

Protein Purification — 375
Chapter 54 in *Practical Skills in Biomolecular Sciences* Third Edition
Rob Reed, David Holmes, Jonathan Weyers and Allan Jones

Enzyme Studies — 383
Chapter 55 in *Practical Skills in Biomolecular Sciences* Third Edition
Rob Reed, David Holmes, Jonathan Weyers and Allan Jones

Membrane Transport Processes — 394
Chapter 56 in *Practical Skills in Biomolecular Sciences* Third Edition
Rob Reed, David Holmes, Jonathan Weyers and Allan Jones

Photosynthesis and Respiration — 401
Chapter 57 in *Practical Skills in Biomolecular Sciences* Third Edition
Rob Reed, David Holmes, Jonathan Weyers and Allan Jones

Genetics

Mendelian Genetics — 415
Chapter 58 in *Practical Skills in Biomolecular Sciences* Third Edition
Rob Reed, David Holmes, Jonathan Weyers and Allan Jones

Bacterial and Phage Genetics — 422
Chapter 59 in *Practical Skills in Biomolecular Sciences* Third Edition
Rob Reed, David Holmes, Jonathan Weyers and Allan Jones

Molecular Genetics I – Fundamental Principles — 431
Chapter 60 in *Practical Skills in Biomolecular Sciences* Third Edition
Rob Reed, David Holmes, Jonathan Weyers and Allan Jones

Molecular Genetics II – PCR and Related Applications — 441
Chapter 61 in *Practical Skills in Biomolecular Sciences* Third Edition
Rob Reed, David Holmes, Jonathan Weyers and Allan Jones

Molecular Genetics III – Genetic Engineering Techniques — 448
Chapter 62 in *Practical Skills in Biomolecular Sciences* Third Edition
Rob Reed, David Holmes, Jonathan Weyers and Allan Jones

Analysis and Presentation of Data

Manipulating and Transforming Raw Data — 457
Chapter 61 in *Practical Skills in Biology* Fourth Edition
Allan Jones, Rob Reed and Jonathan Weyers

Using Graphs — 461
Chapter 62 in *Practical Skills in Biology* Fourth Edition
Allan Jones, Rob Reed and Jonathan Weyers

Presenting Data in Tables 473
Chapter 63 in *Practical Skills in Biology* Fourth Edition
Allan Jones, Rob Reed and Jonathan Weyers

Hints for Solving Numerical Problems 478
Chapter 64 in *Practical Skills in Biology* Fourth Edition
Allan Jones, Rob Reed and Jonathan Weyers

Descriptive Statistics 489
Chapter 65 in *Practical Skills in Biology* Fourth Edition
Allan Jones, Rob Reed and Jonathan Weyers

Choosing and Using Statistical Tests 500
Chapter 66 in *Practical Skills in Biology* Fourth Edition
Allan Jones, Rob Reed and Jonathan Weyers

Chemistry

Procedures in Volumetric Analysis 515
Chapter 21 in *Practical Skills in Chemistry*
by John R. Dean, Alan M. Jones, David Holmes, Rob Reed,
Jonathan Weyers and Allan Jones

Acid-Base Titrations 522
Chapter 22 in *Practical Skills in Chemistry*
by John R. Dean, Alan M. Jones, David Holmes, Rob Reed,
Jonathan Weyers and Allan Jones

Infrared Spectroscopy 525
Chapter 28 in *Practical Skills in Chemistry*
by John R. Dean, Alan M. Jones, David Holmes, Rob Reed,
Jonathan Weyers and Allan Jones

Nuclear Magnetic Resonance Spectroscopy 535
Chapter 29 in *Practical Skills in Chemistry*
by John R. Dean, Alan M. Jones, David Holmes, Rob Reed,
Jonathan Weyers and Allan Jones

Mass Spectrometry 545
Chapter 30 in *Practical Skills in Chemistry*
by John R. Dean, Alan M. Jones, David Holmes, Rob Reed,
Jonathan Weyers and Allan Jones

Index 550

Study and Examination Skills

1 The importance of transferable skills

Skills terminology – *different phrases may be used to describe transferable skills, depending on place or context. These include: 'personal transferable skills' (PTS), 'key skills', 'core skills' and 'competences'.*

This chapter outlines the range of transferable skills and their significance to biomolecular scientists. It also indicates where practical skills fit into this scheme. Having a good understanding of this topic will help you place your work at university in a wider context. You will also gain an insight into the qualities that employers expect you to have developed by the time you graduate. Awareness of these matters will be useful when carrying out personal development planning (PDP) as part of your studies.

The range of transferable skills

Table 1.1 provides a comprehensive listing of university-level transferable skills under six skill categories. There are many possible classifications – and a different one may be used in your institution or field of study. Note particularly that 'study skills', while important, and rightly emphasised at the start of many courses, constitute only a subset of the skills acquired by most university students.

The phrase '*Practical Skills*' in the title of this book indicates that there is a special subset of transferable skills related to work in the laboratory. However, although this text deals primarily with skills and techniques required for laboratory practicals and associated studies, a broader range of material is included. This is because the skills concerned are important, not only in the biosciences but also in the wider world. Examples include time management, evaluating information and communicating effectively.

Using course materials – *study your course handbook and the schedules for each practical session to find out what skills you are expected to develop at each point in the curriculum. Usually the learning objectives/outcomes (p. 24) will describe the skills involved.*

 KEY POINT Biomolecular sciences are essentially practical subjects, and therefore involve highly developed laboratory skills. The importance that your lecturers place on practical skills will probably be evident from the large proportion of curriculum time you will spend on practical work in your course.

The word 'skill' implies much more than the robotic learning of, for example, a laboratory routine. Of course, some of the tasks you will be asked to carry out in practical classes *will* be repetitive. Certain techniques require manual dexterity and attention to detail if accuracy and precision are to be attained, and the necessary competence often requires practice to make perfect. However, a deeper understanding of the context of a technique is important if the skill is to be appreciated fully and then transferred to a new situation. That is why this text is not simply a 'recipe book' of methods and protocols and why it includes background information, tips and worked examples, as well as study exercises to test your understanding.

Example *The skills involved in teamwork cannot be developed without a deeper understanding of the interrelationships involved in successful groups. The context will be different for every group and a flexible approach will always be required, according to the individuals involved and the nature of the task.*

Transferability of skills

Transferable skills are those which allow someone with knowledge, understanding or ability gained in one situation to adapt or extend this for application in a different context. In some cases, the transfer of a skill is immediately obvious. Take, for example, the ability to use a spreadsheet to summarise biological data and create a graph to

Practical Skills

Table 1.1 Transferable skills identified as important in the biosciences. The list has been compiled from several sources, including the UK Quality Assurance Agency for Higher Education Subject Benchmark Statement for the Biosciences and for Biomedical Sciences. Particularly relevant chapters are shown for the skills covered by this book (numbers in **bold coloured** text indicate a deeper, or more extensive, treatment)

Skill category	Examples of skills and competences	Relevant chapters in this textbook
Generic skills for bioscientists	Having an appreciation of the complexity and diversity of life and life processes	11 28 29 33 36 56–59
	Reading and evaluating biological literature with a full and critical understanding	4 8 9
	Ability to communicate a clear and accurate account of a biological topic, both verbally and in writing	14 15 16 17–19
	Applying critical and analytical skills to evaluate evidence regarding theories and hypotheses	9 30
	Using a variety of methods for studying the biosciences	32–62
	Having the ability to think independently, set personal tasks and solve problems	30 31 65
Intellectual skills	Recognising and applying biological theories, concepts and principles	9 30
	Analysing, synthesising and summarising information critically	9 19 63–67
	Obtaining evidence to formulate and test hypotheses; applying knowledge to address familiar and unfamiliar problems	28 29 30 31 67
	Recognising and explaining moral, ethical and legal issues in biological research	20 21 32 33 34
Experimental and observational skills	Carrying out basic laboratory techniques and understanding the principles that underlie them	20 21–30 32 41 43 49 60
	Working in the laboratory safely, responsibly and legally, with due attention to ethical aspects	20 21 31–34
	Designing, planning, conducting and reporting on biological investigations and data arising from them	14 15 18 30 31
	Obtaining, recording, collating and analysing biological data	28–34 35 36 37 63–67
	Carrying out basic techniques relevant to core subjects in biomedical science (biochemistry, molecular genetics, immunology, microbiology)	20–26 27 28–31 32 33 34 35 36 37 38 39 41 42 43 44–48 49 50–59 60 61 62–67
Numeracy, communication and IT skills	Understanding and using data in several forms (e.g. numerical, textual, verbal and graphical)	4 9 63–66
	Communicating in written, verbal, graphical and visual forms	14 15 16 17–19 63 64 65
	Citing and referencing the work of others in an appropriate manner	8 9 19
	Obtaining data, including the concepts behind sampling and sampling errors, calibration and types of error	27 28 30 31 37 65–67
	Processing, interpreting and presenting data, and applying appropriate statistical methods for summarising and analysing data	11 63–65 66 67
	Solving problems with calculators and computers, including the use of tools such as spreadsheets	10 11 12 20 65
	Using computer technology to communicate and as a source of biological information	10 11 12 13
Interpersonal and teamwork skills	Working individually or in teams as appropriate; identifying individual and group goals and acting responsibly and appropriately to achieve them	3
	Recognising and respecting the views and opinions of others	3
	Evaluating your own performance and that of others	3 7
	Appreciating the interdisciplinary nature of contemporary biosciences	1 19
Self-management and professional development skills	Working independently, managing time and organising activities	2 30 31
	Identifying and working towards targets for personal, academic and career development	1 7
	Developing an adaptable and effective approach to study and work (including revision and exam technique)	2 4 5 6

illustrate results. Once the key concepts and commands are learned (Chapter 12), they can be applied to many instances outside the biosciences where this type of output is used. This is not only true for similar data sets, but also in unrelated situations, such as making up a financial balance sheet and creating a pie chart to show sources of expenditure. Similarly, knowing the requirements for good graph drawing and tabulation (Chapters 63 and 64), perhaps practised by hand in earlier work, might help you use spreadsheet commands to make the output suit your needs.

Other cases may be less clear but equally valid. For example, towards the end of your undergraduate studies you may be involved in designing experiments as part of your project work. This task will draw on several skills gained at earlier stages in your course, such as preparing solutions (Chapters 22–25), deciding about numbers of replicates and experimental layout (Chapters 30 and 31) and perhaps carrying out some particular method of observation, measurement or analysis (Chapters 37–62). How and when might you transfer this complex set of skills? In the workplace, it is unlikely that you would be asked to repeat the same process, but in critically evaluating a problem or in planning a complex project for a new employer, you will need to use many of the time management, organisational and analytical skills developed when designing and carrying out experiments. The same applies to information retrieval and evaluation and writing essays and dissertations, when transferred to the task of analysing or writing a business report.

Personal development planning

Many universities have schemes for personal development planning (PDP), which may go under slightly different names such as progress file or professional development plan. You will usually be expected to create a portfolio of evidence on your progress, then reflect on this, and subsequently set yourself plans for the future, including targets and action points. Analysis of your transferable skills profile will probably form part of your PDP. Other aspects commonly included are:

- your aspirations, goals, interests and motivations;
- your learning style or preference (see p. 21);
- your assessment transcript or academic profile information (e.g. record of grades in your modules);
- your developing CV (see p. 38).

Taking part in PDP can help focus your thoughts about your university studies and future career. This is important, as many biosciences degrees do not lead only to a single, specific occupation. The PDP process will introduce you to some new terms and will help you to describe your personality and abilities. This will be useful when constructing your CV and when applying for jobs.

What your future employer will be looking for

At the end of your course, which may seem some time away, you will aim to get a job and start on your chosen career path. You will need to sell yourself to your future employer, firstly in your application form

---

Opportunities to develop and practise skills in your private or social life – you could, for example, practise spreadsheet skills by organising personal or club finances using Microsoft Excel, or teamwork skills within any university clubs or societies you may join (see Chapter 7).

Types of PDP portfolio and their benefits – some PDP schemes are centred on academic and learning skills, while others are more focused on career planning. Some are carried out independently and others in tandem with a personal tutor or advisory system. Some PDP schemes involve creating an online portfolio, while others are primarily paper-based. Each method has specific goals and advantages, but whichever way your scheme operates, maximum benefit will be gained from being fully involved with the process.

Definition

Employability – the 'combination of in-depth subject knowledge, work awareness, subject-specific, generic and career management skills, and personal attributes and attitudes that enable a student to secure suitable employment and perform excellently throughout a career spanning a range of employers and occupations.' *(Higher Education Academy Centre for Bioscience* definition of employability for bioscientists)

and curriculum vitae (Chapter 7), and perhaps later at interview. Companies rarely employ bioscience graduates simply because they know how to carry out a particular lab routine or because they can remember specific facts about their chosen degree subject. Instead, employers tend to look for a range of qualities and transferable skills that together define an attribute known as 'graduateness'. This encompasses, for example, the ability to work in a team, to speak effectively and write clearly about your work, to understand complex data and to manage a project to completion. All of these skills can be developed at different stages during your University studies.

 KEY POINT Factual knowledge is important in degrees with a strong vocational element, but understanding how to find and evaluate information is usually rated more highly by employers than the ability to memorise facts.

Most likely, your future employer(s) will seek someone with a organised yet flexible mind, capable of demonstrating a logical approach to problems – someone who has a range of skills and who can transfer these skills to new situations. Many competing applicants will probably have similar qualifications. If you want the job, you will have to show that your additional skills place you above the other candidates.

Sources for further study

Drew, S. and Bingham, R. (2004) *The Student Skills Guide*, 2nd edn. Gower Publishing Ltd, Aldershot.

McMillan, K. and Weyers, J.D.B. (2006) *The Smarter Student: Skills and Strategies for Success at University*. Pearson Education, Harlow.

Race, P. (1999) *How to Get a Good Degree: Making the Most of Your Time at University*. Open University Press, Buckingham.

Study exercises

1.1 Evaluate your skills. Examine the list of skill topics shown in Table 1.1 (p. 4). Now create a new table with two columns, like the one on next page. The first half of this table should indicate *five* skills you feel confident about and show where you demonstrated this skill (for example, 'working in a team' and 'in a first year group project in molecular biology'). The second half of the table should show *five* skills you do not feel confident about, or you recognise need development (e.g. 'communicating in verbal form'). List these and then list ways in which you think the course material for your current modules will provide opportunities to develop these skills, or what activities you might take to improve them (e.g. 'forming a study group with colleagues').

1.2 Find skills resources. For at least one of the skills in the second half of Table 1.1, check your university's library database to see if there are any texts on that subject. Alternatively, carry out a search for relevant websites (there are many); decide which are useful and 'bookmark' them for future use (Chapter 10).

1.3 Analyse your goals and aspirations. Spend a little time thinking what you hope to gain from university. See if your friends have the same aspirations. Think about and/or discuss how these goals can be achieved, while keeping the necessary balance between university work, paid employment and your social life.

Study exercises (continued)

Skills I feel confident about	Where demonstrated
1.	
2.	
3.	
4.	
5.	
Skills that I could develop	**Opportunities for development**
6.	
7.	
8.	
9.	
10.	

2 Managing your time

One of the most important activities that you can do is to organise your personal and working time effectively. There is a lot to do at university and a common complaint is that there isn't enough time to accomplish everything. In fact, research shows that most people use up a lot of their time without realising it through ineffective study or activities such as extended coffee breaks. Developing your time management skills will help you achieve more in work, rest and play, but it is important to remember that putting time management techniques into practice is an individual matter, requiring a level of self-discipline not unlike that required for dieting. A new system won't always work perfectly straight away, but through time you can develop a system that is effective for you. An inability to organise your time effectively, of course, results in feelings of failure, frustration, guilt and being out of control in your life.

Setting your goals

The first step is to identify clearly what you want to achieve, both in work and in your personal life. We all have a general idea of what we are aiming for, but to be effective, your goals must be clearly identified and priorities allocated. Clear, concise objectives can provide you with a framework in which to make these choices. Try using the 'SMART' approach, in which objectives should be:

- Specific – clear and unambiguous, including what, when, where, how and why.
- Measurable – having quantified targets and benefits to provide an understanding of progress.
- Achievable – being attainable within your resources.
- Realistic – being within your abilities and expectations.
- Timed – stating the time period for completion.

Having identified your goals, you can now move on to answer four very important questions:

1. Where does your time go?
2. Where should your time go?
3. What are your time-wasting activities?
4. What strategies can help you?

Analysing your current activities

The key to successful development of time management is a realistic knowledge of how you currently spend your time. Start by keeping a detailed time log for a typical week (Fig. 2.1), but you will need to be truthful in this process. Once you have completed the log, consider the following questions:

- How many hours do I work in total and how many hours do I use for relaxation?
- What range of activities do I do?
- How long do I spend on each activity?
- What do I spend most of my time doing?

Definition

Time management – a system for controlling and using time as efficiently and as effectively as possible.

Advantages of time management – these include:

- a feeling of much greater control over your activities;
- avoidance of stress;
- improved productivity – achieve more in a shorter period;
- improved performance – work to higher standards because you are in charge;
- increase in time available for non-work matters – work hard, but play hard too.

Example The objective "to spend an extra hour each week on directed study in microbiology next term" fulfils the SMART criteria, in contrast to a general intention "to study more".

Fig. 2.1 Example of how to lay out a time log. Write activities along the top of the page, and divide the day into 15-minute segments as shown. Think beforehand how you will categorise the different things you do, from the mundane (laundry, having a shower, drinking coffee, etc.) to the well-timetabled (tutorial meeting, sports club meeting) and add supplementary notes if required. At the end of each day, place a dot in the relevant column for each activity and sum the dots to give a total at the bottom of the page. You will need to keep a diary like this for at least a week before you see patterns emerging.

Time slots	Activity	Notes
7.00–7.15		
7.15–7.30		
7.30–7.45		
7.45–8.00		
8.00–8.15		
8.15–8.30		
8.30–8.45		
8.45–9.00		
9.00–9.15		

- What do I spend the least amount of my time doing?
- Are my allocations of time in proportion to the importance of my activities?
- How much of my time is ineffectively used, e.g. for uncontrolled socialising or interruptions?

If you wish, you could use a spreadsheet (Chapter 12) to produce graphical summaries of time allocations in different categories as an aid to analysis and management. Divide your time into:

- Committed time – timetabled activities involving your main objectives/ goals.
- Maintenance time – time spent supporting your general life activities (shopping, cleaning, laundry, etc.).
- Discretionary time – time for you to use as you wish, e.g. recreation, sport, hobbies, socialising.

Avoiding time-wasting activities

Look carefully at those tasks that could be identified as time-wasting activities. They include gossiping, over-long breaks, uninvited interruptions and even ineffective study periods. Try to reduce these to a minimum, but do not count on eliminating them entirely. Remember also that some relaxation *should* be programmed into your daily schedule.

Organising your tasks

Having analysed your time usage, you can now use this information, together with your objectives and prioritised goals, to organise your activities, both on a short-term and a long-term basis. Consider using a diary-based system (such as those produced by Filofax, TMI and Day-Timer) that will help you plan ahead and analyse your progress.

Divide your tasks into several categories, such as:

- Urgent – must be done as a top priority and at short notice (e.g. doctor's appointment).

Quality in time management – avoid spending a lot of time doing unproductive studying, e.g. reading a textbook without specific objectives for that reading. Make sure you test your recall of the material, if you are working towards an examination (p. 20).

Being assertive – if friends and colleagues continually interrupt you, find a way of controlling them, before they control you. Indicate clearly on your door that you do not wish to be disturbed and explain why. Otherwise, try to work away from disturbance.

Practical Skills

WEEKLY DIARY   Week beginning:

	Sunday	Monday	Tuesday	Wednesday	Thursday	Friday	Saturday
DATE							
7–8 am		Breakfast	Breakfast	Breakfast	Breakfast	Breakfast	
8–9		Preparation	Preparation	Preparation	Preparation	Preparation	Breakfast
9–10	Breakfast	PE112(L)	PE112(L)	PE112(L)	PE112(L)	BIOL(P)	Travel
10–11	FREE	CHEM(L)	CHEM(L)	CHEM(L)	CHEM(L)	BIOL(P)	WORK
11–12	STUDY	STUDY	STUDY	STUDY	STUDY	BIOL(P)	WORK
12–1 pm	STUDY	BIOL(L)	BIOL(L)	BIOL(L)	BIOL(L)	TUTORIAL	WORK
1–2	Lunch	Lunch	Lunch	Lunch	Lunch	Lunch	Lunch
2–3	(VOLLEY-	CHEM(P)	STUDY	SPORT	PE112(P)	STUDY	WORK
3–4	BALL	CHEM(P)	STUDY	(VOLLEY-	PE112(P)	STUDY	WORK
4–5	MATCH)	CHEM(P)	STUDY	BALL	PE112(P)	SHOPPING	WORK
5–6	FREE	STUDY	STUDY	CLUB)	STUDY	TEA ROTA	WORK
6–7	Tea	Tea	Tea	Tea	Tea	Tea	Tea
7–8	FREE*	STUDY	STUDY	FREE*	STUDY	FREE*	FREE
8–9	FREE*	STUDY	STUDY	FREE*	STUDY	FREE*	FREE
9–10	FREE*	FREE*	STUDY	FREE*	STUDY	FREE*	FREE
Study (h)	2	10	11	4	11	6	0
Other (h)	13	5	4	11	4	9	15

Total study time = 44 h

Fig. 2.2 A weekly diary with example of entries for a first year science student with a Saturday job and active membership of a volleyball club. Note that 'free time' changes to 'study time', e.g. for periods when assessed work is to be produced or during revision for exams. Study time (including attendance at lectures, practicals and tutorials) thus represents between 42 and 50 per cent of the total time.

- Routine – predictable and regular and therefore easily scheduled (e.g. attending lectures or playing sport).
- One-off activities – usually with rather shorter deadlines and which may be of high priority (e.g. a tutorial assignment or seeking advice on a specific issue).
- Long-term tasks – sometimes referred to as 'elephant tasks' that are too large to 'eat' in one go (e.g. learning a language). These are best managed by scheduling frequent small 'bites' to achieve the task over a longer timescale.

You should make a weekly plan (Fig. 2.2) for the routine activities, with gaps for less predictable tasks. This should be supplemented by individual daily checklists, preferably written at the end of the previous working day. Such plans and checklists should be flexible, forming the basis for most of your activities except when exceptional circumstances intervene. The planning must be kept brief, however, and should be scheduled into your activities. Box 2.1 provides tips for effective time management during your studies.

KEY POINT Review each day's plan at the end of the previous day, making such modifications as are required by circumstances, e.g. adding an uncompleted task from the previous day or a new and urgent task.

Use checklists as often as possible – post your lists in places where they are easily and frequently visible, such as in front of your desk. Ticking things off as they are completed gives you a feeling of accomplishment and progress, increasing motivation.

Matching your work to your body's rhythm – everyone has times of day when they feel more alert and able to work. Decide when these times are for you and programme your work accordingly. Plan relaxation events for periods when you tend to be less alert.

Box 2.1 Tips for effective planning and working

- Set guidelines and review expectations regularly.
- Don't procrastinate: don't keep putting off doing things you know are important – they will not go away but they will increase to crisis point.
- Don't be a perfectionist – perfection is paralysing.
- Learn from past experience – review your management system regularly.
- Don't set yourself unrealistic goals and objectives – this will lead to procrastination and feelings of failure.
- Avoid recurring crises – they are telling you that something is not working properly and needs to be changed.
- Learn to concentrate effectively and don't let yourself be distracted by casual interruptions.
- Learn to say 'no' firmly but graciously when appropriate.
- Know your own body rhythms: e.g. are you a morning person or an evening person?
- Learn to recognise the benefits of rest and relaxation at appropriate times.
- Take short but complete breaks from your tasks – come back feeling refreshed in mind and body.
- Work in suitable study areas and keep your own workspace organised.
- Avoid clutter (physical and mental).
- Learn to access and use information effectively (Chapter 9).
- Learn to read and write accurately and quickly (Chapters 4 and 16).

Sources for further study

Anon. *Day-Timer*. Available: http://www.daytimer.co.uk
Last accessed: 01/04/07.
[Website for products of Day-Timers Europe Ltd., Chene Court, Poundwell Street, Modbury, Devon PL21 0QJ]

Anon. *Filofax*. Available: http://www.filofax.co.uk
Last accessed: 01/04/07.
[Website for products of Filofax UK, Unit 3, Victoria Gardens, Burgess Hill, West Sussex, RH15 9NB]

Anon. *TMI Website*. Available: http://www.tmi.co.uk
Last accessed: 01/04/07.
[Website for products of TMI (Time Manager International A/S), 50 High Street, Henley-in-Arden, Solihull, West Midlands B95 5AN]

Mayer, J.L. (1999) *Time Management for Dummies*, 2nd edn. IDG Books Worldwide, Inc., Foster City.

Study exercises

2.1 **Evaluate your time usage.** Compile a spreadsheet to keep a record of your daily activities in 15-minute segments for a week. Analyse this graphically and identify areas for improvement.

2.2 **List your short-, medium- and long-term tasks and allocate priorities.** Produce several lists, one for each of the three timescales, and prioritise each item. Use this list to plan your time management, by scheduling high priority tasks and leave low priority activities to 'fill in' the spare time that you may identify. This task should be done on a regular (monthly) basis to allow for changing situations.

2.3 **Plan an 'elephant' task.** Spend some time planning how to carry out a large or difficult task (learning a language or learning to use a complex computer program) by breaking it down into achievable segments ('bites').

3 Working with others

> **Definitions**
>
> **Team** – a team is not a bunch of people with job titles, but a congregation of individuals, each of whom has a role which is understood by other members. Members of a team seek out certain roles and they perform most effectively in the ones that are most natural to them.
>
> **Team role** – a tendency to behave, contribute and interrelate with others in a particular way.
>
> (both after Belbin, 1993)

It is highly likely that you will be expected to work with fellow students during practicals and study exercises. This might take the form of sharing tasks or casual collaboration through discussion, or it might be formally directed teamwork such as problem-based learning (Box 6.1) or preparing a poster (Chapter 14). Interacting with others can be extremely rewarding and realistically represents the professional world, where teamworking is common. The advantages of working with others include:

- Teamworking is usually synergistic in effect – it often results in better ideas, produced by the interchange of views, and better output, due to the complementary skills of team members.
- Working in teams can provide support for individuals within the team.
- Levels of personal commitment can be enhanced through concern about letting the team down.
- Responsibilities for tasks can be shared.

However, you can also feel both threatened and exposed if teamwork is not managed properly. Some of the main reasons for negative feelings towards working in groups include:

- Reservations about working with strangers – not knowing whether you will be able to form a friendly and productive relationship.
- Worries over rejection – a perception of being unpopular or being chosen last by the group.
- Concerns over levels of personal commitment – these can be enhanced through a desire to perform well, so the team as a whole achieves its target.
- Fear of being held back by others – especially for those who have been successful in individual work already.
- Lack of previous experience – worries about the kinds of personal interactions likely to occur and the team role likely to suit you best.
- Concerns about the outcomes of peer assessment – in particular, whether others will give you a fair mark for your efforts.

> *Peer assessment* – this term applies to marking schemes in which all or a proportion of the marks for a teamwork exercise are allocated by the team members themselves. Read the instructions carefully before embarking on the exercise, so you know which aspects of your work your fellow team members will be assessing. When deciding what marks to allocate yourself, try to be as fair as possible with your marking.

Teamwork skills

Some of the key skills you will need to develop to maximise the success of your teamworking activities include:

- Interpersonal skills. How do you react to new people? Are you able to both listen and communicate easily with them? How do you deal with conflicts and disagreements?
- Delegation/sharing of tasks. The primary advantage of teamwork is the sharing of effort and responsibility. Are you willing/able to do this? It involves trusting your team members. How will you deal with those group members who don't contribute fully?
- Effective listening. Successful listening is a skill that usually needs developing, e.g. during the exchange of ideas within a group.
- Speaking clearly and concisely. Effective communication is a vital part of teamwork, both between team members and when presenting team outcomes to others. Try to develop your communication skills through learning and practice (see Chapter 15).

> *Gaining confidence through experience* – the more you take part in teamwork, the more you know how teams operate and how to make teamwork effective for you.

- Providing constructive criticism. It is all too easy to be negative but only constructive criticism will have a positive effect on interactions with others.

Collaboration for learning

Much collaboration is informal and consists of pairs or groups of individuals getting together to exchange materials and ideas while studying. It may consist of a 'brainstorming' session for a topic or piece of work, or sharing efforts to research a topic. This has much to commend it and is generally encouraged. However, it is vital that this collaborative learning is distinguished from the collaborative writing of assessed documents: the latter is not usually acceptable and, in its most extreme form, is plagiarism, usually with a heavy potential punishment in university assessment systems. Make sure you know what plagiarism is, what unacceptable collaboration is, and how they are treated within your institution.

Studying with others – teaming up with someone else on your course for revision (a 'study buddy') is a potentially valuable activity and may especially suit some types of learners (p. 20). It can help keep your morale high when things get tough. You might consider:

- sharing notes, textbooks and other information;
- going through past papers together, dissecting the questions and planning answers;
- talking to each other about a topic (good for aural learners. p. 21);
- giving tutorials to each other about parts of the course that have not been fully grasped.

KEY POINT Collaboration is inappropriate during the final phase of an assessed piece of work unless you have been directed to produce a group report. Collaboration is encouraged during research and learning activities but the final write-up must normally be your own work. The extreme of producing copycat write-ups is regarded as plagiarism (p. 51) and will be punished accordingly.

The dynamics of teamworking

It is important that team activities are properly structured so that each member knows what is expected of them. Allocation of responsibilities usually requires the clear identification of a leader. Several studies of groups have identified different team roles that derive from differences in personality. You should be aware of such categorisations, both in terms of your own predispositions and those of your fellow team members, as it will help the group to interact more productively. Belbin (1993) identified eight such roles, recently extended to nine, as shown in Table 3.1. Several of the categories shown in Table 3.1 are suitable for a leader, including 'co-ordinator' and 'shaper'.

Web-based resources and support for brainstorming – websites such as http://www.brainstorming.co.uk give further information and practical advice for teamworking.

In formal team situations, your course organiser may deal with these issues; even if they do not, it is important that you are aware of these roles and their potential impact on the success or failure of teamwork. You should try to identify your own 'natural' role: if asked to form a team, bear these different roles in mind during your selection of colleagues and your interactions with them. The ideal team should contain members capable of adopting most of these roles. However, you should also note the following points:

- People will probably best fit one of these roles naturally as a function of their personality and skills.
- Group members may be suited to more than one role.
- In some circumstances, team members may be required to adapt and take a different role from the one that they feel suits them.
- No one role is 'better' than any other. For good teamwork, the group should have a balance of personality types present.
- People may have to adopt multiple roles, especially if the team size is small.

Recording group discussions – make sure you structure meetings (including writing agendas) and note their outcomes (taking minutes and noting action points).

Table 3.1 A summary of the team roles described by Belbin (1993). A good team requires members who are able to undertake appropriate roles at different times. Each role provides important strengths to a team, and its compensatory weaknesses should be accepted within the group framework

Team role	Personality characteristics	Typical function in a team	Strengths	Allowable weaknesses
Co-ordinator	Self-confident, calm and controlled	Leading: causing others to work towards shared goals	Good at spotting others' talents and delegating activities	Often less creative or intellectual than others in the group
Shaper	Strong need for achievement; outgoing; dynamic; highly strung	Leading: generating action within team; imposing shape and pattern to work	Providing drive and realism to group activities	Can be headstrong, emotional and less patient than others
Innovator	Individualistic, serious-minded; often unorthodox	Generating action; imposing shape and pattern to work activities	Creative, innovative and knowledgeable	Tendency to work in isolation; ideas may not always be practical
Monitor–evaluator	Sober, unemotional and prudent	Analysing problems and evaluating ideas	Shrewd judgement	May work slowly; not usually a good motivator
Implementer	Well-organised and self-disciplined, with practical common sense	Doing what needs to be done	Organising abilities and common sense	Lack of flexibility and tendency to resist new ideas
Teamworker	Sociable, mild and sensitive	Being supportive, perceptive and diplomatic; keeping the team going	Good listener; reliable and flexible; promotes team spirit	Not comfortable when leading; may be indecisive
Resource investigator	Extrovert, enthusiastic, curious and communicative	Exploiting opportunities; finding resources; external relations	Quick thinking; good at developing others' ideas	May lose interest rapidly
Completer–finisher	Introvert and anxious; painstaking, orderly and conscientious	Ensuring completion of activity to high standard	Good focus on fulfilling objectives and goals	Obsessive about details; may wish to do all the work to control quality
Specialist	Professional, self-motivated and dedicated	Providing essential skills	Commitment and technical knowledge	Contribute on a narrow aspect of project; tend to be single-minded

KEY POINT In formal teamwork situations, be clear as to how individual contributions are to be identified and recognised. This might require discussion with the course organiser. Make sure that recognition, including assessment, is truly reflective of effort. Failure to ensure that this is the case can lead to disputes and feelings of unfairness.

Your lab partner(s)

Many laboratory sessions in the life sciences involve working in pairs or small groups. In some cases, you may work with the same partner(s) for series of practicals or for a complete module. The relationship you develop as a team is important to your progress, and can enhance your understanding of the material and the grades you obtain. Tips for building a constructive partnership include:

- Introduce yourselves at the first session and take a continuing interest in each other's interests and progress at university.
- At appropriate points, discuss the practical (both theory and tasks) and your understanding of what is expected of you.

- Work jointly to complete the practical effectively, avoiding the situation where either partner dominates the activities and gains most from the practical experience.
- Share tasks according to your strengths, but do this in such a way that one partner can learn new skills and knowledge from the other.
- Make sure you ask questions of each other and communicate any doubts about what you have to do.
- Discuss other aspects of your course, e.g. by comparing notes from lectures or ideas about in-course assessments.
- Consider meeting up outside the practical sessions to study, revise and discuss exams.

Text reference

Belbin, R.M. (1993) *Team Roles at Work*. Butterworth-Heinemann, Oxford.

Source for further study

Belbin, R.M. *The Belbin Website*. Available: http://www.belbin.com/
Last accessed: 01/04/07.

Study exercises

3.1 **Evaluate your 'natural' team role(s).** Using Table 3.1 as a source, decide which team role best fits your personality.

3.2 **Keep a journal during a group activity.** Record your feelings and observations about experiences of working with other students. After the event, review the journal, then draw up a strategy for developing aspects where you feel you might have done better.

3.3 **Reflect upon your teamwork abilities.** Draw up a list of your reactions to previous efforts at collaboration or teamwork and analyse your strengths and weaknesses. How could these interactions have been improved or supported more effectively?

4 Taking notes from lectures and texts

Choose note-taking methods appropriately – the method you choose to take notes might depend on the subject; the lecturer and their style of delivery; and your own preference.

Note-taking is an essential skill that you will require in many different situations, such as:

- listening to lectures and seminars;
- attending meetings and tutorials;
- reading texts and research papers;
- finding information on the World Wide Web.

KEY POINT Good performance in assessments and exams is built on effective learning and revision (Chapters 5 and 6). However, both ultimately depend on the quality of your notes.

Taking notes from lectures

Compare lecture notes with a colleague – looking at different sets of notes for the same lecture may reveal interesting differences in approach, depth and detail.

Taking legible and meaningful lecture notes is essential if you are to make sense of them later, but many students find it difficult when starting their university studies. Begin by noting the date, course, topic and lecturer on the first page of each day's notes. Number every page in case they get mixed up later. The most popular way of taking notes is to write in a linear sequence down the page, emphasising the underlying structure via headings, as in Fig. 16.2. However, the 'pattern' and 'Mind Map' methods (Figs 4.1 and 4.2) have their advocates: experiment, to see which method you prefer.

Adjusting to the styles of your lecturers – recognise that different approaches to lecture delivery demand different approaches to note-taking. For example, if a lecturer seems to tell lots of anecdotes or spend much of the time on examples during a lecture, do not switch off – you still need to be listening carefully to recognise the key take-home messages. Similarly, if a lecture includes a section consisting mainly of images, you should still try to take notes – names of organisms, locations, key features, even quick sketches. These will help prompt your memory when revising. Do not be deterred by lecturers' idiosyncrasies; in every case you still need to focus and take useful notes.

Whatever technique you use, don't try to take down all of the lecturer's words, except when an important definition or example is being given, or when the lecturer has made it clear that he/she is dictating. Listen first, then write. Your goal should be to take down the structure and reasoning behind the lecturer's approach in as few words and phrases as possible. At this stage, follow the lecturer's sequence of delivery. Use headings and leave plenty of space, but don't worry too much about being tidy – it is more important that you get down the appropriate information in a readable form. Use abbreviations to save time. Recognise that you may need to alter your note-taking technique to suit different lecturers' styles.

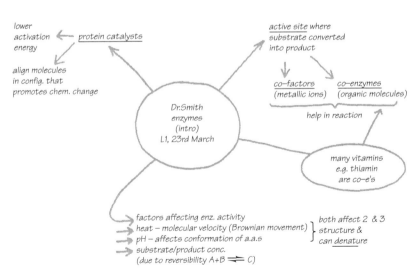

Fig. 4.1 An example of 'pattern' notes, an alternative to the more commonly used 'linear' format. Note the similarity to the 'spider diagram' method of brainstorming ideas (Fig. 16.2).

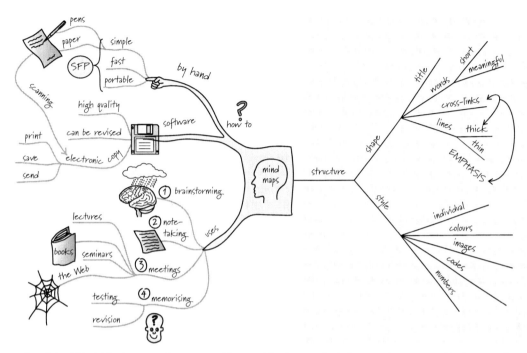

Fig. 4.2 Example of the 'Mind Map' approach to note-taking and 'brainstorming'. Start at the centre with the overall topic title, adding branches and sub-branches for themes and subsidiary topics. 'Basic' maps consist of a branched hierarchy overwritten with key words (e.g. shaded portion). Connections should be indicated with arrows; numbering and abbreviations are encouraged. To aid recall and creativity, Buzan (2006) recommends use of colour, different fonts, 3-dimensional doodles and other forms of emphasis (e.g. non-shaded portion).

Example Commonly used abbreviations include:

$\exists$	there are, there exist(s)
$\therefore$	therefore
$\because$	because
$\propto$	is proportional to
$\rightarrow$	leads to, into
$\leftarrow$	comes from, from
$\rightarrowtail$	involves several processes in a sequence
$1°, 2°$	primary, secondary (etc.)
$\approx, \cong$	approximately, roughly equal to
$=, \neq$	equals, not equal to
$\equiv, \not\equiv$	equivalent, not equivalent to
$<, >$	smaller than, bigger than
$\gg$	much bigger than
[X]	concentration of X
Σ	sum
Δ	change
f	function
#	number
∞	infinity, infinite

You should also make up your own abbreviations relevant to the context, e.g. if a lecturer is talking about photosynthesis, you could write 'PS' instead, etc.

Make sure you note down references to texts and take special care to ensure accuracy of definitions and numerical examples. If the lecturer repeats or otherwise emphasises a point, highlight (e.g. by underlining) or make a margin note of this – it could come in useful when revising. If there is something you don't understand, ask at the end of the lecture, or make an appointment to discuss the matter if there isn't time to deal with it then. Tutorials may provide an additional forum for discussing course topics.

Lectures delivered by PowerPoint or similar presentation programs

Some students make the mistake of thinking that lectures delivered as computer-based presentations with an accompanying handout or Web-resource require little or no effort by way of note-taking. While it is true that you may be freed from the need to copy out large diagrams and the basic text may provide structure, you will still need to adapt and add to the lecturer's points. Much of the important detail and crucial emphasis will still be delivered verbally. Furthermore, if you simply listen passively to the lecture, or worse, try to work from the handout alone, it will be far more difficult to understand and remember the content.

If you are not supplied with handouts, you may be given access to the electronic file, so that you can print out the presentation beforehand, perhaps in the '3 slides per page' format that allows space for notes alongside each slide (Fig. 4.3). Scan through this before the lecture if you can; then, during the presentation, focus on listening to what the lecturer has to say. Note down any extra details, points of emphasis and examples. After lectures, you could also add notes from supplementary reading. The text in presentations can be converted to word processor format if you have access to the electronic file. In PowerPoint, this can be

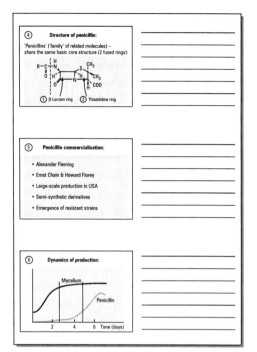

Fig. 4.3 An example of a printout from PowerPoint in 'Handouts (3 slides per page)' format.

achieved from the *Outline View* option on the *View* menu. You can copy and paste text between programs in the normal fashion, then modify font size and colour as appropriate.

'Making up' your notes

As soon as possible after each lecture, work through your notes, tidying them up and adding detail where necessary. Add emphasis to any headings you have made, so that the structure is clearer. If you feel it would be more logical for your purposes, change the order. Compare your notes with material in a textbook and correct any inconsistencies. Make notes from, or copy, any useful material you see in textbooks, ready for revision.

Taking notes from books and journal papers

Scanning and skimming are useful techniques.

Scanning

This involves searching for relevant content. Useful techniques are to:

- decide on key words relevant to your search;
- check that the book or journal title indicates relevance;
- look through the contents page (either paper titles in a journal volume, or chapter titles in a textbook);
- look at the index, if present.

Skimming

This is a valuable way to gain the maximum amount of information in the minimum amount of time, by reading as little of a text as is required. Essentially, the technique (also termed 'surveying') requires you to look at the structure of the text, rather than the detail. In a sense, you are trying to see the writer's original plan and the purpose behind each part of the text. Look through the whole of the piece first, to gain an overview of its scope and structure. Headings provide an obvious clue to structure, if present. Next, look for the 'topic sentence' in each paragraph (p. 103), which is often the first. You might then decide that the paragraph contains a definition that is important to note, or it may contain examples, so may not be worth reading for your purpose.

When you have found relevant material, note-taking fulfils the vital purpose of helping you understand and remember the information. If you simply read it, either directly or from a photocopy, you risk accomplishing neither. The act of paraphrasing (using different words to give the same meaning) makes you think about the meaning and forces you to express this for yourself. It is an important active learning technique. A popular method of skimming and note-taking is called the SQ3R technique (Box. 4.1).

 KEY POINT *Obtaining information and then understanding it are distinct, sequential parts of the process of learning. As discussed in Chapter 5 (Table 5.1), you must be able to do more than recall facts to succeed.*

Methods for finding and evaluating texts and articles are discussed further in Chapters 8 and 9.

Printing PowerPoint slides – *use the 'Black and White' option on the* Print *menu to avoid wasting ink on printing of coloured backgrounds. If you wish to use colour, remember that slides can be difficult to read if printed in small format. Always print a sample page before printing the whole lecture.*

Scanning effectively – *you need to stay focused on your key words, otherwise you may be distracted by apparently interesting but peripheral material.*

Spotting sequences – *writers often number their points (firstly, secondly, thirdly, etc.) and looking for these words in the text can help you skim it quickly.*

Making sure you have all the details of a source – *when taking notes from a text or journal paper: (a) always take full details (Chapter 8); (b) if copying word-for-word make sure you indicate this using quotes and take special care to ensure you do not alter the original wording.*

Box 4.1 The SQ3R technique for skimming texts

Survey Get a quick overview of the contents of the book or chapter, perhaps by rapidly reading the contents page or headings.

Question Ask yourself what the material covers and how precisely it relates to your study objectives.

Read Now read the text, paying attention to the ways it addresses your key questions.

Recall Recite to yourself what has been stated every few paragraphs. Write notes of this if appropriate, paraphrasing the text rather than copying it.

Review Think about what you have read and/or review your notes as a whole. Consider where it all fits in.

Sources for further study

Anon. *Mind Map.* Available: http://en.wikipedia.org/wiki/Mind_mapping. Last accessed: 01/04/07. [An independent review of mind-mapping and its history]

Anon. *FreeMind – Free Mind Mapping Software.* Available: http://freemind.sourceforge.net/wiki/index.php/Main_Page. Last accessed: 01/04/07. [An alternative to pen-and-paper mind mapping]

Buzan, A. (2006) *The Ultimate Book of Mind Maps.* Harper Thorsons, New York.

Morris, S. and Smith, J. (1998) *Understanding Mind Maps . . . in a Week.* [Foreword by Buzan, A.] Hodder and Stoughton, London.

Study exercises

4.1 Experiment with a new note-taking technique. If you haven't tried the pattern or mind-mapping methods (Figs 4.1 and 4.2), carry out a trial to see how they work for you. Research the methods first by consulting appropriate books or websites.

4.2 Carry out a 'spring clean' of your desk area and notes. Make a concerted effort to organise your notes and handouts, investing if necessary in files and folders. This will be especially valuable at the start of a revision period.

4.3 Try out the SQ3R technique. The next time you need to obtain information from a text, compare this method (Box 4.1) with others you may have adopted in the past. Is it faster, and does it aid your ability to recall the information?

5 Learning and revising

There are many different ways of learning and at university you have the freedom to choose which approach to study suits you best. You should tackle this responsibility with an open mind, and be prepared to consider new options. Understanding how you learn best and how you are expected to think about your discipline will help you to improve your approach to study and to understand your branch of bioscience at a deeper level. Adopting active methods of studying and revision that are suited to your personality can make a significant difference to your performance. Your department will publish material that can help too. Taking account of learning outcomes and marking/assessment criteria, for example, can help you focus your revision.

KEY POINT At university, you are expected to set your own agenda for learning. There will be timetabled activities, assessments and exam deadlines, but it is your responsibility to decide how you will study and learn, how you will manage your time, and, ultimately, what you will gain from the experience. You should be willing to challenge yourself academically to discover your full potential.

Learning styles

Significance of learning styles – no one learning style is 'better' than the others; each has its own strengths and weaknesses. However, since many university exams are conducted using 'reading and writing' modes of communication, you may need to find ways of expressing yourself appropriately using the written word (see Box 5.1).

We don't all learn in the same way. Your preferred learning style is simply the one that suits you best for receiving, communicating and understanding information. It therefore involves approaches that will help you learn and perform most effectively. There are many different ways of describing learning styles, and you may be introduced to specific schemes during your studies. While methods and terminology may differ among these approaches, it is important to realise that the important thing is the *process* of analysing your learning style, together with the way you use the information to modify your approach to studying, rather than the specific type of learner that you may identify yourself to be.

Learning styles and teaching styles – there may be a mismatch between your preferred learning style and the corresponding 'teaching style' used by your lecturers, in which case you will need to adapt appropriately (see Box 5.1).

A useful scheme for describing learning styles is the VARK system devised by Fleming (2001). By answering a short online questionnaire, you can 'diagnose' yourself as one of the types shown in Box 5.1, which also summarises important outcomes relating to how information and concepts can be assimilated, learned and expressed. People show different degrees of alignment with these categories; and research indicates that the majority of students are multi-modal learners – that is, falling into more than one category – rather than being only in one grouping. By carrying out an analysis like this, you can become more aware of your personal characteristics and think about whether the methods of studying you currently use are those that are best suited to your needs.

KEY POINT Having a particular learning preference or style does not mean that you are automatically skilled in using methods generally suited to that type of learner. You must work at developing your ability to take in information, study and cope with assessment.

Box 5.1 How to diagnose your learning preferences using the VARK learning styles scheme

Visit www.vark-learn.com to carry out the online diagnostic test, reflect on whether it is a fair description of your preferences, and think about whether you might change the way you study to improve your performance. None of the outcomes should be regarded as prescriptive – you should mix techniques as you see fit and only use methods that you feel comfortable adopting. Adapted with permission from material produced by Fleming (2001).

Learning style Description of learning preferences	Outcomes for your learning, studying and exam technique		
	Advice for taking in information and understanding it	Best methods of studying for effective learning	Ways to cope with exams so you perform better
Visual: *You are interested in colour, layout and design. You probably prefer to learn from visual media or books with diagrams and charts. You tend to add doodles and use highlighters on lecture and revision notes and express ideas and concepts as images.*	Use media incorporating images, diagrams, flow-charts etc. When constructing notes, employ underlining, different colours and highlighters. Use symbols as much as you can, rather than words. Leave plenty of white space in your notes. Experiment with the 'mind map' style of note-taking (p. 16).	Use similar methods to those described in column two. Reduce lecture notes to pictures. Try to construct your own images to aid understanding, then test your learning by redrawing these from memory.	Plan answers diagrammatically. Recall the images and doodles you used in your notes. Use diagrams in your answers (making sure they are numbered and fully labelled). As part of your revision, turn images into words.
Aural: *You prefer discussing subjects and probably like to attend tutorials and listen to lecturers, rather than read textbooks. Your lecture notes may be poor because you would rather listen than take notes.*	Make sure you attend classes, discussions and tutorials. Note and remember the interesting examples, stories, jokes. Leave spaces in your notes for later recall and 'filling'. Discuss topics with a 'study buddy'. Record lectures (with lecturer's permission).	Expand your notes by talking with others and making additional notes from the textbook. Ask others to 'hear' you talk about topics. Read your summarised notes aloud to yourself. Record your vocalised notes and listen to them later.	When writing answers, imagine you are talking to an unseen examiner. Speak your answers inside your head. Listen to your voices and write them down. Practise writing answers to old exam questions.
Read–Write: *You prefer using text in all formats. Your lecture notes are probably good. You tend to like lecturers who use words well and have lots of information in sentences and notes. In note-taking, you may convert diagrams to text and text to bullet points.*	Focus on note-taking. You may prefer the 'linear' style of note-taking (p. 101). Use the following in your notes: lists; headings; glossaries and lists of definitions. Expand you notes by adding further information from handouts, textbooks and library readings	Reduce your notes to lists or headings. Write out and read the lists again and again (silently). Turn actions, diagrams, charts and flowcharts into words. Rewrite the ideas and principles into other words. Organise diagrams and graphs into statements, e.g. 'The trend is . . .'.	Plan and write out exam answers using remembered lists. Arrange your words into hierarchies and points.

(continued)

Box 5.1 (continued)

Kinesthetic: *You tend to recall by remembering real events and lecturers' 'stories'. You probably prefer field excursions and lab work to theory and like lecturers who give real-life examples. Your lecture notes may be weak because the topics did not seem 'concrete' or 'relevant'.*	Focus on examples that illustrate principles. Concentrate on applied aspects and hands-on approaches, but try to understand the theoretical principles that underpin them. When taking in information, use all your senses – sight, touch, taste, smell, hearing.	Put plenty of examples, pictures and photographs into your notes. Use case studies and applications to help with principles and concepts. Talk through your notes with others. Recall your experience of lectures, tutorials, experiments or field trips.	Write practice answers and paragraphs. Recall examples and things you did in the lab or field trip. Role-play the exam situation in your own room.
Multi-modal: *Your preferences fall into two or more of the above categories. You are able to use these different modes as appropriate.*	If you are diagnosed as having two dominant preferences or several equally dominant preferences, read the study strategies above that apply to each of these. You may find it necessary to use more than one strategy for learning and communicating, feeling less secure with only one.		

Example A set of learning objectives taken from an introductory lecture on bacterial cell structure.
After this lecture, you should be able to:
☐ Define the following terms:
   ✓ prokaryote
   ✓ eukaryote
   ✓ envelope
   ✓ fimbriae
   ✓ F pilus
   ✓ plasmid
☐ Draw a labelled diagram to illustrate the principal components of a bacterial cell.
☐ Explain the functions of the major cellular components.
☐ Demonstrate knowledge of the relative magnitude of bacteria and eukaryotic cells, in terms of typical linear dimensions and volumes.
☐ Describe the basic process of cell division and give examples of typical timescales for different bacteria, e.g. *Escherichia coli, Clostridium perfringens, Mycobacterium tuberculosis.*

Thinking about thinking

The thinking processes that students are expected to carry out can be presented in a sequence, starting with shallower thought processes and ending with deeper processes, each of which builds on the previous level (see Table 5.1). The first two categories in this ladder apply to gaining basic knowledge and understanding, important when you first encounter a topic. Processes three to six are those additionally carried out by high-performing university students, with the latter two being especially relevant to final-year students, researchers and professionals. Naturally, the tutors assessing you will want to reward the deepest thinking appropriate for your level of study. This is often signified by the words they use in assessment tasks and marking criteria (column four, Table 5.1, and p. 108), and while this is not an exact process, being more aware of this agenda can help you to gain more from your studies and appreciate what is being demanded of you.

 KEY POINT When considering assessment questions, look carefully at words used in the instructions. These cues can help you identify what depth is expected in your answer. Take special care in multi-part questions, because the first part may require lower-level thinking, while in later parts, marks may be awarded for evidence of deeper thinking.

The role of assessment and feedback in your learning

Your starting point for assessment should be the learning outcomes or objectives for each module, topic or learning activity. You will usually find them in your module handbook. They state in clear terms what your tutors expect you to be able to accomplish after participating in each part and reading around the topic. Also of value will be marking/assessment

Table 5.1 A ladder of thinking processes, moving from shallower thought processes (top of table) to deeper levels of thinking (bottom of table). This table is derived from research by Benjamin Bloom et al. (1956). When considering the cue words in typical question instructions, bear in mind that the precise meaning will always depend on the context. For example, while 'describe' is often associated with relatively simple processes of recall, an instruction like 'describe how the human brain works' demands higher-level understanding. Note also that while a 'cue word' is often given at the start of a question/instruction, this is not universally so.

Thinking processes and description (in approximate order of increasing 'depth')	Example in life sciences	Example of typical question structure, with cue word highlighted	Other cue words used in question instructions
1. **Knowledge (knowing facts)**. If you know information, you can *remember* or *recognise* it. This does not always mean you understand it at a higher level.	You might know the order of bases in a piece of DNA but not understand what this means.	*Describe* the main components of a biological membrane.	• define • list • state • identify
2. **Comprehension**. If you comprehend a fact, you *understand* what it means.	You might know the order of bases in a piece of DNA and understand that they code in triplets for specific amino acids.	*Explain* how membrane components are involved in the accumulation of solutes within living cells.	• contrast • compare • distinguish • interpret
3. **Application**. To apply a fact means that you can *put it to use* in a particular context.	You might be able to take the DNA base sequence and work out the amino acid sequence of the protein for which they code.	Using the Nernst equation, and realistic values for the membrane potential and solute concentrations, *demonstrate* how Na⁺ ions must be actively transported out of the cells of marine organisms.	• calculate • illustrate • solve • show
4. **Analysis**. To analyse information means that you are able to *break it down into parts* and show how these components *fit together*.	You might be able to construct a three-dimensional model of a protein derived from the base sequence.	Drawing on information about membrane structure, *defend* the endosymbiotic theory of eukaryote evolution.	• compare • explain • consider • infer
5. **Synthesis**. To synthesise, you need to be able to *extract relevant facts* from a body of knowledge and use these to *address an issue in a novel way* or *create something new*.	You might be able to work out the function of a protein for which you know the sequence of bases, based on a comparison with other like proteins.	*Devise* an experiment to test the hypothesis that a specific membrane fraction contains a functional ATPase involved in glucose transport.	• design • integrate • test • create
6. **Evaluation**. If you evaluate information, you *arrive at a judgement* based on its importance relative to the topic being addressed.	You might be able to comment on theories about how a protein has evolved, by considering the structure of related proteins and relating this to the taxonomic position of their source species.	*Evaluate* the relative importance of passive and active transport in the accumulation of heavy metal salts by the main groups of soil fungi.	• review • assess • consider • justify

criteria or grade descriptors, which state in general terms what level of attainment is required for your work to reach specific grades. These are more likely to be defined at faculty/college/school/department level and consequently published in appropriate handbooks or websites. Reading learning outcomes and grade descriptors will give you a good idea of what to expect and the level of performance required to reach your personal goals. Relate them to both the material covered (e.g. in lectures and practicals, or online) and past exam papers. Doing this as you study and revise will indicate whether further reading and independent studying is required, and of what type. You will also have a much clearer picture of how you are likely to be assessed.

> **Definitions**
>
> **Learning objectives/outcomes** – statements of the knowledge, understanding or skills that a learner will be able to demonstrate on successful completion of a module, topic or learning activity.
>
> **Formative assessments** – these may be mid-term or mid-semester tests and are often in the same format as later exams. They are intended to give you feedback on your performance. You should use the results to measure your performance against the work you put in, and to find out, either from grades or tutor's comments, how you could do better in future. If you don't understand the reason for your grade, talk to your tutor.
>
> **Summative assessments** – these include end-of-year or end-of-module exams. They inform others about the standard of your work. In continuous or 'in-course' assessment, the summative elements are spread out over the course. Sometimes these exams may involve a formative aspect, if feedback is given.

 KEY POINT Use the learning objectives for your course (normally published in the handbook) as a fundamental part of your revision planning. These indicate what you will be expected to be able to do after taking part in the course, so exam questions are often based on them. Check this by reference to past papers.

There are essentially two types of assessment – formative and summative, although the distinction may not always be clear-cut (see margin). The first way you can learn from formative assessment is to consider the grade you obtained in relation to the work you put in. If this is a disappointment to you, then there must be a mismatch between your understanding of the topic and the marking scheme and that of the marker, or a problem in the writing or presentation of your assignment. This element of feedback is also present in summative assessment.

The second way to learn from formative assessment is through the written feedback and notes on your work. These comments may be cryptic, or scribbled hastily, so if you don't understand or can't read them, ask the tutor who marked the work. Most tutors will be pleased to explain how you could have improved your mark. If you find that the same comments appear frequently, it may be a good idea to seek help from your university's academic support unit. Take along examples of your work and feedback comments, so they can give you the best possible advice. Another suggestion is to ask to see the work of another student who obtained a good mark, and compare it with your own. This will help you judge the standard you should be aiming for.

Preparing for revision and examinations

Before you start revising, find out as much as you can about each exam, including:

- format and duration;
- date and location;
- types of question;
- whether any questions/sections are compulsory;
- whether the questions are internally or externally set or assessed;
- whether the exam is 'open book', and if so, which texts or notes are allowed.

> *Time management when revising* – this is vital to success and is best achieved by creating a revision timetable (Box 5.2).

Your course tutor is likely to give you details of exam structure and timing well beforehand, so that you can plan your revision; the course handbook and past papers (if available) can provide further useful details. Always check that the nature of the exam has not changed before you consult past papers.

Organising and using lecture notes, assignments and practical reports

> *Filing lecture notes* – make sure your notes are kept neatly and in sequence by using a ring binder system. File the notes in lecture or practical sequence, adding any supplementary notes or photocopies alongside.

Given their importance as a source of material for revision, you should have sorted out any deficiencies or omissions in your lecture notes and practical reports at an early stage. For example, you may have missed a lecture or practical due to illness, etc., but the exam is likely to assume attendance throughout the year. Make sure you attend classes whenever possible and keep your notes up to date. Your practical reports and any assignment work will contain specific comments from the teaching staff, indicating where

Box 5.2 How to prepare and use a revision timetable

1. **Make up a grid showing the number of days until your exams are finished.** Divide each day into several sections. If you like revising in large blocks of time, use am, pm and evening slots, but if you prefer shorter periods, divide each of these in two, or use hourly divisions (see also the table in study exercise 5.1).

2. **Write in your non-revision commitments**, including any time off you plan to allocate and physical activity at frequent intervals. Try to have about one-third or a quarter of the time off in any one day. Plan this in relation to your best times for useful work – for example, some people work best in the mornings, while others prefer evenings. If you wish, use a system where your relaxation time is a bonus to be worked for; this may help you motivate yourself.

3. **Decide on how you wish to subdivide your subjects** for revision purposes. This might be among subjects, according to difficulty (with the hardest getting the most time), or within subjects, according to topics. Make sure there is an adequate balance of time among topics and especially that you do not avoid working on the subject(s) you find least interesting or most difficult.

4. **Allocate the work to the different slots available on your timetable.** You should work backwards from the exams, making sure that you cover every exam topic adequately in the period just before each exam. You may wish to colour-code the subjects.

5. **As you revise, mark off the slots completed** – this has a positive psychological effect and will boost your self-confidence.

6. **After the exams, revisit your timetable** and decide whether you would do anything differently next time.

Using feedback from tutors – it is always worth reading any comments on your work as soon as it is returned. If you don't understand the comments, or are unsure about why you might have lost marks in an assignment, ask for an explanation.

marks were lost, corrections, mistakes, inadequacies, etc. Most lecturers are quite happy to discuss such details with students on a one-to-one basis and this information may provide you with clues to the expectations of individual lecturers that may be useful in exams set by the same members of staff. However, you should never 'fish' for specific information on possible exam questions, as this is likely to be counter-productive.

Revision

Begin early, to avoid last-minute panic. Start in earnest several weeks beforehand, and plan your work carefully:

- Prepare a revision timetable – an 'action plan' that gives details of specific topics to be covered (Box 5.2). Plan your revision around when (and where) your examinations are to be held. Try to keep to your timetable. Time management during this period is as important as keeping to time during the exam itself.
- Study the learning objectives/outcomes for each topic (usually published in the course handbook) to get an idea of what lecturers expect from you.
- Use past papers as a guide to the form of exam and the type of question likely to be asked (Box 5.3).

Recognise when your concentration powers are dwindling – take a short break when this happens and return to work refreshed and ready to learn. Remember that 20 minutes is often quoted as a typical limit to full concentration effort.

- Remember to have several short (5 minute) breaks during each hour of revision and a longer break every few hours. In any day, try to work for a maximum of three-quarters of the time.
- Include recreation within your schedule: there is little point in tiring yourself with too much revision, as this is unlikely to be profitable.
- Make your revision as active and interesting as possible (see below): the least productive approach is simply to read and reread your notes.
- Ease back on the revision near the exam: plan your revision to avoid last-minute cramming and overload fatigue.

Box 5.3 How to use past exam papers in your revision

Past exam papers are a valuable resource for targeting your revision.

1. **Find out where the past exam papers are kept.** Copies may be lodged in your department or the library; or they may be accessible online.
2. **Locate and copy relevant papers for your module(s).** Check with your tutor or course handbook that the style of paper will not change for the next set of examinations.
3. **Analyse the design of the exam paper.** Taking into account the length in weeks of your module, and the different lecturers and/or topics for those weeks, note any patterns that emerge. For example, can you translate weeks of lectures/practicals into numbers of questions or sections of the paper? Consider how this might affect your revision plans and exam tactics, taking into account (a) any choices or restrictions offered in the paper, and (b) the different types of questions asked (i.e. multiple choice, short-answer or essay).
4. **Examine carefully the style of questions.** Can you identify the expectations of your lecturers? Can you relate the questions to the learning objectives? How much extra reading do they seem to expect? Are the questions fact-based? Do they require a synthesis based on other knowledge? Can you identify different styles for different lecturers? Consider how the answers to these questions might affect your revision effort and exam strategy.
5. **Practise answering questions.** Perhaps with friends, set up your own mock exam once you have done a fair amount of revision, but not too close to the exams. Use a relevant past exam paper and don't study it beforehand. You need not attempt all of the paper at one sitting. You'll need a quiet room in a place where you will not be interrupted (e.g. a library). Keep close track of time during the mock exam and try to do each question in the length of time you would normally assign to it (see p. 31) – this gives you a feel for the speed of thought and writing required and the scope of answer possible. Mark each other's papers and discuss how each of you interpreted the question and laid out your answers and your individual marking schemes.
6. **Practise writing answer plans and starting answers.** This can save time compared to the 'mock exam' approach. Practice in starting answers can help you get over stalling at the start and wasting valuable time. Writing essay plans gets you used to organising your thoughts quickly and putting your thoughts into a logical sequence.

Aiding recall through effective note-taking – the Mind Map technique (p. 16), when used to organise ideas, is claimed to enhance recall by connecting the material to visual images or linking it to the physical senses.

Active revision

The following techniques may prove useful in devising an active revision strategy:

- 'Distil' your lecture notes to show the main headings and examples. Prepare revision sheets with details for a particular topic on a single sheet of paper, arranged as a numbered checklist. Wall posters are another useful revision aid.
- Confirm that you know the material by testing yourself – take a blank sheet of paper and write down all you know. Check your full notes to see if you missed anything out. If you did, go back immediately to a fresh blank sheet and redo the example. Repeat, as required.
- Memorise definitions and key phrases: definitions can be a useful starting point for many exam answers. Make up lists of relevant facts or definitions associated with particular topics. Test yourself repeatedly on these, or get a friend to do this. Try to remember *how many* facts or definitions you need to know in each case – this will help you recall them all during the exam.
- Use mnemonics and acronyms to commit specific factual information to memory. Sometimes, the dafter they are, the better they seem to work.

Question-spotting – *avoid this risky strategy. Lecturers are aware that this approach may be taken and try to ask questions in an unpredictable manner. You may find that you are unable to answer on unexpected topics which you failed to revise. Moreover, if you have a preconceived idea about what will be asked, you may also fail to grasp the nuances of the exact question set, and thereby fail to provide a focused answer.*

Revision checks – *it is important to test yourself frequently during revision, to ensure that you have retained the information you are revising.*

Final preparations – *try to get a good night's sleep before an exam. Last-minute cramming will be counter-productive if you are too tired during the exam.*

- Use pattern diagrams or mind maps as a means of testing your powers of recall on a particular topic (p. 16).
- Draw diagrams from memory: make sure you can label them fully.
- Try recitation as an alternative to written recall. Talk about your topic to another person, preferably someone in your class. Talk to yourself if necessary. Explaining something out loud is an excellent test of your understanding.
- Associate facts with images or journeys if you find this method works.
- Use a wide variety of approaches to avoid boredom during revision (e.g. record information on audio tape, use cartoons, or any other method, as long as it's not just reading).
- Form a revision group to share ideas and discuss topics with other students.
- Prepare answers to past papers, e.g. write essays or, if time is limited, write essay plans (see Box 5.3).
- If your subject involves numerical calculations, work through representative problems.
- Make up your own questions: the act of putting yourself in the examiner's mind-set by inventing questions can help revision. However, you should not rely on 'question spotting': this is risky!

The evening before your exam should be spent in consolidating material, and checking through summary lists and plans. Avoid introducing new material at this late stage: your aim should be to boost your confidence, putting yourself in the right frame of mind for the exam itself.

Text references

Bloom, B., Englehart, M. Furst, E., Hill, W. and Krathwohl, D. (1956) *Taxonomy of Educational Objectives: The Classification of Educational Goals. Handbook I: Cognitive Domain.* New York, Toronto: Longmans, Green.

Fleming, N.D. (2001) *Teaching and Learning Styles: VARK Strategies.* Neil Fleming, Christchurch.

Fleming, N.D. *VARK: A Guide to Learning Styles.* Available: http://www.vark-learn.com/
Last Accessed 01/04/07.

Sources for further study

Burns, R. (1997) *The Student's Guide to Passing Exams.* Kogan Page, London.

Hamilton, D. (1999) *Passing Exams: A Guide for Maximum Stress and Minimum Stress.* Cassell, London.

Many universities host study skills websites; these can be found using 'study skills', 'revision' or 'exams' as key words in a search engine.

Study exercises

5.1 Draw up a revision timetable. Use the techniques discussed in Box 5.2 to create a revision timetable for your forthcoming exams. You may wish to use or adapt the following arrangement, either on paper or within a spreadsheet.

A revision timetable planner

Date	Morning		Lunch	Afternoon		Tea/Dinner	Evening	
	Session 1	Session 2		Session 1	Session 2		Session 1	Session 2

5.2 Make use of past exam papers. Use the techniques discussed in Box 5.2 to improve your revision strategy: assess their effectiveness in a particular exam, or series of exams.

5.3 Try out new active revision techniques. Try any or all of the methods mentioned on pages 26–8 when revising. Compare notes with a colleague – which seems to be the most successful technique for you and for the topic you are revising?

6 Curriculum options, assessments and exams

> **Definition**
>
> **Transcript** – this is your record of achievement at university. Normally it will consist of details of each module or course you have taken, and an indication of the grade or mark achieved. It will also show your final (honours) classification, that is: first class, upper second class (2.1), lower second class (2.2), third class or unclassified (note: some UK universities do not differentiate second class degrees).

Many universities operate a modular system for their biosciences degree courses. This allows greater flexibility in subject choice and accommodates students studying on different degree paths. Modules also break a subject into discrete, easily assimilated elements. They have the advantage of spreading assessment over the academic year, but they can also tempt you to avoid certain difficult subjects or to feel that you can forget about a topic once the module is finished.

> *KEY POINT You should select your modules with care, mindful of potential degree options and how your transcript and CV will appear to a prospective employer. If you feel you need advice, consult your personal tutor or study advisor.*

As you move between levels of the university system, you will be expected to have passed a certain number of modules, as detailed in the progression criteria. These may be expressed using a credit point system. Students are normally allowed two attempts at each module exam and the resits often take place at the end of the summer vacation. If a student does not pass at the second attempt, they may be asked to 'carry' the subject in a subsequent year, and in severe cases of multiple failure, they may be asked to re-take the whole year or even leave the course. Consequently, it is worth finding out about these aspects of your degree. They are usually published in relevant handbooks.

You are unlikely to have reached this stage in your education without being exposed to the examination process. You may not enjoy being assessed, but you probably want to do well in your course. It is therefore important to understand why and how you are being tested. Identifying and improving the skills required for exam success will allow you to perform to the best of your ability.

> **Aiming high** – your goal should be to perform at your highest possible level and not simply to fulfil the minimum criteria for progression. This will lay sound foundations for your later studies. Remember too that a future employer might ask to see your academic transcript, which will detail all your module grades including any fails/resits, and will not just state your final degree classification.

Assessed coursework

There is a component of assessed coursework in many modules. This often tests specific skills, and may require you to demonstrate thinking at deeper levels (Table 5.1). The common types of coursework assessment are covered at various points in this book:

- practical exercises (Chapters 20–67);
- essays (Chapters 16 and 17);
- numerical problems (Chapter 65);
- data analysis (Chapters 65–67);
- poster and spoken presentations (Chapters 14 and 15);
- literature surveys and reviews (Chapter 19);
- project work (Chapters 18 and 31);
- problem-based learning (Box 6.1).

> **Avoiding plagiarism** – this is a key issue for assessed coursework - see p. 51 for a definition and Chapter 8 for appropriate methods of referring to the ideas and results of others using citation.

At the start of each year or module, read the course handbook or module guide carefully to find out when any assessed work needs to be submitted. Note relevant dates in your diary, and use this information to plan your work. Take special note if deadlines for different modules clash, or if they coincide with social or sporting commitments.

Box 6.1 Problem-based learning (PBL)

In this relatively new teaching method, you are likely to be presented with a 'real world' problem or issue, often working within a team. As you tackle the problem, you will gain factual knowledge, develop skills and exercise critical thinking (Chapter 9). Because there is a direct and relevant context for your work, and because you have to employ active learning techniques, the knowledge and skills you gain are likely to be more readily assimilated and remembered. This approach also more closely mimics workplace practices. PBL usually proceeds as follows:

1. **You are presented with a problem** (e.g. a case study, a hypothetical patient, a topical issue).

2. **You consider what issues and topics you need to research,** by discussion with others if necessary. You may need to identify where relevant resources can be found (Chapters 8–11).

3. **You then need to rank the issues and topics in importance,** allocating tasks to group members, if appropriate. A structured approach is an important aspect of PBL.

4. **Having carried out the necessary research, you should review what information has been obtained.** As a result, new issues may need to be explored and, where appropriate, allocated to group members.

5. **You will be asked to produce an outcome, such as a report, diagnosis, seminar presentation or poster.** An outline structure will be required, and for groups, further allocation of tasks will be needed to accomplish this goal.

If asked to carry out PBL as part of your course, it is important to get off to a good start. At first, the problem may seem unfamiliar. However, once you become involved in the work, you will quickly gain confidence. If working as part of a group, make sure that your group meets as early as possible, that you attend all sessions and that you do the necessary background reading. When working in a team, a degree of self-awareness is necessary regarding your 'natural' role in group situations (Table 3.1). Various methods are used for assessing PBL, including written, oral and poster presentations, and the assessment may involve peer marking.

KEY POINT If, for some valid reason (e.g. illness), you will be late with an assessment, speak to your tutors as soon as possible. They may be able to take extenuating circumstances into account by not applying a marking penalty. They will let you know what paperwork you may be required to submit to support your claim.

Summative exams – general points

Summative exams (p. 24) normally involve you answering questions without being able to consult other students or your notes. Invigilators are present to ensure appropriate conduct, but departmental representatives may be present for some of the exam. Their role is to sort out any subject-related problems, so if you think something is wrong, ask at the earliest opportunity. It is not unknown for parts of questions to be omitted in error, or for double meanings to arise, for example.

Planning

When preparing for an exam, make a checklist of the items you'll need (see p. 35). On the day of the exam, give yourself sufficient time to arrive at the correct room, without the risk of being late. Double-check the times and places of your exams, both well before the exam, and also on arrival. If you arrive at the exam venue early, you can always rectify a mistake if you find you've gone to the wrong place.

Curriculum Options, Assessments and Exams   **31**

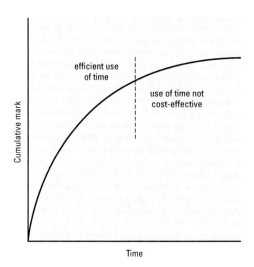

Fig. 6.1 Exam marks as a function of time. The marks awarded in a single answer will follow the law of diminishing returns – it will be far more difficult to achieve the final 25% of the available marks than the initial 25%. Do not spend too long on any one question.

Using the question paper – unless this is specifically forbidden, you should write on the question paper to plan your strategy, keep to time and organise answers.

Checking exam answers – look for:
- errors of fact;
- missing information
- grammatical and spelling errors;
- errors of scale and units;
- errors in calculations.

Adopting different tactics according to the exam – you should adjust your exam strategy (and revision methods) to allow for the differences in question types used in each exam paper.

Tackling the paper
Begin by reading the instructions at the top of the exam paper carefully, so that you do not make any errors based on lack of understanding of the exam structure. Make sure that you know:

- how many questions are set;
- how many must be answered;
- whether the paper is divided into sections;
- whether any parts are compulsory;
- what each question/section is worth, as a proportion of the total mark;
- whether different questions should be answered in different books.

Do not be tempted to spend too long on any one question or section: the return in terms of marks will not justify the loss of time from other questions (see Fig. 6.1). Take the first 10 minutes or so to read the paper and plan your strategy, before you begin writing. Do not be put off by those who begin immediately; it is almost certain they are producing unplanned work of a poor standard.

Underline the key phrases in the instructions, to reinforce their message. Next, read through the set of questions. If there is a choice, decide on those questions to be answered and decide on the order in which you will tackle them. Prepare a timetable which takes into account the amount of time required to complete each section and which reflects the allocation of marks – there is little point in spending one-quarter of the exam period on a question worth only 5 per cent of the total marks. Use the exam paper to mark the sequence in which the questions will be answered and write the finishing times alongside; refer to this timetable during the exam to keep yourself on course.

Reviewing your answers
At the end of the exam, you should allow at least 10 minutes to check through your script. Make sure your name and/or ID number is on each exam book as required and on all other sheets of paper, including graph paper, even if securely attached to your script, as it is in your interest to ensure that your work does not go astray.

 KEY POINT Never leave any exam early. Most exams assess work carried out over several months within a time period of 2–3 hours and there is always something constructive you can do with the remaining time to improve your script.

Special considerations for different types of exam question

Essay questions
Essay questions let examiners test the depth of your comprehension and understanding as well as your recall of facts. Essay questions give you plenty of scope to show what you know. They suit those with a good grasp of principles but who perhaps have less ability to recall individual details.

Before you tackle a particular question, you must be sure of what is required in your answer. Ask yourself 'What is the examiner looking for in this particular question?' and then set about providing a *relevant* answer. Consider each word in the question and highlight, underline or circle the key words. Make sure you know the meaning of the terms

> **Box 6.2 Writing under exam conditions**
>
> Make sure you go into an exam with a strategy for managing the available time.
>
> - **Allocate some time (say 5 per cent of the total) to consider which questions to answer and in which order.**
> - **Share the rest of the time among the questions, according to the marks available.** Aim to optimise the marks obtained. A potentially good answer should be allocated slightly more time than one you don't feel so happy about. However, don't concentrate on any one answer (see Fig. 6.1).
> - **For each question divide the time into planning, writing and revision phases** (see p. 107).
>
> Employ time-saving techniques as much as possible:
>
> - **Use spider diagrams** (Fig. 16.2) **or mind maps** (Fig. 4.2) to organise and plan your answer.
> - **Use diagrams and tables** to save time in making difficult and lengthy explanations, but make sure you refer to each one in the text.
> - **Use standard abbreviations** to save time repeating text but always explain them at the first point of use (e.g. PCR, polymerase chain reaction).
> - **Consider speed of writing and neatness** – especially when selecting the type of pen to use – ballpoint pens are fastest, but they can tend to smudge. You can only gain marks if the examiner can read your script.
> - **Keep your answer simple and to the point**, with clear explanations of your reasoning.
>
> Make sure your answer is relevant.
>
> - **Don't include irrelevant facts** just because you memorised them during revision, as this may do you more harm than good. You must answer the specific question that has been set.
> - **Time taken to write irrelevant material is time lost from another question.**

given in Table 17.1 so that you can provide the appropriate information, where necessary. Spend some time planning your writing (see Chapter 16). Refer back to the question frequently as you write, to confirm that you are keeping to the subject matter. Box 6.2 gives advice on writing essays under exam conditions.

It is usually a good idea to begin with the question that you are most confident about. This will reassure you before tackling more difficult parts of the paper. If you run out of time, write in note form. Examiners are usually understanding, as long as the main components of the question have been addressed and the intended structure of the answer is clear. Common reasons for poor exam answers in essay-style questions are listed in Box 6.3.

Multiple-choice and short-answer questions

Multiple-choice questions (MCQs) and short-answer questions (SAQs) are generally used to test the breadth and detail of your knowledge. The various styles that can be encompassed within the SAQ format allow for more demanding questions than MCQs, which may emphasise memory work and specific factual information.

A good approach for MCQ papers is as follows:

1. First trawl: read through the questions fairly rapidly, noting the 'correct' answer in those you can attempt immediately, perhaps in pencil.
2. Second trawl: go through the paper again, checking your original answers and this time marking up the answer sheet properly.
3. Third trawl: now tackle the difficult questions and those that require longer to answer (e.g. those based on numerical problems).

Penalties for guessing – if there is a penalty for incorrect answers in a multiple choice test, the best strategy is not to answer questions when you know your answer is a complete guess. Depending on the penalty, it may be beneficial to guess if you can narrow the choice down to two options (but beware false or irrelevant alternatives). However, if there are no such penalties, then you should provide an answer to all questions.

Box 6.3 Reasons for poor exam answers to essay-style questions

The following are reasons that lecturers often cite when they give low marks for essay answers:

- **Not answering the exact question set**. Either failing to recognise the specialist terms used in the question, failing to demonstrate an understanding of the terms by not providing definitions, failing to carry out the precise instruction in a question, or failing to address all aspects of the question.
- **Running out of time**. Not matching the time allocated to the extent of the answer. Frequently, this results in spending too long on one question and not enough on the others, or even failing to complete the paper.
- **Not answering all parts** of a multiple part question, or not recognising that one part (perhaps involving more complex ideas) may carry more marks than another.
- **Failing to provide evidence** to support an answer. Forgetting to state the 'obvious' – either basic facts or definitions.
- **Failing to illustrate an answer appropriately**, either by not including a relevant diagram, or by providing a diagram that does not aid communication, or by not including examples.
- **Incomplete answer(s)**. Failing to answer appropriately due to lack of knowledge.
- **Providing irrelevant evidence** to support an answer. 'Waffling' to fill space.
- **Illegible handwriting**.
- **Poor English**, such that facts and ideas are not expressed clearly.
- **Lack of logic** or structure to the answer.
- **Factual errors**, indicating poor note-taking, poor revision or poor recall.
- **Failing to correct obvious mistakes** by re-reading an answer before submitting the script.

At higher levels, the following aspects are especially important:

- **Not providing enough in-depth information**.
- **Providing a descriptive rather than an analytical answer** – focusing on facts, rather than deeper aspects of a topic.
- **Not setting a problem in context**, or not demonstrating a wider understanding of the topic. However, make sure you don't overdo this, or you may risk not answering the question set.
- **Not giving enough evidence of reading around the subject**. Wider reading can be demonstrated by quoting relevant papers and reviews and by giving author names and dates of publication.
- **Not considering both sides of a topic/debate, or not arriving at a conclusion**.

Answer the question as requested – this is true for all questions, but especially important for SAQs. If the question asks for a diagram, make sure you provide one; if it asks for a specified number of aspects of a topic, try to list this number of points; if there are two or more parts, provide appropriate answers to all aspects. This may seem obvious, but many marks are lost for not following instructions.

One reason for adopting this three-phase approach is that you may be prompted to recall facts relevant to questions looked at earlier. You can also spend more time per question on the difficult ones.

When unsure of an answer, the first stage is to rule out options that are clearly absurd or have obviously been placed there to distract you. Next, looking at the remaining options, can you judge between contrasting pairs with alternative answers? Logically, both cannot be correct, so you should see if you can rule one of the pair out. Watch out, however, in case *both* may be irrelevant to the answer. If the question involves a calculation, try to do this independently from the answers, so you are not influenced by them.

In SAQ papers, there may be a choice of questions. Choose your options carefully – it may be better to gain half marks for a correct answer to half a question, than to provide a largely irrelevant answer that apparently covers the whole question but lacks the necessary detail. For the SAQ form of question, few if any marks are given for writing style. Think in 'bullet points' and list the crucial points only. The time for answering SAQ questions may be tight, so get down to work fast, starting with answers that demand remembered facts. Stick to your timetable by moving on to the next question as soon as possible.

Strategically, it is probably better to get part-marks for the full number of questions than good marks for only a few.

Practical and information-processing exams

The prospect of a practical or information-processing exam may cause you more concern than a theory exam. This may be due to a limited experience of practical examinations, or to the fact that practical and observational skills are tested, as well as recall, description and analysis of factual information. Your first thoughts may be that it is not possible to prepare for such exams but, in fact, you can improve your performance by mastering the various practical techniques described in this book.

You may be allowed to take your laboratory reports and other texts into the practical exam. Don't assume that this is a soft option, or that revision is unnecessary: you will not have time to read large sections of your reports or to familiarise yourself with basic principles, etc. The main advantage of 'open book' exams is that you can check specific details of methodology, reducing your reliance on memory, provided you know your way around your practical manual and notes. In all other respects, your revision and preparation for such exams should be similar to theory exams. Make sure you are familiar with all of the practical exercises, including any work carried out in class by your partner (since exams are assessed on individual performance). If necessary, check with the teaching staff to see whether you can be given access to the laboratory, to complete any exercises that you have missed.

At the outset of the practical exam, determine or decide on the order in which you will tackle the questions. A question in the latter half of the paper may need to be started early on in the exam period (e.g. an enzyme assay requiring 2-hour incubation in a 3-hour exam). Such questions are included to test your forward-planning and time-management skills. You may need to make additional decisions on the allocation of material; e.g. if you are given 30 sterile test tubes, there is little value in designing an experiment that uses 25 of these to answer question 1, only to find that you need at least 15 tubes for subsequent questions.

Make sure you explain your choice of apparatus and experimental design. Calculations should be set out in a stepwise manner, so that credit can be given, even if the final answer is incorrect (see p. 471). If there are any questions that rely on recall of factual information and you are unable to remember specific details, make sure that you describe the item fully, so that you gain credit for observational skills. Alternatively, leave a gap and return to the question at a later stage.

Oral exams and interviews

An oral interview is sometimes a part of final degree exams, representing a chance for the external examiner(s) to meet individual students and to test their abilities directly and interactively. In some departments, orals are used to validate the exam standard, or to test students on the borderline between exam grades. Sometimes an interview may form part of an assessment, as with project work or posters. This type of exam is often intimidating – many students say they don't know how to revise for an oral – and many candidates

---

Examples These are principal types of question you are likely to encounter in a practical or information-processing exam:

Manipulative exercises Often based on work carried out during your practical course. Tests dexterity, specific techniques (e.g. sterile technique, p. 203).

'Spot' tests Short questions requiring an identification, or brief descriptive notes on a specific item (e.g. a prepared slide). Tests knowledge of seen material or ability to transfer this to a new example.

Calculations May include the preparation of aqueous solutions at particular concentrations (p. 143) and statistical exercises (p. 491). Tests numeracy.

Data analyses May include the preparation and interpretation of graphs (p. 465) and numerical information, from data either obtained during the exam or provided by the examiner. Tests problem-solving skills.

Preparing specimens for examination with a microscope Tests staining technique and light microscopy technique (p. 163).

Interpreting images Sometimes used when it is not possible to provide living specimens, e.g. in relation to electron microscopy. Can test a variety of skills.

Terminology – the oral exams are sometimes known simply as 'orals' or, borrowing Latin, as 'viva voce' (by the living voice) exams or 'vivas'.

Curriculum Options, Assessments and Exams   35

worry that they will be so nervous they won't be able to do themselves justice.

Preparation is just as important for oral exams as it is for written exams:

- Think about your earlier performances – if the oral follows written papers, it may be that you will be asked about questions you did not do so well on. These topics should be revised thoroughly. Be prepared to say how you would approach the questions if given a second chance.
- Read up a little about the examiner – he or she may focus their questions in their area of expertise.
- Get used to giving spoken answers – it is often difficult to transfer between written and spoken modes. Write down a few questions and get a friend to ask you them, possibly with unscripted follow-up queries.
- Research and think about topical issues in your subject area – some examiners will feel this reflects how interested you are in your subject.

Your conduct during the oral exam is important, too:

- Arrive promptly and wear reasonably smart clothing. Not to do either might be considered disrespectful by the examiner.
- Take your time before answering questions. Even if you think you know the answer immediately, take a little while to check mentally whether you have considered all angles. A considered, logical approach will be more impressive than a quick but ill-considered response.
- Start answers with the basics, then develop into deeper aspects. There may be both surface and deeper aspects to a topic and more credit will be given to a student who mentions the latter.
- When your answer is finished, stop speaking. A short, crisp answer is better than a rambling one.
- If you don't know the answer, say so. To waffle and talk about irrelevant material is more damaging than admitting that you don't know.
- Make sure your answer is balanced. Talk about the evidence and opinions on both sides of a contentious issue.
- Don't disagree violently with the examiner. Politely put your point of view, detailing the evidence behind it. Examiners will be impressed by students who know their own mind and subject area. However, they will expect you to support a position at odds with the conventional viewpoint.
- Finally, be positive and enthusiastic about your topic.

Counteracting anxiety before and during exams

Adverse effects of anxiety need to be overcome by anticipation and preparation well in advance (Box 6.4). Exams, with their tight time limits, are especially stressful for perfectionists. To counteract this tendency, focus on the following points during the exam:

- Don't expect to produce a perfect essay – this won't be possible in the time available.

Allow yourself to relax in oral exams – *external examiners are experienced at putting students at ease. They will start by asking 'simple-to-answer' questions, such as what modules you did, how your project research went, and what your career aspirations are. Look on the external as a friend rather than a foe.*

Creating an exam action list – *knowing that you have prepared well, checked everything on your list and gathered together all you need for an exam will improve your confidence and reduce anxiety. Your list might include:*

- *Verify time, date and place of the exam.*
- *Confirm travel arrangements to exam hall.*
- *Double-check module handbooks and past papers for exam structure.*
- *Think through use of time and exam strategy.*
- *Identify a quiet place near the exam hall to carry out a last-minute check on key knowledge (e.g. formulae, definitions, diagram labels).*
- *Ensure you have all the items you wish to take to the exam, e.g.*
  - *pens, pencils (with sharpener and eraser);*
  - *ruler;*
  - *correction fluid;*
  - *calculator (allowable type), if required;*
  - *sweets and drink, if allowed;*
  - *tissues*
  - *watch or clock;*
  - *ID card*
  - *texts and/or notes, if an open book exam;*
  - *lucky charm/mascot.*
- *Lay out clothes (if exam is early in the morning).*
- *Set alarm and/or ask a friend or family member to check you are awake on time.*

> **Box 6.4 Strategies for combating the symptoms of exam anxiety**
>
> **Sleeplessness** – this is common and does little harm in the short term. Get up, have a snack, do some light reading or other activity, then return to bed. Avoid caffeine (e.g. tea, coffee and cola) for several hours before going to bed.
>
> **Lack of appetite** – again commonplace. Eat what you can, and take sugary sweets into the exam to keep energy levels up in case you become tired.
>
> **Fear of the unknown** – it can be a good idea to visit the exam room beforehand, so you can become familiar with the location. By working through the points given in the exam action list on p. 35, you will be confident that nothing has been left out.
>
> **Worries about timekeeping** – get a *reliable* alarm clock or a new battery for an old one. Arrange for an alarm phone call; ask a friend or relative to make sure you are awake on time. Make reliable travel arrangements, to arrive on time. If your exam is early in the morning, it may be a good idea to get up early for a few days beforehand.
>
> **Blind panic during an exam** – explain how you feel to an invigilator. Ask to go for a supervised walk outside. Do some relaxation exercises (see below), then return to your work. If you are having problems with a specific question, it may be appropriate to speak to the departmental representative at the exam, to check that you are not misinterpreting the question.
>
> **Feeling tense** – shut your eyes, take several slow, deep breaths, do some stretching and relaxing muscle movements. During exams, it can be a good idea to do this between questions, and possibly to have a complete rest for a minute or so. Before exams, try some exercise activity or escape temporarily from your worries by watching TV or a movie.
>
> **Running out of time** – don't panic when the invigilator says 'five minutes left'. It is surprising how much you can write in this time. Write note-style answers or state the areas you would have covered: you may get some credit.

After the exam – try to avoid becoming involved in prolonged analyses with other students over the 'ideal' answers to the questions; after all, it is too late to change anything at this stage. Go for a walk, watch TV for a while, or do something else that helps you relax, so that you are ready to face the next exam with confidence.

- Don't spend too long planning your answer – once you have an outline plan, get started.
- Don't spend too much time on the initial parts of an answer, at the expense of the main message.
- Concentrate on getting all of the basic points across – markers are looking for the main points first, before allocating extra marks for the detail.
- Don't be obsessed with neatness, either in handwriting, or in the diagrams you draw, but make sure your answers are legible.
- Don't worry if you forget something. You can't be expected to know everything. Most marking schemes give a first class grade to work that misses out on up to 30 per cent of the marks available.

 KEY POINT Everyone worries about exams. Anxiety is a perfectly natural feeling. It works to your advantage, as it helps provide motivation and the adrenaline that can help you 'raise your game' on the day.

There is a lot to be said for treating exams as a game. After all, they are artificial situations contrived to ensure that large numbers of candidates can be assessed together, with little risk of cheating. They have conventions and rules, just like games. If you understand the rationale behind them and follow the rules, this will aid your performance.

Sources for further study

Acres, D. (1998) *Passing Exams Without Anxiety: How to Get Organised, be Prepared and Feel Confident of Success.* How to Books, London.

Burns, R. (1997) *The Student's Guide to Passing Exams.* Kogan Page, London.

Hamilton, D. (1999) *Passing Exams: A Guide for Maximum Success and Minimum Stress.* Cassell, London.

Many universities host study skills websites; these can be found using 'study skills', 'revision' or 'exams' as key words in a search engine.

Study exercises

6.1 **Analyse your past performances.** Think back to past exams and any feedback you received from them. How might you improve your performance? Consider ways in which you might approach the forthcoming exam differently. If you have kept past papers and answers to continuous assessment exercises, look at any specific comments your lecturers may have made.

6.2 **Share revision notes with other students.** Make a revision plan (see pp. 10 and 26) and then allocate some time to discussing your revision notes with a colleague. Try to learn from his or her approach. Discuss any issues you do not agree upon.

6.3 **Plan your exam tactics.** Find out from your module handbook or past papers what the format of each paper will be. Confirm this if necessary with staff. Decide how you will tackle each paper, allocating time to each section and to each question within the sections (see p. 107). Write a personal checklist of requirements for the exam (see p. 35).

7 Preparing your curriculum vitae

> **Definition**
>
> **Curriculum vitae** (or CV for short) – a Latin phrase that means 'the course your life has taken'.

Many students only think about their curriculum vitae immediately before applying for a job. Reflecting this, chapters on preparing a CV are usually placed near the end of texts of this type. Putting the chapter near the beginning of this book emphasises the importance of focusing your thoughts on your CV at an early stage in your studies. There are four main reasons why this can be valuable:

1. Considering your CV and how it will look to a future employer will help you think more deeply about the direction and value of your academic studies.
2. Creating a draft CV will prompt you to assess your skills and personal qualities and how these fit into your career aspirations.
3. Your CV can be used as a record of all the relevant things you have done at university and then, later, will help you communicate these to a potential employer.
4. Your developing CV can be used when you apply for vacation or part-time employment.

Personal development planning (PDP) and your CV – many PDP schemes (p. 5) also include an element of career planning that may involve creating a draft or generic CV. The PDP process can help you improve the structure and content of your CV, and the language you use within it.

 KEY POINT Developing your skills and qualities needs to be treated as a long-term project. It makes sense to think early about your career aspirations so that you can make the most of opportunities to build up relevant experience. A good focus for such thoughts is your developing curriculum vitae, so it is useful to work on this from a very early stage.

Skills and personal qualities

Skills (sometimes called competences) are generally what you have learned to do and have improved with practice. Table 1.1 summarises some important skills for bioscientists. This list might seem quite daunting, but your tutors will have designed your courses to give you plenty of opportunities to develop your expertise. Personal qualities, on the other hand, are predominantly innate. Examples include honesty, determination and thoroughness (Table 7.1). These qualities need not remain static, however, and can be developed or changed according to your experiences. By consciously deciding to take on new challenges and responsibilities, not only can you develop your personal qualities, but you can also provide supporting evidence for your CV.

Understanding skills and qualities – it may be helpful to think about how the skills and qualities in Tables 1.1 and 7.1 apply to particular activities during your studies, since this will give them a greater relevance.

Personal qualities and skills are interrelated because your personal qualities can influence the skills you gain. For example, you may become highly proficient at a skill requiring manual dexterity if you are particularly adept with your hands. Being able to transfer your skills is highly important (Chapter 1) – many employers take a long-term view and look for evidence of the adaptability that will allow you to be a flexible employee and one who will continue to develop skills.

Focusing on evidence – it is important to be able to provide specific information that will back up the claims you make under the 'skills and personal qualities' and other sections of your CV. A potential employer will be interested in your level of competence (what you can actually do) and in situations where you have used a skill or demonstrated a particular quality. These aspects can also be mentioned in your covering letter or at interview.

Developing your curriculum vitae

The initial stage involves making an audit of the skills and qualities you already have, and thinking about those you might need to develop. Tables 7.1 and 1.1 could form a basis of this self-appraisal. Assessing your skills may be easier than critically analysing your

Table 7.1 Some positive personal qualities

> Adaptability
> Conscientiousness
> Curiosity
> Determination
> Drive
> Energy
> Enthusiasm
> Fitness and health
> Flexible approach
> Honesty
> Innovation
> Integrity
> Leadership
> Logical approach
> Motivation
> Patience
> Performance under stress
> Perseverance
> Prudence
> Quickness of thought
> Seeing others' viewpoints
> Self-confidence
> Self-discipline
> Sense of purpose
> Shrewd judgement
> Social skills (sociability)
> Taking initiative
> Tenacity
> Tidiness
> Thoroughness
> Tolerance
> Unemotional approach
> Willingness to take on challenges

Seeing yourself as others see you – *you may not recognise all of your personal qualities and you may need someone else to give you an honest appraisal. This could be anyone whose opinion you value: a friend, a member of your family, a tutor, or a careers adviser.*

Setting your own agenda – *you have the capability to widen your experience and to demonstrate relevant personal qualities through both curricular and extracurricular activities.*

Paying attention to the quality of your CV – *your potential employer will regard your CV as an example of your very best work and will not be impressed if it is full of mistakes or badly presented, especially if you claim 'good written communication' as a skill!*

personal characteristics. In judging your qualities, try to take a positive view and avoid being overly modest. It is important to think of your qualities in a specific context, e.g. 'I have shown that I am trustworthy, by acting as treasurer for the Biochemistry Society', as this evidence will form a vital part of your CV and job applications.

If you can identify gaps in your skills, or qualities that you would like to develop, especially in relation to the needs of your intended career, the next step is to think about ways of improving them. This will be reasonably easy in some cases, but may require some creative thinking in others. A relatively simple example would be if you decided to learn a new language or to keep up with one you learned at school. There are likely to be many local college and university courses dealing with foreign languages at many different levels, so it would be a straightforward matter to join one of these. A rather more difficult case might be if you wished to demonstrate 'responsibility', because there are no courses available on this. One route to demonstrate this quality might be to put yourself up for election as an officer in a student society or club; another could be to take a leading role in a relevant activity within your community (e.g. voluntary work such as hospital radio). If you already take part in activities like these, your CV should relate them to this context.

Basic CV structures and their presentation

Box 7.1 illustrates the typical parts of a CV and explains the purpose of each part. Employers are more likely to take notice of a well-organised and well-presented CV, in contrast to one that is difficult to read and assimilate. They will expect it to be concise, complete and accurate. There are many ways of presenting information in a CV, and you will be assessed partly on your choices.

- **Order**. There is some flexibility as to the order in which you can present the different parts (see Box 7.1). A chronological approach within sections helps employers gain a picture of your experience.
- **Personality and 'colour'**. Make your CV different by avoiding standard or dull phrasing. Try not to focus solely on academic aspects: you will probably work in a team, and the social aspects of teamwork will be enhanced by your outside interests. However, make sure that the reader does not get the impression that these interests dominate your life.
- **Style**. Your CV should reflect *your* personality, but not in such a way that it indicates too idiosyncratic an approach. It is probably better to be formal in both language and presentation, as flippant or chatty expressions will not be well received.
- **Neatness**. Producing a well-presented, word-processed CV is very important. Use a laser-quality printer and good-quality paper; avoid poor-quality photocopying at all costs.
- **Layout**. Use headings for different aspects, such as personal details, education, etc. Emphasise words (e.g. with capitals, bold, italics or underlining) sparingly and with the primary aim of making the structure clearer. Remember that careful use of white space is important in design.
- **Grammar and proofreading**. Look at your CV carefully before you submit it, as sloppy errors give a very poor impression. Even if you use a spell-checker, some errors may creep in. Ask someone whom you regard as a reliable proofreader to comment on it (many tutors will do this, if asked in advance).

Box 7.1 The structure and components of a typical CV and covering letter

There is no right or wrong way to write a CV, and no single format applies. It is probably best to avoid software templates and CV 'wizards' as they can create a bland, standardised result, rather than something that demonstrates your individuality.

You should include the following, with appropriate sub-headings, generally in the order given below:

1. **Personal details**. This section *must* include your full name and date of birth, your address (both home and term-time, with dates, if appropriate) and a contact telephone number at each address. If you have an email account, you might also include this. You need only mention sex if your name could be either male or female.

2. **Education**. Choose either chronological order, or reverse chronological order and make sure you take the same approach in all other sections. Give educational institutions and dates (month, year) and provide more detail for your degree course than for your previous education. Remember to mention any prizes, scholarships or other academic achievements. Include your overall mark for the most recent year of your course, if it seems appropriate. Make sure you explain any gap years.

3. **Work experience**. Include all temporary, part-time, full-time or voluntary jobs. Details include dates, employer, job title and major duties involved.

4. **Skills and personal qualities**. Tables 1.1 and 7.1 give examples of the aspects you might include. Emphasise your strengths, and tailor this section to the specific requirements of the post (the 'job description'): for example, the practical skills you have gained during your degree studies if the post is a biological one, but concentrate on generic transferable skills and personal qualities for other jobs. Provide supporting evidence for your statements in all cases.

5. **Interests and activities**. This is an opportunity to bring out the positive aspects of your personality, and explain their relevance to the post you are applying for. Aim to keep this section short, or it may seem that your social life is more important than your education and work experience. Include up to four separate items, and provide sufficient detail to highlight the *positive* aspects of your interests (e.g. positions of responsibility, working with others, communication, etc.). Use sections 4 and 5 to demonstrate that you have the necessary attributes to fulfil the major requirements of the post.

6. **Referees**. Include the names (and titles), job descriptions, full postal addresses, contact telephone numbers and email addresses of two referees (rarely, some employers may ask for three). It is usual to include your personal tutor or course leader at university (who among other things will verify your marks), plus another person – perhaps a current or former employer, or someone who runs a club or society and who knows your personal interests and activities. Unless you have kept in touch with a particular teacher since starting university, it is probably best to choose current contacts, rather than those from your previous education.

Some other points to consider:

- Try to avoid jargon and over-complicated phrases in your CV: aim for direct, active words and phrases (see Box 16.1).

- Most employers will expect your CV to be word-processed (and spell-checked). Errors in style, grammar and presentation will count against you, so be sure to check through your final version (and ask a reliable person to second-check it for you).

- Aim for a *maximum* length of two pages, printed single-sided on A4 paper, using a 'formal' font (e.g. Times Roman or Arial) of no less than 12 point for the main text. Always print onto good quality white paper. Avoid fussy use of colour, borders or fonts.

- Don't try to cram in too much detail. Use a clear and succinct approach with short sentences and lists to improve 'readability' and create structure. Remember that your aim is to catch the eye of your potential employer, who may have many applications to work through.

- It is polite to check that people are willing to act as a referee for you and to provide them with an up-to-date copy of your CV.

Your single-page covering letter should have four major components:

1. **Letterhead**. Include your contact details, the recipient's name and title (if known) and address, plus any job reference number.

2. **Introductory paragraph**. Explain who you are and state the post you are applying for.

3. **Main message**. This is your opportunity to sell yourself to a potential employer, highlighting particular attributes and experience. Keep it to three or four sentences at the most and relate it to the particular skills and qualities demanded in the job or person specification.

4. **Concluding paragraph**. A brief statement that you look forward to hearing the outcome of your application is sufficient.

Finally, add either 'Yours sincerely' (where the recipient's name is known) or 'Yours faithfully' (in a letter beginning 'Dear Sir or Madam') and then end with your signature.

- **Relevance**. If you can, slant your CV towards the job description and the qualifications required (see below). Make sure you provide evidence to back up your assertions about skills, qualities and experience.
- **Accuracy and completeness**. Check that all your dates tally; otherwise, you will seem careless. It is better to be honest about your grades and (say) a period of unemployment, than to cover this up or omit details that an employer will want to know. They may be suspicious if you leave things out.

Adjusting your CV

You should fine-tune your CV for each post. Employers frequently use a 'person specification' to define the skills and qualities demanded in a job, often under headings such as 'essential' and 'desirable'. This will help you decide whether to apply for a position and it assists the selection panel to filter the applicants. Highlight relevant qualifications as early in your CV as possible. Be selective – don't include every detail about yourself. Emphasise relevant parts and leave out irrelevant details, according to the job. Similarly, your letter of application is not merely a formal document but is also an opportunity for persuasion (Box 7.1). You can use it to state your ambitions and highlight particular qualifications and experience. However, don't go over the top – always keep the letter to a single page.

Creating a generic CV – as you may apply for several jobs, it is useful to construct a CV in electronic format (e.g. as a Word file) which includes all information of potential relevance. This can then be modified to fit each post. Having a prepared CV on file will reduce the work each time you apply, while modifying this will help you focus on relevant skills and attributes for the particular job.

 KEY POINT A well-constructed and relevant CV won't necessarily guarantee you a job, but it may well get you onto the short list for interview. A poor-quality CV is a sure route to failure.

Sources for further study

Anon. (2000) *How to Write a Curriculum Vitae*. University of London Careers Service, London.

Anon. *Doctorjob.com Website*. Available: http://doctorjob.com
Last accessed: 01/04/07.

Anon. *Graduate Prospects Website*. Available: http://www.prospects.ac.uk/cms/ShowPage/Home_page/p!eLaXi
Last accessed: 01/04/07.
[Higher Education Careers Services Unit, containing good examples of CVs]

Anon. *Applying for Jobs*. Available: http://www.hero.ac.uk/uk/studying/careers_and_life long_learning/applying_for_jobs285.cfm
Last accessed: 01/04/07.

Study exercises

7.1 Evaluate your personal attributes. Using Table 7.1, list *five* qualities that you would use to best describe yourself, and cite the evidence you might give to a potential employer to convince them that this was the case. List *five* attributes you could develop, then indicate how you might do this.

7.2 Create a generic CV. Drawing on your school record of achievement, or any CV already prepared, e.g. for a part-time job, create a word-processed generic CV. Save the file in an appropriate (computer) folder and make a back-up copy. Print out a copy for filing. Periodically update the word-processed version. If appropriate, save different versions to be used in different contexts (e.g. when applying for a vacation job).

7.3 Think about your future career and ask for advice. Make an appointment with one of the advisors in your university's careers service. Ask about career options for graduates with your intended degree, or determine what qualifications or module options might be appropriate for occupations that interest you.

Communicating Information

14 Organising a poster display

Learning from others – look at the various types of posters around your university and elsewhere; the best examples will be visual, not textual, with a clear structure that helps get the key messages across.

A scientific poster is a visual display of the results of an investigation, usually mounted on a rectangular board. Posters are used in undergraduate courses to display project results or assignment work, and at scientific meetings to communicate research findings.

In a written report you can include a reasonable amount of specific detail and the reader can go back and reread difficult passages. However, if a poster is long-winded or contains too much detail, your reader is likely to lose interest.

> KEY POINT A poster session is like a competition – you are competing for the attention of people in a room. Because you need to attract and hold the attention of your audience, make your poster as interesting as possible. Think of it as an advertisement for your work and you will not go far wrong.

Preliminaries

Before considering the content of your poster, you should find out:

- the linear dimensions of your poster area, typically up to 1.5 m wide by 1.0 m high;
- the composition of the poster board and the method of attachment, whether drawing pins, Velcro tape, or some other form of adhesive; and whether these will be provided – in any case, it is safer to bring your own;
- the time(s) when the poster should be set up and when you should attend;
- the room where the poster session will be held.

Design

Plan your poster with your audience in mind, as this will dictate the appropriate level for your presentation. Aim to make your poster accessible to a broad audience. Since a poster is a *visual* display, you must pay particular attention to the presentation of information: work that may have taken hours to prepare can be ruined in a few minutes by the ill-considered arrangement of items (Fig. 14.1). Begin by making a draft sketch of the major elements of your poster. It is worth discussing your intended design with someone else, as constructive advice at the draft stage will save a lot of time and effort when you prepare the final version (or consult Simmonds and Reynolds, 1994).

Layout

One approach is to divide the poster into several smaller areas, perhaps six or eight in all, and prepare each as a separate item on a piece of card. Alternatively, you can produce a single large poster on one sheet of paper or card and store it inside a protective cardboard tube. However, a single large poster may bend and crease, making it difficult to flatten out. In addition, photographs and text attached to the backing sheet may work loose; a large printed poster with embedded images is an alternative approach.

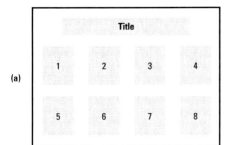

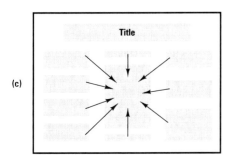

Fig. 14.1 Poster design. (a) An uninspiring design: sub-units of equal area, reading left to right, are not recommended. (b) This design is more interesting and the text will be easier to read (column format). (c) An alternative approach, with a central focus and arrows/tapes to guide the reader.

Presenting a poster at a formal conference – *it can be useful to include your photograph for identification purposes, e.g. in the top right-hand corner of the poster.*

Making up your poster – *text and graphics printed on good-quality paper can be glued directly onto a contrasting mounting card: use photographic spray mountant or Pritt rather than liquid glue. Trim carefully using a guillotine to give equal margins, parallel with the paper. Photographs should be scanned into an electronic format, or placed in a window mount to avoid the tendency for their corners to curl. Another approach is to trim pages or photographs to their correct size, then encapsulate in plastic film: this gives a highly professional finish and is easy to transport.*

Producing composite material for posters – *PowerPoint is generally more useful than Word when you wish to include text, graphics and/or images on the same page. It is possible to use PowerPoint to produce a complete poster (Box 14.1), although it can be expensive to have this printed out commercially to A1 or A0 size.*

Subdividing your poster means that each smaller area can be prepared on a separate piece of paper or card, of A4 size or slightly larger, making transport and storage easier. It also breaks the reading matter up into smaller pieces, looking less formidable to a potential reader. By using pieces of card of different colours you can provide emphasis for key aspects, or link text with figures or photographs.

You will need to guide your reader through the poster and headings/sub-headings will help with this aspect. It may be appropriate to use either a numbering system, with large, clear numbers at the top of each piece of card, or a system of arrows (or thin tapes) to link sections within the poster (see Fig. 14.1). Make sure that the relationship is clear and that the arrows or tapes do not cross.

Title

Your chosen title should be concise (no more than eight words), specific and interesting, to encourage people to read the poster. Make the title large and bold – it should run across the top of your poster, in letters at least 4 cm high, so that it can be read from the other side of the room. Coloured spirit-based marker and block capitals drawn with a ruler work well, as long as your writing is readable and neat (the colour can be used to add emphasis). Alternatively, you can print out each word in large font, using a word processor. Details of authors, together with their addresses (if appropriate), should be given, usually across the top of the poster in somewhat smaller lettering than the title.

Text

Write in short sentences and avoid verbosity. Keep your poster as visual as possible and make effective use of the spaces between the blocks of text. Your final text should be double-spaced and should have a minimum capital letter height of 8 mm (minimum font size 36 point), preferably greater, so that the poster can be read at a distance of 1 m. One method of obtaining text of the required size is to photo-enlarge standard typescript (using a good-quality photocopier), or use a high-quality (laser) printer. It is best to avoid continuous use of text in capitals, since it slows reading and makes the text less interesting to the reader. Also avoid italic, 'balloon' or decorative styles of lettering.

 KEY POINT Keep text to a minimum – aim to have a maximum of 500 words in your poster.

Subtitles and headings

These should have a capital letter height of 15–20 mm, and should be restricted to two or three words. They can be produced by word processor, photo-enlargement, by stencilling or by hand, using pencilled guidelines (but make sure that no pencil marks are visible on your finished poster).

Colour

Consider the overall visual effect of your chosen display, including the relationship between your text, diagrams and the backing board. Colour can be used to highlight key aspects of your poster. However, it is very easy to ruin a poster by the inappropriate choice and application of colour. Careful use of two, or at most three, complementary colours and shades will be easier on the eye and should aid comprehension. Colour can be used to link the text with the visual images (e.g. by picking out a colour in a photograph and using the same colour on the mounting board for the accompanying text). For PowerPoint posters, careful choice of colours for the various elements will enhance the final product (Box 14.1). Use coloured inks or water-based paints to provide colour in diagrams and figures, as felt pens rarely give satisfactory results.

Content

The typical format is that of a scientific report (see Box 18.1), i.e. with the same headings, but with a considerably reduced content. Keep references within the text to a minimum – interested parties can always ask you for further information. Also note that most posters have a summary/conclusions section at the end, rather than an abstract.

Introduction

This should give the reader background information on the broad field of study and the aims of your own work. It is vital that this section is as interesting as possible, to capture the interest of your audience. It is often worth listing your objectives as a series of numbered points.

> **Presenting at a scientific meeting** – never be tempted to spend the minimum amount of time converting a piece of scientific writing into poster format – the least interesting posters are those where the author simply displays pages from a written communication (e.g. a journal article) on a poster board.

> **Designing the materials and methods section** – photographs or diagrams of apparatus can help to break up the text of this section and provide visual interest. It is sometimes worth preparing this section in a smaller typeface.

Materials and methods

Keep this short, and describe only the principal techniques used. You might mention any special techniques, or problems of general interest.

Results

Don't present your raw data: use data reduction wherever possible, i.e. figures and simple statistical comparisons. Graphs, diagrams, histograms and pie charts give clear visual images of trends and relationships and should be used in place of data tables (see p. 455). Final copies of all figures should be produced so that the numbers can be read from a distance of 1 m. Each should have a concise title and legend, so that it is self-contained: if appropriate, a series of numbered points can be used to link a diagram with the accompanying text. Where symbols are used, provide a key on each graph (symbol size should be at least 5 mm). Avoid using graphs straight from a written version, e.g. a project report, textbook or a paper, without considering whether they need modification to meet your requirements.

> **Keeping graphs and diagrams simple** – avoid composite graphs with different scales for the same axis, or with several trend lines (use a maximum of three trend lines per graph).

Conclusions

This is where many readers will begin, and they may go no further unless you make this section sufficiently interesting. This part needs to be the strongest section of your poster, summarising the main points. Refer to your figures here to draw the reader into the main part of your poster. A slightly larger or bolder typeface may add emphasis, though too many different typefaces can look messy. For the reference list, a smaller font can be used.

> **Listing your conclusions** – a series of numbered points is a useful approach, if your findings fit this pattern.

The poster session

A poster display session may be organised as part of the assessment of your coursework, and this usually means those held at scientific meetings and

> **Consider providing a handout** – this is a useful way to summarise the main points of your poster, so that your readers have a permanent record of the information you have presented.

Box 14.1 How to create a poster using PowerPoint

Software such as PowerPoint can be used to produce a high-quality poster, providing you have access to a good colour printer. However, you should avoid the standard templates available on the Web as they encourage unnecessary uniformity and stifle creativity, leading to a less satisfying end result. The following steps give practical advice on creating a poster as a single PowerPoint slide:

1. **Sketch out your plans.** Decide on the main poster elements (images, graphs, tables and text sections) and their relationship with each other and draw out a one-page 'storyboard' (see Fig. 14.1). Think about colours for background, text and graphics (use two or three complementary colours): dark text on a light background is clearer (high contrast), and uses less ink when printing. Also consider how you will link the elements in sequence, to guide readers through your 'story'.

2. **Get your material ready.** Collect together individual files for pictures, figures and tables. Make any required adjustments to images, graphs or tables before you import them into your poster.

3. **Create a new/blank slide.** Open PowerPoint and from *File > New* select *Blank presentation*. Next, choose the *Blank Presentation* option for *New Presentation > New option*. Then use the *File > Page setup* menu to select either *Landscape* or *Portrait* orientation and to set the correct page size (use *Width* and *Height* commands, or select a standard size such as A4, A3, A2, etc.). Right-click on the slide and select *Ruler* and *Guides* (to help position elements within the slide – the horizontal and vertical guidelines can be dragged to different positions at later stages, as required) and also select an appropriate *Background* colour. In general, avoid setting a picture as your background as this tends to detract from the content of the poster. Before going further, save your work. Repeat this frequently and in more than one location (e.g. hard drive and USB memory stick).

4. **Add graphics.** For images, use the drop-down *Insert* menu, select *Picture, From File* and browse to *Insert* the correct file. The *Insert, Object* command performs a similar function for Excel charts (graphs). Alternatively, use the copy-and-paste functions of complementary software. Once inserted, resize using the *sizing handles* in one of the corners (for photographs, take care not to alter one dimension relative to the other, or the image will be distorted). To reposition, put the mouse pointer over the image, left-click and hold, then drag to new location. While the *Drawing* toolbar offers standard shapes and other useful features, you should avoid clipart (jaded and over-used) and poor-quality images from the Web (always use the highest resolution possible) – if you do not have your final images, use blank text boxes to show their position within the draft poster.

5. **Add text.** Use either the *Drawing* toolbar to select a *Text box* and place this on your slide, then either type in your text (use the *Enter* key to provide line spacing within the box) or copy-and-paste text from a word-processed file. You will need to consider the font size for the printed poster (e.g. for an A0 poster (size 1189 × 841 mm), a printed font size of 24 point is appropriate for the main text, with larger fonts for headings and titles. If you find things difficult to read on-screen, use the *Zoom* function (either select a larger percentage in the *Zoom box* on the standard toolbar, or hold down the *Ctrl* key and use the mouse wheel to scroll up (*zoom*) or down (*reduce*). Use a separate text box for each element of your poster and don't be tempted to type too much text into each box – write in succinct phrases, using bullet points and numbered lists to keep text concise (aim for no more than 50 words per text box). Select appropriate font styles and colours using the *Format > Font* menu. For a background colour or surrounding line, right-click and use the *Format Text Box* command (line thickness and colour can then be altered using the *Drawing* toolbar). Present supplementary text elements in a smaller font. For example, details of methodology, references cited.

6. **Add boxes, lines and/or arrows** to link elements of the poster and guide the reader (see Fig. 14.1). These features are available from the *Drawing* toolbar. Note that new inserts are overlaid over older inserts – if this proves to be a problem, select the relevant item and use the *Draw > Order* functions to change its relative position.

7. **Review your poster.** Get feedback from another student, or your tutor, e.g. on a small printed version, or use a projector to view your poster without printing (adjust the distance between projector and screen to give the correct size).

8. **Revise and edit your poster.** Revisit your work and remove as much unnecessary text as possible. Delete any component that is not essential to the message of the poster. Keep graphs simple and clear (p. 455 gives further advice). 'White space' is important in providing structure.

9. **Print the final version.** Use a high-resolution colour printer (this may be costly, so you should wait until you are sure that no further changes are needed).

Coping with questions in assessed poster sessions – you should expect to be asked questions about your poster, and to explain details of figures, methods, etc.

conferences. Staff and fellow students (delegates at conferences) will mill around, looking at the posters and chatting to their authors, who are usually expected to be in attendance. If you stand at the side of your poster throughout the session you are likely to discourage some readers, who may not wish to become involved in a detailed conversation about the poster. Stand nearby. Find something to do – talk to someone else, or browse among the other posters, but remain aware of people reading your poster and be ready to answer any queries they may raise. Do not be too discouraged if you aren't asked lots of questions: remember, the poster is meant to be a self-contained, visual story, without need for further explanation.

A poster display will never feel like an oral presentation, where the nervousness beforehand is replaced by a combination of satisfaction and relief as you unwind after the event. However, it can be a very satisfying means of communication, particularly if you follow these guidelines.

Text reference

Simmonds, D. and Reynolds, L. (1994) *Data Presentation and Visual Literacy in Medicine and Science.* Butterworth-Heinemann, London.

Sources for further study

Alley, M. (2003) *The Craft of Scientific Presentations: Critical Steps to Succeed and Critical Errors to Avoid.* Springer-Verlag, New York.

Briscoe, M.H. (2000) *Preparing Scientific Illustrations: A Guide to Better Posters, Presentations and Publications,* 2nd edn. Springer-Verlag, New York.

Davis, M.F. (2005) *Scientific Papers and Presentations,* 2nd edn. Academic Press, New York.

Gosling, P.J. (1999) *Scientist's Guide to Poster Presentations.* Kluwer, New York.

Hess, G. and Liegel, L. *Creating Effective Poster Presentations.* Available: http://www.ncsu.edu/project/posters.
Last accessed 01/04/07.

Study exercises

14.1 Design a poster. Working with one or more partners from your year group, decide on a suitable poster topic (perhaps something linked to your current teaching programme). Working individually, make an outline plan of the major elements of the poster, with appropriate sub-headings and a brief indication of the content and relative size of each element (including figures, diagrams and images). Exchange draft plans with your partners and arrange a session where you can discuss their merits and disadvantages.

14.2 Prepare a checklist for assessing the quality of a poster presentation. After reading through this chapter, prepare a 10-point checklist of assessment criteria under the heading 'What makes a good poster presentation?' Compare your list with the one that we have provided (p. 509) – do you agree with our criteria, or do you prefer your own list (and can you justify your preferences)?

14.3 Evaluate the posters in your university. Most universities have a wide range of academic posters on display. Some may cover general topics (e.g. course structures), while others may deal with specific research topics (e.g. poster presentations from past conferences). Consider their good and bad features (if you wish to make this a group exercise, you might compare your evaluation with that of other students, in a group discussion session).

15 Giving a spoken presentation

Opportunities for practising speaking skills – these include:

- answering lecturers' questions;
- contributing in tutorials;
- talking to informal groups;
- giving your views at formal (committee) meetings;
- demonstrating or explaining to other students, e.g. during a practical class;
- asking questions in lectures/seminars;
- answering an examiner's questions in an oral exam.

Learning from experience – use your own experience of good and bad lecturers to shape your performance. Some of the more common errors include:

- speaking too quickly;
- reading from notes or from slides and ignoring the audience;
- inexpressive, impersonal or indistinct speech;
- distracting mannerisms;
- poorly structured material with little emphasis on key information;
- factual information too complex and detailed;
- too few or too many visual aids.

Testing the room – if possible, try to rehearse your talk in the room in which it will be presented. This will help you to make allowance for layout of equipment, lighting, acoustics and sight lines that might affect the way you deliver your talk. It will also put you more at ease on the day, because of the familiarity of the surroundings.

Most students feel very nervous about giving talks. This is natural, since very few people are sufficiently confident and outgoing that they look forward to speaking in public. Additionally, the technical nature of the subject matter may give you cause for concern, especially if you feel that some members of the audience have a greater knowledge than you have, e.g. your tutors. However, this is a fundamental method of scientific communication and an important transferable skill, therefore it forms an important component of many courses.

The comments in this chapter apply equally to informal talks, e.g. those based on assignments and project work, and to more formal conference presentations. It is hoped that the advice and guidance given below will encourage you to make the most of your opportunities for public speaking, but there is no substitute for practice. Do not expect to find all of the answers from this, or any other, book. Rehearse, and learn from your own experience.

 KEY POINT *The three 'Rs' of successful public speaking are: reflect – give sufficient thought to all aspects of your presentation, particularly at the planning stage; rehearse – to improve your delivery; revise – modify the content and style of your material in response to your own ideas and to the comments of others.*

Preparation

Preliminary information

Begin by marshalling the details needed to plan your presentation, including:

- the duration of the talk;
- whether time for questions is included;
- the size and location of the room;
- the projection/lighting facilities provided, and whether pointers or similar aids are available.

It is especially important to find out whether the room has the necessary equipment for digital projection (e.g. PC, projector and screen, black-out curtains or blinds, appropriate lighting) or overhead projection before you prepare your audio-visual aids. If you concentrate only on the spoken part of your presentation at this stage, you are inviting trouble later on. Have a look around the room and try out the equipment at the earliest opportunity, so that you are able to use the lights, projector, etc. with confidence. For digital projection systems, check that you can load/present your material. Box 15.1 gives practical advice on the use of PowerPoint.

Audio-visual aids

If you plan to use overhead transparencies, find out whether your department has facilities for their preparation, whether these facilities are available for your use, and the cost of materials. Adopt the following guidelines:

- Keep text to a minimum: present only the key points, with up to 20 words per slide/transparency.

Box 15.1 Tips on preparing and using PowerPoint slides in a spoken presentation

Microsoft PowerPoint can be used to produce high-quality visual aids, assuming a computer and digital projector are available in the room where you intend to speak. The presentation is produced as a series of electronic 'slides' onto which you can insert images, diagrams and text. When creating your slides, bear the following points in mind:

- **Plan the structure of your presentation.** Decide on the main topic areas and sketch out your ideas on paper. Think about what material you will need (e.g. pictures, graphs) and what colours to use for background and text.
- **Choose slide layouts according to purpose.** Once PowerPoint is running, from the *Insert* menu select *New Slide > Choose an Autolayout*. You can then add material to each new slide to suit your requirements.
- **Select your background with care.** Many of the pre-set background templates available within the *Format* menu (*Apply design template* option) are best avoided, since they are over-used and fussy, diverting attention from the content of the slides. Conversely, flat, dull backgrounds may seem uninteresting, while brightly coloured backgrounds can be garish and distracting. Choose whether to present your text as a light-coloured font on a dark background (more restful but perhaps less engaging if the room is dark) or a dark-coloured font on a light background (more lively).
- **Use visual images throughout.** Remember the familiar maxim 'a picture is worth ten thousand words'. A presentation composed entirely of text-based slides will be uninteresting: adding images and diagrams will brighten up your talk considerably (use the *Insert* menu, *Picture* option). Images can be taken with a digital camera, scanned in from a printed version or copied and pasted from the Web, but you should take care not to break copyright regulations. 'Clipart' is copyright-free, but should be used sparingly as most people will have seen the images before and they are rarely wholly relevant. Diagrams can be made from components created using the *Drawing* toolbar while graphs and tables can be imported from other programs, e.g. Excel (Box 14.1 gives further specific practical advice on adding graphics, saving files, etc.).
- **Keep text to a minimum.** Aim for no more than 20 words on a single slide (e.g. four/five lines containing a few words per line). Use headings and sub-headings to structure your talk: write only key words or phrases as 'prompts' to remind you to cover a particular point during your talk – never be tempted to type whole sentences as you will then be reduced to reading these from the screen during your presentation, which is boring.
- **Use a large, clear font style.** Use the *Slide Master* option within the *View* menu to set the default font to a non-serif style such as Arial, or Comic Sans MS, and an appropriate colour. Default fonts for headings and bullet points are intentionally large, for clarity. Do not reduce these to anything less than 28 point font size (preferably larger) to cram in more words: if you have too much material, create a new slide and divide up the information.
- **Animate your material.** The *Slide Show* menu provides a *Custom Animation* function that enables you to introduce the various elements within a slide, e.g. text can be made to *Appear* one line at a time, to prevent the audience from reading ahead and help maintain their attention.
- **Don't overdo the special effects.** PowerPoint has a wide range of features that allow complex slide transitions and animations, additional sounds, etc. but these quickly become irritating to an audience unless they have a specific purpose within your presentation.
- **Always edit your slides before use.** Check through your slides and cut out any unnecessary words, adjust the layout and animation. Remember the maxim 'less is more' – avoid too much text; too many bullet points; too many distracting visual effects or sounds.

When presenting your talk:

- **Work out the basic procedures beforehand.** Practise, to make sure that you know how to move forwards and backwards in your slideshow, turn the screen on and off, hide the mouse pointer, etc.
- **Don't forget to engage your audience.** Despite the technical gadgetry, *you* need to play an active role in the presentation, as explained elsewhere in this chapter.
- **Don't go too fast.** Sometimes, new users tend to deliver their material too quickly: try to speak at a normal pace and practise beforehand.
- **Consider whether to provide a handout.** PowerPoint has several options, including some that provide space for notes (e.g. Fig. 4.3). However, a handout should not be your default option, as there is a cost involved.

Using audio-visual aids – don't let the equipment and computer gadgetry distract you from the essential rules of good speaking (Box 15.2): remember that *you* are the presenter.

- Make sure the text is readable: try out your material beforehand.
- Use several simpler figures rather than a single complex graph.
- Avoid too much colour on overhead transparencies: blue and black are easier to read than red or green.
- Don't mix slides and transparencies as this is often distracting.
- Use spirit-based pens for transparencies: use alcohol for corrections.
- Transparencies can be produced from typewritten or printed text using a photocopier, often giving a better product than pens. Note that you must use special heat-resistant acetate sheets for photocopying.

Electronic presentation software (e.g. PowerPoint) can replace these specialist requirements, as long as the necessary facilities are available for your talk (see below).

Audience

You should consider your audience at the earliest stage, since they will determine the appropriate level for your presentation. If you are talking to fellow students you may be able to assume a common level of background knowledge. In contrast, a research lecture given to your department, or a paper at a meeting of a scientific society, will be presented to an audience from a broader range of backgrounds. An oral presentation is not the place for a complex discussion of specialised information: build up your talk from a low level. The speed at which this can be done will vary according to your audience. As long as you are not boring or patronizing, you can cover basic information without losing the attention of the more knowledgeable members in your audience.

Pitching your talk at the right level – the general rule should be: 'do not over-estimate the background knowledge of your audience'. This sometimes happens in student presentations, where fears about the presence of 'experts' can encourage the speaker to include too much detail, overloading the audience with facts.

Content

While the specific details in your talk will be for you to decide, most spoken presentations share some common features of structure, as described below.

Introductory remarks

It is vital to capture the attention of your audience at the outset. Consequently, you must make sure your opening comments are strong, otherwise your audience will lose interest before you reach the main message. Remember it takes a sentence or two for an audience to establish a relationship with a new speaker. Your opening sentence should be some form of preamble and should not contain any key information. For a formal lecture, you might begin with 'Thank you for that introduction. My talk today is about ...' then restate the title and acknowledge other contributors, etc. You might show a transparency or slide with the title printed on it, or an introductory photograph, if appropriate. This should provide the necessary settling-in period.

Getting the introduction right – a good idea is to have an initial slide giving your details and the title of your talk, and a second slide telling the audience how your presentation will be structured. Make eye contact with all sections of the audience during the introduction.

After these preliminaries, you should introduce your topic. Begin your story on a strong note – avoid timid or apologetic phrases.

Opening remarks are unlikely to occupy more than 10% of the talk. However, because of their significance, you might reasonably spend up to 25% of your preparation time on them.

What to cover in your introduction – You should:

- explain the structure of your talk;
- set out your aims and objectives;
- explain your approach to the topic.

Allowing time for slides – as a rough guide you should allow at least two minutes per illustration, although some diagrams may need longer, depending on content. Make a note of the halfway point, to help you check timing/pace.

 KEY POINT *Make sure you have practised your opening remarks, so that you can deliver the material in a flowing style, with less chance of mistakes.*

The main message

This section should include the bulk of your experimental results or literature findings, depending on the type of presentation. Keep details of methods to the minimum needed to explain your data. This is *not* the place for a detailed description of equipment and experimental protocol (unless it is a talk about methodology). Results should be presented in an easily digested format.

 KEY POINT *Do not expect your audience to cope with large amounts of data; use a maximum of six numbers per slide. Remember that graphs and diagrams are usually better than tables of raw data, since the audience will be able to see the visual trends and relationships in your data (p. 455).*

Present summary statistics (Chapter 66) rather than individual results. Show the final results of any analyses in terms of the statistics calculated, and their significance (p. 485), rather than dwelling on details of the procedures used. Figures should not be crowded with unnecessary detail. Every diagram should have a concise title and the symbols and trend lines should be clearly labelled, with an explanatory key where necessary. When presenting graphical data (Chapter 63) always 'introduce' each graph by stating the units for each axis and describing the relationship for each trend line or data set.

KEY POINT *Use summary slides at regular intervals, to maintain the flow of the presentation and to emphasise the main points.*

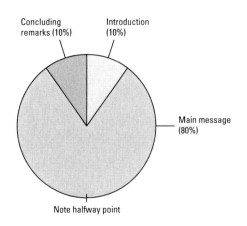

Fig. 15.1 Pie chart showing time allocation for a typical presentation.

Take the audience through your story step-by-step at a reasonable pace. Try not to rush the delivery of your main message due to nervousness. Avoid complex, convoluted story-lines – one of the most distracting things you can do is to fumble backwards through PowerPoint slides or overhead transparencies. If you need to use the same diagram or graph more than once then you should make two (or more) copies. In a presentation of experimental results, you should discuss each point as it is raised, in contrast to written text where the results and discussion may be in separate sections. The main message typically occupies approximately 80% of the time allocated to an oral presentation (Fig. 15.1).

Concluding remarks

Having captured the interest of your audience in the introduction and given them the details of your story in the middle section, you must now bring your talk to a conclusion. Do not end weakly, e.g. by running out of steam on the last slide. Provide your audience with a clear 'take-home message', by returning to the key points in your presentation. It is often appropriate to prepare a slide or overhead transparency listing your main conclusions as a numbered series.

Final remarks – make sure you give the audience sufficient time to assimilate your final slide: some of them may wish to write down the key points. Alternatively, you might provide a handout, with a brief outline of the aims of your study and the major conclusions.

Box 15.2 Hints on spoken presentations

In planning the delivery of your talk, bear the following aspects in mind:

- **Using notes.** Many accomplished speakers use abbreviated notes for guidance, rather than reading word-for-word from a prepared script. When writing your talk:
  (a) **Consider preparing your first draft as a full script**: write in spoken English and keep the text simple, to avoid a formal, impersonal style. Your aim should be to *talk* to your audience, not to *read* to them.
  (b) **If necessary, use note-cards with key words and phrases**: it is best to avoid using a full script in the final presentation. As you rehearse and your confidence improves, a set of note-cards may be an appropriate format. Mark the position of slides/key points, etc.: each note-card should contain details of structure as well as content. Your notes should be written/printed in text large enough to be read easily during the presentation (also check that the lecture room has a lectern light or you may have problems reading your notes if the lights are dimmed). Each note-card or sheet should be clearly numbered, so that you do not lose your place.
  (c) **Decide on the layout of your talk**: give each sub-division a heading in your notes, so that your audience is made aware of the structure.
  (d) **Memorise your introductory/closing remarks**: you may prefer to rely on a full written version for these sections, in case your memory fails, or if you suffer 'stage fright'.
  (e) **Using PowerPoint** (Box 15.1): here, you can either use the 'notes' option (*View> Notes Page*), or you may even prefer to dispense with notes entirely, since the slides will help structure your talk, acting as an *aide-memoire* for your material.
- **Work on your timing.** It is essential that your talk is the right length and the correct pace:
  (a) **Rehearse your presentation**: ask a friend to listen and to comment constructively on those parts that were difficult to follow, to improve your performance.
  (b) **Use 'split times' to pace yourself**: following an initial run-through, add the times at which you should arrive at the key points of your talk to your notes. These timing marks will help you keep to time during the final presentation.
  (c) **Avoid looking at your wristwatch when speaking**; this sends a negative signal to the audience. Use a wall clock (where available), or take off your watch and put it beside your notes so that you can glance at it without distracting the audience.
- **Consider your image.** Make sure that the image you project is appropriate for the occasion:
  (a) **Think about what to wear**: aim to be respectable without 'dressing up', otherwise your message may be diminished.
  (b) **Maintain a good posture**: it will help your voice projection if you stand upright, rather than slouching, or leaning over a lectern.
  (c) **Deliver your material with expression**: project your voice towards the audience at the back of the room and make sure you look round to make eye contact with all sections of the audience. Arm movements and subdued body language will help maintain the interest of your audience. However, you should avoid extreme gestures (it may work for some TV personalities but it is not recommended for the beginner).
  (d) **Try to identify and control any repetitive mannerisms**: repeated 'empty' words/phrases, fidgeting with pens, keys, etc. will distract your audience. Note-cards held in your hand give you something to focus on, while laser pointers will show up any nervous hand tremors. Practising in front of a mirror may help.
- **Think about questions.** Once again, the best approach is to prepare beforehand:
  (a) **Consider what questions are likely to come up, and prepare brief answers.** However, do not be afraid to say 'I don't know': your audience will appreciate honesty rather than vacillation if you don't have an answer for a particular question.
  (b) **If no questions are asked, you might pose a question yourself** and then ask for opinions from the audience: if you use this approach, you should be prepared to comment briefly if your audience has no suggestions, to avoid the presentation ending in an embarrassing silence.

Signal the end of your talk by saying 'finally ...', 'in conclusion ...', or a similar comment and then finish speaking after that sentence. Your audience will lose interest if you extend your closing remarks beyond this point. You may add a simple end phrase (for example, 'thank you') as you put your notes into your folder, but do not say 'that's all folks!', or make any similar offhand remark. Finish as strongly and as clearly as you started. Box 15.2 gives further advice on presentation.

Sources for further study

Alley, M. (2003) *The Craft of Scientific Presentations: Critical Steps to Succeed and Critical Errors to Avoid.* Springer-Verlag, New York.

Capp, C.C. and Capp, G.R. (1989) *Basic Oral Communication*, 5th edn. Prentice Hall, Harlow.

Matthews, C. and Marino, J. (1999) *Professional Interactions: Oral Communication Skills of Science, Technology and Medicine.* Pearson, Harlow.

Radel, J. *Oral Presentations.* Available: http://www.biology.eku.edu/RITCHISO/oralpres.html Last accessed: 01/04/07.

Study exercises

15.1 Prepare a checklist for assessing the quality of an oral presentation. After reading through this chapter, prepare a 10-point checklist of assessment criteria under the heading 'What makes a good oral presentation?'. Compare your list with the one that we have provided (p. 509) – do you agree with our criteria, or do you prefer your checklist? (Can you justify your preferences?)

15.2 Evaluate the presentation styles of other speakers. There are many opportunities to assess the streng-ths and weaknesses of academic 'public speakers', including your lecturers, seminar speakers, presenters of TV documentaries, etc. Decide in advance how you are going to tackle the evalua-tion (e.g. with a quantitative marking scheme, or a less formal procedure).

15.3 Rehearse a talk and get feedback on your performance. There are a number of approaches you might take, including: (i) recording and reviewing your presentation using a digital movie camera; or (ii) giving your talk to a small group of fellow students and asking them to provide constructive feedback.

16 General aspects of scientific writing

Written communication is an essential component of all sciences. Most courses include writing exercises in which you will learn to describe ideas and results accurately, succinctly and in an appropriate style and format. The following features/aspects are common to all forms of scientific writing.

Time management – *practical advice is given in Chapter 2.*

Organising your time

Making a timetable at the outset helps ensure that you give each stage adequate attention and complete the work on time (e.g. Fig. 16.1). To create and use a timetable:

1. Break down the task into stages.
2. Decide on the proportion of the total time each stage should take.
3. Set realistic deadlines for completing each stage, allowing some time for slippage.
4. Refer to your timetable frequently as you work: if you fail to meet one of your deadlines, make a serious effort to catch up as soon as possible.

KEY POINT *The appropriate allocation of your time to reading, planning, writing and revising will differ according to the task in hand (see Chapters 17–19).*

Monday:	morning	Lectures (University)
	afternoon	Practical (University)
	evening	Initial analysis and brainstorming (Home)
Tuesday:	morning	Lectures (University)
	afternoon	Locate sources (Library)
	evening	Background reading (Library)
Wednesday:	morning	Background reading (Library)
	afternoon	Squash (Sports hall)
	evening	Planning (Home)
Thursday:	morning	Lectures (University)
	afternoon	Additional reading (Library)
	evening	Prepare outline (Library)
Friday:	morning	Lab class (University)
	afternoon	Write first draft (Home)
	evening	Write first draft (Home)
Saturday:	morning	Shopping (Town)
	afternoon	Review first draft (Home)
	evening	Revise first draft (Home)
Sunday:	morning	Free
	afternoon	Produce final copy (Home)
	evening	Proof read and print essay (Home)
Monday:	morning	Final read-through and check Submit essay (deadline midday)

Fig. 16.1 Example timetable for writing a short essay.

Organising your information and ideas

Before you write, you need to gather and/or think about relevant material (Chapters 8 and 9). You must then decide:

- what needs to be included and what doesn't;
- in what order it should appear.

Start by jotting down headings for everything of potential relevance to the topic (this is sometimes called 'brainstorming'). A spider diagram (Fig. 16.2) or a Mind Map (Fig. 4.2) will help you organise these ideas. The next stage is to create an outline of your text (Fig. 16.3). Outlines are valuable because they:

- force you to think about and plan the structure;
- provide a checklist so nothing is missed out;
- ensure the material is balanced in content and length;
- help you organise figures and tables by showing where they will be used.

KEY POINT *A suitable structure is essential to the narrative of your writing, and should be carefully considered at the outset.*

Creating an outline – *an informal outline can be made simply by indicating the order of sections on a spider diagram (as in Fig. 16.2).*

In an essay or review, the structure of your writing should help the reader to assimilate and understand your main points. Subdivisions of the topic could simply be related to the physical nature of the subject matter (e.g. levels of organisation of a protein) and should proceed logically (e.g. primary structure then secondary, etc.). A chronological approach is good for evaluation of past work (e.g. the development of

General Aspects of Scientific Writing

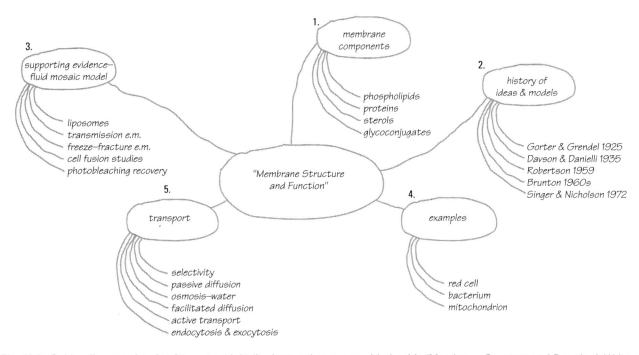

Fig. 16.2 Spider diagram showing how you might 'brainstorm' an essay with the title 'Membrane Structure and Function'. Write out the essay title in full to form the spider's body, and as you think of possible content place headings around this to form its legs. Decide which headings are relevant and which are not and use arrows to note connections between subjects. This may influence your choice of order and may help to make your writing flow because the links between paragraphs will be natural. You can make an informal outline directly on a spider diagram by adding numbers indicating a sequence of paragraphs (as shown). This method is best when you must work quickly, as with an essay written under exam conditions.

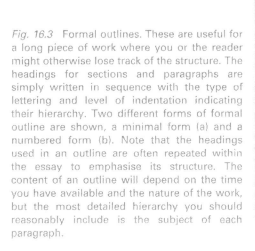

Fig. 16.3 Formal outlines. These are useful for a long piece of work where you or the reader might otherwise lose track of the structure. The headings for sections and paragraphs are simply written in sequence with the type of lettering and level of indentation indicating their hierarchy. Two different forms of formal outline are shown, a minimal form (a) and a numbered form (b). Note that the headings used in an outline are often repeated within the essay to emphasise its structure. The content of an outline will depend on the time you have available and the nature of the work, but the most detailed hierarchy you should reasonably include is the subject of each paragraph.

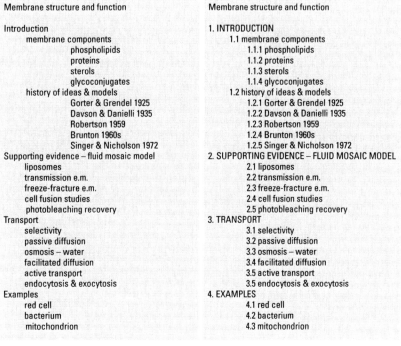

the concept of DNA as the hereditary material), whereas a step-by-step comparison might be best for certain exam questions (e.g. 'Discuss the differences between prokaryotes and eukaryotes'). There is little choice about structure for practical and project reports (see p. 111).

Writing

Adopting a scientific style

Your main aim in developing a scientific style should be to get your message across directly and unambiguously. While you can try to achieve this through a set of 'rules' (see Box 16.1), you may find other requirements driving your writing in a contradictory direction. For instance, the need to be accurate and complete may result in text littered with technical terms, and the flow may be continually interrupted by references to the literature. The need to be succinct also affects style and readability through the use of, for example, stacked noun-adjectives (e.g. 'restriction fragment length polymorphism') and acronyms (e.g. 'RFLP'). Finally, style is very much a matter of taste and each tutor, examiner, supervisor or editor will have pet loves and hates which you may have to accommodate. Different assignments will need different styles; Box 16.2 gives further details.

Developing technique

Writing is a skill that can be improved, but not instantly. You should analyse your deficiencies with the help of feedback from your tutors, be prepared to change work habits (e.g. start planning your work more carefully), and willing to learn from some of the excellent texts that are available on scientific writing (p. 106).

Improving your writing skills – you need to take a long-term view if you wish to improve this aspect of your work. An essential preliminary is to invest in and make full use of a personal reference library (see Box 16.3).

Getting started

A common problem is 'writer's block' – inactivity or stalling brought on by a variety of causes. If blocked, ask yourself these questions:

- Are you comfortable with your surroundings? Make sure you are seated comfortably at a reasonably clear desk and have minimised the possibility of interruptions and distractions.
- Are you trying to write too soon? Have you clarified your thoughts on the subject? Have you done enough preliminary reading?
- Are you happy with the underlying structure of your work? If you haven't made an outline, try this. If you are unhappy because you can't think of a particular detail at the planning stage, just start writing – it is more likely to come to you while you are thinking of something else.

Talking about your work – discussing your topic with a friend or colleague might bring out ideas or reveal deficiencies in your knowledge.

- Are you trying to be too clever? Your first sentence doesn't have to be earth-shattering in content or particularly smart in style. A short statement of fact or a definition is fine. If there will be time for revision, first get your ideas down on paper and then revise grammar, content and order later.
- Do you really need to start writing at the beginning? Try writing the opening remarks after a more straightforward part. For example, with reports of laboratory work, the Materials and Methods section may be the easiest place to start.

Writing with a word processor – use the dynamic/interactive features of the word processor (Chapter 13) to help you get started: first make notes on structure and content, then expand these to form a first draft and finally revise/improve the text.

- Are you too tired to work? Don't try to 'sweat it out' by writing for long periods at a stretch: stop frequently for a rest.

Box 16.1 How to achieve a clear, readable style

Words and phrases

- Choose short, clear words and phrases rather than long ones: e.g. use 'build' rather than 'fabricate'; 'now' rather than 'at the present time'. At certain times, technical terms must be used for precision, but don't use jargon if you don't have to.
- Don't worry too much about repeating words, especially when introducing an alternative might subtly alter your meaning.
- Where appropriate, use the first person to describe your actions ('We decided to'; 'I conclude that'), but not if this is specifically discouraged by your supervisor or department.
- Favour active forms of writing ('the observer completed the survey in ten minutes') rather than a passive style ('the survey was completed by the observer in ten minutes').
- Use tenses consistently. Past tense is always used for materials and methods ('samples were taken from...') and for reviewing past work ('Smith (1990) concluded that...'). The present tense is used when describing data ('Fig. 1 shows...'), for generalisations ('Most authorities agree that...') and conclusions ('To conclude, ...').
- Use statements in parentheses sparingly – they disrupt the reader's attention to your central theme.
- Avoid clichés and colloquialisms – they are usually inappropriate in a scientific context.

Punctuation

- Try to use a variety of types of punctuation, to make the text more interesting to read.
- Decide whether you wish to use 'closed' punctuation (frequent commas at the end of clauses) or 'open' punctuation (less frequent punctuation) and be consistent.
- Don't link two sentences with a comma. Use a full stop, this is an example of what *not* to do.
- Pay special attention to apostrophes, using the following rules:
  - To indicate possession, use an apostrophe before an 's' for a singular word (e.g. the rat's temperature was ...') and after the s for a plural word ending in s (e.g. the rats' temperatures were = the temperatures of the rats were). If the word has a special plural (e.g. woman → women) then use the apostrophe before the s (the women's temperatures were ...).
  - When contracting words, use an apostrophe (e.g. do not = don't; it's = it is), but remember that contractions are generally *not* used in formal scientific writing.
  - Do *not* use an apostrophe for 'its' as the possessive form of 'it' (e.g. 'the university and its surroundings') Note that 'it's' is reserved for 'it is'. This is an exception to the general rule and a very common mistake.
  - Never use an apostrophe to indicate plurals. Even for abbreviations, the accepted style is now to omit the apostrophe for the plural (e.g. write 'the ELISAs were').

Sentences

- Don't make them overlong or complicated.
- Introduce variety in structure and length.
- If unhappy with the structure of a sentence, try chopping it into a series of shorter sentences.

Paragraphs

- Get the paragraph length right – five sentences or so. Do *not* submit an essay that consists of a single paragraph, nor one that contains single sentence paragraphs.
- Make sure each paragraph is logical, dealing with a single topic or theme.
- Take care with the first sentence in a paragraph (the 'topic' sentence); this introduces the theme of the paragraph. Further sentences should then develop this theme, e.g. by providing supporting information, examples or contrasting cases.
- Use 'linking' words or phrases to maintain the flow of the text within a paragraph (e.g. 'for example'; 'in contrast'; 'however'; 'on the other hand').
- Make your text more readable by adopting modern layout style. The first paragraph in any section of text is usually *not* indented, but following paragraphs may be (by the equivalent of three character spaces). In addition, the space between paragraphs should be slightly larger than the space between lines. Follow departmental guidelines if these specify a format.
- Group paragraphs in sections under appropriate headings and sub-headings to reinforce the structure underlying your writing.
- Think carefully about the first and last paragraphs in any piece of writing: these are often the most important as they respectively set the aims and report the conclusions.

Note: If you're not sure what is meant by any of the terms used here, consult a guide on writing (see sources for further study).

Box 16.2 Using appropriate writing styles for different purposes (with examples)

Note that courses tend to move from assignments that are predominantly descriptive in the early years to a more analytical approach towards the final year (see Chapter 5). Also, different styles may be required in different sections of a write-up, e.g. descriptive for introductory historical aspects, becoming more analytical in later sections.

Descriptive writing

This is the most straightforward style, providing factual information on a particular subject and is most appropriate:

- in essays where you are asked to 'describe' or 'explain' (p. 108)
- when describing the results of a practical exercise, e.g.: 'The experiment shown in Figure 1 confirmed that enzyme activity was strongly influenced by temperature, as the rate observed at 37°C was more than double that seen at 20°C.'

However, in literature reviews and essays where you are asked to 'discuss' (p. 108) a particular topic, the descriptive approach is mostly inappropriate, as in the following example, where a large amount of specific information from a single scientific paper has been used, without any attempt to highlight the most important points:

'In a study carried out between July and October 2002, a total of 225 sputum samples from patients attending 25 different clinics in England and Wales were screened. Bacteria were isolated from 67.6% of these samples, with 47.42% of the samples giving *Pseudomonas aeruginosa*, 11.76% *Burkholderia cepacia* and 8.59% *Stentrophomonas maltophilia* (Grey and Gray, 2003).'

In the most extreme examples, whole paragraphs or pages of essays may be based on descriptive factual detail from a single source, often with a single citation at the end of the material, as above: such essays often score low marks in essays where evidence of deeper thinking is required (Chapter 5).

Comparative writing

This technique is an important component of academic writing, and it will be important to develop your comparative writing skills as you progress through your course. Its applications include:

- answering essay questions and assignments of the 'compare and contrast' type (p. 108)
- comparing your results with previously published work in the Discussion section of a practical report.

To use this style, first decide on those aspects you wish to compare and then consider the material (e.g. different literature sources) from these aspects – in what ways do they agree or disagree with each other? One approach is to compare/contrast a different aspect in each paragraph. At a practical level, you can use 'linking' words and phrases to help orientate your reader, as you move between aspects where there is agreement and disagreement. These include, for agreement: 'in both cases'; 'in agreement with'; 'is also shown by the study of'; 'similarly'; 'in the same way', and for disagreement: 'however'; 'although'; 'in contrast to'; 'on the other hand'; 'which differs from'. The comparative style is fairly straightforward, once you have decided on the aspects to be compared. The following brief example compares two different studies using this style:

'While Grey and Gray (2003) reported that *Pseudomonas aeruginosa* was present in 47.4% of 225 UK sputum samples, Black and White (2006) showed that 89.1% of sputum samples from 2592 patients were positive for this bacterium.'

Comparative text typically makes use of two or more references per paragraph.

Analytical writing

Typically, this is the most appropriate form of writing for:

- a review of scientific literature on a particular topic;
- an essay where you are asked to 'discuss' (p. 108) different aspects of a particular topic;
- evaluating a number of different published sources within the Discussion section of a final-year project dissertation.

By considering the significance of the information provided in the various sources you have read, you will be able to take a more critical approach. Your writing should evaluate the importance of the material in the context of your topic (see also Chapter 9). In analytical writing, you need to demonstrate critical thinking (p. 55) and personal input about the topic in a well-structured text that provides clear messages, presented in a logical order and demonstrating synthesis from a number of sources by appropriate use of citations (p. 47). Detailed information and relevant examples are used only to explain or develop a particular aspect, and not simply as 'padding' to bulk up the essay, as in the following example:

'*Pseudomonas aeruginosa* is often isolated from sputum samples of cystic fibrosis patients: a short-term UK study with a relatively small sample size (225 patients) isolated this bacterium from around half of all samples (Grey and Gray, 2003), while a longer-term study with a far larger sample size (2592 patients) gave an isolation rate of almost 90% (Black and White, 2006).'

Analytical writing is based on a broad range of sources, typically with several citations per paragraph.

Box 16.3 How to improve your writing ability by consulting a personal reference library

Using dictionaries

We all know that a dictionary helps with spelling and definitions, but how many of us use one effectively? You should:

- Keep a dictionary beside you when writing and always use it if in any doubt about spelling or definitions.
- Use it to prepare a list of words which you have difficulty in spelling: apart from speeding up the checking process, the act of writing out the words helps commit them to memory.
- Use it to write out a personal glossary of terms. This can help you memorise definitions. From time to time, test yourself.

Not all dictionaries are the same! Ask your tutor or supervisor whether he/she has a preference and why. Try out the *Oxford Advanced Learner's Dictionary*, which is particularly useful because it gives examples of use of all words and helps with grammar, e.g. by indicating which prepositions to use with verbs. Dictionaries of biology tend to be variable in quality, possibly because the subject is so wide and new terms are continually being coined. *Henderson's Dictionary of Biological Terms* is a useful example.

Using a thesaurus

A thesaurus contains lists of words of similar meaning grouped thematically; words of opposite meaning always appear nearby.

- Use a thesaurus to find a more precise and appropriate word to fit your meaning, but check definitions of unfamiliar words with a dictionary.
- Use it to find a word or phrase 'on the tip of your tongue' by looking up a word of similar meaning.
- Use it to increase your vocabulary.

Roget's Thesaurus is the standard. Collins also publishes a combined dictionary and thesaurus.

Using guides for written English

These provide help with the use of words.

- Use guides to solve grammatical problems such as when to use 'shall' or 'will', 'which' or 'that', 'effect' or 'affect', 'can' or 'may', etc.
- Use them for help with the paragraph concept and the correct use of punctuation.
- Use them to learn how to structure writing for different tasks.

Revising your text – to improve clarity and shorten your text, 'distil' each sentence by taking away unnecessary words and 'condense' words or phrases by choosing a shorter alternative.

Learning from others – ask a colleague to read through your draft and comment on its content and overall structure.

Revising your text

Wholesale revision of your first draft is strongly advised for all writing, apart from in exams. When using a word processor, this can be a simple process. Where possible, schedule your writing so you can leave the first draft to 'settle' for at least a couple of days. When you return to it fresh, you will see more easily where improvements can be made. Try the following structured revision process, each stage being covered in a separate scan of your text:

1. Examine content. Have you included everything you need to? Is all the material relevant?
2. Check the grammar and spelling. Can you spot any 'howlers'?
3. Focus on clarity. Is the text clear and unambiguous? Does each sentence really say what you want it to say?
4. Be succinct. What could be missed out without spoiling the essence of your work? It might help to imagine an editor has set you the target of reducing the text by 15%.
5. Improve style. Could the text read better? Consider the sentence and paragraph structure and the way your text develops to its conclusion.

Common errors

These include:

- Problems over singular and plural words ("a bacteria is"; "the results shows")
- Verbose text ("One definition that can be employed in this situation is given in the following sentence.")
- Misconstructed sentences ("Health and safety regulations should be made aware of . . .")
- Misuse of punctuation, especially commas and apostrophes (for examples, see Box 16.1).
- Poorly-constructed paragraphs (for advice, see Box 16.1).

Sources for further study

Burchfield, R.W. (ed.) (2004) *Fowler's Modern English Usage*, revised 3rd edn. Oxford University Press, New York.

Clark, R. *The English Style Book. A Guide to the Writing of Scholarly English.* Available: http://www.litency.com/stylebook/stylebook.php Last accessed: 01/04/07.

Kane, T.S. (1994) *The New Oxford Guide to Writing.* Oxford University Press, New York.
[This is excellent for the basics of English – it covers grammar, usage and the construction of sentences and paragraphs]

Lindsay, D. (1995) *A Guide to Scientific Writing*, 2nd edn. Longman, Harlow.

McMillan, K.M. and Weyers, J.D.B. (2006) *The Smarter Student: Skills and Strategies for Success at University.* Pearson, Harlow.

Partridge, E. (1978) *You Have a Point There.* Routledge, London.
[This covers punctuation in a very readable manner]

Pechnick, J.A. and Lamb, B.C. (1994) *How to Write about Biology.* Harper Collins, London.

Tichy, H.J. (1988) *Effective Writing for Engineers, Managers and Scientists*, 2nd edn. Wiley, New York.
[This is strong on scientific style and clarity in writing]

Study exercises

16.1 'Brainstorm' an essay title. Pair up with a partner in your class. Together, pick a suitable essay title from a past exam paper. Using the spider diagram or another technique, individually 'brainstorm' the title. Meet afterwards, compare your ideas, and discuss their relative merits and disadvantages.

16.2 Improve your writing technique. From the following checklist, identify the *three* weakest aspects of your writing, either in your own opinion or from essay/assignment feedback:

- grammar;
- paragraph organisation;
- presentation of work;
- punctuation;
- scientific style;
- sentence structure/variety;
- spelling;
- structure and flow;
- vocabulary.

Now either borrow a book from a library or buy a book that deals with your weakest aspects of writing. Read the relevant chapters or sections and for each aspect write down some tips that should help you in future.

16.3 Improve your spelling and vocabulary with two lists. Create a pair of lists and pin these up beside your desk. One should be entitled *Spelling Mistakes* and the other *New Words*. Now, whenever you make a mistake in spelling or have to look up how to spell a word in a dictionary, add the problem word to your spelling list, showing where you made the mistake. Also, whenever you come across a word whose meaning is unclear to you, look it up in a dictionary and write the word and its meaning in the 'new words' vocabulary list.

17 Writing essays

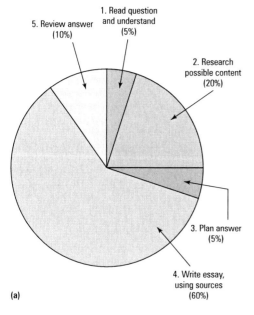

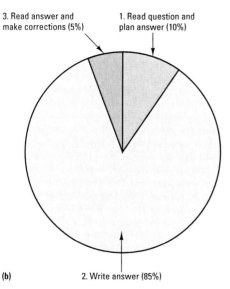

Fig. 17.1 Typical division of time for an essay written as part of an in-course assessment (a) or under exam conditions (b).

Considering essay content – *it is rarely enough simply to lay down facts for the reader – you must analyse them and comment on their significance (see p. 104).*

The function of an essay is to show how much you understand about a topic and how well you can organise and express your knowledge.

Organising your time

The way you should divide your time when producing an essay depends on whether you are writing it for in-course assessment or under exam conditions (Fig. 17.1). Essays written over a long period with access to books and other resources will probably involve a research element, not only before the planning phase but also when writing (Fig. 17.1a). For exams, it is assumed that you have revised appropriately (Chapter 5) and essentially have all the information at your fingertips. To keep things uncomplicated, the time allocated for each essay should be divided into three components – planning, writing and reviewing (Fig. 17.1b), and you should adopt time-saving techniques whenever possible (Box 6.2).

Making a plan for your essay

Dissect the meaning of the essay question or title

Read the title very carefully and think about the topic before starting to write. Consider the definitions of each of the important nouns (this can help in approaching the introductory section). Also think about the meaning of the verb(s) used and try to follow each instruction precisely (see Table 17.1). Don't get side-tracked because you know something about one word or phrase in the title: consider the whole title and all its ramifications. If there are two or more parts to the question, make sure you give adequate attention to each part.

Consider possible content and examples

Research content using the methods described in Chapters 8 and 9. If you have time to read several sources, consider their content in relation to the essay title. Can you spot different approaches to the same subject? Which do you prefer as a means of treating the topic in relation to your title? Which examples are most relevant to your case, and why?

Construct an outline

Every essay should have a structure related to its title.

 KEY POINT Most marks for essays are lost because the written material is badly organised or is irrelevant. An essay plan, by definition, creates order and, if thought about carefully, should ensure relevance.

Your plan should be written down (but scored through later if written in an exam book). Think about an essay's content in three parts:

1. The introductory section, in which you should include definitions and some background information on the context of the topic being considered. You should also tell your reader how you plan to approach the subject.
2. The middle of the essay, where you develop your answer and provide relevant examples. Decide whether a broad analytical approach is appropriate or whether the essay should contain more factual detail.

Ten Golden Rules for essay writing – these are framed for in-course assessments (p. 29), though many are also relevant to exams (see also Box 6.2).

1. Read the question carefully, and decide exactly what the assessor wants you to achieve in your answer.
2. Make sure you understand the question by considering all aspects – discuss your approach with colleagues or a tutor.
3. Carry out the necessary research (using books, journals, the Web), taking appropriate notes. Gain an overview of the topic before getting involved with the details.
4. Always plan your work in outline before you start writing. Check that your plan covers the main points and that it flows logically.
5. Introduce your essay by showing that you understand the topic and stating how you intend to approach it.
6. As you write the main content, ensure it is relevant by continually looking back at the question.
7. Use headings and sub-headings to organise and structure your essay.
8. Support your statements with relevant examples, diagrams and references where appropriate.
9. Conclude by summarising the key points of the topic, indicating the present state of knowledge, what we still need to find out and how this might be achieved.
10. Always review your essay before submitting it. Check grammar and spelling and confirm that you have answered *all* aspects of the question.

Table 17.1 Instructions often used in essay questions and their meanings. When more than one instruction is given (e.g. compare and contrast; describe and explain), make sure you carry out *both* or you may lose a large proportion of the available marks (see also Table 5.1)

Account for:	give the reasons for
Analyse:	examine in depth and describe the main characteristics of
Assess:	weigh up the elements of and arrive at a conclusion about
Comment:	give an opinion on and provide evidence for your views
Compare:	bring out the similarities between
Contrast:	bring out dissimilarities between
Criticise:	judge the worth of (give both positive and negative aspects)
Define:	explain the exact meaning of
Describe:	use words and diagrams to illustrate
Discuss:	provide evidence or opinions about, arriving at a balanced conclusion
Enumerate:	list in outline form
Evaluate:	weigh up or appraise; find a numerical value for
Explain:	make the meaning of something clear
Illustrate:	use diagrams or examples to make clear
Interpret:	express in simple terms, providing a judgement
Justify:	show that an idea or statement is correct
List:	provide an itemised series of statements about
Outline:	describe the essential parts only, stressing the classification
Prove:	establish the truth of
Relate:	show the connection between
Review:	examine critically, perhaps concentrating on the stages in the development of an idea or method
State:	express clearly
Summarise:	without illustrations, provide a brief account of
Trace:	describe a sequence of events from a defined point of origin

3. The conclusion, which you can make quite short. You should use this part to summarise and draw together the components of the essay, without merely repeating previous phrases. You might mention such things as: the broader significance of the topic; its future; its relevance to other important areas of biology. Always try to mention both sides of any debate you have touched on, but beware of 'sitting on the fence'.

KEY POINT *Use paragraphs to make the essay's structure obvious. Emphasise them with headings and sub-headings unless the material beneath the headings would be too short or trivial.*

Start writing

- Never lose track of the importance of content and its relevance. Repeatedly ask yourself: 'Am I really answering this question?' Never waffle just to increase the length of an essay. Quality, rather than quantity, is important.
- Illustrate your answer appropriately. Use examples to make your points clear, but remember that too many similar examples can stifle the flow of an essay. Use diagrams where a written description would be difficult or take too long. Use tables to condense information.

Using diagrams – give a title and legend to each diagram so that it makes sense in isolation and point out in the text when the reader should consult it (e.g. 'as shown in Fig. 1 ...' or 'as can be seen in the accompanying diagram, ...').

- Take care with your handwriting. You can't get marks if your writing is illegible. Try to cultivate an open form of handwriting, making the individual letters large and distinct. If there is time, make out a rough draft from which a tidy version can be copied.

Reviewing your answer

Learning from lecturers' and tutors' comments – ask for further explanations if you don't understand a comment or why an essay was less successful than you thought it should have been.

Make sure that you leave enough time to:
- re-read the question to check that you have answered all points;
- re-read your essay to check for errors in punctuation, spelling and content. Make any corrections obvious. In an exam, don't panic if you suddenly realise you've missed a large chunk out as the reader can be redirected to a supplementary paragraph if necessary.

Sources for further study

Anon. *Yahoo! Directory: Writing > Essays and Research Papers*. Available: http://dir.yahoo.com/Social_Science/Communications/Writing/Essays_and_Research_Papers
Last accessed: 01/04/07.
[An extensive directory of web resources]

Anon. (2004) *Essay and Report Writing Skills*. Open University, Milton Keynes.

Good, S. and Jensen, B. (1995) *The Student's Only Survival Guide to Essay Writing*. Orca Book Publishers, Victoria, BC.

Study exercises

17.1 Practise dissecting essay titles. Use past exam papers, or make up questions based on learning objectives for your course and your lecture notes. Take each essay title and carefully 'dissect' the wording, working out exactly what you think the assessor expects you to do (see e.g. Table 17.1).

17.2 Write essay plans under self-imposed time limits. Continuing from study exercise 17.1, outline plans for essays from a past exam paper. Allow yourself a maximum of 5 minutes per outline. Within this time your main goal is to create an essay plan. To do this, you may need to 'brainstorm' the topic. Alternatively, if you allocate 10 minutes per essay, you may be able to provide more details, e.g. list the examples you could describe.

17.3 Practise reviewing your work carefully. For the next assignment you write, review it fully as part of the writing process. This will require you to finish the first draft about one week before the hand-in date, e.g. by setting yourself an earlier deadline than the submission date. This exercise is best done with a word-processed essay. Don't worry if it is a little over the word limit at this stage.

(a) Print out a copy of the essay. Do not look at it for at least two days after finishing this version.
(b) Review 1: spelling, grammar and sense. Read through the draft critically (try to imagine it had been written by someone else) and correct any obvious errors that strike you. Does the text make sense? Do sentences/paragraphs flow smoothly?
(c) Review 2: structure and relevance. Consider again the structure of the essay, asking yourself whether you have really answered the question that was set (see study exercise 17.1). Are all the parts in the right order? Is anything missed out? Have you followed precisely the instruction(s) in the title? Are the different parts of the essay linked together well?
(d) Review 3: shorten and improve style. Check the word count. Shorten the essay if required. Look critically at phrasing and, even if the essay is within the word limit, ask yourself whether any of the words are unnecessary or whether the text could be made more concise, more precise or more apt.

18 Reporting practical and project work

Typical structure of scientific reports – this usually follows the 'IMRaD' acronym: **I**ntroduction, **M**aterials and Methods, **R**esults **a**nd **D**iscussion.

Practical reports, project reports, theses and scientific papers differ greatly in depth, scope and size, but they all have the same basic structure. Some variation is permitted, however (see Box 18.1), and you should always follow the advice or rules provided by your department.

Additional parts may be specified: for project reports, dissertations and theses, a Title page is often required and a List of Figures and Tables as part of the Contents section. When work is submitted for certain degrees, you may need to include certain declarations and statements made by the student and supervisor. In scientific papers, a list of Key Words is often added following the Abstract: this information may be combined with words in the title for computer cross-referencing systems.

KEY POINT Department, school or faculty regulations may specify a precise format for producing your report or thesis. Obtain a copy of these rules at an early stage and follow them closely, to avoid losing marks.

Practical and project reports

These are exercises designed to make you think more deeply about your experiments and to practise and test the skills necessary for writing up research work. Special features are:

- Introductory material is generally short and, unless otherwise specified, should outline the aims of the experiment(s) with a minimum of background material.
- Materials and methods may be provided by your supervisor for practical reports. If you make changes to this, you should state clearly what you did. With project work, your lab notebook (see p. 176) should provide the basis for writing this section.
- Great attention in assessment will be paid to presentation and analysis of data. Take special care over graphs (see p. 464 for further advice). Make sure your conclusions are justified by the evidence you present.

Options for discussing data – the main optional variants of the general structure include combining Results and Discussion into a single section and adding a separate Conclusions section.

- The main advantage of a joint Results and Discussion section is that you can link together different experiments, perhaps explaining why a particular result led to a new hypothesis and the next experiment. However, a combined Results and Discussion section may contravene your department's regulations, so you should check before using this approach.
- The main advantage of having a separate Conclusions section is to draw together and emphasise the chief points arising from your work, when these may have been 'buried' in an extensive Discussion section.

Theses and dissertations

These are submitted as part of the examination for a degree following an extended period of research. They act to place on record full details about your experimental work and will normally only be read by those with a direct interest in it – your examiners or colleagues. Note the following:

- You are allowed scope to expand on your findings and to include detail that might otherwise be omitted in a scientific paper.
- You may have problems with the volume of information that has to be organised. One method of coping with this is to divide your thesis into chapters, each having the standard format (as in Box 18.1). A General Introduction can be given at the start and a General Discussion at the end. Discuss this with your supervisor.

Oral assessments – there may be an oral exam (viva voce) associated with the submission of a thesis or dissertation. The primary aim of the examiners will be to ensure that you understand what you did and why you did it.

Reporting Practical and Project Work

Box 18.1 The structure of reports of experimental work

Undergraduate practical and project reports are generally modelled on this arrangement or a close variant of it, because this is the structure used for nearly all research papers and theses. The more common variations include Results and Discussion combined into a single section and Conclusions appearing separately as a series of points arising from the work. In scientific papers, a list of Key Words (for computer cross-referencing systems) may be included following the Abstract. Acknowledgements may appear after the Contents section, rather than near the end. Department or faculty regulations for producing theses and reports may specify a precise format; they often require a Title page to be inserted at the start and a List of Figures and Tables as part of the Contents section, and may specify declarations and statements to be made by the student and supervisor.

Part (in order)	Contents/purpose	Checklist for reviewing content
Title	Explains what the project was about	Does it explain what the text is about succinctly?
Authors plus their institutions	Explains who did the work and where; also where they can be contacted now	Are all the details correct?
Abstract/Summary	Synopsis of methods, results and conclusion of work described. Allows the reader to grasp quickly the essence of the work	Does it explain why the work was done? Does it outline the whole of your work and your findings?
List of Contents	Shows the organisation of the text (not required for short papers)	Are all the sections covered? Are the page numbers correct?
Abbreviations	Lists all the abbreviations used (but not those of SI, chemical elements or standard biochemical terms)	Have they all been explained? Are they all in the accepted form? Are they in alphabetical order?
Introduction	Orientates the reader, explains why the work has been done and its context in the literature, why the methods used were chosen, why the experimental organisms were chosen. Indicates the central hypothesis behind the experiments	Does it provide enough background information and cite all the relevant references? Is it of the correct depth for the readership? Have all the technical terms been defined? Have you explained why you investigated the problem? Have you outlined your aims and objectives? Have you explained your methodological approach? Have you stated your hypothesis?
Materials and Methods	Explains how the work was done. Should contain sufficient detail to allow another competent worker to repeat the work	Is each experiment covered and have you avoided unnecessary duplication? Is there sufficient detail to allow repetition of the work? Are proper scientific names and authorities given for all organisms? Have you explained where you got them from? Are the correct names, sources and grades given for all chemicals?
Results	Displays and describes the data obtained. Should be presented in a form which is easily assimilated (graphs rather than tables, small tables rather than large ones)	Is the sequence of experiments logical? Are the parts adequately linked? Are the data presented in the clearest possible way? Have SI units been used properly throughout? Has adequate statistical analysis been carried out? Is all the material relevant? Are the figures and tables all numbered in the order of their appearance? Are their titles appropriate? Do the figure and table legends provide all the information necessary to interpret the data without reference to the text? Have you presented the same data more than once?
Discussion/ Conclusions	Discusses the results: their meaning, their importance; compares the results with those of others; suggests what to do next	Have you explained the significance of the results? Have you compared your data with other published work? Are your conclusions justified by the data presented?
Acknowledgements	Gives credit to those who helped carry out the work	Have you listed everyone that helped, including any grant-awarding bodies?
Literature Cited (Bibliography)	Lists all references cited in appropriate format: provides enough information to allow the reader to find the reference in a library	Do all the references in the text appear on the list? Do all the listed references appear in the text? Do the years of publications and authors match? Are the journal details complete and in the correct format? Is the list in alphabetical order, or correct numerical order?

Steps in the production of a practical report or thesis

Choose the experiments you wish to describe and decide how best to present them

Try to start this process before your lab work ends, because at the stage of reviewing your experiments, a gap may become apparent (e.g. a missing control) and you might still have time to rectify the deficiency. Irrelevant material should be ruthlessly eliminated, at the same time bearing in mind that negative results can be extremely important (see p. 188). Use as many different forms of data presentation as are appropriate, but avoid presenting the same data in more than one form. Relegate large tables of primary data to an appendix and summarise the important points within the main text (with a cross-reference to the appendix). Make sure that the experiments you describe are representative: always state the number of times they were repeated and how consistent your findings were.

Choosing between graphs and tables – graphs are generally easier for the reader to assimilate, while tables can be used to condense a lot of data into a small space.

Make up plans or outlines for the component parts

The overall structure of practical and project reports is well defined (see Box 18.1), but individual parts will need to be organised as with any other form of writing (see Chapter 16).

Repeating your experiments – remember, if you do an experiment twice, you have repeated it only once.

Write

The Materials and Methods section is often the easiest to write once you have decided what to report. Remember to use the past tense and do not allow results or discussion to creep in. The Results section is the next easiest as it should only involve description. At this stage, you may benefit from jotting down ideas for the Discussion – this may be the hardest part to compose as you need an overview both of your own work and of the relevant literature. It is also liable to become wordy, so try hard to make it succinct. The Introduction shouldn't be too difficult if you have fully understood the aims of the experiments. Write the Abstract and complete the list of references at the end. To assist with the latter, it is a good idea as you write to jot down the references you use or to pull out their cards from your index system.

Presenting your results – remember that the order of results presented in a report need not correspond with the order in which you carried out the experiments: you are expected to rearrange them to provide a logical sequence of findings.

Revise the text

Once your first draft is complete, try to answer all the questions given in Box 18.1. Show your work to your supervisors and learn from their comments. Let a friend or colleague who is unfamiliar with your subject read your text; they may be able to pinpoint obscure wording and show where information or explanation is missing. If writing a thesis, double-check that you are adhering to your institution's thesis regulations.

Using the correct tense – always use the past tense to describe the methodology used in your work, since it is now complete. Use the present tense only for generalisations and conclusions.

Prepare the final version

Markers appreciate neatly produced work but a well-presented document will not disguise poor science! If using a word processor, print the final version with the best printer available. Make sure figures are clear and in the correct size and format.

Submit your work

Your department will specify when to submit a thesis or project report, so plan your work carefully to meet this deadline or you may lose marks. Tell your supervisor early of any circumstances that may cause delay. And check to see whether any forms are required for late submission, or evidence of extenuating circumstances.

Definition

Peer review – the process of evaluation and review of a colleague's work. In scientific communication, a paper is reviewed by two or more expert reviewers for comments on quality and significance as a key component of the validation procedure.

Producing a scientific paper

Scientific papers are the means by which research findings are communicated to others. Peer-reviewed papers are published in journals; each covers a well-defined subject area and publishes details of the format they expect.

KEY POINT Peer review is an important component of the process of scientific publication; only those papers whose worth is confirmed by the peer-review process will be published.

It would be very unusual for an undergraduate to submit a paper on his or her own – this would normally be done in collaboration with your project supervisor, and only then if your research has satisfied appropriate criteria. However, it is important to understand the process whereby a paper comes into being (Box 18.2), as this can help you understand and interpret the primary literature.

Box 18.2 Steps in producing a scientific paper

Scientific papers are the lifeblood of any science and it is a major landmark in your scientific career to publish your first paper. The main steps in doing this should include the following:

Assessing potential content

The work must be of an appropriate standard to be published and should be 'new, true and meaningful'. Therefore, before starting, the authors need to review their work critically under these headings. The material included in a scientific paper will generally be a subset of the total work done during a project, so it must be carefully selected for relevance to a clear central hypothesis – if the authors won't prune, the referees and editors of the journal certainly will.

Choosing a journal

There are thousands of journals covering biology and each covers a specific area (which may change through time). The main factors in deciding on an appropriate journal are the range of subjects it covers, the quality of its content and the number and geographical distribution of its readers. The choice of journal always dictates the format of a paper since authors must follow to the letter the journal's 'Instructions to Authors'.

Deciding on authorship

In multi-author papers, a contentious issue is often who should appear as an author and in what order they should be cited. Where authors make an equal contribution, an alphabetical order of names may be used. Otherwise, each author should have made a substantial contribution to the paper and should be prepared to defend it in public. Ideally, the order of appearance will reflect the amount of work done rather than seniority. This may not always happen in practice!

Writing

The paper's format will be similar to that shown in Box 18.1 and the process of writing will include outlining, reviewing, etc., as discussed elsewhere in this chapter. Figures must be finished to an appropriate standard and this may involve preparing photographs or digital images of them.

Submitting

When completed, copies of the paper are submitted to the editor of the chosen journal with a simple covering letter. A delay of one to two months usually follows while the manuscript is sent to two or more anonymous referees who will be asked by the editor to check that the paper is novel, scientifically correct and that its length is justified.

Responding to referees' comments

The editor will send on the referees' comments to the authors, who will then have a chance to respond. The editor will decide on the basis of the comments and replies to them whether the paper should be published. Sometimes quite heated correspondence can result if the authors and referees disagree.

Checking proofs and waiting for publication

If a paper is accepted, it will be sent off to the typesetters. The next the authors see of it is the proofs (first printed version in style of journal), which have to be corrected carefully for errors and returned. Eventually, the paper will appear in print, but a delay of six months following acceptance is not unusual. Nowadays, papers are often available electronically, *via* the Web, in PDF format - see p. 49 for advice on how to cite 'online early' papers using the DOI system.

Sources for further study

Berry, R. (2004) *The Research Project: How to Write it*, 5th edn. Routledge, London.

Davis, M. (2005) *Scientific Papers and Presentations*. Academic Press, London.

Day, R.A. and Gastel, B. (2006) *How to Write and Publish a Scientific Paper*, 6th edn. Cambridge University Press, Cambridge.

Lobban, C.S. and Schefter, M. (1992) *Successful Lab Reports: A Manual for Science Students*. Cambridge University Press, Cambridge.

Luck, M. (1999) *Your Student Research Project*. Gower, London.

Luey, B. (2002) *Handbook for Academic Authors*, 4th edn. Cambridge University Press, Cambridge.

Matthews, J.R., Bowen, J.M. and Matthews, R. (2000) *Successful Science Writing: a Step-by-Step Guide for the Biological and Medical Sciences*, 2nd edn. Cambridge University Press, Cambridge

Valiela, I. (2001) *Doing Science: Design, Analysis and Communication of Scientific Research*. Oxford University Press, Oxford.
[Covers scientific communication, graphical presentations and aspects of statistics]

Study exercises

18.1 Write a formal 'Materials and Methods' section. Adopting the style of a research paper (i.e. past tense, all relevant detail reported such that a competent colleague could repeat your work), write out the Materials and Methods for a practical you have recently carried out. Ask a colleague or tutor to comment on what you have written.

18.2 Describe a set of results in words. Again adopting the style of a research paper, write a paragraph describing the results contained in a particular table or graph. Ask a colleague or tutor to comment on your description, to identify what is missing or unclear.

18.3 Write an abstract for a paper. Pair up with a colleague. Each of you should independently choose a different research paper in a current journal. Copy the paper, but mask over the abstract section, having first counted the words used. Swap papers. Now, working to the same number of words as in the original, read the paper and provide an abstract of its contents. Then compare this with the real abstract. Compare your abstracts.

19 Writing literature surveys and reviews

The literature survey or review is a specialised form of essay which summarises and reviews the evidence and concepts concerning a particular area of research.

> **KEY POINT** *A literature review should not be a simple recitation of facts. The best reviews are those which analyse information rather than simply describe it.*

Making up a timetable

Figure 19.1 illustrates how you might divide up your time for writing a literature survey. There are many subdivisions in this chart because of the size of the task: in general, for lengthy tasks, it is best to divide up the work into manageable chunks. Note also that proportionately less time is allocated to writing itself than with an essay. In a literature survey, make sure that you spend adequate time on research and revision.

Selecting a topic

You may have no choice in the topic to be covered, but if you do, carry out your selection as a three-stage process:

1. Identify a broad subject area that interests you.
2. Find and read relevant literature in that area. Try to gain a broad impression of the field from books and general review articles. Discuss your ideas with your supervisor.
3. Select a relevant and concise title. The wording should be considered very carefully as it will define the content expected by the reader. A narrow subject area will cut down on the amount of literature you will be expected to review, but will also restrict the scope of the conclusions you can make (and *vice versa* for a wide subject area).

Scanning the literature and organising your references

You will need to carry out a thorough investigation of the literature before you start to write. The key problems are as follows:

- Getting an initial toehold in the literature. Seek help from your supervisor, who may be willing to supply a few key papers to get you started. Hints on expanding your collection of references are given on p. 45.
- Assessing the relevance and value of each article. This is the essence of writing a review, but it is difficult unless you already have a good understanding of the field. Try reading earlier reviews in your area and discussing the topic with your supervisor or other academic staff.
- Clarifying your thoughts. Subdividing the main topic and assigning your references to these smaller subject areas may help you gain a better overview of the literature.

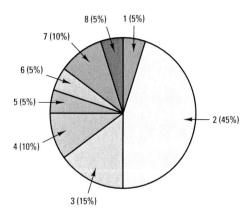

Fig. 19.1 Pie chart showing how you might allocate time for a literature survey:
1. select a topic;
2. scan the literature;
3. plan the review;
4. write first draft;
5. leave to settle;
6. prepare a structured review of text;
7. write final draft;
8. produce top copy.

Creating a glossary – one barrier to developing an understanding of a new topic is the jargon used. To overcome this, create your own glossary. You may wish to cross-reference a range of sources to ensure the definitions are reliable and context-specific. Remember to note your sources in case you wish to use the definition within your review.

Using index cards – these can help you organise large numbers of references. Write key points and author information on each card – this helps when considering where the reference fits into the literature. Arrange the cards in subject piles, eliminating irrelevant ones. Order the cards in the sequence you wish to write in.

Defining terms – the introduction is a good place to explain the meaning of the key terms used in your survey or review.

Balancing opposing views – even if you favour one side of a disagreement in the literature, your review should provide a balanced and fair description of all the published views of the topic. Having done this, if you do wish to state a preference, give reasons for your opinion.

Making citations – a review of literature poses stylistic problems because of the need to cite large numbers of papers; in the Annual Review series this is overcome by using numbered references (see p. 48).

Deciding on structure and content

The general structure and content of a literature survey are described below.

The *Annual Review* series (available in most university libraries) provides good examples of appropriate style for reviews of the biosciences.

Introduction

The introduction should give the general background to the research area, concentrating on its development and importance. You should also make a statement about the scope of your survey; as well as defining the subject matter to be discussed, you may wish to restrict the period being considered.

Main body of text

The review itself should discuss the published work in the selected field and may be subdivided into appropriate sections. Within each portion of a review, the approach is usually chronological, with appropriate linking phrases (e.g. 'Following on from this, ...'; 'Meanwhile, Bloggs (2002) tackled the problem from a different angle ...'). However, a good review is much more than a chronological list of work done. It should:

- allow the reader to obtain an overall view of the current state of the research area, identifying the key areas where knowledge is advancing;
- show how techniques are developing and discuss the benefits and disadvantages of using particular organisms or experimental systems;
- assess the relative worth of different types of evidence – this is the most important aspect (see Chapter 9). Do not be intimidated from taking a critical approach as the conclusions you may read in the primary literature aren't always correct;
- indicate where there is conflict in findings or theories, suggesting if possible which side has the stronger case;
- indicate gaps in current knowledge.

You do not need to wait until you have read all the sources available to you before starting to write the main body. Word processors allow you to modify and move pieces of text at any point and it will be useful to write paragraphs about key sources, or groups of related papers, as you read them. Try to create a general plan for your review as soon as possible. Place your draft sections of text under an appropriate set of subheadings that reflects your plan, but be prepared to rearrange these and re-title or re-order sections as you proceed. Not only will working in this way help to clarify your thoughts, but it may help you avoid a last-minute rush of writing near to the submission date.

Conclusions

The conclusions should draw together the threads of the preceding parts and point the way forward, perhaps listing areas of ignorance or where the application of new techniques may lead to advances.

References, etc.

The References or Literature Cited section should provide full details of all papers referred to in the text (see p. 48). The regulations for your department may also specify a format and position for the title page, list of contents, acknowledgements, etc.

Source for further study

Rudner, L.M. and Schafer, W.D. (1999) How to write a scholarly research report. *Practical Assessment, Research & Evaluation*, **6** (13).
Available: http://pareonline.net/getvn.asp?v=6&n=13
Last accessed: 01/04/07.

Study exercises

19.1 Summarise the main differences between a review and a scientific paper. From the many subject areas in the *Annual Review* series (find *via* your library's periodical indexing system), pick one that matches your subject interests, and within this find a review that seems relevant or interesting. Read the review and write down *five* ways in which the writing style and content differ from those seen in primary scientific papers.

19.2 Gather a collection of primary sources for a topic. From the journal section of the library, select an interesting scientific paper published about 5–10 years ago. First, work *back* from the references cited by that paper: can you identify from the text or the article titles which are the most important and relevant to the topic? List *five* of these, using the proper conventions for citing articles in a reference list (see Chapter 8). Note that each of these papers will also cite other articles, always going back in time. Now using the Science Citation Index or a similar system (e.g. the Web of Science website at http://wos.mimas.ac.uk or Google Scholar at http://scholar.google.com/), work *forward* and find out who has cited your selected article in the time since its publication. Again list the *five* most important articles found.

19.3 Write a synopsis of a review. Again using one of the *Annual Review* series as a source, allow yourself just *five* single-sentence bullet points to summarise the key points reported in a particular review.

Fundamental Laboratory Techniques

20 Your approach to practical work

Developing practical skills – these will include:

- designing experiments
- observing and measuring
- recording data
- analysing and interpreting data
- reporting/presenting.

All knowledge and theory in science have originated from practical observation and experimentation: this is equally true for disciplines as diverse as microscopy and molecular genetics. Practical work is an important part of most courses and often accounts for a significant proportion of the assessment marks. The abilities developed in practical classes will continue to be useful throughout your course and beyond, some within science and others in any career you choose (see Chapter 1).

Being prepared

 KEY POINT   You will get the most out of practicals if you prepare well in advance. Do not go into a practical session assuming that everything will be provided, without any input on your part.

The main points to remember are:

- Read any handouts in advance: make sure you understand the purpose of the practical and the particular skills involved. Does the practical relate to, or expand upon, a current topic in your lectures? Is there any additional preparatory reading that will help?
- Take along appropriate textbooks, to explain aspects in the practical.
- Consider what safety hazards might be involved, and any precautions you might need to take, before you begin (p. 126).
- Listen carefully to any introductory guidance and note any important points: adjust your schedule/handout as necessary.
- During the practical session, organise your bench space – make sure your lab book is adjacent to, but not within, your working area. You will often find it easiest to keep clean items of glassware, etc. on one side of your working space, with used equipment on the other side.
- Write up your work as soon as possible, and submit it on time, or you may lose marks.
- Catch up on any work you have missed as soon as possible – preferably, before the next practical session.

Using textbooks in the lab – take this book along to the relevant classes, so that you can make full use of the information during the practical sessions.

 SAFETY NOTE Mobile phones – these should never be used in a lab class, as there is a risk of contamination from hazardous substances. Always switch off your mobile phone before entering a laboratory. Conversely, they are an extremely useful accessory for fieldwork.

Ethical and legal aspects

You will need to consider the ethical and legal implications of biological work at several points during your studies:

- Safe working means following a code of safe practice, supported by legislation, alongside a moral obligation to avoid harm to yourself and others, as discussed in Chapter 21.
- Any laboratory work that involves working with animals must be carefully considered.
- Fieldwork can have a legal aspect, e.g. in relation to specimen collection, since legislation may protect particular habitats and/or species. Ethical and human aspects that must all be considered before any work is carried out include: possible damage to fauna and flora; safety of others within your group; implications of your work to others, e.g. landowners, or other users.

Getting to grips with bioethics – in addition to any moral implications of your lab practicals and field work, you may have the opportunity to address broader issues within your course (see Box 19.1). Professional scientists should always consider the consequences of their work, and it is therefore important that you develop your appreciation of these issues alongside your academic studies.

Box 20.1 Bioethics

Contemporary bioscience degree programmes place increasing emphasis on the ethical and social impacts of scientific advances, and on the need for scientists engaged in potentially controversial work to communicate their ideas to the general public. Bioscience research raises many moral and legal dilemmas, requiring difficult choices to be made (e.g. animal testing of medical products), and students are likely to be asked to reflect on bioethical topics, e.g. in group discussions and debates on current issues such as:

- environmental ethics (e.g. use of genetically modified plants, hunting of endangered species);
- animal ethics (e.g. factory farming, transgenic animals, xenotransplantation);
- medical and social ethics (e.g. human cloning; embryo research, genetic testing).

In discussing such topics (sometimes referred to as ethical, legal and social issues, ELSI), you will find that there is rarely a 'right' or 'wrong' answer, and it is important to be able to consider these issues in a logical manner, and to provide a reasoned argument in support of a particular viewpoint. Gaining experience in such debates should also help you understand some of the issues linked to the public understanding of science, and how these can be addressed (see: http://www.copus.org.uk/pubs_guides.html). While a full exposition is beyond the scope of this book, the following provides a framework of principles for considering particular topics:

- Beneficence – the obligation to do good (for example, if it is possible to prevent suffering by a particular course of action, then it should be carried out).
- Non-maleficence – the duty to cause no harm (contained within the Hippocratic Oath of medical practitioners).
- Justice – the obligation to treat all people fairly and impartially (for example, lack of discrimination between people on the grounds of race or sex).
- Autonomy – the duty to allow an individual to make their own choices, without constraints (this principle underlies the notion of informed consent in medical research).
- Respect – the need to show due regard for others (for example, by taking into account the rights and beliefs of all people equally).
- Rationality – the notion that a particular action or choice should be based on reason and logic (many scientists would argue that the scientific method is an example of rationality).
- Precautionary principle – the notion that it is better not to carry out an action if there is any risk of harm (for example, in deciding that the risks of building nuclear power stations outweigh their potential benefits).

Understanding the various theories of ethics may also help you formulate your ideas. These include:

- Utilitarianism – the notion that it is ethical to choose the action that produces the greatest good for the greatest number.
- Deontology – a theory that states than a particular action is either intrinsically good (right) or bad (wrong). According to deontological theory, decisions should be based on the actions themselves, rather than on their consequences.
- Virtue theory – the notion that making decisions according to established virtues (e.g. honesty, wisdom, justice) will lead to ethically valid choices.
- Objectivism – the theory that what is right and wrong is intrinsic and applies equally to all people, places and times (the alternative is that morality is subjective, being dependent on the views of each individual).

The principles and issues of bioethics are considered in a number of websites, including:

- introductory bioethics (e.g. http://www.accessexcellence.org/RC/AB/IE/);
- bioethics resources (e.g. http://bioethics.od.nih.gov/, http://www.ethicsweb.ca/resources/ and http://www.beep.ac.uk/content/index.php/);
- ethics discussion groups (e.g. http://www.beep.ac.uk/discuss/index.php?c=6 and http://www-hsc.usc.edu/~mbernste/).

The following introductory texts give further information and guidance:
Bryant et al. (2005),
Gert et al. (2006) and
Mepham (2005).

Basic requirements

Recording practical results

An A4 loose-leaf ring binder offers flexibility, since you can insert laboratory handouts or lined and graph paper at appropriate points. The danger of losing one or more pages from a loose-leaf system is the main drawback. Bound books avoid this problem, although those containing alternating lined/graph or lined/blank pages tend to be wasteful – it is often better to paste sheets of graph paper into a bound book, as required.

A good quality HB pencil or propelling pencil is recommended for recording your raw data, making diagrams, etc. as mistakes are easily corrected. Buy a black, spirit-based (permanent) marker for labelling experimental glassware, Petri plates, etc. Fibre-tipped fine line drawing/lettering pens are useful for preparing final versions of hand-drawn graphs and diagrams for assessment purposes. Use a see-through ruler (with an undamaged edge) for graph drawing, so that you can see data points and information below the ruler as you draw.

Calculators

These range from basic machines with no pre-programmed functions and only one memory, to sophisticated programmable portable computers with many memories. The following may be helpful when using a calculator:

- Power sources. Choose a battery-powered machine, rather than a mains-operated or solar-powered type. You will need one with basic mathematical/scientific operations, including powers, logarithms (p. 476), roots and parentheses (brackets), together with statistical functions such as sample means and standard deviations (Chapter 66).
- Mode of operation. The older operating system used by e.g. Hewlett-Packard calculators is known as the reverse Polish notation: to calculate the sum of two numbers, the sequence is 2 [enter] 4 + and the answer 6 is displayed. The more usual method of calculating this equation is as $2 + 4 =$, which is the system used by the majority of modern calculators. Most newcomers find the latter approach to be more straightforward. Spend some time finding out how a calculator operates, e.g. does it have true algebraic logic ($\sqrt{}$ then number, rather than number then $\sqrt{}$)? How does it deal with (and display) scientific notation and logarithms (p. 476)?
- Display. Some calculators will display an entire mathematical operation (e.g. '$2 + 4 = 6$'), while others simply display the last number/operation. The former type may offer advantages in tracing errors.
- Complexity. In the early stages, it is usually better to avoid the more complex machines, full of impressive-looking, but often unused, pre-programmed functions – go for more memory, parentheses or statistical functions rather than engineering or mathematical constants. Programmable calculators may be worth considering for more advanced studies. However, it is important to note that such calculators are often unacceptable for exams.

Presenting more advanced practical work

In some practical reports and in project work, you may need to use more sophisticated presentation equipment. Computer-based graphics packages can be useful – choose easily read fonts such as Arial or Helvetica for posters and consider the layout and content carefully. Alternatively, you

---

Presenting results – while you don't need to be a graphic designer to produce work of a satisfactory standard, presentation and layout are important and you will lose marks for poorly presented work. Chapter 63 gives further practical advice.

Using calculators for numerical problems – Chapter 65 gives further advice.

Using inexpensive calculators – many unsophisticated calculators have a restricted display for exponential numbers and do not show the 'power of 10', e.g. displaying 2.4×10^{-5} as 2.4^{-05}, or $2.4E-05$, or even $2.4-05$.

Presenting graphs and diagrams – ensure these are large enough to be easily read: a common error is to present graphs or diagrams that are too small, with poorly chosen scales (see p. 457).

Printing on acetates – *standard overhead transparencies are not suitable for use in laser printers or photocopiers: you need to make sure that you use the correct type.*

could use fine line drawing pens and dry-transfer lettering/symbols, although this can be more time-consuming than computer-based systems, e.g. using Microsoft Excel (p. 458).

To prepare overhead transparencies for spoken presentations, you can use spirit-based markers and acetate sheets. An alternative approach is to photocopy onto special acetates, or print directly from a computer-based package, using a laser printer and special acetates, or use a PC-based presentation package (e.g. PowerPoint) and a suitable projector. Further advice on content and presentation is given in Chapter 15.

Text references

Bryant, J., Baggott le Velle, L. and Searle, J. (2005) *Introduction to Bioethics*. Wiley, Chichester.

Gert, B., Culver, C.M. and Clouser, K.D. (2006) *Bioethics: a Systematic Approach*, 2nd edn. Oxford University Press, Oxford.

Mepham, B. (2005) *Bioethics: an Introduction for the Biosciences*. Oxford University Press, Oxford.

Sources for further study

Barnard, C.J., Gilbert, F.S. and MacGregor, P.K. (2001) *Asking Questions in Biology: Key Skills for Practical Assessments and Project Work*, 2nd edn. Prentice Hall, Harlow.

Howell, J.H., Sale, W.F. and Callahan, D. (2000) *Life Choices: A Hastings Center Introduction to Bioethics*, 2nd edn. Georgetown University Press, Washington, DC.

Mier-Jedrzejowicz, W.A.C. (1999) *A Guide to HP Handheld Calculators and Computers*. Wilson-Barnett, Tustin. [Provides further guidance on the use of Hewlett-Packard calculators (reverse Polish notation)]

Singer, P. and Kuhse, H. (2006) *Bioethics: an Anthology*. Blackwell, Oxford.

Study exercises

20.1 Consider the value of practical work. Spend a few minutes thinking about the purpose of practical work within a specific part of your course (e.g. a particular first year module) and then write a list of the six most important points. Compare your list with the generic list that we have provided on p. 510, which is based on our experience as lecturers – does it differ much from your list, which is drawn up from a student perspective?

20.2 Make a list of items required for a particular practical exercise. This exercise is likely to be most useful if you can relate it to an appropriate practical session on your course. However, we have given a model list for a tissue dissection as an example.

20.3 Check your calculator skills. Carry out the following mathematical operations, using either a hand-held calculator or a PC with appropriate 'calculator' software.

(a) $5 \times (2 + 6)$

(b) $(8.3 \div [6.4 - 1.9]) \times 24$ (to four significant figures)

(c) $(1 \div 32) \times (5 \div 8)$ (to three significant figures)

(d) $1.2 \times 10^5 + 4.0 \times 10^4$ in scientific notation (see p. 476)

(e) $3.4 \times 10^{-2} - 2.7 \times 10^{-3}$ in 'normal' notation (i.e. conventional notation, not scientific format) and to three decimal places.

(See also the numerical exercises in Chapter 65.)

21 Health and safety

*Health and safety legislation – in the UK, the **Health & Safety at Work, etc. Act 1974** provides the main legal framework for health and safety. The **Control of Substances Hazardous to Health (COSHH) Regulations 2002** impose specific legal requirements for risk assessment wherever hazardous chemicals or biological agents are used, with Approved Codes of Practice for the control of hazardous substances, carcinogens and biological agents, including pathogenic microbes.*

Health and safety law requires academic institutions to provide a working environment that is safe and without risk to health. Where appropriate, training and information on safe working practices must be provided. Students and staff must take reasonable care to ensure the health and safety of themselves and of others, and must not misuse any safety equipment.

KEY POINT All practical work must be carried out with safety in mind, to minimise the risk of harm to yourself and to others – safety is everyone's responsibility.

Risk assessment

The most widespread approach to safe working practice involves the use of risk assessment, which aims to establish:

1. The intrinsic chemical, biological and physical hazards, together with any maximum exposure limits (MELs) or occupational exposure standards (OESs), where appropriate. Chemical manufacturers provide data sheets listing the hazards associated with particular chemical compounds, while pathogenic (disease-causing) microbes are categorised according to their ability to cause illness (p. 205).
2. The risks involved, by taking into account the amount of substance to be used, the way in which it will be used and the possible routes of entry into the body (Fig. 21.1).

Definitions

Hazard – the ability of a substance or biological agent to cause harm.

Risk – the likelihood that a substance or biological agent might be harmful under specific circumstances.

Distinguishing between hazard and risk – one of the hazards associated with water is drowning. However, the risk of drowning in a few drops of water is negligible.

KEY POINT It is important to distinguish between the intrinsic hazards of a particular substance and the risks involved in its use in a particular exercise.

3. The persons at risk, and the ways in which they might be exposed to hazardous substances, including accidental exposure (spillage).
4. The steps required to prevent or control exposure. Ideally, a non-hazardous or less hazardous alternative should be used. If this is not feasible, adequate control measures must be used, e.g. a fume cupboard or other containment system. Personal protective equipment (e.g. lab coats, safety glasses) must be used in addition to such containment measures. A safe means of disposal will be required.

The outcome of the risk assessment process must be recorded and appropriate safety information must be passed on to those at risk. For most practical classes, risk assessments will have been carried out in advance by the person in charge: the information necessary to minimise the risks to students may be given in the practical schedule. Make sure you know how your department provides such information and that you have read the appropriate material before you begin your practical work. You should also pay close attention to the person in charge at the beginning of the practical session, as they may emphasise the major hazards and risks. In project work, you will need to be involved in the risk assessment process along with your supervisor, before you carry out any practical work.

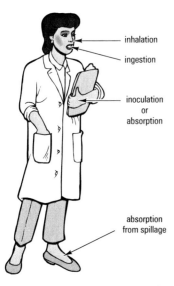

Fig. 21.1 Major routes of entry of harmful substances into the body.

Fig. 21.2 Warning labels for specific chemical hazards. These appear on suppliers' containers and on tape used to label working vessels.

In addition to specific risk assessments, most institutions will have a safety handbook, giving general details of safe working practices, together with the names and telephone numbers of safety personnel, first aiders, hospitals, etc. Make sure you read this and follow any instructions given.

Basic rules for laboratory work

- Make sure you know what to do in case of fire, including exit routes, how to raise the alarm, and where to gather on leaving the building. Remember that the most important consideration at all times is human safety: do not attempt to fight a fire unless it is safe to do so.
- All laboratories display notices telling you where to find the first aid kit and who to contact in case of accident or emergency. Report all accidents, even those appearing insignificant – your department will have a formal recording procedure to comply with safety legislation.
- Wear appropriate protective clothing at all times – a clean lab coat (buttoned up), plus safety glasses if there is any risk to the eyes.
- Never smoke, eat or drink in any laboratory, because of the risks of contamination by inhalation or ingestion (Fig. 21.1).
- Never mouth pipette any liquid. Use a pipette filler (see p. 128) or, if appropriate, a pipettor (pp. 129–30).
- Take care when handling glassware – see p. 132 for details.
- Know the warning symbols for specific chemical hazards (see Fig. 21.2).
- Use a fume cupboard for hazardous chemicals. Make sure that it is working and then open the front only as far as is necessary: many fume cupboards are marked with a maximum opening.
- Always use the minimum quantity of any hazardous materials.
- Work in a logical, tidy manner and minimise risks by thinking ahead.
- Always clear up at the end of each session. This is an important aspect of safety, encouraging a responsible attitude towards laboratory work.
- Dispose of waste in appropriate containers. Most labs will have bins for sharps, glassware, hazardous solutions and radioactive waste.

Genetic engineering and molecular genetics

In the UK, the Genetically Modified Organisms (Contained Use) Regulations 2000 define the legal requirements for risk assessment and for the notification of work involving genetic manipulation.

The Advisory Committee on Genetic Manipulation provides guidance on the use of genetically manipulated organisms.

The Health and Safety Executive (HSE) has specific responsibility for the operation of these regulations and is the regulatory authority for genetic manipulation in the UK.

Additional legal constraints apply to practical work involving genetic manipulation. A specific risk assessment must be carried out for any experiment where a cell or organism is modified by genetic engineering techniques (Chapter 62) involving the insertion of DNA into a cell or organism in which it does not normally occur. Before any practical work can be carried out, it must be authorised by the establishment's genetic manipulation safety committee and notified to the relevant authority. Such work must be carried out with appropriate containment, to prevent the accidental release of genetically modified organisms into the environment.

Practicals in molecular genetics will involve some of the techniques of genetic manipulation. Typically, these will be examples of 'self-cloning' where recombinant DNA molecules are constructed from fragments of DNA which naturally occur in that organism, for example, the transformation of laboratory strains of *Escherichia coli* using a pUC plasmid (p. 448).

Sources for further study

American Chemical Society (2003) *Safety in Academic Chemistry Laboratories*, 7th edn. Available: http://membership.acs.org/c/ccs/pub_3.htm SACL_Students.pdf
Last accessed: 01/04/07.

Anon. *BUBL Link – Biochemical Safety*. Available: http://bubl.ac.uk/link/b/biochemicalsafety.htm
Last accessed: 01/04/07.
[Provides a single-step link to a number of databases giving information on chemical and microbiological hazards]

Anon. *Howard Hughes Medical Institute On-line Safety Course*. Available: http://www.hhmi.org/about/labsafe/safescience.html
Last accessed: 01/04/07.
[This site, designed for laboratory workers, enables you to test your knowledge and understanding of laboratory safety]

Furr, A.K. (2000) *CRC Handbook of Laboratory Safety*. CRC Press, Boca Raton, FL.

Health and Safety Executive (2005) *Control of Substances Hazardous to Health Regulations*. HSE, London.

Study exercises

21.1 Test your knowledge of safe working procedures. After reading the appropriate sections of this book, can you remember the following:

(a) The four main steps involved in the process of risk assessment?

(b) The major routes of entry of harmful substances into the body?

(c) The warning labels for the major chemical hazard symbols (either describe them or draw them from memory)?

(d) The international symbol for a biohazard?

(e) The international symbol for radioactivity?

21.2 Locate relevant health and safety features in a laboratory. Find each of the following in one of the laboratories used as part of your course (draw a simple location map, if this seems appropriate):

(a) fire exit(s);

(b) fire-fighting equipment;

(c) first aid kit;

(d) 'sharps' container;

(e) container for disposal of broken glassware;

(f) eye wash station (where appropriate).

21.3 Investigate the health and safety procedures in operation at your university. Can you find out the following?

(a) your university's procedure in case of fire;

(b) the colour coding for fire extinguishers available in your department and their recommendations for use;

(c) the accident reporting procedure used in your department;

(d) your department's Code of Safe Practice relating to a specific aspect of bioscience, e.g. working with micro-organisms.

21.4 Carry out risk assessments for specific chemical hazards. Look up the hazards associated with the use of the following chemicals and list the appropriate protective measures required to minimise risk during use in a lab class:

(a) formaldehyde solution, used as a preservative for animal tissue, to be used for microscopic examination (Chapter 27);

(b) acetone, e.g. for use as a solvent for the quantitative analysis of plant pigments (Chapter 41);

(c) sodium hydroxide, used in solid form to prepare a dilute solution to be used for pH adjustment (Chapter 25).

22 Working with liquids

 SAFETY NOTE *Using hazardous liquids and solutions (flammable, corrosive, toxic, etc.) – make sure the liquid is properly contained and work out how to deal with any spillages before you begin.*

Measuring and dispensing liquids

The equipment you should choose to measure out liquids depends on the volumes being dispensed, the accuracy required and the number of times the job must be done (Table 22.1).

Table 22.1 Criteria for choosing a method for measuring out a liquid

Method	Best volume range	Accuracy	Usefulness for repetitive measurement
Pasteur pipette	30 μl to 2 ml	Low	Convenient
Conical flask/beaker	25–5000 ml	Very low	Convenient
Measuring cylinder	5–2000 ml	Medium	Convenient
Volumetric flask	5–2000 ml	High	Convenient
Burette	1–100 ml	High	Convenient
Glass pipette	1–100 ml	High	Convenient
Mechanical pipettor	5–1000 μl	High*	Convenient
Syringe	0.5–20 μl	Medium**	Convenient
Microsyringe	0.5–50 μl	High	Convenient
Weighing	Any (depends on accuracy of balance)	Very high	Inconvenient

* If correctly calibrated and used properly (see p. 130).
** Accuracy depends on width of barrel: large volumes are less accurate.

Certain liquids may cause problems:

- High viscosity liquids are difficult to dispense: allow time for all the liquid to transfer.
- Organic solvents may evaporate rapidly, making measurements inaccurate: work quickly; seal containers without delay.
- Solutions prone to frothing (e.g. protein and detergent solutions) are difficult to measure and dispense: avoid forming bubbles due to overagitation; do not transfer quickly.
- Suspensions (e.g. cell cultures) may sediment: thoroughly mix them before dispensing.

Reading any volumetric scale – make sure your eye is level with the bottom of the liquid's meniscus and take the reading from this point.

Pasteur pipettes

Hold correctly during use (Fig. 22.1) – keep the pipette vertical, with the middle fingers gripping the barrel while the thumb and index finger provide controlled pressure on the bulb. Squeeze gently to dispense individual drops. To avoid the risk of cross-contamination, take care not to draw up solution into the bulb or to lie the pipette on its side. Alternatively, use a plastic disposable 'Pastette'.

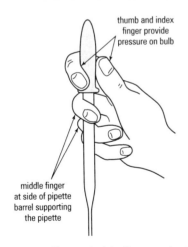

Fig. 22.1 How to hold a Pasteur pipette.

 SAFETY NOTE *Pasteur pipettes should be used with care for hazardous solutions: remove the tip from the solution before fully releasing pressure on the bulb – the air taken up helps prevent spillage.*

Measuring cylinders and volumetric flasks

These must be used on a level surface so that the scale is horizontal; you should first fill with solution until just below the desired mark; then fill slowly (e.g. using a Pasteur pipette) until the meniscus is level with the mark. Allow time for the solution to run down the walls of the vessel.

Working with Liquids

Burettes

Burettes should be mounted vertically on a clamp stand – don't overtighten the clamp. First ensure the tap is closed and fill the body with solution using a funnel. Open the tap and allow some liquid to fill the tubing below the tap before first use. Take a meniscus reading, noting the value in your notebook. Dispense the solution *via* the tap and measure the new meniscus reading. The volume dispensed is the difference between the two readings. Titrations are usually performed on a magnetic stirrer.

Pipettes

These come in various designs, including graduated and bulb (volumetric) pipettes (Fig. 22.2). Take care to check the volume scale before use: some empty from full volume to zero, others from zero to full volume; some scales refer to the shoulder of the tip, others to the tip either by gravity or after blowing out.

 KEY POINT *For safety reasons, it is never permissible to mouth pipette – various aids are available such as the Pi-pump.*

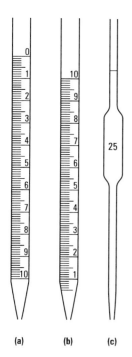

Fig. 22.2 Glass pipettes – graduated pipette, reading from zero to shoulder (a); graduated pipette, reading from maximum to tip, by gravity (b); bulb (volumetric) pipette, showing volume (calibration mark to tip, by gravity) above the bulb (c).

Pipettors (autopipettors)

These come in two basic types:

- Air displacement pipettors. For routine work with dilute aqueous solutions. One of the most widely used examples is the Gilson Pipetman (Fig. 22.3). Box 22.1 gives details on its use.
- Positive displacement pipettors. For non-standard applications, including dispensing viscous, dense or volatile liquids, or certain procedures in molecular genetics, e.g. the PCR (p. 439), where an air displacement pipettor might create aerosols, leading to errors.

Air displacement and positive displacement pipettors may be:

- Fixed volume: capable of delivering a single factory-set volume.
- Adjustable: where the volume is determined by the operator across a particular range of values.
- Pre-set: movable between a limited number of values.
- Multi-channel: able to deliver several replicate volumes at the same time.

Whichever type you use, you must ensure that you understand the operating principles of the volume scale and the method for changing the volume delivered – some pipettors are easily misread.

A pipettor must be fitted with the correct disposable tip before use: each manufacturer produces different tips to fit particular models. Specialised tips are available for particular applications e.g. PCR (p. 439).

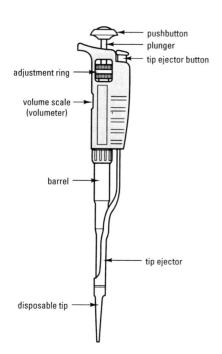

Fig. 22.3 A pipettor – the Gilson Pipetman.

Syringes

Syringes should be used by placing the tip of the needle in the solution and drawing the plunger up slowly to the required point on the scale. Check the barrel to make sure no air bubbles have been drawn up. Expel slowly and touch the syringe on the edge of the vessel to remove any liquid adhering to the end of the needle. Microsyringes should always be cleaned before and after use by repeatedly drawing up and expelling

Box 22.1 Using a pipettor to deliver accurate, reproducible volumes of liquid

A pipettor can be used to dispense volumes with accuracy and precision, by following this stepwise procedure:

1. **Select a pipettor that operates over the appropriate range.** Most adjustable pipettors are accurate only over a particular working range and should not be used to deliver volumes below the manufacturer's specifications (minimum volume is usually 10–20% of maximum value). Do not attempt to set the volume above the maximum limit, or the pipettor may be damaged.
2. **Set the volume to be delivered.** In some pipettors, you 'dial up' the required volume. Types like the Gilson Pipetman have a system where the scale (or 'volumeter') consists of three numbers, read from top to bottom of the barrel, and adjusted using the black knurled adjustment ring (Fig. 22.3). This number gives the first three digits of the volume scale and thus can only be understood by establishing the maximum volume of the Pipetman, as shown on the push-button on the end of the plunger (Fig. 22.3). The following examples illustrate the principle for two common sizes of Pipetman:

   P1000 Pipetman (maximum volume 1000 µl) if you dial up

   | 1 |
   | 0 |
   | 0 |

   the volume is set at 1000 µl

   P20 Pipetman (maximum volume 20 µl) if you dial up

   | 1 |
   | 0 |
   | 0 |

   the volume is set at 10.0 µl

   Note: The Pipetman scale is *not* a percentage one.

3. **Fit a new disposable tip to the end of the barrel.** Make sure that it is the appropriate type for your pipettor and that it is correctly fitted. Press the tip on firmly using a slight twisting motion – if not, you will take up less than the set volume and liquid will drip from the tip during use. Tips are often supplied in boxes, for ease of use: if sterility is important, make sure you use appropriate sterile technique at all times (p. 203). *Never, ever, try to use a pipettor without its disposable tip.*
4. **Check your delivery.** Confirm that the pipettor delivers the correct volume by dispensing volumes of distilled water and weighing on a balance, assuming $1 \text{ mg} = 1\,\mu l = 1 \text{ mm}^3$. The value should be within 1% of the selected volume. For small volumes, measure several 'squirts' together, e.g. 20 'squirts' of $5\,\mu l = 100 \text{ mg}$. If the pipettor is inaccurate (p. 175) giving a biased result (e.g. delivering significantly more or less than the volume set), you can make a temporary correction by adjusting the volumeter scale down or up accordingly (the volume *delivered* is more important than the value *displayed* on the volumeter), or have the pipettor recalibrated. If the pipettor is imprecise (p. 175), delivering a variable amount of liquid each time, it may need to be serviced. After calibration, fit a clean (sterile) tip if necessary.
5. **Draw up the appropriate volume.** Holding the pipettor *vertically*, press down on the plunger/push-button until a resistance (spring-loaded stop) is met. Then place the end of the tip in the liquid. Keeping your thumb on the plunger/push-button, release the pressure slowly and evenly: watch the liquid being drawn up into the tip, to confirm that no air bubbles are present. Wait a second or so, to confirm that the liquid has been taken up, then withdraw the end of the tip from the liquid. Inexperienced users often have problems caused by drawing up the liquid too quickly/carelessly. If you accidentally draw liquid into the barrel, seek assistance from your demonstrator or supervisor as the barrel will need to be cleaned before further use.
6. **Make a quick visual check on the liquid in the tip.** Does the volume seem reasonable? (e.g. a $100\,\mu l$ volume should occupy approximately half the volume of a P200 tip). The liquid will remain in the tip, without dripping, as long as the tip is fitted correctly and the pipettor is not tilted too far from a vertical position.
7. **Deliver the liquid.** Place the end of the tip against the wall of the vessel at a slight angle (10–15° from vertical) and press the plunger/push-button slowly and smoothly to the first (spring-loaded) stop. Wait a second or two, to allow any residual liquid to run down the inside of the tip, then press again to the final stop, dispensing any remaining liquid. Remove from the vessel with the plunger/push-button still depressed.
8. **Eject the tip.** Press the tip ejector button if present (Fig. 22.3). If the tip is contaminated, eject directly into an appropriate container, e.g. a beaker of disinfectant, for microbiological work, or a labelled container for hazardous solutions (p. 126). For repeat delivery, fit a new tip if necessary and begin again at step 5 above. Always make sure that the tip is ejected before putting a pipettor on the bench.

pure solvent. The dead space in the syringe needle can occupy up to 4% of the nominal syringe volume. A way of avoiding such problems is to fill the dead space with an inert substance (e.g. silicone oil) after sampling. Alternatively use a syringe where the plunger occupies the needle space (small volumes only).

Balances

These can be used to weigh accurately (p. 140) how much liquid you have dispensed. Convert mass to volume using the equation:

$$\text{mass}/\text{density} = \text{volume} \qquad [22.1]$$

Example Weighing 9 g of liquid with a density of 1.2 g ml^{-1} will give a volume of $9 \div 1.2 = 7.5$ ml

Densities of common solvents can be found in Lide (2006). You will also need to know the liquid's temperature, as density is temperature dependent.

Holding and storing liquids

Test tubes

These are used for colour tests, small-scale reactions, holding cultures, etc. The tube can be sterilised by heating (p. 203) and maintained in this state with a cap or cotton wool plug.

Beakers

Working with beakers and flasks – remember that volume graduations, where present, are often inaccurate and should be used only where approximations will suffice.

Beakers are used for general purposes, e.g. heating a solvent while the solute dissolves, carrying out a titration, etc.

Conical (Erlenmeyer) flasks

These are used for storage of solutions: their wide base makes them stable, while their small mouth reduces evaporation and is easily sealed.

Bottles and vials

Storing light-sensitive chemicals – use a coloured vessel or wrap aluminium foil around a clear vessel.

These are used when the solution needs to be sealed for safety, sterility or to prevent evaporation or oxidation. They usually have a screw top or ground glass stopper to prevent evaporation and contamination. Many types are available, including 'bijou', 'McCartney', 'universal' and 'Winkler'.

Storing an aqueous solution containing organic constituents – unless this has been sterilised or is toxic, microbes will start growing, so store for short periods in a refrigerator: older solutions may not give reliable results.

You should clearly label all stored solutions (see p. 139), including relevant hazard information, preferably marking with hazard warning tape (p. 126). Seal vessels in an appropriate way, e.g. using a stopper or a sealing film such as Parafilm or Nescofilm to prevent evaporation. To avoid degradation store your solution in a fridge, but allow it to reach room temperature before use.

Creating specialised apparatus

Glassware systems incorporating ground glass connections such as Quickfit are useful for setting up combinations of standard glass components, e.g. for chemical reactions. In project work, you may need to adapt standard forms of glassware for a special need. A glassblowing service (often available in chemistry departments) can make special items to order.

SAFETY NOTE Special cleaning of glass – for an acid wash use dilute acid, e.g. 100 mmol l^{-1} (100 mol m^{-3}) HCl. Rinse thoroughly at least three times with distilled or deionised water. Glassware that must be exceptionally clean (e.g. for a micronutrient study) should be washed in a chromic acid bath, but this involves toxic and corrosive chemicals and should only be used under supervision.

Choosing between glass and plastic

Bear in mind the following points:

- Reactivity. Plastic vessels often distort at relatively low temperatures; they may be flammable, may dissolve in certain organic solvents and

Table 22.2 Spectral cutoff values for glass and plastics (λ_{50} = wavelength at which transmission of EMR is reduced to 50%)

Material	λ_{50} (nm)
Routine glassware	340
Pyrex glass	292
Polycarbonate	396
Acrylic	342
Polyester	318
Quartz	220

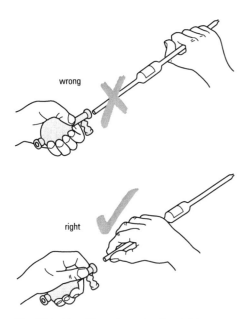

Fig. 22.4 Handling glass pipettes tubing.

may be affected by prolonged exposure to ultraviolet (UV) light. Some plasticisers may leach from vessels and have been shown to have biological activity. Glass may adsorb ions and other molecules and then leach them into solutions, especially in alkaline conditions. Pyrex glass is stronger than ordinary soda glass and can withstand temperatures up to 500 °C.

- Rigidity and resilience. Plastic vessels are not recommended where volume is critical as they may distort through time: use class A volumetric glassware for accurate work, e.g. preparing solutions (Chapter 24). Glass vessels are more easily broken than plastic, which is particularly important for centrifugation (see p. 301).
- Opacity. Both glass and plastic absorb light in the UV range of the EMR spectrum (Table 22.2). Quartz should be used where this is important, e.g. in cuvettes for UV spectrophotometry (see p. 280).
- Disposability. Plastic items may be cheap enough to make them disposable, an advantage where there is a risk of chemical or microbial contamination.

Cleaning glass and plastic

Take care to avoid the possibility of contamination arising from prior use of chemicals or inadequate rinsing following washing. A thorough rinse with distilled or deionised water immediately before use will remove dust and other deposits and is good practice in quantitative work, but ensure that the rinsing solution is not left in the vessel. 'Strong' basic detergents (e.g. Pyroneg) are good for solubilising acidic deposits. If there is a risk of basic deposits remaining, use an acid wash. If there is a risk of contamination from organic deposits, a rinse with Analar grade ethanol is recommended. Glassware can be disinfected by washing with a sodium hypochlorite bleach such as Chloros, with sodium metabisulphite or a blended commercial product such as Virkon – dilute as recommended before use and rinse thoroughly with sterile water after use. Alternatively, to sterilise glassware, heat to at least 121 °C for 15 min in an autoclave or 160 °C for 3 h in an oven.

Box 22.2 Safe working with glass

Many minor accidents in the laboratory are due to lack of care with glassware. You should follow these general precautions:

- **Always wear safety glasses when there is any risk of glass breakage** – e.g. when using low pressures, or heating solutions.
- **Take care when attaching tubing to glass tubes and when putting glass tubes into bungs** – always hold the tubing and glassware close together, as shown in Fig. 22.4 , and wear thick gloves when appropriate.
- **Use a 'soft' Bunsen flame when heating glassware** – this avoids creating a hot spot, where cracks may start: always use tongs or special heat-resistant gloves when handling hot glassware (never use a rolled-up paper towel).
- **Do not use chipped or cracked glassware** – it may break under very slight strain and should be disposed of in the broken glassware bin.
- **Never carry large glass bottles/flasks by their necks** – support them with a hand underneath or, better still, carry them in a basket.
- **Do not force bungs too firmly into bottles** – they can be extremely difficult to remove. If you need a tight seal, use a screw-top bottle with a rubber or plastic seal.
- **Dispose of broken glass thoroughly and carefully** – use disposable paper towels and wear thick gloves. Always put pieces of broken glass in the correct bin.

Text reference

Lide, D.R. (ed.) (2006) *CRC Handbook of Chemistry and Physics*, 87th edn. CRC Press, Boca Raton.

Sources for further study

Anon. *CHEMnet BASE*. Available http://www.chemnetbase.com Last accessed: 01/04/07.
[Online access to the *Handbook of Chemistry and Physics*]

Boyer, R. (2000) *Modern Experimental Biochemistry*, 3rd edn. Benjamin Cummings, San Francisco.

Henrikson, C., Byrd, L.C. and Hunter, N.W. (2005) *A Laboratory Manual for General, Organic and Biochemistry*, 5th edn. McGraw-Hill, New York.

Seidman, L.A. and Moore, C.J. (2000) *Basic Laboratory Methods for Biotechnology: Textbook and Laboratory Reference*. Prentice Hall, New Jersey.

Study exercises

22.1 Decide on the appropriate methods and equipment for the following procedures:

(a) Preparing one litre of ethanol at approximately 70% v/v in water for use as a general-purpose reagent.

(b) Adding 10 μl of a sample to the well of an agarose gel during a molecular biology procedure.

(c) Preparing a calibration standard of 100 ml of DNA, to contain 200.0 μg ml^{-1}, for spectrophotometry.

(d) Carrying out a titration curve for a buffer solution.

22.2 Write a protocol for calibrating and using a pipettor. After reading this chapter, prepare a detailed stepwise protocol explaining how to use a pipettor to deliver a specific volume, say of 500 μl (e.g. using a Gilson Pipetman, or an alternative if your department does not use this type). Ask another student to evaluate your protocol and provide you with written feedback – either simply by reading through your protocol, or by trying it out with a pipettor as part of a class exercise (check with a member of staff before you attempt this in a laboratory).

22.3 Determine the accuracy and precision of a pipette. Using the following data for three different models of pipettor, determine which pipettor is most *accurate* and which is most *precise* (check p. 175 if you are unsure of the definitions of these two terms). All three pipettors were set to deliver 1000 μl (1.000 ml) and 10 repetitive measurements of the weight of the volume of water in grams delivered were made using a three-place balance:

Model A pipettor: 0.986; 0.971; 0.993; 0.964; 0.983; 0.996; 0.977; 0.969; 0.982; 0.974

Model B pipettor: 1.013; 1.011; 1.010; 1.009; 1.011; 1.010; 1.011; 1.009; 1.011; 1.012

Model C pipettor: 0.985; 1.022; 1.051; 1.067; 0.973; 0.982; 0.894; 1.045; 1.062; 0.928

In your answer, you should support your conclusions with appropriate numerical (statistical) evidence (see Chapter 66 for appropriate measures of location and dispersion).

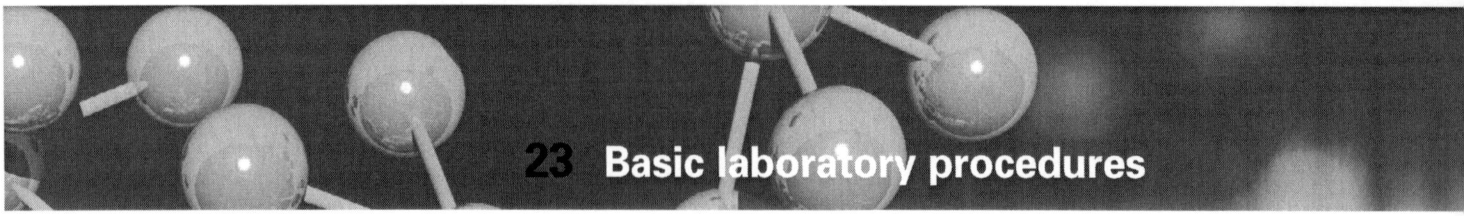

23 Basic laboratory procedures

Using chemicals responsibly – be considerate to others: always return store room chemicals promptly to the correct place. Report when supplies are getting low to the person who looks after storage/ordering. If you empty an aspirator or wash bottle, fill it up from the appropriate source.

Finding out about chemicals – The Merck Index (O'Neil et al., 2006: Fig. 23.1) and the CRC Handbook of Chemistry and Physics (Lide, 2006) are useful sources of information on the physical and biological properties of chemicals, including melting and boiling points, solubility, toxicity, etc.

8671. Sodium Chloride. [7647-14-5] Salt; common salt. ClNa; mol wt 58.44. Cl 60.67%, Na 39.34%. NaCl. The article of commerce is also known as **table salt**, **rock salt** or **sea salt**. Occurs in nature as the mineral **halite**. Produced by mining (rock salt), by evaporation of brine from underground salt deposits and from sea water by solar evaporation: *Faith, Keyes & Clark's Industrial Chemicals*, F. A. Lowenheim, M. K. Moran, Eds. (Wiley-Interscience, New York, 4th ed., 1975) pp 722-730. Toxicity studies: E. M. Boyd, M. N. Shanas, *Arch. Int. Pharmacodyn.* **144**, 86 (1963). Comprehensive monograph: D. W. Kaufmann, *Sodium Chloride*, ACS Monograph Series no. **145** (Reinhold, New York, 1960) 743 pp.

Cubic, white crystals, granules, or powder; colorless and transparent or translucent when in large crystals. d 2.17. The salt of commerce usually contains some calcium and magnesium chlorides which absorb moisture and make it cake. mp 804° and begins to volatilize at a little above this temp. One gram dissolves in 2.8 ml water at 25°, in 2.6 ml boiling water, in 10 ml glycerol; very slightly sol in alcohol. Its soly in water is decreased by HCl. Almost insol in concd HCl. Its aq soln is neutral. pH: 6.7-7.3. d of satd aq soln at 25° is 1.202. A 23% aq soln of sodium chloride freezes at −20.5°C (5°F). LD$_{50}$ orally in rats: 3.75 ±0.43 g/kg (Boyd, Shanas).

Note: **Blusalt**, a brand of sodium chloride contg trace amounts of cobalt, iodine, iron, copper, manganese, zinc is used in farm animals.

Caution: Not generally considered poisonous. Accidental substitution of NaCl for lactose in baby formulas has caused fatal poisoning.

USE: Natural salt is the source of chlorine and of sodium as well as of all, or practically all, their compds, e.g., hydrochloric acid, chlorates, sodium carbonate, hydroxide, etc.; for preserving foods; manuf soap, dyes—to salt them out; in freezing mixtures; for dyeing and printing fabrics, glazing pottery, curing hides; metallurgy of tin and other metals.

THERAP CAT: Electrolyte replenisher; emetic; topical anti-inflammatory.

THERAP CAT (VET): Essential nutrient factor. May be given orally as emetic, stomachic, laxative or to stimulate thirst (prevention of calculi). Intravenously as isotonic solution to raise blood volume, to combat dehydration. Locally as wound irrigant, rectal douche.

Fig. 23.1 Example of typical *Merck Index* entry above showing type of information given for each chemical. From *The Merck Index: An Encyclopedia of Chemicals, Drugs and Biologicals*, Thirteenth edition, O'Neil, M.J., Smith, A., Heckelman, P.E. and Obenchain, J.R. Jr. (eds). Merck & Co, Inc., Whitehouse Station, NJ, USA, 2001. Reproduced with permission. Copyright © 2001 by Merck & Co, Inc. All rights reserved.

Using chemicals

 Safety aspects

In practical classes, the person in charge has a responsibility to inform you of any hazards associated with the use of chemicals. For routine practical procedures, a risk assessment (p. 125) will have been carried out by a member of staff and relevant safety information will be included in the practical schedule: an example is shown in Table 23.1.

In project work, your first duty when using an unfamiliar chemical is to find out about its properties, especially those relating to safety. Your department must provide the relevant information to allow you to do this. If your supervisor has filled out the form, read it carefully before signing. Box 23.1 gives further advice.

 KEY POINT Before you use any chemical you must find out whether safety precautions need to be taken, and complete the appropriate forms confirming that you appreciate the risks involved.

Selection

Chemicals are supplied in various degrees of purity and this is always stated on the manufacturer's containers. Suppliers differ in the names given to the grades and there is no conformity in purity standards. Very pure chemicals cost more, sometimes a lot more, and should only be used if the situation demands. If you need to order a chemical, your department will have a defined procedure for doing this.

Preparing solutions

Solutions are usually prepared with respect to their molar concentrations (e.g. mmol l^{-1}, or mol m^{-3}), or mass concentrations (e.g. g l^{-1}, or kg m^{-3}):

Table 23.1 Representative risk assessment information for a practical exercise in molecular biology, involving the isolation of DNA

Substance	Hazards	Comments
Sodium dodecyl sulphate (SDS)	Irritant Toxic	Wear gloves
Sodium hydroxide (NaOH)	Highly corrosive Severe irritant	Wear gloves
Isopropanol	Highly flammable Irritant/corrosive Potential carcinogen	No naked flames Wear gloves
Phenol	Highly toxic Causes skin burns Potential carcinogen	Use in fume hood Wear gloves
Chloroform	Volatile and toxic Irritant/corrosive Potential carcinogen	Use in fume hood Wear gloves

Box 23.1 Safe working with chemicals

You should always treat chemicals as potentially dangerous, following these general precautions:

- **Do not use any chemical until you have considered the risks involved** – for lab classes, you should carefully read all hazard and risk information provided *before* you start work. In project work, you may need to be involved in the risk assessment process with your supervisor.
- **Wear a laboratory coat at all times** – the coat should be fully fastened and cleaned appropriately, should any chemical compound be spilled on it. Closed-toe footwear will protect your feet should any spillages occur.
- **Make sure you know where the safety apparatus is kept before you begin working** – this apparatus includes eye bath, fire extinguishers and blanket, first aid kit.
- **Wear safety glasses and gloves when working with toxic, irritant or corrosive chemicals, and for any substances where the hazards are not yet fully characterised** – make sure you understand the hazard warning signs (p. 126) along with any specific hazard coding system used in your department.

Carry out procedures with solid material in a fume cupboard.

- **Use aids such as pipette fillers to minimise the risk of contact with hazardous solutions** – these aids are further detailed on p. 129.
- **Never smoke, eat, drink or chew gum in a lab where chemicals are handled** – this will minimise the risk of ingestion.
- **Label all solutions appropriately** – use the appropriate hazard warning information (see p. 126/203/275).
- **Report all spillages of chemicals/solutions** – make sure that spillages are cleaned up properly.
- **Store hazardous chemicals only in the appropriate locations** – for example, a spark-proof fridge is required for flammable liquids; acids and solvents should not be stored together.
- **Dispose of chemicals in the correct manner** – if unsure, ask a member of staff (do not assume that it is safe to use the lab waste bin or the sink for disposal).
- **Wash hands after any direct contact with chemicals or biochemical material** – always wash your hands at the end of a lab session.

Examples
Using Eqn 23.1, 25 g of a substance dissolved in 400 ml of water would have a mass concentration of
$25 \div 400 = 0.0625 \text{ g ml}^{-1}$
$(\equiv 62.5 \text{ mg ml}^{-1} \equiv 62.5 \text{ g l}^{-1})$

Using Eqn 23.1, 0.4 mol of a substance dissolved in 0.5 litres of water would have a molar concentration of
$0.4 \div 0.5 = 0.8 \text{ mol l}^{-1} (\equiv 800 \text{ mmol l}^{-1})$.

Solving solubility problems – if your chemical does not dissolve after a reasonable time:

- check the limits of solubility for your compound (see Merck Index, O'Neil et al. 2006);
- check the pH of the solution – solubility often changes with pH, e.g. you may be able to dissolve the compound in an acidic or an alkaline solution.

both can be regarded as an amount of *substance* per unit volume of *solution*, in accordance with the relationship:

$$\text{Concentration} = \frac{\text{amount}}{\text{volume}} \qquad [23.1]$$

The most important aspect of Eqn [23.1] is to recognise clearly the units involved, and to prepare the solution accordingly: for molar concentrations, you will need the relative molecular mass of the compound, so that you can determine the mass of substance required. Further advice on concentrations and interconversion of units is given in Box 24.1.

Box 23.2 shows the steps involved in making up a solution. The concentration you require is likely to be defined by a protocol you are following, and the grade of chemical and supplier may also be specified. Success may depend on using the same source and quality, e.g. with enzyme work. To avoid waste, think carefully about the volume of solution you require, though it is always a good idea to err on the high side because you may spill some or make a mistake when dispensing it. Try to choose one of the standard volumes for vessels, as this will make measuring out easier.

Use distilled or deionised water to make up aqueous solutions and stir to make sure all the chemical is dissolved. Magnetic stirrers are the most convenient means of doing this: carefully drop a clean magnetic stirrer bar ('flea') in the beaker, avoiding splashing; place the beaker centrally on the stirrer plate, switch on the stirrer and gradually increase the speed of stirring. When the crystals or powder have completely dissolved, switch off and retrieve the flea with a magnet or another flea. Take care not to contaminate your solution when you do this, and rinse the flea with distilled water.

Box 23.2 How to make up an aqueous solution of known concentration from solid material

1. **Find out or decide the concentration of chemical required** and the degree of purity necessary.
2. **Decide on the volume of solution required.**
3. **Find out the relative molecular mass of the chemical (M_r).** This is the sum of the atomic (elemental) masses of the component elements and can be found on the container. If the chemical is hydrated, i.e. has water molecules associated with it, these must be included when calculating the mass required.
4. **Work out the mass of chemical that will give the concentration desired in the volume required.**
   Suppose your procedure requires you to prepare 250 ml of 0.1 mol l^{-1} NaCl.
   (a) Begin by expressing all volumes in the same units, either millilitres or litres (e.g. 250 ml as 0.25 litres).
   (b) Calculate the number of moles required from Eqn [23.1]: 0.1 = amount (mol) ÷ 0.25.
   By rearrangement, the required number of moles is thus 0.1 × 0.25 = 0.025 mol.
   (c) Convert from mol to g by multiplying by the relative molecular mass (M_r for NaCl = 58.44)
   (d) Therefore, you need to make up 0.025 × 58.44 = 1.461 g to 250 ml of solution, using distilled water.

   In some instances, it may be easier to work in SI units, though you must be careful when using exponential numbers (p. 475).
   Suppose your protocol states that you need 100 ml of 10 mmol l^{-1} KCl.
   (a) Start by converting this to 100×10^{-6} m^3 of 10 mol m^{-3} KCl.
   (b) The required number of mol is thus $(100 \times 10^{-6}) \times (10) = 10^3$.
   (c) Each mol of KCl weighs 72.56 g (M_r, the relative molecular mass).
   (d) Therefore you need to make up 72.56 × 10^{-3} g = 72.56 mg KCl to 100×10^{-6} m^3 (100 ml) with distilled water.
   See Box 24.1 for additional information.

5. **Weigh out the required mass of chemical to an appropriate accuracy.** If the mass is too small to weigh to the desired degree of accuracy, consider the following options:
   (a) Make up a greater volume of solution.
   (b) Make up a stock solution which can be diluted at a later stage (p. 134).
   (c) Weigh the mass first, and calculate what volume to make the solution up to afterwards using Eqn [23.1].
6. **Add the chemical to a beaker or conical flask then add a little less water than the final amount required.** If some of the chemical sticks to the paper, foil or weighing boat, use some of the water to wash it off.
7. **Stir and, if necessary, heat the solution to ensure all the chemical dissolves.** You can determine when this has happened visually by observing the disappearance of the crystals or powder.
8. **If required, check and adjust the pH of the solution when cool** (see p. 153).
9. **Make up the solution to the desired volume.** If the concentration needs to be accurate, use a class A volumetric flask; if a high degree of accuracy is not required, use a measuring cylinder (class B).
   (a) Pour the solution from the beaker into the measuring vessel using a funnel to avoid spillage.
   (b) Make up the volume so that the meniscus comes up to the appropriate measurement line (p. 128). For accurate work, rinse out the original vessel and use this liquid to make up the volume.
10. **Transfer the solution to a reagent bottle or a conical flask and label the vessel clearly.**

'Obstinate' solutions may require heating, but do this only if you know that the chemical will not be damaged at the temperature used. Use a stirrer-heater to keep the solution mixed as you heat it. Allow the solution to cool before you measure volume or pH as these change with temperature.

Stock solutions

Stock solutions are valuable when making up a range of solutions containing different concentrations of a reagent or if the solutions have some common ingredients. They also save work if the same solution is used over a prolonged period (e.g. a nutrient solution). The stock solution is more concentrated than the final requirement and is diluted as

Table 23.2 Use of stock solutions. Suppose you need a set of solutions 10 ml in volume containing differing concentrations of KCl, with and without reagent Q. You decide to make up a stock of KCl at twice the maximum required concentration (50 mmol l^{-1} = 50 mol m^{-3}) and a stock of reagent Q at twice its required concentration. The table shows how you might use these stocks to make up the media you require. Note that the total volumes of stock you require can be calculated from the table (end column)

Stock solutions	Volume of stock required to make required solutions (ml)						Total volume of stock required (ml)
	No KCl plus Q	No KCl minus Q	15 mmol l^{-1} KCl plus Q	15 mmol l^{-1} KCl minus Q	25 mmol l^{-1} KCl plus Q	25 mmol l^{-1} KCl minus Q	
50 mmol l^{-1} KCl	0	0	3	3	5	5	16
[reagent Q] × 2	5	0	5	0	5	0	15
Water	5	10	2	7	0	5	29
Total	10	10	10	10	10	10	60

appropriate when the final solutions are made up. The principle is best illustrated with an example (Table 23.2).

Preparing dilutions

Making a single dilution

You may need to dilute a stock solution to give a particular mass concentration, or molar concentration. Use the following procedure:

1. Transfer an accurate volume of stock solution to a volumetric flask, using appropriate equipment (Table 22.1).
2. Make up to the calibration mark with solvent – add the last few drops from a pipette or solvent bottle, until the meniscus is level with the calibration mark.
3. Mix thoroughly, either by repeated inversion (holding the stopper firmly) or by prolonged stirring, using a magnetic stirrer. Make sure you add the magnetic flea *after* the volume adjustment step.

Making a dilution – use the relationship $[C_1]V_1 = [C_2]V_2$ to determine volume or concentration (see Box 24.1).

Removing a magnetic flea from a volumetric flask – use a strong magnet to bring the flea to the top of the flask, to avoid contamination during removal.

For routine work using dilute aqueous solutions where the highest degree of accuracy is not required, it may be acceptable to substitute test tubes or conical flasks for volumetric flasks. In such cases, you would calculate the volumes of stock solution and diluent required, with the assumption that the final volume is determined by the sum of the individual volumes of stock and diluent used (e.g. Table 23.2). Thus, a twofold dilution would be prepared using 1 volume of stock solution and 1 volume of diluent. The dilution factor is obtained from the ratio of the initial concentration of the stock solution and the final concentration of the diluted solution. The dilution factor can be used to determine the volumes of stock and diluent required in a particular instance. For example, suppose you wanted to prepare 100 ml of a solution of NaCl at 0.2 mol l^{-1}. Using a stock solution containing 4.0 mol l^{-1} NaCl, the dilution factor is $0.2 \div 4.0 = 0.05 = 1/20$ (a twentyfold dilution). Therefore, the amount of stock solution required is 1/20th of 100 ml = 5 ml and the amount of diluent needed is 19/20ths of 100 ml = 95 ml.

Using the correct volumes for dilutions – it is important to distinguish between the volumes of the various liquids: a one-in-ten dilution is obtained using 1 volume of stock solution plus 9 volumes of diluent (1 + 9 = 10). Note that when this is shown as a ratio, it may represent the initial and final volumes (e.g. 1:10) or, sometimes, the volumes of stock solution and diluent (e.g. 1:9).

Preparing a dilution series

Dilution series are used in a wide range of procedures, including the preparation of standard curves for calibration of analytical instruments (Chapter 37), and in microbiology (Chapter 32) and immunoassay

Using diluents – *various liquids are used, including distilled or deionised water, salt solutions, buffers, Ringer's solution (p. 223), etc., according to the specific requirements of the procedure.*

(Chapter 38), where a range of dilutions of a particular sample is often required. A variety of different approaches can be used:

Linear dilution series Here, the concentrations are separated by an equal amount, e.g. a series containing protein at 0, 0.2, 0.4, 0.6, 0.8, 1.0 µg ml^{-1}. Such a dilution series might be used to prepare a calibration curve for spectrophotometric assay of protein concentration (Box 50.1), or an enzyme assay (p. 381). Use $[C_1]V_1 = [C_2]V_2$ (p. 145) to determine the amount of stock solution required for each member of the series, with the volume of diluent being determined by subtraction.

Logarithmic dilution series Here, the concentrations are separated by a constant proportion, often referred to as the step interval. This type of serial dilution is useful when a broad range of concentrations is required, e.g. for titration of biologically active substances (p. 238), making a plate count of a suspension of microbes (p. 206), or when a process is logarithmically related to concentration.

The most common examples are:

- Doubling dilutions – where each concentration is half that of the previous one (twofold step interval, $\log_2$ dilution series). First, make up the most concentrated solution at twice the volume required. Measure out half of this volume into a vessel containing the same volume of diluent, mix thoroughly and repeat, for as many doubling dilutions as are required. The concentrations obtained will be 1/2, 1/4, 1/8, 1/16, etc. times the original (i.e. the dilutions will be two, four, eight and sixteenfold, etc.).
- Decimal dilutions – where each concentration is one-tenth that of the previous one (tenfold step interval, $\log_{10}$ dilution series). First, make up the most concentrated solution required, with at least a 10% excess. Measure out one-tenth of the volume required into a vessel containing nine times as much diluent, mix thoroughly and repeat. The concentrations obtained will be 1/10, 1/100, 1/1000, etc. times the original (i.e. dilutions of 10^{-1}, 10^{-2}, 10^{-3}, etc.). To calculate the actual concentration of solute, multiply by the appropriate dilution factor.

When preparing serial doubling or decimal dilutions, it is often easiest to add the appropriate amount of diluent to several vessels beforehand, as shown in the worked example in Figure 23.2. When preparing a dilution series, it is essential that all volumes are dispensed accurately, e.g. using calibrated pipettors (p. 130), otherwise any inaccuracies will be compounded, leading to gross errors in the most dilute solutions.

Harmonic dilution series Here, the concentrations in the series take the values of the reciprocals of successive whole numbers, e.g. 1, 1/2, 1/3, 1/4, 1/5, etc. The individual dilutions are simply achieved by a stepwise increase in the volume of diluent in successive vessels, e.g. by adding 0, 1, 2, 3, 4 and 5 times the volume of diluent to a set of test tubes, then adding a constant unit volume of stock solution to each vessel. Although there is no dilution transfer error between individual dilutions, the main disadvantage is that the series is non-linear, with a step interval that becomes progressively smaller as the series is extended.

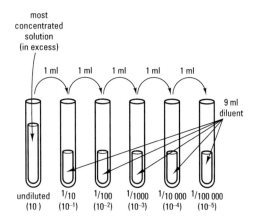

Fig. 23.2 Preparation of a dilution series. The example shown is a decimal dilution series, down to 1/100 000 (10^{-5}) of the solution in the first (left-hand) tube. Note that all solutions must be mixed thoroughly before transferring the volume to the next in the series. In microbiology and cell culture, sterile solutions and appropriate aseptic technique will be required (p. 203).

Preparing a dilution series using pipettes or pipettors – use a fresh pipette or disposable tip for each dilution, to prevent carry-over of solutions.

Solutions must be thoroughly mixed before measuring out volumes for the next dilution. Use a fresh measuring vessel for each dilution to avoid contamination, or wash your vessel thoroughly between dilutions. Clearly label the vessel containing each dilution when it is made: it is easy to get confused! When deciding on the volumes required, allow for the aliquot removed when making up the next member in the series. Remember to discard any excess from the last in the series if volumes are critical.

Mixing solutions and suspensions

Various devices may be used, including:

- Magnetic stirrers and fleas. Magnetic fleas come in a range of shapes and sizes, and some stirrers have integral heaters. During use, stirrer speed may increase as the instrument warms up.
- Vortex mixers. For vigorous mixing of small volumes of solution, e.g. when preparing a dilution series in test tubes. Take care when adjusting the mixing speed – if the setting is too low, the test tube will vibrate rather than creating a vortex, giving inadequate mixing. If the setting is too high, the test tube may slip from your hand.
- Orbital shakers and shaking water baths. These are used to provide controlled mixing at a particular temperature, e.g. for long-term incubation and cell growth studies (p. 233).
- Bottle rollers. For cell culture work, ensuring gentle, continuous mixing.

 SAFETY NOTE Using a vortex mixer with open or capped test tubes – do not vortex too vigorously or liquid will spill from the top of the tube, creating a contamination risk.

Storing chemicals and solutions

Labile chemicals may be stored in a fridge or freezer. Take special care when using chemicals that have been stored at low temperature: the container and its contents must be warmed up to room temperature before use, otherwise water vapour will condense on the chemical. This may render any weighing you do meaningless and it could ruin the chemical. Other chemicals may need to be kept in a desiccator, especially if they are deliquescent (water-absorbing).

 SAFETY NOTE Cleaning up chemical spillages – you must always clean up any spillages of chemicals, as you are the only person who knows the risks from the spilled material.

 KEY POINT Label all stored chemicals clearly with the following information: the chemical name (if a solution, state solute(s), concentration(s) and pH if measured), plus any relevant hazard warning information, the date made up, and your name.

Separating components of mixtures and solutions

Particulate solids (e.g. soils) can be separated on the basis of size using sieves. These are available in stacking forms which fit on automatic shakers. Sieves with the largest pores are placed at the top and the assembly is shaken for a fixed time until the sample separates. Suspensions of solids in liquids may be separated out by centrifugation (see p. 297) or filtration. Various forms of filter paper are available having different porosities and purities. Vacuum-assisted filtration speeds up the process and is best carried out with a filter funnel attached to a filter flask. Filtration through pre-sterilised membranes with very small pores (e.g. the Millipore type) is an excellent method of sterilising small volumes of solution. Solvents can be removed from solutes by heating, using rotary film evaporation under low pressure and, for water, by freeze drying. The last two are especially useful for heat-labile solutes – refer to the manufacturers' specific instructions for use.

Using balances

Electronic balances with digital readouts are now favoured over mechanical types: they are easy to read and their self-taring feature means the mass of the weighing boat or container can be subtracted automatically before weighing an object. The most common type offers accuracy down to 1 mg over the range 1 mg to 160 g, which is suitable for most biological applications.

To operate a standard self-taring balance:

1. Check that it is level, using the adjustable feet to centre the bubble in the spirit level (usually at the back of the machine). For accurate work, make sure a draught shield is on the balance.
2. Place an empty vessel in the middle of the balance pan and allow the reading to stabilise. *If the object is larger than the pan, take care that no part rests on the body of the balance or the draught shield as this will invalidate the reading.* Press the tare bar to bring the reading to zero.
3. Place the chemical or object carefully in the vessel (powdered chemicals should be dispensed with a suitably sized clean spatula). Take care to avoid spillages.
4. Allow the reading to stabilise and make a note of the value.
5. If you add excess chemical, take great care when removing it. Switch off if you need to clean any deposit accidentally left on or around the balance.

Larger masses should be weighed on a top-loading balance to an appropriate degree of accuracy. Take care to note the limits for the balance: while most have devices to protect against overloading, you may damage the mechanism. In the field, spring or battery-operated balances may be preferred. Try to find a place out of the wind to use them. For extremely small masses, there are electrical balances that can weigh down to 1 µg, but these are very delicate and must be used under supervision.

Measuring and controlling temperature

Heating samples

Care is required when heating samples – there is a danger of fire whenever organic material is heated and a danger of scalding from heated liquids. Safety glasses should always be worn. Use a thermostatically controlled electric stirrer-heater if possible. If using a Bunsen burner, keep the flame well away from yourself and your clothing (tie back long hair). Use a non-flammable mat beneath a Bunsen to protect the bench. Switch off when no longer required. To light a Bunsen, close the air hole first, then apply a lit match or lighter. Open the air hole if you need a hotter, more concentrated flame: the hottest part of the flame is just above the apex of the blue cone in its centre.

Ovens and drying cabinets may be used to dry specimens or glassware. They are normally thermostatically controlled. If drying organic material for dry weight measurement, do so at about 80 °C to avoid caramelising the sample. Always state the actual temperature used as this affects results. Check that all water has been driven off by weighing until a constant mass is reached.

Weighing – never weigh anything directly onto a balance's pan: you may contaminate it for other users. Use a weighing boat or a slip of aluminium foil. Otherwise, choose a suitable vessel like a beaker, conical flask or aluminium tray.

Deciding on which balance to use – select a balance that weighs to an appropriate number of decimal places. For example, you should use a top-loading balance weighing to one decimal place for less accurate work. Note that a weight of 6.4 g on such a balance may represent a true value of between 6.350 g and 6.449 g (to three decimal places).

Using a balance – it is poor technique to use a large container to weigh out a small amount of a chemical: you are attempting to make accurate measurements of a small difference between two large numbers. Instead, use a small weighing container.

SAFETY NOTE Heating/cooling glass vessels – take care if heating or cooling glass vessels rapidly as they may break when heat stressed. Freezing aqueous solutions in thin-walled glass vessels is risky because ice expansion may break the glass.

Using thermometers – some are calibrated for use in air, others require partial immersion in liquid and others total immersion – check before use.

SAFETY NOTE If a mercury thermometer is broken, report the spillage, as mercury is a poison.

Example Water vapour can be removed by passing gas over dehydrated $CaCO_3$, and CO_2 may be removed by bubbling through KOH solution.

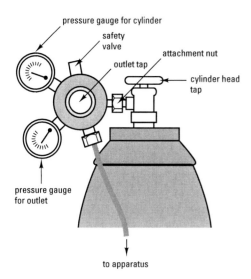

Fig. 23.3 Parts of a cylinder head regulator. The regulator is normally attached by tightening the attachment nut clockwise; the exception is with cylinders of hydrogen, where the special regulator is tightened *anticlockwise* to avoid the chance of this potentially explosive gas being incorrectly used.

Basic Laboratory Procedures

Cooling samples and specimens

Fridges and freezers are used for storing stock solutions and chemicals that would either break down or become contaminated at room temperature. Normal fridge and freezer temperatures are about 4 °C and −15 °C respectively. Ice baths can be used when reactants must be kept close to 0 °C. Most bioscience departments will have a machine which provides flaked ice for use in these baths. If common salt is mixed with ice, temperatures below 0 °C can be achieved. A mixture of ethanol and solid CO_2 will provide a temperature of −72 °C if required. To freeze a specimen quickly, immerse in liquid nitrogen (−196 °C) using tongs and wearing an apron and thick gloves, as splashes will damage your skin. Always work in a well-ventilated room.

Maintaining constant temperature

Thermostatically controlled temperature rooms and incubators can be used to maintain temperature at a desired level. Always check with a thermometer or thermograph that the thermostat is accurate enough for your study. To achieve a controlled temperature on a smaller scale, e.g. for an oxygen electrode (pp. 339–341), use a water bath. These usually incorporate heating elements, a circulating mechanism and a thermostat. Baths for sub-ambient temperatures have a cooling element.

Controlling atmospheric conditions

Gas composition

The atmosphere may be 'scrubbed' of certain gases by passing through a U-tube or Dreschel bottle containing an appropriate chemical or solution.

For accurate control of gas concentrations, use cylinders of pure gas; the contents can be mixed to give specified concentrations by controlling individual flow rates. The cylinder head regulator (Fig. 23.3) allows you to control the pressure (and hence flow rate) of gas; adjust using the controls on the regulator or with spanners of appropriate size. Before use, ensure the regulator outlet tap is off (turn anticlockwise), then switch on at the cylinder (turn clockwise) – the cylinder dial will give you the pressure reading for the cylinder contents. Now switch on at the regulator outlet (turn clockwise) and adjust to desired pressure/flow setting. To switch off, carry out the above directions in reverse order.

To control dissolved gas composition in liquids, either 'degas' under vacuum or bubble with another gas, e.g. when preparing oxygen-free liquids for HPLC (p. 305) by bubbling with nitrogen.

Pressure

Many forms of pump are used to pressurise or provide partial vacuum, usually to force gas or liquid movement. Each has specific instructions for use. Many laboratories are supplied with 'vacuum' (suction) and pressurised air lines that are useful for procedures such as vacuum-assisted filtration. Make sure you switch off the taps after use. Take special care with glass items kept at very low or high pressures. These should be contained within a metal cage to minimise the risk of injury (e.g. a vacuum desiccator, as shown in Fig. 23.4).

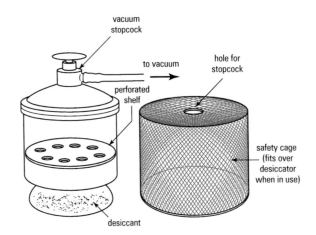

Fig. 23.4 Vacuum desiccator (with mesh safety cage).

Measuring time

Many experiments and observations need to be carefully timed. Large-faced stopclocks allow you to set and follow 'experimental time' and remove the potential difficulties in calculating this from 'real time' on a watch or clock. Some timers incorporate an alarm which you can set to warn when readings or operations must be carried out; 24-h timers are available for controlling light and temperature regimes.

Using a timer – always set the alarm before the critical time, so that you have adequate time to react.

Text references

Lide, D.R. (ed.) (2006) *CRC Handbook of Chemistry and Physics*, 87th edn. CRC Press, Boca Raton.

O'Neil, M.J., Smith, A. and Heckelman, P.E. (2006) *The Merck Index: An Encyclopedia of Chemicals, Drugs and Biologicals*, 14th edn. Merck & Co., Inc., Whitehouse Station.

Sources for further study

Anon. *CHEMnet BASE*. Available:
http://www.chemnetbase.com
Last accessed: 01/04/07.
[Online access to the *Handbook of Chemistry and Physics*.]

Jack, C.R. (1995) *Basic Biochemical Laboratory Procedures and Computing*. Oxford University Press, Oxford.

Seidman, L.A. and Moore, C.J. (2000) *Basic Laboratory Methods for Biotechnology: Textbook and Laboratory Reference*. Prentice Hall, New Jersey.

Study exercises

23.1 **Practise the calculations involved in preparing specific volumes of aqueous solutions (see also study exercise 24.1).** What mass of substance would be required to prepare each of the following (answer in each case to three significant figures):
(a) 100 ml of NaCl at 50 mmol l^{-1} (M_r of NaCl = 58.44)?
(b) 250 ml of mannitol at 0.10 mol l^{-1} (M_r of mannitol = 182.17)?
(c) 200 ml of a bovine serum albumin solution at 800 μg ml^{-1}?
(d) 0.5 litres of MgCl$_2$ prepared from the hexahydrate salt at 22.5 mmol l^{-1} (M_r of MgCl$_2$·6H$_2$O = 203.30)?
(e) 400 ml of DNA at 20 ng μl^{-1}?

Study exercises (continued)

23.2 Practise the calculations involved in preparing dilutions (answer in each case to three significant figures).
(a) If you added 1.0 ml of an aqueous solution of NaCl at 0.4 mol l^{-1} to 9.0 ml of water, what would be the final concentration of NaCl in mmol l^{-1}?
(b) If you added 25 ml of an aqueous solution of DNA at 10 µg ml^{-1} to a 500 ml volumetric flask and made it up to the specified volume with water, what would be the final concentration of DNA, in ng ml^{-1}?
(c) If you added 10 µl of an aqueous solution of sucrose at 200 mmol l^{-1} to a 250 ml volumetric flask and made it up to the specified volume with water, what would be the final concentration of sucrose, in nmol ml^{-1}?
(d) How would you prepare 250 ml of KCl at a final concentration of 20.0 mmol l^{-1} from a solution containing KCl at 0.2 mol l^{-1}?
(e) How would you prepare 1×10^{-3} m^3 of glucose at a final concentration of 50 µmol m^{-3} from a stock solution containing glucose at 20.0 g m^{-3} (M_r of glucose = 180.16)?

23.3 Practise the calculations involved in using stock solutions. Suppose you had the following stock solutions: NaCl 100.0 mmol l^{-1}; KCl 200.0 mmol l^{-1}; CaCl$_2$ 160.0 mmol l^{-1}; glucose 5.0 mmol l^{-1}. Calculate the volumes of each stock solution and the volume of water required to prepare each of the following (answer in each case to three significant figures):
(a) 1.0 ml of a solution containing only KCl at 10.0 mmol l^{-1}
(b) 50 ml of a solution containing NaCl at 2.5 mmol l^{-1} and glucose at 0.5 mmol l^{-1}
(c) 100 ml of a solution containing NaCl at 5.0 mmol l^{-1}, KCl at 2.5 mmol l^{-1}, CaCl$_2$ at 40.0 mmol l^{-1} and glucose at 0.25 mmol l^{-1}
(d) 10 ml of a solution containing NaCl, CaCl$_2$ and KCl, all at 20.0 µmol ml^{-1}
(e) 25 ml of a solution containing CaCl$_2$ at 8.0 mmol l^{-1}, KCl at 50.0 mmol l^{-1} and glucose at 20.0 nmol ml^{-1}.

24 Principles of solution chemistry

SAFETY NOTE Working with solutions – many solutes and solvents used in biosciences are potentially toxic, corrosive, oxidising or flammable, and they may also be carcinogenic (see p. 126). Further, there is a risk of accident involving the vessels used when preparing, storing and dispensing solutions (see Box 22.2).

Definitions

Electrolyte – a substance that dissociates, either fully or partially, in water to give two or more ions.

Relative atomic mass (A_r) – the mass of an atom relative to $^{12}C = 12$.

Relative molecular mass (M_r) – the mass of a compound's formula unit relative to $^{12}C = 12$.

Mole (of a substance) – the equivalent in mass to relative molecular mass in grams.

Expressing solute concentrations – you should use SI units wherever possible. However, you are likely to meet non-SI concentrations and you must be able to deal with these units too.

Example A 1.0 molar solution of NaCl would contain 58.44 g NaCl (the relative molecular mass) per litre of solution.

A solution is a homogeneous liquid, formed by the addition of solutes to a solvent (usually water in biological systems). The behaviour of solutions is determined by the type of solutes involved and by their proportions, relative to the solvent. Many laboratory exercises involve calculation of concentrations, e.g. when preparing an experimental solution at a particular concentration (p. 136), or when expressing data in terms of solute concentration (p. 135). Make sure that you understand the basic principles set out in this chapter before you tackle such exercises.

Solutes can affect the properties of solutions in several ways, including:

Electrolytic dissociation

This occurs where individual molecules of an electrolyte dissociate to give charged particles (ions). For a strong electrolyte, e.g. NaCl, dissociation is essentially complete. In contrast, a weak electrolyte, e.g. acetic acid, will be only partly dissociated, depending upon the pH and temperature of the solution (pp. 152–3).

Osmotic effects

These are the result of solute particles lowering the effective concentration of the solvent (water). These effects are particularly relevant to biological systems since membranes are far more permeable to water than to most solutes. Water moves across biological membranes from the solution with the higher effective water concentration to that with the lower effective water concentration (osmosis).

Ideal/non-ideal behaviour

This occurs because solutions of real substances do not necessarily conform to the theoretical relationships predicted for dilute solutions of so-called ideal solutes. It is often necessary to take account of the non-ideal behaviour of real solutions, especially at high solute concentrations (see Lide, 2006, and Robinson and Stokes, 2002 for appropriate data).

Concentration

In SI units (p. 183), the concentration of a solute in a solution is expressed in $mol\,m^{-3}$, which is convenient for most biological purposes. The concentration of a solute is usually symbolised by square brackets, e.g. [NaCl]. Details of how to prepare a solution using SI and non-SI units are given in Box 23.2.

A number of alternative ways of expressing the relative amounts of solute and solvent are in general use, and you may come across these terms in your practical work or in the literature:

Molarity

This is the term used to denote molar concentration, $[C]$, expressed as moles of solute per litre volume of solution ($mol\,l^{-1}$). This non-SI term continues to find widespread usage, in part because of the familiarity of working scientists with the term, but also because laboratory glassware is calibrated in millilitres and litres, making the preparation of molar and

Box 24.1 Useful procedures for calculations involving molar concentrations

1. **Preparing a solution of defined molarity.** For a solute of known relative molecular mass, M_r, the following relationship can be applied:

$$[C] = \frac{\text{mass of solute/relative molecular mass}}{\text{volume of solution}} \quad [24.1]$$

So, if you wanted to make up 200 ml (0.2 l) of an aqueous solution of NaCl (M_r 58.44) at a concentration of 500 mmol l^{-1} (0.5 mol l^{-1}), you could calculate the amount of NaCl required by inserting these values into Eqn [24.1]:

$$0.5 = \frac{\text{mass of solute}/58.44}{0.2}$$

which can be rearranged to

$$\text{mass of solute} = 0.5 \times 0.2 \times 58.44 = 5.844 \, \text{g}$$

The same relationship can be used to calculate the concentration of a solution containing a known amount of a solute, e.g. if 21.1 g of NaCl were made up to a volume of 100 ml (0.1 l), this would give

$$[\text{NaCl}] = \frac{21.1/58.44}{0.1} = 3.61 \, \text{mol l}^{-1}$$

2. **Dilutions and concentrations.** The following relationship is very useful if you are diluting (or concentrating) a solution:

$$[C_1]V_1 = [C_2]V_2 \quad [24.2]$$

where $[C_1]$ and $[C_2]$ are the initial and final concentrations, while V_1 and V_2 are their respective volumes: each pair must be expressed in the same units. Thus, if you wanted to dilute 200 ml of 0.5 mol l^{-1} NaCl to give a final molarity of 0.1 mol l^{-1}, then, by substitution into Eqn [24.2]:

$$0.5 \times 200 = 0.1 \times V_2$$

Thus $V_2 = 1\,000$ ml (in other words, you would have to add water to 200 ml of 0.5 mol l^{-1} NaCl to give a final volume of 1 000 ml to obtain a 0.1 mol l^{-1} solution).

3. **Interconversion.** A simple way of interconverting amounts and volumes of any particular solution is to divide the amount and volume by a factor of 10^3: thus a molar solution of a substance contains 1 mol l^{-1}, which is equivalent to 1 mmol ml^{-1}, or 1 µmol µl^{-1}, or 1 nmol nl^{-1}, etc. You may find this technique useful when calculating the amount of substance present in a small volume of solution of known concentration, e.g. to calculate the amount of NaCl present in 50 µl of a solution with a concentration (molarity) of 0.5 mol l^{-1} NaCl:

(a) this is equivalent to 0.5 µmol µl^{-1};

(b) therefore 50 µl will contain 50×0.5 µmol = 25 µmol.

Alternatively, you may prefer to convert to primary SI units, for ease of calculation (see Box 23.1).

The 'unitary method' (p. 477) is an alternative approach to these calculations.

---

millimolar solutions relatively straightforward. However, the symbols in common use for molar (M) and millimolar (mM) solutions are at odds with the SI system and many people now prefer to use mol l^{-1} and mmol l^{-1} respectively, to avoid confusion. Box 24.1 gives details of some useful approaches to calculations involving molarities.

Molality

Example A 0.5 molal solution of NaCl would contain $58.44 \times 0.5 = 29.22$ g NaCl per kg of water.

This is used to express the concentration of solute relative to the *mass* of solvent, i.e. mol kg^{-1}. Molality is a temperature-independent means of expressing solute concentration, rarely used except when the osmotic properties of a solution are of interest (p. 148).

Per cent composition (% w/w)

Example A 5% w/w sucrose solution contains 5 g sucrose and 95 g water (= 95 ml water, assuming a density of 1 g ml^{-1}) to give 100 g of solution.

This is the solute mass (in g) per 100 g solution. The advantage of this expression is the ease with which a solution can be prepared, since it simply requires each component to be pre-weighed (for water, a volumetric measurement may be used, e.g. using a measuring cylinder) and then mixed together. Similar terms are parts per thousand (‰), e.g. mg g^{-1}, and parts per million (ppm), e.g. µg g^{-1} (see below).

Per cent concentration (% w/v and % v/v)

For solutes added in solid form, this is the number of grams of solute per 100 ml solution. This is more commonly used than per cent composition, since solutions can be accurately prepared by weighing out the required amount of solute and then making this up to a known volume using a volumetric flask. The equivalent expression for liquid solutes is % v/v.

The principal use of mass/mass or mass/volume terms (including $g\,l^{-1}$) is for solutes whose relative molecular mass is unknown (e.g. cellular proteins), or for mixtures of certain classes of substance (e.g. total salt in sea water). You should *never* use the per cent term without specifying how the solution was prepared, i.e. by using the qualifier w/w, w/v or v/v. For mass concentrations, it is simpler to use mass per unit volume, e.g. $mg\,l^{-1}$, $\mu g\,\mu l^{-1}$, etc.

> **Example** A 5% w/v sucrose solution contains 5 g sucrose in 100 ml of solution. A 5% v/v glycerol solution would contain 5 ml glycerol in 100 ml of solution.
> Note that when water is the solvent this is often not specified in the expression, e.g. a 20% v/v ethanol solution contains 20% ethanol made up to 100 ml of solution using water.

Parts per million (ppm) and parts per billion (ppb) concentration

Ppm is a non-SI weight per volume (w/v) concentration term commonly used in quantitative analysis such as flame photometry, atomic absorption spectroscopy and gas chromatography, where low concentrations of solutes are analysed. The term ppm is equivalent to the expression of concentration as $\mu g\,ml^{-1}$ ($10^{-6}\,g\,ml^{-1}$) and a 1.0 ppm solution of a substance will have a concentration of $1.0\,\mu g\,ml^{-1}$ ($1.0 \times 10^{-6}\,g\,ml^{-1}$).

Parts per billion (ppb) is an extension of this concentration term as $ng\,ml^{-1}$ ($10^{-9}\,g\,ml^{-1}$) and is commonly used to express concentrations of very dilute solutions. For example, the allowable concentration of arsenic in water is 0.05 ppm, more conveniently expressed as 50 ppb. Also note that ppm and ppb are sometimes used as a weight per weight (w/w) term, e.g. $\mu g\,g^{-1}$ and $ng\,g^{-1}$ respectively.

> **Alternative expressions** – parts per thousand (‰) w/v can be used as an alternative to $g\,l^{-1}$

> **Example** The concentration of a NaCl solution is stated as 3 ppm. This is equivalent to $3\,\mu g\,ml^{-1}$ ($3\,mg\,l^{-1}$). The relative molecular mass of NaCl is $58.44\,g\,mol^{-1}$, so the solution has a concentration of $3 \times 10^{-6} \div 58.44\,mol\,ml^{-1} = 5.13 \times 10^{-8}\,mol\,ml^{-1} = 0.0513\,\mu mol\,ml^{-1} = 51.3\,\mu mol\,l^{-1}$.

Activity (a)

This is a term used to describe the *effective* concentration of a solute. In dilute solutions, solutes can be considered to behave according to ideal (thermodynamic) principles, i.e. they will have an effective concentration equivalent to the actual concentration. However, in concentrated solutions ($\geqslant 500\,mol\,m^{-3}$), the behaviour of solutes is often non-ideal, and their effective concentration (activity) will be less than the actual concentration $[C]$. The ratio between the effective concentration and the actual concentration is called the activity coefficient (γ) where

$$\gamma = \frac{a}{[C]} \qquad [24.3]$$

Equation [24.3] can be used for SI units ($mol\,m^{-3}$), molarity ($mol\,l^{-1}$) or molality ($mol\,kg^{-1}$). In all cases, γ is a dimensionless term, since a and $[C]$ are expressed in the same units. The activity coefficient of a solute is effectively unity in dilute solution, decreasing as the solute concentration increases (Table 24.1). At high concentrations of certain ionic solutes, γ may increase to become greater than unity.

Table 24.1 Activity coefficient of NaCl solutions as a function of molality. Data from Robinson and Stokes (2002)

Molality	Activity coefficient at 25 °C
0.1	0.778
0.5	0.681
1.0	0.657
2.0	0.668
4.0	0.783
6.0	0.986

> **KEY POINT** Activity is often the correct expression for theoretical relationships involving solute concentration (e.g. where a property of the solution is dependent on concentration). However, for most practical purposes, it is possible to use the actual concentration of a solute rather than the activity, since the difference between the two terms can be ignored for dilute solutions.

Example A solution of NaCl with a molality of 0.5 mol kg^{-1} has an activity coefficient of 0.681 at 25 °C and a molal activity of $0.5 \times 0.681 = 0.340$ mol kg^{-1}.

The particular use of the term 'water activity' is considered below, since it is based on the mole fraction of solvent, rather than the effective concentration of solute.

Equivalent mass (equivalent weight)

Equivalence and normality are outdated terms, although you may come across them in older texts. They apply to certain solutes whose reactions involve the transfer of charged ions, e.g. acids and alkalis (which may be involved in H$^+$ or OH$^-$ transfer), and electrolytes (which form cations and anions that may take part in further reactions). These two terms take into account the valency of the charged solutes. Thus the equivalent mass of an ion is its relative molecular mass divided by its valency (ignoring the sign), expressed in grams per equivalent (eq) according to the relationship:

Examples For carbonate ions (CO$_3^{2-}$), with a relative molecular mass of 60.00 and a valency of 2, the equivalent mass is $60.00/2 = 30.00$ g eq^{-1}.
For sulphuric acid (H$_2$SO$_4$, relative molecular mass 98.08), where 2 hydrogen ions are available, the equivalent mass is $98.08/2 = 49.04$ g eq^{-1}.

$$\text{equivalent mass} = \frac{\text{relative molecular mass}}{\text{valency}} \quad [24.4]$$

For acids and alkalis, the equivalent mass is the mass of substance that will provide 1 mol of either H$^+$ or OH$^-$ ions in a reaction, obtained by dividing the molecular mass by the number of available ions (n), using n instead of valency as the denominator in Eqn [24.4].

Normality

Example A 0.5 N solution of sulphuric acid would contain $0.5 \times 49.04 = 24.52$ g l^{-1}.

A 1 normal solution (1 N) is one that contains one equivalent mass of a substance per litre of solution. The general formula is:

$$\text{normality} = \frac{\text{mass of substance per litre}}{\text{equivalent mass}} \quad [24.5]$$

Osmolarity

Example Under ideal conditions, 1 mol of NaCl dissolved in water would give 1 mol of Na$^+$ ions and 1 mol of Cl$^-$ ions, equivalent to a theoretical osmolarity of 2 osmol l^{-1}.

This non-SI expression is used to describe the number of moles of osmotically active solute particles per litre of solution (osmol l^{-1}). The need for such a term arises because some molecules dissociate to give more than one osmotically active particle in aqueous solution.

Osmolality

This term describes the number of moles of osmotically active solute particles per unit mass of solvent (osmol kg^{-1}). For an ideal solute, the osmolality can be determined by multiplying the molality by n, the number of solute particles produced in solution (e.g. for NaCl, $n = 2$). However, for real (i.e. non-ideal) solutes, a correction factor (the osmotic coefficient, ϕ) is used:

Example A 1.0 mol kg^{-1} solution of NaCl has an osmotic coefficient of 0.936 at 25 °C and an osmolality of $1.0 \times 2 \times 0.936 = 1.872$ osmol kg^{-1}.

$$\text{osmolality} = \text{molality} \times n \times \phi \quad [24.6]$$

If necessary, the osmotic coefficients of a particular solute can be obtained from tables (e.g. Table 24.2): non-ideal behaviour means that ϕ may have values >1 at high concentrations. Alternatively, the osmolality of a solution can be measured using an osmometer.

Table 24.2 Osmotic coefficients of NaCl solutions as a function of molality. Data from Robinson and Stokes (2002)

Molality	Osmotic coefficient at 25 °C
0.1	0.932
0.5	0.921
1.0	0.936
2.0	0.983
4.0	1.116
6.0	1.271

Colligative properties and their use in osmometry

Several properties vary in direct proportion to the effective number of osmotically active solute particles per unit mass of solvent and can be used

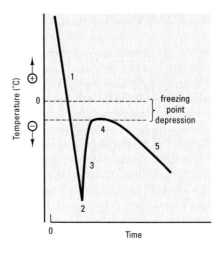

Fig. 24.1 Temperature responses of a cryoscopic osmometer. The response can be subdivided into:
1. initial supercooling
2. initiation of crystallisation
3. crystallisation/freezing
4. plateau, at the freezing point
5. slow temperature decrease.

to determine the osmolality of a solution. These colligative properties include freezing point, boiling point and vapour pressure.

An osmometer is an instrument which measures the osmolality of a solution, usually by determining the freezing point depression of the solution in relation to pure water, a technique known as cryoscopic osmometry. A small amount of sample is cooled rapidly and then brought to the freezing point (Fig. 24.1), which is measured by a temperature-sensitive thermistor probe calibrated in $mosmol\,kg^{-1}$. An alternative method is used in vapour pressure osmometry, which measures the relative decrease in the vapour pressure produced in the gas phase when a small sample of the solution is equilibrated within a chamber.

Using an osmometer – it is vital that the sample holder and probe are clean, otherwise small droplets of the previous sample may be carried over, leading to inaccurate measurement.

Osmotic properties of solutions

Several interrelated terms can be used to describe the osmotic status of a solution. In addition to osmolality, you may come across the following:

Osmotic pressure

This is based on the concept of a membrane permeable to water, but not to solute molecules. For example, if a sucrose solution is placed on one side and pure water on the other, then a passive driving force will be created and water will diffuse across the membrane into the sucrose solution, since the effective water concentration in the sucrose solution will be lower (see Fig. 24.2). The tendency for water to diffuse into the sucrose solution could be counteracted by applying a hydrostatic pressure equivalent to the passive driving force. Thus, the osmotic pressure of a solution is the excess hydrostatic pressure required to prevent the net flow of water into a vessel containing the solution. The SI unit of osmotic pressure is the pascal, Pa ($= kg\,m^{-1}\,s^{-2}$). Older sources may use atmospheres, or bars, and conversion factors are given in Table 29.1. Osmotic pressure and osmolality can be interconverted using the expression $1\,osmol\,kg^{-1} = 2.479\,MPa$ at 25 °C.

Example A $1.0\,mol\,kg^{-1}$ solution of NaCl at 25 °C has an osmolality of $1.872\,osmol\,kg^{-1}$ and an osmotic pressure of $1.872 \times 2.479 = 4.641\,MPa$.

The use of osmotic pressure has been criticised as misleading, since a solution does not exhibit an 'osmotic pressure' unless it is placed on the other side of a selectively permeable membrane from pure water.

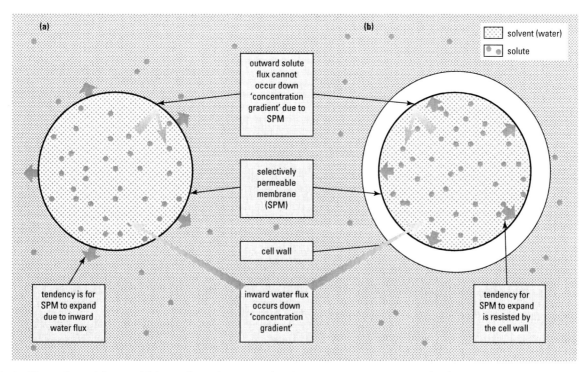

Fig. 24.2. Illustration of forces driving solvent (water) and solute movement across a selectively-permeable membrane (SPM). Energetically, both solutes and solvents tend to move down their respective 'concentration gradient' (strictly, down their chemical potential gradient). However, solute molecules cannot leave the model cells illustrated because they cannot pass through the SPM. In the situation illustrated in (a), water will tend to move from outside the cell to within because the solute molecules have effectively 'diluted' the water within the cell (illustrated by the density of point shading), creating a gradient in 'concentration' and because this molecule is able to pass through the SPM, the result will be an expansion of this model cell (short arrows). The osmotic pressure is the (theoretical) pressure that would need to be applied to prevent this. If the model cell were surrounded by a cell wall, as in (b), this would resist expansion, leading to internal pressurisation (turgor pressure, p. 150).

Water activity (a_w)

This is a term often used to describe the osmotic behaviour of microbial cells. It is a measure of the relative proportion of water in a solution, expressed in terms of its mole fraction, i.e. the ratio of the number of moles of water (n_w) to the total number of moles of all substances (i.e. water and solutes) in solution (n_t), taking into account the molal activity coefficient of the solvent, water (i.e. γ_w):

$$a_w = \gamma_w \frac{n_w}{n_t} \quad [24.7]$$

The water activity of pure water is unity, decreasing as solutes are added. One disadvantage of a_w is the limited change which occurs in response to a change in solute concentration: a 1.0 mol kg^{-1} solution of NaCl has a water activity of 0.967 (Table 24.3).

Osmolality, osmotic pressure and water activity are measurements based solely on the osmotic properties of a solution, with no regard for any other driving forces, e.g. hydrostatic and gravitational forces. In circumstances where such other forces are important, you will need to measure a variable that takes into account these aspects of water status, namely water potential.

Table 24.3 Water activity (a_w) of NaCl solutions as a function of molality. Data from Robinson and Stokes (2002)

Molality	a_w
0.1	0.997
0.5	0.984
1.0	0.967
2.0	0.932
4.0	0.852
6.0	0.760

Examples A 1.0 mol kg^{-1} solution of NaCl has a (negative) water potential of -4.641 MPa.

Pure water at 0.2 MPa pressure (about 0.1 MPa above atmospheric pressure) has a (positive) water potential of 0.1 MPa.

Water potential (hydraulic potential) and its applications

Water potential, Ψ_w, is the most appropriate measure of osmotic status in many areas of the biosciences. It is a term derived from the chemical potential of water. It expresses the difference between the chemical potential of water in the test system and that of pure water under standard conditions and has units of pressure (i.e. Pa). It is a more appropriate term than osmotic pressure because it is based on sound theoretical principles and because it can be used to predict the direction of passive movement of water, since water will flow down a gradient of chemical potential (i.e. osmosis occurs from a solution with a higher water potential to one with a lower water potential). Pure water at 20 °C and at 0.1 MPa pressure (i.e. $\approx$ atmospheric) has a water potential of zero. The addition of solutes will lower the water potential (i.e. make it negative), while the application of pressure, e.g. from hydrostatic or gravitational forces, will raise it (i.e. make it positive).

Often, the two principal components of water potential are referred to as the solute potential, or osmotic potential (Ψ_s, sometimes symbolised as Ψ_π or π) and the hydrostatic pressure potential (Ψ_p) respectively. For a solution at atmospheric pressure, the water potential is due solely to the presence of osmotically active solute molecules (osmotic potential) and may be calculated from the measured osmolality (osmol kg^{-1}) at 25 °C, using the relationship:

$$\Psi_w (\text{MPa}) = \Psi_s (\text{MPa}) = -2.479 \times \text{osmolality} \qquad [24.8]$$

For aquatic microbial cells, e.g. algae, fungi and bacteria, equilibrated in their growth medium at atmospheric pressure, the water potential of the external medium will be equal to the cellular water potential ('isotonic') and the latter can be derived from the measured osmolality of the medium (Eqn [24.8]) by osmometry (p. 142). The water potential of such cells can be subdivided into two major parts, the cell solute potential (Ψ_s) and the cell turgor pressure (Ψ_p) as follows:

$$\Psi_w = \Psi_s + \Psi_p \qquad [24.9]$$

To calculate the relative contribution of the osmotic and pressure terms in Eqn [24.9], an estimate of the internal osmolality is required, e.g. by measuring the freezing point depression of expressed intracellular fluid. Once you have values for Ψ_w and Ψ_s, the turgor pressure can be calculated by substitution into Eqn [24.9].

Measuring water potential – Eqn [24.9] ignores the effects of gravitational forces – for systems where gravitational effects are important an additional term is required (Nobel, 2005).

For terrestrial plant cells, the water potential may be determined directly using a vapour pressure osmometer, by placing a sample of the material within the osmometer chamber and allowing it to equilibrate. If Ψ_s of expressed sap is then measured, Ψ_p can be determined from Eqn [24.9].

The van't Hoff relationship can be used to estimate Ψ_s, by summation of the osmotic potentials due to the major solutes, determined from their concentrations, as:

$$\Psi_s = -RTn\phi[C] \qquad [24.10]$$

where RT is the product of the universal gas constant and absolute temperature (2 479 J mol^{-1} at 25 °C), n and ϕ are as previously defined and $[C]$ is expressed in SI terms as mol m^{-3}.

Text references

Lide, D.R. (ed.) (2006) *CRC Handbook of Chemistry and Physics*, 87th edn. CRC Press, Boca Raton.

Nobel, P.S. (2005) *Physicochemical and Environmental Plant Physiology*, 2nd edn. Academic Press, New York.

Robinson, R.A. and Stokes, R.H. (2002) *Electrolyte Solutions*. Dover Publications, New York.

Sources for further study

Burtis, C.A. and Ashwood, E.R. (2001) *Fundamentals of Clinical Chemistry*, 5th edn. Saunders, Philadelphia.

Chapman, C. (1998) *Basic Chemistry for Biology*. McGraw-Hill, New York.

O'Neil, M.J., Smith, A. and Heckelman, P.E. (2006) *The Merck Index: An Encyclopedia of Chemicals, Drugs and Biologicals*, 14th edn. Merck & Co., Inc., Whitehouse Station.

Seidman, L.A. and Moore, C.J. (2000) *Basic Laboratory Methods for Biotechnology: Textbook and Laboratory Reference*. Prentice Hall, New Jersey.

Study exercises

24.1 Practise calculations involving molar concentrations (see also study exercises 23.1 and 23.2). What mass of substance would be required to prepare each of the following aqueous solutions (answer in grams, to three decimal places in each case):
  (a) 1 litre of NaCl at a concentration of 1 molar? (M_r of NaCl = 58.44.)
  (b) 250 ml of $CaCl_2$ at 100 mmol l^{-1}? (M_r of $CaCl_2$ = 110.99.)
  (c) 2.5 l of mannitol at 10 nmol μl^{-1}? (M_r of mannitol = 182.17.)
  (d) 400 ml of KCl at 5% w/v?
  (e) 250 ml of glucose at 2.50 mol m^{-3}? (M_r of glucose = 180.16.)

24.2 Practise expressing concentrations in different ways. Express all answers to three significant figures:
  (a) What is 5 g l^{-1} sucrose, expressed in terms of molarity? (M_r of sucrose = 342.3.)
  (b) What is 1.0 mol m^{-3} NaCl, expressed in g l^{-1}? (M_r of NaCl = 58.44.)
  (c) What is 5% v/v ethanol, expressed in terms of molarity? (M_r of ethanol = 46.06 and density of ethanol at 25°C = 0.789 g ml^{-1}.)
  (d) What is 150 mmol l^{-1} glucose, expressed in terms of per cent concentration (% w/v)? (M_r of glucose = 180.16.)
  (e) What is a 1.0 molal solution of KCl, expressed as per cent composition (% w/w)? (M_r of KCl = 74.55.)

24.3 Calculate osmolality and osmotic potentials. Answer to three significant figures in all cases.
  (a) Assuming NaCl, KCl and $CaCl_2$ behave according to ideal thermodynamic principles, what would be the predicted osmolality of a solution containing:
   (i) NaCl alone, at 50 mmol kg^{-1}?
   (ii) KCl at 200 mmol kg^{-1} and $CaCl_2$ at 40 mmol kg^{-1}?
   (iii) NaCl at 100 mmol kg^{-1}, KCl at 60 mmol kg^{-1} and $CaCl_2$ at 75 mmol kg^{-1}?
  (b) What is the predicted osmotic pressure and osmotic potential of each of the solutions in (a) at 25°C?

25 pH and buffer solutions

Definitions

Acid – a compound that acts as a proton donor in aqueous solution.

Base – a compound that acts as a proton acceptor in aqueous solution.

Conjugate pair – an acid together with its corresponding base.

Alkali – a compound that liberates hydroxyl ions when it dissociates. Since hydroxyl ions are strongly basic, this will reduce the proton concentration.

Ampholyte – a compound that can act as both an acid and a base. Water is an ampholyte since it may dissociate to give a proton and a hydroxyl ion (amphoteric behaviour).

 SAFETY NOTE Working with strong acids or alkalis – these can be highly corrosive; rinse with plenty of water, if spilled.

Table 25.1 Effects of temperature on the ion product of water (K_w), H^+ ion concentration and pH at neutrality. Values calculated from Lide (2006)

Temp. (°C)	K_w (mol^2 l^{-2})	[H$^+$] at neutrality (nmol l^{-1})	pH at neutrality
0	0.11×10^{-14}	33.9	7.47
4	0.17×10^{-14}	40.7	7.39
10	0.29×10^{-14}	53.7	7.27
20	0.68×10^{-14}	83.2	7.08
25	1.01×10^{-14}	100.4	7.00
30	1.47×10^{-14}	120.2	6.92
37	2.39×10^{-14}	154.9	6.81
45	4.02×10^{-14}	199.5	6.70

Example Human blood plasma has a typical H$^+$ concentration of approximately 0.4×10^{-7} mol l^{-1} ($= 10^{-7.4}$ mol l^{-1}), giving a pH of 7.4.

pH is a measure of the amount of hydrogen ions (H$^+$) in a solution: this affects the solubility of many substances and the activity of most biological systems from individual molecules to whole organisms. It is usual to think of aqueous solutions as containing H$^+$ ions (protons), though protons actually exist in their hydrated form as hydronium ions (H$_3$O$^+$). The proton concentration of an aqueous solution [H$^+$] is affected by several factors:

- Ionisation (dissociation) of water, which liberates protons and hydroxyl ions in equal quantities, according to the reversible relationship:

$$H_2O \rightleftharpoons H^+ + OH^- \qquad [25.1]$$

- Dissociation of acids, according to the equation:

$$H\text{–}A \rightleftharpoons H^+ + A^- \qquad [25.2]$$

where H–A represents the acid and A$^-$ is the corresponding conjugate base. The dissociation of an acid in water will increase the amount of protons, reducing the amount of hydroxyl ions as water molecules are formed (Eqn [25.1]). The addition of a base (usually, as its salt) to water will decrease the amount of H$^+$, due to the formation of the conjugate acid (Eqn [25.2]).

Dissociation of alkalis, according to the relationship:

$$X\text{–}OH \rightleftharpoons X^+ + OH^- \qquad [25.3]$$

where X–OH represents the undissociated alkali. Since the dissociation of water is reversible (Eqn [25.1]), in an aqueous solution the production of hydroxyl ions will effectively act to 'mop up' protons, lowering the proton concentration.

Many compounds act as acids, bases or alkalis: those which are almost completely ionised in solution are usually called strong acids or bases, while weak acids or bases are only slightly ionised in solution.

In an aqueous solution, most of the water molecules are not ionised. In fact, the extent of ionisation of pure water is constant at any given temperature and is usually expressed in terms of the ion product (or ionisation constant) of water, K_w:

$$K_w = [H^+][OH^-] \qquad [25.4]$$

where [H$^+$] and [OH$^-$] represent the molar concentration (strictly, the activity) of protons and hydroxyl ions in solution, expressed as mol l^{-1}. At 25 °C, the ion product of pure water (Table 25.1) is 10^{-14} mol^2 l^{-2} (i.e. 10^{-8} mol^2 m^{-6}). This means that the concentration of protons in solution will be 10^{-7} mol l^{-1} (10^{-4} mol m^{-3}), with an equivalent concentration of hydroxyl ions (Eqn [25.1]). Since these values are very low and involve negative powers of 10, it is customary to use the pH scale, where:

$$pH = -\log_{10}[H^+] \qquad [25.5]$$

and [H$^+$] is the proton activity in mol l^{-1} (see p. 146).

KEY POINT *While pH is strictly the negative logarithm (to the base 10) of H^+ activity, in practice H^+ concentration in $mol\,l^{-1}$ (equivalent to $kmol\,m^{-3}$ in SI terminology) is most often used in place of activity since the two are virtually the same, given the limited dissociation of H_2O. The pH scale is not SI: nevertheless, it continues to be used widely in biological science.*

The value where an equal amount of H^+ and OH^- ions are present is termed neutrality: at 25 °C the pH of pure water at neutrality is 7.0. At this temperature, pH values below 7.0 are acidic while values above 7.0 are alkaline.

Always remember that the pH scale is a logarithmic one, not a linear one: a solution with a pH of 3.0 is not twice as acidic as a solution of pH 6.0, but one thousand times as acidic (i.e. contains 1000 times the amount of H^+ ions). Therefore, you may need to convert pH values into proton concentrations before you carry out mathematical manipulations (see Box 66.2). For similar reasons, it is important that pH change is expressed in terms of the original and final pH values, rather than simply quoting the difference between the values: a pH change of 0.1 has little meaning unless the initial or final pH is known.

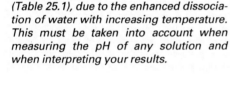

Measuring pH – the pH of a neutral solution changes with temperature (Table 25.1), due to the enhanced dissociation of water with increasing temperature. This must be taken into account when measuring the pH of any solution and when interpreting your results.

SAFETY NOTE *Preparing a dilute acid solution using concentrated acid – always slowly add the concentrated acid to water, not the reverse, since the strongly exothermic process can trigger a violent reaction with water.*

SAFETY NOTE *Preparing an alkali solution – typically, the alkali will be in solid form (e.g. NaOH) and addition to water will rapidly raise the temperature of the solution: use only heat-resistant glassware, cooled with water, if necessary.*

Measuring pH

pH electrodes

Accurate pH measurements can be made using a pH electrode, coupled to a pH meter. The pH electrode is usually a combination electrode, comprising two separate systems: an H^+-sensitive glass electrode and a reference electrode which is unaffected by H^+ ion concentration (Fig. 25.1). When this is immersed in a solution, a pH-dependent voltage between the two electrodes can be measured using a potentiometer. In most cases, the pH electrode assembly (containing the glass and reference electrodes) is connected to a separate pH meter by a cable, although some hand-held instruments (pH probes) have the electrodes and meter within the same assembly, often using an H^+-sensitive field effect transistor in place of a glass electrode, to improve durability and portability.

Box 25.1 gives details of the steps involved in making a pH measurement with a glass pH electrode and meter.

pH indicator dyes

These compounds (usually weak acids) change colour in a pH-dependent manner. They may be added in small amounts to a solution, or they can be used in paper strip form. Each indicator dye usually changes colour over a restricted pH range, typically 1–2 pH units (Table 25.2): universal indicator dyes/papers make use of a combination of individual dyes to measure a wider pH range. Dyes are not suitable for accurate pH measurement as they are affected by other components of the solution including oxidising and reducing agents and salts. However, they are useful for:

- estimating the approximate pH of a solution;
- determining a change in pH, for example at the end-point of a titration or the production of acids during bacterial metabolism (pp. 217–18);
- establishing the approximate pH of intracellular compartments, for example the use of neutral red as a 'vital' stain (p. 167).

Table 25.2 Properties of some pH indicator dyes

Dye	Acid–base colour change	Useful pH range
Thymol blue (acid)	red–yellow	1.2–6.8
Bromophenol blue	yellow–blue	1.2–6.8
Congo red	blue–red	3.0–5.2
Bromocresol green	yellow–blue	3.8–5.4
Resazurin	orange–violet	3.8–6.5
Methyl red	red–yellow	4.3–6.1
Litmus	red–blue	4.5–8.3
Bromocresol purple	yellow–purple	5.8–6.8
Bromothymol blue	yellow–blue	6.0–7.6
Neutral red	red–yellow	6.8–8.0
Phenol red	yellow–red	6.8–8.2
Thymol blue (alkaline)	yellow–blue	8.0–9.6
Phenolphthalein	none–red	8.3–10.0

Box 25.1 Using a glass pH electrode and meter to measure the pH of a solution

The following procedure should be used whenever you make a pH measurement: consult the manufacturer's handbook for specific information, where necessary. Do not be tempted to miss out any of the steps detailed below, particularly those relating to the effects of temperature, or your measurements are likely to be inaccurate.

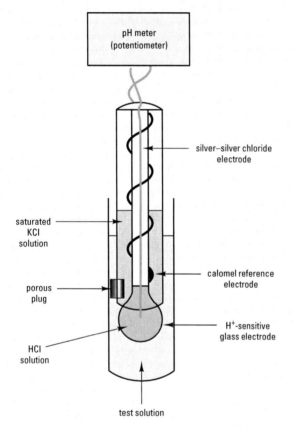

Fig. 25.1 Measurement of pH using a combination pH electrode and meter. The electrical potential difference recorded by the potentiometer is directly proportional to the pH of the test solution.

1. **Stir the test solution thoroughly before you make any measurement:** it is often best to use a magnetic stirrer. Leave the solution for sufficient time to allow equilibration at lab temperature.

2. **Record the temperature of every solution you use,** including all calibration standards and samples, since this will affect K_w, neutrality and pH.

3. **Set the temperature compensator on the meter to the appropriate value.** This control makes an allowance for the effect of temperature on the electrical potential difference recorded by the meter: it does *not* allow for the other temperature-dependent effects mentioned elsewhere. Basic instruments have no temperature compensator, and should only be used at a specified temperature, either 20 °C or 25 °C, otherwise they will not give an accurate measurement. More sophisticated systems have automatic temperature compensation.

4. **Rinse the electrode assembly with distilled water** and gently dab off the excess water onto a clean tissue: check for visible damage or contamination of the glass electrode (consult a member of staff if the glass is broken or dirty). Also check that the solution within the glass assembly is covering the metal electrode.

5. **Calibrate the instrument:** set the meter to 'pH' mode, if appropriate, and then place the electrode assembly in a standard solution of known pH, usually pH 7.00. This solution may be supplied as a liquid, or may be prepared by dissolving a measured amount of a calibration standard in water: calibration standards are often provided in tablet form, to be dissolved in water to give a particular volume of solution. Adjust the calibration control to give the correct reading. Remember that your calibration standards will only give the specified pH at a particular temperature, usually either 20 °C or 25 °C. If you are working at a different temperature, you must establish the actual pH of your calibration standards, either from the supplier, or from literature information.

6. **Remove the electrode assembly from the calibration solution and rinse again with distilled water:** dab off the excess water. Basic instruments have no further calibration steps (single-point calibration), while the more refined pH meters have additional calibration procedures.

   If you are using a basic instrument, you should check that your apparatus is accurate over the appropriate pH range by measuring the pH of another standard whose pH is close to that expected for the test solution. If the standard does not give the expected reading, the instrument is not functioning correctly: consult a member of staff.

   If you are using an instrument with a slope control function, this will allow you to correct for any deviation in electrical potential from that predicted by the theoretical relationship (at 25 °C, a change in pH of 1.00 unit should result in a change

Box 25.1 (continued)

in electrical potential of 59.16 mV) by performing a two-point calibration. Having calibrated the instrument at pH 7.00, immerse in a second standard at the same temperature as that of the first standard, usually buffered to either pH 4.00 or pH 9.00, depending upon the expected pH of your samples. Adjust the slope control until the exact value of the second standard is achieved (Fig. 25.2). A pH electrode and meter calibrated using the two-point method will give accurate readings over the pH range from 3 to 11: laboratory pH electrodes are not accurate outside this range, since the theoretical relationship between electrical potential and pH is no longer valid.

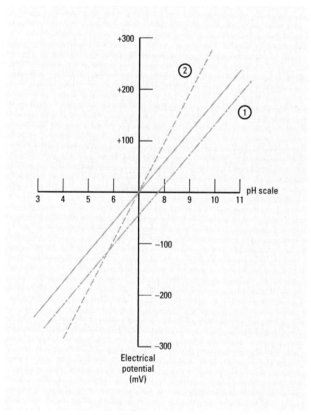

Fig. 25.2 The relationship between electrical potential and pH. The solid line shows the response of a calibrated electrode while the other plots are for instruments requiring calibration: 1 has the correct slope but incorrect isopotential point (calibration control adjustment is needed); 2 has the correct isopotential point but incorrect slope (slope control adjustment is needed).

7. **Once the instrument is calibrated, measure the pH of your solution(s)**, making sure that the electrode assembly is rinsed thoroughly between measurements. You should be particularly aware of this requirement if your solutions contain organic biological material, e.g. soil, tissue fluids, protein solutions, etc., since these may adhere to the glass electrode and affect the calibration of your instrument. If your electrode becomes contaminated during use, check with a member of staff before cleaning: avoid touching the surface of the glass electrode with abrasive material. Allow sufficient time for the pH reading to stabilise in each solution before taking a measurement: for unbuffered solutions, this may take several minutes, so do not take inaccurate pH readings due to impatience.

8. **After use, the electrode assembly must not be allowed to dry out.** Most pH electrodes should be stored in a neutral solution of KCl, either by suspending the assembly in a small beaker, or by using an electrode cap filled with the appropriate solution (typically 1.0 mol l^{-1} KCl buffered at pH 7.0). However, many labs simply use distilled water as a storage solution, leading to loss of ions from the interior of the electrode assembly. In practice, this means that pH electrodes stored in distilled water will take far longer to give a stable reading than those stored in KCl.

9. **Switch the meter to zero (where appropriate), but do not turn off the power:** pH meters give more stable readings if they are left on during normal working hours.

   Problems (and solutions) include: inaccurate and/or unstable pH readings caused by cross-contamination (rinse electrode assembly with distilled water and blot dry between measurements); development of a protein film on the surface of the electrode (soak in 1% w/v pepsin in 0.1 mol l^{-1} HCl for at least an hour); deposition of organic or inorganic contaminants on the glass bulb (use an organic solvent, such as acetone, or a solution of 0.1 mol l^{-1} disodium ethylenediamine-tetraacetic acid, respectively); drying out of the internal reference solutions (drain, flush and refill with fresh solution, then allow to equilibrate in 0.1 mol l^{-1} HCl for at least an hour); cracks or chips to the surface of the glass bulb (use a replacement electrode).

Buffers

Rather than simply measuring the pH of a solution, you may wish to *control* the pH, e.g. in metabolic experiments, or in a growth medium for cell culture (p. 233). In fact, you should consider whether you need to control pH in any experiment involving a biological system, whether whole organisms, isolated cells, subcellular components or biomolecules. One of the most effective ways to control pH is to use a buffer solution.

A buffer solution is usually a mixture of a weak acid and its conjugate base. Added protons will be neutralised by the anionic base while a reduction in protons, e.g. due to the addition of hydroxyl ions, will be counterbalanced by dissociation of the acid (Eqn [25.2]); thus the conjugate pair acts as a 'buffer' to pH change. The innate resistance of most biological fluids to pH change is due to the presence of cellular constituents that act as buffers, e.g. proteins, which have a large number of weakly acidic and basic groups in their amino acid side chains.

Buffer capacity and the effects of pH

The extent of resistance to pH change is called the buffer capacity of a solution. The buffer capacity is measured experimentally at a particular pH by titration against a strong acid or alkali: the resultant curve will be strongly sigmoidal, with a plateau where the buffer capacity is greatest (Fig. 25.3). The mid-point of the plateau represents the pH where equal quantities of acid and conjugate base are present, and is given the symbol pK_a, which refers to the negative logarithm (to the base 10) of the acid dissociation constant, K_a, where

$$K_a = \frac{[H^+][A^-]}{[HA]} \qquad [25.6]$$

By rearranging Eqn [25.6] and taking negative logarithms, we obtain:

$$pH = pK_a + \log_{10}\frac{[A^-]}{[HA]} \qquad [25.7]$$

This relationship is known as the Henderson–Hasselbalch equation and it shows that the pH will be equal to the pK_a when the ratio of conjugate base to acid is unity, since the final term in Eqn [25.7] will be zero. Consequently, pK_a is an important factor in determining buffer capacity at a particular pH. In practical terms, a buffer solution will work most effectively at pH values about one unit either side of the pK_a.

Selecting an appropriate buffer

When selecting a buffer, you should be aware of certain limitations to their use. Citric acid and phosphate buffers readily form insoluble complexes with divalent cations, while phosphate can also act as a substrate, activator or inhibitor of certain enzymes. Both of these buffers contain biologically significant quantities of cations, e.g. Na^+ or K^+. TRIS (Table 25.3) is often toxic to biological systems: due to its high lipid solubility it can penetrate membranes, uncoupling electron transport reactions in whole cells and isolated organelles. In addition, it is markedly affected by temperature, with a tenfold increase in H^+ concentration from 4 °C to 37 °C. A number of zwitterionic molecules (having both positive

---

Definition

Buffer solution – one which resists a change in H^+ concentration (pH) on addition of acid or alkali.

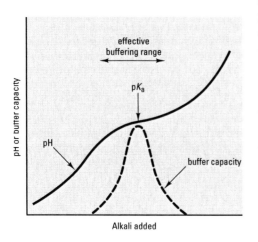

Fig. 25.3 Theoretical pH titration curve for a buffer solution. pH change is lowest and buffer capacity is greatest at the pK_a of the buffer solution.

Choosing a buffer – an ideal buffer for biological purposes would possess the following characteristics:

- impermeability to biological membranes;
- biological stability and lack of interference with metabolic and biological processes;
- lack of significant absorption of ultraviolet or visible light;
- lack of formation of insoluble complexes with cations;
- minimal effect of ionic composition or salt concentration;
- limited pH change in response to temperature.

Table 25.3 pK_a values at 25 °C and M_r of some acids and bases (upper section) and some large organic zwitterions (lower section) commonly used in buffer solutions. For polyprotic acids, where more than one proton may dissociate, the pK_a values are given for each ionisation step. Only the trivial acronyms of the larger molecules are provided: their full names can be obtained from the catalogues of most chemical suppliers.

Acid or base	pK_a value(s)	M_r
Acetic acid	4.8	60.1
Boric acid	9.2	61.8
Citric acid	3.1, 4.8, 5.4	191.2
Glycylglycine	3.1, 8.2	132.1
Phosphoric acid	2.1, 7.1, 12.3	98.0
Phthalic acid	2.9, 5.5	166.1
Succinic acid	4.2, 5.6	118.1
TRIS (base)*	8.3	121.1
CAPS (free acid)	10.4	221.3
CHES (free acid)	9.3	207.3
HEPES (free acid)	7.5	238.3
MES (free acid)	6.1	213.2
MOPS (free acid)	7.2	209.3
PIPES (free acid)	6.8	302.4
TAPS (free acid)	8.4	243.3
TRICINE (free acid)	8.1	179.2

*Note that this compound is hygroscopic and should be stored in a desiccator; also see text regarding its biological toxicity (p. 156).

Table 25.4 Preparation of sodium phosphate buffer solutions for use at 25 °C. Prepare separate stock solutions of (a) disodium hydrogen phosphate (M_r = 141.96) and (b) sodium dihydrogen phosphate (M_r = 156.01), both at 200 mol m^{-3}. Buffer solutions (at 100 mol m^{-3}) are then prepared at the required pH by mixing together the volume of each stock solution shown in the table, then diluting to a final volume of 100 ml using distilled or deionised water

Required pH (at 25 °C)	Volume of stock (a) Na$_2$HPO$_4$ (ml)	Volume of stock (b) NaH$_2$PO$_4$ (ml)
6.0	6.2	43.8
6.2	9.3	40.7
6.4	13.3	36.7
6.6	18.8	31.2
6.8	24.5	25.5
7.0	30.5	19.5
7.2	36.0	14.0
7.4	40.5	9.5
7.6	43.5	6.5
7.8	45.8	4.2
8.0	47.4	2.6

and negative groups) have been introduced to overcome some of the disadvantages of traditional buffers. These newer compounds are often referred to as 'Good buffers', to acknowledge the work of Dr N.E. Good: HEPES is one of the most useful zwitterionic buffers, with a pK_a of 7.5 at 25 °C.

These zwitterionic substances are usually added to water as the free acid: the solution must then be adjusted to the correct pH with a strong alkali, usually NaOH or KOH. Alternatively, they may be used as their sodium or potassium salts, adjusted to the correct pH with a strong acid, e.g. HCl. Consequently, you may need to consider what effects such changes in ion concentration may have in a solution where zwitterions are used as buffers. In addition, zwitterionic buffers can interfere with protein determinations (e.g. Lowry method, p. 354).

Figure 25.4 shows a number of traditional and zwitterionic buffers and their effective pH ranges. When selecting one of these buffers, aim for a pK_a which is in the direction of the expected pH change (Table 25.3). For example, HEPES buffer would be a better choice of buffer than PIPES for use at pH 7.2 for experimental systems where a pH increase is anticipated, while PIPES would be a better choice where acidification is expected.

Preparation of buffer solutions

Having selected an appropriate buffer, you will need to make up your solution to give the desired pH. You will need to consider two factors:

- The ratio of acid and conjugate base required to give the correct pH.
- The amount of buffering required; buffer capacity depends upon the absolute quantities of acid and base, as well as their relative proportions.

In most instances, buffer solutions are prepared to contain between 10 mmol l^{-1} and 200 mmol l^{-1} of the conjugate pair. While it is possible to calculate the quantities required from first principles using the Henderson–Hasselbalch equation (Eqn [25.7]), there are sources that tabulate the amount of substance required to give a particular volume of solution with a specific pH value for a range of buffers (e.g. Anon., 2007; Radiometer, 2006). For traditional buffers, it is customary to mix stock solutions of acidic and basic components in the correct proportions to give the required pH (Table 25.4). For zwitterionic acids, the usual procedure is to add the compound to

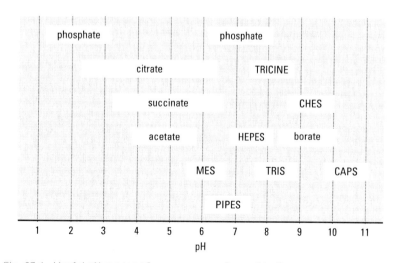

Fig. 25.4 Useful pH ranges of some commonly used buffers.

water, then bring the solution to the required pH by adding a specific amount of strong alkali or acid (obtained from tables). Alternatively, the required pH can be obtained by dropwise addition of alkali or acid, using a meter to check the pH, until the correct value is reached. When preparing solutions of zwitterionic buffers, the acid may be relatively insoluble. Do not wait for it to dissolve fully before adding alkali to change the pH – the addition of alkali will help bring the acid into solution (but make sure it has all dissolved before the desired pH is reached).

Remember that buffer solutions will only work effectively if they have sufficient buffering capacity to resist the change in pH expected during the course of the experiment. Thus a weak solution of HEPES (e.g. $10\,\text{mmol l}^{-1}$, adjusted to pH 7.0 with NaOH) will not be able to buffer the growth medium of a dense suspension of cells for more than a few minutes.

Finally, when preparing a buffer solution based on tabulated information, always confirm the pH with a pH meter before use.

Text references and sources for further study

Anon. (2007) *BioBasics Technical Library: Phosphate Buffer Table*.
Available: http://www.sigmaaldrich.com/Area_of_Interest/Research_Essentials/Biochemicals/Key_Resources/Technical_Library.html
Last accessed: 01/04/07.

Galster, H. (1991) *pH Measurement: Fundamentals, Methods, Applications, Instrumentation*. Wiley, New York.

Lide, D.R. (ed.) (2006) *CRC Handbook of Chemistry and Physics*, 87th edn. CRC Press, Boca Raton.

Radiometer (2006) *pH Theory and Practice: a Radiometer Analytical Guide*.
Available: http://www.radiometer-analytical.com/all_resource_centre.asp?code=112
Last accessed: 01/04/07.

Rilbe, H. (1996) *pH and Buffer Theory – a New Approach*. Wiley, New York.

Study exercises

25.1 Practise interconverting pH values and proton concentrations. Express all answers to three significant figures.
 (a) What is pH 7.4 expressed as [H^+] in mol l^{-1}?
 (b) What is pH 4.1 expressed as [H^+] in mol m^{-3}?
 (c) What is the pH of a solution containing H^+ at $2 \times 10^{-5}\,\text{mol l}^{-1}$?
 (d) What is the pH of a solution containing H^+ at $10^{-12.5}\,\text{mol l}^{-1}$?
 (e) What is the pH of a solution containing H^+ at $2.8 \times 10^{-5}\,\text{mol m}^{-3}$?

25.2 Decide on a suitable buffer to use. In the following instances, choose a buffer that would be suitable:
 (a) Maintaining the pH at 8.5 during an enzyme assay of a cell-free extract at 25 °C.
 (b) Keeping a stable pH of 6.5 in an experiment to measure the uptake of radiolabelled glucose by a dense suspension of *E. coli*.
 (c) Carrying out an assay of photosynthetic activity at pH 7.2 at temperatures of 10 °C, 20 °C and 30 °C.
 (d) Stabilising pH at 5.5 during enzyme extraction, in a solution where you intend to measure total protein concentration at a later stage.

25.3 Practise using the Henderson–Hasselbalch equation. What are the relative proportions of deprotonated (A^-) and protonated (HA) forms of each substance at the following pH values:
 (a) acetic acid ($pK_a = 4.8$) for use in an experiment at pH 3.8;
 (b) boric acid ($pK_a = 9.2$) for use in an experiment at pH 9.5;
 (c) HEPES ($pK_a = 7.5$) for use in an experiment at pH 8.1.

The Investigative Approach

25 The principles of measurement

> **Definition**
>
> **Variable** – any characteristic or property that can take one of a range of values (contrast this definition with that for a **parameter**, which is a numerical constant in any particular instance).

The term data (singular = datum, or data value) refers to items of information, and you will use different types of data from a wide range of sources during your practical work. Consequently, it is important to appreciate the underlying features of data collection and measurement.

Variables

Biological variables (Fig. 25.1) can be classified as follows:

Quantitative variables

These are characteristics whose differing states can be described by means of a number. They are of two basic types:

- Continuous variables, such as length; these are usually measured against a numerical scale. Theoretically, they can take any value on the measurement scale. In practice, the number of significant figures of a measurement is directly related to the precision of your measuring system; for example, dimensions measured with Vernier calipers will provide readings of greater precision than a millimetre ruler (p. 134).
- Discontinuous (discrete) variables, such as the number of eggs in a nest; these are always obtained by counting and therefore the data values must be whole numbers (integers). There are no intermediate values – for example, you never find 1.25 eggs in a nest.

> **Working with discontinuous variables** – note that while the original data values must be integers, derived data and statistical values do not have to be whole numbers. Thus, it is perfectly acceptable to express the mean number of children per family as 2.4.

Ranked variables

These provide data that can be listed in order of magnitude (i.e. ranked). A familiar example is the abundance of an organism in a sample, which is often expressed as a series of ranks, e.g. rare = 1, occasional = 2, frequent = 3, common = 4, and abundant = 5. When such data are given numerical ranks, rather than descriptive terms, they are sometimes called 'semi-quantitative data'. Note that the difference in magnitude between ranks need not be consistent. For example, regardless of whether there was a one-year or a five-year gap between offspring in a family, their rank in order of birth would be the same.

Qualitative variables (attributes)

These are non-numerical and descriptive; they have no order of preference, and therefore are not measured on a numerical scale nor ranked in order of magnitude, but are described in terms of categories. Examples include viability (i.e. dead or alive) and shape (e.g. round, flat, elongated, etc.).

Variables may be independent or dependent. Usually, the variable under the control of the experimenter (e.g. time) is the independent variable, while the variable being measured is the dependent variable (p. 185). Sometimes it is not appropriate to describe variables in this way, and they are then referred to as interdependent variables (e.g. the length and breadth of an organism).

The majority of data values are recorded as direct measurements, readings or counts, but there is an important group, called derived (or computed), that results from calculations based on two or more data values, e.g. ratios, percentages, indices and rates.

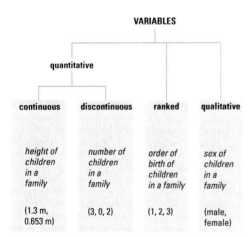

Fig. 25.1 Examples of the different types of variables as used to describe some characteristics of families

Measurement scales

Variables may be measured on different types of scale:

- Nominal scale: this classifies objects into categories based on a descriptive characteristic. It is the only scale suitable for qualitative data.
- Ordinal scale: this classifies by rank. There is a logical order in any number scale used.
- Interval scale: this is used for quantitative variables. Numbers on an equal-unit scale are related to an arbitrary zero point.
- Ratio scale: this is similar to the interval scale, except that the zero point now represents an absence of that character (i.e. it is an absolute zero). In contrast to the interval scale, the ratio of two values is meaningful (e.g. a temperature of 200K is twice that of 100K).

The measurement scale is important in determining the mathematical and statistical methods used to analyse your data. Table 25.1 presents a summary of the important properties of these scales. Note that you may be able to measure a characteristic in more than one way, or you may be able to convert data collected in one form to a different form. For instance, you might measure light in terms of the photon flux density (p. 343) between particular wavelengths of the EMR spectrum (ratio scale), or simply as 'blue' or 'red' (nominal scale); you could find out the dates of birth of individuals (interval scale) but then use this information to rank them in order of birth (ordinal scale). Where there are no other constraints, you should use a ratio scale to

> **Examples** A **nominal scale** for temperature is not feasible, since the relevant descriptive terms can be ranked in order of magnitude.
>
> An **ordinal scale** for temperature measurement might use descriptive terms, ranked in ascending order, e.g. cold = 1, cool = 2, warm = 3, hot = 4.
>
> The **Celsius scale** is an interval scale for temperature measurement, since the arbitrary zero corresponds to the freezing point of water (0 °C).
>
> The **Kelvin scale** is a ratio scale for temperature measurement since 0 K represents a temperature of absolute zero (for information, the freezing point of water is 273.15 K on this scale).

Table 25.1 Some important features of scales of measurement

	Measurement scale			
	Nominal	Ordinal	Interval	Ratio
Type of variable	Qualitative (Ranked)* (Quantitative)*	Ranked (Quantitative)*	Quantitative	Quantitative
Examples	Species Sex Colour	Abundance scales Reproductive condition Optical assessment of colour development	Fahrenheit temperature scale Date (BC/AD)	Kelvin temperature scale Weight Length Response time Most physical measurements
Mathematical properties	Identity	Identity Magnitude	Identity Magnitude Equal intervals	Identity Magnitude Equal intervals True zero point
Mathematical operations possible on data	None	Rank	Rank Addition Subtraction	Rank Addition Subtraction Multiplication Division
Typical statistics used	Only those based on frequency of counts made: contingency tables, frequency distributions, etc. Chi-square test	Non-parametric methods, sign tests. Mann–Whitney U-test	Almost all types of test, t-test, analysis of variance (ANOVA), etc. (check distribution before using, Chapter 66)	Almost all types of test, t-test, ANOVA, etc. (check distribution before using, Chapter 66)

*In some instances (see text for examples).

measure a quantitative variable, since this will allow you to use the broadest range of mathematical and statistical procedures (Table 25.1).

Accuracy and precision

Accuracy is the closeness of a measured or derived data value to its true value, while precision is the closeness of repeated measurements to each other (Fig. 25.2). A balance with a fault in it (i.e. a bias, see below) could give precise (i.e. very repeatable) but inaccurate (i.e. untrue) results. Unless there is bias in a measuring system, precision will lead to accuracy and it is precision that is generally the most important practical consideration, if there is no reason to suspect bias. You can investigate the precision of any measuring system by repeated measurements of individual samples.

Absolute accuracy and precision are impossible to achieve, due to both the limitations of measuring systems for continuous quantitative data and the fact that you are usually working with incomplete data sets (samples, p. 179). It is particularly important to avoid spurious accuracy in the presentation of results; include only those digits that the accuracy of the measuring system implies (p. 408). This type of error is common when changing units (e.g. inches to metres) and in derived data, especially when calculators give results to a large number of decimal places.

Bias (systematic error) and consistency

Bias is a systematic or non-random distortion and is one of the most troublesome difficulties in using numerical data. Biases may be associated with incorrectly calibrated instruments, e.g. a faulty pipettor, or with experimental manipulations, e.g. shrinkage during the preservation of a specimen. Bias in measurement can also be subjective, or personal, e.g. an experimenter's preconceived ideas about an 'expected' result.

Bias can be minimised by using a carefully standardised procedure, with fully calibrated instruments. You can investigate bias in 'trial runs' by measuring a single variable in several different ways, to see whether the same result is obtained.

If a personal bias is possible, 'blind' measurements should be made where the identity of individual samples is not known to the operator, e.g. using a coding system.

Measurement error

All measurements are subject to error, but the dangers of misinterpretation are reduced by recognising and understanding the likely sources of error and by adopting appropriate protocols and calculation procedures.

A common source of measurement error is carelessness, e.g. reading a scale in the wrong direction or parallax errors. This can be reduced greatly by careful recording and may be detected by repeating the measurement. Other errors arise from faulty or inaccurate equipment, but even a perfectly functioning machine has distinct limits to the accuracy and precision of its measurements. These limits are often quoted in manufacturers' specifications and are applicable when an instrument is new; however, you should allow for some deterioration with age. Further errors are introduced when the subject being studied is open to influences outside your control. Resolving such problems requires appropriate experimental design (Chapter 31) and sampling procedures (Chapter 30).

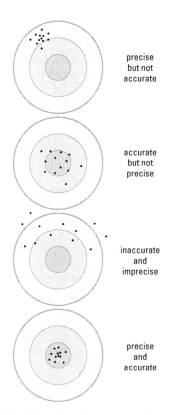

Fig. 25.2 'Target' diagrams illustrating precision and accuracy

Minimising errors – determine early in your study what the dominant errors are likely to be and concentrate your time and effort on reducing these.

Working with derived data – special effort should be made to reduce measurement errors because their effects can be magnified when differences, ratios, indices or rates are calculated.

One major influence virtually impossible to eliminate is the effect of the investigation itself: even putting a thermometer in a liquid may change the temperature of the liquid. The very act of measurement may give rise to a confounding variable (p. 186) as discussed in Chapter 31.

Sources for further study

Anon. *Measurement*. Available: http://wikipedia.org/wiki/Measurement
Last accessed: 09/04/07.

Erikson, B.H. and Nosanchuk, T.A. (1992) *Understanding Data*, 2nd edn. Open University Press, Milton Keynes.
[A text aimed at social science students but with clear explanations of issues that are generic, including information on analysis of data.]

Friedrich, G.W. *Basic Principles of Measurement. Methods of Inquiry*. Available: http://www.scils.rutgers.edu/~gusf/measurement.html
Last accessed: 09/04/07.
[Course notes covering diverse aspects of enquiry.]

National Instruments *Measurement Encyclopedia*. Available: http://zone.ni.com/devzone/nidzgloss.nsf/glossary/
Last accessed: 09/04/07.

Study exercises

25.1 Classify variables. Decide on the type of variables used for the following measures, indicating whether they are quantitative or qualitative, continuous or discontinuous, and the type of scale that would be used.

(a) Number of organisms in a population.
(b) Length of individuals in a population.
(c) Colour of flowers.
(d) Species present in a sample.
(e) Date of a sample.
(f) Reproductive condition of an animal.

25.2 Assess errors and accuracy of a set of measurements. Assume that you are asked to measure the length, breadth and height of each of a sample of 20 limpet shells (see figure) using (a) a ruler and (b) a pair of Vernier calipers (p. 134). Identify the sources of error likely to be present in your measurements using each tool and the precision to which you would be able to quote your data. Devise a protocol for taking these measurements that would help you to minimise the errors and maximise the accuracy and precision.

25.3 Investigate types of errors. A student weighed a set of standard masses on two electronic balances and obtained the readings shown in the table below. Explain these results in terms of the type of error involved in each case.

Lateral (a) and ventral (b) views of a limpet to show measurement variables length (L), breadth (B) and height (H). Note also the uneven edge of the shell.

Comparison of weights of masses on two balances

	Standard mass (g)				
	10	25	50	100	250
Reading (balance A)	10.050	25.049	50.051	100.048	250.052
Reading (balance B)	10.004	25.011	50.021	100.039	250.102

27 Making observations

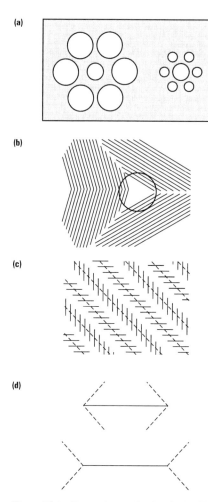

Fig. 27.1 Examples of 'optical illusions' caused by problems of perception. Image (a) shows how the sizes of adjacent objects can distort a simple comparison of size: the central circles in either hexagonal pattern are the same size. Image (b) reveals how shapes can be distorted by adjacent linear objects: the inner shape is a perfect circle. Image (c) illustrates how directional cues can lead to confusion: the dashed lines are parallel. Image (d) shows how adjacent shapes can make comparison of simple linear dimensions difficult: the two solid lines are the same length. In all cases, the correct perception can only be confirmed with a measurement aid such as a ruler or compass.

Observer effects – remember that your presence, or the act of observation itself, may influence the event you are observing. Take appropriate precautions, like using a hide when observing animal behaviour.

Observations provide the basic information leading to the formulation of hypotheses, an essential step in the scientific method (see Fig. 31.1). Observations are obtained either directly by our senses or indirectly through the use of instruments that extend our senses and may be either:

- Qualitative: described by words or terms rather than by numbers and including subjective descriptions in terms of variables such as colour, shape and smell; often recorded using photographs and drawings.
- Quantitative: numerical values derived from counts or measurements of a variable (see Chapter 25), frequently requiring use of some kind of instrument.

 KEY POINT Although qualitative and quantitative observations are useful in biology, you should try to make numerical counts or measurements wherever possible, as this allows you to define your observations more rigorously and make objective comparisons using statistical tools.

Factors influencing the quality of observations

Perception

Observation is highly dependent upon the perception of the observer (Fig. 27.1). Perception involves both visual and intuitive processes, so your interpretation of what you see is very dependent upon what you already know or have seen before. Thus, two persons observing the same event or object may 'see' it differently, a good example of bias. This is frequently true in microscopy where experience is an important factor in interpretation.

When you start biology, your knowledge base will be limited and your experience restricted. Practical training in observation provides the opportunity to develop both aspects of your skills in a process that is effectively a positive feedback loop – the more you know/see as a result of practice, the better will your observations become.

Precision and error

Obviously very important for interpretive accuracy, with both human and non-human components. These are dealt with in Chapter 25.

Artefacts

These are artificial features introduced usually during some treatment process such as chemical fixation prior to microscopic examination. They may be included in the interpretive process if their presence is not recognised – again, prior experience and knowledge are important factors in spotting artefacts (see Chapter 44 and especially Fig. 44.1 in relation to microscopy).

Developing observational skills

You must develop your knowledge and observational skills to benefit properly from your practical work. The only way to acquire these skills is through extensive practice.

Preparation – thorough theoretical groundwork before a practical class or examination is vitally important for improving the quality of your observations (see Fig. 27.1).

Perceptual illusions – beware deceptions such as holes appearing as bumps in photographs, misinterpretations due to lack of information on scale and other simple but well-known tricks of vision (see Fig. 27.1).

Making counts – acetate sheets can be used as overlays for photographs, drawings and other images, then as each object is counted it can be marked off using a water-based marker pen. Remember to mark identification points in case the sheet slips!

Make sure your observations are:

- relevant, i.e. directed towards a clearly defined objective;
- accurate, i.e. related to a scale whenever possible;
- repeatable, i.e. as error free (precise) as possible.

Much biological work attempts to relate structure to function, often through careful analysis of structure at different levels. One of the best ways to develop observational skills is by making accurate drawings or diagrams, forcing you to look more carefully than is usual (see Chapter 28). An important observational skill to develop is the interpretation of two-dimensional images, such as sections through plant/animal material and photographs, in terms of the three-dimensional forms from which they are derived. This requires a clear understanding of the nature of the image in terms of both scale and orientation (see Chapter 44).

Counting

Counting is an observational skill that requires practice to become both accurate and efficient. It is easy to make errors or lose count when working with large numbers of objects. Use a counting aid whenever human error might be significant. There are many such aids such as tally counters, tally charts and specialised counting devices like colony counters. It is important to avoid counting items twice. For example, when counting microbial colonies on Petri plates, each colony can be marked off as it is counted on the base of the plate using a spirit-based marker pen.

Another valuable technique is to use a grid system to organise the counting procedure. This has become formalised in equipment such as the haemocytometer used for counting blood cells (see Box 46.1, p. 275). Remember that you must decide on a protocol for sampling, particularly with regard to the direction of counting within the grid and for dealing with boundary overlaps to prevent double counting at the edges of the grid squares (see Fig. 30.4, p. 181).

Observation during examinations

Making appropriate observations during practical examinations often causes difficulty, particularly when qualitative observations are needed, e.g. when asked to classify a specimen, giving reasons. Answering such questions clearly requires biological knowledge but also requires a strategy to provide the relevant observations. Thus for the above example, observations relevant to determining each taxonomic level should be made and recorded: set out your observations in a logical sequence so that you show the examiner how you arrived at your conclusion.

A similar approach is needed when asked to make observations related to structure or function. For example, if comments on a method of locomotion are required, do not make observations on irrelevant structures. You will obtain maximum marks only if your answers are concise and relevant (Chapter 6).

Text references and sources for further study

Akins, K. (ed.) (1996) *Perception.* Vancouver Studies in Cognitive Science, Vol. 5. Oxford University Press, Oxford.
[A detailed examination of the problems of perception from a psychological perspective.]

Edwards, B. (2000). *The New Drawing on the Right Side of your Brain.* Souvenir Press, London.

Raulin, M.L. *Naturalistic Observation and Case-Study Research.* Available: http://www.abacon.com/graziano/ch06/index.htm
Last accessed 09/04/07.

Study exercises

27.1 Assess the difficulties of making counts of large numbers. The dots in the area shown in the figure below represent bacterial colonies on a Petri dish. To appreciate the difficulties of counting numbers accurately, make three rapid independent attempts at counting the dots by eye (without using any aids). Now subdivide the area, such that the number of colonies per subdivision is more easily counted (i.e. is a number between 1 and 20). Again carry out three further estimates using these subdivisions, summing the counts obtained each time. Compare the results using both methods – which is the more precise and which the more accurate?

27.2 Extrapolate from 2-D to 3-D. Each of the diagrams on the next page represents a series of nine transverse sections taken at regular intervals through an organism or part of an organism. Use these diagrams to construct a 3-D representation of the structure in each case.

27.3 Trick your brain into letting you do better drawings. One of the problems with making drawings is that the brain thinks it 'knows' how things should look. This may cause distortion in diagrams. One way to produce realistic drawings of relatively complex objects is to turn the object upside down, then try to draw it. Because the brain no longer recognises the object, it focuses on the relationships between spaces and lines and allows you to make a reasonable copy (Edwards, 2000). Try this out by taking a photograph of an object that you would normally find difficult to draw, e.g. a person's face or an animal, and using this technique to copy it.

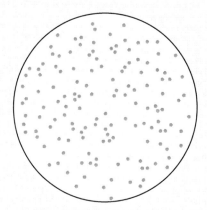

Bacterial colonies on a Petri dish

(continued)

Study exercises (continued)

Sections taken at regular intervals through organisms or parts of organisms

32 Making notes of practical work

When carrying out advanced lab work or research projects, you will need to master the important skill of managing data and observations and learn how to keep a record of your studies in a lab book. This is important for the following reasons:

- An accurate and neat record helps when using information later, perhaps for exam purposes or when writing a report.
- It allows you to practise important skills such as scientific writing, drawing diagrams, preparing graphs and tables and interpreting results.
- Analysing and writing up your data as you go along prevents a backlog at the end of your study time.
- You can show your work to a future employer to prove you have developed the skills necessary for writing up properly; in industry, this is vital so that others in your team can interpret and develop your work.

Understanding what's expected – especially when taking notes for a lab-based practical, pay special attention to the aims and learning objectives (pp. 22–4) of the session, as these will indicate the sorts of notes you should be taking, including content and diagrams, and the ways in which you should present these for assessment.

KEY POINT A good set of lab notes should:

- outline the purpose of your experiment or observation;
- set down all the information required to describe your materials and methods;
- record all relevant information about your results or observations and provide a visual representation of the data;
- note your immediate conclusions and suggestions for further experiments.

Collecting and recording primary data

Individual observations (e.g. laboratory temperature) can be recorded in the text of your notes, but tables are the most convenient way to collect large amounts of information. When preparing a table for data collection, you should:

Recording primary data – never be tempted to jot down data on scraps of paper: you are likely to lose them, or to forget what individual values mean.

1. Use a concise title or a numbered code for cross-referencing.
2. Decide on the number of variables to be measured and their relationship with each other and lay out the table appropriately:
   (a) The first column of your table should show values of the independent (controlled) variable, with subsequent columns for the individual (measured) values for each replicate or sample.
   (b) If several variables are measured for the same organism or sample, each should be given a row.
   (c) In time-course studies, put the replicates as columns grouped according to treatment, with the rows relating to different times.
3. Make sure the arrangement reflects the order in which the values will be collected. Your table should be designed to make the recording process as straightforward as possible, to minimise the possibility of mistakes. For final presentation, a different arrangement may be best (Chapter 63).
4. Consider whether additional columns are required for subsequent calculations. Create a separate column for each mathematical manipulation, so the step-by-step calculations are clearly visible. Use a computer spreadsheet (pp. 68–73) if you are manipulating lots of data.

Designing a table for data collection – use a spreadsheet or the table-creating facility in a word processor to create your table. This will allow you to reorganise it easily if required. Make sure there is sufficient space in each column for the values – if in doubt, err on the generous side.

Identifying your notes – *always put a date and time on each of your primary record sheets. You may also wish to add your name and details of the type of observation or experiment.*

5. Use a pencil to record data so that mistakes can be easily corrected.
6. Take sufficient time to record quantitative data unambiguously – use large clear numbers, making sure that individual numerals cannot be confused.
7. Record numerical data to an appropriate number of significant figures, reflecting the accuracy and precision of your measurement (p. 408). Do not round off data values, as this might affect the subsequent analysis of your data.
8. Record discrete or grouped data as a tally chart (see p. 383), each row showing the possible values or classes of the variable. Provided that tally marks are of consistent size and spacing, this method has the advantage of providing an 'instant' frequency distribution chart.
9. Prepare duplicated recording tables if your experiments or observations will be repeated.
10. Explain any unusual data values or observations in a footnote. Don't rely on your memory.

Recording details of project work

The recommended system is one where you make a dual record.

Primary record

The primary record is made at the bench or in the field. In this, you must concentrate on the detail of materials, methods and results. Include information that would not be used elsewhere, but that might prove useful in error tracing: for example, if you note how a solution was made up (exact volumes and weights used rather than concentration alone), this could reveal whether a miscalculation had been the cause of a rogue result. Note the origin, type and state of the chemicals and organism(s) used. Make rough diagrams to show the arrangement of replicates, equipment, etc. If you are forced to use loose paper to record data, make sure each sheet is dated and taped to your lab book, collected in a ring binder, or attached together with a treasury tag. The same applies to traces, printouts and graphs.

The basic order of the primary record should mirror that of a research report (see p. 105), including: the title and date; brief introduction; comprehensive materials and methods; the data and short conclusions.

Choosing a notebook for primary recording – *a spiral-bound notebook is good for making a primary record – it lies conveniently open on the bench and provides a simple method of dealing with major mistakes!*

Secondary record

You should make a secondary record concurrently or later in a bound book and it ought to be neater, in both organisation and presentation. This book will be used when discussing results with your supervisor, and when writing up a report or thesis, and may be part of your course assessment. Although these notes should retain the essential features of the primary record, they should be more concise and the emphasis should move towards analysis of the experiment. Outline the aims more carefully at the start and link the experiment to others in a series (e.g. 'Following the results of Expt D24, I decided to test whether...'). You should present data in an easily digested form, e.g. as tables of means or as summary graphs. Use appropriate statistical tests (Chapter 66) to support your analysis of the results. The choice of a bound book ensures that data are not easily lost.

Choosing a book for secondary recording – *a hard-backed A4-size lined book is good because you will not lose pages. Graphs, printouts, etc., can be stuck in, as required.*

Formal aspects of keeping a record – *the diary aspect of the record can be used to establish precedence (e.g. for patentable research where it can be important to 'minute' where and when an idea arose and whose it was); for error tracing (e.g. you might be able to find patterns in the work affecting the results); or even for justifying your activities to a supervisor.*

Analysing data as soon as possible – *always analyse and think about data immediately after collecting them as this may influence your subsequent activities.*

- *A graphical indication of what has happened can be particularly valuable.*
- *Carry out statistical analyses before moving on to the next experiment because apparent differences among treatments may not turn out to be statistically significant when tested.*
- *Write down any conclusions you make while analysing your data: sometimes those that seem obvious at the time of doing the work are forgotten when the time comes to write up a report or thesis.*
- *Note ideas for further studies as they occur to you – these may prove valuable later. Even if your experiment appears to be a failure, suggestions as to the likely causes might prove useful.*

Points to note

The dual method of recording deals with the inevitable untidiness of notes taken at the bench or in the field; these often have to be made rapidly, in awkward positions and in a generally complex environment. Writing a second, neater version forces you to consider again details that might have been overlooked in the primary record and provides a duplicate in case of loss or damage.

If you find it difficult to decide on the amount of detail required in Materials and Methods, the basic ground rule is to record enough information to allow a reasonably competent scientist to repeat your work exactly. You must tread a line between the extremes of pedantic, irrelevant detail and the omission of information essential for a proper interpretation of the data – better perhaps to err on the side of extra detail to begin with. An experienced worker can help you decide which subtle shifts in technique are important (e.g. batch numbers for an important chemical, or when a new stock solution is made up and used). Many important scientific advances have been made because of careful observation and record taking or because coincident data were recorded that did not seem of immediate value.

When creating a primary record, take care not to lose any of the information content of the data: for instance, if you only write down means and not individual values, this may affect your ability to carry out subsequent statistical analyses.

There are numerous ways to reduce the labour of keeping a record. Don't repeat Materials and Methods for a series of similar experiments; use devices such as 'method as for Expt B4'. A photocopy might suffice if the method is derived from a text or article (check with supervisor). To save time, make up and copy a checklist in which details such as chemical batch numbers can be entered.

Special requirements for fieldwork

The main problems you will encounter in the field are the effects of the weather while taking a primary record and the distance you might be from a suitable place to make a neat secondary record. Wind, rain and cold temperatures are not conducive to neat note-taking and you should be prepared for the worst possible conditions at all times. Make sure your clothing allows you to feel comfortable while recording data.

The simplest method of protecting a field notebook is to enclose it in a clear polythene bag large enough for you to take notes inside (Fig. 32.1). Alternatively, you could use a clipboard with a waterproof cover to shield your notes or a special notebook with a waterproof cover. When selecting a field notebook, choose a small size – the dimensions of outside pockets may dictate the upper size limit.

If recording results and observations outdoors:

- Use a pencil as ink pens such as ballpoints smudge in wet conditions, are temperamental in the cold and may not work at awkward angles. Don't forget to take a sharpener.
- Prepare well to enhance the speed and quality of your field note-taking – the date and site details can be written down before setting out and tables can be made out ready for data entry.

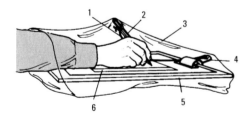

Fig. 32.1 Simple but effective method for keeping notes dry in the field

1. string attaching pencil to clip
2. pencil (not pen) writes on damp paper
3. transparent plastic bag
4. bulldog clip
5. hardboard (at least 31 × 22 cm for A4 paper)
6. record sheet.

Selecting a waterproof notebook – notebooks with waterproof paper are available (for example the 'Aquascribe' brand supplied by Hawkins and Mainwaring Ltd, Newark, Notts). These are relatively expensive, but you can re-use them by rubbing out.

- Transcribe field notes to a duplicate primary record at your base each time you return there. There is a very real risk of your losing or damaging a field notebook. Also, poor weather may prevent full note-taking and the necessary extra details should be written up while fresh in your memory.
- Consider using a voice recorder rather than a notebook, in which case voice transcription into written form should also take place while your memory is fresh in case the sound quality is poor.
- Use photographs to set data in context, when appropriate. Develop photographs as soon as possible to check their suitability. Consider using a digital camera when the suitability of a record must be guaranteed at the time of the visit (see Chapter 48).

Field data may be logged automatically, stored temporarily in the instrument's electronic memory, and downloaded to a portable computer ('data logger') when convenient. The information is then transferred to a data bank back at base. If you are using this system, make back up copies of each period's data as soon as possible – if the recording instrument's memory is cleared or overwritten after reading there may be no recourse if the logging machine fails.

Using communal records

If working with a research team, you may need to use their communal databases. These avoid duplication of effort and ensure uniformity in techniques. You will be expected to use the databases carefully and to contribute to them properly. They might include:

- a shared notebook of common techniques (e.g. how to make up media or solutions);
- a set of simplified step-by-step instructions for use of equipment. Manuals are often complex and poorly written and it may help to redraft them, incorporating any differences in procedure adopted by the group;
- an alphabetical list of suppliers of equipment and consumables (perhaps held on a card-index system);
- a list of chemicals required by the group and where they are stored;
- the risk-assessment sheets for dangerous procedures (p. 119);
- the record book detailing the use of radioisotopes and their disposal.

SAFETY NOTE *Maintaining and consulting communal lab records* – these activities may form a part of the safety requirements for working in a laboratory.

Sources for further study

Anon. *Suggestions for Keeping Laboratory Notebooks.* Available: http://otl.stanford.edu/inventors/resources/labnotebooks.html
Last accessed 09/04/07.
[A Stanford University website that looks at the laboratory notebook from the patenting perspective.]

Kanare, H.M. (1985) *Writing the Laboratory Notebook.* American Chemical Society, Washington, DC.

Pechenik, J.A.A. (2001) *Short Guide to Writing About Biology*, 4th edn. Longman, Harlow.
[Chapter 8 is particularly relevant.]

Slingsby, D. and Cook, C. (1986) *Practical Ecology.* Macmillan Education, London.

Study exercises

32.1 Design a primary data collection sheet for a behavioural study. Imagine you wish to describe the feeding behaviour of birds at their nest. You wish to study the nature of the food, the relative participation of both parents, and the frequency of visits during the daytime with a view to analysis using a spreadsheet.

32.2 Outline the advantages and problems that would be associated with using a tape recorder to record data. Consider the behavioural study described in Study exercise 32.1 as an example.

32.3 Design a secondary record sheet for the collection and analysis of a set of count data for the number of bacterial colonies developed in each of ten replicates for each of five different treatments. Assume that you need to calculate means, variances, etc., and then to compare the results for each treatment.

29 SI units and their use

Dimensionless measurements – some quantities can be expressed as dimensionless ratios or logarithms (e.g. pH), and in these cases you do not need to use a qualifying unit.

When describing a measurement, you normally state both a number and a unit (e.g. 'the length is 1.85 metres'). The number expresses the ratio of the measured quantity to a fixed standard, while the unit identifies that standard measure or dimension. Clearly, a single unified system of units is essential for efficient communication of such data within the scientific community. The Système International d'Unités (SI) is the internationally ratified form of the metre-kilogram-second system of measurement and represents the accepted scientific convention for measurements of physical quantities.

Another important reason for adopting consistent units is to simplify complex calculations where you may be dealing with several measured quantities (see p. 471). Although the rules of the SI are complex and the scale of the base units is sometimes inconvenient, to gain the full benefits of the system you should observe its conventions strictly.

The description of measurements in SI involves:

- seven base units and two supplementary units, each having a specified abbreviation or symbol (Table 29.1);
- derived units, obtained from combinations of base and supplementary units, which may also be given special symbols (Table 29.2);
- a set of prefixes to denote multiplication factors of 10^3, used for convenience to express multiples or fractions of units (Table 29.3).

Table 29.1 The base and supplementary SI units

Measured quantity	Name of SI unit	Symbol
Base units		
Length	metre	m
Mass	kilogram	kg
Amount of substance	mole	mol
Time	second	s
Electric current	ampere	A
Temperature	kelvin	K
Luminous intensity	candela	cd
Supplementary units		
Plane angle	radian	rad
Solid angle	steradian	sr

Table 29.2 Some important derived SI units

Measured quantity	Name of unit	Symbol	Definition in base units	Alternative in derived units
Energy	joule	J	$m^2\,kg\,s^{-2}$	N m
Force	newton	N	$m\,kg\,s^{-2}$	$J\,m^{-1}$
Pressure	pascal	Pa	$kg\,m^{-1}\,s^{-2}$	$N\,m^{-2}$
Power	watt	W	$m^2\,kg\,s^{-3}$	$J\,s^{-1}$
Electric charge	coulomb	C	A s	$J\,V^{-1}$
Electric potential difference	volt	V	$m^2\,kg\,A^{-1}\,s^{-3}$	$J\,C^{-1}$
Electric resistance	ohm	Ω	$m^2\,kg\,A^{-2}\,s^{-3}$	$V\,A^{-1}$
Electric conductance	siemens	S	$s^3\,A^2\,kg^{-1}\,m^{-2}$	$A\,V^{-1}$ or Ω^{-1}
Electric capacitance	farad	F	$s^4\,A^2\,kg^{-1}\,m^{-2}$	$C\,V^{-1}$
Luminous flux	lumen	lm	cd sr	
Illumination	lux	lx	$cd\,sr\,m^{-2}$	$lm\,m^{-2}$
Frequency	hertz	Hz	s^{-1}	
Radioactivity	becquerel	Bq	s^{-1}	
Enzyme activity	katal	kat	mol substrate s^{-1}	

Table 29.3 Prefixes used in the SI

Multiple	Prefix	Symbol	Multiple	Prefix	Symbol
10^{-3}	milli	m	10^3	kilo	k
10^{-6}	micro	μ	10^6	mega	M
10^{-9}	nano	n	10^9	giga	G
10^{-12}	pico	p	10^{12}	tera	T
10^{-15}	femto	f	10^{15}	peta	P
10^{-18}	atto	a	10^{18}	exa	E
10^{-21}	zepto	z	10^{21}	zetta	Z
10^{-24}	yocto	y	10^{24}	yotta	Y

Example $10\,\mu g$ is correct, while $10\mu g$, $10\ \mu\ g$, and $10\mu\ g$ are incorrect. 2.6 mol is right, but 2.6 mols is wrong.

Recommendations for describing measurements in SI units

Basic format

- Express each measurement as a number separated from its units by a space. If a prefix is required, no space is left between the prefix and the unit it refers to. Symbols for units are only written in their singular

form and do not require full stops to show that they are abbreviated or that they are being multiplied together.
- Give symbols and prefixes appropriate upper or lower case initial letters as this may define their meaning. Upper case symbols are named after persons but when written out in full they are not given initial capital letters.

Example n stands for nano and N for newtons.

- Show the decimal sign as a full point on the line. Some metric countries continue to use the comma for this purpose and you may come across this in the literature: commas should not therefore be used to separate groups of thousands. In numbers that contain many significant figures, you should separate multiples of 10^3 by spaces rather than commas.

Example 1 982 963.192 309 kg (perhaps better expressed as 1.982 963 192 309 Gg).

Compound expressions for derived units

- Take care to separate symbols in compound expressions by a space to avoid the potential for confusion with prefixes. Note, for example, that 200 m s (metre-seconds) is different from 200 ms (milliseconds).
- Express compound units using negative powers rather than a solidus (/): for example, write $mol\,m^{-3}$ rather than mol/m^3. The solidus is reserved for separating a descriptive label from its units (see p. 455).
- Use parentheses to enclose expressions being raised to a power if this avoids confusion: for example, a photosynthetic rate might be given in $mol\,CO_2\,(mol\,photons)^{-1}\,s^{-1}$.
- Where there is a choice, select relevant (natural) combinations of derived and base units: e.g. you might choose units of $Pa\,m^{-1}$ to describe a hydrostatic pressure gradient rather than $kg\,m^{-2}\,s^{-2}$, even though these units are equivalent and the measurements are numerically the same.

Use of prefixes

- Use prefixes to denote multiples of 10^3 (Table 29.3) so that numbers are kept between 0.1 and 1000.
- Treat a combination of a prefix and a symbol as a single symbol. Thus, when a modified unit is raised to a power, this refers to the whole unit including the prefix.

Examples

- 10 μm is preferred to 0.000 01 m or 0.010 mm.
- $1\,mm^2 = 10^{-6}\,m^2$ (not one-thousandth of a square metre).
- $1\,dm^3$ (1 litre) is more properly expressed in SI as $1 \times 10^{-3}\,m^3$.
- The mass of a neutrino is 10^{-36} kg.
- State as $MW\,m^{-2}$ rather than $W\,mm^{-2}$.

- Avoid the prefixes deci (d) for 10^{-1}, centi (c) for 10^{-2}, deca (da) for 10 and hecto (h) for 100 as they are not strictly SI.
- Express very large or small numbers as a number between 1 and 10 multiplied by a power of 10 if they are outside the range of prefixes shown in Table 29.3.
- Do not use prefixes in the middle of derived units: they should be attached only to a unit in the numerator (the exception is in the unit for mass, kg).

KEY POINT For the foreseeable future, you will need to make conversions from other units to SI units, as much of the literature quotes data using imperial, c.g.s. or other systems. You will need to recognise these units and find the conversion factors required. Examples relevant to biology are given in Box 29.1. Table 29.4 provides values of some important physical constants in SI units.

Box 29.1 Conversion factors between some redundant units and the SI

Quantity	SI unit/symbol	Old unit/symbol	Multiply number in old unit by this factor for equivalent in SI unit*	Multiply number in SI unit by this factor for equivalent in old unit*
Area	square metre/m^2	acre	4.04686×10^3	0.247105×10^{-3}
		hectare/ha	10×10^3	0.1×10^{-3}
		square foot/ft^2	0.092903	10.7639
		square inch/in^2	645.16×10^{-9}	1.55000×10^6
		square yard/yd^2	0.836127	1.19599
Angle	radian/rad	degree/°	17.4532×10^{-3}	57.2958
Energy	joule/J	erg	0.1×10^{-6}	10×10^6
		kilowatt hour/kWh	3.6×10^6	0.277778×10^{-6}
		calorie/cal	4.1868	0.2388
Length	metre/m	Ångstrom/Å	0.1×10^{-9}	10×10^9
		foot/ft	0.3048	3.28084
		inch/in	25.4×10^{-3}	39.3701
		mile	1.60934×10^3	0.621373×10^{-3}
		yard/yd	0.9144	1.09361
Mass	kilogram/kg	ounce/oz	28.3495×10^{-3}	35.2740
		pound/lb	0.453592	2.20462
		stone	6.35029	0.157473
		hundredweight/cwt	50.8024	19.6841×10^{-3}
		ton (UK)	1.01605×10^3	0.984203×10^{-3}
Pressure	pascal/Pa	atmosphere/atm	101325	9.86923×10^{-6}
		bar/b	100000	10×10^{-6}
		millimetre of mercury/mmHg	133.322	7.50064×10^{-3}
		torr/Torr	133.322	7.50064×10^{-3}
Radioactivity	becquerel/Bq	curie/Ci	37×10^9	27.0270×10^{-12}
Temperature	kelvin/K	centigrade (Celsius) degree/°C	°C + 273.15	K − 273.15
		Fahrenheit degree/°F	(°F + 459.67) × 5/9	(K × 9/5) − 459.67
Volume	cubic metre/m^3	cubic foot/ft^3	0.0283168	35.3147
		cubic inch/in^3	16.3871×10^{-6}	61.0236×10^3
		cubic yard/yd^3	0.764555	1.30795
		UK pint/pt	0.568261×10^{-3}	1759.75
		US pint/liq pt	0.473176×10^{-3}	2113.38
		UK gallon/gal	4.54609×10^{-3}	219.969
		US gallon/gal	3.78541×10^{-3}	264.172

*In the case of temperature measurements, use formulae shown.

Table 29.4 Some physical constants in SI terms

Physical constant	Symbol	Value and units
Avogadro's constant	N_A	6.022174×10^{23} mol^{-1}
Boltzmann's constant	k	1.380626×10^{-23} $J\,K^{-1}$
Charge of electron	e	1.602192×10^{-19} C
Gas constant	R	8.31443 $J\,K^{-1}\,mol^{-1}$
Faraday's constant	F	9.648675×10^4 $C\,mol^{-1}$
Molar volume of ideal gas at STP	V_0	0.022414 $m^3\,mol^{-1}$
Speed of light in vacuo	c	2.997924×10^8 $m\,s^{-1}$
Planck's constant	h	6.626205×10^{-34} $J\,s$

Some implications of SI in bioscience

Volume

The SI unit of volume is the cubic metre, m^3, which is rather large for practical purposes. The litre (l) and the millilitre (ml) are technically obsolete, but are widely used and glassware is still calibrated using them. Note also that the US spelling is liter. In some instances you may find litre given the symbol L, rather than l to avoid confusion with the number 1 and with I, e.g. in certain font styles.

> **Using units for volume** – in this book, we use l and ml where you would normally find equipment calibrated in that way, but use SI units where this simplifies calculations. In formal scientific writing, constructions such as $1 \times 10^{-6} m^3$ ($= 1 ml$) and $1 mm^3$ ($= 1 \mu l$) may be used.

Mass

The SI unit for mass is the kilogram (kg) rather than the gram (g): this is unusual because the base unit has a prefix applied.

Amount of substance

You should use the mole (mol, i.e. Avogadro's constant, see Table 29.4) to express very large numbers. The mole gives the number of atoms in the atomic mass, a convenient constant. Always specify the elementary unit referred to in other situations (e.g. mol photons $m^{-2} s^{-1}$).

> **Expressing enzyme activity** – the derived SI unit is the katal (kat) which is the amount of enzyme that will transform 1 mol of substrate in 1 s (see Chapter 55).

Concentration

The SI unit of concentration, $mol\,m^{-3}$, is quite convenient for biological systems. It is equivalent to the non-SI term 'millimolar' ($mM \equiv mmol\,l^{-1}$) while 'molar' ($M \equiv mol\,l^{-1}$) becomes $kmol\,m^{-3}$. Note that the symbol M in the SI is reserved for mega and hence should not be used for concentrations. If the solvent is not specified, then it is assumed to be water (see Chapter 24).

> **Converting between concentration units** – being able to express concentrations in different units (pp. 144–7) is important as this skill is frequently used when following instructions and interpreting data.

Time

In general, use the second (s) when reporting physical quantities having a time element (e.g. give photosynthetic rates in mol $CO_2\,m^{-2}\,s^{-1}$). Hours (h), days (d) and years should be used if seconds are clearly absurd (e.g. samples were taken over a 5-year period). Note, however, that you may have to convert these units to seconds when doing calculations.

Temperature

The SI unit is the kelvin, K. The degree Celsius scale has units of the same magnitude, °C, but starts at 273.15 K, the melting point of ice at STP. Temperature is similar to time in that the Celsius scale is in widespread use, but note that conversions to K may be required for calculations. Note also that you *must not* use the degree sign (°) with K and that this symbol must be in upper case to avoid confusion with k for kilo; however, you *should* retain the degree sign with °C to avoid confusion with the coulomb, C.

> **Definition**
>
> **STP** – standard temperature and pressure = 293.15 K and 0.101 325 MPa.

Light

While the first six base units in Table 29.1 have standards of high precision, the SI base unit for luminous intensity, the candela (cd) and the derived units lm and lx (Table 29.2), are defined in 'human' terms. They are, in fact, based on the spectral responses of the eyes of 52 American GIs measured in

1923! Clearly, few organisms 'see' light in the same way as this sample of humans. Also, light sources differ in their spectral quality. For these reasons, it is better to use expressions based on energy or photon content (e.g. $W\,m^{-2}$ or mol photons $m^{-2}\,s^{-1}$) in studies other than those on human vision. Ideally you should specify the photon wavelength spectrum involved (see Chapter 40).

Sources for further study

Institute of Biology (1997) *Biological Nomenclature: Recommendations on Terms, Units and Symbols*, 2nd edn. Institute of Biology, London.

National Institute of Standards and Technology *The NIST Reference on Constants, Units and Uncertainty: International System of Units (SI)*. Available: http://physics.nist.gov/cuu/Units/
Last accessed: 01/04/07.

Pennycuick, C.J. (1988) *Conversion Factors: SI Units and Many Others: Over 2100 Conversion Factors for Biologists and Mechanical Engineers Arranged in 21 Quick-Reference Tables*. The University of Chicago Press, Chicago.

Rowlett, R. *How Many? A Dictionary of Units of Measurement*. Available: http://www.unc.edu/~rowlett/units
Last accessed: 01/04/07.

Tapson, F. *A Dictionary of Units*. Available: http://www.ex.ac.uk/cimt/dictunit/dictunit.htm
Last accessed: 01/04/07.

Study exercises

29.1 Practise converting between units. Using Box 29.1 as a source, convert the following amounts into the units shown. Give your answers to three significant figures.
  (a) 101 000 Pa into atmospheres
  (b) One square yard into square millimetres
  (c) One UK pint into millilitres
  (d) 37 °C into kelvin
  (e) 11 stone 6 pounds into kilograms.

29.2 Practise using prefixes appropriately. Simplify the following number/unit combinations using an appropriate prefix so that the number component lies between 0.1 and 1000.
  (a) 10 000 mm
  (b) 0.015 ml
  (c) 5×10^9 J
  (d) 65 000 m s^{-1}
  (e) 0.000 000 000 1 g

29.3 Check units in an equation. The Hagan–Poiseuille equation describes water flux in smooth cylindrical pipes assuming laminar flow. This equation can be expressed as:

$$J_v = \frac{r^2 \times \delta P}{8\eta \times \delta x}$$

where

  r = the radius of the cylindrical pipe (m);
  η ('eta') = the viscosity of the liquid (Pa s);
  δP = the pressure difference across the ends of the pipe (Pa);
  δx = the length of the pipe (m).

Verify, by putting relevant units in place of the variables in this equation and simplifying the resulting relationship, that appropriate units for J_v are m s^{-1} (mean flow rate).

30 Scientific method and design of experiments

Science is a body of knowledge based on observation and experiment. Biological scientists attempt to explain life in terms of theories and hypotheses. They make predictions from these hypotheses and test them by experiment or further observations. The philosophy and sociology that underlie this process are complex topics (see e.g. Chalmers, 1999) and any brief description must involve simplifications.

Figure 30.1 models the scientific process you are most likely to be involved in – testing hypotheses on a small scale. These represent the sorts of explanation that can give rise to predictions which can be tested by an experiment or a series of observations. For example, you might put forward the hypothesis that the rate of K^+ efflux from a particular cell type is dependent on the intracellular concentration of calcium ions. This might then lead to a prediction that the application of a substance known to decrease the intracellular concentration of calcium ions, would reduce K^+ efflux from the cells. An experiment could be set up to test this hypothesis and the results would either confirm or falsify the hypothesis.

If confirmed, a hypothesis is retained with greater confidence. If falsified, it is either rejected outright as false, or modified and retested. Alternatively, it might be decided that the experiment was not a valid test of the hypothesis (perhaps because it was later found that the applied substance could not penetrate the cell membrane to a presumed site of action).

Nearly all scientific research deals with the testing of small-scale hypotheses. These hypotheses operate within a theoretical framework that has proven to be successful (i.e. is confirmed by many experiments and is consistently predictive). This operating model or 'paradigm' is not changed readily, and even if a result appears that seems to challenge the conventional view, it would not be overturned immediately. The conflicting result would be 'shelved' until an explanation was found after further investigation. In the example used above, a relevant paradigm could be the notion that life processes are ultimately chemical in nature.

Although changes in paradigms are rare, they are important, and the scientists who recognise them become famous. For example, a 'paradigm shift' can be said to have occurred when Darwin's ideas about natural selection replaced special creation as an explanation for the origin of species. Generally, however, results from hypothesis-testing tend to support and develop ('articulate') the paradigm, enhancing its relevance and strengthening its status. Thus, research in the area of population genetics has developed and refined Darwinism.

Where do ideas for small-scale hypotheses come from? They arise from one or more thought processes on the part of a scientist:

- analogy with other systems;
- recognition of a pattern;
- recognition of departure from a pattern;
- invention of new analytical methods;
- development of a mathematical model;
- intuition;
- imagination.

> **Definitions**
>
> There are many interpretations of the following terms. For the purposes of this chapter, the following will be used.
>
> **Paradigm** – theoretical framework so successful and well-confirmed that most research is carried out within its context and doesn't challenge it – even significant difficulties can be 'shelved' in favour of its retention.
>
> **Theory** – a collection of hypotheses that covers a range of natural phenomena – a 'larger-scale' idea than a hypothesis. Note that a theory may be 'hypothetical', in the sense that it is a tentative explanation.
>
> **Hypothesis** – an explanation tested in a specific experiment or by a set of observations. Tends to involve a 'small scale' idea.
>
> **(Scientific) Law** – this concept can be summarised as an equation (law) that provides a succinct encapsulation of a system, often in the form of a mathematical relationship. The term is often used in the physical sciences (e.g. 'Beer's Law', p. 280).

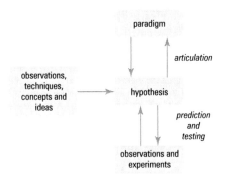

Fig. 30.1 A model of scientific method as used when testing hypotheses on a small scale. Hypotheses can arise as a result of various thought-processes on the part of the scientist, and are consistent with the overlying paradigm. Each hypothesis is testable by experiment or observation, leading to its confirmation or rejection. Confirmed hypotheses act to strengthen the status of the paradigm, but rejected ones do not immediately result in the paradigm's replacement.

Deciding whether to accept or reject a hypothesis – this is sometimes clear-cut, as in some areas of genetics, where experiments can be set up to result in a binary outcome (Chapter 58). In many other cases, the existence of 'biological variation' means that statistical techniques need to be employed (Chapters 66 and 67; Box 58.2).

Recently, it has been recognised that the process of science is not an entirely objective one. For instance, the choice of analogy which led to a new hypothesis might well be subjective, depending on past knowledge or understanding. Also, science is a social activity, where researchers put forward and defend viewpoints against those who hold an opposing view; where groups may work together towards a common goal; and where effort may depend on externally dictated financial opportunities and constraints. As with any other human activity, science is bound to involve an element of subjectivity.

No hypothesis can ever be rejected with certainty. Statistics allow us to quantify as vanishingly small the probability of an erroneous conclusion, but we are nevertheless left in the position of never being 100% certain that we have rejected all relevant alternative hypotheses, nor 100% certain that our decision to reject some alternative hypotheses was correct. However, despite these problems, experimental science has yielded and continues to yield many important findings.

 KEY POINT *The fallibility of scientific 'facts' is essential to grasp. No explanation can ever be 100% certain as it is always possible for a new alternative hypothesis to be generated. Our understanding of biology changes all the time as new observations and methods force old hypotheses to be retested.*

Definition

Mathematical model – an algebraic summary of the relationship between the variables in a system.

Quantitative hypotheses involve a mathematical description of the biological system. They can be formulated concisely by mathematical models. Formulating models is often useful because it forces deeper thought about mechanisms and encourages simplification of the system. A mathematical model:

- is inherently testable through experiment;
- identifies areas where information is lacking or uncertain;
- encapsulates many observations;
- allows you to predict the behaviour of the system.

Remember, however, that assumptions and simplifications required to create a model may result in it being unrealistic. Further, the results obtained from any model are only as good as the information put into it.

The terminology of experimentation

In many experiments, the aim is to provide evidence for causality. If x causes y, we expect, repeatably, to find that a change in x results in a change in y. Hence, the ideal experiment of this kind involves measurement of y, the dependent (measured) variable, at one or more values of x, the independent variable, and subsequent demonstration of some relationship between them. Experiments therefore involve comparisons of the results of treatments – changes in the independent variable as applied to an experimental subject. The change is engineered by the experimenter under controlled conditions.

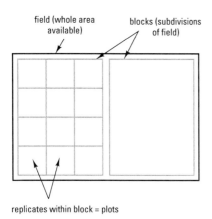

Fig. 30.2 Terminology and physical arrangement of elements in an experiment. Each block should contain the complete range of treatments (treatments may be replicated more than once in each block).

Subjects given the same treatment are known as replicates (they may be called plots). A block is a grouping of replicates or plots. The blocks are contained in a field, i.e. the whole area (or time) available for the experiment (Fig. 30.2). These terms originated from the statistical analysis of agricultural experiments, but they are now used across all areas of bioscience.

Why you need to control variables in experiments

Interpretation of experiments is seldom clear-cut because uncontrolled variables always change when treatments are given.

Confounding variables

These increase or decrease systematically as the independent variable increases or decreases. Their effects are known as systematic variation. This form of variation can be disentangled from that caused directly by treatments by incorporating appropriate controls in the experiment. A control is really just another treatment where a potentially confounding variable is adjusted so that its effects, if any, can be taken into account. The results from a control may therefore allow an alternative hypothesis to be rejected. There are often many potential controls for any experiment.

The consequence of systematic variation is that you can never be certain that the treatment, and the treatment alone, has caused an observed result. By careful design, you can, however, 'minimise the uncertainty' involved in your conclusion. Methods available include:

- Ensuring, through experimental design, that the independent variable is the only major factor that changes in any treatment.
- Incorporating appropriate controls to show that potential confounding variables have little or no effect.
- Selecting experimental subjects randomly to cancel out systematic variation arising from biased selection.
- Matching or pairing individuals among treatments so that differences in response due to their initial status are eliminated.
- Arranging subjects and treatments randomly so that responses to systematic differences in conditions do not influence the results.
- Ensuring that experimental conditions are uniform so that responses to systematic differences in conditions are minimised. When attempting this, beware 'edge effects' where subjects on the periphery of the layout receive substantially different conditions from those in the centre.

Nuisance variables

These are uncontrolled variables which cause differences in the value of y independently of the value of x, resulting in random variation. Experimental biology is characterised by the high number of nuisance variables that are found and their relatively great influence on results: biological data tend to have large errors. To reduce and assess the consequences of nuisance variables:

- incorporate replicates to allow random variation to be quantified;
- choose subjects that are as similar as possible;
- control random fluctuations in environmental conditions.

Constraints on experimental design

Box 30.1 outlines the important stages in designing an experiment. At an early stage, you should find out how resources may constrain the design. For example, limits may be set by availability of subjects, cost of treatment, availability of a chemical or bench space. Logistics may be a factor (e.g. time taken to record or analyse data).

Example Suppose you wish to investigate the effect of a metal ion on the growth of a culture. If you add the metal as a salt to the culture and then measure the growth, you will immediately introduce at least two confounding variables, compared with a control that has no salt added. Firstly, you will introduce an anion that may also affect growth in its own right, or in combination with the metal ion; secondly, you will alter the osmotic potential of the medium (see p. 150). Both of these effects could be tested using appropriate controls.

Reducing edge effects – one approach is to incorporate a 'buffer zone' of untreated subjects around the experiment proper.

Evaluating design constraints – a good way to do this is by processing an individual subject through the experimental procedures – a 'preliminary run' can help to identify potential difficulties.

Box 30.1 Checklist for designing and performing an experiment

1. **Preliminaries**

   (a) **Read background material** and decide on a subject area to investigate.

   (b) **Formulate a simple hypothesis to test.** It is preferable to have a clear answer to one question than to be uncertain about several questions.

   (c) **Decide which dependent variable you are going to measure and how.** Is it relevant to the problem? Can you measure it accurately, precisely and without bias?

   (d) **Think about and plan the statistical analysis of your results.** Will this affect your design?

2. **Designing**

   (a) **Find out the limitations on your resources.**

   (b) **Choose treatments that alter the minimum of confounding variables.**

   (c) **Incorporate as many effective controls as possible.**

   (d) **Keep the number of replicates as high as is feasible.**

   (e) **Ensure that the same number of replicates is present in each treatment.**

   (f) **Use effective randomisation and blocking arrangements.**

3. **Planning**

   (a) **List all the materials you will need.** Order any chemicals and make up solutions; grow, collect or breed the experimental subjects you require; check equipment is available.

   (b) **Organise space and/or time** in which to do the experiment.

   (c) **Account for the time taken to apply treatments and record results.** Make out a timesheet if things will be hectic.

4. **Carrying out the experiment**

   (a) **Record the results and make careful notes of everything you do** (see p. 176). Make additional observations to those planned if interesting things happen.

   (b) **Repeat the experiment** if time and resources allow.

5. **Analysing**

   (a) **Graph data as soon as possible** (during the experiment if you can). This will allow you to visualise what has happened and make adjustments to the design (e.g. timing of measurements).

   (b) **Carry out the planned statistical analysis.**

   (c) **Jot down conclusions and new hypotheses** arising from the experiment.

Your equipment or facilities may affect design because you cannot regulate conditions as well as you might desire. For example, you may be unable to ensure that temperature and lighting are equal over an experiment laid out in a glasshouse, or you may have to accept a great deal of initial variability if your subjects are collected from the wild. This problem is especially acute for experiments carried out under field conditions.

Using replicates

Replicate results show how variable the response is within treatments. They allow you to compare the differences among treatments in the context of the variability within treatments – you can do this *via* statistical tests such as analysis of variance (Chapter 67). Larger sample sizes tend to increase the precision of estimates of parameters and increase the chances of showing a significant difference between treatments if one exists. For statistical reasons (weighting, ease of calculation, fitting data to certain tests) it is often best to keep the number of replicates even. Remember that the degree of independence of replicates is important: sub-samples cannot act as replicate samples – they tell you about variability in the measurement method but not in the quantity being measured.

Deciding the number of replicates in each treatment – try to maximise the number of replicates in each treatment within the constraints of time and resources available.

If the total number of replicates available for an experiment is limited by resources, you may need to compromise between the number of treatments and the number of replicates per treatment. Statistics can help here, for it is possible to work out the minimum number of replicates you would need to show a certain difference between pairs of means (say 10%) at a specified level of significance (say $P = 0.05$). For this, you need to obtain a prior estimate of variability within treatments (see Sokal and Rohlf, 1994).

Randomisation of treatments

The two aspects of randomisation you must consider are:

- positioning of treatments within experimental blocks;
- allocation of treatments to the experimental subjects.

For relatively simple experiments, you can adopt a completely randomised design; here, the position and treatment assigned to any subject are defined randomly. You can draw lots, use a random number generator on a calculator, or use the random number tables which can be found in most books of statistical tables (see Box 30.2).

Box 30.2 How to use random number tables to assign subjects to positions and treatments

This is one method of many that could be used. It requires two sets of n random numbers – where n is the total number of subjects used.

1. **Number the subjects in any arbitrary order** but in such a way that you know which is which (i.e. mark or tag them).

2. **Decide how treatments will be assigned**, e.g. first five subjects selected treatment A; second five – treatment B, etc.

3. **Use the first set of random numbers in the sequence obtained to identify subjects and allocate them to treatment groups** in order of selection as decided in (2).

4. **Map the positions for subjects in the block or field. Assign numbers to these positions using the second set of random numbers**, working through the positions in some arbitrary order, e.g. top left to bottom right.

5. **Match the original numbers given to subjects with the position numbers.**

To obtain a sequence of random numbers:

1. **Decide on the range of random numbers you need.**

2. **Decide how you wish to sample the random number tables** (e.g. row-by-row and top to bottom) and your starting point.

3. **Moving in the selected manner, read the sequence of numbers until you come to a group that fits your needs** (e.g. in the sequence 978186, 18 represents a number between 1 and 20). Write this down and continue sampling until you get a new number. If a number is repeated, ignore it. Small numbers need to have the appropriate number of zeros preceding (e.g. 5 = 05 for a range in the tens, 21 = 021 for a range in the hundreds).

4. **When you come to the last number required, you don't need to sample any more:** simply write it down.

Example: You find the following random number sequence in a table and wish to select numbers between 1 and 10 from it.

```
9059146823    4862925166    1063260345
1277423810    9948040676    6430247598
8357945137    2490145183    5946242208
6588812379    2325701558    3260726568
```

Working left to right and top to bottom, the order of numbers found is 5, 10, 3, 9, 4, 6, 2, 1, 8, 7 as indicated by coloured type. If the table is sampled by working row-by-row right to left from bottom to top, the order is 6, 10, 7, 2, 9, 3, 4, 8, 1, 5.

Example If you knew that soil type varied in a graded fashion across a field, you might arrange blocks to be long thin rectangles at right angles to the gradient to ensure conditions within the block were as even as possible.

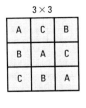

Fig. 30.3 Examples of Latin square arrangements for 3 and 4 treatments. Letters indicate treatments; the number of possible arrangements for each size of square increases greatly as the size increases.

Example A Latin square format is used for agar bioassays of antibiotics and vitamins (p. 238). Replicates are arranged so that gradients across the plate are cancelled out.

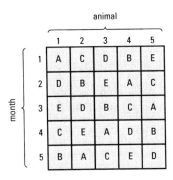

Fig. 30.4 Example of how to use a Latin square design to arrange sequential treatments. The experimenter wishes to test the effect of drugs A–E on weight gain, but only has five animals available. Each animal is fed on a control diet for the first 3 weeks of each month, then on a control diet plus drug for the last week. Weights are taken at start and finish of each treatment. Each animal receives all treatments.

A completely randomised layout has the advantage of simplicity but cannot show how confounding variables alter in space or time. This information can be obtained if you use a blocked design in which the degree of randomisation is restricted. Here, the experimental space or time is divided into blocks, each of which accommodates the complete set of treatments (Fig. 30.2). When analysed appropriately, the results for the blocks can be compared to test for differences in the confounding variables and these effects can be separated out from the effects of the treatments. The size and shape (or timing) of the block you choose is important: besides being able to accommodate the number of replicates desired, the suspected confounding variable should be relatively uniform within the block.

A Latin square is a method of placing treatments so that they appear in a balanced fashion within a square block or field. Treatments appear once in each column and row (see Fig. 30.3), so the effects of confounding variables can be 'cancelled out' in two directions at right angles to each other. This is effective if there is a smooth gradient in some confounding variable over the field. It is less useful if the variable has a patchy distribution, where a randomised block design might be better.

Latin square designs are useful in serial experiments where different treatments are given to the same subjects in a sequence (e.g. Fig. 30.4). A disadvantage of Latin squares is the fact that the number of plots is equal to the number of replicates, so increases in the number of replicates can only be made by the use of further Latin squares.

Pairing and matching subjects

The paired comparison is a special case of blocking used to reduce systematic variation when there are two treatments. Examples of its use are:

- 'Before and after' comparison. Here, the pairing removes variability arising from the initial state of the subjects, e.g. weight gain of mice on a diet, where the weight gain may depend on the initial weight.
- Application of a treatment and control to parts of the same subject or to closely related subjects. This allows comparison without complications arising from different origin of subjects, e.g. drug or placebo given to sibling rats, virus-containing or control solution swabbed on left or right halves of a leaf.
- Application of treatment and control under shared conditions. This allows comparison without complications arising from different environments of subjects, e.g. rats in a cage, plants in a pot.

Matched samples represent a restriction on randomisation where you make a balanced selection of subjects for treatments on the basis of some attribute or attributes that may influence results, e.g. age, sex, prior history. The effect of matching should be to 'cancel out' the unwanted source(s) of variation. Disadvantages include the subjective element in choice of character(s) to be balanced, inexact matching of quantitative characteristics, the time matching takes and possible wastage of unmatched subjects.

When analysed statistically, both paired comparisons and matched samples can show up differences between treatments that might otherwise be rejected on the basis of a fully randomised design, but note that the statistical analysis may be different.

Multifactorial experiments

The simplest experiments are those in which one treatment (factor) is applied at a time to the subjects. This approach is likely to give clear-cut answers, but it could be criticised for lacking realism. In particular, it cannot take account of interactions among two or more conditions that are likely to occur in real life. A multifactorial experiment (Fig. 30.5) is an attempt to do this; the interactions among treatments can be analysed by specialised statistics.

Multifactorial experiments are economical on resources because of 'hidden replication'. This arises when two or more treatments are given to a subject because the result acts statistically as a replicate for each treatment. Choice of relevant treatments to combine is important in multifactorial experiments; for instance, an interaction may be present at certain concentrations of a chemical but not at others (perhaps because the response is saturated). It is also important that the measurement scale for the response is consistent, otherwise spurious interactions may occur. Beware when planning a multifactorial experiment that the numbers of replicates do not get out of hand: you may have to restrict the treatments to 'plus' or 'minus' the factor of interest (as in Fig. 30.5).

Repetition of experiments

Even if your experiment is well designed and analysed, only limited conclusions can be made. Firstly, what you can say is valid for a particular place and time, with a particular investigator, experimental subject and method of applying treatments. Secondly, if your results were significant at the 5% level of probability (p. 493), there is still an approximately one-in-twenty chance that the results did arise by chance. To guard against these possibilities, it is important that experiments are repeated. Ideally, this would be done by an independent scientist with independent materials. However, it makes sense to repeat work yourself so that you can have full confidence in your conclusions. Many scientists recommend that experiments are done three times in total, but this may not be possible in undergraduate practical classes or project work.

Definition

Interaction – where the effect of treatments given together is greater or less than the sum of their individual effects.

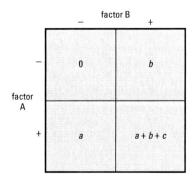

Fig. 30.5 Design of a simple multifactorial experiment. Factors A and B have effects a and b when applied alone. When both are applied together, the effect is denoted by $a + b + c$.

- If $c = 0$, there is no interaction (e.g. $2 + 2 + c = 4$).
- If c is positive, there is a positive interaction (synergism) between A and B (e.g. $2 + 2 + c = 5$).
- If c is negative, there is a negative interaction (antagonism) between A and B (e.g. $2 + 2 + c = 3$).

Reporting results – it is good practice to report how many times your experiments were repeated (in Materials and Methods); in the Results section, you should add a statement saying that the illustrated experiment is representative, or one explaining any differences between results obtained.

Text references

Chalmers, A.F. (1999) *What is This Thing Called Science?* 3rd edn. Open University Press, Buckingham.

Sokal, R.R. and Rohlf, F.J. (1994) *Biometry*, 3rd edn. W.H. Freeman, San Francisco.

Sources for further study

Heath, D. (1995) *An Introduction to Experimental Design and Statistics for Biology*. UCL Press, London.

Maber, J. (1999) *Data Analysis for Biomolecular Sciences*. Longman, Harlow.

Quinn, G.P. and Keough, M.J. (2002) *Experimental Design and Data Analysis for Biologists*. Cambridge University Press, Cambridge.

Study exercises

30.1 Consider the application of a Latin square design. Treatments W, X, Y and Z are to be applied to potted plants in a glasshouse where the researcher suspects there may be a slight gradation in temperature and light over an oblong bench that can hold a total of 48 pots. You decide to use an experimental design consisting of three 4×4 blocks of plants, each arranged in a different Latin square design (see below). Assign treatments to the locations in the diagram below. Explain why this design will help to eliminate the effects of the confounding variables.

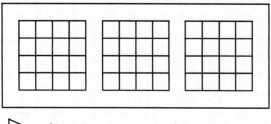

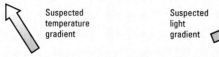

Layout for glasshouse experiment

30.2 Generate random numbers. Produce a list of 20 random whole numbers between 1 and 5 using a spreadsheet. If using MS Excel, investigate the RAND() and INT functions. The RAND() function produces a random number between 0 and 1, so you will need to multiply by a constant factor to scale your final output appropriately. Copy your test formula(e) to several cells to test empirically whether it works.

30.3 Investigate possible interactions. Treatments A, B and C involve tests of three different nutrients on proliferation of cells in a culture of carrot cells. Analyse the results in the table below to determine whether any interactions have occurred between the four possible combinations of treatments. No complex statistical analysis is required, simply a calculation of mean experimental effect (treatment minus controls) in each combined treatment and a comparison with the data for the relevant treatments on their own (in other words, assume that observed differences reflect true underlying differences). Classify the results as 'no interaction', 'antagonism' or 'synergism'.

Results of cell proliferation experiment

Treatment	Replicate (growth of callus in g)				
	1	2	3	4	5
Control	4.8	5.2	5.2	4.8	5.0
A	6.5	7.3	7.0	7.1	7.1
B	7.7	8.3	8.5	7.5	8.0
C	10.7	10.0	9.9	9.8	9.6
A+B	6.6	7.2	6.8	7.2	7.2
A+C	11.9	11.7	12.0	12.0	12.4
B+C	17.3	16.8	17.1	16.6	17.2
A+B+C	12.8	12.9	13.1	13.3	12.9

31 Project work

Research projects are an important component of the final-year syllabus for most degree programmes in the biosciences, while shorter projects may also be carried out during courses in earlier years. Project work can be extremely rewarding, although it does present a number of challenges. The assessment of your project is likely to contribute significantly to your degree grade, so all aspects of this work should be approached in a thorough manner.

Obtaining ethical approval – if any aspect of your project involves work with human or animal subjects, then you must obtain the necessary ethical clearance before you begin; consult your department's ethical committee for details.

Deciding on a topic to study

Assuming you have a choice, this important decision should be researched carefully. Make appointments to visit possible supervisors and ask them for advice on topics that you find interesting. Use library texts and research papers to obtain further background information. Perhaps the most important criterion is whether the topic will sustain your interest over the whole period of the project. Other things to look for include:

- Opportunities to learn new skills. Ideally, you should attempt to gain experience and skills that you might be able to 'sell' to a potential employer.
- Ease of obtaining valid results. An ideal project provides a means to obtain 'guaranteed' data for your report, but also the chance to extend knowledge by doing genuinely novel research.
- Assistance. What help will be available to you during the project? A busy lab with many research students might provide a supportive environment should your potential supervisor be too busy to meet you often; on the other hand, a smaller lab may provide the opportunity for more personal interaction with your supervisor.
- Impact. Your project may result in publishable data: discuss this with your prospective supervisor.

Using the Internet as an information source for project work – since many university departments and research groups have home pages on the World Wide Web, searches using relevant key words may indicate where research in your area is currently being carried out. Academics usually respond positively to emailed questions about their area of expertise.

Asking around – one of the best sources of information about supervisors, laboratories and projects is past students. Some of the postgraduates in your department may be products of your own system and they could provide an alternative source of advice.

Planning your work

As with any lengthy exercise, planning is required to make the best use of the time allocated (p. 10). This is true on a daily basis as well as over the entire period of the project. It is especially important not to underestimate the time it will take to write and produce your thesis (see below). If you wish to benefit from feedback given by your supervisor, you should aim to have drafts in his/her hands in good time. Since a large proportion of marks will be allocated to the report, you should not rush its production.

If your department requires you to write an interim report, look on this as an opportunity to clarify your thoughts and get some of the time-consuming preparative work out of the way. If not, you should set your own deadlines for producing drafts of the introduction, materials and methods section, etc.

Liaising with your supervisor(s) – this is essential if your work is to proceed efficiently. Specific meetings may be timetabled, e.g. to discuss a term's progress, review your work plan or consider a draft introduction. Most supervisors also have an 'open-door' policy, allowing you to air current problems. Prepare well for all meetings: have a list of questions ready before the meeting; provide results in an easily digestible form (but take your lab notebook along); be clear about your future plans for work.

 KEY POINT Project work can be very time-consuming. Try not to neglect other aspects of your course – make sure your lecture notes are up-to-date and collect relevant supporting information as you go along.

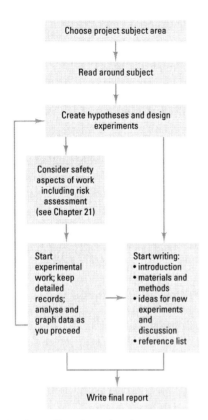

Fig. 31.1 Flowchart showing a recommended sequence of events in carrying out an undergraduate research project.

Sources of further information for project work:

- experimental design checklist (Box 30.1);
- advice on recording results (p. 176);
- describing and analysing numerical data (pp. 480–491);
- checklist for presenting your final report (Box 18.1).

Getting started

Figure 31.1 is a flowchart illustrating how a project might proceed; at the start, don't spend too long reading the literature and working out a lengthy programme of research. Get stuck in and do an experiment. There's no substitute for 'getting your hands dirty' for stimulating new ideas:

- even a 'failed' experiment will provide some useful information which may allow you to create a new or modified hypothesis;
- pilot experiments may point out deficiencies in experimental technique that will need to be rectified;
- the experience will help you create a realistic plan of work.

Designing experiments or sampling procedures

Design of experiments and sampling procedure is covered in Chapter 30. Avoid being too ambitious at the start of your work. It is generally best to work with a simple hypothesis and design your experiments or sampling around this. A small pilot experiment or test sample will highlight potential stumbling blocks including resource limitations, whether in materials, or time, or both.

Working in a laboratory environment

During your time as a project student, you are effectively a guest in your supervisor's laboratory.

- Be considerate – keep your 'area' tidy and offer to do your share of lab duties such as calibrating the pH meter, replenishing stock solutions, distilled water, etc., maintaining cultures, tending plants or animals.
- Use instruments carefully – they could be worth more than you think. Careless use may invalidate calibration settings and ruin other people's work as well as your own.
- Do your homework on techniques you intend to use – there's less chance of making costly mistakes if you have a good background understanding of the methods you will be using.
- Always seek advice if you are unsure of what you are doing.

 KEY POINT It is essential that you follow all the safety rules applying to the laboratory or field site. Make sure you know all relevant procedures – if in doubt, ask beforehand.

Keeping notes and analysing your results

Tidy record keeping is often associated with good research, and you should follow the advice and hints given in Chapter 28 (p. 180). Try to keep copies of all files relating to your project. As you obtain results, you should always calculate, analyse and graph data as soon as you can (see Fig. 31.1). This can reveal aspects that may not be obvious in numerical or readout form. Don't be worried by negative results – these can sometimes be as useful as positive results if they allow you to eliminate hypotheses – and don't be too dispirited if things do not work first time. Thomas Edison's maxim 'Genius is one per cent inspiration and ninety-nine per cent perspiration' certainly applies to research work.

Writing your project report

The structure of scientific reports is dealt with in Chapter 18. The following advice concerns methods of accumulating relevant information.

Introduction This is a big piece of writing that can be very time-consuming. Therefore, the more work you can do on it early on the better. Allocate some time at the start for library work (without neglecting your lab work), to build up a database of references (Chapter 8). Photocopying can be expensive, but you will find it valuable to have copies of key reviews and references handy when writing away from the library. Discuss proposals for content and structure with your supervisor to ensure your effort is relevant. Leave space at the end for a section on aims and objectives. This is important to orientate readers (including assessors), but you may prefer to finalise the content after the results have been analysed.

Materials and Methods Note as many details as possible *when doing the experiment or making observations*. Don't rely on your memory or hope that the information will still be available when you come to write up. Even if it is, chasing these details can waste valuable time.

Brushing up on your IT skills – word processors and spreadsheets are extremely useful when producing a thesis. Chapters 12 and 13 detail key features of these programs. You might benefit from attending courses on the relevant programs or studying manuals or texts so that you can use them more efficiently.

Results Show your supervisor graphed and tabulated versions of your data promptly. These can easily be produced using a spreadsheet (p. 74), but you should seek your supervisor's advice on whether the design and print quality is appropriate to be included in your report. You may wish to use a specialist graphics program to produce publishable-quality graphs and charts: allow some time for learning its idiosyncrasies. If you are producing a poster for assessment (Chapter 14), be sure to mock up the design well in advance. Similarly, think ahead about your needs for any seminar or poster you will present.

Using drawings and photographs – these can provide valuable records of sampling sites or experimental set-ups and could be useful in your report. Plan ahead and do the relevant work at the time of carrying out your research rather than afterwards.

Discussion Because this comes at the end of your report, and some parts can only be written after you have all the results in place, the temptation is to leave the discussion to last. This means that it might be rushed – not a good idea because of the weight attached by assessors to your analysis of data and thoughts about future experiments. It will help greatly if you keep notes of aims, conclusions and ideas for future work as you go along (Fig. 31.1). Another useful tip is to make notes of comparable data and conclusions from the literature as you read papers and reviews.

Acknowledgements Make a special place in your notebook for noting all those who have helped you carry out the work for use when writing this section of the report.

Using a word processor to record your ideas – remember that you can note down your thoughts and any other important information relevant to the Results and Discussion sections of your project in a file that can then form the basis of your first draft (p. 81); that way, you won't forget to include these points in your final report.

References Because of the complex formats involved (p. 48), these can be tricky to type. To save time, process them in batches as you go along – bibliographic software (e.g. Endnote) can help with organisation of references (p. 47).

 KEY POINT Make sure you are absolutely certain about the deadline for submitting your report, and work towards submitting a day or so before it. If you leave things until the last moment, you may find access to printers, photocopiers and binding machines is difficult.

Sources for further study

Barnard, C., Gilbert, F. and McGregor, P. (2001) *Asking Questions in Biology: Key Skills for Practical Assessments and Project Work*, 2nd edn. Prentice Hall, Harlow.

Ebel, H.F., Bliefert, C. and Russey, W.E. (2004) *The Art of Scientific Writing: from Student Reports to Professional Publications in Chemistry and Related Fields*, 2nd edn. Wiley-VCH, Weinheim, Germany.
[Covers record-keeping, notebooks, report writing and dissertations]

Jardine, F.H. (1994) *How to do your Student Project in Chemistry*. CRC Press, Boca Raton.
[Also contains general information relevant to bioscience projects]

Luck, M. (1999) *Your Student Research Project*. Gower, London.

Marshall, P. (1997) *Research Methods: How to Design and Conduct a Successful Project*. How To Books, Plymouth.

Ruxton, G. and Palgrave, N. (2006) *Experimental Design for the Life Sciences*. Oxford University Press, Oxford.

Sweetnam, D. (2004) *Writing your Dissertation: the Bestselling Guide to Planning, Preparing and Presenting First-Class Work*, 3rd edn. How To Books, Oxford.

Study exercises

Note: These exercises assume that you have started a research project, or are about to start one, as part of your studies.

31.1 **Prepare a project plan.** Make a formal plan for your research project, incorporating any milestones dictated by your department, such as interim reports and final submission dates. Discuss your plan with your supervisor and incorporate his or her comments. Refer back to the plan frequently during your project, to see how well you are meeting your deadlines.

31.2 **Resolve to write up your work as you go along.** Each time you complete an experiment or observation, write up the materials and methods, analyse the data and draw up the graphs as soon as you can. While you may reject some of your work at a later stage, you may wish to modify it; this will spread out the majority of the effort and allow time for critical thinking close to the final submission date.

31.3 **Devise a computer database for keeping details of your references.** Keeping these records up-to-date will save you a lot of time when writing up. You will need to decide on an appropriate referencing format, or find out about that followed by your department (see Chapter 8).

28 Drawing and diagrams

Drawing has an important place in biological teaching because of its role in developing observation skills. You need to look at a specimen very carefully to be able to draw it accurately, while labelling a diagram forces you to think about the component structures and their positions. If your observation of a specimen is poor, so too will be your diagram.

Strictly, a *drawing* is a detailed and accurate representation of a specimen, requiring no previous biological knowledge. This level of artwork is never required for normal practical work. A *diagram*, on the other hand, needs to be accurate in its general proportions, but is otherwise very stylised, showing only the most important features. Biological knowledge is required to select items for inclusion and to decide what detail to ignore. Diagrams are often called figures in formal scientific writing, but may sometimes be referred to loosely as drawings, since the above distinction is frequently ignored.

Figure numbering in reports – if producing several diagrams, graphs, etc., number them consecutively as Figure 1, Figure 2, etc., and use this system to refer to them in the text.

KEY POINT *You may not feel confident about being able to produce quality artwork in practicals, especially when the time allowed is limited. However, the requirements of biological drawing are not as demanding as you might think and the skills required can be learned. By following the guidelines and techniques explained below, most students should be able to produce good diagrams.*

The main types of figure

Cell diagrams

In a cell diagram, your aim is to show accurately the details of the individual cells in a tissue. You would normally draw a cell diagram from a specimen viewed by light microscopy at a magnification of 200× or more. Your diagram should be detailed, but need not comprise more than a few cells, especially if they are all similar (Fig. 28.1(a)). You may be asked to draw only one complete cell, plus *part* of any neighbouring cells, so that the interrelationship between cells can be seen. Use labels to note any structures that are stained, and what this means in terms of cell chemistry. For cells with thin walls, you may need to exaggerate the wall thickness to show it on your diagram. If this is necessary, add an explanatory note to your diagram (see Fig. 28.1(a)).

Tissue diagrams or maps

The purpose of a tissue diagram is to show how the tissues are placed in a section through a whole organism or one of its organs (compare Figs 28.1(b) and 28.1(c)). You should not include any cellular detail unless specifically instructed, and shading or hatching should be kept to a minimum. The main difficulty you will encounter is deciding where to draw the boundary line between tissues: cell differentiation is rarely discrete, so cells at the boundary may show characteristics of both tissues. A certain amount of background knowledge and interpretive skill is required for this.

Definitions

When drawing cell diagrams, tissue maps and body plans, it might help to bear in mind the following definitions:

Cells – the 'building blocks' of life. Always separated from their environment and other cells by a membrane ± a cell wall. Typical dimensions: 10 to 100 μm. Examples: leaf mesophyll cell; kidney nephron cell; yeast cell.

Tissues – collections of cells having similar structure and function. From one cell thick to many cell layers. Examples: palisade mesophyll in a leaf; renal cortex in the kidney.

Organs – structures consisting of several tissues working as a group to perform a specific function. Examples: leaf; kidney.

Practical Skills

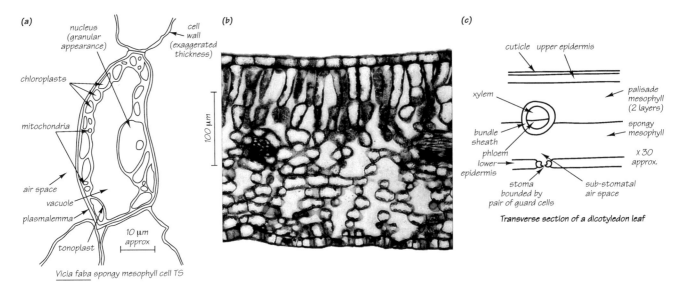

Fig. 28.1 Cell and tissue diagrams compared with a photomicrograph. (a) Cell diagram showing representative spongy mesophyll cells from an electron microscope image of a leaf section; (b) a photomicrograph of a transverse section through a leaf (light microscopy). Different scales are used, but in both cases they are indicated by a bar (see p. 259). Note how the cell-wall structure is clarified in (a) by slightly exaggerating its thickness. In a light microscope section less detail of organelles would be visible. (c) Tissue diagram of part of a TS leaf section. Note that this diagram is uncluttered by detail and the individual tissues are not shaded – the scale is indicated by a magnification factor relevant to the actual size of the diagram (reduced here) (see p. 258).

Choosing what to include in a morphological diagram – since the essence of this type of drawing is to show only the important features, it is worth making a list of the items to be included, or to highlight these where mentioned in your practical schedule. Don't forget, however, to add enough anatomical detail, e.g. of a body plan, to place outline features in the appropriate context. Always add a scale and effective labelling. Note that if you wish to show fine detail, consider doing this on a separate higher-scale diagram.

Morphological diagrams and body plans

In a morphological diagram, the objective is to provide a lifelike representation indicating the main surface features (Fig. 28.2). For a body plan, your aim is to show the relationships between segments, organs or other body parts, often following a dissection (Fig. 28.3). In both cases, shading should be avoided, unless it serves to highlight a particular feature. You should fully label both types of diagram and add notes where relevant. The main problem you are likely to encounter is keeping the different parts in proportion to each other – this can be solved by using construction lines and frames (see below).

 KEY POINT Remember that in drawing some 'artistic licence' is possible, allowing you to merge features seen on different specimens, or show, for example, the inside or underside of one of the parts.

Continuous tone drawings are more refined versions of the above where shading is used to provide a pictorial representation of the item of interest, providing fine detail and realism (Fig. 28.4). These drawings are generally only used for presentation work and are not normally expected from students. It can be argued that a properly taken photograph (Chapter 48) is often more appropriate and less time-consuming.

Apparatus diagrams

Here, your aim is to portray the components of some experimental set-up as a diagram (Fig. 28.5). Note that these figures are normally drawn as a section rather than a perspective drawing. Also, you may be more concerned with the relationship between parts than with showing them

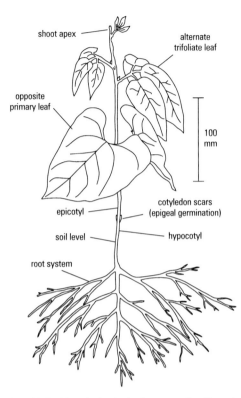

Fig. 28.2 Morphological diagram of a French bean seedling, *Phaseolus vulgaris* L. Note the lack of shading compared with Fig. 28.4.

Fig. 28.4 Continuous tone drawing of a French bean seedling, *Phaseolus vulgaris* L. Compare with Fig. 28.2.

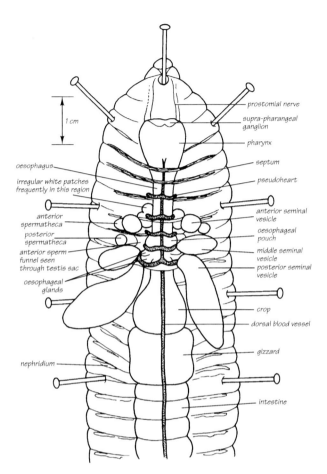

Fig. 28.3 Example of a body plan. This is a diagram of a general dissection of an earthworm, *Lumbricus terrestris* L., dorsal view. Note the lack of shading and the use of labels and annotations to identify clearly what has been drawn.

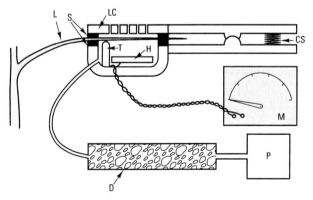

Fig. 28.5 Two-dimensional apparatus diagram of a diffusion porometer (not drawn to uniform scale). Note the use of letters to simplify labelling – these should be explained in the figure legend (e.g. L = leaf, P = pump).

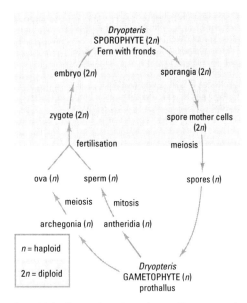

Fig. 28.6 Example of a chart: diagrammatic life cycle of the fern *Dryopteris filix-mas*. Note the use of shading and chromosome number to denote haploid and diploid phases of the life cycle, and how a key is used to explain this.

to a uniform scale. For example, in Fig. 28.5, the leaf clamp (LC) is exaggerated in size compared with other components so that its internal detail can be shown.

Charts

The main purpose of a chart is to organise information (e.g. Fig. 28.6). You can use charts to communicate complex ideas, procedures or lists of facts by simplifying, grouping and appropriate layout. In biology, they are particularly useful for illustrating life cycles, metabolic pathways and organisational hierarchies. Flowcharts (e.g. Fig. 33.1) are a specialised form. To be effective, charts must be logically organised. A good chart should clarify the parts and their relationships and its presentation should be simple, clear and visually pleasing. Make several rough sketches with different arrangements before deciding on the final version. You should use appropriate words or symbols to denote the components and link them with lines or arrows to show sequences or interrelationships. Computer software packages can be used to enhance the quality of presentation.

Graphs and histograms

These are used to display numerical information (data) in a form that is easily assimilated. Chapter 62 covers the main types of graphs and how they should be constructed.

Steps towards drawing a good diagram

To produce good figures, both planning and careful execution are needed (Box 28.1).

Planning

The first stage of any drawing is to decide exactly what to draw – this may seem obvious, but until you have focused your thoughts, you will not be able to decide on the answers to the following questions:

- What is the purpose of the drawing?
- What type of drawing is required?
- What should go into it?
- What magnification or reduction is required?

Box 28.1 Checklist for making a good diagram

1. **Decide exactly what you are going to draw and why.**
2. **Decide how large the diagram should be.**
3. **Decide where you are going to place the diagram on the page.**
4. **Start drawing:**
   (a) Draw what you see, not what you expect to see.
   (b) Use carefully measured construction lines to provide the correct proportions.
   (c) Avoid shading.
   (d) Avoid excess detail, especially in tissue diagrams.
   (e) Use conventions where appropriate e.g. cut edges represented by a double line.
5. **Label the drawing/diagram carefully and comprehensively.**
6. **Give the diagram a title, scale and legend:** include organism, classification, part drawn, orientation, stain(s), magnification, etc.

Once these decisions are made, you can determine the position and size of your diagram. Your diagram should be as large as possible, but remember to leave space for legends and labels.

Materials

Most diagrams for practicals are drawn in pencil, to allow corrections to be made. Propelling pencils are valuable for ensuring constant line thickness but they do not allow you the flexibility to vary line thickness as you can by changing the angle of an ordinary pencil. If you prefer to use an ordinary pencil, sharpen it frequently. Invest in a good-quality eraser – those of poor quality tend to smudge badly – and frequently clean its working surface on a spare piece of paper. Always use plain paper for drawing, and if you are asked to supply your own, make sure it is of good quality. Use pen and ink to create line drawings for illustration purposes in posters, project reports, etc. Such diagrams should be in black and white only. Computer drawing programs can also provide good-quality output suitable for these tasks.

Producing labels for diagrams in formal reports – *create the text using a word processor and high-quality printer. This can then be stuck on your diagram using e.g. Pritt Stick, taking care to keep all the text parallel. If the diagram is now photocopied or photographed with high contrast, the use of different pieces of paper cannot be detected. Alternatively, scan the diagram and add labels using a suitable program such as PowerPoint or Word.*

Constructing a diagram

1. Draw a faint rectangle in pencil to show the figure boundaries.
2. Draw very faint 'construction lines' using a 2H pencil with a sharp point to get the basic proportions and outlines correct before progressing. These should be erased once the basic drawing is complete. To lay in construction lines, use a ruler or pair of dividers to determine the actual proportions of the object to be drawn and then, using these dimensions, construct a scaled frame to allow further important reference points to be located (Fig. 28.7).
3. Draw the main outlines faintly with your 2H pencil. When satisfied, go over the lines with a sharp HB or 2B pencil. Draw firm, continuous lines, not hesitant, scratchy ones and make sure that junctions between lines are properly drawn. If you need to distinguish between different regions within your drawing, use hatching or stippling (but not shading), and avoid drawing unnecessary detail such as large numbers of cells (see Fig. 28.8). Always draw what you see, rather than what you think you should see, and seek advice at an early stage if you can't see a particular structure, or if you think your specimen is atypical. When drawing a specimen that is symmetrical or that contains repeated forms, it will save time if you draw its outline, and only provide detail in one of the replicated elements.
4. Complete your drawing by adding labels. This requires you to interpret your observations and helps you to remember what the structures look like. Careful and accurate labelling is as important as the drawing itself. It should be done clearly and neatly using either radiating or horizontal lines ending in arrowheads or large dots to indicate exact label references. The lines should not cross. Labels should be written clearly in one orientation, so that they can be read without needing to turn the paper. Annotations (short explanatory notes in brackets below the labels) are strongly recommended – regard them as notes to yourself about what you have seen, or what has been pointed out by tutors. For practical work, use a pencil for labelling in case your demonstrator or tutor corrects your work.

Using construction lines – *these are vital for producing well-proportioned drawings.*

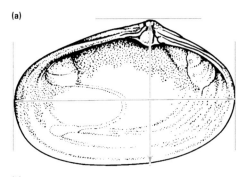

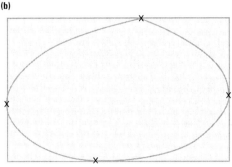

Fig. 28.7 How to draw an object in proportion: (a) determine linear dimensions; (b) construct a frame and outline, using reference points (x) determined from scaled measurements of the original specimen.

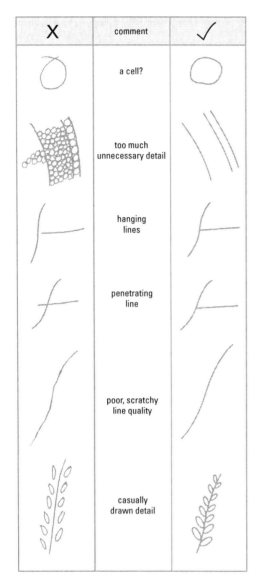

Fig. 28.8 Examples of common errors in biological drawings. Most of these mistakes are due to lack of care or attention to detail – easily solved!

5. Add a title, a scale or magnification factor (see Chapter 44), and a legend. The legend should provide all relevant information, including:
   (a) the binomial Latin name for the organism and a taxonomic classification, if appropriate;
   (b) details of the preparation of the subject, e.g. TS, whole mount, ventral dissection, etc., and any stains used.

Drawing from the microscope

Begin by positioning the paper beside your drawing hand and use the 'opposite' eye for examining the specimen; thus for a right-handed person, the paper is placed on the right of the microscope and you use your left eye. With a binocular microscope, use only one of the eyepieces; if you keep both eyes open, it is possible with practice to learn to draw and see the page with one eye while observing the specimen with the other. For specimens that need to be drawn very accurately in project work, projection devices such as the *camera lucida* may be required.

Avoiding mistakes

There are four main categories of error in student diagrams:

- Incorrect positioning and proportions – solve these problems by following steps 1–3 above.
- Forgetting to add a title, scale or a full set of labels – use a checklist like that provided in Box 28.1 to ensure your diagram is complete.
- Untidiness in presentation – avoid this type of error by using the correct materials as discussed above and by taking care – untidiness is frequently due to lack of attention to detail, as illustrated in Fig. 28.8.
- Biological inaccuracies – this kind of mistake is the most important, and will lose you most marks. Avoiding these errors requires preparation before the practical, so that you know more about what you are drawing and, for instance, have a good idea which parts are which *before* you start. Try to focus clearly on the objectives of the practical, and listen carefully to any tips given by your tutors, which may relate to the particular specimen(s) available on that day rather than to those in your notes or texts.

Finally, it is important to realise that you cannot expect biological drawing skills to develop overnight. This skill, like any other, requires much practice. Try to learn from any feedback your tutor may provide, and if your marks are consistently low without explanation, seek advice.

Sources for further study

Jepson, M. (1942) *Biological Drawings with Notes, Parts 1 and 2*, 5th edn. John Murray, London.
[Although this book is old, you may find it in your library. Provides examples of labelled biological drawings in an exemplary style.]

Leslie, C.W. (1984) *The Art of Field Sketching: A Naturalist's Sketchbook*. Prentice-Hall, Englewood Cliffs, NJ.

Sodt, J. *Botanical Illustration: A Selected Bibliography*. Available: http://www.library.wwu.edu/ref/subjguides/botill.htm

Last accessed 09/04/07.
[A wide-ranging source of information about drawing plant life. Although the focus of this site is on botanical illustration, it contains many useful references.]

Zweifel, F.W. (1988) *A Handbook of Biological Illustration*, 2nd edn. University of Chicago Press, Chicago.

Study exercises

28.1 Practise making well-proportioned drawings. Choose a picture of an animal or plant from a biology textbook or a printout from the Web. Using the techniques described on pp. 172–3 and Fig 28.7, make your own drawing based on this image.

28.2 Devise a flowchart. Present the instructions given in one of the Boxes in this text (for example, Boxes 39.1, 46.1 and 52.1) in the form of a flowchart.

28.3 Compare and contrast the main types of biological diagrams. Complete the following table.

Comparison of the main types of biological diagrams.

Feature	Cell diagram	Tissue diagram	Morphological diagram
Should the diagram carry a scale?			
What might be an appropriate scale in m?			
Approximate number of cells visible			
Are organelles visible?			
Should shading be used?			
Should labelling be used?			

(Type of diagram spans Cell diagram, Tissue diagram, Morphological diagram columns.)

29 Basic fieldwork procedures

Fieldwork is often one of the most enjoyable and rewarding aspects of undergraduate courses in biology, although it can also be arduous, especially in bad weather. The aims of fieldwork include:

- introducing a broad range of living organisms, with consideration of their roles in natural communities;
- developing skills in field techniques, with particular reference to sampling strategies and methods, and to the measurement of environmental variables;
- providing experience in the use of identification keys and fostering diagnostic skills;
- developing skills in data handling, analysis, interpretation and presentation, together with the writing of scientific reports;
- experiencing teamwork over an extended period;
- broadening environmental awareness.

To make the most of your time and effort while carrying out field investigations, you will need clear objectives and adequate preparation. You can minimise problems by carefully thinking your activities through beforehand.

KEY POINT Field studies are often time-restricted, and access to equipment and apparatus may be limited, so anticipation is the key to successful and safe fieldwork.

Typical coursework assessments for field excursions – these include:

- For specimens you have seen, recalling the names, diagnostic features, habitat, growth form, behaviour and physiology, etc.
- Making a diagrammatic representation of the site visited, e.g. a labelled map or transect.
- Interpreting key features of a photograph of the area visited.
- Interpreting data representing aspects of the location, based on the knowledge you have gained about the organisms and their environment.

Preventing rain damage to printed information in the field – use laminated sheets for keys, maps and any other items you wish to protect.

Field excursions

You are likely to make short visits to field sites during the early part of your course. These excursions provide an opportunity to gain an overview of a particular environment in a single visit, e.g. an afternoon excursion to a marine rocky shore. Note the following points:

- Carry out the recommended background reading – you will get much more out of the visit if you read any specific handouts or notes in advance. If you plan to take these into the field, you can put them in separate polythene pockets – if necessary, you can make these fully waterproof by sealing the top opening with adhesive tape. It might also be useful to carry out some general background reading on the area to be visited. You should pay particular attention to any risk assessment information and safety guidance notes.
- Make sure you are well-prepared for the weather and the terrain – listen to the weather forecast and take appropriate clothing to keep you warm and dry. Just because your visit is short, you should not dismiss this aspect of the trip. It is better to take extra clothing that you do not use than to leave it at home. Match your footwear with the environment, e.g. wear waterproof boots, not trainers, on a visit to a rocky shore. Brightly coloured clothing is a good idea, since this will be easier to spot if you become lost, or separated from the main group. Take account of the exposure of the site – in bright sunshine,

you may need a hat and sun protection cream, but in chilly and windy conditions a hat, gloves and a warm coat would be required. A large pair of domestic rubber gloves can be worn over woollen gloves for work in cold and wet conditions, e.g. sampling a lake in winter. Students without appropriate clothing may be unable to participate in all or some of the field activities.

- During the excursion, keep up with the field leader – you will need to be close enough to hear any commentary. Take notes as you go, to get maximum value from the visit: your course assessment may test recall and understanding and this may be your only opportunity to visit this location with a knowledgeable 'guide'. Remember to bring a suitable notebook and pencil for making notes and recording data – for further details, see Chapter 32. Rewrite your field notes soon after returning from the excursion, adding additional detail, as required.
- Act responsibly at all times – consider your own safety and that of others in the group, and minimise the impact of your visit on the environment, e.g. keep to paths, where appropriate. Follow at all times any specific safety instructions you are given during the visit.

 SAFETY NOTE Anticipate possible hazards – for example, there may be a significant risk of slipping on seaweed-covered rocks, on wet grass or on loose scree slopes. Also consider the risks involved in handling plants and animals (poisons, allergies, bites, stings, etc.) – if in doubt, ask your instructor before picking up any unknown specimen.

Planning and preparation for field-based project work

In an individual or group project involving a fieldwork component, you are likely to carry out a more involved and detailed study, in comparison to a field excursion: Box 29.1 gives details of the key questions to consider.

 KEY POINT Always construct a checklist (e.g. Table 29.1) of the required fieldwork equipment, paying attention to the smallest detail.

Making a preliminary inspection of a field site

It is unlikely that you will be able to answer the questions in Box 29.1 if you or your instructor have not visited the field area in advance. The location, for example, may be remote, rugged or dangerous; it may be intertidal or subject to problems of access; the area may present social hazards (e.g. in some urban areas) or be culturally sensitive (e.g. nature reserves or archaeologically important sites). All fieldwork programmes, therefore, should be preceded by a 'scoping' or orientation study involving a preliminary inspection of the area. Make sure that the organisms that you wish to study are present and accessible. Any problems relating to access problems should be resolved at this stage and permissions obtained, where appropriate. Always make sure that landowners know who you are, where you are from, what you intend to do, and when you propose to be on their land. Be open and honest with them and involve them in your project.

 KEY POINT Never enter private land without permission: this is potentially trespassing. After your fieldwork, always thank the landowner for their help and support.

Table 29.1 A checklist of generic equipment for fieldwork

- Notebook and pencil
- Datasheets
- Indelible marker pens
- Camera
- Food and drink (plus emergency rations)
- First aid equipment
- Safety (protective) gear
- Whistle
- Two-way radio or mobile phone
- Watch or clock
- Torch
- Multi-purpose knife (e.g. Swiss Army knife)
- Maps and compass or GPS (position locating device)
- Hand lens
- Specialised measuring equipment
- Sampling equipment (including spares)
- Specimen storage materials and labels
- Rucksack or other carrying devices
- Laminated guides and plans
- Specific items such as field guides, biological keys, etc.
- Survival bag (e.g. for longer trips, or in remote locations)
- Buoyancy aid and line (e.g. for trips near deep water)

Box 29.1 Questions you should consider before carrying out a field project

- **What are the aims and objectives of the study?** Your fieldwork must be realistic in its scope and purpose; most people overestimate what can be done in a given time period, especially if the weather is poor. Recognise your own limitations and those of others who might be involved and do not overstretch yourself.

- **How long and how many periods of fieldwork are required?** If your work requires more than one visit, what time interval will be needed between visits? This may affect the logistics and cost of the work.

- **What are the safety implications of the work?** It is always essential to take full account of safety issues, and you should read and take note of the basic rules in Chapter 20. Make sure that a risk assessment has been carried out before any field-based project work is carried out: the major hazards should be considered, and any steps required to minimise risk must be identified in advance (see p. 119).

- **Am I likely to encounter any difficult conditions or environments?** Plans should always be based upon a preliminary site inspection and you should always discuss your intentions with more experienced fieldworkers.

- **What samples are needed and how will they be collected?** The number, frequency, nature and spatial or temporal distribution of samples must be consistent with the purpose of the fieldwork and may require considerable planning in advance. Too many samples can be as much of a problem as too few, so determine the minimum required sample size with statistical evaluation in mind. Chapter 30 considers aspects of sampling strategy. The sampling or environmental measurement and recording protocol that you choose will influence your equipment requirements.

- **How should samples be stored for transport to the laboratory?** If the work involves collection of samples or specimens, make sure you include appropriate storage vessels and a means of labelling them (e.g. a spirit-based marker). You may also need to identify temporary storage locations in the field – note that samples may be ruined by inappropriate field storage (e.g. inactivation of bacteria by brief exposure to bright sunlight) and this damage may not be obvious on return to the laboratory. Water samples may require refrigeration to minimise changes in biological composition. Remember too that large numbers of water, sediment or soil samples can be very heavy; make sure that you have considered the best way to transport them. Chapters 34, 35 and 36 give further advice on collecting and preserving biological specimens.

- **What equipment will be required?** You should compile a detailed list well in advance, and then check that everything is available and in working order.

- **What transport will be needed for personnel, equipment and samples?** You may need to arrange for a vehicle or a boat of appropriate size. On land, choose a vehicle appropriate for the terrain. In shallow water, use only an inflatable dinghy or shallow-draught boat. In deep water, make sure that the boat can withstand any possible sea conditions. Take account of the capacity of your means of transport – to overload a vehicle or boat is a risk to safety.

SAFETY NOTE Remember your responsibilities during fieldwork – although your activities will normally be organised in consultation with a member of staff, this does not free you from a personal responsibility for your own and the group's safety.

No sampling or survey work is carried out at the inspection stage. However, the field area should be walked over and examined. Appropriate notes should be taken, and, where applicable, the site should be photographed, so that the strategy for subsequent sampling can be determined.

Working safely and responsibly

Always wear appropriate clothing and safety equipment and let someone else know where you will be working and when you expect to return. Do not then alter this plan without informing that person. Remember, too,

SAFETY NOTE *Safe and responsible fieldwork also means:*
- *Do not drink alcohol or take drugs (other than prescribed medication) in the field.*
- *Do not annoy or antagonise the local population.*
- *Leave gates as you find them.*
- *Do not drop litter.*
- *Avoid fire risks, especially when working in wooded areas, sand dunes or heathland.*
- *Do not frighten or disturb livestock or domestic animals.*
- *Do not attempt to carry too much. Keep rucksack loads below 14 kg. Make several trips rather than risk exhaustion or injury by trying to carry too great a load.*
- *You should also observe the general safety points discussed in Chapter 20.*

that what may be a comfortable environment for one person may not be so for another, depending on their outdoor experience and interests. Do not attempt to work from a boat or climb a rock face without prior training. If your work requires specialist support you must seek assistance from an appropriately qualified individual (e.g. a qualified rock climber or a scuba diver).

Sources for further study

Anon. (1995) *Field Work Code of Practice.* CVCP, London.

Jones, A.M., Duck, R., Reed, R. and Weyers, J.D.B. (2000). *Practical Skills in Environmental Science.* Prentice Hall, Harlow.

Nichols, D. (1999) *Safety in Biological Fieldwork. Guidance Notes for Codes of Practice*, 4th edn. Institute of Biology, London.

Watts, S. and Halliwell, L. (eds) (1996) *Essential Environmental Science: Methods and Techniques.* Routledge, London.

Study exercises

29.1 **Check your knowledge of generic fieldwork requirements.** What items are missing from the following list of generic fieldwork items? Notebook and pencil; datasheets; camera; food and drink (plus emergency rations); safety (protective) gear; whistle, watch or clock; torch; multi-purpose knife; map(s) and compass/GPS device; hand lens; specialised measuring equipment; specimen storage materials and labels; rucksack or other carrying devices; laminated guides and plans; field guides.

29.2 **List specific items that would be needed to carry out an ecological investigation.** This could be, for example, of the fauna and flora of a marine rocky shore or of plants in a meadow. Make a note of why each item is required and any problems that might be associated with its use.

29.3 **Consider the hazards of a particular field study.** List the potential hazards of carrying out a survey of woodlice in an urban estate environment.

30 Samples and sampling

When carrying out research, it is unlikely that you can observe or measure every individual in the population of organisms in which you are interested. In practice, statistics obtained from a subset (or sample) are used to estimate relevant parameters for the biological population. Samples consist of data values for a particular variable (e.g. length), each recorded from an individual sampling unit (e.g. a limpet) in a sample of n units (e.g. $n = 50$ limpets) taken from the population under investigation (e.g. those limpets on a particular rocky shore). The term 'replicate' can be applied either to the measurement (i.e. repeated measurement or observation taken at the same time or location) or the actual sampling unit.

Symbols are used to represent each type of sample statistic: these are given Roman-character symbols, e.g. $\overline{Y}$ for the sample mean, while the equivalent population parameter is given a Greek symbol, e.g. μ for the population mean. When estimating population parameters from sample statistics, the sample size is important, larger sample sizes allowing greater statistical confidence. However, the optimum sample size is normally a balance between statistical and practical considerations.

This chapter deals mainly with sampling in fieldwork where natural populations are to be observed under undisturbed conditions; however, the same principles may also apply in a laboratory context (see Chapter 31).

Specifying the population being sampled

At the outset, it is important to provide a complete description of the biological population being sampled. Failure to do this will make your results difficult to interpret or to compare with other observations, including your own. When populations are to be compared, ideally only the variable under consideration should differ if maximum significance is to be placed upon the analysis.

Deciding on a sampling strategy

Selecting a sample involves the formulation of rules and methods (the sampling protocol) by which some members of the population are included in the sample. The chosen sample is then measured using defined procedures to obtain relevant data. Finally, the information so obtained is processed to calculate appropriate statistics. Good sampling design takes into account all of these procedures, and it should relate to the objectives of your investigation by providing valid estimates of features of the population(s) of interest.

Truly representative samples should be:

- taken at random, or in a manner that ensures that every member of the population has an equal chance of being selected;
- large enough to provide sufficient precision in estimation of population characteristics;
- unbiased by the sampling procedure or equipment.

Definitions

Biological population – all those individuals within a specified time or space about which inferences are to be made, specified according to some biological definition (perhaps related to life history, growth stage, or sex), and normally investigated at a particular location and time.

Parameter – a numerical constant or mathematical function used to describe a particular population (e.g. the mean height of 18-year-old females).

Statistic – an estimate of a parameter obtained from a sample (e.g. the height of 18-year-old females based on those in your class).

Choosing relevant population factors to specify – these should include:

- geographical location;
- type of habitat;
- date and time of sampling;
- age, sex, physiological condition and health of sampled organisms;
- other details relevant to your work, e.g. an index of pollution.

This information might apply to all members of the population or from a matrix of data associated with each replicate.

Selecting specimens – do not include data for which the appropriate population specification is unavailable.

Samples and Sampling

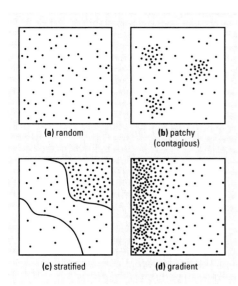

Fig. 30.1 Types of distributions

Definitions

Homogeneous – evenly distributed.

Patchy – showing clustered (contagious) distribution (e.g. numbers of parasites within hosts).

Gradient – a distribution that varies smoothly over the sampling area.

Stratified – showing a distribution with discrete levels or strata (e.g. algae on a rocky shore)

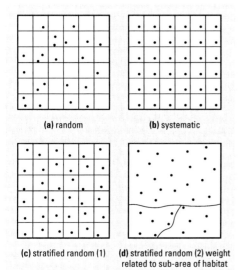

Fig. 30.2 Basic methods of sampling

These requirements may well conflict and there is rarely any unique best answer to a sampling problem. You should, however, take great care to minimise bias, or population parameters inferred from your samples will be unrealistic and this may invalidate your work and its conclusions.

 KEY POINT *Sampling involves choices on the part of the investigator. A 'sampling strategy' should allow you to obtain reliable and useful information about your particular population(s), while using your resources efficiently.*

The sampling protocol

You should decide on a sampling protocol before any investigation proceeds. The main aspects to be determined are: the position of samples; the size and shape of the sampling area; and the number of sampling units in each sample. Before this can be done, however, information is required about the likely distribution of organisms. This can be even (homogeneous), patchy (contagious), stratified (homogeneous within sub-areas) or present as a gradient (Fig. 30.1). You might decide which type applies from a pilot study, published research or by analogy with other systems.

Locating your samples

The positions where sampling takes place can be determined either randomly, systematically or in some stratified manner (Fig. 30.2).

In simple random sampling (Fig. 30.2(a)), a 2-D coordinate grid is superimposed on the area to be investigated. The required number of grid reference data pairs is then obtained using random numbers (p. 188) and samples taken at these points. Every organism in the population thus has an equal chance of selection, but the area may not be covered evenly. This method is best if the distribution of organisms is homogeneous.

Systematic sampling (Fig. 30.2(b)) involves selecting the location of the first sampling position at random and then taking samples at fixed distances from this. This method has the advantage of simplicity and it is often used where the intention is to map data. The disadvantages are firstly, that the results can be biased if the interval between sampling positions coincides with some periodic distribution of the population, and secondly, that there is no reliable method of estimating the standard error of the sample mean.

Stratified random sampling may be preferable if you wish to avoid these disadvantages yet still ensure that each part of the area is represented (e.g. where you suspect there to be stratified micro-habitats, as in a tidal seashore). The area is divided into sub-areas within which random sampling is carried out. These can either be constant in size (Fig. 30.2(c)) or related to known features in the sampling area (Fig. 30.2(d)). If the latter, strata are normally sampled in proportion to their area. 'Weighting' is the general term applied to sampling procedures that allow the calculated statistics to represent the population better by accounting for differences in the distribution of the chosen character. You can analyse data from different strata by a one-way analysis of variance (see Chapter 66).

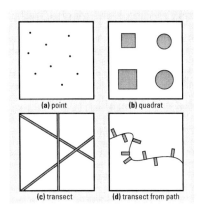

Fig. 30.3 Methods of positioning samples

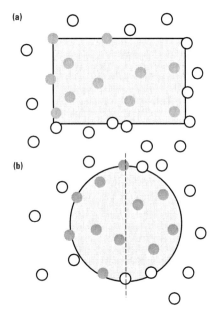

Fig. 30.4 Examples of sampling protocols for reducing edge effects. Filled circles represent objects to be counted or measured, open circles those to be ignored. (a) Rectangular area. All objects touching the top and left-hand sides including the top left-hand corner are included, as well as those clearly within the perimeter. (b) Circular area. All objects clearly within the perimeter are included as well as those touching the perimeter on the left-hand side. In the bottom half, those that would touch both the perimeter and the imaginary plane of symmetry are ignored.

The dimensions of the sampling area

The chief options are:

- point sampling (Fig. 30.3(a));
- quadrat sampling (Fig. 30.3(b));
- transect (traverse) sampling (Fig. 30.3(c)).

Quadrats are usually either circular or square. A circular quadrat has the advantage that its position can be marked as a single (central) point and the area defined by use of a tape measure, whereas a square quadrat may require marking at each corner. Transects are generally used when it is difficult to move through the site to position quadrats. If a defined path is present, transects taken at right angles to the path (Fig. 30.3(d)) will save time in reaching the sites.

The problems involved are best illustrated by reference to the selection of a quadrat for fieldwork sampling. Note first that the maximum number of independent sampling units (quadrats) is equal to the area occupied by the population under study (total theoretical sampling area) divided by the area of the sampling units (quadrat area).

The distribution and size of the organisms must be considered: it is obvious that you would require different-sized quadrats for trees in a forest than for daisies on a lawn. When the distribution is truly random, then all quadrat sizes are equally effective for estimating population parameters (assuming the total number of individuals sampled is equal). If the distribution is patchy, a smaller quadrat size may be more effective than a larger one: too large an area might obscure the true nature of the clumped distribution. Alternatively, you may wish to exclude a patchy distribution from your investigation and should thus choose a relatively large sampling area. If the distribution is stratified or graded, then sampling area is generally less important than sampling position.

Small sample areas have the advantage that more small samples can usually be taken for the same amount of labour. This may result in increased precision and many small areas will cover a wider range of the habitat than few large ones, so the catch can be more representative. However, sampling error at the edge of quadrats is proportionately greater as sample area diminishes, increasing as the scale of the quadrat and the sampled item become closer. To avoid such effects, you need to establish a protocol for dealing with items that overlap the edge of the quadrats (Fig. 30.4). These protocols are also valid when sampling objects in e.g. microscope fields.

Number of sampling units per sample

When small numbers of sampling units are used, this can lead to imprecise estimates of population parameters because the values of the sample statistics will be susceptible to the effects of random variation – this is especially true if the underlying spatial distribution is patchy, as is often the case. You may then be unable to demonstrate statistically that there are differences between the populations. On the other hand, measuring very large numbers of replicates may represent an impractical workload.

To estimate appropriate numbers of sampling units, you can use data from a pilot study to work out the probability of detecting a specified difference in the measured variable at a specified confidence level (see Sokal

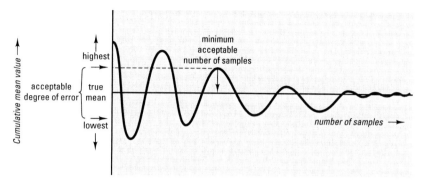

Fig. 30.5 Determination of the number of sampling units required

and Rohlf, 1994). However, the formulae involved are complex. A simpler method, usually performed as a pilot study, is as follows:

1. Take five sampling units at random and calculate the arithmetic mean of the measured variable for this sample.
2. Take five more units and calculate the mean for the ten units you have now collected.
3. Continue sampling in five-unit steps and plot the cumulative mean value against the number of samples. When the mean fluctuates within acceptable limits, say ±5 per cent, a suitable number of sampling units has been reached (Fig. 30.5).

The sequence of events in creating a sampling strategy is shown in Fig. 30.6.

Sampling in time

Sampling in time presents a different set of problems. If examining a phenomenon that fluctuates regularly (e.g. with a period governed by day and night, or high and low tide), then the frequency of sampling has to be determined with that periodicity in mind. In fact, you should consider the possible existence of periodic phenomena, even if they are not immediately obvious. If the phenomenon you are investigating is likely to change through time by some complex function such as the logarithmic decay in pollutant concentration, then your sampling intervals should be spaced according to an appropriate logarithmic series.

Subsampling

If you do not wish to sample the whole of a quadrat, perhaps because the density of sampling units is too high and it would take too long, then you can employ subsampling by studying a defined part of the quadrat – also known as two-stage sampling. If studying the density of organisms, this can simplify counting: assuming the organisms are randomly distributed, and only a small proportion of the population is sampled in each subunit, then the counts should follow a Poisson distribution (see Chapter 66). If this distribution is confirmed, only single subsamples need to be counted to estimate the total numbers in each quadrat. The precision of the estimate then depends on the size of the count.

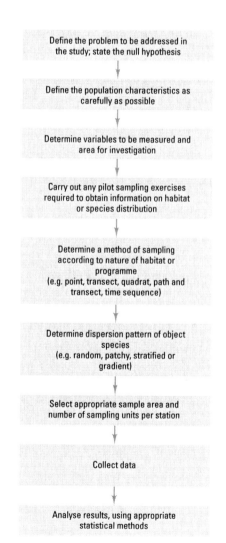

Fig. 30.6 Flowchart outlining decisions required for field sampling studies

Text reference

Sokal, R.R. and Rohlf, F.J. (1994) *Biometry*, 3rd edn. W.H. Freeman, San Francisco.

Sources for further study

Baker, J.M. and Wolff, W.J. (1987) *Biological Surveys of Estuaries and Coasts*. Cambridge University Press, Cambridge.
[Includes chapters on planning biological surveys and safety as well as numerous sections on sampling specific habitats.]

Keith, L.H. (1991) *Environmental Sampling and Analysis: a Practical Guide*. Lewis Publishers Inc., Michigan.
[Advanced text that is particularly good on how to devise protocols.]

New, T.R. (1998) *Invertebrate Surveys for Conservation*. Oxford University Press, Oxford.
[Excellent source for most aspects of sampling.]

Slingsby, D. and Cook, C. (1986) *Practical Ecology*. Macmillan Education, London.
[Contains a general introduction to sampling and other aspects of fieldwork.]

Study exercises

30.1 Calculate the density of subsampled copepods. A sample of plankton was collected using a 5 litre water bottle sampler. It was filtered upon collection through a 200 μm mesh net and the contents retained and preserved. Because the resulting volume of material was too large to count all the individuals present, the material was subsampled by taking ten replicate 1 per cent aliquots. The numbers of copepods collected for each subsample determined using a counting chamber are given in the table below. From these data, calculate the mean density of copepods as (a) density per litre and (b) density per cubic metre of water.

Sub-samples of copepods

1	2	3	4	5	6	7	8	9	10
234	255	342	276	288	295	324	301	299	273

30.2 Devise a protocol for counting limpets on a rocky platform using some form of quadrat. Take into account the size of the organism (up to 50 mm long), their normal density (up to 50 per square metre) and the significance of edge effects at the periphery of your sampling device. Assume that the limpets are randomly distributed.

30.3 Calculate the number of samples required for a given level of accuracy. The table below presents data on the effect of number of samples on the estimate of the mean for a population of limpets. Calculate the minimum number of samples that would be necessary to provide a measure of the mean (sample mean) that is within 10 per cent of the true (population) mean (which is 50).

Estimated mean after taking different numbers of samples

Number of samples	1	3	5	10	15	20	25	30	35	40
Estimated mean (limpets m^{-2})	90	62	42.8	56.1	43.3	54.8	52.7	48.9	50.6	50

Obtaining and Identifying Specimens

34 Collecting animals and plants

If you are required to collect material you may need to use a formal sampling procedure (see Chapter 30) or simple qualitative collecting. Your choice of equipment will depend upon your objectives. Some of the main reasons for collecting include:

- Obtaining specimens for subsequent laboratory experimentation or observation: this requires collection of living material while causing minimum stress and damage to the organisms.
- Making estimates of population and community parameters: this can be destructive (requiring killing of specimens) or non-destructive, depending upon the objectives of the study and any requirements for subsequent laboratory work, e.g. sorting and identification.
- Gathering specimens for museum-type collections: here, the main objective is to obtain undamaged and representative specimens, usually in a preserved form (see Chapter 35).
- Collecting for subsequent chemical/biochemical analysis: this may require a formal protocol. This objective requires care in both the method of collection and subsequent storage to avoid inducing chemical changes that are artefacts of the collection and storage processes. Deep freezing is usually the preferred storage method when subsequent chemical analysis is likely.

The main 'rules' for collecting are:

- Collect only enough for your purposes.
- Treat animals with respect at all times: do not cause unnecessary stress or suffering. There are formal rules for the handling of many vertebrate species but the same attitude should be taken towards all living organisms.
- Minimise damage and stress during collection and transport: stressed organisms are of little use for realistic experimentation.
- Be aware of the limitations and bias of your collecting equipment: this is particularly important for formal sampling procedures where collecting devices almost always have such problems. This may require specific testing for your particular usage.
- Keep good records of collection details.

KEY POINT *Check any legislation relating to the species or habitats you are intending to use. Obtain any permits required and strictly obey any regulations.*

Equipment for collecting

There is an immense variety of collecting and sampling devices available, see e.g. Fig. 34.1. In general, the collection of remote and/or animal samples presents the greatest problems. The more remote the operator is

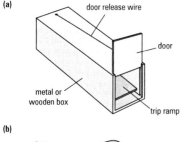
(a)
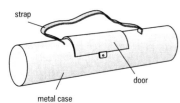
(b)
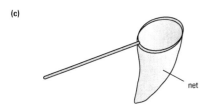
(c)
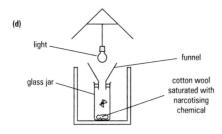
(d)

(e)
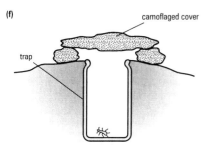
(f)

Fig. 34.1 Examples of devices used for collecting and sampling. (a) Small vertebrate trap – the animal walks up the ramp towards bait and releases the trap door by a simple trip mechanism. (b) Botanist's vasculum – used in the field for collecting and protecting plant, fungal and animal specimens. (c) Sweep net – used for catching flying insects or aquatic specimens. (d) Light trap – used for catching night-flying insects. (e) 'Pooter' – device for collecting small animals such as insects by suction. (f) Pitfall trap – animal falls into camouflaged trap and cannot escape.

from the point of sampling, the more difficult is the evaluation of the quality of the sample in terms of its representativeness. Animal collection can be difficult because of factors such as mobility and complications introduced when allowing for avoidance behaviour.

Collecting formal samples

Here the objective is to obtain specimens that both qualitatively and quantitatively represent some well-defined habitat(s). Some of the more obvious practical considerations are:

- Are there 'edge effects' associated with the sampling method? Because of such effects you may need to adopt a special protocol, e.g. to determine whether or not a specimen falling on the boundary is included or excluded from the sample (p. 181).
- Will the sample be uniform? For example, remote sampling of marine sediments of different texture using grabs often results in 'bites' of different depth being taken.
- Does the method sample all components of the biota equally well? This may not be important as long as it adequately samples those components in which you are interested.
- How accurately can the location of the sample be defined, especially with regard to other samples intended as replicates? This is particularly important in remote sampling.
- Is the size of the sampling unit adequate for the size and distribution of the object, species or communities being sampled? In remote sampling, this can often be a problem and frequently the method used is a compromise.
- Can interspecies interactions affect the integrity of the sample after collection but before processing? Factors such as predation can be prevented by using an appropriate chemical fixative immediately upon collection (see Chapter 35). Freezing is a poor option here since, upon thawing, many animal and plant tissues tend to disintegrate and the specimens are in poor condition for subsequent identification, etc.

General collecting strategies

For qualitative collecting, where a representative sample of the population or habitat is not the objective, the effectiveness of equipment may be of less significance. Here, suitability for capturing living and undamaged specimens may be the principal criterion in choosing your equipment. For mobile animals this often involves nets and traps of various kinds, combined with narcotising agents such as smoke for insects (see Chapter 35). If specimens can be killed before capture, then spray biocides can be a useful aid. Some useful general points about all collections are:

- Think ahead and be prepared. Your collecting equipment must be appropriate for the task.
- Keep a good record of collection details: this is particularly important if the collection is for museum or herbarium purposes.
- Keep collected plants in a humid atmosphere to ensure good condition. The vasculum originally designed for this purpose has largely been replaced by the polythene bag in most circumstances, except where mechanical damage is likely.

Definition

Edge effect – in this context, any phenomenon associated with the sampling procedure at the edge of a sampling device.

Defining location – for greatest precision, use the coordinates obtained from a large-scale map; for less precise measurements, coordinates given by satellite-based or radio-based position-fixing may be appropriate.

Recording information – the following details should be recorded upon collection: date, time, location (grid ref.), habitat details, collecting technique, preservation technique.

SAFETY NOTE Safety issues associated with collecting – risks from the organisms themselves include toxicity of some plant specimens; allergic reactions to stings and bites; venomous bites and stings; and viral and bacterial infections from bites and excreta. Risks associated with the collecting environment and equipment include rough terrain, sharp edges and water hazards. Always follow the safety rules of your department and any more specific advice produced locally. Before you collect in a new environment, make sure you consult an experienced worker who can alert you to potential problems.

- Keep all living animals in conditions as similar as possible to the environment from which they were collected. Aquatic specimens are usually particularly temperature-sensitive and should be kept in a Thermos flask or 'Coolbox' to prevent rapid temperature changes.
- Rigid containers are better than plastic bags for most purposes since they help to prevent damage to the specimens. Plastic containers should be chosen in preference to glass for most purposes. Make sure that the container seals properly to avoid the loss of specimens and to prevent loss of water.

Sources for further study

Anon. *An Introduction to Collecting Plants*.
Available: http://www.anbg.gov.au/cpbr/herbarium/collecting/index.html
Last accessed: 09/04/07.
[Centre for Plant Biodiversity Research website with links and information relating to collecting plants. Based upon Australian habitats; lots of useful information and suggestions.]

Census of Marine Life *Investigating Marine Life. How do Scientists Collect Organisms?* http://www.coml.org/edu/tech/collect/col1.htm
Last accessed: 09/04/07.
[Well-illustrated guide to marine sampling techniques.]

Lincoln, R.J. and Sheals, J.G. (1979) *Invertebrate Animals: Collection and Preservation*. Cambridge University Press, Cambridge.
[An excellent methodology source for all invertebrate groups. Note that some of the chemicals recommended have been superseded by other, usually less toxic, equivalents.]

Wolberg, D. and Reinard, P. (1997) *Collecting the Natural World: Legal Requirements and Personal Liability for Collecting Plants, Animals, Rocks, Minerals and Fossils*. Geoscience Press, Phoenix.
[Deals with US regulations; gives a good idea of what needs to be considered.]

Study exercises

34.1 **Specify collection equipment.** What equipment would you use to make a collection of the following organisms and how would you transport them to the laboratory in a living condition?
   (a) *Daphnia* (planktonic crustacean)
   (b) earthworms
   (c) bluebells
   (d) ants

34.2 **Specify preservation methods.** How would you preserve organisms (a)–(d) in Study exercise 34.1 for subsequent examination?

34.3 **Specify labelling details.** What items would you include on the labels for preserved specimens of organisms (a)–(d) in Study exercise 34.1?

35 Fixing and preserving animals and plants

Fixation is a chemical process that stops autolysis and stabilises protein and other major components of tissues so that during subsequent processing, the tissues retain as fully as possible the form they had in life. Preservation allows material to be stored indefinitely by destroying any bacteria, fungi, etc., that could degrade the specimen. Preservation and fixation used to be synonymous in that most of the commonly used preservatives also had a fixative action. This has changed with the introduction of 'phenoxetols'. These are good preservatives but since they do not arrest autolysis, their use must be preceded by treatment with a true fixative: for that reason, they are called post-fixation preservatives.

Fixing specimens for histological or cytological work – the procedures and chemicals used are critical to success. Consult specialist texts such as Kiernan (1999) or materials and methods sections of relevant research papers.

KEY POINT *Take care when fixing specimens – remember, anything that will fix your specimen is also capable of fixing you. Be very careful in your handling of these materials and obey all safety precautions.*

Your choice of fixative will depend upon the material to be fixed and the purpose for which it is being fixed. Some of the most common fixatives (Table 35.1) may be used alone although more often they are used in mixtures, the object being to combine the virtues of the various ingredients.

Narcotisation

Narcotisation (relaxation) is usually advisable for animals since many are highly contractile and assume grossly distorted postures if placed straight into fixative; it is also more humane to narcotise before killing. Failure to narcotise may result in contortion, rupture of the body wall or evisceration and reduce the scientific value of the specimen. The need for narcotisation varies with the type of organism and the objectives of the study: some of the most widely used narcotising agents are given in Table 35.2.

Acting humanely – always narcotise animals before carrying out any procedures that might cause pain.

Table 35.1 Some of the most widely used fixatives/preservatives and their properties. Note that there are significant safety issues with nearly all these substances (see column 4).

Substance	Fixation	Usage	Notes
Formaldehyde	+	4% v/v	Comes as 40% v/v solution (= formalin): normal dilution 1 + 9. Make up with sea water for marine specimens; buffer for calcareous specimens. Use in a fume cupboard – health hazard
Ethanol	+	70% v/v aqueous	Highly volatile; inflammable; containers must be well sealed to prevent evaporation; causes shrinkage and decolorisation as well as loss of lipids
Acetic acid	+	In mixtures	Pungent vapour
Picric acid	+	In mixtures	Risk of explosion; detonates readily on contact with some metals. Not recommended for routine student use (significant health risk)
Mercuric chloride	+	In mixtures	Extremely poisonous; corrosive to metal implements; tissue will contain mercuric salt deposits
Osmium tetroxide	+	1% v/v in buffer or vapour	Both fluid and vapour highly toxic; use in fume cupboard only. Excellent for cytological detail but is expensive and can only be used for very small specimens due to poor penetration speed. Vapour good for protozoa
Propylene phenoxetol	–	1–2% v/v aqueous	Relatively expensive but innocuous and effective preservative: needs pre-fixation stage
Glutaraldehyde	+	2–4% v/v in buffer	Highly toxic and severe skin irritant. Must be used cold

Table 35.2 Narcotising agents and their characteristics

Agent	Usage	Notes
Cold (chilling)	Cold-blooded animals	Effective form of relaxing many animals such as tropical and sub-tropical invertebrates
Heat (warming)	Slow heat	Works for some animals. Start from ambient but keep time period as short as possible
Magnesium sulphate or	7% w/v in water	Quite effective for many invertebrates but beware of osmotic problems if made up in sea water: keep exposure times fairly short (1–2 h)
Magnesium chloride	20% w/v in water	
Menthol crystals	Float on water	Slow but effective for many aquatic animals
Chloral hydrate	1% w/v in sea water	General narcotising agent
Ethanol	10% v/v dropwise	Slow and rather tedious process for all but very small specimens
MS-222 (Tricaine)	Use as 0.05% w/v aqueous solution	Good for marine and freshwater fish: very rapid effect (15 s–1 min)
Chloretone	0.1–0.5% w/v in water	General narcotising agent
Ethyl acetate	Vapour	Effective for most insects (kills as well); highly volatile
Ether	Vapour	Effective for vertebrates
Chloroform	Vapour	Effective for vertebrates

 SAFETY NOTE **Keep safety in mind** – many fixatives and preservatives are toxic and a risk assessment should be made before they are used. Always use safety equipment when handling these potentially toxic chemicals, including goggles, buttoned lab coat and latex gloves.

Selecting your fixative/preservative

The main factors you must consider in selecting the preservative/fixative suitable for your materials are:

- Speed of penetration, which determines the size of object that can be fixed. Some fixatives penetrate very slowly (e.g. osmium salts) and only very small pieces of tissue can be fixed. Others such as formaldehyde penetrate very rapidly and allow fixation of relatively large specimens, e.g. whole animals. It may be necessary to inject fixatives into the body cavities of large specimens to ensure adequate fixation.
- Shrinkage: fixatives such as ethanol and mercuric chloride cause tissues to shrink, thus distorting them significantly. The addition of glacial acetic acid is frequently used to reduce this problem.
- Hardening: ethanol is particularly liable to cause tissues to harden on prolonged exposure to concentrations above 70% v/v. For whole specimens, the addition of 3% v/v glycerol to 70% v/v alcohol reduces this, making it a useful medium for specimen storage (not to be used for histological studies). The time in hardening solutions must be carefully controlled for histological material.
- Decolorisation: for whole specimens, the loss of colour may be detrimental and solvents such as ethanol must be used with care since they readily remove many pigments.
- Osmotic problems: distortion of tissues by osmotic movements of water can be rapid and serious. Some fixatives are best made up in sea water if they are to be used for marine specimens to avoid osmotic swelling of tissues. Others such as osmium salts and glutaraldehyde usually need to be made up in a buffer solution isotonic with the tissues.
- Decalcification: acidic fixatives such as unbuffered formalin and those containing picric acid or acetic acid readily dissolve calcareous

structures. For histological preparations this may be desirable, but for whole specimens it is highly undesirable. The acidic properties of formaldehyde can be overcome by neutralising with calcium salts.
- Other chemical reactions: fixatives containing mercuric chloride usually result in the deposition of mercuric precipitates in the tissues. For histological preparations, these must be removed by thorough washing and if necessary by post-treating with iodised alcohol. There is no good reason for using mercuric chloride fixatives other than for histological work and the benefits must be weighed against the dangers and difficulties. Note also that you must not use metal implements with such fixatives as they are extremely corrosive – use only wooden or glass implements.

 SAFETY NOTE Working with mercuric chloride – remember that mercury is a cumulative poison. Take care to avoid contact or ingestion during use.

Preservation

Wet preservation

This is used mainly for animals and is usually preceded by narcotisation. The process of fixation is then comparatively straightforward: sometimes it is necessary to arrange the body and appendages of the animal using tapes, elastic bands, etc. before fixation begins (see Fig. 35.1). The solutions most commonly used for wet preservation have been either:

- a 5–10% v/v aqueous solution of formalin (=40% v/v aqueous formaldehyde), neutralised with calcium carbonate or some other agent to prevent any acidity in the solution resulting in the slow dissolution of calcareous structures (decalcification); or
- 50–70% v/v ethanol.

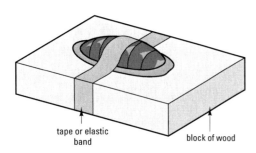

Fig. 35.1 Method for constraining animal specimens liable to curl up or flex during wet preservation

Formaldehyde is cheap and non-inflammable but tends to stiffen and harden tissues on prolonged exposure: safety considerations make it particularly problematical to use because of its noxious vapour. Ethanol is inflammable, highly volatile and tends to cause shrinkage and decolorisation in soft-bodied animals: pass them slowly through a graded series of concentrations to minimise this problem. Industrial methylated spirit is quite suitable for preservation if ethanol is unavailable.

A 1–2% v/v aqueous solution of propylene phenoxetol has been widely used for vertebrate and invertebrate preservation provided it is preceded by the use of a fixative: it preserves the natural colour well and leaves the material pliable. It is comparatively expensive but is non-flammable and non-volatile.

 SAFETY NOTE Working with formaldehyde – always use formaldehyde in a fume cupboard as it is not only an excellent fixative but a significant health hazard.

Wet preservation of plant material is relatively straightforward and has the advantage of preserving the form of the specimen providing it is done carefully. Herbaceous plants should be preserved in a solution with an alcohol base, while for succulents a formaldehyde base is better. Algae, fungi, ferns, lichens and seed plants can be stored wet in formalin acetic acid (FAA) solution: this is made from formalin, acetic acid and ethanol in the ratio 85:10:5. To retain the green coloration of ferns and seed plants, add a crystal of copper sulphate to the solution and incubate the specimens for 3–10 days, then transfer to ordinary FAA solution. Colours of flowers cannot easily be preserved using wet preservatives.

For marine specimens, make up a stock of 1 part propylene glycol with 1 part formalin. Use this in the ratio of 1 part stock to 8 parts sea water for fixation and 1 part stock to 9 parts sea water for subsequent preservation. This markedly reduces the formaldehyde concentration without loss of the fixative/preservative effect and has been shown to be satisfactory even for specimens with calcareous components.

Dry preservation

This is used mainly for vertebrate taxidermy, and for arthropod and plant material, although the development of freeze-drying has made this procedure applicable to almost any type of small or medium-sized organism. Arthropods and vascular plants, with their hard exoskeletons and vascular tissue respectively, are most easily dealt with since the skeletons prevent collapse and loss of form upon drying. Similarly, some types of fungal fruiting bodies such as 'bracket fungi' can be preserved in a dry state.

Flowering plants, or parts of plants, mosses and liverworts, should be dried in a plant press (Fig. 35.2) between sheets of absorbent paper. Interpose a sheet of muslin or greaseproof paper between the paper and the specimen to prevent the specimen adhering to the paper. Great care must be exercised in the cleaning and arrangement of the plant so that essential features are not obscured. Change the drying sheets daily until dry. Drying by heating the press is particularly good for preserving colour: use a low-temperature oven or place on a radiator.

Algae are usually dried in a press after floating them on to a sheet of cartridge paper that is immersed in sea water: this allows careful arrangement of the often delicate algal fronds (see Fig. 35.3). Once suitably arranged, the process is essentially the same as for flowering plants although regular changes of the absorbent paper layers are necessary and the drying procedure usually takes several days. Fungi are best preserved by drying without any attempt at pressing.

Storage

Careful maintenance of specimens after the fixation/preservation processes is very important. Specimens may be stored for long periods provided they are protected from pests such as mites: this may require the use of chemicals such as naphthalene in air-tight containers. Seeds are frequently stored after air-drying only, but they have a finite period of viability.

The storage conditions for all dried materials are critical; these include moderate to low temperatures and very low humidity. Storage of wet material should be in appropriate, vapour-tight vessels: even then the containers will need checking and topping up over time, particularly when volatile preservatives are used. Labelling must be comprehensive and should contain information on the fixation and preservation processes as well as ecological and taxonomic details.

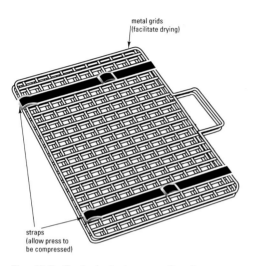

Fig. 35.2 Typical plant press. Specimens are placed between absorbent paper sheets within the metal grids, then the straps are tightened.

Using a plant press – be sure to label each specimen carefully in a way that ensures that the label cannot be lost during paper changes.

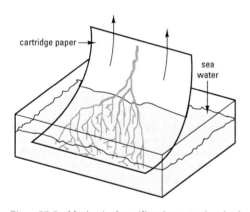

Fig. 35.3 Method for 'floating out' algal specimens. Arrows show direction of lifting

Text reference

Kiernan, J.A. (1999) *Histological and Histochemical Methods: Theory and Practice*, 3rd edn. Scion Publishing Ltd., Bloxham.

Sources for further study

Centre for Plant Biodiversity Research *Plant Collection Procedures and Specimen Preservation.* Available: http://www.anbg.gov.au/cpbr/herbarium/collecting/collection-procedures.html
Last accessed 09/04/07

Lincoln, R.J. and Sheals, J.G. (1979) *Invertebrate Animals: Collection and Preservation.* Cambridge University Press, Cambridge.

Lo Bianco, S. *Classic Resources: The Methods Employed at the Naples Zoological Station for the Preservation of Marine Animals (Trans. E. O. Hovey).* Available: http://www.mbl.edu/BiologicalBulletin/CLASSICS/Russell/Russell-pp11-50.html#Preparations
Last accessed 09/04/07.
[Web reprint of a 'classic' resource.]

National Museum of National History, Smithsonian Institution. *Algae Research: Collection and Preservation.* Available: http://www.nmnh.si.edu/botany/projects/algae/collpres.htm
Last accessed 09/04/07.

Smalldon, G. and Lee, E.W. (1979) *A Synopsis of Methods for the Narcotisation of Marine Invertebrates.* Royal Scottish Museum, Edinburgh.

Study exercises

35.1 **Write a protocol for collection, preservation and fixation of an animal specimen.** This could be for any animal, but an illustrative answer is given for an earthworm specimen that is to be processed for microscopic sectioning.

35.2 **Write a protocol for the collection and preservation of a plant specimen.** This could be for any plant, but an illustrative answer is given for a marine seaweed to be used as a reference specimen for future identification purposes.

35.3 **State the difficulties encountered when fixing materials for microscopic sectioning.** Make a list of the safety problems associated with the use of five fixatives used in microscopy.

37 Naming and classifying organisms

> **Definitions**
>
> **Systematics** – the study of the diversity of living organisms and of the evolutionary relationships between them.
>
> **Classification (taxonomy)** – the study of the theory and methods of organisation of taxa and, therefore, a part of systematics.
>
> **Taxon (plural taxa)** – an assemblage of organisms sharing some basic features.
>
> **Nomenclature** – the allocation of names to taxa.
>
> **Identification** – the placing of organisms into taxa (see Chapters 38 and 39).

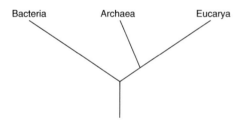

Fig. 37.1 Simplified diagram of the three major domains of the Tree of Life (based on the rRNA sequencing work of Carl Woese).

> **Definition**
>
> **Saprotroph** – a heterotrophic organism that feeds on dead organic matter. (Other nutritional terms are defined on p. 271).

The use of scientific names is fundamental to all aspects of biological science since it aims to provide a system of identification that is precise, fixed and of universal application. Without such a system, comparative studies would be impossible.

There are two possible bases for such classification:

- Phenetic taxonomy, which involves grouping on the basis of phenotypic similarity, frequently using complex statistical techniques to obtain objective measures of similarity. The characters used have been largely morphological and anatomical, but biochemical, cytological and other characters are increasingly used, especially for microbes (p. 219) where structural characters are few.
- Phylogenetic (=phyletic) taxonomy, which involves grouping on the basis of presumed evolutionary, and therefore genetic, relationships.

These two systems are often broadly similar in outcome, since closely related organisms are usually fairly similar to each other and because judgements of evolutionary relationships are usually themselves based upon similarities. The situation is made more complex by phenomena such as convergent and divergent evolution; phyletic classifications are also liable to subjective bias.

New classifications have been proposed on the basis of studies of biomolecules, e.g. rRNA sequences, that are robust and objective. One such arrangement divides organisms into three major domains: Bacteria, Archaea and Eucarya (Fig. 37.1).

The hierarchical system

Another approach is to recognise six kingdoms if the viruses are included:

- Virales: the viruses. Very simple in structure, obligate parasites of prokaryotic or eukaryotic cells.
- Prokaryotae (Monera): the bacteria and cyanobacteria. Prokaryotic, non-nucleate and lacking membrane-bound organelles, such as: mitochondria, plastids, endoplasmic reticulum, Golgi apparatus.
- Protista: includes algae, protozoa and slime moulds. Eukaryotic and mainly unicellular. A heterogeneous group.
- Plantae: the plants. Eukaryotic, walled, mainly multicellular and usually photoautotrophs.
- Fungi: eukaryotic, usually syncytial, walled and saprotrophic.
- Animalia: the animals. Eukaryotic, multicellular and heterotrophs.

Six other levels of taxa are generally accepted: phylum, class, order, family, genus and species, although in botany and microbiology, division is used instead of phylum.

 KEY POINT Whereas the levels of taxa above genus are rather subjective and vary among authorities, the use of genus and species names are governed by strict, internationally agreed conventions called Codes of Nomenclature.

There are three major Codes of Nomenclature, the Botanical, Zoological and Bacteriological Codes, which operate on similar but not identical principles (Lapage et al., 1992; Wakeham-Dawson et al., 1999; Greuter et al., 2000). Fungi are covered by the International Code of Botanical Nomenclature despite the fact that they are not plants.

The basis of classification

No simple definition of a species is possible, but there are two generally used definitions:

- A group of organisms capable of interbreeding and producing fertile offspring – this, however, excludes all asexual, parthenogenetic and apomictic forms.
- A group of organisms showing a close similarity in phenotypic characteristics – this would include morphological, anatomical, biochemical, ecological and life history characters.

DNA-based definition of a species – in microbiology, members of a bacterial species can be characterised by having a DNA similarity of ⩾70 per cent.

 KEY POINT *The basic unit of classification is the species, which represents a group of recognizably similar individuals, clearly distinct from other such groups.*

When species are compared, groups of species may show a number of features in common; they are then arranged into larger groupings known as genera (singular genus). This process can be repeated at each taxonomic stage to form a hierarchical system of classification whose different levels are known as taxonomic ranks. The number of levels in this system is arbitrary and based upon practical experience – the seven levels normally used have been found to be sufficient to accommodate the majority of the variation observed in nature.

Example When the cockle *Cardium edule* was renamed *Cerastoderma edule*, it was commonly referred to in textbooks as *Cerastoderma (Cardium) edule*: this is strictly not correct practice but can be helpful for non-specialists.

When a generic or specific name is changed as a result of further study, the former name becomes a synonym; you should always try to use the latest name. Where a generic name has been changed recently, the old name is occasionally given in parentheses to allow easy reference to the old name. There are taxonomic reference works available for each discipline or subdiscipline, such as the *Flora Europea* (for plants) and the *Plymouth Marine Fauna* (for British marine animals); these often provide the current versions of a name and often its synonyms.

Nomenclature in practice

Writing taxonomic names – always underline or italicise generic and species names to avoid confusion: thus bacillus is a descriptive term for rod-shaped bacteria, while <u>Bacillus</u> is a generic name.

The scientific name of an organism is effectively a symbol or cipher that removes the need for repeated use of descriptions. It normally comprises two words and is, therefore, called a binomial term. The name of the genus is followed by a second term that identifies the species, e.g. *Quercus robur*, *Apis mellifera*, *Pseudomonas aeruginosa* or *Saccharomyces cerevisae* (Table 37.1).

Table 37.1 Example of taxonomies for a plant, an animal and a bacterium

Common name	English oak	Honey-bee	Pseudomonas	Baker's yeast
Kingdom	Plantae	Animalia	Monera	Fungi
Phylum/Division	Anthophyta	Arthropoda	Gracilicutes	Ascomycota
Class	Dicotyledonae	Insecta	Scotobacteria	Incertae sedis[1]
Order	Fagales	Hymenoptera	Pseudomonadales	Saccharomycetales
Family	Fagaceae	Apidae	Pseudomonadaceae	Saccharomycetaceae
Genus	*Quercus*	*Apis*	*Pseudomonas*	*Saccharomyces*
Species	*Q. robur*	*A. mellifera*	*P. aeruginosa*	*S. cerevisiae*

[1] Meaning = 'uncertain taxonomic position'

> **Box 37.1 Basic rules for the writing of taxonomic names**
>
> - **Names of the seven levels of taxa should take lower-case initial letters,** e.g. class Mollusca or kingdom Fungi.
> - **The Latin forms of all taxon names except the specific name take initial capital letters,** e.g. 'the Arthropoda...' but anglicised versions do not, e.g. 'the arthropods...'.
> - **The names of the higher taxa are all plural,** hence 'the Mollusca are...' while the singular of the anglicised version is used for a single member of that taxon, hence 'a mollusc is...'.
> - **The binomial system gives each species two terms,** the first being the generic name and the second the specific name, which must never be used by itself. The genus and species names are distinguished from the rest of the text either:
>   (a) by being underlined (when handwritten), e.g. Patella vulgata; or
>   (b) by being set in italics (in print or on a word processor), e.g. *Patella vulgata*.
> - **The generic name is singular and always takes an initial capital letter.** If you use the generic name this implies that the point being made is a generic characteristic unless the specific name is present. Write the generic name in full when first used in a text, e.g. *Patella vulgata*, but subsequent references can be abbreviated to its initial letter, e.g. *P. vulgata*, unless this will result in confusion with another genus also being considered.
> - **The abbreviation 'sp.' should be used in place of the specific name if a single unspecified species of a genus is being referred to,** e.g. *Patella* sp.; it is not underlined or italicised. If more than one unspecified species is meant, then the correct form is 'spp.', e.g. *Patella* spp.
> - **Common names should not normally be written with a capital letter,** e.g. limpet.
> - **The name of each species should be followed by the authority:** on first usage in formal reports and in titles, the name or names of the person(s) to whom that name is attributed and the date of that description should be quoted. These names may sometimes be abbreviated, e.g. L. for Linnaeus and standard abbreviations must be used. If the species was first described under its current generic name, the authority's name, often in abbreviated form is added, e.g. *Quercus robur* L. If, however, the species was first described under a different genus, the name of the author of the original description is presented in parentheses, e.g. *Escherichia coli* (Migula) Castellani and Chalmers. Note that in zoology, the date the description was published is also included, e.g. *Ischnochiton kermadecensis* Iredale, 1914. The use of authorities should be confined to formal papers, final year project reports, etc.; they would not normally be used in practical reports, short assignments or examinations.

Common names are often interesting, but totally unsatisfactory for use in biological nomenclature because of the lack of consistency in their use: the Codes of Nomenclature were established to prevent the ambiguities associated with informal/common names.

The Codes require that all scientific names are either Latin or treated as Latin, written in the Latin alphabet and subject to the rules of Latin grammar. Consequently, you must be very precise in your use of such names. In some cases, the Codes stipulate a standardised ending for the names of all taxa of a given taxonomic rank, e.g. names of all animal families must end in -idae while plant, fungal and bacterial families end in -aceae. When used in a formal scientific context, you should follow the specific name by the authority on which that name is based, i.e. the name of the person describing that species and the date of the description.

Box 37.1 summarises the basic rules for writing taxonomic names.

Taxa below the rank of species

Some use is made of taxa below the rank of species. Within zoology, this is confined to the term subspecies, so the names of subspecies have three components, e.g. *Mus musculus domesticus*, no rank term being necessary.

Example The full, formal name of baker's yeast is *Saccharomyces cerevisiae* Meyen ex Hansen 1883. After first use, this would be abbreviated to *S. cerevisiae*.

Applications of bacterial typing – typing methods are used widely in epidemiological studies, e.g. tracing a particular type of bacterium responsible for a food-poisoning outbreak, or tracking the development of a particular type of antibiotic-resistant bacterium in a hospital.

In bacteriology, the use of subspecies is again acceptable although a word indicating rank is usually inserted, e.g. *Bacillus subtilis* subsp. *niger*. The Bacteriological Code considers ranks from the level of subspecies up to, and including, class: the use of the term variety is discouraged, as the term is synonymous with subspecies. However, other terms are in widespread use for taxa below the species level, especially in medical microbiology and plant pathology, when a particular strain of bacterium has been identified (pp. 231–2). Subspecies identification is often referred to as typing and the following terms apply:

- biovar, or biotype: subdivided according to biochemical characteristics;
- serovar or serotype: subdivided by serological methods, using antibodies (see Chapter 51);
- pathovar: subdivided according to pathogenicity (ability to cause disease);
- phagovar or phage type: subdivided according to susceptibility to particular viruses.

Many micro-organisms are now referred to by their generic and specific names followed by a culture collection reference number, e.g. *Bacillus subtilis* NCTC 10400, where NCTC stands for the National Collection of Type Cultures and 10400 is the reference number of that strain in the collection.

Understanding microbiological terminology – the term strain *is widely used, particularly in the context of the practice of lodging microbiological strains with culture collections, while the term* isolate *is often used for a pure culture derived from a natural (wild) population. Cell lines (Chapter 47) are often given code names and/or reference numbers.*

In botany, several categories below the rank of species are recognised and a term of rank is used before the name, e.g. *Salix repens* var. *fusca*: the term var. is short for the Latin word *varietas* and is subordinate to the term subspecies in the Botanical Code. The term cultivar (cv.) is an important modern term frequently used in experimental work and refers to cultivated varieties of plants.

The special case of viruses

The classification and nomenclature of viruses is less advanced than for cellular organisms and the current nomenclature has been arrived at on a piecemeal, ad hoc basis. The International Committee for Virus Taxonomy proposed a unified classification system, dividing viruses into 50 families on the basis of:

- host preference;
- nucleic acid type (i.e. DNA or RNA);
- whether the nucleic acid is single- or double-stranded;
- the presence or absence of a surrounding envelope.

Virus family names end in -viridae and genus names in -virus. (Note that these names are *not* latinised and the genus–species binomial is not now approved). However, this system has not yet been adopted universally and many viruses are still referred to by their trivial names or by code names (sigla), e.g. the bacterial viruses ϕX174, T4, etc. Many of the names used reflect the diseases caused by the virus. Often, a three-letter abbreviation is used, e.g. HIV (for human immunodeficiency virus), TMV (for tobacco mosaic virus).

Example The virus that causes tobacco mosaic disease belongs to the genus *Tobamovirus* and can be referred to as tobacco mosaic tobamovirus, tobacco mosaic virus or TMV.

Text references

Greuter, W., et al. (eds) (2000) *International Code of Botanical Nomenclature (St Louis Code)*. Koeltz Scientific Books, Königstein.

Lapage, et al. (eds) (1992) *International Code of Nomenclature of Bacteria (1990 revision)*. American Society for Microbiology, Washington, DC.

Wakeham-Dawson, A., et al. (eds) (1999) *International Code of Zoological Nomenclature*, 4th edn. The International Trust for Zoological Nomenclature, London.

Sources for further study

Anon. *International Code of Botanical Nomenclature (St Louis Code)*. Available: http://www.bgbm.fu-berlin.de/iapt/nomenclature/code/SaintLouis/0000St.Luistitle.htm Last accessed 09/04/07.
[Web version of Greuter et al. (2000).]

Boone, D.R. *Naming a New Procaryotic Taxon*. Available: http://methanogens.pdx.edu/naming.html Last accessed 09/04/07.
[Succinct indication of procedures for naming a new prokaryote species.]

Buchen-Osmond, C. (2002) *Universal Virus Database of the International Committee on Taxonomy of Viruses*. Available: http://www.ncbi.nlm.nih.gov/ICTVdb/ Last accessed 09/04/07.
[Includes a catalogue of approved virus names, image gallery, descriptions, etc.]

Fauquet, C.M., Mayo, M.A., Maniloff, J., Desselberger, U. and Ball, L.A. (2005) *Virus Taxonomy. Eighth Report of the International Committee on the Taxonomy of Viruses*. Elsevier, London.

Institute of Biology (2000) *Biological Nomenclature: Recommendations on Terms, Units and Symbols*. Institute of Biology, London.
[Includes sections on taxonomy and classification of organisms.]

Jeffrey, C. (1989) *Biological Nomenclature*, 3rd edn. Edward Arnold, London.

Study exercises

37.1 Research full classifications. Provide the full classification of the following species, laid out as in the table in Study exercise 37.2.
(a) The limpet *Patella vulgata*
(b) The Great White Shark
(c) The Giant Redwood
(d) The earthworm *Lumbricus terrestris*.

37.2 What is wrong with the hierarchical classifications given in the table below?

Examples of hierarchical classifications

Nautilus	Edible mushroom	E. coli bacterium
Animalia	Fungi	Monera (Bacteria)
Mollusca	Basidiomycota	Proteobacteria
Cephalopoda	Basidiomycetes	Enterobacteriales
Nautilidae	Agaricales	Gamma subdivision
Nautilida	Agaricaceae	Enterobacteriaceae
Nautilus pompilius	*Agaricus bisporus*	*Escherichia coli*

37.3 Compare a variety of current textbooks and Internet sites to discover alternative classification schemes at kingdom level. Make your own notes regarding the *evidence* used to support the alternative kingdom classifications.

38 Identifying plants and animals

Fig. 38.1 Research scientists examining a rocky shore habitat at Orkney, Scotland, and making counts of the different species present, using quadrats. This image illustrates the problems facing a field biologist in identifying plants and animals. What are the species present here? How would you start to identify a specimen, say a mollusc or an alga, from this environment, if its identity were unknown to you? In the absence of specialised knowledge, the task is made far easier by using a flora or fauna with a key that will guide you to descriptions of candidate species.

Out-of-date guides – taxonomists frequently change the names of taxa and update their classification. An up-to-date identification requires an up-to-date guide.

Observing rare specimens – take special care not to collect, disturb or destroy rare plants and animals in the course of your observations.

The normal way to identify plants or animals is to use identification guides. These consist of two parts:

1. Written and pictorial descriptions of organisms, which you compare with your unknown specimen to aid in its identification. Good descriptions direct you to the crucial diagnostic features for the relevant taxon, explain the range of variability found and point out biological and ecological characteristics of importance.
2. Keys, which help you find the likely description for your specimen rapidly and simply. Most keys are arranged to present you with a series of choices, usually dichotomous (dividing in two). The paired statements of each 'couplet' are framed to be contrasting and mutually exclusive. Each choice you make narrows down the possibilities for your specimen until you find the appropriate description.

The authors of identification guides assume that you have a live or preserved specimen to hand and the means to observe it closely and measure it (see Fig. 38.1). The terminology in guides is designed to combine precision with brevity. Guides for animal species are called faunas and those for plant species are called floras.

 KEY POINT To use an identification guide properly, you need to know enough of the vocabulary to understand the choices presented to you, but all identification guides provide both a glossary and a list of abbreviations to help with this.

The best identification guides are those that lead you in the simplest way to a correct identification. If you need to choose one for your area of interest, think about the following questions:

- What degree of prior knowledge is assumed? Some guides are written for novices, whereas others assume an expert's command of terminology. If tempted to go for the former type, consider whether it will always be suitable for your needs.
- What is the scope of the guide? Guides may be restricted in the taxa they consider or in the geographical region that is covered; this will suit you if your interests are similarly narrow. However, if your interests are wide, the relevant guide may be so large as to be unwieldy in the field.
- How well is the guide illustrated? Good-quality illustrations enhance the ease of use of a guide – features can be shown pictorially that might involve an off-putting specialised vocabulary to describe. Accurately coloured illustrations can be helpful, but note that colour can be a variable character: look for good line diagrams that highlight the critical diagnostic points.
- Is the guide divided into parts? Good guides are arranged in short parts dealing with different levels of the taxonomic hierarchy. This speeds up identification by allowing you to skip initial material when you have a fair idea of the specimen's identity.
- Do you like the style of the key? As discussed below, there are several ways in which a key can be presented, one of which may suit you more than the others. If you can't actually test out a key yourself on

real specimens, the next best thing is to ask for the opinion of someone who has used it.

Types of key

Bracketed keys

Here, numbered pairs of adjacent lines in the key present you with a choice and either 'send' you to a new couplet or provide the tentative identification (Box 38.1).

Box 38.1 Example of a bracketed key

Part of key to ragworts (Fig. 38.2), modified after Stace (1997):

1. Ligules <8 mm or 0; capitula cylindrical, about 2× as long as wide 2
   Ligules >8 mm; capitula bell-shaped in flower, about 1.5× as long as wide. 3
2. Ligules usually 0; achenes ⩽ 2.5 mm   *Senecio vulgaris*
   Ligules usually present; achenes <3 mm   *Senecio cambrensis*

If there are no ligules on your specimen or they are less than 8 mm in length, you should proceed to choice 2, but if they are present and greater than 8 mm in length, you should proceed to choice 3.

In this case, the choice at 2 is sufficient to pin down the species; sometimes quite early in a key a distinctive characteristic may allow the specimen to be 'identified' to species level, while for the other options the specimen's identity remains open. Note the use of more than one comparison in each couplet to provide confirmation.

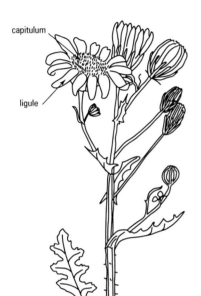

Fig. 38.2 Ragwort flower (*Senecio* spp.). Adapted from figure from *New Flora of the British Isles*, 2nd edition, Cambridge University Press, (Stace, C.A., 1997).

Indented keys

In this method, the pairs of choices are indented and given the same number. They are separated by other choices further down the sequence. Having made a choice, you look at the next couplet below which will be one indent level further in. When a choice is sufficiently distinctive, the tentative identity of your specimen will be given (Box 38.2).

Box 38.2 Example of an indented key

Part of key for common species of true bumblebee (Fig. 38.3), modified after Prys-Jones and Corbet (1987):

1. Thorax with black area(s)
   2. Thorax all black
      3. Pollen baskets with red hairs   *Bombus rudarius*
      3. Pollen baskets with black hairs   *Bombus lapidarius*
   2. Thorax black with yellow or brown patches
      4. Tail white, buff or brown
         5. Scutellum black   *Bombus terrestris*
         5. Scutellum yellow   *Bombus hortorum*
      4. Tail red or orange   *Bombus pratorum*
1. Thorax without any black   *Bombus pascuorum*

If your bumblebee has a black thorax with yellow patches, proceed to choice 4; if its tail is brown, carry on to choice 5, etc.

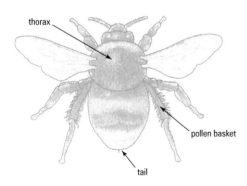

Fig. 38.3 Bumblebee (*Bombus* spp., ...)

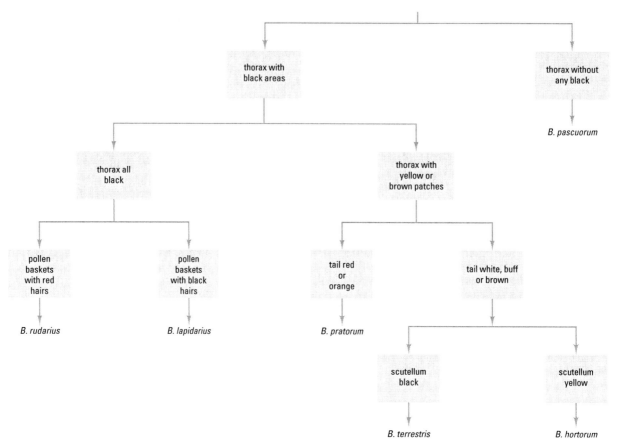

Fig. 38.4 A key for bumblebees (*Bombus* spp.) laid out in the form of a flowchart.

 KEY POINT *Bracketed keys have the advantage that they keep the couplets close together for ready comparison, but indented keys show the relationships between taxa more clearly and allow you to backtrack more easily if an error has been made.*

Flowchart keys

In this form of key, the choices are laid out in the form of a flowchart (Fig. 38.4), which allows easier cross-checking of options but is only feasible where there are a small number of choices. To use this type of key, follow the arrows after making each choice in the sequence; this will lead you on to another choice and eventually to the tentative identification.

Multiaccess keys

These allow you to choose the characters used in the key according to the state of your specimen (Box 38.3). They are useful in situations where:

- important characters are difficult to observe;
- characters are likely to be misinterpreted;
- a single character would be unreliable in isolation;
- a part is missing or seems abnormal.

Another way of presenting a multi-access key is in the form of a table. For instance, the taxa to be distinguished could make up the rows

Box 38.3 Example of a multi-access key

Part of key to species of willowherbs, modified after Stace (1997):

Stigma 4-lobed	A
Stigma club-shaped	B
Seeds minutely uniformly papillose	C
Seeds with longitudinal papillose ridges	D
Stems erect or erect at apex	E
Stems trailing on ground	F

ACE	Petals 10–16 mm, purplish pink	*Epilobium hirsutum*
	Petals 5–9 mm, paler	*Epilobium parviflorum*
BCE	Petioles 4–15 mm, plant perennating by rosettes	*Epilobium roseum*
	Petioles ⩽ 4 mm, plant perennating by stolons ending in tight bud	*Epilobium palustre*

If your specimen had a 4-lobed stigma and erect stems, but no seeds were available to examine, you could 'identify' it as either *E. hirsutum* or *E. parviflorum* and distinguish between these choices on the basis of petal size and colour.

---

of the table and the relevant characteristics the columns (see Table 39.1). Like the flowchart, this type of key is limited to a small number of choices.

Computerised keys

These simplify the initial stages of identification by providing a series of menu-like choices for the user. They can rapidly provide a list of tentative identifications on even a few positive choices from these menus, ranking these in likelihood of being correct. You can then work down the list, comparing your specimen to a description of the proposed species (see below). At present, computerised keys are more likely to be used in a laboratory or field station than in the field.

Advice for using keys

- Note down the route taken (i.e. the numbering system for the decision tree): this makes it easier to trace back your path through the key.
- At each step, read the full description for both choices before arriving at a decision about which one to take.
- Never guess if you do not know the precise meaning of the terms used – consult the key's glossary and list of abbreviations. Where measurements are required, use a ruler – do not guess sizes.
- If features are very small, use an appropriate lens to inspect them clearly.
- If the key is a multipart one, look carefully at the descriptions for higher levels of taxa before progressing to the species key: this not only acts as a check that you are correct up to this stage, but may also provide definitions of useful terminology.
- If both of a pair of choices seem reasonable, try out each route – one will usually prove to be unsuitable at a later stage.

Problems of identification – *sometimes your best attempts to identify the specimen will be confronted by the existence of sexual dimorphism or polymorphism, juvenile and adult phases (e.g. gametophyte and sporophyte phases for certain plants), local forms, non-native taxa, etc. A good guide will point out these problems where they occur.*

KEY POINT *When you arrive at the end of a key's path*, do not simply accept this as a reliable identification of your specimen. *Compare your specimen with the full description of the species.*

Comparing specimens with descriptions

If the specimen doesn't fit the description properly, follow the instructions outlined below:

- Compare the specimen with neighbouring descriptions: in a well-organised guide, those of similar species will be together.
- Go back along the path of the key and re-examine each decision you have made. Try going down the alternative route for any that might have been questionable.
- Check to see whether you inadvertently went down the wrong pathway even though you made the correct diagnosis.
- Bear in mind the possibility that your specimen is not typical. A good key will use characteristics that are constant, but biological variation will often throw up an oddity to confuse you. Try to obtain another specimen, preferably not genetically related to the original.
- Consider the possibility that it could be outside its normal geographical range or even new to science.

The ultimate check on an identification is a comparison of the specimen with an authentically named specimen in a museum or herbarium. The ultimate comparison would be with the type specimen, the specimen used when the species or subspecies was first described and named. If this is the only specimen collected by the author(s) who named the species, it is called a holotype. Other 'type' specimens include:

- paratypes – those other than the type specimen also used by the author(s) at the time of the original description;
- syntypes – a collection of specimens used for the original collection, but from which no one specimen was defined as the type specimen;
- lectotype – a particular syntype subsequently chosen and designated through publication to act as the type specimen.

Text references

Prys-Jones, O.E. and Corbet, S.A. (1987) *Naturalist's Handbook 6. Bumblebees.* Cambridge University Press, Cambridge.

Stace, C.A. (1997) *New Flora of the British Isles*, 2nd edn. Cambridge University Press, Cambridge.

Sources for further study

Schmidt, D. *Flora and Fauna Field Guides.* Available: http://gateway.library.uiuc.edu/bix/fieldguides/flora.htm Last accessed 09/04/07.
[Listing of floras and faunas across the world; also classified by type of organism, e.g. mammals of Central and South America.]

Thompson Scientific *BiologyBrowser.* Available: http://www.biologybrowser.org/ Last accessed 09/04/07.
[Includes a search engine for information about specific organisms.]

University of Illinois *Flora and Fauna Field Guides.* Available: http://www.library.uiuc.edu/bix/fieldguides/flora.htm Last accessed 09/04/07.
[A listing of guides by geographical region.]

Study exercises

38.1 Devise a key. The table below gives details of some invertebrate phyla from freshwater habitats, together with illustrations of 'representative' animals. The taxa are presented in no particular order. Use this information to construct a dichotomous key to help with preliminary identification of specimens. Ask a colleague to test your key by correctly assigning the animals to phyla from their pictures alone. Refine the key if problems are encountered during testing.

Members of some invertebrate phyla

Taxonomic division and description	Representative animal
Phylum Mollusca, Class Bivalvia Soft bodied with hard shell in two parts	Zebra mussel, *Dreissena polymorpha*
Phylum Arthropoda Sub-phylum Arachnida Hard bodied with four pairs of jointed legs	Water spider, *Argyroneta aquatica*
Phylum Mollusca, Class Gastropoda Soft bodied with hard shell in one part	River snail, *Viviparus viviparus*
Phylum Platyhelminthes Soft bodied, free moving, flattened and unsegmented, without legs	Flatworm, *Dugesia lugubris*
Phylum Arthropoda Sub-phylum Insecta Hard bodied with three pairs of jointed legs	Great diving beetle, *Dytiscus marginalis*

(continued)

Study exercises (continued)

Taxonomic division and description	Representative animal
Phylum Annelida **Class Oligochaetae** Soft bodied, free moving, divided into 15 or more segments (without head capsule or leg-like appendages)	 Water worms, various sp.
Phylum Nematoda Soft bodied, free-moving, thread-like and unsegmented, without legs	 Nematodes, various sp.
Phylum Arthropoda **Sub-phylum Crustacea** Hard bodied, with four or more pairs of jointed legs	 Freshwater shrimp, *Crangonyx pseudogracilis*
Phylum Coelenterata **Class Hydrozoa** Soft bodied, surface-attached, unsegmented and tube-like, without legs	 e.g. *Hydra* sp.

38.2 Use a standard flora or fauna. Visit your university library and take out one of the standard keys for the local flora or fauna. Find a specimen plant or animal whose identity you do not know (take care not to kill or destroy any protected specimens) and use the key and its descriptions to identify it as best you can. Check your identification with a tutor.

38.3 Search for Web-based identification keys and test them. Hypertext (p. 58) lends itself well to dichotomous or multiple choices, taking you to new pages via links based on appropriate options. Search for a Web-based identification key using a search engine and try it out.

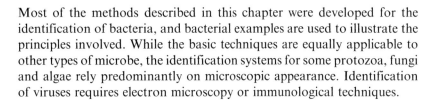

39 Identifying microbes

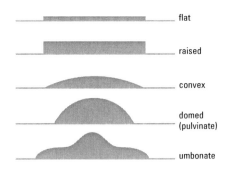

Fig. 39.1 Colony elevations (cross-sectional profile)

Most of the methods described in this chapter were developed for the identification of bacteria, and bacterial examples are used to illustrate the principles involved. While the basic techniques are equally applicable to other types of microbe, the identification systems for some protozoa, fungi and algae rely predominantly on microscopic appearance. Identification of viruses requires electron microscopy or immunological techniques.

 KEY POINT Identification of bacteria is often based on a combination of a number of different features, including growth characteristics, microscopic examination, physiological or biochemical characterisation, and, where necessary, immunological tests.

Direct observation

Once a microbe has been isolated (Chapter 36) and cultured in the laboratory (Chapter 46), the visual appearance of individual colonies on the surface of a solidified medium may provide useful information. Bacteria typically produce smooth, glistening colonies, varying in diameter from <1 mm to >1 cm. Actinomycete colonies are often <1 cm, with a shrivelled, powdery surface. Filamentous fungi usually grow as large, spreading mycelia with a matt appearance and are identified by microscopy, using the morphological characteristics of their reproductive structures. Yeasts produce small, glistening colonies; identification usually involves microscopy, combined with physiological and biochemical tests similar to those used for bacteria.

Colony characteristics

When measuring colony size, choose a typical colony, well spaced from any others, as colony size is affected by competition for nutrients. The characteristics of a microbial colony on a particular medium include:

- Size: some bacteria produce punctiform colonies, with a diameter of less than 1 mm, while motile bacteria may spread over the entire plate.
- Form: colonies may be circular, irregular, lenticular (spindle-shaped) or filamentous.
- Elevation: colonies may be flat, raised, convex, etc. (see Fig. 39.1).
- Margin: the edge of a colony may be entire (smooth) or more distinctive, e.g. undulate, lobate or filamentous (Fig. 39.2).
- Consistency: colonies may be viscous (or mucoid), butyrous (of similar consistency to butter) or friable (dry and granular), etc.
- Colour: some bacteria produce characteristic pigments. A few pigments are fluorescent under UV light.
- Optical properties: colonies may be translucent or opaque.
- Haemolytic reactions on blood agar: many pathogenic bacteria produce characteristic zones of haemolysis. Alpha haemolysis is a partial breakdown of the haemoglobin from the erythrocytes, producing a green zone around the colony, while beta haemolysis is the complete destruction of haemoglobin, producing a clear zone.

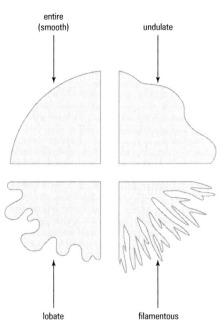

Fig. 39.2 Colony margins (surface view)

 SAFETY NOTE Never attempt to smell mould cultures, because of the risk of inhaling large numbers of spores.

- Odour: some actinomycetes and cyanobacteria produce earthy odours, while certain bacteria and yeasts produce fruity or 'off' odours. However, odour is not a reliable characteristic in bacterial identification.

Microscopic examination

Bacteria are usually observed using an oil immersion objective at a total magnification of ×1000 (p. 255).

Motility

Using the hanging drop technique – place a drop of bacterial suspension on a coverslip and invert over a cavity slide so that the drop does not make contact with the slide: motile aerobes are best observed at the edge of the droplet, where oxygen is most abundant.

Wet mounts can be prepared by placing a small drop of bacterial suspension on a clean, degreased slide, adding a coverslip and examining the film without delay. For aerobes, areas near air bubbles or by the edge of the coverslip give best results, whereas anaerobes show greatest motility in the centre of the preparation, with rapid loss of motility due to oxygen toxicity.

Prepare wet mounts using young cultures in exponential growth in a liquid medium (p. 272). It is best to work with cultures grown at 20 or 25 °C, since those grown at 37 °C may not be actively motile on cooling to room temperature. It is essential to distinguish between the following:

Assessing motility – if you have not seen bacterial motility before, it is worth comparing your unknown bacterium with a positive and a negative control.

- True motility, due to the presence of flagella: bacteria dart around the field of view, changing direction in zigzag, tumbling movements.
- Brownian motion: non-motile bacteria show a localised, vibratory, random motion, due to bombardment of bacterial cells by molecules in the solution.
- Passive motion, due to currents within the suspension: all cells will be swept in the same direction at a similar rate of movement.
- Gliding motility: a slower, intermittent movement, parallel to the longitudinal axis of the cell, requiring contact with a solid surface.

Cell shape

Using cell shape in microbial identification – many bacteria are pleomorphic, varying in size and shape according to the growth conditions and the age of the culture: thus, other characteristics are required for identification.

Bacteria are subdivided into the following groups:

- Cocci (singular, coccus): spherical, or almost spherical, cells, sometimes growing in pairs (diplococci), chains or clumps.
- Rods: straight, cylindrical cells of variable length, with flattened, tapered or rounded ends; sometimes termed bacilli. Short rods are sometimes called cocco-bacilli.
- Curved rods: the curvature varies according to the organism, from short curved rods, sometimes tapered at one end, to spiral shapes.
- Branched filaments: characteristic of actinomycete bacteria.

Gram staining

 SAFETY NOTE The Gram-staining procedure involves toxic dyes and flammable solvents: avoid skin contact and extinguish any naked flames (e.g. Bunsens).

This is the most important differential staining technique in bacteriology (Box 39.1 gives details). It enables us to divide bacteria into two distinct groups, Gram-positive and Gram-negative, according to a particular staining procedure (the technique is given a capital letter, since it is named after its originator, H.C. Gram). The basis of the staining reaction is the different structure of the cell walls of Gram-positive and Gram-negative bacteria. Heat fixation of air-dried bacteria causes some shrinkage, but cells retain their shape: to measure cell dimensions use a chemical fixative (p. 246).

Assessing the Gram status of an unknown bacterium – if a pure culture gives both Gram-positive and Gram-negative cells, identical in size and shape, it can be regarded as a Gram-positive organism that is demonstrating Gram-variability.

Gram staining should be carried out using light smears of young, active cultures, since older cultures may give variable results. In particular, certain Gram-positive bacteria may stain Gram-negative if older cultures are used.

Box 39.1 Preparation of a heat-fixed, Gram-stained smear

Preparation of a heat-fixed smear

The following procedure will provide you with a thin film of bacteria on a microscope slide, for staining.

1. **Take a clean microscope slide and pass it through a Bunsen flame twice**, to ensure it is free of grease. Allow to cool.
2. **Using a sterile inoculating loop, place a single drop of water in the centre of the slide and then mix in a small amount of sample** from a single bacterial colony with the drop, until the suspension is slightly turbid. Smear the suspension over the central area of the slide, to form a thin film. For liquid cultures, use a single drop of culture fluid, spread in a similar manner.
3. **Allow to air-dry at room temperature**, or high above a Bunsen flame: air-drying must proceed gently, or the cells will shrink and become distorted.
4. **Fix the air-dried film by passage through a Bunsen flame.** Using a slide holder or forceps, pass the slide, film side up, rapidly through the hottest part of the flame (just above the blue cone). The temperature of the slide should be just too hot for comfort on the back of your hand: note that you must not overheat the slide or you may burn yourself (you will also ruin the preparation).
5. **Allow to cool**: the smear is now ready for staining.

Gram-staining procedure

The version given here is a modification of the Hucker method, since acetone is used to decolorise the smear. Note that some of the staining solutions used are flammable, especially the acetone decolorising solvent: you must make sure that all Bunsens are turned off during staining. The procedure should be carried out with the slides suspended over a sink, using a staining rack.

1. **Flood a heat-fixed smear with 2% w/v crystal violet in 20% v/v ethanol:water** and leave for 1 min.
2. **Pour off the crystal violet and rinse briefly with tap water. Flood with Gram's iodine** (2 g KI and 1 g I_2 in 300 ml water) for 1 min.
3. **Rinse briefly with tap water** and leave the tap running gently.
4. **Tilt the slide and decolorise with acetone** for 2–3 s: acetone should be added dropwise to the slide until no colour appears in the effluent. This step is critical, since acetone is a powerful decolorising solvent and must not be left in contact with the slide for too long.
5. **Immediately immerse the smear in a gentle stream of tap water**, to remove the acetone.
6. **Pour off the water and counterstain for 10–15 s using 2.5% w/v safranin** in 95% v/v ethanol:water.
7. **Pour off the counterstain, rinse briefly with tap water, then dry the smear** by blotting gently with absorbent paper: all traces of water must be removed before the stained smear is examined microscopically.
8. **Place a small drop of immersion oil on the stained smear: examine directly** (without a coverslip) using an oil-immersion objective (p. 255).

Gram-positive bacteria retain the crystal violet (primary stain) and appear purple while Gram-negative bacteria are decolorised by acetone and counterstained by the safranin, appearing pink or red when viewed microscopically.

Other decolorising solvents are sometimes used, including ethanol:water, ethyl ether:acetone and acetone:alcohol mixtures. The time of decolorisation must be adjusted, depending upon the strength of the solvents used, e.g. 95% v/v ethanol:water is less powerful than acetone, requiring around 30 s to decolorise a smear.

This Gram-variability is due to autolytic changes in the cell wall of Gram-positive bacteria. Developing spores are often visible as unstained areas within older vegetative cells of *Bacillus* and *Clostridium*. Other stains are required to demonstrate spores, capsules or flagella (p. 249).

Basic laboratory tests

At least two simple biochemical tests are usually performed:

Oxidase test

This identifies cytochrome *c* oxidase, an enzyme found in obligate aerobic bacteria. Soak a small piece of filter paper in a fresh solution of 1% (w/v) N-N-N′-N′-tetramethyl-p-phenylenediamine dihydrochloride on a clean microscope slide. Rub a small amount from the surface of a young, active

Performing the oxidase test – never use a nichrome wire loop, as this will react with the oxidase reagent, giving a false positive result.

colony on to the filter paper using a glass rod, a *plastic* loop or a wooden applicator stick: a purple-blue colour within 10 s is a positive result.

Catalase test

This identifies catalase, an enzyme found in obligate aerobes and in most facultative anaerobes, which catalyses the breakdown of hydrogen peroxide into water and oxygen ($2 H_2O_2 \rightarrow 2H_2O + O_2$). Transfer a small sample of your unknown bacterium on to a coverslip using a disposable plastic loop or glass rod. Invert on to a drop of hydrogen peroxide: the appearance of bubbles within 30 s is a positive reaction. This method minimises the dangers from aerosols formed when gas bubbles burst.

> SAFETY NOTE The catalase and oxidase reagents are irritants and could be harmful if swallowed. Avoid skin contact and ingestion.

The oxidase and catalase tests effectively allow us to subdivide bacteria on the basis of their oxygen requirements, without using agar shake cultures (p. 213) and overnight incubation, since, for the most part:

- obligate aerobes will be oxidase and catalase positive;
- facultative anaerobes will be oxidase negative and catalase positive;
- microaerophilic bacteria, aerotolerant anaerobes and strict (obligate) anaerobes will be oxidase and catalase negative – the latter group will grow only under anaerobic conditions (p. 213).

Avoiding false negatives – ensure you use sufficient material during oxidase and catalase testing, otherwise you may obtain a false negative result: a clearly visible 'clump' of bacteria should be used.

Once you have reached this stage (colony characteristics, motility, shape, Gram reaction, oxidase and catalase status) it may be possible to make a tentative identification, at least for certain Gram-positive bacteria, at the generic level. To identify Gram-negative bacteria, particularly the oxidase-negative, catalase-positive rods, further tests are required.

Identification tables: further laboratory tests

Bacteria are asexual organisms and strains of the same species may give different results for individual biochemical/physiological tests. This variation is allowed for in identification tables (multi-access keys, p. 223), based on the results of a large number of tests. Identification tables are often used for particular subgroups of bacteria, after Gram staining and basic laboratory tests have been performed: an example is shown in Table 39.1.

Carbohydrate utilisation tests and isolation media – many diagnostic agar-based media incorporate one or more specific carbohydrates and pH indicator dyes, thereby providing additional information as part of the isolation procedure (p. 232).

Table 39.1 Identification table for selected Gram-negative rods

Bacterium	Biochemical test								
	1	2	3	4	5	6	7	8	9
Escherichia coli	v	+	−	+	−	v	v	+	−
Proteus mirabilis	−	−	v	−	+	−	+	−	+
Morganella morganii	−	−	−	−	+	−	+	+	−
Vibrio parahaemolyticus	−	+	v	−	−	+	+	+	−
Salmonella spp.	−	+	v	−	−	+	+	−	+

Key to biochemical tests and symbols:
1. sucrose utilisation
2. mannitol utilisation
3. citrate utilisation
4. β-galactosidase activity
5. urease activity
6. lysine decarboxylase activity
7. ornithine decarboxylase activity
8. indole production
9. H_2S production
+, >90% of strains tested positive
−, <10% of strains tested positive
v, 10–90% of strains tested positive

A large number of specific biochemical and physiological tests are used in bacterial identification including:

- Carbohydrate utilisation tests. Some bacteria can use a particular carbohydrate as a carbon and energy source. Acidic end products can be identified using a pH indicator dye (p. 147) while CO_2 is detected in liquid culture using an inverted small test tube (Durham tube, Fig. 39.3). Aerobic breakdown (via respiration) is termed oxidation, and anaerobic breakdown is known as fermentation. Identification tables usually incorporate tests for several different carbohydrates, e.g. Table 39.1.
- Enzyme tests. Most of these incorporate a substance that changes colour if the enzyme is present, e.g. a pH indicator, or a chromogenic substrate.
- Tests for specific end products of metabolism, e.g. the production of indole due to the metabolic breakdown of the amino acid tryptophan, or H_2S from sulphur-containing amino acids.

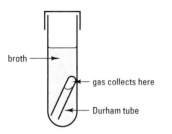

Fig. 39.3 Durham tube in carbohydrate utilisation broth. Air within the Durham tube is replaced by broth during the autoclaving procedure.

Molecular approaches to microbial identification – *several novel methods of detection and identification are based on nucleic acid techniques, including use of the PCR and Southern blotting to detect particular microbes (pp. 335–6).*

Identification kits

Some biochemical tests are now supplied in kit form, e.g. the API 20E system incorporates 20 tests within a sterile plastic strip (Fig. 39.4). After inoculation and overnight incubation, the results of the tests are converted into a seven-digit code, for comparison with known bacteria using either a reference book (the Analytical Profile Index), or a computer program. While kit identification systems save time and labour, they are more expensive and less flexible than conventional biochemical tests.

Immunological tests

Tests used in diagnostic microbiology include:

- Agglutination tests: based on the reaction between specific antibodies and a particular bacterium (p. 310). These tests are particularly useful for subdividing biochemically similar bacteria.
- Fluorescent antibody tests: the reaction between a labelled antibody and a particular bacterium can be visualised using UV microscopy. The direct fluorescent antibody test uses fluorescein isothiocyanate as the label.
- Enzyme-linked immunoassay tests using antibodies labelled with a particular enzyme, e.g. the double antibody sandwich ELISA or competitive ELISA tests (p. 313).

While such tests can give specific and accurate confirmation of the identity of a bacterium under controlled laboratory conditions, they are often too expensive and time-consuming for routine identification purposes, especially when large numbers of tests are required.

Fig. 39.4 Example of a bacterial identification kit (API 20E) property of bioMérieux S.A/Andrea Bannuscher.

Typing methods

The identification of bacteria at subspecies level is known as typing: this is usually done in a specialist laboratory, e.g. as part of an epidemiological study to establish the source of an infection. Various methods are used:

- Antigen typing or serotyping is based on immunological tests.
- Phage typing is based on the susceptibility of different strains to certain bacterial viruses (phages).
- Biotyping is based on biochemical differences between different strains e.g. enzyme profiles or antibiotic resistance screening ('antibiograms').
- Bacteriocin typing: bacteriocins are proteins released by bacteria that inhibit the growth of other members of the same species.

Practical applications of bacterial typing – E. coli O157:H7 is a serotype of this bacterium that is capable of causing severe human disease: it can be identified on the basis of an agglutination reaction with an appropriate antiserum. Typing is also discussed on p. 225.

Sources for further study

Barrow, G.I. and Feltham, R.K.A. (1993) *Cowan and Steel's Manual for the Identification of Medical Bacteria*, 3rd edn. Cambridge University Press, Cambridge.
[A standard reference work for the identification of clinically-important bacteria.]

Cullimore, D.R. (2000) *Practical Atlas for Bacterial Identification*. CRC Press, Boca Raton, Florida.
[Detailed coverage of practical methods applicable to a wide range of bacteria.]

Fisher, F., Cook, N.B., Fisher, F.W. and Kaszczuk, S. (1998) *Fundamentals of Diagnostic Mycology*. Saunders, Philadelphia.
[Covers identification techniques for medically important fungi.]

Fox, A. (ed.) *Journal of Microbiological Methods*, Elsevier, London (available through Science Direct at: http://www.sciencedirect.com/)
[Provides information on novel developments in all aspects of microbiological methods, including microbial culture/isolation and molecular approaches.]

Holt, J.G. (1994) *Bergey's Manual of Determinative Bacteriology*, 9th edn. Williams & Wilkins, Baltimore.

Macfaddin, J.F. (2000) *Biochemical Tests for the Identification of Medical Bacteria*, 3rd edn. Williams & Wilkins, Philadelphia.
[Explains the operating principles underlying most of the biochemical tests in routine use in diagnostic bacteriology.]

Study exercises

39.1 Describe the colonial characteristics of selected microbes. Either research the features of the following microbes (e.g. via the Web), or look at well-isolated individual colonies in the laboratory, following overnight growth on a suitable medium:

(a) *E. coli*;
(b) *Pseudomonas aerogenes*;
(c) *Staphylococcus aureus*;
(d) *Streptococcus pneumoniae*;
(e) *Bacillus subtilis*;
(f) *Saccharomyces cerevisiae*.

39.2 Find out how some of the tests used in microbial identification work. Research the operating principles that underpin the following tests, using microbiological textbooks or the Web, and prepare brief notes explaining how each of the following tests works:

(a) oxidase test;
(b) indole test;
(c) β-galactosidase test;
(d) urease test;
(e) lysine decarboxylase test;
(f) H_2S production.

39.3 Identify the following oxidase-negative, catalase-positive, Gram-negative rod-shaped bacteria, using Table 39.1.

(a) Positive for citrate, urease, ornithine decarboxylase and H_2S only.
(b) Positive for sucrose, mannitol, β-galactosidase, lysine decarboxylase and indole only.
(c) Positive for mannitol, citrate, lysine decarboxylase, ornithine decarboxylase and H_2S only.
(d) Positive for urease, ornithine decarboxylase and indole only.
(e) Positive for mannitol, β-galactosidase and indole only.

40 The purpose and practice of dissection

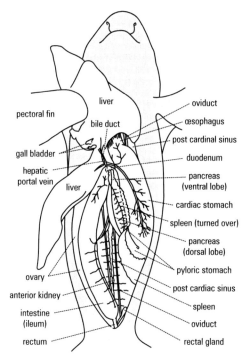

Fig. 40.1 Labelled diagram of abdomen of young female dogfish. Good drawings are essential to get the most out of dissections (see Chapter 28).

Dissection in zoology involves the display or removal of parts of any dead animal, whereas vivisection is an operation on a living animal. Dissection is usually associated with teaching animal structure, but the skills involved are used widely within the life sciences for:

- investigating anatomy and morphology;
- making physiological preparations of nerves, muscles and other organs;
- investigating parasites in various body organs;
- removing specific body organs/tissues for chemical analysis;
- investigating reproductive status;
- removing organs/tissues for histological/histochemical investigation;
- manipulating living material, as in grafting processes and preparing diagrams to show selected features (Fig. 40.1).

There is considerable debate about the ethics of using animals for dissection (p. 116) but this must be distinguished from that of vivisection. It is part of a complex and emotive issue, but one viewpoint is that dissection is required for effective zoology teaching: experience of dissection at an *appropriate* stage of the curriculum is both enlightening and teaches an essential technical skill. However, if dissection is definitely not for you, it is best to discover this as early as possible in your career.

 KEY POINT The primary objectives of dissection are:
- *personal exploration of animal structure and function;*
- *development of manipulative skills;*
- *production of reports based upon personal observation.*

The ground rules of dissection

There are some important rules to be considered if dissection is to be an acceptable procedure in zoological teaching.

Humane treatment

The use of any animal, whatever its level of organisation, for experimentation or dissection must be a considered act with due regard for humane treatment and killing (see Chapter 19). Remember there are very specific regulations for the use of vertebrate species and some higher invertebrates: check with your supervisor.

Maximum benefit

Any animal should be used for as many investigations as possible.

Preparation

Your responsibility – because of the ethical aspect to dissection, you have a particular responsibility to attend to and make careful and effective use of a dissection specimen.

Prepare for the exercise by ensuring that practical schedules and relevant texts are consulted before attempting a dissection. If you have to miss a practical, inform the organiser in advance to prevent animals being killed or prepared unnecessarily.

Types of dissections

Dissection can be carried out at three levels of sophistication, related directly to the size of the organism or structure being investigated. Gross dissection of large organisms requires equipment very different from that used for 'normal' dissection in that large knives replace scalpels, etc. Normal-scale dissection involves creatures from a dog down to an earthworm, and requires equipment typical of commercial dissection kits/instruments. Fine-scale dissection for the removal or display of organs, glands, etc., from small animals usually requires only mounted needles, fractured glass edges for cutting and a dissecting microscope or magnifier. Fine-scale dissection requires extensive practice, and the preparation of material (relaxation and fixation, p. 206) is much more critical than for larger specimens.

Equipment

> **SAFETY NOTE** Taking care with dissection equipment – sharp blades should be used with extreme caution and disposed of immediately after use in a sharps container.

Table 40.1 lists basic equipment required for normal dissection: some comments on use are given below (see also Fig. 40.2). Commercial dissection kits often contain inappropriate components and you should buy your instruments individually from specialist suppliers, if possible. Equipment for fine-scale dissections can be assembled easily to your own specifications.

Table 40.1 List of basic equipment recommended for dissection.

Quantity	Description
1	All-metal scalpel, stainless steel, 45 mm blade
1 each	Swann–Morton scalpel handles, sizes 3 and 4
1 each	Swann–Morton blades, packets of nos. 10, 11, 12, 15, 22 and 24
2	Dissecting needles, straight, stainless steel
2	Blunt seekers, stainless steel: metal handles
1	Fine forceps, blunt points, stainless steel, 112 mm length
1	Coarse forceps, blunt points, stainless steel, 112 mm length
1	Coarse scissors, open shanks, straight points, stainless steel, 150 mm length
1	Fine scissors, open shanks, straight fine points, stainless steel, 110 mm length
1	Section lifter

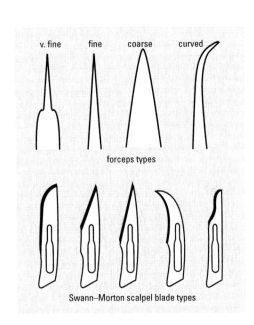

Fig. 40.2 Variations in design of dissection equipment

Scalpels

There are two basic types of scalpels, fixed blade and replaceable blade. Buy at least one 45-mm blade length solid forged scalpel, preferably made of stainless steel: use this for coarser cutting procedures as it can be resharpened using an oil-stone. The Swann–Morton scalpel comprises a handle and a disposable blade: blades come in a variety of shapes and sizes (Fig. 40.2), curved-edge ones being the most useful. Blades fit only specific handle sizes so make sure that they match.

Forceps

Buy at least one each of coarse and fine stainless steel forceps. The latter are very delicate and you should check that the points meet precisely before purchasing. Look after them very carefully as the points are easily damaged: use only on soft tissues. Use large and small forceps for general purposes but keep fine ones for delicate work only.

Dissecting scissors

Buy two pairs of stainless steel, pointed scissors, one medium-large for coarse work and cutting small bones and one fine pair for delicate work. Points and edges are easily damaged and require professional sharpening when repair is needed, so use carefully. A pair of bone cutters is optional but do *not* use scissors to cut large bones.

The Purpose and Practice of Dissection

Dissecting (mounted) needles
Buy at least two with metal handles and protect the points; use them for dissecting membranes in areas where damage is acceptable. Needles have many non-dissection functions associated with other fine manipulative techniques.

Box 40.1 Basic stages of an animal dissection

The basic sequence of steps is outlined in Fig. 40.3.

1. **The animal should have been killed as humanely and as recently as possible**: this will probably be done for you by the class supervisor or technician according to the rules appropriate to the type of animal. The method of killing should be chosen carefully to keep the specimen relaxed. If preserved material is used, wash out excess fixative thoroughly before dissecting.

2. **Orient the specimen carefully**; determine the dorsal/ventral, anterior/posterior or other oral/aboral axes (see Fig. 44.4) and work out the correct orientation of the specimen for dissection. Invertebrates are usually dissected from the dorsal surface (the nerve cord being ventral in position) and vertebrates from the ventral surface (the nerve cord being dorsal in position): however, special objectives may require a different orientation.

3. **Open the body cavity carefully.** This is usually done using forceps to lift the skin away from underlying organs while using a fine pair of scissors to make an initial opening; the scissors are then used to extend this opening in anterior–posterior and lateral directions until the skin flaps can be pinned back. Pin out by placing tension on the skin flaps, breaking down any restricting membranes using a seeker.

4. **Subsequent dissection procedure depends upon your objectives.** To display the system being investigated as clearly and neatly as possible requires:

    (a) identification of organs, blood vessels and nerves initially visible;

    (b) separation of these structures from the membranes that hold them in place. This is best achieved using a blunt instrument such as a seeker to avoid damage;

    (c) removal or displacement of organs that obscure parts of the system you wish to display. Displacement using pins to hold the organ in position is the preferred option when possible;

    (d) if your objective is to display a blood system or a nervous system, you must follow individual vessels/nerves from their point of origin to their destination organ; do this using a blunt seeker to remove covering membranes by working carefully along the structure. Do not pull sideways during this procedure as this often results in breakages; such tissues are usually much stronger in the direction of their length than when pulled laterally.

    (e) tidy up loose pieces and wash away residual blood, etc.; use more pins to finalise the display of the system and then make notes and drawings as necessary (see e.g. Figs 28.3 and 40.1).

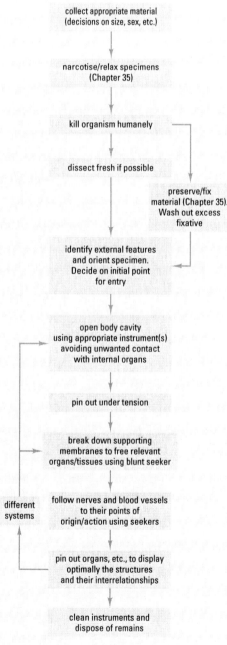

Fig. 40.3 Flowchart for dissection

Box 40.2 Tips for improving dissection technique

The following tips should help you develop your dissection technique:

- **Use a blunt seeker and controlled tension on the tissue to remove the connective tissue** that binds tissues and organs together: use scalpels sparingly.
- **Insert pins/awls obliquely** so that they do not interfere with further dissection. In segmented animals such as earthworms, fine pins can be used to mark the position of specific segments.
- **Dissect most invertebrates under water:** the water buoys up the tissues/organs and assists dissection. Change the water if it becomes clouded but do not allow flowing water to run directly on to the specimen as it will damage delicate structures.
- **Dissect vertebrates and larger invertebrates in air** on a dissection board, using cotton-wool swabs to remove excess blood and other body fluids.
- **Keep tissues under tension while dissecting,** but avoid damaging them with forceps/fingers.
- **Cut away from the organs when using scalpels or scissors** and keep scissor points away from structures you are attempting to free. This is essential when opening the body cavities of small animals.
- **Use appropriately sized equipment:** e.g. do not attempt to use large scissors to open an earthworm. Never use delicate instruments for coarse work.
- *Never* **remove anything until you know what you are removing.**
- **Dissect along structures and not across them,** particularly for tubular structures such as nerves and blood vessels.
- **Use fresh material whenever possible.** Fixatives make tissues more brittle and inelastic: alcohol storage tends to harden skin, muscle and connective tissue.
- **Keep your instruments clean and sharp:** you can't dissect with blunt or dirty equipment. Dry instruments after washing and wipe with an oily cloth. Cut nothing but tissues with scalpels and scissors. Do not sharpen pencils with scalpels or stick mounted needles into the bench or the dissecting board.
- **Be hygienic.** Wear rubber gloves if you have cuts or lesions on your hands. Wash hands and equipment thoroughly when finished. Dispose of animal remains carefully as instructed.

Dissecting seekers

These have blunt points and should have metal handles. Use for breaking down membranes holding delicate organs together. Again, they have multiple uses.

Recommended accessory equipment

A teat (Pasteur) pipette is valuable for washing delicate organs and tissues. A camel-hair brush is useful for removing material from delicate structures. Pins and awls are usually provided by your department and are used for fixing the dissection specimen to the board (awls) or wax dish (pins): small pins are invaluable for pinning organs aside for display purposes. For measurement, and as an aid to drawing, a pair of dividers is very useful, as is a small steel ruler.

Carrying out the dissection

The basic sequence of steps in a typical dissection is shown in Box 40.1 and Fig. 40.3, while Box 40.2 gives tips to help you improve your dissection technique.

Sources for further study

About Inc. *About Biology: Online dissections*. Available: http://biology.about.com/od/onlinedissections/Online_Dissections.htm
Last accessed 09/04/07
[A compendium of Web-based dissections.]

Fishbeck, D. and Sebastini, A. (2001) *Comparative Anatomy: Manual of Vertebrate Dissection*. Morton Publishing Co., Englewood, NJ.

Kardong, K.V. and Zalisko, E.J. (1998) *Comparative Vertebrate Anatomy: A Laboratory Dissection Guide*. McGraw-Hill, Boston.

Morgan, M. *Earthworm Dissection*. Available: http://www.microscopy-uk.org.uk/mag/articles/worm.html
Last accessed 09/04/07.
[A simple, descriptive account of an earthworm dissection.]

Walker, W.F. and Homberger, D. (1997) *Anatomy and Dissection of the Rat*, 3rd edn. W.H. Freeman, New York.

Whitehouse, R.H. and Grove, A.J. (1947) *The Dissection of the Crayfish*. University Tutorial Press, London.
[Part of a classic series including manuals on frog, rabbit, dogfish, earthworm and cockroach. Includes lots of information on how to carry out detailed dissections.]

Study exercises

40.1 Make a list of situations where dissection is used. There are many areas of zoology (and botany) where dissection technique is required. List up to ten situations where you would require to use some form of dissection technique to acquire material to work with, or to carry out manipulations before or after experiments.

40.2 Decide on appropriate dissection instruments. What instruments should you use for the following operations:

(a) following a nerve or blood vessel from source to ending?

(b) opening up the body cavity of the earthworm?

(c) removing or severing muscle tissue in the body wall of a vertebrate?

40.3 Research general rules of dissection. What is the general rule for deciding on which side a general dissection should be commenced: (a) for invertebrates; and (b) for vertebrates?

42 Preparing specimens for light microscopy

 SAFETY NOTE The fixatives, solvents, embedding media and stains used in the preparation of specimens for light microscopy should all be assumed to be toxic, and treated accordingly. Take care to follow the safety advice provided by laboratory staff and tutors.

Preparative techniques are crucial to successful microscopical investigation because the chemical and physical processes involved have the potential for making the material difficult to work with and for producing artefacts. The basic steps (outlined in Fig. 41.3) are similar in most cases, but the exact details (e.g. timing, chemicals used and their concentrations) differ according to the material being examined and the purpose of the investigation. It is usually best to follow a recipe that has worked in the past for your material (see Grimstone and Skaer, 1972, Kiernan, 1999).

Chemical fixation

The main purpose of fixation is to preserve material in a lifelike manner. The process of fixation for microscopy is much more critical than for whole specimens (see Chapter 35) and only small pieces of tissue should be used. The fixation solutions used for microscopy are intended to:

- penetrate rapidly to prevent post-mortem changes in the cells;
- coagulate the cell contents into insoluble substances;
- protect tissues against shrinkage and distortion during subsequent processing;
- allow cell parts to become selectively and clearly visible when stained.

Fixing specimens – tissues must be fixed as soon as possible after death. If this is not possible, specimens should be stored at low temperature (4°C) and for as short a time as possible.

Fixative solutions are usually mixtures of chemicals selected for their combined properties (see Chapter 35). Your choice of fixative from the numerous recipes available in reference texts will depend upon both the type of investigation and the nature of the material.

 KEY POINT Poor fixation can produce artefacts, particularly where coagulant fixatives are used.

When using a fixative for microscopy, observe the following points:

- Use fresh solutions: some of the fluids are unstable and do not keep well. Do not reuse fixative.
- Always use plenty of fixing fluid compared with the volume of material to be fixed (not less than a 10:1 fixative:sample volume ratio).
- Avoid underfixation or overfixation: in general the optimum time will be a function of several factors including:
  (a) The penetration capacity of the fixative.
  (b) The size of the piece of tissue: this should always be small and have as large a surface:volume ratio as possible.
  (c) The type of tissue to be fixed: uniform tissues fix more quickly than complex tissues, where one component may form a barrier to others. The presence of chitin usually means a slow rate of penetration. Tissues filled with air can be difficult to submerge and infiltrate; this can be overcome by fixation in a partial vacuum.
  (d) The temperature: increased temperature results in increased penetration rate, but also tends to make tissue brittle.
- Wash the specimen thoroughly after fixation: residues of fixative can interfere with subsequent processes. The washing may be in water or another appropriate solution.

Decalcifying specimens – this may be necessary if calcareous structures remain after fixation: this is usually done using a 5–10 per cent solution of EDTA followed by thorough washing.

Dehydration and clearing

A high water content in tissues will usually hinder subsequent processing so they must be dehydrated. This is done with an organic solvent using a series of solutions graded from pure water to pure solvent. Ethanol/water mixtures are often used in histology. Dehydration must be carried out carefully, using prescribed time schedules to avoid distortion and hardening. Protect delicate specimens by mixing solutions in the container rather than by transferring the specimen as this is when damage will occur.

Allowing time for dehydration – *never be tempted to rush, because incomplete dehydration will do more than anything else to ruin a preparation.*

The chemicals used for dehydration are not usually soluble in the embedding medium and tissues must therefore be infiltrated with an intermediate fluid, miscible with the waxes and resins normally used: such fluids make the tissues transparent and are termed clearing agents. The most widely used ones are hydrocarbons such as xylene; many are volatile, pose significant health risks, and tend to harden tissues rapidly. Clearing oils such as clove oil and cedarwood oil are safer, but slower in action. Terpineol is useful since it does not require such complete dehydration and can be used straight from 90% v/v ethanol. Note that every trace of oil must be removed during the infiltration (embedding) stage, or specimens will not embed properly.

Embedding and sectioning

Embedding involves infiltrating the specimen with a medium that will solidify and support it when sectioned. The specimen is either passed through a series of gradually increasing concentrations of the embedding material (e.g. wax or epoxy resin) dissolved in the dehydrating or clearing agent, or placed straight into pure embedding agent. Several changes of embedding material are made to remove the last traces of solvent, then the embedding agent is solidified by cooling or by a polymerisation treatment. Precise protocols can be found in specialist books (e.g. Kiernan, 1999). The specimen must be oriented carefully during the solidification process to aid section-cutting in known planes.

 SAFETY NOTE *The sharp blades use in hand sectioning are an obvious hazard. Identify beforehand the location of the nearest first aid kit. Take care to carry out such procedures with care and away from others who may inadvertently brush against you.*

The aim of sectioning is to provide a thin slice through the tissues suitable for observing cellular details at maximum resolution. After trimming, blocks of embedded material are sectioned using a microtome. Section-cutting techniques are complex and require time for familiarisation. Sections are attached to slides after flattening the sections by a flotation procedure carried out either directly on the glass slide or in a water bath. The sections are attached to the slide by coating the latter with a very thin layer of albumen before drying them down on to this sticky surface.

For some procedures, specimens are frozen in isopentane, freon or dry-ice/ethanol mixtures and then maintained below −20 °C until sectioned. Sectioning is performed in a cold chamber (cryostat) where there is a microtome; sections are removed directly to glass slides. If storage is necessary, slides are placed over a desiccant in a refrigerator, but this is possible only for short periods.

Hand-cut sections can be made through certain relatively stiff materials, notably stems and roots of herbaceous plants:

1. Grasp the object firmly between your index finger and thumb.
2. Brace your elbows against your ribs to steady your hands.
3. Rest the side of a fresh and sharp razor blade on the index finger holding the specimen (Fig. 42.1).

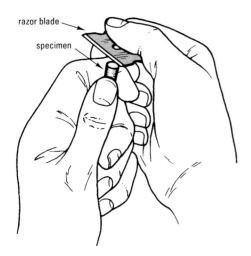

Fig. 42.1 Preparation of a hand-cut section

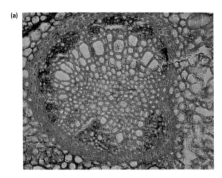

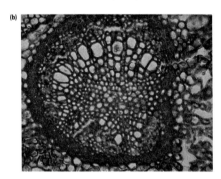

Fig. 42.2 Simulation of the introduction of contrast into a biological specimen by staining. Leaf midrib transverse section showing in greyscale the effects 'before' (a) and 'after' (b) staining xylem tissue red with phloroglucinol. In colour, the effects of staining are even more distinct.

Definitions

Metachromic stains – have the capacity to stain different structures different colours.

Orthochromic stains – never change colour whatever they stain.

Mordants – chemicals (salts and hydroxides of divalent and trivalent metals) that increase the efficiency of stains usually by forming complexes with the stain.

Counterstains – stains that apply a background colour to contrast with stained structures.

Negative staining – where the background is stained rather than the structure of interest.

4. Pull the blade towards your body so that you cut the specimen with a slicing action.
5. Repeat this action quickly, pushing the object slowly upwards with your thumb.
6. Float off the cut sections on to a watch glass of water.

This is a relatively tricky procedure, requiring practice. You may find it helps to lubricate the razor and object by wetting them before cutting. Thin material, such as a leaf, is best sectioned when supported between the halves of a longitudinally split cylinder of pith or fresh carrot. Always cut lots of sections, because only a small proportion will be thin enough to use – the best ones are often wedge-shaped, tapering off to thin edges.

Staining

The purpose of staining in microscopy is to:

- add contrast to the image;
- identify chemical components of interest;
- locate particular tissues, cells or organelles.

This is achieved in different ways for different types of microscopy. In standard light microscopy, contrast is achieved by staining the structure of interest with a coloured dye (e.g. Fig. 42.2); in UV microscopy, contrast is obtained using fluorescent stains. Physico-chemical properties of the stain cause it to attach to certain structures preferentially or be taken up across cell membranes.

Stains for light microscopy are categorised according to the charge on the dye molecule. Stains like haematoxylin, whose coloured part is a cation (i.e. basic dyes), stain acidic, anionic substances like nucleic acids: such structures are termed basophilic. Stains like eosin, whose coloured part is an anion (i.e. acid dyes), stain basic, cationic substances: such structures are termed acidophilic. Acid dyes tend to stain all tissue components, especially at low pH, and are much used as counterstains. Staining is progressive if it results in some structures taking up the dye preferentially. Staining is regressive if it involves initial over-staining followed by decolorisation (differentiation) of those structures that do not bind the dye tightly (e.g. Gram staining, p. 229).

Certain 'vital' stains (e.g. neutral red) are used to determine cell viability or the pH of cell compartments such as plant vacuoles. 'Mortal' stains (e.g. Evans' blue) are excluded from living cells but diffuse into dead ones and are used to assay cell mortality.

Stains and staining procedures

There is a huge range of stains for light microscopy but the features of those used commonly for cytology in botany, microbiology and zoology are given in Table 42.1. Consult appropriate texts for full details of (a) how to make up stains and (b) the protocol to use. Results depend on technique: follow the recommended procedures carefully. There are many stains used in histochemistry for identifying various classes of macromolecules such as DNA, RNA, proteins, lipids, carbohydrates (chitin, cellulose, starch, callose, pectins, glycogen) and heteropolymers (e.g. lipopolysaccharides, peptido-glycans, proteoglycans). Consult specialist texts for methods (e.g. Grimstone and Skaer, 1972; Horobin and Kiernan, 2002).

Table 42.1 A selection of stains for light microscopy of sections

	Stain	What it stains	Comments
Plant cells	Chlorazol black	Cell walls: black Nuclei: black, yellow or green Suberin: amber	The solvent used (70 per cent ethanol in water or water alone) affects colours developed
	Neutral red	Living cells: pink (pH < 7)	A 'vital' stain used to determine cell viability or to visualise plant protoplasts in plasmolysis experiments; best used at neutral external pH
	Phloroglucinol/HCl	Lignified cell walls: red	Care is required because the acid may damage microscope lenses
	Ruthenium red	Pectins: red	Shows up the middle lamella
	Safranin + Fast green	Nuclei, chromosomes, cuticle and lignin: red Other components: green	Stain in safranin first, then counterstain with fast green (light green will substitute). A differentiation step is required
	Toluidine blue	Lignified cell walls: blue Cellulose cell walls: purple	Best to apply dilute and allow progressive staining to occur
Fungi and bacteria	Giemsa	Bacterial chromosome: purple Bacterial cytoplasm: colourless	Also used in zoology to stain protozoa
	Gram	Gram-positive bacteria: violet/purple Gram-negative bacteria: red/pink Yeasts: violet/purple	See p. 230 for procedure
	Gray	Bacterial flagella: red	Uses toxic chemicals: mercuric chloride and formaldehyde. Leifson's stain is an alternative
	Lactophenol cotton blue	Fungal cytoplasm: blue (hyphal wall unstained)	Shrinkage may occur
	Nigrosin or India ink	Background: grey–black	Negative stains for visualisation of capsules: requires a very thin film
	Shaeffer and Fulton	Bacterial endospores: green Vegetative cells: pink/red	Malachite green is primary stain, heated for 5 min. Counterstained with safranin
	Ziehl–Neelsen	Actinomycetes: red Bacterial endospores: red Other microbes: blue	Requires heat treatment of fuchsin primary stain, decolorisation with ethanol–HCl and a methylene blue counterstain (acid-fast structures remain red)
Animal cells	Azure A/eosin B	Nuclei, RNA: blue Basophilic cells: blue–violet Most other cells: pale blue Muscle cells: pink Necrosing cells: pink Cartilage matrix: red–violet Bone: pink Red blood cells: orange–red Mucins: green–blue/blue–violet	Used in pathology – shows up bacteria as blue; must be fresh; care required over pH: Mann's methyl blue/eosin gives similar results
	Chlorazol black	Chitin: greenish-black Nuclei: black, yellow or green Glycogen: pink or red	Solvent (70% v/v ethanol in water or water alone) affects colours formed
	Iron haematoxylin	Nuclei, chromosomes and red blood cells: black Other structures: grey or blue–black	Good for resolving fine detail; iron alum used as mordant before haematoxylin to differentiate
	Mallory	Nuclei: red Nucleoli: yellow Collagen, mucus: blue Red blood cells: yellow Cytoplasm: pink or yellow	Simple, one-stage stain; fades within a year; not to be used with osmium-containing fixatives. Heidenhain's azan gives similar results but does not fade. Cason's one-step Mallory is a rapidly applied stain that is particularly good for connective tissue
	Masson's trichrome	Collagen, mucus: green Cytoplasm: orange or pink	Used as a counterstain after, for example, iron haematoxylin, which will have stained nuclei black. Not to be used after osmium fixation
	Mayer's (haemalum and eosin; 'H&E')	Nuclei: blue/purple Cytoplasm: pink	Alum used as mordant for haematoxylin; eosin is the counterstain. To show up collagen, use van Gieson's stain as counterstain

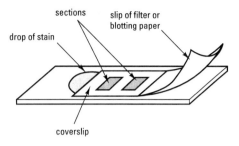

Fig. 42.3 How to irrigate a section with stain by drawing it through with a piece of filter paper.

Transferring sections – a camel-hair paintbrush is useful for transferring sections between staining solutions and on to a slide.

Using coverslips – all wet specimens for microscopic examination must be covered with a coverslip to protect the objective lens from water, oil, stains and dirt.

Using mountants – most mountants require that all water is removed from the section by transfer through increasing concentrations of ethanol until 100 per cent ethanol is reached.

Stains for light microscopy are normally applied by one of four methods:

- Floating the sections on the stain.
- Applying the stain to a smear fixed on to a slide, e.g. when staining bacteria and blood.
- Drawing the stain through under a coverslip as shown in Fig. 42.3.
- Immersing slides with sections attached into a staining trough. This is best for bulk staining.

Most stains need to act in an aqueous medium, so sections that have been embedded in wax must be rehydrated before staining. The wax is dissolved, e.g. in Histo-clear for 1–3 min, the Histo-clear replaced by 100 per cent ethanol (1 min) then 70% (v/v) ethanol : water for 1 min., finally transferring to distilled water. Fresh sections of plants or heat-fixed smears of microbes generally require no pretreatment before staining if they are not to be retained after examination.

Mounting sections

Wet mounts

These are used for observing fresh specimens. The following steps are involved:

1. Isolate the specimen.
2. Place the specimen in a small droplet of the relevant fluid (fresh water, sea water, etc.) on a microscope slide.
3. Gently lower a coverslip on to the droplet, using forceps or two needles and avoiding bubbles.
4. Remove any excess water on or around the coverslip with absorbent paper.

Entire specimens can be examined under the light microscope providing they are small enough to be mounted on a glass slide. They may be mounted in cavity slides or by using ring mounts.

Temporary mounts

These essentially involve wet mounting in a mountant with a short useful life, e.g. for identification purposes. It may be desirable to clear the specimen first and a dual-purpose substance such as lactophenol, which will clear from 70% (v/v) ethanol, is recommended.

Permanent mounts

These protect sections during examination and allow storage without deterioration. A permanent mount involves sealing your section under a coverslip in a mountant. The mountants used are clear resins dissolved in a slowly evaporating solvent. A good mountant has a similar refractive index to the tissue being mounted, remains clear through time, is chemically inert and will harden quickly. Natural resins like Canada balsam take a long time to dry, are variable in quality and tend to colour up and crack in time. The newer synthetic resins and plastics such as DPX

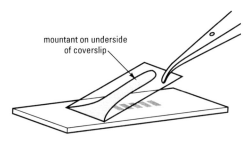

Fig. 42.4 How to lower a coverslip painted with mountant on its underside on to sections on a slide.

Making squash preparations – these may be required for any type of mount. The smallest specimens can be squashed after mounting by applying gentle pressure on the coverslip with your forceps. Larger specimens can be squashed between two slides before fixing and mounting – this ensures that higher pressures are applied evenly.

mountant are superior: they dry quickly, are available in a range of refractive indices, and do not yellow with age. For tissue components soluble in organic solvents, aqueous mounting media based on, for example, gelatine or glycerol should be used. Find out which mountant is recommended for your particular sections.

The recommended procedure when mounting sections on a slide is:

1. Apply a little mountant to a coverslip of appropriate size.
2. Turn the coverslip over and place on its edge to one side of the sections as in Fig. 42.4.
3. Lower the coverslip slowly down on to the sections so as to displace all the air and sandwich the sections between the slide and the coverslip.
4. Press firmly from the centre outwards to distribute the mounting medium evenly.
5. Allow the solvent to evaporate – best results come from slow drying when time allows, but many synthetic mountants will tolerate brief heating when speed is essential.

Text references and sources for further study

Davidson, M.W. and Abramowitz, M. *Molecular Expressions. Exploring the World of Optics and Microscopy*. Available: http://micro.magnet.fsu.edu/index.html
Last accessed 09/04/07.
[Covers many areas of basic knowledge underlying microscopy, including preparative procedures.]

Grimstone, A.V. and Skaer, R.J. (1972) *A Guidebook to Microscopical Methods*. Cambridge University Press, Cambridge.

Horobin, R.W. and Kiernan, J.A. (eds) (2002) *Conn's Biological Stains. A Handbook of Dyes, Stains and Fluorochromes for use in biology and Medicine*, 10th edn. Bios Scientific Publishers, Oxford.

Kiernan, J.A. (1999) *Histological and Histochemical Methods: Theory and Practice*, 3rd edn. Scion Publishing, Bloxham.

Study exercises

42.1 Select appropriate stains. From Table 42.1, identify a stain you could use to help indicate the presence of the following: (a) glycogen in a liver section; (b) woody (lignified) cells in a plant stem section; (c) a fungal pathogen in a leaf section; (d) living cells in an onion epidermal peel; (e) mucus in a lung section; (f) bacteria in a food sample.

42.2 Explain the processes involved in preparing a section for light microscopy. Prepare one-sentence summaries of the reasons for carrying out the following procedures: (a) fixation; (b) dehydration; (c) embedding; (d) sectioning; (e) staining; (f) permanent mounting.

42.3 Investigate toxicity and safety aspects of some of the staining reagents used for light microscopy. Many chemicals used in staining are toxic or otherwise dangerous. As examples, investigate the specific hazards associated with the use of the following stains: (a) chlorazol black; (b) neutral red; (c) malachite green; (d) safranin.

Handling Cells and Tissues

32 Sterile technique and microbial culture

Sterile technique (aseptic technique) is the name given to the procedures used in cell culture. While the same general principles apply to all cell types, you are most likely to learn the basic procedures using bacteria and most of the examples given in this section refer to bacterial culture.

Sterile technique serves two main purposes:

1. To prevent accidental contamination of laboratory cultures due to microbes from external sources, e.g. skin, clothing or the surrounding environment.
2. To prevent microbial contamination of laboratory workers, in this instance you and your fellow students.

Achieving a sterile state – you should assume that all items of laboratory equipment have contaminating microbes on their surfaces, unless they have been destroyed by some form of sterilisation. Such items will only remain sterile if they do not come into contact with the non-sterile environment.

KEY POINT All *microbial cell cultures should be treated as if they contained potentially harmful organisms. Sterile technique forms an important part of safety procedures, and must be followed whenever cell cultures are handled in the laboratory.*

Care is required because:

- You may accidentally isolate a harmful microbe as a contaminant when culturing a relatively harmless strain.
- Some individuals are more susceptible to infection and disease than others – not everyone exposed to a particular microbe will become ill.
- Laboratory culture involves purifying and growing large numbers of microbial cells – this represents a greater risk than small numbers of the original microbe.
- A microbe may change its characteristics, perhaps as a result of gene exchange or mutation.

The international biohazard symbol, shown in Figure 32.1, is used to indicate a significant risk due to a pathogenic microbe (p. 206).

Fig. 32.1 International symbol for a biohazard. Usually red on a yellow background, or black on a red background.

Sterilisation procedures

Given the ubiquity of microbes, the only way to achieve a sterile state is by their destruction or removal. Several methods can be used to achieve this objective:

Heat treatment

This is the most widespread form of sterilisation and is used in several basic laboratory procedures including the following:

- **Red heat sterilisation.** Achieved by heating metal inoculating loops, forceps, needles, etc., in a Bunsen flame (Fig. 32.2). This is a simple and effective form of sterilisation as no microbe will survive even a brief exposure to a naked flame. Flame sterilisation using alcohol is used for glass rods and spreaders (see below).
- **Dry heat sterilisation.** Here, a hot air oven is used at a temperature of at least 160 °C for at least 2 h. This method is used for the routine sterilisation of laboratory glassware. Dry heat procedures are of little value for items requiring repeated sterilisation during use.

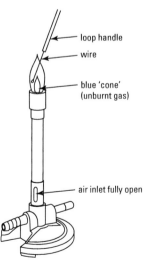

Fig. 32.2 'Flaming' a wire loop. Keep the loop in the hottest part of the Bunsen flame (just outside the blue 'cone') until the wire is red hot.

Fig. 32.3 Autoclave tape – the lower sample is untreated while the upper sample (with dark diagonal lines) has been autoclaved.

Using a sterile filter – *most filters are supplied as pre-sterilised items. Make sure you follow a procedure that does not contaminate the filter on removal from its protective wrapping.*

 SAFETY NOTE When working with biocides, take care to avoid skin contact or ingestion, as most are toxic and irritant. If contact does occur, rinse with plenty of water.

Using molten agar – *a water bath (at 45–50°C) can be used to keep an agar-based medium in its molten state after autoclaving. Always dry the outside of the container on removal from the water bath, to reduce the risk of contamination from microbes in the water, e.g. during pour plating (p. 207).*

- Moist heat sterilisation. This is the method of choice for many laboratory items, including most fluids, apart from heat-sensitive media. It is also used to decontaminate liquid media and glassware after use. The laboratory autoclave is used for these purposes. Typically, most items will be sterile after 15 min at 121 °C, although large items may require a longer period. The rapid killing action results from the latent heat of condensation of the pressurised steam, released on contact with cool materials in the autoclave. While special heat-sensitive tape (Fig. 32.3) is sometimes used to check that the autoclave is operating correctly, a better approach is to use spores of *Bacillus stearothermophilus*.

Radiation

Many disposable plastic items used in microbiology and cell biology are sterilised by exposure to UV or ionising radiation. They are supplied commercially in sterile packages, ready for use. Ultraviolet radiation has limited use in the laboratory, while ionising radiation (e.g. γ-rays) requires industrial facilities and cannot be operated on a laboratory scale.

Filtration

Heat-labile solutions (e.g. complex macromolecules, including proteins, antibiotics, serum) are particularly suited to this form of sterilisation. The filters come in a variety of shapes, sizes and materials, usually with a pore size of either 0.2 or 0.45 µm. The filtration apparatus and associated equipment are usually sterilised by autoclaving, or by dry heat. Passage of liquid through a sterile filter of pore size 0.2 µm into a sterile vessel is usually sufficient to remove bacteria but not viruses, so filtered liquids are not necessarily virus-free.

Chemical agents

These are known as disinfectants, or biocides, and are most often used for the disposal of contaminated items following laboratory use, e.g. glass slides and pipettes. They are also used to treat spillages. The term 'disinfection' implies destruction of disease-causing bacterial cells, although spores and viruses may not always be destroyed. Remember that disinfectants require time to exert their killing effect – any spillage should be covered with an appropriate disinfectant and left for at least 10 min before mopping up.

Use of laboratory equipment

Working area

One of the most important aspects of good sterile technique is to keep your working area as clean and tidy as possible. Start by clearing all items from your working surface, wipe the bench down with disinfectant and then arrange the items you need for a particular procedure so that they are close at hand, leaving a clear working space in the centre of your bench.

Media

Cells may be cultured in either a liquid medium (broth), or a solidified medium (p. 232). The gelling agent used in most solidified media is agar, a complex polysaccharide from red algae that produces a stiff transparent gel when used at 1–2% (w/v). Agar is used because it is relatively resistant to degradation by most bacteria and because of its rheological properties – an agar medium melts at 98 °C, remaining solid at all temperatures used for routine laboratory culture. However, once melted it does not solidify again until the temperature falls to about 44 °C. This means that heat-sensitive constituents (e.g. vitamins, blood, cells) can be added aseptically to the medium after autoclaving.

Inoculating loops

The initial isolation and subsequent transfer of microbes between containers can be achieved using a sterile inoculating loop. Most teaching laboratories use nichrome wire loops in a metal handle. A wire loop can be repeatedly sterilised by heating the wire, loop downwards and almost vertical, in the hottest part of a Bunsen flame until the whole wire becomes red hot (Fig. 32.2). Then the loop is removed from the flame to minimise heat transfer to the handle. After cooling for 8–10 s (without touching any other object) it is ready for use.

When resterilising a contaminated wire loop in a Bunsen flame after use, do not heat the loop too rapidly, as the sample may spatter, creating an aerosol: it is better to soak the loop for a few minutes in disinfectant than to risk heating a fully charged (contaminated) inoculating loop.

Working with plastic disposable loops – these are used in many research laboratories: pre-sterilised and suitable for single use, they avoid the hazards of naked flames and the risk of aerosol formation during heating. Discard into disinfectant solution after use.

Using a Bunsen burner to reduce airborne contamination – working close to the updraught created by a Bunsen flame reduces the likelihood of particles falling from the air into an open vessel.

Using glass pipettes – these are plugged with cotton wool at the top before being autoclaved inside a metal can. Flame the open end of the can on removal of a pipette, to prevent contamination of the remaining pipettes. Autopipettors and sterile disposable tips (p. 129) offer an alternative approach.

Containers

There is a risk of contamination whenever a sterile bottle, flask or test tube is opened. One method that reduces the chance of airborne contamination is quickly to pass the open mouth of the glass vessel through a flame. This destroys any microbes on the outer surfaces nearest to the mouth of the vessel. In addition, by heating the air within the neck of the vessel, an outwardly directed air flow is established, reducing the likelihood of microbial contamination.

It is general practice to flame the mouth of each vessel immediately after opening and then repeat the procedure just before replacing the top. Caps, lids and cotton wool plugs must not be placed on the bench during flaming and sampling: they should be removed and held using the smallest finger of one hand, to minimise the risk of contamination. This also leaves the remaining fingers free to carry out other manipulations. With practice, it is possible to remove the tops from two tubes, flame each tube and transfer material from one to the other while holding one top in each hand.

Laminar flow cabinets

These are designed to prevent airborne contamination, e.g. when preparing media or subculturing microbes or tissue cultures. Sterile air is produced by passage through a high efficiency particulate air (HEPA) filter: this is then directed over the working area, either horizontally (towards the operator) or downwards. The operator handles specimens, media, etc. through an opening at the front of the cabinet. Note that standard laminar flow cabinets do *not* protect the worker from contamination and must not be used with pathogenic microbes: special safety cabinets and laboratories are used for work with ACDP hazard group 3 and 4 microbes (Table 32.1) and for samples that might contain such pathogens.

Table 32.1 Classification of microbes on the basis of hazard. The following categories are recommended by the UK Advisory Committee on Dangerous Pathogens (ACDP)

Hazard group	Comments
1	Unlikely to cause human disease
2	May cause disease: possible hazard to laboratory workers, minimal hazard to community
3	May cause severe disease: may be a serious hazard to laboratory workers, may spread to community
4	Causes severe disease: is a serious hazard to laboratory workers, high risk to community

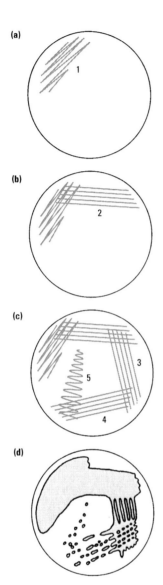

Fig. 32.4 Preparation of a streak plate for single colonies. (a) Using a sterile metal loop, take a small sample of the material to be streaked. Distribute the sample over a small sector of the plate (area 1), then flame the loop and allow to cool (approximately 8–10 s). (b) Make several small streaks from the initial sector into the adjacent sector (area 2), taking care not to allow the streaks to overlap. Flame the loop and allow to cool. (c) Repeat the procedure for areas 3 and 4, resterilising the loop between each step. Finally, make a single, long streak, as shown for area 5. (d) The expected result after incubation at the appropriate temperature (e.g. 37 °C for 24 h): each step should have diluted the inoculum, giving individual colonies within one or more sectors on the plate. Further subculture of an individual colony should give a pure (clonal) culture.

 KEY POINT *The most obvious risks when handling microbial cultures are those due to ingestion or entry via a cut in the skin – all cuts should be covered with a plaster or disposable plastic gloves. A less obvious source of hazard is the formation of aerosols of liquid droplets from microbial suspensions, with the risk of inhalation, or surface contamination of other objects.*

Microbiological hazards

The following steps will minimise the risk of aerosol formation:

- Use stoppered tubes when shaking, centrifuging or mixing microbial suspensions.
- Pour solutions gently, keeping the difference in height to a minimum.
- Discharge pipettes onto the side of the container.

Other general rules which apply in all laboratories include:

- Take care with sharp instruments, including needles and glass Pasteur pipettes.
- Do not pour waste cultures down the sink – they must be autoclaved.
- Put other contaminated items (e.g. slides, pipettes) into disinfectant after use.
- Wipe down your bench with disinfectant when practical work is complete.
- Always wash your hands before leaving the laboratory.

Plating methods

Many culture methods make use of a solidified medium within a Petri plate. A variety of techniques can be used to transfer and distribute the organisms prior to incubation. The three most important procedures are described below.

Streak dilution plate

Streaking a plate for single colonies is one of the most important basic skills in microbiology, since it is used in the initial isolation of a cell culture and in maintaining stock cultures, where a streak dilution plate with single colonies all of the same type confirms the purity of the strain. A sterile inoculating loop is used to streak the organisms over the surface of the medium, thereby diluting the sample (Fig. 32.4). The aim is to achieve single colonies at some point on the plate: ideally, such colonies are derived from single cells (e.g. in the case of unicellular bacteria, animal and plant cell lines) or from groups of cells of the same species (in filamentous or colonial forms). Single colonies, containing cells of a single species and derived from a single parental cell, form the basis of most pure culture methods (p. 211).

Note the following:

- Keep the lid of the Petri plate as close to the base as possible to reduce the risk of aerial contamination.
- Allow the loop to glide over the surface of the medium. Hold the handle at the balance point (near the centre) and use light, sweeping movements, as the agar surface is easily damaged and torn.
- Work quickly, but carefully. Do not breathe directly onto the exposed agar surface and replace the lid as soon as possible.

Spread plate

This method is used with cells in suspension, either in a liquid growth medium or in an appropriate sterile diluent. It is one method of quantifying the number of viable cells (or colony-forming units) in a sample, after appropriate dilution (p. 137).

An L-shaped glass spreader is sterilised by dipping the end of the spreader in a beaker containing a small amount of 70% v/v alcohol, allowing the excess to drain from the spreader and then igniting the remainder in a Bunsen flame. After cooling, the spreader is used to distribute a known volume of cell suspension across the plate (Fig. 32.5). *There is a significant fire risk associated with this technique*, so take care not to ignite the alcohol in the beaker, e.g. by returning an overheated glass rod to the beaker. The alcohol will burn with a pale blue flame that may be difficult to see, but will readily ignite other materials (e.g. a laboratory coat). Another source of risk comes from small droplets of flaming alcohol shed by an overloaded spreader onto the bench and this is why you *must* drain excess alcohol from the spreader *before* flaming. Some laboratories now provide plastic disposable spreaders for student use, to avoid the risk of fire.

Pour plate

This procedure also uses cells in suspension, but requires molten agar medium, usually in screw-capped bottles containing sufficient medium to prepare a single Petri plate (i.e. 15–20 ml), maintained in a water bath at 45–50 °C. A known volume of cell suspension is mixed with this molten agar, distributing the cells throughout the medium. This is then poured without delay into an empty sterile Petri plate and incubated, giving widely spaced colonies (Fig. 32.6). Furthermore, as most of the colonies are formed within the medium, they are far smaller than those of the surface streak method, allowing higher cell numbers to be counted (e.g. over 100 colonies per plate): some workers pour a thin layer of molten agar onto the surface of a pour plate after it has set, to ensure that no surface colonies are produced. Most bacteria and fungi are not killed by brief exposure to temperatures of 45–50 °C, though the procedure may be more damaging to microbes from low temperature conditions, e.g. psychrophilic bacteria.

One disadvantage of the pour plate method is that the typical colony morphology seen in surface-grown cultures will not be observed for those colonies that develop within the agar medium. A further disadvantage is that some of the suspension will be left behind in the screw-capped bottle. This can be avoided by transferring the suspension to the Petri plate, adding the molten agar, then swirling the plate to mix the two liquids. However, even when the plate is swirled repeatedly and in several directions, the liquids are not mixed as evenly as in the former procedure.

KEY POINT When working with molten agar, keep tubes and bottles of molten agar in a water bath until you are ready to use them, as they will begin to set within a couple of minutes at room temperature.

Working with phages

Bacterial viruses ('bacteriophages', or simply 'phages') are often used to illustrate the general principles involved in the detection and enumeration of viruses. They also have a role in genome mapping of bacteria. Individual

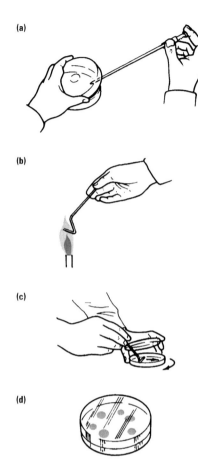

Fig. 32.5 Preparation of a spread plate. (a) Transfer a small volume of cell suspension (0.05–0.5 ml) to the surface of a solidified medium in a Petri plate. (b) Flame sterilise a glass spreader and allow to cool (8–10 s). (c) Distribute the liquid over the surface of the plate using the sterile spreader. Make sure of an even coverage by rotating the plate as you spread; allow the liquid to be absorbed into the agar medium. Incubate under suitable conditions. (d) After incubation, the microbial colonies should be distributed over the surface of the plate.

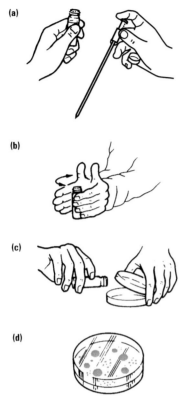

Fig. 32.6 Preparation of a pour plate. (a) Add a known volume of cell suspension (0.05–1.0 ml) to a small bottle of molten agar medium from a 45 °C water bath. (b) Mix thoroughly, by rotating between the palms of the hands: do not shake or this will cause frothing of the medium. (c) Pour the mixture into an empty, sterile Petri plate and allow to set. Incubate under suitable conditions. (d) After incubation, the microbial colonies will be distributed throughout the medium: any cells deposited at the surface will give larger, spreading colonies.

phage particles (virions) are too small to be seen by light microscopy, but are detected by their effects on susceptible hostcells:

- Virulent phages will infect and replicate within actively growing host cells, causing cell lysis and releasing new infective phages – this 'lytic cycle' takes ≈30 min for T-even phages of *E. coli*, e.g. T4.
- Temperate phages are a specialised group, capable either of lytic growth or an alternative cycle, termed lysogeny – the phage becomes latent within a host cell (lysogen), typically by insertion of its genetic information into the host cell genome, becoming a 'prophage'. At a later stage, termed induction, the prophage may enter the lytic cycle. A widely used example is λ phage of *E. coli*.

The lytic cycle can be used to detect and quantify the number of phages in a sample. A known volume of sample is mixed with susceptible bacterial cells in molten soft agar medium (45–50 °C), then poured on top of a plate of the same medium, creating a thin layer of 'top agar'. The upper layer contains only half the normal amount of agar, to allow phages to diffuse through the medium and attach to susceptible cells. On incubation, the bacteria will grow throughout the agar to produce a homogeneous 'lawn' of cells, except in those parts of the plate where a phage particle has infected and lysed the cells to create a clear area, termed a plaque (Fig. 32.7). Each plaque is due to a single functional phage (i.e. a plaque-forming unit, or PFU). A count of the number of plaques can be used to give the number of phages in a particular sample (e.g. as $PFU\,ml^{-1}$), with appropriate correction for dilution and the volume of sample counted in an analogous manner to a bacterial plate count (p. 237). When counting plaques in phage assays you should view them against a black background to make them easier to see: mark each plaque with a spirit-based marker to ensure an accurate count. Temperate phages often produce cloudy plaques, because many of the infected cells will be lysogenised rather than lysed, creating turbidity within the plaque. Samples of material from within the plaque can be used to subculture the phage for further study, perhaps in a broth culture where the phages will cause widespread cell lysis and a decrease in turbidity. Alternatively, phages can be stored by adding chloroform to aqueous suspensions – this will prevent contamination by cellular micro-organisms. A similar approach can be used to detect and count animal or human viruses, using a monolayer of a susceptible animal or human cells.

Electron microscopy (EM, p. 159) provides an alternative approach to the detection of viruses, avoiding the requirement for culture of infected

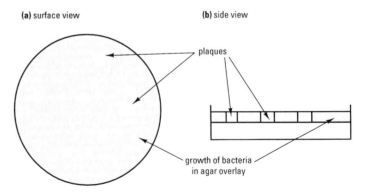

Fig. 32.7 Phage plaques in a 'lawn' of susceptible host bacterium.

host cells, and giving a faster result. However, it requires access to specialised equipment and expertise. EM counts are often higher than culture-based methods, for similar reasons to those described for bacteria (p. 237).

Labelling your plates and cultures

Labelling Petri plates – the following information should be recorded on the base:

- date
- growth medium
- your name or initials
- brief details of the organism, experimental treatment, etc.

Petri plates should always be labelled on the *base*, rather than the lid, using a permanent marker. Restrict your labelling to the outermost region of the plate to avoid problems when counting colonies, assessing growth, etc. After labelling, Petri plates usually are incubated upside-down in a temperature-controlled incubator (often at 37 °C) for an appropriate period (usually 18–72 h). Plates are also usually kept upside-down on the lab bench – following incubation, the base (containing medium and microbes) can then be lifted from the lid and examined.

Sources for further study

Anon. (1995) *Advisory Committee on Dangerous Pathogens: Categorisation of Biological Agents According to Hazard and Categories of Containment*, 4th edn. HSE Books, London.

Anon. (2004) *Advisory Committee on Dangerous Pathogens: Approved List of Biological Agents*. Available: http://www.hse.gov.uk/pubns/misc208.pdf Last accessed: 01/04/07.

Bartelt, M.A. (1999) *Diagnostic Bacteriology: A Study Guide*. Davis, New York.

Collins, C.H., Lyne, P.M., Grange, J.M. and Falkingham, J. (2004) *Collins and Lyne's Microbiological Methods*, 8th edn. Hodder Arnold, London.

Hawkey. P. and Lewis, D. (2004) *Medical Bacteriology: a Practical Approach*, 2nd edn. Oxford University Press, Oxford.

Isenberg, H.D. (1998) *Essential Procedures for Clinical Microbiology*. American Society for Microbiology, Washington, DC.

Rhodes, P.M. and Stanbury, P.F. (1997) *Applied Microbial Physiology: a Practical Approach*. Oxford University Press, Oxford.

Study exercises

32.1 Check that you understand how to label a plate culture. List the key aspects and then compare your list against the one provided on p. 514.

32.2 Decide on the best method of sterilisation. What would be the most appropriate method of sterilisation for the following items?
(a) A box of 100 plastic tips to be used with a pipettor.
(b) A 50 ml batch of blood, for use in 5% v/v blood agar plates.
(c) A 1 litre batch of MacConkey agar.
(d) Ten 5 ml glass pipettes.
(e) A microbiological wire, used for 'stab' cultures.
(f) A 10 ml sample of a heat-sensitive solution of an antibiotic, to be used as a component of a selective isolation medium.

32.3 Find out the biohazard classification (UK ACDP categorisation – Table 32.1) for the following microbes:
(a) *Salmonella typhimurium* (b) *Leptospira interrogans* (c) *Shigella dysenteriae* (type 1) (d) *E. coli* K12 (e) *E. coli* O157 (f) Human immunodeficiency virus (HIV) (g) *Cryptococcus neoformans* var. *neoformans* (h) *Mycobacterium tuberculosis* (i) *Lactobacillus plantarum* (j) Marburg virus.

32.4 Consider the advantages and disadvantages of spread plating and pour plating methods. Having read through this chapter, list up to six pros/cons of each plating method and compare your answers either with the list on pp. 514–15, or with those of other students, as a group exercise.

33 Isolating, identifying and naming microbes

Micro-organisms have a broad range of applications in the life sciences:

- Microbes are widely used as model systems, as they are often easier to study under controlled laboratory conditions than 'higher' organisms.
- Bacteria and fungi are used as sources of particular biomolecules, e.g. for the characterisation of a specific enzyme *in vitro*.
- Pathogenic microbes are studied at the molecular level, to find new methods of identification using biochemical 'markers' and to investigate the molecular basis of their pathogenicity.
- Microbes are used in biotechnology and industrial microbiology as sources of particular biomolecules, e.g. production of blood-clotting factors using yeast cells.

In your practical classes, you are likely to gain experience of a wide range of laboratory exercises involving microbes, particularly bacteria such as *Escherichia coli* (*E. coli*), using the procedures of sterile technique described in Chapter 32.

Alternatives to traditional culture-based methods – microbes can now be investigated by molecular methods, including the amplification of specific nucleic acid sequences by PCR (p. 441), and by immunoassay, e.g. ELISA (p. 261).

Sampling microbes

Microbes can be studied by taking samples for analysis, usually in one of the following ways:

- by direct examination of individual cells of a particular microbe, e.g. using fluorescence microscopy (p. 160);
- by isolating/purifying a particular species or related individuals of a taxonomic group, e.g. the faecal indicator bacterium *E. coli* in sea water;
- by studying microbial processes, rather than individual microbes, either *in situ* or in the laboratory.

Avoiding contamination during sampling – always remember that you are the most important source of contamination of field samples: components of the oral or skin microflora are the most likely contaminants.

Sampling techniques include the use of swabs, Sellotape strips and agar contact methods for sampling surfaces, bottles for aquatic habitats, plastic bags and corers for soils and sediments. A wide range of complex apparatus is available for accurately sampling water or soil at particular depths.

KEY POINT An important feature of all microbiological sampling protocols is that the sampling apparatus must be sterile; strict aseptic technique must be used throughout the sampling process (see Chapter 32).

SAFETY NOTE When working with newly isolated microbes, you should always treat them as potentially harmful (p. 203) until they have been identified.

The sampling method must minimise the chance of contamination with microbes from other sources, especially the exterior of the sampling apparatus and the operator. For example, if you are sampling an aquatic habitat, stand downstream of the sampling site. A portable Bunsen burner or spirit lamp can be used to assist sterile technique during field sampling, e.g. while flaming a loop (p. 203). Alternatively, use disposable single-use plastic loops.

Process the sample as quickly as possible to minimise any changes in microbiological status. As a general guideline, many procedures require samples to be analysed within 6 h of collection. Changes in aeration, pH and water content may occur after collection. Some microbes are more susceptible to such effects, e.g. anaerobic bacteria may not survive if the sample is exposed

Isolating, Identifying and Naming Microbes

Sub-sampling *– to minimise the effects of changes in temperature, aeration and water status during transportation, a primary sample may be returned to the laboratory, where the working sample (sub-sample) is then taken (e.g. from the centre of a large block of soil).*

to air. Sunlight can also inactivate bacteria; samples should be shielded from direct sunlight during collection and transport to the laboratory.

Soil and water samples are often kept cool (at 0–5 °C) during transport to the laboratory. In contrast, some microbes adapted to grow in association with warm-blooded animals may be damaged by low temperatures. An alternative approach is to keep the sample near the ambient sampling temperature using an insulated vessel (e.g. Thermos flask).

Isolating a particular microbe

Several different approaches may be used to obtain microbes in pure culture. The choice of method will depend upon the microbe to be isolated: some organisms are relatively easy to isolate, while others require more involved procedures.

Separation methods

Obtaining a pure culture *– if a single colony from a primary isolation medium is used to prepare a streak dilution plate and all the colonies on the second plate appear identical, then a pure culture has been established. Otherwise, you cannot assume that your culture is pure and you should repeat the subculture until you have a pure culture.*

Most microbial isolation procedures involve some form of separation to obtain individual microbial cells. The most common approach is to use an agar-based medium for primary isolation, with streak dilution, spread plating or pour plating to produce single colonies, each derived from a single type of microbe (pp. 206–7). It is often necessary to dilute samples before isolation, so that a small number of individual microbial cells are transferred to the growth medium. Strict serial dilution (pp. 137–8) of a known amount of sample is needed for quantitative work.

 KEY POINT *If your aim is to isolate a particular microbe, perhaps for further investigation, you will need to subculture individual colonies from the primary isolation plate to establish a pure culture, also known as an axenic culture.*

Pure cultures of most microbes can be maintained indefinitely, using sterile technique and microbial culture methods (Chapter 32).

Other separation techniques include:

Using a sonicator *– minimise heat damage with short treatment 'bursts' (typically up to 1 min), cooling the sample between bursts, e.g. using ice.*

- Dilution to extinction. This involves diluting the sample to such an extent that only one or two microbes are present per millilitre: small volumes of this dilution are then transferred to a liquid growth medium (broth). After incubation, most of the tubes will show no growth, but some tubes will show growth, having been inoculated with a single viable microbe at the outset. This should give a pure culture, though it is wasteful of resources.
- Sonication/homogenisation. This is useful for separating individual microbial cells from each other and from inert particles, prior to isolation. However, some decrease in viability is likely.
- Filtration. This can be useful where the number of microbes is low. Samples can be passed through a sterile cellulose ester filter (pore size 0.2 μm), which is then incubated on the surface of an appropriate solidified medium. Sieving and filtration techniques (p. 139) are often used in soil microbiology to subdivide a sample on the basis of particle size.
- Micromanipulation. It may be possible to separate a microbe from contaminants using a micropipette and dissecting microscope. The microbe can then be transferred to an appropriate growth medium, to give a pure culture. However, this is rarely an easy task for the novice.

Definitions

Psychrophile – a microbe with an optimum temperature for growth of <20 °C (literally 'cold-loving').

Psychrotroph – a microbe with an optimum temperature for growth of ⩾20 °C, but capable of growing at lower temperature, typically 0–5 °C (literally 'cold-feeding').

Thermophile – a microbe with an optimum growth temperature of >45 °C (literally 'heat-loving').

Mesophile – a microbe with an optimum growth temperature of 20–45 °C (literally 'middle-loving').

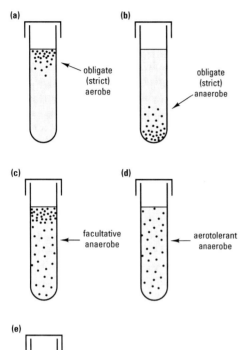

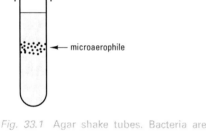

Fig. 33.1 Agar shake tubes. Bacteria are suspended in molten agar at 45–50 °C and allowed to cool. The growth pattern after incubation reflects the atmospheric (oxygen) requirements of the bacterium.

Definitions

Facultative anaerobe – a microbe that grows by aerobic respiration when oxygen is present, switching to fermentation under anaerobic conditions.

Aerotolerant anaerobe – a microbe that grows by fermentation, but which is insensitive to air/oxygen (in contrast to strict anaerobes, which are typically killed by exposure to air/oxygen).

Capnophile – a microbe that thrives in the presence of high levels of atmospheric carbon dioxide.

- Motility. Phototactic microbes (including photosynthetic flagellates and motile cyanobacteria) will move towards a light source; heterotrophic flagellate bacteria will move through a filter of appropriate pore size into a nutrient solution, or away from unfavourable conditions (chemotaxis).

Selective and enrichment methods

Laboratory incubation under selective/enrichment conditions will allow particular microbes to be isolated in pure culture. Selective and enrichment techniques can be considered together, since they both enhance the growth of a particular microbe when compared with its competitors and they are often combined in specific media (Box 33.1).

The fundamental difference between selective and enrichment techniques is that the former use growth conditions unfavourable for competitors while the latter provide improved growth conditions for the chosen microbes.

> **KEY POINT** Selective methods are based on the use of physico-chemical conditions that will permit the growth of a particular group of microbes while inhibiting others. Enrichment techniques encourage the growth of certain bacteria, usually by providing additional nutrients in the growth medium.

Methods based on specific physical conditions include:

- Temperature. Psychrophilic and psychrotrophic microbes can be isolated by incubating the growth medium at 4 °C, while thermophilic microbes require temperatures above 45 °C for isolation. Short-term heat treatment of samples can be used to select for endospore-forming bacteria, e.g. 70–80 °C for 5–15 min, prior to isolation.
- Atmosphere. Many eukaryotic microbes are obligate aerobes, requiring an adequate supply of oxygen to grow. Bacteria vary in their responses to oxygen: obligate anaerobes are the most demanding, growing only under anaerobic conditions (e.g. in an anaerobic cabinet or jar). Oxygen requirements can be determined using the agar shake tube method as part of the isolation procedure (Fig. 33.1). Some pathogenic bacteria grow best in an atmosphere with a reduced oxygen status and increased CO_2 concentration: such carboxyphilic bacteria (capnophiles) are grown in an incubator where the gas composition can be adjusted.
- Centrifugation. This can be used to separate buoyant microbes from their non-buoyant counterparts – on centrifugation, such organisms will collect at the surface while the remaining microbes will sediment. Alternatively, density gradient methods may be used (p. 298). Centrifugation can be combined with repeated washing, to separate microbes from contaminants.
- Ultraviolet irradiation. Some microbes are tolerant of UV treatment and can be selected by exposing samples to UV light. However, the survivors may show a greater rate of mutation.
- Illumination. Samples may be enriched for cyanobacteria and microalgae by incubation under a suitable light regime. For dilute samples, where the number of photosynthetic microbes is too low to give the sample any visible green coloration, there is a risk of photoinhibition and loss of viability if the irradiance is too high. Such samples need shading during initial growth.

Box 33.1 Differential media for bacterial isolation: an example

MacConkey agar is both a selective and a differential medium, useful for the isolation and identification of intestinal Gram-negative bacteria. Each component in the medium has a particular role:

- **Peptone:** (a meat digest) provides a rich source of complex organic nutrients, to support the growth of non-exacting bacteria.
- **Bile salts:** toxic to most microbes apart from those growing in the intestinal tract (selective agent).
- **Lactose:** present as an additional, specific carbon source (enrichment agent).
- **Neutral red:** a pH indicator dye, to show the decrease in pH which accompanies the breakdown of lactose (differential agent).
- **Crystal violet:** selectively inhibits the growth of Gram-positive bacteria. (This component is only present in certain formulations of MacConkey agar.)

Any intestinal Gram-negative bacterium capable of fermenting lactose will grow on MacConkey agar to produce large red–purple colonies, the red colouration being due to the neutral red indicator under acidic conditions while the purple colouration, often accompanied by a metallic sheen, is due to the precipitation of bile salts and crystal violet at low pH.

In contrast, enteric Gram-negative bacteria unable to metabolise lactose will give colonies with no obvious pigmentation. This differential medium has been particularly useful in medical microbiology, since many enteric bacteria are unable to ferment lactose (e.g. *Salmonella*, *Shigella*) while others metabolise this carbohydrate (e.g. *Escherichia coli*, *Klebsiella* spp.). Colonial morphology (Fig. 33.2) on such a medium can give an experienced bacteriologist important clues to the identity of an organism, e.g. capsulate *Klebsiella* spp. characteristically produce large, convex, mucoid colonies with a weak pink colouration, due to the fermentation of lactose, while *E. coli* produces smaller, flattened colonies with a stronger red colouration and a metallic sheen.

Chemical methods form the mainstay of bacteriological isolation techniques and various media have been developed for the isolation of specific groups of bacteria. The chemicals involved can be subdivided into the following groups:

- Nutrients that encourage the growth of certain microbes: including the addition of a particular carbon source, or specific inorganic nutrients.
- Selectively toxic substances: for example, salt-tolerant, Gram-positive cocci can be grown in a medium containing 7.5% w/v NaCl, which prevents the growth of most common heterotrophic bacteria. Several media include dyes as selective agents, particularly against Gram-positive bacteria.
- Antibiotics: for example, the use of antibacterial agents (e.g. penicillin, streptomycin, chloramphenicol) in media designed to isolate fungi, or the use of antifungal agents (e.g. cycloheximide, nystatin) in bacterial media. Some antibacterial agents show a narrow spectrum of toxicity and these can be incorporated into selective isolation media for resistant bacteria, e.g. metronidazole for anaerobic bacteria.
- Substances that affect the pH of the medium: for example, the use of alkaline peptone water at pH 8.6 for the isolation of *Vibrio* spp.

Example Cellulolytic bacteria can be isolated from soil or water using a growth medium that contains cellulose as the major source of carbon.

Example Slow-growing *Legionella pneumophila* can be isolated from water samples using media containing the antibiotics vancomycin, polymyxin and cycloheximide to suppress other microbes.

 KEY POINT Note that subcultures from a primary isolation medium must be grown in a non-selective medium, to confirm the purity of the isolate.

Many of the selective and enrichment media used in bacteriology are able to distinguish between different types of bacteria: such media are termed differential media or diagnostic media and they are often used in the preliminary stages of an identification procedure. Box 33.1 gives details of the constituents of MacConkey medium, a selective, differential

Table 33.1 Selective agents in bacteriological media

Substance	Selective for
Azide salts	*Enterococcus* spp.
Bile salts	Intestinal bacteria
Brilliant green	Gram-negative bacteria
Gentian violet	Gram-negative bacteria
Lauryl sulphate	Gram-negative bacteria
Methyl violet	*Vibrio* spp.
Malachite green	*Mycobacterium*
Polymyxin	*Bacillus* spp.
Sodium selenite	*Salmonella* spp.
Sodium chloride	Halotolerant bacteria *Staphylococcus aureus*
Sodium tetrathionate	*Salmonella* spp.
Tergitol/surfactant	Intestinal bacteria
Trypan blue	*Streptococcus* spp.

medium used in clinical microbiology (e.g. for the isolation of certain faecal bacteria), while Table 33.1 gives details of other selective agents.

Further details on methods can be found in Collins *et al.* (2004) or Hurst *et al.* (2002). Note that isolation procedures for a particular microbe often combine several of the techniques described above. For instance, a protocol for isolating a food-poisoning bacterium from a foodstuff might involve:

1. Homogenisation of a known amount of sample in a suitable diluent.
2. Serial decimal dilution.
3. Separation procedures using spread or pour plates to quantify the number of bacteria of a particular type present in the foodstuff and provide a plate count (p. 237).
4. Selective/enrichment procedures, e.g. specific media/temperatures/atmospheric conditions, depending on the bacteria to be isolated.
5. Confirmation of identity: any organism growing on a primary isolation medium would require subculture and further tests, to confirm the preliminary identification as discussed below.

Identifying a particular microbe

Most of the methods described in this chapter were developed for the identification of bacteria, and bacterial examples are used to illustrate the principles involved. While the basic techniques are equally applicable to other types of microbe, the identification systems for some protozoa, fungi and algae rely predominantly on microscopic appearance. Identification of viruses requires electron microscopy or immunological techniques.

KEY POINT Identification of bacteria is often based on a combination of a number of different features, including growth characteristics, microscopic examination, physiological or biochemical characterisation, and, where necessary, immunological tests.

Direct observation

Once a microbe has been isolated and cultured in the laboratory (Chapter 32), the visual appearance of individual colonies on the surface of a solidified medium may provide useful information. Bacteria typically produce smooth, glistening colonies, varying in diameter from <1 mm to >1 cm. Actinomycete colonies are often <1 cm, with a shrivelled, powdery surface. Filamentous fungi usually grow as large, spreading mycelia with a matt appearance and are identified by microscopy, using the morphological characteristics of their reproductive structures. Yeasts produce small, glistening colonies: identification usually involves microscopy, combined with physiological and biochemical tests similar to those used for bacteria.

Colony characteristics

When measuring colony size, choose a typical colony, well spaced from any others as colony size is affected by competition for nutrients. The characteristics of a microbial colony on a particular medium include:

- Size: some bacteria produce punctiform colonies, with a diameter of less than 1 mm, while motile bacteria may spread over the entire plate.

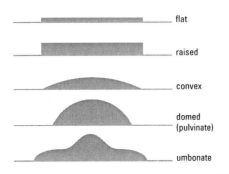

Fig. 33.2 Colony elevations (cross-sectional profile).

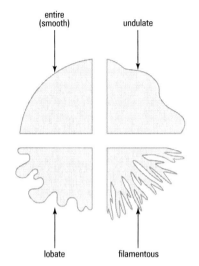

Fig. 33.3 Colony margins (surface view).

 SAFETY NOTE *Working with mould cultures – never attempt to smell mould cultures, because of the risk of inhaling large numbers of spores.*

Using cell shape in microbial identification – many bacteria are pleomorphic, varying in size and shape according to the growth conditions and the age of the culture: thus, other characteristics are required for identification.

 SAFETY NOTE *The Gram staining procedure involves toxic dyes and flammable solvents: avoid skin contact and extinguish any naked flames (e.g. nearby Bunsens).*

Assessing the Gram status of an unknown bacterium – if a pure culture gives both Gram-positive and Gram-negative cells, identical in size and shape, it can be regarded as a Gram-positive organism that is demonstrating Gram-variability.

- Form: colonies may be circular, irregular, lenticular (spindle-shaped) or filamentous.
- Elevation: colonies may be flat, raised, convex, etc. (see Fig. 33.2).
- Margin: the edge of a colony may be entire (smooth) or more distinctive, e.g. undulate, lobate or filamentous (Fig. 33.3).
- Consistency: colonies may be viscous (or mucoid), butyrous (of similar consistency to butter) or friable (dry and granular), etc.
- Colour: some bacteria produce characteristic pigments. A few pigments are fluorescent under UV light.
- Optical properties: colonies may be translucent or opaque.
- Haemolytic reactions on blood agar: many pathogenic bacteria produce characteristic zones of haemolysis. Alpha haemolysis is a partial breakdown of the haemoglobin from the erythrocytes, producing a green zone around the colony, while beta haemolysis is the complete destruction of haemoglobin, producing a clear zone.
- Odour: some actinomycetes and cyanobacteria produce earthy odours, while certain bacteria and yeasts produce fruity or 'off' odours. However, odour is not a reliable characteristic in bacterial identification and smelling plates creates a risk of inhaling microbes (p. 206).

Microscopic examination – cell shape

Bacteria are usually observed using an oil immersion objective at a total magnification of ×1000 (p. 166). Bacteria are subdivided into the following groups, based on their cell shape:

- Cocci (singular, coccus): spherical, or almost spherical, cells, sometimes growing in pairs (diplococci), chains or clumps.
- Rods: straight, cylindrical cells of variable length with flattened, tapered or rounded ends – sometimes termed bacilli. Short rods are sometimes called cocco-bacilli.
- Curved rods: the curvature varies according to the organism, from short curved rods, sometimes tapered at one end, to spiral shapes.
- Branched filaments: characteristic of actinomycete bacteria.

Gram staining

This is the most important differential staining technique in bacteriology (Box 33.2 gives details). It enables us to divide bacteria into two distinct groups, Gram-positive and Gram-negative, according to a particular staining procedure (the technique is given a capital letter, since it is named after its originator, H.C. Gram). The basis of the staining reaction is the different structure of the cell walls of Gram-positive and Gram-negative bacteria. Heat fixation of air-dried bacteria causes some shrinkage, but cells retain their shape: to measure cell dimensions use a chemical fixative.

Gram staining should be carried out using light smears of young, active cultures, since older cultures may give variable results. In particular, certain Gram-positive bacteria may stain Gram-negative if older cultures are used. This Gram-variability is due to autolytic changes in the cell wall of Gram-positive bacteria. Developing spores are often visible as unstained areas within older vegetative cells of *Bacillus* and *Clostridium*. Other stains are required to demonstrate spores, capsules or flagella (p. 167).

Box 33.2 Preparation of a heat-fixed, Gram-stained smear

Preparation of a heat-fixed smear
The following procedure will provide you with a thin film of bacteria on a microscope slide, for staining.

1. **Take a clean microscope slide and pass it through a Bunsen flame twice** to ensure it is free of grease. Allow to cool.
2. **Using a sterile inoculating loop, place a single drop of water in the centre of the slide and then mix in a small amount of sample** from a single bacterial colony with the drop, until the suspension is slightly turbid. Smear the suspension over the central area of the slide, to form a thin film. For liquid cultures, use a single drop of culture fluid, spread in a similar manner.
3. **Allow to air-dry at room temperature**, or high above a Bunsen flame: air-drying must proceed gently or the cells will shrink and become distorted.
4. **Fix the air-dried film by passage through a Bunsen flame.** Using a slide holder or forceps, pass the slide, film side up, rapidly through the hottest part of the flame (just above the blue cone). The temperature of the slide should be just too hot for comfort on the back of your hand: note that you must not overheat the slide or you may burn yourself (you will also ruin the preparation).
5. **Allow to cool**: the smear is now ready for staining.

Gram-staining procedure
The version given here is a modification of the Hucker method, since acetone is used to decolourise the smear. Note that some of the staining solutions used are flammable, especially the acetone decolourising solvent: you must make sure that all Bunsens are turned off during staining. The procedure should be carried out with the slides suspended over a sink, using a staining rack.

1. **Flood a heat-fixed smear with 2% w/v crystal violet in 20% v/v ethanol:water** and leave for 1 min.
2. **Pour off the crystal violet and rinse briefly with tap water. Flood with Gram's iodine** (2 g KI and 1 g I_2 in 300 ml water) for 1 min.
3. **Rinse briefly with tap water** and leave the tap running gently.
4. **Tilt the slide and decolourise with acetone** for 2–3 s: acetone should be added dropwise to the slide until no colour appears in the effluent. This step is critical, since acetone is a powerful decolourising solvent and must not be left in contact with the slide for too long.
5. **Immediately immerse the smear in a gentle stream of tap water** to remove the acetone.
6. **Pour off the water and counterstain for 10–15 s using 2.5% w/v safranin** in 95% v/v ethanol:water.
7. **Pour off the counterstain, rinse briefly with tap water, then dry the smear** by blotting gently with absorbent paper: all traces of water must be removed before the stained smear is examined microscopically.
8. **Place a small drop of immersion oil on the stained smear: examine directly** (without a coverslip) using an oil-immersion objective (p. 166).

Gram-positive bacteria retain the crystal violet (primary stain) and appear purple, while Gram-negative bacteria are decolourised by acetone and counterstained by the safranin, appearing pink or red when viewed microscopically.

Other decolourising solvents are sometimes used, including ethanol:water, ethyl ether:acetone and acetone:alcohol mixtures. The time of decolourisation must be adjusted, depending upon the strength of the solvents used, e.g. 95% v/v ethanol:water is less powerful than acetone, requiring around 30 s to decolourise a smear.

Using the hanging drop technique – place a drop of bacterial suspension on a coverslip and invert over a cavity slide so that the drop does not make contact with the slide: motile aerobes are best observed at the edge of the droplet, where oxygen is most abundant.

Motility
Wet mounts can be prepared by placing a small drop of bacterial suspension on a clean, degreased slide, adding a coverslip and examining the film by light microscopy without delay. For aerobes, areas near air bubbles or by the edge of the coverslip give best results, while anaerobes show greatest motility in the centre of the preparation, with rapid loss of motility due to oxygen toxicity.

Prepare wet mounts using young cultures in exponential growth in a liquid medium (p. 233). It is best to work with cultures grown at 20 or 25 °C, since those grown at 37 °C may not be actively motile on cooling to room temperature. It is essential to distinguish between the following:

- True motility, due to the presence of flagella: bacteria dart around the field of view, changing direction in zigzag, tumbling movements.

Assessing motility – if you have not seen bacterial motility before, it is worth comparing your unknown bacterium to a positive and a negative control.

- Brownian motion: non-motile bacteria show a localised, vibratory, random motion, due to bombardment of bacterial cells by molecules in the solution.
- Passive motion, due to currents within the suspension: all cells will be swept in the same direction at a similar rate of movement.
- Gliding motility: a slower, intermittent movement, parallel to the longitudinal axis of the cell, requiring contact with a solid surface.

Basic laboratory tests

At least two simple biochemical tests are usually performed:

1. Oxidase test This identifies cytochrome c oxidase, an enzyme found in obligate aerobic bacteria. Soak a small piece of filter paper in a fresh solution of 1% (w/v) N-N-N'-N'-tetramethyl-p-phenylenediamine dihydrochloride on a clean microscope slide. Rub a small amount from the surface of a young, active colony onto the filter paper using a glass rod, a *plastic* loop or a wooden applicator stick: a purple-blue colour within 10 s is a positive result.

Performing the oxidase test – never use a nichrome wire loop, as this will react with the oxidase reagent, giving a false positive result.

2. Catalase test This identifies catalase, an enzyme found in obligate aerobes and in most facultative anaerobes, which catalyses the breakdown of hydrogen peroxide into water and oxygen ($2H_2O_2 \rightarrow 2H_2O + O_2$). Transfer a small sample of your unknown bacterium onto a coverslip using a disposable plastic loop or glass rod. Invert onto a drop of hydrogen peroxide: the appearance of bubbles within 30 s is a positive reaction. This method minimises the dangers from aerosols formed when gas bubbles burst.

 SAFETY NOTE *The catalase and oxidase reagents are irritants and could be harmful if swallowed. Avoid skin contact and ingestion.*

The oxidase and catalase tests effectively allow us to subdivide bacteria on the basis of their oxygen requirements, without using agar shake cultures (p. 212) and overnight incubation, since, for the most part:

- obligate aerobes will be oxidase and catalase positive;
- facultative anaerobes are generally oxidase negative and catalase positive;
- microaerophilic bacteria, aerotolerant anaerobes and strict (obligate) anaerobes will be oxidase and catalase negative – the latter group will grow only under anaerobic conditions (p. 212).

Avoiding false negatives – ensure you use sufficient material during oxidase and catalase testing, otherwise you may obtain a false negative result: a clearly visible 'clump' of bacteria should be used.

Once you have reached this stage (colony characteristics, motility, shape, Gram reaction, oxidase and catalase status) it may be possible to make a tentative identification, at least for certain Gram-positive bacteria, at the generic level. To identify Gram-negative bacteria, particularly the oxidase-negative, catalase-positive rods, further tests are required.

Identification tables: further laboratory tests

Bacteria are asexual organisms and strains of the same species may give different results for individual biochemical/physiological tests. This variation is allowed for in identification tables, based on the results of a large number of tests. Identification tables are often used for particular subgroups of bacteria, after Gram staining and basic laboratory tests have been performed: an example is shown in Table 33.2.

Carbohydrate utilisation tests and isolation media – many diagnostic agar-based media incorporate one or more specific carbohydrates and pH indicator dyes, thereby providing additional information as part of the isolation procedure (Box 33.1).

Table 33.2 Identification table for selected Gram-negative rods

Bacterium	Biochemical test								
	1	2	3	4	5	6	7	8	9
Escherichia coli	v	+	−	+	−	v	v	+	−
Proteus mirabilis	−	−	v	−	+	−	+	−	+
Morganella morganii	−	−	−	−	+	−	+	+	−
Vibrio parahaemolyticus	−	+	v	−	−	+	+	+	−
Salmonella spp.	−	+	v	−	−	+	+	−	+

Key to biochemical tests and symbols
1. sucrose utilisation
2. mannitol utilisation
3. citrate utilisation
4. β-galactosidase activity
5. urease activity
6. lysine decarboxylase activity
7. ornithine decarboxylase activity
8. indole production
9. H_2S production
+, >90% of strains tested positive
−, <10% of strains tested positive
v, 10–90% of strains tested positive

A large number of specific biochemical and physiological tests are used in bacterial identification, including:

- Carbohydrate utilisation tests. Some bacteria can use a particular carbohydrate as a carbon and energy source. Acidic end-products can be identified using a pH indicator dye (p. 153) while CO_2 is detected in liquid culture using an inverted small test tube (Durham tube, Fig. 33.4). Aerobic breakdown (via respiration) is termed oxidation while anaerobic breakdown is known as fermentation. Identification tables usually incorporate tests for several different carbohydrates, e.g. Table 33.2.
- Enzyme tests. Most of these incorporate a substance which changes colour if the enzyme is present, e.g. a pH indicator, or a chromogenic or fluorogenic substrate (p. 382).
- Tests for specific end-products of metabolism, e.g. the production of indole due to the metabolic breakdown of the amino acid tryptophan, or H_2S from sulphur-containing amino acids.

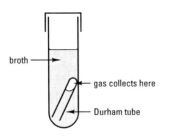

Fig. 33.4 Durham tube in carbohydrate utilisation broth. Air within the Durham tube is replaced by broth during the autoclaving procedure.

Identification kits

Some biochemical tests are now supplied in kit form, e.g. the API 20E system incorporates 20 tests within a sterile plastic strip (Fig. 33.5). After inoculation and overnight incubation, the results of the tests are converted into a seven-digit code, for comparison with known bacteria using either a reference book (the Analytical Profile Index), or a computer program. While kit identification systems save time and labour, they are more expensive and less flexible than conventional biochemical tests.

Molecular approaches to microbial identification – several novel methods of detection and identification are based on nucleic acid techniques, including use of the PCR and Southern blotting to detect particular microbes (pp. 433–4).

Fig. 33.5 Example of a bacterial identification kit (API 20E). Property of bioMérieux S.A./ photographer: Andrea Bannuscher.

Immunological tests

Tests used in diagnostic microbiology include:

- Agglutination tests: based on the reaction between specific antibodies and a particular bacterium (p. 258). These tests are particularly useful for subdividing biochemically similar bacteria.
- Fluorescent antibody tests: the reaction between a labelled antibody and a particular bacterium can be visualised using UV microscopy. The direct fluorescent antibody test uses fluorescein isothiocyanate as the label.
- Enzyme-linked immunoassay tests using antibodies labelled with a particular enzyme, e.g. the double antibody sandwich ELISA or competitive ELISA tests (p. 261).

While such tests can give specific and accurate confirmation of the identity of a bacterium under controlled laboratory conditions, they are often too expensive and time-consuming for routine identification purposes, especially when large numbers of tests are required.

Typing methods

The identification of bacteria at subspecies level is known as typing: this is usually done in a specialist laboratory, e.g. as part of an epidemiological study to establish the source of an infection. Various methods are used:

- Antigen typing or serotyping is based on immunological tests.
- Phage typing is based on the susceptibility of different strains to certain bacterial viruses (phages).
- Biotyping is based on biochemical differences between different strains, e.g. enzyme profiles or antibiotic resistance screening ('antibiograms').
- Bacteriocin typing: bacteriocins are proteins released by bacteria which inhibit the growth of other members of the same species.

Naming microbes

The use of scientific names is fundamental to all aspects of biological science since it aims to provide a system of identification which is precise, fixed and of universal application. Without such a system, comparative studies would be impossible.

There are two possible bases for such classification:

- Phenetic taxonomy, which involves grouping on the basis of phenotypic similarity, frequently using complex statistical techniques to obtain objective measures of similarity. The characters used have been largely morphological and anatomical, but biochemical, cytological and other characters are increasingly used, especially for microbes, where structural characteristics are few.
- Phylogenetic (= phyletic) taxonomy, which involves grouping on the basis of presumed evolutionary, and therefore genetic, relationships.

These two systems are often broadly similar in outcome, since closely related organisms are usually fairly similar to each other and because judgements of evolutionary relationships are usually themselves based upon similarities. The situation is made more complex by phenomena such as convergent and divergent evolution; phylogenetic classifications are also liable to subjective bias.

Practical applications of bacterial typing – E. coli *O157:H7* is a serotype of this bacterium that is capable of causing severe human disease: it can be identified on the basis of an agglutination reaction with an appropriate antiserum. Other applications include tracking the development and spread of particular types of antibiotic-resistant bacteria in hospitals and the community.

Definitions

Systematics – the study of the diversity of living organisms and of the evolutionary relationships between them.

Classification (taxonomy) – the study of the theory and methods of organisation of taxa and therefore, a part of systematics.

Taxon (plural taxa) – an assemblage of organisms sharing some basic features.

Nomenclature – the allocation of names to taxa.

Identification – the placing of organisms into taxa.

The basis of classification

No single, simple definition of a species is possible, but there are two generally used definitions:

- A group of organisms capable of interbreeding and producing fertile offspring – this, however, excludes all asexual organisms, such as bacteria and some other microbes.
- A group of organisms showing a close similarity in phenotypic characteristics – this would include morphological, anatomical, biochemical, ecological and life history characters or, increasingly, similarities in nucleic acid sequence.

KEY POINT The basic unit of classification is the species, which represents a group of recognisably similar individuals, clearly distinct from other such groups.

DNA-based definition of a species – in bacteriology, members of a single species can be characterised by sharing at least 70% DNA, based on nucleic acid hybridisation studies (p. 433).

When species are compared, groups of species may show a number of features in common; they are then arranged into larger groupings known as genera (singular, genus). Cellular microbes (e.g. bacteria, archaea, fungi, protozoa and algae) are given two Latin terms to identify their genus and species (a Latin 'binomial'), for example *Escherichia coli*. All scientific names of organisms are either Latin or are treated as Latin, written in the Latin alphabet and subject to the rules of Latin grammar. Consequently, you must be very precise in your use of such names. When used in a formal scientific context, the specific name can be followed by the authority on which the name is based, i.e. the name of the person describing that species, and the year in which it was first described. For example, the full, formal name of baker's yeast is *Saccharomyces cerevisiae* Meyen ex Hansen 1883. Where specified, additional terms may follow the species name to indicate the type or subspecies, e.g. *Shigella dysenteriae* type 1. Many micro-organisms are now referred to by their generic and specific names followed by a culture collection reference number, e.g. *Bacillus subtilis* NCTC 10400, where NCTC stands for the National Collection of Type Cultures and 10400 is the reference number of that strain in the collection.

Writing taxonomic names – always underline (handwritten text) or italicise (word-processed text) genus and species names to avoid confusion: thus bacillus is a descriptive term for a rod-shaped bacterium, while <u>Bacillus</u> is a generic name.

After first use in a text, the genus name may be abbreviated to a single letter, e.g. *E. coli*, as long as this does not cause confusion with other genera; for example, with *Escherichia coli* and *Enterococcus faecalis*, these could be abbreviated to *Esch. coli* and *Ent. faecalis*, to avoid confusion. Where the species name is unknown, the (non-italicised) abbreviation 'sp.' (singular) or 'spp.' (plural) should be used, e.g. *Enterococcus* sp. denotes a single unknown species of the genus *Enterococcus*.

Understanding microbiological terms – the term *strain* is widely used, particularly in the context of the practice of lodging microbiological strains with culture collections, while the term *isolate* is often used for a pure culture derived from a natural (wild) population. Cell lines (Chapter 34) are often given code names and/or reference numbers.

KEY POINT While the use of taxa above those of the genus are somewhat subjective and may vary between different authorities, the use of genus and species names are governed by strict internationally agreed conventions called Codes of Nomenclature (e.g. Lapage, 1992 for bacteria).

Understanding the hierarchical taxonomic system – the taxonomic groups, in decreasing level, are: kingdom, phylum or division, class, order, family, genus, species. An example of a full hierarchical classification would be: Bacteria (monera); Gracilicutes; Scotobacteria; Pseudomonadales; Pseudomonadaceae; <u>Pseudomonas</u>; <u>Pseudomonas aeruginosa</u>. Note that the names of orders normally end in '-ales' while family names usually end in '-aceae' and that all names of taxa apart from species begin with a capital letter.

The classification and nomenclature of viruses are less advanced than for cellular organisms and the current nomenclature has been arrived at on a piecemeal, *ad hoc* basis. The International Committee for Virus Taxonomy proposed a unified classification system, dividing viruses into 50 families on the basis of: host preference, nucleic acid type

Example The virus that causes tobacco mosaic disease belongs to the *Tobamovirus* group and can be referred to as tobacco mosaic tobamovirus, tobacco mosaic virus or TMV.

(i.e. DNA or RNA), whether the nucleic acid is single or double stranded and the presence or absence of a surrounding envelope. Many viruses are still referred to by trivial names or by code-names (sigla), e.g. the bacterial viruses ϕX174, T4, etc. Many of the names used reflect the diseases caused by the virus. Often, a three-letter abbreviation is used, e.g. HIV (for human immunodeficiency virus), TMV (for tobacco mosaic virus).

Text references

Collins, C.H., Lyne, P.M. and Grange, J.M. (2004) *Microbiological Methods*, 8th edn. Hodder-Arnold, London.

Hurst, C.J., Crawford, R.L. and Knudsen, G.R. (2002) *Manual of Environmental Microbiology*, 2nd edn. American Society for Microbiology, Washington, DC.

Lapage, S.P. (1992) *International Code of Nomenclature for Bacteria: 1990 revision*. American Society for Microbiology, Washington, DC.

Sources for further study

Atlas, R.M. (2004) *Handbook of Microbiological Media*, 3rd edn. CRC Press, Boca Raton.
[Gives details of culture media for a broad range of microbes]

Barrow, G.I. and Feltham, R.K.A. (1993) *Cowan and Steel's Manual for the Identification of Medical Bacteria*, 3rd edn. Cambridge University Press, Cambridge.
[A standard reference work for the identification of clinically important bacteria]

Boone, D.R. *Naming a New Procaryotic Taxon*. Available: http://methanogens.pdx.edu/naming.html Last accessed: 01/04/07.
[Succinct indication of procedures for naming a new prokaryote species]

Buchen-Osmond, C. (2002) Universal Virus Database of the International Committee on Taxonomy of Viruses. Available: http://www.ncbi.nlm.nih.gov/ICTVdb/. Last accessed: 01/04/07.

Cullimore, D.R. (2000) *Practical Atlas for Bacterial Identification*. CRC Press, Boca Raton.
[Detailed coverage of practical methods applicable to a wide range of bacteria]

Eaton, A.D., Clesceri, L.S., Rice, E.W. and Greenberg, A.E. (2005) *Standard Methods for Examination of Water and Wastewater*, 21st edn. American Public Health Association, Washington, DC.
[Gives standard US protocols for a range of indicator bacteria]

Fauquet, C.M., Mayo, M.A., Maniloff, J., Desselberger, U. and Ball, L.A. (2005) *Virus Taxonomy. Eighth Report of the International Committee on the Taxonomy of Viruses*. Elsevier, London.

Fisher, F., Cook, N.B., Fisher, F.W. and Kaszczuk, S. (1998) *Fundamentals of Diagnostic Mycology*. Saunders, Philadelphia.
[Covers identification techniques for medically important fungi]

Fox, A. (ed.) *Journal of Microbiological Methods*. Elsevier, London (available through Science Direct at: http://www.sciencedirect.com/)
[Provides information on novel developments in all aspects of microbiological methods, including microbial culture/isolation and molecular approaches]

Holt, J.G. (1994) *Bergey's Manual of Determinative Bacteriology*, 9th edn. Williams and Wilkins, Baltimore.

Institute of Biology (2000) *Biological Nomenclature: Recommendations on Terms, Units and Symbols*. Institute of Biology, London.
[Includes sections on taxonomy and classification of organisms]

Jeffrey, C. (1989) *Biological Nomenclature*, 3rd edn. Edward Arnold, London.

Levett, P.N. (ed.) (1991) *Anaerobic Microbiology: A Practical Approach*. IRL Press, Oxford.
[Provides details of the methods used to isolate and culture anaerobic micro-organisms]

Macfaddin, J.F. (2000) *Biochemical Tests for the Identification of Medical Bacteria*, 3rd edn. Williams and Wilkins, Philadelphia.
[Explains the operating principles underlying most of the biochemical tests in routine use in diagnostic bacteriology]

Pepper, I.L. and Gerba, C.P. (2005) *Environmental Microbiology: A Laboratory Manual*, 2nd edn. Academic Press, New York.

Rochelle, P.A. (2001) *Environmental Molecular Microbiology: Protocols and Applications*. Horizon Scientific Press, Norwich.
[Covers molecular approaches to the detection and identification of microbes in environmental samples]

Study exercises

33.1 Plan a collection/sampling strategy for a target group of microbes. What approaches might you take in the following instances?
  (a) Collecting representatives of the normal skin microflora.
  (b) Sampling psychrotrophic anaerobes from the subsurface mud of an estuary.
  (c) Collecting a sample of sea water to enumerate faecal indicator bacteria.
  (d) Mapping the microflora on the surface of a leaf.

33.2 Decide on an appropriate isolation procedure for a particular microbe. How might you isolate the following microbes from a sample?
  (a) Photosynthetic flagellate algae in a sample of pond water.
  (b) Bacteria growing as a biofilm on the surface of sand particles.
  (c) Faecal streptococci (enterococci) present at low density (<1 per ml) in a sample of river water.
  (d) A distinctively shaped bacterium present at very low numbers in a sample containing a high number of other microbes of different shape but with similar nutritional requirements.

33.3 Investigate the selective basis of microbiological media. Using textbooks on bacteriological methods or the Web, research how each of the following media operates, in terms of their selective and diagnostic (differential) features:
  (a) Mannitol salt agar for *Staphylococcus aureus*.
  (b) Membrane lauryl sulphate broth for coliforms and *E. coli*.
  (c) Slanetz and Bartley medium for faecal streptococci (enterococci).
  (d) Mannitol pyruvate egg yolk polymyxin (MPYP) medium for *Bacillus cereus*.

33.4 Describe the colonial characteristics of selected microbes. Either research the features of the following microbes (e.g. *via* the Web), or look at well-isolated individual colonies in the laboratory, following overnight growth on a suitable medium:
  (a) *Escherichia coli*;
  (b) *Pseudomonas aerogenes*;
  (c) *Staphylococcus aureus*;
  (d) *Streptococcus pneumoniae*;
  (e) *Bacillus subtilis*;
  (f) *Saccharomyces cerevisiae*.

33.5 Find out how some of the tests used in microbial identification work. Research the operating principles that underpin the following tests, using microbiological textbooks or the Web, and prepare brief notes explaining how each of the tests works:
  (a) oxidase test;
  (b) indole test;
  (c) β-galactosidase test;
  (d) urease test;
  (e) lysine decarboxylase test;
  (f) H_2S production.

33.6 Identify the following oxidase-negative, catalase-positive, Gram-negative rod-shaped bacteria, using Table 33.2.
  (a) Positive for citrate, urease, ornithine decarboxylase and H_2S only.
  (b) Positive for sucrose, mannitol, β-galactosidase, lysine decarboxylase and indole only.
  (c) Positive for mannitol, citrate, lysine decarboxylase, ornithine decarboxylase and H_2S only.
  (d) Positive for urease, ornithine decarboxylase and indole only.
  (e) Positive for mannitol, β-galactosidase and indole only.

34 Working with animal and plant tissues and cells

> **Definitions**
>
> **In vivo** – occurring within a living organism (literally 'in life').
>
> **In vitro** – occurring outside a living organism, in an artificial environment (literally 'in glass').
>
> **In silico** – occurring within a computer system (literally 'in silicon').

While the aim of many studies is to isolate, quantify or characterise individual molecules from a biological system, e.g. the purification of a particular enzyme (p. 373), this is not always the most appropriate course of action. Depending on the purpose of the investigation, it may be more relevant to study the functioning of biomolecules within more complex systems, in order to understand their role in a particular biological process. At one extreme this may be carried out *in vivo*, using whole multicellular organisms (e.g. individual animals or plants), while the other extreme is represented by *in vitro* studies, using subcellular 'cell-free' extracts. Between these two extremes, a range of tissue and cell culture techniques offers some of the biological complexity of the intact organism combined with a degree of experimental control that may not be obtainable *in vivo*. The ethics and costs of whole animal experimentation have provided an additional stimulus to the development of *in vitro* methods, e.g. toxicity tests using mammalian cell culture rather than laboratory animals, while developments in molecular genetics have led to further applications of cell and tissue culture.

Animal tissues and organs

Physiological experiments are carried out using either whole organisms, or a range of animal organs and tissues, including heart, liver, muscle, etc.

> **Working with vertebrate animals and their organs/tissues** – remember that procedures must be consistent with the law, i.e. in the UK, the Animals Scientific Procedures Act 1986.

 KEY POINT A major practical consideration is that the tissue should be studied as soon as possible after the death of the animal, typically under laboratory conditions that mimic the *in vivo* environment as closely as possible.

In most instances, the experiments are relatively short term (<24 h) and the aim is to maintain the tissue in a physiological state similar to that within the living organism. For metabolic studies, the whole organ or a tissue slice (typically 1–10 mm thick) will be bathed in an appropriate perfusion fluid, supplied either by gravity or by peristaltic pump. Practical considerations include:

> **Using human tissues** – in the UK, the Human Tissue Act 2004 applies and work is licensed and overseen by the Human Tissue Authority. Note that established cell lines (p. 225) are excluded from the regulations.

- Inorganic solute requirements – the chemical composition of the perfusion fluid is usually chosen to reflect the major inorganic ion requirements of the tissue. For short-term studies, a number of so-called 'physiological salt solutions' may be used, e.g. Ringer's solution, one formulation of which is given in Table 34.1.
- Oxygen requirements – it may be necessary to increase the O_2 content of the perfusion fluid by bubbling with air, in order to meet the oxygen demand of the innermost parts of the tissue. However, this can lead to oxygen toxicity in the outermost parts and an alternative approach is simply to increase the rate of perfusion.
- Physicochemical conditions – including temperature (usually controlled to $\pm 1\,°C$ of normal body temperature), water status (the perfusion fluid and the tissue should be isotonic, p. 150), pH and buffering capacity (e.g. some perfusion fluids have elevated an $NaHCO_3$ concentration, to mimic the buffering capacity of mammalian serum).

Table 34.1 Composition of Ringer's solution (simplified formulation, for amphibians, etc., pH 7.4–7.6)

Compound	Amount per litre (g)
NaCl	6.0
KCl	0.075
$CaCl_2$	0.1
$NaHCO_3$	0.1

Definitions

Apoplasm – that part of the plant body outside the symplasm.

Light compensation point – the amount of photosynthetically active radiation (PAR) where photosynthetic CO_2 uptake is balanced by CO_2 production due to respiration and photorespiration (p. 400).

Plasmodesmata – transverse connections through the cell wall, linking the cytoplasm of adjacent plant cells and creating a symplasm.

Working with plant tissues and organs – the ethical problems associated with animal and human tissues are avoided, and there is a decreased risk of infection to the laboratory worker; practical manipulations are often carried out in horizontal laminar-flow cabinets (p. 205).

Table 34.2 Components of Long Ashton medium (nitrate version)

Stock solution: mass of component required per litre of solution (g)	Volume of stock solution to make 1 litre of medium (ml)
Major nutrients	
KNO_3: 50.60	8
$Ca(NO_3)_2$: 80.25	8
$MgSO_4 \cdot 7H_2O$: 46.00	8
$NaH_2PO_4 \cdot 2H_2O$: 52.00	4
Micronutrients	
FeKEDTA: 3.30	5
$MnSO_4 \cdot 4H_2O$: 2.23	1
$ZnSO_4 \cdot 7H_2O$: 0.29	1
$CuSO_4 \cdot 5H_2O$: 0.25	1
H_3BO_3: 3.10	1
$Na_2MoO_4 \cdot 2H_2O$: 0.12	1
NaCl: 5.85	1
$CoSO_4 \cdot 7H_2O$: 0.056	1

Effects of humans on plants – remember that your exhaled breath will be nearly saturated with water vapour and will contain CO_2 at 3–4% v/v, some 100 times more concentrated than atmospheric CO_2: your breath can thus affect rates of transpiration and photosynthesis.

- Organic nutrient requirements – for longer-term studies, suitable nutrients will be required: these may be chemically defined additives, e.g. vitamins, amino acids, proteins, etc., or biological fluids such as plasma or serum. Glucose is often added as a carbon and energy source (Freshney, 2005).

Plant tissues and organs

Individual plant components (e.g. leaves, leaf slices and epidermal strips) can be isolated from the main plant body for study under controlled conditions. Since photosynthetic plant parts are autotrophic, they may be maintained *in vitro* for longer than animal organs, given adequate light and CO_2. However, most plant cells are joined via plasmodesmata, and the separation of such connections when the component is removed from the plant often leads to death when these connections are broken.

 KEY POINT *Plants show wound responses that may affect the metabolic processes under study. For these reasons, the most suitable systems for longer-term studies are often whole organs, e.g. whole leaves or entire root systems.*

Water culture (hydroponics), e.g. in Long Ashton medium (Table 34.2), is an alternative approach, offering greater control over the root environment (Dodds and Robert, 1995).

It is best to use vigorous, healthy stock plants and to follow a well-established procedure, taking account of the following:

- Sterility – strict attention to sterile technique can be essential to the success of many longer-term experiments. Decontamination of plant organs may be especially difficult where specimens are obtained from soil: to achieve this, use a surface wash with disinfectant (e.g. 10% w/v sodium hypochlorite), followed by several rinses with sterile water.
- Gaseous environment – in general, the experimental system should be well ventilated. Actively photosynthetic tissues will rapidly deplete the atmospheric CO_2 in a closed vessel: plant parts may also produce physiologically active gases, such as ethylene, especially at wound sites. Turgor loss may occur in isolated plant parts unless a high humidity is maintained.
- Nutrition – plant tissues may benefit from a supply of inorganic ions, including K^+, SO_4^{2-}, etc. and may require certain vitamins, micronutrients and plant hormones for prolonged studies.
- Physicochemical conditions – light is the most important environmental requirement for green plant parts. At atmospheric CO_2 concentrations, the light compensation point is about 5–10 µmol photons PAR $m^{-2} s^{-1}$ (p. 278) and photosynthesis is usually saturated between 500 and 2000 µmol photons PAR $m^{-2} s^{-1}$, depending on the plant type. Light quality and photoperiod (daylength) are also important: fluorescent tubes and incandescent bulbs that mimic the photosynthetic spectrum of sunlight are available (see p. 277), while the photoperiod can be controlled using a timer. The water potential of aqueous media can be adjusted using an impermeant osmoticum such as mannitol, and pH values are often kept close to those of the apoplasm ($\approx$pH 6) using appropriate buffers, if necessary (p. 156).

Definitions

Cell line – a cell culture derived by passage of a primary culture.

Clone – a population of cells derived from a single original cell, i.e. sharing the same genotype.

Confluence/confluent growth – merging of individual cells to form a continuous layer.

Continuous cell line – a culture with the capacity for unlimited multiplication *in vitro*. Sometimes termed an established cell line.

Finite cell line – a culture with a limited capacity for growth *in vitro* (maximum number of cell doublings).

Hybridoma – a hybrid cell produced by fusion of a tumour (myeloma) cell and an antibody-producing B lymphocyte (p. 257). A cell line derived from a single hybridoma will produce a single antibody type (monoclonal).

Immortalisation – conversion of a finite cell line into a continuous cell line.

Passage – an alternative term for subculture.

Primary culture – a cell culture derived from tissue or organ fragments (explants). Primary culture ends on first subculture.

Senescence – the end-point in the limited lifespan of a finite cell line, characterised by the lack of proliferation. Conversion to an established (continuous) cell line requires 'escape' from senescence.

Stem cell – an undifferentiated animal cell that can generate new stem cells through mitosis, or can develop and differentiate into a particular specialised cell type.

Transformation – a permanent alteration in the growth characteristics of a finite cell line that may include (i) changes in morphology, (ii) an increased growth rate and/or (iii) the acquisition of an infinite lifespan, often termed immortalisation. Transformation may be spontaneous or may be induced by chemical agents or viruses, and often involves a change in chromosome number.

Cell and tissue cultures

Many of the basic principles involved in culturing animal and plant cells are broadly similar to those described for microbial cell culture (p. 203). Definitions for several key terms are given in the margin.

Applications of cell and tissue culture

The main uses of animal and plant cell culture systems include:

- Experimental model systems in biochemistry, pharmacology and physiology: cell culture offers certain advantages over whole organism studies, with greater control over environmental conditions and biological variability. The use of genetically defined clones of cells may simplify the analysis of experimental data. Conversely, results obtained with specialised cell-based systems might be unrepresentative of a broader range of cell types and may be more difficult to interpret in terms of the whole organism.
- Studies of the growth requirements of particular cells: including studies of the positive effects of growth factors or growth-promoting substances, and the negative effects of xenobiotics or cytotoxic compounds. The use of cell culture in bioassays and mutagenicity testing is considered in Chapter 35.
- Studies of cell development and differentiation: including aspects of the cell cycle and gene expression. Cell cultures retaining their ability to differentiate *in vitro* are particularly interesting to researchers, while the lack of differentiation and unlimited growth of many animal cell lines makes them useful models of tumour development.
- Pathological studies: including culture of foetal cells for karyotyping, detection of genetic abnormalities, e.g. trisomy, etc.
- Genetic manipulation: cell culture techniques have played an essential role in the development of molecular biology, including the production of transgenic animals and plants by techniques such as transfection, etc.
- Biotechnology: including the industrial production of therapeutic proteins, vaccines and monoclonal antibodies using large-scale hybridoma culture techniques similar to those used in microbiology (pp. 233–5).
- Stem cell technology: increasingly important in biomedical research and development.

KEY POINT One of the main differences between organ and tissue incubation techniques and those used in cell/tissue culture is that the former aim simply to maintain metabolic and physiological activity for a limited period, while the latter provide conditions suitable for cell growth, division and development in vitro *over an extended timescale, from a few days to several months.*

Animal cell culture systems

These may be established either from whole organisms (e.g. chick embryo), discrete organs (e.g. rat liver) or from blood (e.g. lymphocytes), typically using *mild* enzymic tissue disruption techniques, where necessary.

KEY POINT While, in theory, it is possible to culture nucleated cells from virtually any source, in practice the highest rates of success are most often achieved with young, actively growing tissues.

SAFETY NOTE Working with tissue cultures safely – do not confuse laminar flow hoods (p. 205) with biosafety cabinets, as they perform completely different functions. A horizontal laminar flow hood is designed to minimise contamination of the culture, rather than the worker.

SAFETY NOTE When sterilising during tissue culture procedures, (Box 34.1), take care when using 70% v/v alcohol near a naked flame (e.g. Bunsen or spirit lamp) – it is easily ignited and burns with a weakly visible flame.

Example Dulbecco's modified Eagle's medium contains Ca^{2+}, Fe^{3+}, Mg^{2+}, K^+, Na^+, Cl^-, SO_4^{2-}, PO_4^{2-}, glucose, 20 amino acids, 10 vitamins, inositol and glutathione. Foetal calf serum is usually added at up to 20% v/v.

Using antibiotics in cell culture – a typical antimicrobial supplement might include antibacterial agents, e.g. penicillin and streptomycin, an antifungal agent, e.g. griseofulvin, and an antimycoplasmal agent, e.g. gentamicin. Use sparingly to treat contamination and discontinue once the contaminant has been eradicated.

The principal considerations in animal cell culture are:

- Safety: it is important to be aware of the potential dangers of infection from cell cultures. Although avian and rodent cells present a reduced risk of disease transmission compared to human cells, all animal cell cultures must be regarded as a potential source of pathogenic microbes, and appropriate sterile technique (Box 34.1) must be used at all times (see also p. 203). Work involving human tissues and cell cultures must be carried out in a biosafety cabinet by trained, experienced personnel – you are unlikely to gain practical experience with such cultures in the early stages of your course. For other cell cultures, you will need to follow the code of safe practice of your department – consult your Departmental Safety Officer if you have any doubts about safe working procedures.
- Whether to use a primary culture or a cell line: freshly isolated cells are more likely to reflect the biochemical activities of cells *in vivo*, though they will have a limited lifespan in culture, requiring repeated isolation for longer-term projects. Continuous cell lines are more easily cultured and offer the advantage that their growth requirements in culture may be known in some detail, especially for the more widely used cell lines (e.g. BHK, HeLa).
- The requirements for a solid substratum: some cells must be attached to a solid surface in order to grow. Anchorage dependence is a typical feature of primary cultures and finite cell lines – such cultures show density-dependent growth inhibition once the cells have formed a confluent monolayer on the surface of the substratum. An alternative approach is to grow such cells on a particulate support using 'microcarrier' beads. In contrast, many continuous cell lines can be maintained in suspension culture, as individual cells or aggregates.
- The physicochemical conditions, including pH (typically 7.2–7.5) and buffering capacity, osmolality (usually $300 \pm 20\,\mathrm{mosmol\,kg^{-1}}$) and temperature (e.g. 35–37 °C for mammalian cells).
- The requirements of the culture medium: these will include the provision of inorganic ions (as a balanced salt solution), a carbon/energy source plus other organic nutrients, and in some cases a supplement containing antimicrobial agents to counter the risks of contamination. For example, Dulbecco's modified Eagle's medium is used for many mammalian cell types that grow as adherent monolayers, while suspension cultures of continuous cell lines can be maintained using less stringent media. To support growth, the basal medium is usually supplemented with serum (usually foetal calf serum, at up to 20% v/v), or a chemically defined serum-like supplement containing a mixture of proteins, polypeptides, hormones, lipids and trace components. The *in vitro* level of CO_2 and O_2 must also be considered: many cell cultures are buffered using bicarbonate, and must be maintained in an atmosphere of elevated CO_2, either in a sealed culture vessel or in a CO_2 incubator, to maintain pH balance. In some cases, a pH-sensitive dye (e.g. phenol red, p. 153) may be incorporated into the growth medium, to provide a visual check on pH status during growth – a colour change indicates acidification of the medium and the growth medium should be renewed.
- The equipment required: this may include a laminar flow hood or biosafety cabinet to reduce the possibility of microbial contamination,

Box 34.1 Sterile technique and its application to animal and plant cell culture

While Chapter 32 gives general advice on the basic principles of sterile technique with microbial cultures, animal and plant cell cultures need additional precautions due to the complexity of some of the procedures and the likelihood of rapid overgrowth of any contaminating microbes. For routine work on the open bench, you should be aware of the following points:

1. **Consider personal clothing and hygiene:** long hair should be tied back, or retained by a net/cap. Avoid pendulous earrings and similar items. Wash your hands at the outset, to remove any loose skin flakes: surgical gloves may be worn, though there is loss of tactile sensitivity and comfort. Hands/gloves should be swabbed with 70% v/v alcohol to further reduce the risk of contamination. If you have a cold/cough, consider wearing a face mask.
2. **Work in a designated quiet area:** there should be no air currents (avoid open windows) and no 'through traffic' or other activity that might give rise to contamination (e.g. microbes should not be cultured in the same location). The term 'quiet area' also highlights that no talking should occur, to reduce the risk of aerosol contamination from the oral microflora.
3. **Organise you work surface:** clear everything from the bench, swab with 70% v/v alcohol, then position all items around a central area, so you can reach every item without reaching across anything. A well-organised workspace reduces the risk of accidental contact between sterile and non-sterile items. Swab bottles and flasks with 70% v/v alcohol before positioning them at the edge of your workspace. Also, swab your work surface between procedures and at the end of the session.
4. **Work close to a Bunsen flame,** positioned centrally within your workspace where convection currents create an upward airflow, reducing the likelihood of particles falling from the air into an open vessel. Always flame the tops of glass bottles (p. 205), but not plastic items, for 2-3 seconds both *before* and *after* opening and closing, rotating the bottles during exposure to the flame. Flame glass pipettes, as described on p. 205.
5. **Tilt flasks and bottles during use:** uncapped culture flasks and media bottles are best kept at a shallow angle, to minimise the risk of airborne contamination.
6. **Work without delay, but do not hurry:** always keep in mind that the air contains contaminant microbes and the longer you leave a vessel open to the air, the more likely it is that it will become contaminated.

When preparing to work in a laminar flow cabinet, the following additional aspects should be considered:

1. **Note whether the airflow is horizontal or vertical:** horizontal flow gives greatest protection to the work area and provides the most stable air flow, while vertical flow cabinets reduce exposure of the operator. A biosafety cabinet should be used for work with potentially hazardous cultures (e.g. human cells/tissues, or with any cells known to be infected with a virus).
2. **Prepare the cabinet:** switch on the cabinet and leave running for at least 10 minutes, then swab the work surface and other interior surfaces with 70% v/v alcohol. Swab the outsides of bottles, flasks, etc. before bringing them into the cabinet. Arrange items around a crescent-shaped central work area.
3. **Carry out your work with due regard for the airflow within the cabinet:** use of a Bunsen burner is often discouraged, as it disrupts the correct airflow, creates heat and can ignite flammable items within the restricted interior of the cabinet. Always try to keep your hands/arms further away from the sterile airstream than the items within the cabinet, e.g. avoid working *behind* an open vessel in a horizontal flow cabinet and *above* an open vessel in a vertical flow cabinet.
4. **Take particular care when using pipettes:** a laminar flow hood will have a restricted interior volume, and it is easy for the novice to touch a pipette tip on the hood, or on another item during use. When removing a pipette from its packaging and connecting to a pipette filler (p. 129) you should point the tip of the pipette away from the user and into the air flow, holding it well above the graduated scale markings to avoid contamination. The pipette and filler are then held horizontally until required: as this can be a tricky procedure for the novice, you should not take a pipette out of its container until you are ready to use it.
5. **Work carefully with open bottles and flasks:** caps can be either placed top-down on the work surface, or held in a crooked finger (Fig. 32.6a). Since the airflow is sterile, there is less need to work swiftly when compared to the open bench, but you should still re-cap all containers as soon as you have finished a particular procedure.

Definitions

Callus – an aggregation of undifferentiated plant cells in culture.

Embryogenic callus – tissue with the capacity to differentiate under defined laboratory conditions, typically in response to plant growth regulators in the medium.

Explant – a fragment of tissue used to initiate a culture (the term is also used in animal culture).

Protoplasts – cells lacking their cell walls.

Sphaeroplasts – cells with attached fragments of their cell walls: osmotically sensitive.

Totipotency – the ability (of any plant cell) to de-differentiate and redifferentiate into any of the cell types found in the mature plant.

suitable culture vessels (typically presterilised, disposable polystyrene dishes, bottles and flasks, treated to create a negatively charged, hydrophilic surface), a supply of high purity water (typically distilled, deionised and carbon filtered), a suitable incubator with temperature control of $\pm 0.5\,°C$ or better, often with CO_2 control and mechanical mixing, and an inverted microscope to examine adherent cell monolayers during growth.

The successive stages of isolation of animal components are shown in Fig. 34.1a. Box 34.1 gives advice on key aspects of sterile technique in cell culture; additional guidance on the principal practical procedures used in animal cell culture is given in Box 34.2.

Plant tissue and cell culture systems

Plant tissue cultures can be established by growing explants of sterilised tissue on the surface of an agar-based growth medium to give a callus of undifferentiated cells. Initial sterilisation is usually achieved by incubation for 15–20 min in 10% w/v sodium hypochlorite.

 KEY POINT *For most plants, cell cultures can be established from a broad range of tissue types, reflecting the totipotency of many plant cells.*

Embryogenic callus may be induced to differentiate, forming tissues and organs on a medium containing appropriate plant hormones: in many

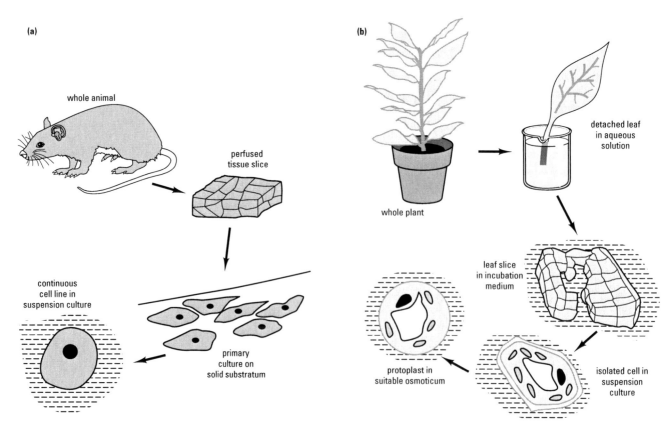

Fig. 34.1 Isolation of animal (a) and plant (b) components for *in vitro* study (note decreasing scale, from organisms to cells).

Box 34.2 Practical procedures in animal cell culture

The following activities are most likely to be encountered during the initial stages of your training:

- **Sterile technique (aseptic technique):** your first exercises are likely to involve transferring sterile liquids between vessels using pipettes, preparing solutions/media and making dilutions.
- **Preparing cells for initial culture:** if you are using a cryopreserved (frozen) vial containing a cell line, it will need to be thawed as rapidly as possible, to minimise the chance of intracellular ice crystal damage – thaw by immersion in water at 37°C (if the cells were initially stored in liquid nitrogen, thaw in a lidded container as there is a risk of explosion). The thawed culture should then be diluted slowly with new growth medium. If you are preparing a primary culture, you will first have to disrupt the tissue (Chapter 36 gives further details of methods available).
- **Checking for contamination:** this can be carried out by visual inspection, looking for: (i) a cloudy medium, sometimes with a film of surface growth; (ii) a change in pH; typically a sharp decrease due to bacterial contamination causes a colour change of the indicator dye (e.g. phenol red turns yellow). Alternatively, examine cultures microscopically, or subculture for specific microbes (Chapter 33).
- **Observing cells:** cells growing as an adherent monolayer can be observed by inverted microscope (×10 or ×40 objective), looking for typical cell morphology, or cytopathic effects (e.g. rounding off and detachment, irregular shape, or internal vacuolation and granulation), as well as any evidence of microbial growth. Microscopy will also tell you when such cells are reaching confluent growth (p. 225), requiring subculture.
- **Counting cells:** while rough estimates of the growth stage and cell number may be made from direct microscopic examination of cultures, a haemocytometer is required for a fully quantitative count, as detailed in Box 35.1. Adherent cells would first be detached (e.g. using 0.25% w/v trypsin) and counted in suspension. Other quantitative approaches include electronic counting (p. 235), and flow cytometry where other cellular activities can be assessed simultaneously. Alternatively, cellular constituents e.g. DNA (p. 370) or protein (p. 353) can be assayed.
- **Checking viability:** this is usually carried out by assessing membrane integrity using either dye exclusion (1–2 min in 0.4% w/v trypan blue or 1.0% w/v naphthalene black) or, less frequently, dye uptake (10 min in 0.001% w/v diacetyl fluorescein or 0.5% w/v neutral red), with viability expressed in terms of the percentage of cells unstained or stained, respectively. When combined with haemocytometry (p. 236), the absolute numbers of live and dead cells can be estimated.
- **Feeding an adherent culture:** cells growing as a monolayer may need to have their growth medium replenished before they are ready for subculture, e.g. due to a drop in pH of the medium, or a deterioration in cell morphology. The original medium should be removed, usually by aspiration using a sterile pipette and suction line, and a replacement volume of fresh medium then added (the new medium should be pre-warmed to 37°C).
- **Subculturing cells ('passaging'):** typically this is carried out when adherent cells reach confluence, or when a suspension culture almost reaches the maximum population density in stationary phase (p. 233). Cells growing as a monolayer must first be detached from their substrate, typically using a small amount of 0.25% w/v trypsin for up to 15 min (it is important to dilute the trypsin following detachment, to prevent further digestion). Cells in suspension culture can simply be diluted in an appropriate volume of new medium. For both types of cell culture, the reseeding density is important, since over-diluted cultures will grow poorly. For primary cultures, low dilution factors (often termed 'split ratios') of between 1:2 and 1:8 are commonly used whereas some continuous cell lines can be diluted by up to 1:100, to give a cell density of $10^4–10^5$ per ml.
- **Harvesting cells:** gentle centrifugation (at 80–100 g for 10 min) is sufficient to pellet most animal cells, leaving the spent medium as a supernatant, which should be either aspirated, or removed using a pipettor (p. 129). Monolayer cultures would be centrifuged following trypsinisation (see above).
- **Freezing cells (cryopreservation):** cells are first suspended at high density (typically around 10^6 per ml) in a suitable freezing solution, usually growth medium or serum supplemented with a cryoprotectant such as 10% v/v DMSO or 20–30% v/v glycerol, dispensed into small vials ('ampoules'/'cryotubes'), and then frozen slowly (ideally, at around 1°C min^{-1}): one approach is to wrap the ampoules in cotton wool inside a small cardboard tube, then place them in an insulated container in a −70°C freezer for ⩾ 6 h, then transfer to a liquid nitrogen storage facility. An alternative approach is to use a proprietary system, e.g. 'Mr Frosty' (Nalgene), to provide a controlled rate of cooling.

cases, these cultures will develop to form plantlets that can be grown on to mature plants, or encapsulated to produce so-called 'artificial seeds'. This approach can be used to study the conditions necessary for differentiation and development, or to propagate rare plants and other valuable stock (e.g. virus-free stock, or genetically altered plants). Callus derived from anthers can be used to provide haploid cell cultures and haploid plants – these are often useful for experimental genetics and breeding purposes.

Plant cell suspension cultures (Fig. 34.1b) are usually obtained by transferring fragments of actively growing callus to a 'shake' flask containing liquid medium in an orbital incubator: gentle agitation causes fragmentation of the callus tissue, to give a suspension culture that will contain individual plant cells and cell aggregates. This culture can be maintained by repeated subculturing of material from the upper layers of the liquid, encouraging the growth of small aggregates. However, in contrast to animal cell suspension cultures, it is rare for plant suspension cultures to be entirely unicellular, due to the presence of plasmodesmata and plant cell walls. Suspension cultures often require a minimum inoculum size on subculture – an inoculum volume of 10% v/v may be necessary to ensure successful subculture. It may also be helpful to add a small amount of 'conditioned' medium from a previous culture. Commercial-scale suspension cultures are used to produce certain plant pigments and secondary metabolites such as flavourings and high value pharmaceutical compounds.

The growth media used for callus and suspension cultures are more complex than those required for intact plant organs and tissues, with organic nutrient supplements in addition to a balanced salt solution. Such organic supplements often include a major carbon and energy source, e.g. sucrose, plus various vitamins and growth regulators (typically, at least one auxin and a cytokinin), together with undefined components such as yeast extract and hydrolysed casein in some instances. Otherwise, the techniques are broadly similar to those described for microbial systems (pp. 203–208).

Plant protoplasts

Some experimental procedures using plant cells require the enzymatic removal of their cell walls, creating protoplasts that can be manipulated *in vitro*, then maintained under conditions that allow the regeneration of cell walls, with subsequent growth and differentiation to give genetically modified plants. Protoplast isolation often involves pretreatment in a concentrated osmoticum (e.g. sucrose or mannitol, at 300–500 mmol l^{-1}) to plasmolyse the cells, weakening the linkage between cell wall and plasma membrane. Enzymatic treatment is often prolonged, taking several hours in a suitable mixture of enzymes, e.g. cellulase and Macerozyme. The resulting material can be sieved through fine nylon mesh, centrifuged at low speed and then re-suspended, to remove subcellular debris and cell aggregates. The protoplast preparation can be checked for fragments of cell wall using a suitable stain, e.g. 0.1% w/v calcofluor white and a fluorescence microscope.

The fusion of protoplasts from different plants can be used to produce a somatic hybrid: this process can be used to circumvent inter-species reproductive barriers, creating novel plants. Protoplast fusion can be induced by chemical 'fusogens', e.g. using polyethylene glycol (PEG) at

Using an orbital incubator – continued rapid agitation (>100 r.p.m.) maintains a homogeneous suspension and encourages gas exchange between the culture medium and the atmosphere.

Example Murashige and Skoog's medium contains a balanced mixture of the principal inorganic ions, plus 7 trace element compounds, 3 vitamins, inositol, glycine and sucrose (at 30 g l^{-1}) as the major carbon source.

Working with protoplasts – remember that all solutions must contain a suitable osmoticum, to prevent bursting. Mannitol is widely used at ≈400 mmol l^{-1}.

Assessing cell and protoplast viability – cytochemical techniques are often used, e.g. exclusion of the mortal stain Trypan blue or Evan's blue. The fluorogenic vital stain fluorescein diacetate is an alternative: it is cleaved by esterases within living cells, liberating fluorescein and giving green–yellow fluorescent cells when viewed by UV fluorescence microscopy.

high concentration, or by electrofusion (incubation under low alternating current to encourage aggregation, then brief exposure to a high voltage electrical field – typically $1000\,V\,cm^{-1}$ for 1–2 ms – causing protoplast fusion). Similar techniques can be used with animal cells. The successive stages of isolation of plant components are illustrated in Fig. 34.1b.

Text references

Dodds, J.H. and Robert, L.W. (1995) *Experiments in Plant Tissue Culture*, 3rd edn. Cambridge University Press, Cambridge.

Freshney, R.I. (2005) *Culture of Animal Cells: A Manual of Basic Technique*, 5th edn. Wiley, New York.

Sources for further study

Anon. (2006) *Fundamental Techniques in Cell Culture ... a Laboratory Handbook*. Sigma-Aldrich/European Collection of Cell Cultures. Available at: http://www.sigmaaldrich.com/Area_of_Interest/Life_Science/Cell_Culture/Key_Resources/Cell_Culture_Manual.html. Last accessed: 01/04/07.

Butler, M. (2003) *Animal Cell Culture and Technology: The Basics*, 2nd edn. Garland Science Press, Oxford.

Doyle, A., Griffiths, J.B. and Newell, D.G. (1998) *Cell and Tissue Culture: Laboratory Procedures*. Wiley, New York.

Evans, D., Coleman, J. and Kearns, A. (2003) *Plant Cell Culture: the Basics*. Garland Science, Oxford.

Hall, R.D. (1999) *Plant Cell Culture Protocols. Methods in Molecular Biology Series*. Humana Press, New Jersey.

Helgason, C.D. (2004) *Basic Cell Culture Protocols*, 3rd edn. Humana Press, New Jersey.

Loyola-Vargas, V.M. and Vazques-Flota, F. (2005) *Plant Cell Culture Protocols*, 2nd edn. Humana Press, New Jersey.

Masters, J. (2000) *Animal Cell Culture: a Practical Approach*, 3rd edn. Oxford University Press, Oxford.

Razdan, M.K. (2003) *Introduction to Plant Tissue Culture*, 2nd edn. Intercept, Andover.

Study exercises

34.1 **Calculate the osmolarity of Ringer's solution from its individual constituents.** Use the data in Table 34.1 and the information in Chapter 24, and assume that each constituent behaves according to ideal thermodynamic principles. Note the following M_r values: NaCl 58.44; KCl 74.55; $CaCl_2$ 110.99; $NaHCO_3$ 84.01, required to convert the data in Table 34.1 to molar concentrations (see p. 136 if you are unsure about this conversion). Give your final answer in $mosmol\,kg^{-1}$, to one decimal place.

34.2 **Find out about the availability of cell cultures using the Web.** Research the culture collections of particular countries via their websites. How useful is each site in terms of features such as: ease of use; access to catalogues; information on individual strains, e.g. culture media; details of costs?

34.3 **Research the origins of the code names used for individual cell lines.** In many instances, the code names given to individual cell lines are derived from their original source. Use the Web to find out the origin of the following descriptors for cell lines:

(a) HeLa; (b) Vero; (c) BHK; (d) CHO; (e) TBY-2.

35 Culture systems and growth measurement

> **Definitions**
>
> **Heterotroph** – an organism that uses complex organic carbon compounds as a source of carbon and energy.
>
> **Photoautotroph** – an organism that uses light as a source of energy and CO_2 as a carbon source (photosynthetic metabolism).
>
> **Chemoautotroph** – an organism that acquires energy from the oxidation of simple inorganic compounds, fixing CO_2 as a source of carbon (chemosynthetic metabolism).

Microbial, animal and plant cell culture methods are based on the same general principles, requiring:

- a pure culture (also known as an axenic culture), perhaps isolated as part of an earlier procedure, or from a culture collection;
- a suitable nutrient medium to provide the necessary components for growth. This medium must be sterilised before use;
- satisfactory growth conditions including temperature, pH, atmospheric requirements, ionic and osmotic conditions;
- sterile technique (p. 203) to maintain the culture in pure form.

Heterotrophic animal cells, fungi and many bacteria require appropriate organic compounds as sources of carbon and energy. Non-exacting bacteria can utilise a wide range of compounds and they are often grown in media containing complex natural substances (including meat extract, yeast extract, soil, blood). Animal cells have more stringent growth requirements (p. 225).

Photoautotrophic bacteria, cyanobacteria and algae are grown in a mineral medium containing inorganic ions including chelated iron, with a light source and CO_2 supply. Plant cells may require additional vitamins and hormones (p. 226). For chemoautotrophic bacteria, the light source is replaced by a suitable inorganic energy source, e.g. H_2S for sulphur-oxidising bacteria, NH_3/NH_4^+ for nitrifying bacteria, etc.

Growth on solidified media

Many organisms can be cultured on an agar-based medium (p. 204).

> **KEY POINT** *An important benefit of agar-based culture systems is that an individual cell inoculated onto the surface can develop to form a visible colony: this is the basis of most microbial isolation and purification methods, including the streak dilution, spread plate and pour plate procedures (pp. 206–7).*

Animal cells are often grown as an adherent monolayer on the surface of a plastic or glass culture vessel (Fig. 35.1), rather than on an agar-based medium (p. 206).

Several types of culture vessel are used:

- Petri plates (Petri dishes): usually the presterilised, disposable plastic type, providing a large surface area for growth.
- Glass bottles or test tubes: these provide sufficient depth of agar medium for prolonged growth of bacterial and fungal cultures, avoiding problems of dehydration and salt crystallisation. Inoculate aerobes on the surface and anaerobes by stabbing down the centre, into the base (stab culture).
- Flat-sided bottles: these are used for animal cell culture, to provide an increased surface area for attachment and allow growth of cells as a surface monolayer. Usually plastic and disposable (Fig. 35.1).

The dynamics of growth are usually studied in liquid culture, apart from certain rapidly growing filamentous fungi, where increases in colony diameter can be measured accurately, e.g. using Vernier callipers.

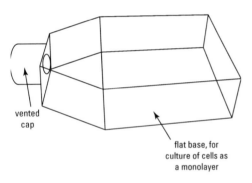

Fig. 35.1 Plastic flask for animal cell culture – this design provides a large surface area for growth of an adherent monolayer of cells.

Harvesting bacteria from an agar plate – colonies can be harvested using a sterile loop, providing large numbers of cells without the need for centrifugation. The cells are relatively free from components of the growth medium; this is useful if the medium contains substances that interfere with subsequent procedures.

Subculturing – when subculturing microbes from a colony on an agar medium, take your sample from the growing edge, so that viable cells are transferred.

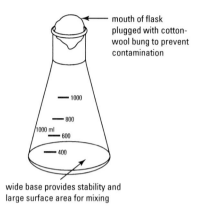

Fig. 35.2 Conical (Erlenmeyer) flask.

Growth in liquid media

Many cells, apart from primary cultures of animal cells, can be grown as a homogeneous unicellular suspension in a suitable liquid medium, where growth is usually considered in terms of cell number (population growth) rather than cell size. Most liquid culture systems need agitation, to ensure adequate mixing and to keep the cells in suspension. An Erlenmeyer flask of 100–2000 ml capacity (Fig. 35.2) can be used to grow a batch culture on an orbital shaker, operating at 20–250 cycles per minute. For aerobic organisms, the surface area of such a culture should be as large as possible: restrict the volume of medium to not more than 20% of the flask volume. Larger cultures may need to be gassed with sterile air and mixed using a magnetic stirrer rather than an orbital shaker. The simplest method of air sterilisation is filtration, using glass wool, non-absorbent cotton wool or a commercial filter unit of appropriate pore size (usually 0.2 μm). Air is introduced via a sparger (a glass tube with many small holes, so that small bubbles are produced) near the bottom of the culture vessel to increase the surface area and enhance gas exchange. More complex systems have baffles and paddles to further improve mixing and gas exchange.

Liquid culture systems may be subdivided under two broad headings:

Batch culture

This is the most common approach for routine liquid culture. Cells are inoculated into a sterile vessel containing a fixed amount of growth medium. Your choice of vessel will depend upon the volume of culture required: larger-scale vessels (e.g. 1 litre and above) are often called 'fermenters' or 'bioreactors', particularly in biotechnology. Growth within the vessel usually follows a predictable S-shaped (sinusoidal) curve when plotted in log–linear format (Fig. 35.3), divided into four components:

1. Lag phase: the initial period when no increase in cell number is seen. The larger the inoculum of active cells the shorter the lag phase will be, provided the cells are transferred from similar growth conditions.
2. Log phase, or exponential phase: where cells are growing at their maximum rate. This may be quantified by the specific growth rate (μ), where:

$$\mu = \frac{2.303 (\log N_x - \log N_0)}{(t_x - t_0)} \quad [35.1]$$

where N_0 is the initial number of cells at time t_0 and N_x is the number of cells at time t_x. For times specified in hours, μ is expressed as h^{-1}.

Prokaryotes grow by binary fission while eukaryotes grow by mitotic cell division; in both cases each cell divides to give two identical offspring. Consequently, the doubling time or generation time (g, or T_2) is:

$$g = \frac{0.301 (t_x - t_0)}{\log N_x - \log N_0} \quad [35.2]$$

Cells grow at different rates, with doubling times ranging from under 20 min for some bacteria to 24 h or more for animal and plant cells. Exponential phase cells are often used in laboratory experiments, since growth and metabolism are nearly uniform.

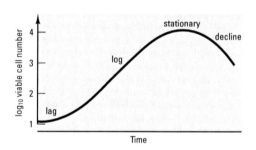

Fig. 35.3 Population growth curve for cells in batch culture (liquid medium).

Example Suppose you counted 2×10^3 cells ($\log_{10} = 3.30$) per unit volume at t_0 and 6.3×10^4 cells ($\log_{10} = 4.80$) after 2 h (t_x).

Substitution into Eqn [35.1] gives $[2.303(4.8 - 3.3)] \div 2 = 1.727 \, h^{-1}$ (or $0.0288 \, min^{-1}$).

Substituting the same values into Eqn [35.2] gives:
$[0.301 \times 2] \div [4.8 - 3.3] = 0.40 \, h$ (or 24 min).

Working with logarithms – note that there is no log value for zero, so you cannot plot zero on a log–linear growth curve or on a death curve.

3. Stationary phase: growth decreases as nutrients are depleted and waste products accumulate. Any increase in cell number is offset by death. This phase is usually termed the 'plateau' in animal cell culture.
4. Decline phase, or death phase: this is the result of prolonged starvation and toxicity, unless the cells are subcultured. Like growth, death often shows an exponential relationship with time, which can be characterised by a rate (specific death rate), equivalent to that used to express growth or, more often, as the decimal reduction time (d, or T_{90}), the time required to reduce the population by 90%:

$$d = \frac{t_x - t_0}{\log N_x - \log N_0} \quad [35.3]$$

Example Suppose you counted 5.2×10^5 cells ($\log 10 = 5.716$) per unit volume at t_0 and 3.7×10^3 cells ($\log 10 = 3.568$) after 60 min (t_x). Substitution into Eqn [35.3] gives $60 \div [2.148] = 27.9$ min. To the nearest minute, this gives a value for d of 28 min.

Some cells undergo rapid autolysis at the end of the stationary period while others show a slower decline.

Batch culture methods can be used to maintain stocks of particular organisms; cells are subcultured onto fresh medium before they enter the decline phase. However, primary cultures of animal cells have a finite life unless transformed to give a continuous cell line, capable of indefinite growth (p. 225).

Continuous culture

This is a method of maintaining cells in exponential growth for an extended period by continuously adding fresh growth medium to a culture vessel of fixed capacity. The new medium replaces nutrients and displaces some of the culture, diluting the remaining cells and allowing further growth.

After inoculating the vessel, the culture is allowed to grow for a short time as a batch culture, until a suitable population size is reached. Then medium is pumped into the vessel: the system is usually set up so that any increase in cell number due to growth will be offset by an equivalent loss due to dilution, i.e. the cell number within the vessel is maintained at a steady state. The cells will be growing at a particular rate (μ), counterbalanced by dilution at an equivalent rate (D):

$$D = \frac{\text{flow rate}}{\text{vessel volume}} \quad [35.4]$$

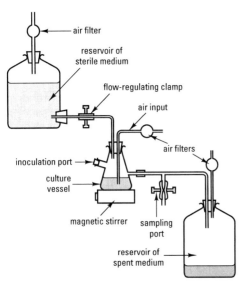

Fig. 35.4 Example of a two-dimensional lab equipment diagram of components of a chemostat.

where D is expressed per unit time (e.g. h^{-1}). In a chemostat, the growth rate is limited by the availability of some nutrient in the inflowing medium, usually either carbon or nitrogen (see Fig. 35.4). In a turbidostat, the input of medium is controlled by the turbidity of the culture, measured using a photocell. A turbidostat is more complex than a chemostat, with additional equipment and controls.

To determine the specific growth rate (μ) of a continuous culture:

1. Measure the flow of medium through the vessel over a known time interval (e.g. connect a sterile measuring cylinder or similar volumetric device to the outlet), to calculate the flow rate.
2. Divide the flow rate by the vessel volume (Eqn [35.4]) to give the dilution rate (D).
3. This equals the specific growth rate, since $D = \mu$ at steady state.

Example Suppose a continuous culture system of 2000 ml volume had a flow of 600 ml over a period of 40 min (flow rate $600 \div 40 = 15$ ml min^{-1}. Substitution into Eqn [35.4] gives a dilution rate D of $15 \div 2000 = 0.00075$ min^{-1} or $0.00075 \times 60 = 0.45$ h^{-1}.

Example Suppose you wanted to convert a doubling time of 20 min to a specific growth rate. Rearrangement of Eqn [35.5] gives $\mu = 0.693 \div 20 = 0.03465$ min^{-1} ($=2.08$ h^{-1}). For the example given on page 234, with a growth rate of 0.45 h^{-1}, substitution into Eqn [35.5] gives 0.693/0.45 = 1.54 h, (approx. 92 min).

4. If you want to know the doubling time (g), calculate using the relationship:

$$g = \frac{0.693}{\mu} \qquad [35.5]$$

(Note that Eqn [35.5] also applies to exponential phase cells in batch culture and is useful for interconverting g and μ.)

Continuous culture systems are more complex to set up than batch cultures. They are prone to contamination, having additional vessels for fresh medium and waste culture: strict aseptic technique is necessary when the medium reservoir is replaced, and during sampling and harvesting. However, they offer several advantages over batch cultures, including the following:

- The physiological state of the cells is more clearly defined, since actively growing cells at the same stage of growth are provided over an extended time period. This is useful for biochemical and physiological studies.
- Monitoring and control can be automated and computerised.
- Modelling can be carried out for biotechnology/fermentation technology.

Measuring growth in cell cultures

The most widely used methods of measuring growth are based on cell number:

Direct microscopic counts

One of the simplest methods is to count the cells in a known volume of medium using a microscope and a counting chamber or haemocytometer (Box 35.1). While this gives a rapid assessment of the total cell number, it does not discriminate between living and dead cells. It is also time-consuming as a large number of cells must be counted for accurate measurement. It may be difficult to distinguish individual cells, e.g. for cells growing as clumps.

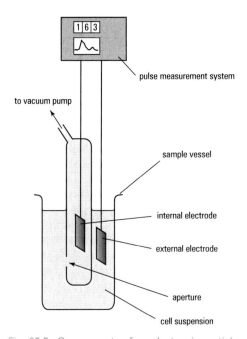

Fig. 35.5 Components of an electronic particle counter. During operation, the cell suspension is drawn through the aperture by the vacuum, creating a 'pulse' of resistance between the two electrodes as each cell passes through the aperture.

Electronic particle counters

These instruments can be used to give a direct (total) count of a suspension of microbial cells. The Coulter counter detects particles due to change in electrical resistance when they pass through a small aperture in a glass tube (Fig. 35.5). It gives a rapid count based on a larger number of cells than direct microscopy. It is well-suited for repeat measurements or large sample numbers and can be linked to a microcomputer for data processing. If correctly calibrated, the counter can also measure cell sizes. A major limitation of electronic counters is the lack of discrimination between living cells, dead cells, cell clumps and other particles (e.g. dust). In addition, the instrument must be set up and calibrated by trained personnel. Flow cytometry is a more specialised alternative, since particles can be sorted as well as counted.

Culture-based counting methods

A variety of culture-based techniques can be used to determine the number of microbes in a sample. A major assumption of such methods is

Alternative approaches to measuring growth – these include biomass, dry weight, turbidity, absorbance or any major cellular component, e.g. protein, nucleic acid, ATP, etc.

Box 35.1 How to use a counting chamber or haemocytometer

A counting chamber is a specially designed slide containing a chamber of known depth with a grid etched onto its lower surface. When a flat coverslip is placed over the chamber, the depth is uniform. Use as follows:

1. **Place the special coverslip over the chamber.** Press the edges firmly, to ensure that the coverslip makes contact with the surface of the slide, but take care that you do not break the slide or coverslip by using too much force. When correctly positioned, you should be able to see interference rings (Newton's rings) at the edge of the coverslip.

2. **Add a small amount of your cell suspension to fill the central space above the grid.** Place on the microscope stage and allow the cells to settle (2–3 min).

3. **Examine the grid microscopically,** using the ×10 objective lens first, since the counting chamber is far thicker than a standard microscope slide. Then switch to the ×40 objective: take care not to scratch the surface of the objective lens, as the special coverslip is thicker than a normal coverslip. For a dense culture, the small squares are used, while the larger squares are used for dilute suspensions. You may need to dilute your suspension if it contains more than 30 cells per small square.

4. **Count the number of cells in several squares:** at least 600 cells should be counted for accurate measurements. Include those cells that cross the upper and left-hand boundaries, but not those that cross the lower and right-hand rulings. A hand tally may be used to aid counting. Motile cells must be immobilised prior to counting (e.g. by killing with a suitable biocide).

5. **Divide the total number of cells (C) by the number of squares counted (S),** to give the mean cell count per square.

6. **Determine the volume (in ml) of liquid corresponding to a single square (V),** e.g. a Petroff–Hausser chamber has small squares of linear dimension 0.2 mm, giving an area of 0.04 mm^2; since the depth of the chamber is 0.02 mm, the volume is $0.04 \times 0.02 = 0.0008$ mm^3; as there are 1000 mm^3 in 1 ml, the volume of a small square is 8×10^{-7} ml; similarly, the volume of a large square (equal to 25 small squares) is 2×10^{-5} ml. Note that other types of counting chamber will have different volumes: check the manufacturer's instructions. For example, the improved Neubauer chamber (Fig. 35.6) has small squares of volume 0.00025 mm$^3 = 2.5 \times 10^{-7}$ ml.

Fig. 35.6 Haemocytometer grid (Improved Neubauer rulings) viewed microscopically. The large square (delimited by triple etched lines) has a volume of 1/250 mm^3 (0.004 mm^3 = 4 µl) while each small square (16 contained within the large square) has a volume of 1/4000 mm^3 (0.00025 mm^3 = 0.25 µl). Note that the boundary line for squares delimited by triple-etched lines is the *middle* line, so this line must be used when counting (count cells straddling the top and left-hand gridlines and ignore those straddling the bottom and right-hand gridlines).

7. **Calculate the cell number per ml by dividing the mean cell count per square by the volume of a single square (in ml).**

8. **Remember to take account of any dilution of your original suspension** in your final calculation by multiplying by the reciprocal of the dilution (M), e.g. if you counted a 1 in 20 dilution of your sample, multiply by 20, or if you diluted to 10^{-5}, multiply by 10^5.

The complete equation for calculating the total microscopic count is:

$$\text{Total cell count (per ml)} = (C \div S \div V) \times M \quad [35.6]$$

e.g. if the mean cell count for a 100-fold dilution of a cell suspension, counted using a Petroff–Hausser chamber, was 12.4 cells in 10 small squares, the total count would be

$$(12.4 \div 10 \div 8 \times 10^{-7}) \times 10^2 = 1.55 \times 10^8 \text{ ml}^{-1}$$

A simpler, less accurate approach is to use a known volume of sample under a coverslip of known area on a standard glass slide, counting the number of cells per field of view using a calibrated microscope of known field diameter, then multiplying up to give the cell number per ml.

Box 35.2 How to make a plate count of bacteria using an agar-based medium

1. **Prepare serial decimal dilutions of the sample in a sterile diluent (pp. 137–8).** The most widely used diluents are 0.1% w/v peptone water or 0.9% w/v NaCl, buffered at pH 7.3. Take care that you mix each dilution before making the next one. For soil, food or other solid samples, make the initial decimal dilution by taking 1 g of sample and making this up to 10 ml using a suitable diluent. Gentle shaking or homogenisation may be required for organisms growing in clumps. The number of decimal dilutions required for a particular sample will be governed by your expected count: dilute until the expected number of viable cells is around 100–1000 ml^{-1}.

2. **Transfer an appropriate volume (e.g. 0.05–0.5 ml) of the lowest dilution to an agar plate** using either the spread plate method or the pour plate procedure (p. 207). At least two, and preferably more, replicate plates should be prepared for each sample. You may also wish to prepare plates for more than one dilution, if you are unsure of the expected number of viable cells.

3. **Incubate under suitable conditions for 18–72 h, then count the number of colonies on each replicate plate at the most appropriate dilution.** The most accurate results will be obtained for plates containing 30–300 colonies. Mark the base of the plate with a spirit-based pen each time you count a colony. Determine the mean colony count per plate at this dilution (C).

4. **Calculate the colony count per ml of that particular dilution** by dividing by the volume (in ml) of liquid transferred to each plate (V).

5. **Now calculate the count per ml of the original sample** by multiplying by the reciprocal of the dilution: this is the multiplication factor (M); e.g. for a dilution of 10^{-3}, the multiplication factor would be 10^3. For soil, food or other solid samples, the count should be expressed per g of sample.

The complete equation for calculating the viable count is:

$$\text{Count per ml (or per g)} = (C \div V) \times M \qquad [35.7]$$

e.g. for a sample with a mean colony count of 5.5 colonies per plate for a volume of 0.05 ml at a dilution of 10^{-7}, the count would be:

$$(5.5 \div 0.05) \times 10^7 = 1.1 \times 10^9 \text{ CFU ml}^{-1}$$

The count should be reported as colony-forming units (CFU) per ml, rather than as cells per ml, since a colony may be the product of more than one cell, particularly in filamentous microbes or in organisms with a tendency to aggregate. You should also be aware of the problems associated with counts of zero – these are best recorded as '<1', and you should then apply the appropriate correction factors for dilution and volume to obtain the detection limit. For example, a zero count (<1) of 100 μl of a five-fold dilution gives a detection limit of $(<1 \div 0.1) \times 5 = <50$ CFU ml^{-1}.

Definition

CFU – colony-forming unit: a cell or group of cells giving rise to a single colony on a solidified medium.

---

that, under suitable conditions, an individual viable microbial cell will be able to multiply and grow to give a visible change in the growth medium, i.e. a colony on an agar-based medium, or turbidity ('cloudiness') in a liquid medium. You are most likely to gain practical experience using bacterial cultures, counted by one or more of the following methods:

- Spread or pour plate methods ('plate counts', p. 207). The most widespread approach is to transfer a suitable amount of the sample to an agar medium, incubate under appropriate conditions and then count the resulting colonies (Box 35.2).
- Membrane filtration. For bacterial samples where the expected cell number is lower than 10 CFU ml^{-1}, pass the sample through a sterile filter (pore size 0.2 or 0.45 μm). The filter is then incubated on a suitable medium until colonies are produced, giving a count by dividing the mean colony count per filter by the volume of sample filtered.
- Multiple tube count, or most probable number (MPN). A bacteriological technique where the sample is diluted and known volumes are transferred to several tubes of liquid medium (typically,

Alternative approaches in plate counting – when large numbers of samples have to be counted, a single agar plate can be divided into segments and a single droplet of each dilution placed into the appropriate segment ('Miles and Misra' droplet counting).

Counting injured or stressed microbes – a resuscitation stage may be required, to allow cells to grow under selective conditions, p. 212.

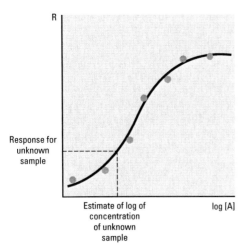

Fig. 35.7 Typical bioassay response curve, showing estimation of an unknown sample. Closed circles represent responses with standard samples. R = response; A = analyte.

Examples The following are typical bioassays:

- measuring the amount of a plant growth substance in a plant extract using the coleoptile straight growth assay;
- measuring the effect of a drug on the dilation of a person's pupils;
- measuring the mutagenic properties of a chemical compound in the Ames test (see Box 35.3).

five tubes at three volumes), chosen so that there is a low probability of the smallest volumes containing a viable cell. After incubation, the number of tubes showing growth (turbidity) is compared to tabulated values to give the most probable number (MPN per ml).

The principal advantage of culture-based counting procedures is that dead cells will not be counted. However, for such techniques the incubation conditions and media used may not allow growth of all cells, underestimating the true viable count. Further problems are caused by cell clumping and dilution errors. In addition, such methods require sterile apparatus and media and the incubation period is lengthy before results are obtained. An alternative approach is to use direct microscopy, combined with 'vital' or 'mortal' staining. For example, the direct epifluorescence technique (DEFT) uses acridine orange and UV epifluorescence microscopy to separate living and dead bacteria, while neutral red is a vital stain used for plant cells. Chapter 27 gives examples of vital/mortal stains for other cell types. A further approach is to use DNA-based methods, such as the polymerase chain reaction (PCR, p. 439).

Bioassays and their applications

A bioassay is a method of quantifying a chemical substance (analyte) by measuring its effect on a biological system under controlled conditions. The hypothetical underlying phenomena are summarised by the relationship:

$$A + Rec \rightleftharpoons ARec \rightarrow R \qquad [35.8]$$

where A is the analyte, Rec the receptor, ARec the analyte–receptor complex and R the response. This relation is analogous to the formation of product from an enzyme–substrate complex and, using similar mathematical arguments to those of enzyme kinetics (p. 386), it can be shown that the expected relationship between [A] and rate of response is hyperbolic (sigmoidal in a log–linear plot like Fig. 35.7). This pattern of response is usually observed in practice if a wide enough range of [A] is tested.

To carry out the assay, the response elicited by the unknown sample is compared to the response obtained for differing concentrations of the substance, as shown in Fig. 35.7. When fitting a curve to standard points and estimating unknowns, the available methods, in order of increasing accuracy, are:

1. fitting by eye;
2. using linear regression on a restricted 'quasi-linear' portion of the assay curve;
3. linearisation followed by regression (e.g. by probit transformation);
4. non-linear regression (e.g. to the Morgan–Mercer–Flodin equation).

In general, bioassay techniques have more potential faults than physicochemical assay techniques. These may include the following:

- A greater level of variability: error in the estimate of the unknown compound will result because no two organisms will respond in exactly the same way. Assay curves vary through time, and because they are non-linear, a full standard curve is required each time the assay is carried out.

- Lack of chemical information: bioassays provide information about *biological* activity; they say little about the chemical structure of an unknown compound. The presence of a specific compound may need to be confirmed by a physicochemical method (e.g. mass spectrometry).
- Possibility of interference: while many bioassays are very specific, it is possible that different chemicals in the extract may influence the results.

Box 35.3 Mutagenicity testing using the Ames test – an example of a widely used bioassay

Chemical carcinogens can be identified by the formation of tumours in laboratory animals exposed to the compound under controlled conditions. However, such animal bioassays are time-consuming and expensive. Dr B.N. Ames and co-workers have shown that most carcinogens are also mutagens, i.e. they will induce mutational changes in DNA. The Ames test makes use of this correlation to provide a simple, rapid and inexpensive bioassay for the initial screening of potential carcinogens. The test makes use of particular strains of *Salmonella typhimurium* with the following characteristics:

- histidine auxotrophy – the tester strains are unable to grow on a minimal medium without added histidine: this characteristic is the result of specific mutational changes to the DNA of these strains, including base substitutions and frame shifts;
- increased cell envelope permeability, to permit access of the test compound to the cell interior;
- defects in excision repair systems and enhanced error-prone repair systems, to reduce the likelihood of DNA repair after treatment with a potential mutagen.

When grown in the presence of a chemical mutagen, the bacteria may revert to prototrophy as a result of back mutations that restore the wild-type phenotype: such revertants grow independently of external histidine and are able to form colonies on minimal medium, unlike the original test strains. The extent of reversion can be used to assess the mutagenic potential of a particular chemical compound. Since many chemicals must be activated *in vivo*, the test incorporates a rat liver homogenate (so-called 'S-9 activator', containing microsomal enzymes) to simulate the metabolic events within the liver. The tester strains are mixed with S-9 activator and a small amount of molten soft agar, then poured as a thin agar overlay (top agar) on a minimal medium plate. The top agar layer contains a very small amount of histidine, to allow the tester strains to divide a few times and express any mutational changes (i.e. prototrophy).

The test can be performed in one of two ways:

1. **Spot test:** a concentrated drop of the test compound is placed at the centre of the plate, either directly on the agar surface, or on a small filter paper disc. The test compound will diffuse into the agar and revertants appear as a ring of colonies around the site of inoculation, as shown in Fig. 35.8. The distance of the ring of colonies from this site provides a measure of the toxicity of the compound, while the number of colonies within the ring gives an indication of the relative mutagenicity of the test substance. The spot test is often carried out in a simplified form, without added S-9 activator, as a rapid preliminary test prior to quantitative analysis.

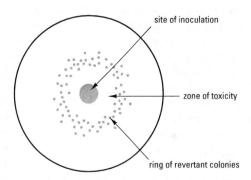

Fig. 35.8 Typical outcome of spot test (Ames test).

2. **Agar incorporation test:** known amounts of the test compound are mixed separately with molten top agar and the other constituents and poured onto separate minimal agar plates. After incubation for 48 h, revertants will appear as evenly dispersed colonies throughout the agar overlay. The number of colonies reflects the relative mutagenicity of the test compounds, with a direct relationship between colony count and the amount of mutagen. Agar incorporation tests can be used to generate dose–response curves similar to that shown in Fig. 35.7. The simplicity, sensitivity and reproducibility of the Ames test has resulted in its widespread use for screening potential carcinogens in many countries, though it is not an infallible test for carcinogenicity.

 SAFETY NOTE Correct handling procedures must be followed at all times, as the test substances may be carcinogenic – testers should wear gloves and avoid skin contact or ingestion.

Despite these problems, bioassays are still much used. They are 'low-tech' and generally cheap to set up. They often allow detection at very low concentrations. Bioassays also provide the means to assess the biological activity of chemicals and to study changes in sensitivity to a chemical, which physicochemical techniques cannot do. Changes in sensitivity may be evident in the shape of the dose–response curve and its position on the concentration axis.

Bioassays can involve responses of whole organisms or parts of organisms. 'Isolated' responding systems (e.g. excised tissues or cells, Chapter 34) decrease the possibility of interference from other parts of the organism. Disadvantages include disruption of nutrition and wound damage during excision. Isolation can continue down to the molecular level, as in immunoassays (Chapter 38).

Bioassays are the basis for characterising the efficacy of drugs and the toxicity of chemicals. Here, response is often treated as a quantal (all-or-nothing) event. The E_{50} is defined as that concentration of a compound causing 50% of the organisms to respond. Where death is the observed response, the LD_{50} describes the concentration of a chemical that would cause 50% of the test organisms to die within a specified period under a specified set of conditions. Box 35.3 presents details of the Ames test, a widely used bioassay used to assess the mutagenicity of chemicals.

Legal use of bioassays – in the UK, where bioassays involving 'higher' (vertebrate) animals are controlled by the Animals Scientific Procedures Act (1986), they can only be carried out under the direct supervision of a scientist licensed by the Home Office.

Considerations when setting up a bioassay

- The response should be easily measured and as metabolically 'close' to the initial binding event as possible.
- The experimental conditions should mimic the *in vivo* environment.
- The standards should be chemically identical to the compound being measured and spread over the expected concentration range being tested.
- The samples should be purified if interfering compounds are present and diluted so the response will be on the 'linear' portion of the assay curve.

To check for interference, the bioassay may be standardised against another method (preferably physicochemical). Related compounds known to be present in the analyte solution should be shown to have minimal activity in the bioassay. If an interfering compound is present, this may show up if a known amount of standard is added to sample vials – the result will not be the sum of independently determined results for the standard and sample.

Sources for further study

Ball, A.S. (1997) *Bacterial Cell Culture: Essential Data*. Wiley, New York.

Cann, A.J. (1999) *Virus Culture: a Practical Approach*. Oxford University Press, Oxford.

Cartledge, T.G. (ed.) (1992) *In Vitro Cultivation of Microorganisms*. Butterworth-Heinemann, Oxford.
[Includes in-depth coverage of the mathematical principles underlying growth in batch and continuous culture]

Hewitt, W. and Vincent, S. (1989) *Theory and Application of Microbiological Assay*. Academic Press, New York.

Jennings, D.H. and Isaac, S. (1995) *Microbial Culture*. Bios, Oxford.

Rhodes, P.M. and Stanbury, P.F. (1997) *Applied Microbial Physiology: A Practical Approach*. Oxford University Press, Oxford.
[Provides details of growth requirements and culture methods applicable to a range of different microbes]

Salanki, J., Jeffrey, D. and Hughes, G.M. (1994) *Biological Monitoring of the Environment: A Manual of Methods*. CAB, Wallingford.

Study exercises

35.1 Calculate the specific growth rate and doubling time of cells in culture. What are the specific growth rates and doubling times of the following? (Give all answers to three significant figures.)
  (a) A broth culture of the yeast *Saccharomyces cerevisiae* growing in exponential phase and containing 5.2×10^4 cells at 10 am and 3.4×10^6 cells at 7 pm.
  (b) A log phase culture of *E. coli* containing 3.0×10^4 CFU ml^{-1} at the start of the experiment and 6.7×10^7 CFU ml^{-1} at the end of the experiment, 200 minutes later.
  (c) An actively growing culture of *Bacillus subtilis* containing 32 bacteria in 25 squares of a haemocytometer chamber at 13.00 hours and 250 bacteria in 25 squares of the same haemocytometer chamber at 15.30 hours.

35.2 Practise the calculations involved in using a haemocytometer to make a direct microscopic count. Express your answers to three significant figures in all cases.
  (a) The mean cell count of a yeast suspension per small square of an improved Neubauer counting chamber was 6.42. If the volume of a small square is 2.5×10^{-7} ml, what is the cell count per ml?
  (b) The following counts were obtained for bacteria in 20 individual small squares of a Petroff–Hauser counting chamber: 26, 36, 42, 35, 27, 16, 29, 50, 24, 43, 41, 35, 18, 36, 33, 47, 25, 46, 32, 57. If the volume of each small square is 8×10^{-7} ml, what is the cell count per ml?
  (c) A 10^{-2} dilution of a dense suspension of *E. coli* was examined microscopically using an improved Neubauer counting chamber, giving a total of 78 cells in a total of 25 small squares. If the volume of each small square is 2.5×10^{-7} ml what is the cell count per ml of the original (undiluted) suspension?
  (d) A dilute suspension of yeast cells was concentrated tenfold by centrifuging 10 ml and resuspending the pellet in 1 ml. The mean cell count of the concentrated yeast suspension per large square of a Petroff–Hauser counting chamber was 6.8. If the volume of the large square is 2×10^{-5} ml, what is the cell count per ml of the original suspension?

35.3 Practise the calculations involved in making a plate count. Express your answers to three significant figures.
  (a) The mean spread plate count for 100 μl of a 10^{-5} dilution of a culture of *E. coli* was 54.4 CFU. What is the mean plate count per ml of the original suspension?
  (b) Three replicate plates of nutrient agar were each spread with 200 μl of a 10^{-3} dilution of a bacterial suspension, giving colony counts of 34, 40 and 37 after 24 h incubation. What is the mean plate count per ml of the original suspension?
  (c) A 20-fold dilution of a yeast suspension was used to prepare four replicate pour plates, each containing 500 μl of this dilution and giving counts of 211, 186, 194 and 202 after incubation. What is the mean plate count per ml of the original suspension?
  (d) A sample of 50 g of raw seafood was homogenised and diluted to 2% w/v in sterile saline solution. Three replicate pour plates were prepared using 1 ml of the diluted sample, giving counts for *E. coli* of 35, 41 and 32. What is the average count of *E. coli* per 100 g of raw seafood?
  (e) Duplicate samples of 250 ml of river water were filtered through separate sterile membranes of pore size 0.2 μm. The membranes were then transferred to the surface of an agar-based medium and incubated for 48 h at 37 °C, giving colony counts of 45 and 32 respectively. What is the mean count per 100 ml of river water?
  (f) A sample of bottled drinking water was processed by serial decimal dilution and 500 μl samples of each dilution were pour plated and incubated at 22 °C for 72 h. The colony counts obtained for the three replicate plates of the 10^{-1} dilution were 28, 32 and 39. Does this water meet the EU regulations for mineral water, which specify a maximum average plate count at 22 °C of 100 CFU ml^{-1}?

35.4 Use bioassay data to estimate the amount of analyte in unknown samples. Bioassay data obtained by exposing water fleas (*Daphnia* sp.) to a purified algal toxin are shown in the table on the next page.
  (a) Graph the data and use the graph to estimate the toxin contents of two samples obtained from two lochs which gave (i) a heart rate of 130 beats per minute and (ii) a heart rate of 25 beats per minute.
  (b) Which one of the two estimates would be the least reliable and why?
  (c) How could you improve the reliability of this estimate, if you had more of the sample?

(continued)

Study exercises (continued)

(d) What unmeasured factors might limit the usefulness of this assay?

Responses of *Daphnia* to toxin

Toxin concentration (mmol l^{-1})	Mean heart rate of *Daphnia* sp. (beats per minute)
1.00×10^{-3}	20
3.16×10^{-4}	23
1.00×10^{-4}	26
3.16×10^{-5}	50
1.00×10^{-5}	180
3.16×10^{-6}	192
1.00×10^{-6}	195

35.5 Use agar diffusion bioassay data to quantify a growth inhibitor in a sample. Agar diffusion bioassay is a simple means of estimating the amount of an inhibitory substance in a test sample, by adding the substance, in aqueous solution, to wells cut in a Petri plate of agar medium seeded with a microbe sensitive to the substance and then incubating the plate at an appropriate growth temperature. Diffusion of the inhibitory substance gives a zone of growth inhibition proportional to $\log_{10}$ of the concentration of substance. In the following example and table (below), the amount of the antibiotic nisin in a test sample can be determined by comparing the degree of growth inhibition of *Bacillus subtilis* for the test solution and a series of standards containing known concentrations of nisin. Prepare a calibration curve and determine the nisin concentration of solution 5 (unknown). Express your answer in $\mu g\, ml^{-1}$, to one decimal place.

Growth inhibition of *B. subtilis* in solutions containing different amounts of nisin

Solution	Nisin content ($\mu g\, ml^{-1}$)	Zone of growth inhibition (mm)
1	40	26.5
2	20	23
3	10	19.5
4	5	17
5	unknown (test solution)	22.5

36 Homogenisation and fractionation of cells and tissues

> **Definitions**
>
> **Disruption** – a process involving structural damage to cells and tissues, to an extent where the biomolecule or complex of interest is released.
>
> **Homogenisation** – a process where cells and tissues are broken into fragments small enough to create a uniform, stable emulsion (a homogenate).
>
> **Homogeniser** – a general term for any equipment used to disrupt or homogenise cells and tissues.

Most biological molecules and subcellular complexes must be isolated from their source material in order to be studied in detail. Unless the biomolecule is already a component of an aqueous medium (e.g. plasma, tissue exudate), the first step will be to disrupt the structure of the appropriate cells or tissues. Following disruption and/or homogenisation, the *in vitro* environment of the homogenate (disrupted tissue) will be very different from that of the intact cell, and it is important that the integrity of the biomolecule or subcellular complex is preserved as far as possible during the isolation procedure.

> **KEY POINT** Disruption may be achieved by chemical, physical or mechanical procedures – the rigour of the technique(s) required will depend on the intracellular location of the molecule and the nature of the source material.

Types of cell and their susceptibility to disruption

Cell walls are a major obstacle to disruption in many organisms and the technique used for a given application must take into account this aspect. With animal cells, there is no cell wall and the plasma membrane and cytoskeleton are relatively weak: unicellular suspensions of animal cells (e.g. blood cells) can be disrupted by gentle techniques. Animal cells within tissues are more difficult to disrupt, due to the presence of connective tissue. Muscle requires vigorous techniques, as a result of additional contractile proteins. For plant cells, the plasma membrane is surrounded by a cellulose cell wall, sometimes with additional components, e.g. lignin or waxes. As a result, large shear forces are often required to disrupt plant cells.

Most Gram-positive and Gram-negative bacteria are surrounded by a rigid, protective cell wall. Peptidoglycan is a major structural component and this may be degraded by the enzyme lysozyme, particularly in certain Gram-positive organisms. In contrast, the outer membrane of Gram-negative bacteria (Fig. 36.1b) protects against lysozyme: however, prior treatment of Gram-negative bacteria with EDTA destabilises the outer membrane by removing Ca^{2+}, making the cell sensitive to lysozyme. In an osmotically buffered solution, the cells will be converted to sphaeroplasts (p. 230), since components the Gram-negative envelope will remain after lysozyme treatment. Note also that some Gram-positive bacteria may be insensitive to lysozyme, due either to a modified peptidoglycan structure or to additional protective outer layers (proteinaceous S-layers: Fig. 36.1a).

The cell walls of filamentous fungi and yeasts are very robust, containing up to 90% polysaccharide (e.g. chitin, mannan), with embedded protein microfibrils – they are often difficult to disrupt, requiring enzymic digestion and/or mechanical homogenisation. Hopkins (1991) gives details of the various systems available for homogenisation and disruption of biological material.

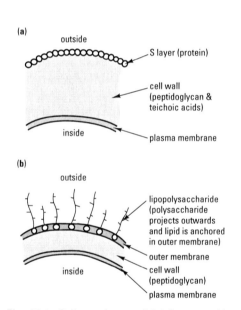

Fig. 36.1 Cell envelopes of (a) Gram-positive and (b) Gram-negative bacteria.

Homogenising media

The solution used for homogenisation serves several purposes, since it acts as a solvent or suspension medium for the released components; it serves as a cooling medium (since many biological macromolecules are denatured by heat); and it contains various reagents that may help to preserve the biological integrity of components. A typical medium will contain:

- Buffer. This replaces the intracellular buffer systems and is needed to prevent pH changes that might denature proteins. TRIS and phosphate buffers are often used for 'physiological' pH values (7.0–8.0); alternatives include various zwitterionic buffers (p. 156).
- Inorganic salts. The intracellular ionic strength is quite high, so KCl and NaCl are often included to maintain the ionic strength of the homogenate. However, the total concentration of inorganic salts should be kept below $100\,\text{mmol}\,\text{l}^{-1}$, to avoid thickening of the homogenate due to solubilisation of structural proteins.
- EDTA. This chelates divalent cations and removes metal ions (e.g. Cu^{2+}, Pb^{2+}, Hg^{2+}) that inactivate proteins by binding to thiol groups. In addition, it removes Ca^{2+} which could activate certain proteases, nucleases and lipases in the homogenate.
- Sucrose. This can be used to prevent osmotic lysis of organelles (e.g. mitochondria, lysosomes), and stabilises proteins from hydrophobic intracellular environments by reducing the polarity of the aqueous medium.
- Mg^{2+}. This helps to preserve the integrity of membrane systems by counteracting the fixed negative charges of membrane phospholipids.
- Protease inhibitors (e.g. phenylmethanesulphonylfluoride, PMSF; L-*trans*-epoxysuccinylleucylamido-(4-guanodino)-butane, E-64; leupeptin). These protect solubilised proteins from digestion by intracellular proteases, mainly released from lysosomes on disruption of the cell. Lysosomal proteases have acid pH optima, another reason for maintaining the pH of the medium close to neutrality.
- Reducing agents (e.g. 2-mercaptoethanol, dithiothreitol, cysteine at $\approx 1\,\text{mmol}\,\text{l}^{-1}$). These reagents prevent oxidation of certain proteins, particularly those with free thiol groups that may be oxidised to disulphide bonds when released from the cell under aerobic conditions.
- Detergents (e.g. Triton X-100, SDS). These cause dissociation of proteins and lipoproteins from the cell membrane, aiding the release of membrane-bound and intracellular components.

Methods of disrupting cells and tissues

Prior to disruption, animal tissues will need to be freed of any visible fat deposits and connective components, and then randomly sliced with fine scissors or a scalpel. Any fibrous and vascular tissue should be removed from plant material. Disruption can be achieved by mechanical and non-mechanical means: the principal applications of various methods are described in Table 36.1 for the major cell and tissue types.

Non-mechanical methods

- Osmotic shock: cells are first placed in a hypertonic solution of high osmolality (p. 147), e.g. 20% w/v sucrose, leading to loss of water. On dilution of this solution (e.g. by addition of water or transfer to a hypotonic solution), the cells will burst due to water influx. Osmotic shock treatment is effective for wall-less cells, sphaeroplasts (p. 230) or protoplasts.

Using chelating agents – EDTA (ethylenediaminetetraacetic acid) and EGTA (ethylenebis(oxethylenenitrilo)-tetraacetic acid). If Mg^{2+} is an important component of the medium, use EGTA, which does not chelate Mg^{2+}.

Extracting proteins from plants – carry out all post-homogenisation procedures as quickly as possible, because plant cell vacuoles may contain phenols that will inactivate proteins when released.

Definitions

Isotonic – a medium with the same water potential (p. 150) as the cells or tissues.

Hypertonic – a medium with a more negative water potential, compared with the cells or medium.

Hypotonic – a medium with a less negative water potential, compared with the cells or medium.

Table 36.1 Summary of techniques for the disruption of tissues and cells – note that safety glasses should be worn for all procedures

Technique	Suitability	Comments
Non-mechanical methods		
Osmotic shock	Animal soft tissues Some plant cells	Small scale only
Freeze/thaw	Animal soft tissues Some bacteria	Time consuming; small scale; closed system – suitable for pathogens with appropriate safety measures; some enzymes are cold-labile
Lytic enzymes, e.g. lipases; proteases pectinase; cellulase	Animal cells Plant cells	Mild and selective; small scale; expensive; enzymes must be removed once lysis is complete
Lysozyme	Some bacteria	Gram-negative bacteria must be pretreated with EDTA. Suitable for some organisms resistant to mechanical disruption.
Mechanical methods		
Pestle and mortar + abrasives	Tough tissues	Not suitable for delicate tissues
Ball mills + glass beads	Bacteria and fungi	May cause organelle damage in eukaryotes
Blenders and rotor-stators	Plant and animal tissues	Ineffective for microbes
Homogenisers (glass and Teflon)	Soft, delicate tissues e.g. white blood cells, liver	Glass may shatter – wear safety glasses during use
Solid extrusion (Hughes press)	Tough plant material; bacteria; yeasts	Small scale
Liquid extrusion (French pressure cell)	Microbial cells	Small scale
Ultrasonication	Microbial cells	Cooling required; small scale; may cause damage to organelles, especially in eukaryotic cells.

- Freezing and thawing: causing leakage of intracellular material, following cell wall and membrane damage and internal disruption due to ice crystal formation.
- Lytic enzymes: damaging the cell wall and/or plasma membrane. Cells can then be disrupted by osmotic shock or gentle mechanical treatment.

Avoiding protein denaturation during homogenisation – excessive frothing of the homogenate indicates denaturation of proteins (think of whipping egg whites for meringue).

Mechanical methods

All mechanical procedures for cell disruption generate heat, and this may denature proteins. Therefore it is very important to cool the starting material, the homogenising medium, and, if possible, the homogeniser itself (to ≈4 °C). The homogenisation should be carried out in short bursts, and the homogenate should be cooled in an ice bath between each burst. Cooling will also reduce the activity of any degradative enzymes in the homogenate. Ideally, carry out homogenisation in a walk-in cooler (a 'cold room') typically at 4–10°C.

Equipment commonly used includes:

- Mixers and blenders. These are similar to domestic liquidisers, with a static vessel and rotating blades. The Waring blender is widely used: it has a stainless steel vessel that will stay cool if pre-chilled. The vessel and blades are designed to maximise turbulence, both disrupting and homogenising tissues and cells.

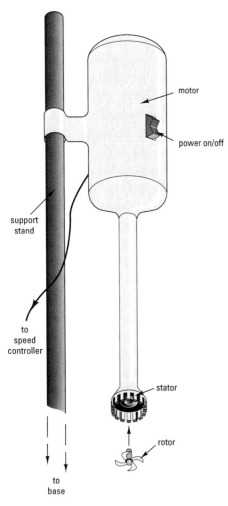

Fig. 36.2 Components of a rotor-stator homogeniser.

Fig. 36.3 Ground-glass homogeniser.

- Ball mills (e.g. Retch mixer mill, Mickle mill). These devices contain glass beads that vibrate and collide with each other and with tissues/cells, leading to disruption.
- Liquid extrusion devices (e.g. French pressure cell). Cells are forced from a vessel to the outside, through a very narrow orifice at high pressures ($\approx$100 MPa). The resulting pressure changes are a powerful means of disrupting cells.
- Solid extrusion (e.g. Hughes press). Here, a frozen cell paste is forced through a narrow orifice, where the shear forces and the abrasive properties of the ice crystals cause cell disruption.
- Rotor-stators (e.g. Ultra-turrax homogeniser). These have a rotor (a set of stainless steel blades) and a stator (a slotted stainless steel cylinder) at the tip of a stainless steel shaft, immersed in the homogenising medium: the arrangement is illustrated in Figure 36.2. The high speed of the rotor blades causes material in the homogenising fluid to be sucked into the dispersing head, where it is pressed radially through the slots in the stator. Along with the cutting action of the rotor blades, the material is subjected to very high shear and thrust and the resulting turbulence in the gap between rotor and stator gives effective mixing. The vigour of the homogenisation process can be altered by varying the rotor speed setting. Various sizes of rotor-stator are available, with typical diameters in the range 8–65 mm: the smaller sizes are particularly useful for small-scale preparations.
- Sonicators. Ultrasonic waves are transmitted to an aqueous suspension of cells via a metal probe. The ultrasound creates bubbles within the liquid and these produce shock waves when they collapse. Successful disruption depends on the correct choice of power and incubation time, together with pH, temperature and ionic strength of the suspension medium, often obtained by trial and error. You can reduce the effects of heating during ultrasonication by using short 'bursts' of power (10–30 s), with rests of 30–60 s in between, and by keeping your cell suspension on ice during disruption. An ultrasonic water bath provides a more gentle means of disrupting certain types of cells, e.g. some bacterial and animal cells.
- Homogenisers. These involve the reciprocating movement of a ground glass or Teflon pestle within a glass tube (Fig. 36.3). Cells are forced against the walls of the tube, releasing their contents. For glass pestles, the tubes also have ground glass homogenising surfaces and may have an overflow chamber. The homogeniser can either be hand operated (e.g. Dounce), or motorised (e.g. Potter-Elvejham). The clearance between the pestle and the tube (range 0.05–0.5 mm) must be chosen to suit the particular application.

Cell fractionation and the isolation of organelles

The fractionation and separation of organelles from a cell homogenate by differential centrifugation are described in Chapter 43. Particular organelles can be obtained by appropriate choice of source tissue and homogenisation method, as illustrated in Table 36.2 for the major types of organelle.

Table 36.2 Isolation and fractionation procedures for various organelles

Stage	Nuclei	Mitochondria	Microsomes	Chloroplasts*
Source	Thymus tissue, which has little cytoplasm, giving high yields.	Beef heart, with fat and connective tissue removed, then cubed and minced. Keep at pH 7.5 using TRIS buffer.	Rat liver, stored overnight to reduce glycogen content.	Spinach leaves, de-ribbed and cut into 1 cm strips.
Pretreatment	Rinse with buffered physiological saline. Suspend in homogenising medium.	Suspend in 2× volume of ice-cold homogenising medium. Squeeze through muslin.	Chop finely with scissors and wash in 2× volume of homogenising medium.	Rinse, then suspend in 3× volume of prechilled homogenising medium.
Homogenising medium	250 mmol l^{-1} sucrose; 10 mmol l^{-1} TRIS/HCl buffer, pH 7.6; 5 mmol l^{-1} MgCl$_2$; 0.2–0.5% v/v Triton X-100	250 mmol l^{-1} sucrose; 10 mmol l^{-1} TRIS/HCl buffer pH 7.7, containing 1 mmol l^{-1} succinic acid and 0.2 mmol l^{-1} EDTA.	250 mmol l^{-1} sucrose; 50 mmol l^{-1} TRIS/HCl buffer, pH 7.5; 25 mmol l^{-1} KCl; 5 mmol l^{-1} MgCl$_2$.	400 mmol l^{-1} sucrose; 25 mmol l^{-1} HEPES/NaOH buffer at pH 7.6; 2 mmol l^{-1} EDTA.
Homogenisation	Waring blender, low speed, 3 min.	Bring to pH 7.8 using 2 mol l^{-1} TRIS base: Waring blender, high speed, 15 s: check and adjust pH to 7.8 and repeat blending step, 5 s.	Potter-Elvejham glass homogeniser with a Teflon pestle – 3 × 5 min at 800 rpm.	Prechilled Waring blender or rotor-stator.
Filtration/ centrifugation	Filter through gauze. Spin at 2000 g for 10 min; discard supernatant. Repeat the homogenisation, filtration and centrifugation stages to improve purity of organelles.	Spin at 1200 g for 20 min: filter supernatant through muslin: centrifuge at 26 000 g for 15 min. Remove and discard upper (lighter) layer of pellet. Resuspend lower layer and rehomogenise (2 × 5 s). Centrifuge at 26 000 g for 15 min.	Centrifuge at 680 g for 10 min – discard pellet; centrifuge at 10 000 g for 10 min – discard pellet; centrifuge at 100 000 g for 60 min – retain pellet. Resuspend in buffer, pH 8.0 and re-centrifuge at 100 000 g for 60 min.	Pass through several layers of muslin (wear gloves); centrifuge at 2500 g for 60 s; resuspend pellet in buffer, pH 7.6; recentrifuge at 2500 g for 60 s. The colour of the supernatants gives a visual indication of chloroplast damage (e.g. if green).
Before use	Resuspend in homogenisation medium without Triton X-100.	Resuspend pellet in buffer, pH 7.8 and either use immediately, or store at −20 °C overnight.	Resuspend in buffer solution at pH 8.0.	Resuspend pellet in appropriate incubation medium, containing sucrose, e.g. for CO$_2$/O$_2$ studies, p. 400.

*An alternative approach is to use plant protoplasts (p. 230) as the starting material, releasing the chloroplasts by gentle lysis – diluting the medium with water.

Text reference

Hopkins, T. (1991) Physical and chemical cell disruption for the recovery of intracellular proteins. In: *Purification and Analysis of Recombinant Proteins* (ed R. Seetharam and S.K. Sharma). pp. 57–84. Marcel Dekker, New York.

Sources for further study

Anon. *Ultrasonic Homogenizers*. Available: http://www.biologics-inc.com/ultrasonic_homogenizers.htm Last accessed: 01/04/07.
[One of many websites dealing with commercial products]

Boyer, R. (2000) *Modern Experimental Biochemistry*, 3rd edn. Benjamin Cummings, San Francisco.
[Also covers centrifugation, spectroscopy, etc.]

Boyer, R. (2005) *Biochemistry Laboratory: Modern Theory and Techniques*. Benjamin Cummings, New York.

Graham, J.M. and Rickwood, D. (1997) *Subcellular Fractionation: A Practical Approach*. Oxford University Press, Oxford.

BioSpec Products Laboratory Cell Disrupters. Available: http://www.biospec.com/Lab%20Cell%20Disrupters%20Review.htm

Last accessed: 01/04/07.
[Provides a brief overview of methods]

Wilson, K. and Walker, J. (eds) (2005) *Principles and Techniques of Biochemistry and Molecular Biology*, 6th edn. Cambridge University Press, Cambridge.
[Also covers a range of other topics, including spectroscopy and chromatography]

Study exercises

36.1 Consider suitable disruption procedures for particular types of cell or tissue. After reading this chapter, identify the major problems associated with disrupting each of the following, and list suitable practical procedures to overcome these problems:
(a) Gram-negative bacteria;
(b) skeletal muscle;
(c) plant tissue.

36.2 Select a suitable disruption technique for the preparation of intracellular components. How might the procedure used to prepare a soluble cytoplasmic protein from fresh liver differ from that required for the isolation of intact organelles?

36.3 Test your understanding of the function of homogenisation medium components. The following list provides a number of possible components that might be used in a homogenisation medium:
(a) sucrose;
(b) low ionic strength buffer;
(c) dithiothreitol;
(d) detergent (e.g. Triton X-100).
Select one of these components to include in a medium for preparing:
   (i) a membrane protein from liver tissue;
   (ii) a mitochondrial suspension from liver tissue;
   (iii) an oxygen-sensitive enzyme from pancreatic tissue;
   (iv) a protein from the sarcolemma (cytoplasm) of skeletal muscle.
Give a brief explanation for your selection in each case.

Analytical Techniques

37 Calibration and its application to quantitative analysis

There are many instances where it is necessary to measure the quantity of a test substance using a calibrated procedure. You are most likely to encounter this approach in one or more of the following practical exercises:

Understanding quantitative measurement – Chapters 28 and 49 contain details of the basic principles of valid measurement, while Chapters 38–48 deal with some of the specific analytical techniques used in biomolecular science. The use of internal standards is covered on p. 318.

- Quantitative spectrophotometric assay of biomolecules (Chapter 41).
- Flame or atomic absorption spectroscopic analysis of metal ions in biological solutions (p. 286).
- Using a chromogenic or fluorogenic substrate to determine the activity of an enzyme (p. 399).
- Quantitative chromatographic analysis, e.g. GC or HPLC (Chapter 45).
- Using a bioassay system (p. 238) to quantify a test substance: examples include immunodiffusion (Chapter 38) and radioimmunoassay (p. 260).

KEY POINT *In most instances, calibration involves the establishment of a relationship between the measured response (the 'signal') and one or more 'standards' containing a known amount of substance.*

In some instances, you can measure a signal due to an inherent property of the substance, e.g. the absorption of UV light by nucleic acids (p. 370), whereas in other cases you will need to react it with another substance to see the result (e.g. molecular weight measurements of DNA fragments after electrophoresis, visualised using ethidium bromide (p. 371), or to produce a measurable response (e.g. the reaction of copper(II) ions and peptide bonds in the biuret assay for proteins, p. 353).

Calibrating laboratory apparatus – this is important in relation to validation of equipment, e.g. when determining the accuracy and precision of a pipettor by the weighing method: see p. 130.

The different types of calibration curve

By preparing a set of solutions (termed 'standards'), each containing either (i) a known *amount* or (ii) a specific *concentration* of the substance, and then measuring the response of each standard solution, the underlying relationship can be established in graphical form as a 'calibration curve', or 'standard curve'. This can then be used to determine either (i) the amount or (ii) the concentration of the substance in one or more test samples. Alternatively, the response can be expressed solely in mathematical terms: an example of this approach is the determination of chlorophyll pigments in plant extracts by measuring absorption at particular wavelengths, and then applying a formula based on previous (published) measurements for purified pigments (p. 282).

There are various types of standard curve: in the simplest cases, the relationship between signal and substance will be linear, or nearly so, and the calibration will be represented best by a straight-line graph (see Box 37.1). In some instances (Fig. 37.1a), you will need to transform either the x values or y values in order to produce a linear graph (e.g. in radial immunodiffusion bioassay, where the y values are squared, p. 259. In other instances, the straight-line relationship may only hold up to a certain value (the 'linear dynamic range') and beyond this point the graph may curve (e.g. in quantitative spectrophotometry, the Beer–Lambert relationship often becomes invalid at high absorbance, giving

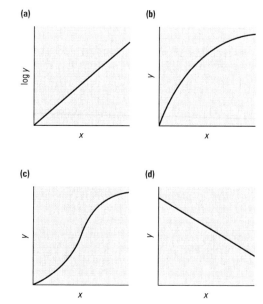

Fig. 37.1 Calibration curves: (a) log-linear; (b) curvilinear; (c) sigmoid, or S-shaped; (d) inverse.

a curve, Fig. 37.1b). Some calibration curves are sigmoid (Fig. 37.1c). Finally, the signal may *decrease* in response to an increase in the substance (Fig. 37.1d), e.g. radioimmunoassay, where an inverse sigmoid calibration curve is obtained (p. 260). In some practical classes, you may be told that the relationship is expected to be linear, curvilinear, or

Box 37.1 The stages involved in preparing and using a calibration curve

1. **Decide on an appropriate test method** – for example, in a project you may need to research the best approach to the analysis of a particular metabolite in your biological material.

2. **Select either (a) amount or (b) concentration, and an appropriate range and number of standards** – in practical classes, this may be given in your schedule, along with detailed instructions on how to make up the standard solutions. In other cases, you may be expected to work this out from first principles (Chapters 22–24 give worked examples) – aim to have evenly spaced values along the *x* axis.

3. **Prepare your standards very carefully** – due attention to detail is required: for example, you should ensure that you check the calibration of pipettors beforehand, using the weighing method (p. 130). Don't forget the 'zero standard' plus any other controls required, e.g. to test for interference due to other chemical substances. Your standards should cover the range of values expected in your test samples.

4. **Assay the standards and the unknown (test) samples** – preferably all at the same time, to avoid introducing error due to changes in the sensitivity or drift in the zero setting of the instrument with time. It is a good idea to measure all of your standard solutions at the outset, and then measure your test solutions, checking that the 'zero standard' and 'top standard' give the same values after, say, every six test measurements. If the re-measured standards do not fall within a reasonable margin of the previous value, then you will have to go back and recalibrate the instrument, and repeat the last six test measurements. If your test samples lie outside the range of your standards, you may need to repeat the assay using diluted test samples (extrapolation of your curve may not be valid, see p. 462).

5. **Draw the standard curve, or determine the underlying relationship** – Figure 37.2 gives an example of a typical linear calibration curve, where the spectrophotometric absorbance of a series of standard solutions is related to the amount of

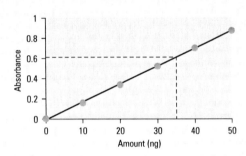

Fig. 37.2 Typical calibration curve for spectrophotometric analysis.

substance. When using a spreadsheet (Box 37.2) or a graphics package (Chapter 12), it is often appropriate to use Model I linear regression (p. 501) to produce a linear trend line (also termed the 'line of best fit') and you can then quote the value of r^2, which is a measure of the 'fit' of the measurements to the line (see p. 501). However, you should take care not to use a linear plot when the underlying relationship is clearly non-linear (Fig. 63.12) and you must consider whether the assumptions of the regression analysis are valid (e.g. for transformed data).

6. **Determine the amount or concentration in each unknown sample** – either by reading the appropriate value from the calibration curve, or by using the underlying mathematical relationship, i.e. $y = a + bx$ (pp. 476–7). Make sure you draw any horizontal and vertical construction lines very carefully – students often lose marks unnecessarily by submitting poorly drawn construction lines within practical reports.

7. **Correct for dilution or concentration, where appropriate** – for example, if you diluted each test sample by 10-fold, then you would need to multiply by 10 to determine the value for the undiluted test sample. As another example, if you assayed 0.2 ml of test sample, you would need to multiply the value obtained from the calibration curve by 5, to give the value per ml.

8. **Quote your test results to an appropriate number of significant figures** – this should reflect the accuracy of the method used (see pp. 472–4), not the size of your calculator's display.

Practical considerations

Amount or concentration?

This first step is often the most confusing for new students. It is vital that you understand the difference between *amount* of substance (e.g. mg, ng), and *concentration* (the amount of substance per unit volume, e.g. $mmol\,l^{-1}$, $mol\,m^{-3}$, % w/v) before you begin your practical work.

KEY POINT *Essentially, you have to choose whether to work in terms of either (i) the total amount of substance in your assay vessel (e.g. test tube or cuvette) or (ii) the final concentration of the substance in your assay vessel, which is independent of the volume used.*

The interconversion of amount and concentration is covered in more detail on p. 135. Either way, this is usually plotted on the x (horizontal) axis and the measured response on the y (vertical) axis.

Example A test tube containing 8 ml of water plus 2 ml of 1% w/v NaCl ($M_r = 58.44$) would have a *mass concentration* of 0.2% w/v NaCl (a five-fold dilution of the original NaCl solution), which can also be expressed as $2\,g\,l^{-1}$; in terms of *molar concentration* (p. 144), this would be equivalent to $2 \div 58.44 = 0.0342\,mol\,l^{-1}$ (to three significant figures). Expressed in terms of the *amount* of NaCl in the test tube, this would be 0.02 g in *mass*, or $0.02 \div 58.44 = 3.42 \times 10^{-4}$ mol (342 μmol) in *moles*.

Choice of standards

In your early practical classes, you may be provided with a stock solution (pp. 136–7), from which you then have to prepare a specified number of standard solutions. In such cases, you will need to understand how to use dilutions to achieve the required amounts or concentrations (pp. 137–8). In later work and projects, you may need to prepare your standards from chemical reagents in solid form, where the important considerations are purity and solubility (p. 135). For professional analysis (e.g. in forensic science or clinical biochemistry), it is often important to be able to trace the original standard or stock solution back to national or international standards (Chapter 49) or to certified reference materials.

Organisations providing national/international standards – these include:

- Laboratory of the Government Chemist (http://www.lgc.co.uk/) for UK standards and European Reference Materials.
- National Institute for Science and Technology (http://www.cstl.nist.gov/), for biochemical/chemical standards in the US.
- Institute for Reference Materials and Measurements (http://www.irmm.jrc.be/) for European standards, including BCR and ERM materials.
- OIE Biological Standards Commission (http://www.oie.int/bsc/eng/en_bsc.htm), for international animal standards.

How many standards are required?

This may be given in your practical schedule, or you may have to decide what is appropriate (e.g. in project work and research). If the form and working range of the standard curve are known in advance, this may influence your choice – for example, linear calibration curves can be established with fewer standards than curvilinear relationships. In some instances, analytical instruments can be calibrated using a single standard solution, often termed a 'calibrator'. Replication of each standard solution is a good idea, since it will give you some information on the variability involved in preparing and assaying the standards. Consider whether you should plot mean values on your standard curve, or whether it is better to plot the individual values (if one value appears to be well off the line, you are likely to have made an error, and you may need to check and repeat).

Preparing your standards

It is extremely important to take the greatest care to measure out all chemicals and liquids very accurately, to achieve the best possible standard curve. The grade of volumetric flask used and temperature of the solution also affect accuracy (grade A apparatus is best). You may also consider what other additives might be required in your standard solutions.

Plotting a standard curve – do not force your calibration line to pass through zero if it clearly does not. There is no reason to assume that the zero value is any more accurate than any other reading you have made.

Box 37.2 How to use a spreadsheet (Microsoft Excel) to produce a linear regression plot

Two approaches are possible: either using the trendline feature or using the regression analysis function within the data analysis tool pack. Both make the assumption that the criteria for Model I linear regression are met (see p. 501). In the example shown below, the following simple data set has been used:

Amount (ng)	Absorbance
0	0.00
10	0.19
20	0.37
30	0.56
40	0.63
50	0.78

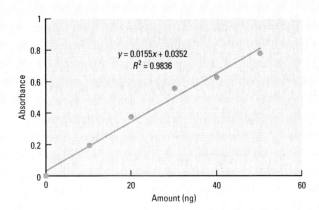

Fig. 37.3 Calibration curve showing line of best fit and details of linear regression equation.

Using the *Trendline* feature. This quick method provides a line of best fit on an Excel chart and can also provide set of equation values for predictive purposes.

1. **Create a graph (*chart*) of your data.** Enter the data in two columns within your spreadsheet, select the data array (highlight using left mouse button) and then, using the *Chart Wizard* icon, select *Chart type > XY (Scatter)*.

2. **Add a trend line.** Right-click on any of the data points on your graph, and select the *Add Trendline* menu. Choose the *Linear* trend line option, but do not click OK at this stage. Rather, from the *Options* menu, select: (i) *Display equation on chart* and (ii) *Display R-squared value on chart*. Now click OK. The equation (shown in the form $y = bx + a$) gives the slope (b) and intercept (a) of the line of best fit, while the *R*-squared value (coefficient of determination, p. 501) gives the proportional fit to the line (the closer this value is to 1, the better the fit of the data to the line).

3. **Modify the graph to improve its effectiveness.** For a graph that is to be used elsewhere (e.g. in a lab write-up or project report), adjust the display to remove the default background and gridlines and change the symbol shape (see Box 63.2 for more advice and examples). Right-click on the trend line and use the *Format Trendline > Custom* menu to adjust the *Weight* of the line to make it thinner. Drag and move the equation panel if you would like to alter its location on the chart, or delete it, having noted the values. Fig. 37.3 shows a calibration curve produced in this way for the data presented above.

4. **Use the regression equation to estimate unknown (test) samples.** By rearranging the equation for a straight line and substituting a particular *y*-value, you can predict the amount/concentration of substance (*x*-value) in a test sample. This is more precise than simply reading the values from the graph using construction lines (Box 37.1). If you are carrying out multiple calculations, the appropriate equation, $x = (y - a)/b$, can be entered into a spreadsheet, for convenience.

Using the *Regression Analysis* tool requires the 'data analysis tool pack' to be loaded beforehand (using *Tools > Add-Ins > Analysis ToolPak*) and provides summary output that contains details of slope, intercept) and coefficient of determination along with an analysis of variance (ANOVA) table (Chapter 67 gives further details of these aspects, including an example of output in study exercise 67.4).

Dealing with interfering substances – one approach is to use the method of 'standard additions', where the standards all contain a fixed additional amount of the sample (see e.g. Dean, 1997). Internal standards (p. 318) can also be used to detect such problems.

For example, do your test samples have high levels of potentially interfering substances, and should these also be added to your standards? Also consider what controls and blank solutions to prepare.

 KEY POINT The validity of your standard curve depends upon careful preparation of standards, especially in relation to accurate dispensing of the volumes of any stock solution and diluting liquid – the results for your test samples can only be as good as your standard curve.

Preparing the calibration curve and determining the amount of the unknown (test) sample(s)

This is described in stepwise fashion in Box 37.1. Check you understand the requirements of graph drawing, especially in relation to plotted curves (p. 457) and the mathematics of straight-line graphs (p. 476). Spreadsheet programs such as Microsoft Excel can be used to produce a regression line for a straight line calibration plot (p. 252). Examples of how to do this are provided in Box 37.2.

Text reference

Dean, J.D. (1997) *Atomic Absorption and Plasma Spectroscopy*, 2nd edn. Wiley, Chichester.
[Chapter 1 deals with calibration, and covers the principle of standard additions]

Sources for further study

Brown, P.J. (1994) *Measurement, Regression, and Calibration*. Clarendon Press, Oxford.
[Covers advanced methods, including curve fitting and multivariate methods]

Mark, H. (1991) *Principles and Practice of Spectroscopic Calibration*. Wiley, New York.

Miller, J.N. and Miller, J.C. (2005) *Statistics and Chemometrics for Analytical Chemistry*, 5th edn. Prentice Hall, Harlow.
[Gives detailed coverage of calibration methods and the validity of analytical measurements]

Study exercises

37.1 Determine unknowns from a hand-drawn calibration curve. The following data are for a set of calibration standards for Zn, measured by atomic absorption spectrophotometry.

Absorbance measurements for a series of standard solutions containing different amounts of zinc

Zinc concentration ($\mu g\, ml^{-1}$)	Absorbance
0	0.000
1	0.082
2	0.174
3	0.257
4	0.340
5	0.408
6	0.463
7	0.511
8	0.543
9	0.561
10	0.575

Draw a calibration curve by hand using graph paper and estimate the concentration of zinc in the following water samples:

(a) an undiluted sample, giving an absorbance of 0.157;
(b) a 20-fold dilution, giving an absorbance of 0.304;
(c) a fivefold dilution, giving an absorbance of 0.550.

Give your answer to three significant figures in each case.

37.2 Determine unknowns from a calibration curve produced in Excel. The following data are for a set of calibration standards for protein content, measured by the Lowry (Folin–Ciocalteau) method (p. 354).

Absorbance measurements for a series of protein standards

Protein (mg per tube)	Absorbance (A_{600})
0.00	0.000
0.02	0.161
0.04	0.284
0.06	0.438
0.08	0.572
0.10	0.762

(continued)

Study exercises (continued)

Using PC-based software (e.g. Excel), fit a trend line (linear regression) and determine the protein content of solutions with the following absorbances:

(a) 0.225   (b) 0.465   (c) 0.682

Give your answers to three significant figures in each case.

37.3 **Identify the errors in a calibration curve.** The figure shows a calibration curve of the type that might be submitted in a practical write-up. List the errors and compare your observations with the list given on p. 519.

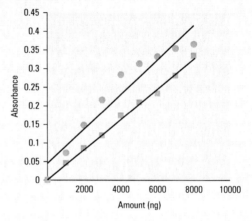

Calibration graph

38 Immunological methods

> **Definitions**
>
> **Antibody** – a protein produced in response to an antigen (an *anti*body-*gen*erating foreign macromolecule).
>
> **Epitope** – a site on the antigen that determines its interaction with a particular antibody.
>
> **Hapten** – a substance that contains at least one epitope, but is too small to induce antibody formation unless it is linked to a macromolecule.
>
> **Ligand** – a molecule or chemical group that binds to a particular site on another molecule.

Antibodies are an important component of the immune system, which protects animals against certain diseases (see Delves *et al.*, 2006). They are produced by B lymphocytes in response to foreign macromolecules (antigens). A particular antibody will bind to a site on a specific antigen, forming an antigen–antibody complex (immune complex). Immunological bioassays use the specificity of this interaction for:

- identifying macromolecules, cellular components or whole cells;
- quantifying a particular substance.

Antibody structure

An antibody is a complex globular protein, or immunoglobulin (Ig). While there are several types, IgG is the major soluble antibody in vertebrates and is used in most immunological assays. Its main features are:

- Shape: IgG is a Y-shaped molecule (Fig. 38.1), with two antigen-binding sites.
- Specificity: variation in amino acid composition at the antigen-binding sites explains the specificity of the antigen–antibody interaction.
- Flexibility: each IgG molecule can interact with epitopes which are different distances apart, including those on different antigen molecules.
- Labelling: regions other than the antigen-binding sites can be labelled, e.g. using a radioisotope or an enzyme with fluorogenic or chromogenic detection (p. 382).

KEY POINT The presence of two antigen-binding sites on a single flexible antibody molecule is relevant to many immunological assays, especially the agglutination and precipitation reactions.

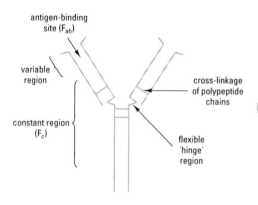

Fig. 38.1 Diagrammatic representation of IgG (antibody).

Antibody production

Polyclonal antibodies

These are commonly used at undergraduate level. They are produced by repeated injection of antigen into a laboratory animal. After a suitable period (3–4 weeks) blood is removed and allowed to clot, leaving a liquid phase (polyclonal antiserum) containing many different IgG antibodies, resulting in:

- cross-reaction with other antigens or haptens;
- batch variation, as individual animals produce slightly different antibodies in response to the same antigen;
- non-specificity, as the antiserum will contain many other antibodies.

> **Producing polyclonal antibodies** – in the UK, this is controlled by government regulations, since it involves vertebrate animals: personnel must be licensed by the Home Office and must operate in accordance with the **Animals Scientific Procedures Act (1986)** and with the **Code of Practice for the Housing and Care of Animals (2005)**.

Standardisation of polyclonal antisera therefore is difficult. You may need to assess the amount of cross-reaction, interbatch variation or non-specific binding using appropriate controls, assayed at the same time as the test samples.

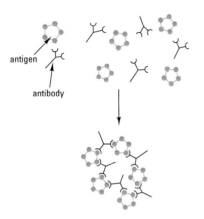

Fig. 38.2 Formation of an antigen–antibody complex (immune complex).

Monoclonal antibodies

These are specific to a single epitope and are produced from individual clones of cells (hybridomas, p. 225), grown using cell culture techniques (p. 232). Such cultures provide a stable source of antibodies of known, uniform specificity. While monoclonal antibodies are likely to be used increasingly in future years, polyclonal antisera are currently employed for many routine immunological assays.

Agglutination tests

When antibodies interact with a suspension of a particulate antigen, e.g. cells or latex particles, the formation of immune complexes (Fig. 38.2) causes visible clumping, termed agglutination. A positive haemagglutination reaction gives an even 'carpet' of red cells over the base of the tube, while a negative reaction gives a tightly packed 'button' of red cells at the bottom of the tube. Agglutination tests are used in several ways:

- Microbial identification: at the species or subspecies level (serotyping), e.g. mixing an unknown bacterium with the appropriate antiserum will cause the cells to agglutinate.
- Latex agglutination (bound antigens): by coating soluble (non-particulate) antigens onto microscopic latex spheres, their reaction with a particular antibody can be visualised.
- Latex agglutination (bound antibodies): antibodies can be bound to latex microspheres, leaving their antigen-binding sites free to react with soluble antigen.
- Haemagglutination: red blood cells can be used as agglutinating particles. However, in some instances, such reactions do not involve antibody interactions (e.g. some animal viruses may haemagglutinate unmodified red blood cells).

Precipitin tests

Immune complexes of antibodies and soluble antigens (or haptens) usually settle out of solution as a visible precipitate: this is termed a precipitin test, or precipitation test. The formation of visible immune complexes in agglutination and precipitation reactions only occurs if antibody and antigen are present in an optimal ratio (Fig. 38.3). It is important to appreciate the shape of this curve: cross-linkage is maximal in the zone of equivalence, decreasing if either component is present in excess. The quantitative precipitin test can be used to measure the antibody content of a solution (Hay *et al.*, 2002 give details). Visual assessment of precipitation reactions forms the basis of several other techniques, described below.

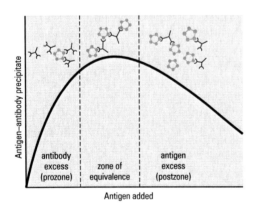

Fig. 38.3 Precipitation curve for an antigen titrated against a fixed amount of antibody.

Immunodiffusion assays

These techniques are easier to perform and interpret than the quantitative precipitin test. Precipitation of antibody and antigen occurs within an agarose gel, giving a visible line corresponding to the zone of equivalence (Fig. 38.3). Details of the main techniques are given in Box 38.1.

Box 38.1 How to carry out immunodiffusion assays

The two most widespread approaches are (i) single radial immunodiffusion (Mancini technique) and (ii) double diffusion immunoassay (Ouchterlony technique).

Single radial immunodiffusion (RID) (Mancini technique)
This is used to quantify the amount of antigen in a test solution, as follows:

1. **Prepare an agarose gel** (1.5% w/v), containing a fixed amount of antibody: allow to set on a glass slide or plate, on a level surface.
2. **Cut several circular wells in the gel.** These should be of a fixed size between 2 and 4 mm in diameter (see Fig. 38.4a). Cut your wells carefully – they should have straight sides and the agarose must not be torn or lifted from the glass plate. All wells should be filled to the top, with a flat meniscus, to ensure identical diffusion characteristics. Non-circular precipitin rings, resulting from poor technique, should not be included in your analysis.
3. **Add a known amount of the antigen or test solution to each well.**
4. **Incubate on a level surface at room temperature in a moist chamber:** diffusion of antigen into the gel produces a precipitin ring. This is usually measured after 2–7 days, depending on the molecular mass of the antigen.
5. **Examine the plates against a black background** (with side illumination), or stain using a protein dye (e.g. Coomassie blue).
6. **Measure the diameter of the precipitin ring**, e.g. using Vernier callipers.
7. **Prepare a calibration curve** (Chapter 37) from the samples containing known amounts of antigen (Fig. 38.4b): the squared diameter of the precipitin ring is directly proportional to the amount of antigen in the well.
8. **Quantify the amount of antigen in your test solutions** using a calibration curve prepared from standards assayed at the same time.

Double diffusion immunoassay (Ouchterlony technique)
This technique is widely used to detect particular antigens in a test solution, or to look for cross-reaction between different antigens.

1. **Prepare an agarose gel** (1.5% w/v) on a level glass slide or plate: allow to set.
2. **Cut several circular wells in the gel.**
3. **Add test solutions of antigen or polyclonal antiserum to adjacent wells.** Both solutions diffuse outwards, forming visible precipitin lines where antigen and corresponding antibody are present in optimal ratio (Fig. 38.5).

The various reactions between antigen and antiserum are:

- **Identity:** two wells containing the same antigen, or antigens with identical epitopes, will give a fused precipitin line (identical interaction between the antiserum and the test antigens, Fig. 38.5a).
- **Non-identity:** where the antiserum contains antibodies to two different antigens, each with its own distinct epitopes, giving two precipitin lines which intersect without any interaction (no cross-reaction, Fig. 38.5b).
- **Partial identity:** where two antigens have at least one epitope in common, but where other epitopes are present, giving a fused precipitin line with a spur (cross-reaction, Fig. 38.5c).

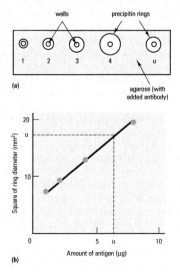

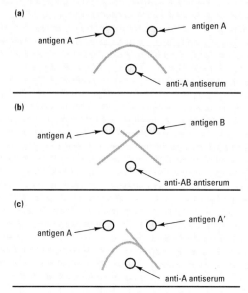

Fig. 38.4 Single radial immunodiffusion (RID). (a) Assay: four standards are shown (wells 1 to 4, each one double the strength of the previous standard), and an unknown (u), run at the same time. (b) Calibration curve. The unknown contains 6.25 μg of antigen. Note the non-zero intercept of the calibration curve, corresponding to the square of the well diameter: do not force such calibration lines through the origin.

Fig. 38.5 Precipitin reactions in double diffusion immunoassay: (a) identity; (b) non-identity; (c) partial identity.

Immunoelectrophoretic assays

These methods combine the precipitin reaction with electrophoretic migration, providing sensitive, rapid assays with increased separation and resolution. The principal techniques are:

Cross-over electrophoresis (counter-current electrophoresis)
Similar to the Ouchterlony technique, since antigen and antibody are in separate wells. However, the movement of antigen and antibody towards each other is driven by a voltage gradient (p. 320): most antigens migrate towards the anode, while IgG migrates towards the cathode. This method is faster and more sensitive than double immunodiffusion, taking 15–20 min to reach completion.

Quantitative immunoelectrophoresis (Laurell rocket immunoelectrophoresis)
Similar to RID, as the antibody is incorporated into an agarose plate while the antigen is placed in a well. However, a voltage gradient moves the antigen into the gel, usually towards the anode, while the antibody moves towards the cathode, giving a sharply peaked, rocket-shaped precipitin line, once equivalence is reached (within 2–10 h). The height of each rocket shape at equivalence is directly proportional to the amount of antigen added to each well. A calibration curve for samples containing a known amount of antigen can be used to quantify the amount of antigen present in test samples (Fig. 38.6).

Radioimmunological methods

These methods use radioisotopes to detect and quantify the antigen–antibody interaction, giving improved sensitivity over agglutination and precipitation methods. The principal techniques are:

Radioimmunoassay (RIA)
This is based on competition between a radioactively labelled antigen (or hapten) and an unlabelled antigen for the binding sites on a limited amount of antibody. The quantity of antigen in a test solution can be determined using a known amount of radiolabelled antigen and a fixed amount of antibody (Fig. 38.7). As with other immunoassay methods, it is important to perform appropriate controls, to screen for potentially interfering compounds. The basic procedure for RIA is:

1. Add appropriate volumes of a sample to a series of small test tubes. Prepare a further set of tubes containing known quantities of the substance to be assayed to provide a standard curve.
2. Add a known amount of radiolabelled antigen (or hapten) to each tube (sample and standard).
3. Add a fixed amount of antibody to each tube (the antibody must be present in limited quantity).
4. Leave at constant temperature for a fixed time (usually 24 h), to allow antigen–antibody complexes to form.
5. Precipitate the antibody and bound antigen using saturated ammonium sulphate, followed by centrifugation.
6. Determine the radioactivity of the supernatant or the precipitate (pp. 269–71).

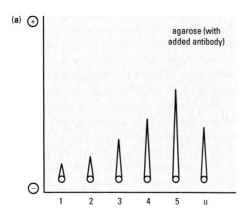
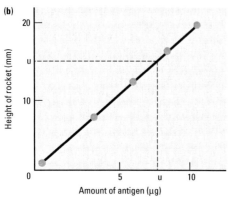

Fig. 38.6 Laurell rocket immunoelectrophoresis. (a) Assay: precipitin rockets are formed by electrophoresis of five standards of increasing concentration (wells 1 to 5) and an unknown (u). (b) Calibration curve: the unknown sample contains 7.7 μg of antigen.

Immunological Methods

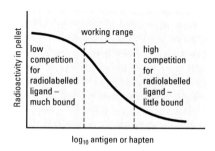

Fig. 38.7 Radioimmunoassay calibration curve. Note that the assay is insensitive at very low and very high antigen levels.

7. Prepare a calibration curve of radioactivity against $\log_{10}$ antigen (Fig. 38.7). The curve is most accurate in the central region, so adjust the amount of antigen in your test sample to fall within this range.

Note the following:

- You must be registered to work with radioactivity: check with the Departmental Safety Officer, if necessary (p. 273).
- Measure all volumes as accurately as possible as the end result depends on the volumetric quantities of unlabelled (sample) antigen, radio-labelled antigen and antibody: an error in any of these reagents will invalidate the assay.
- Incorporate replicates, so that errors can be quantified.
- Seek your supervisor's advice about fitting a curve to your data: this can be a complex process.

Immunoradiometric assay (IRMA)

This technique uses radiolabelled antibody, rather than antigen, for direct measurement of the amount of antigen (or hapten) in a sample. Most immunoradiometric assays are similar to the double antibody sandwich method described below, except that the second antibody is labelled using a radioisotope. The important advantages over RIA are:

- linear relationship between amount of radioactivity and test antigen;
- wider working range for test substance;
- improved stability/longer shelf-life.

Enzyme immunoassays (EIA)

These techniques are also known as enzyme-linked immunosorbent assays (ELISA). They combine the specificity of the antibody–antigen interaction with the sensitivity of enzyme assays using either an antibody or an antigen conjugated (linked) to an enzyme at a site which does not affect the activity of either component. The enzyme is measured by adding an appropriate chromogenic substrate (p. 382), which yields a coloured product. Enzymes offer the following advantages over radioisotopic labels:

- Increased sensitivity: a single enzyme molecule can produce many product molecules, amplifying the signal.
- Simplified assay: enzyme assays are usually easier than radioisotope assays (p. 381).
- Improved stability of reagents: components are generally more stable than their radiolabelled counterparts, giving them a longer shelf-life.
- No radiological hazard: no requirement for specialised containment/disposal facilities.
- Automation is straightforward: using disposable microtitre plates and an optical scanner.

The principal techniques are:

Double antibody sandwich ELISA

This is used to detect specific antigens, involving a three-component complex between a capture antibody linked to a solid support, the antigen, and a second, enzyme-linked antibody (Fig. 38.8). This can be used to detect a particular antigen, e.g. a virus in a clinical sample, or to quantify the amount of antigen.

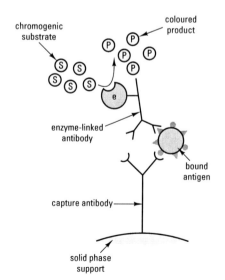

Fig. 38.8 Double antibody sandwich ELISA.

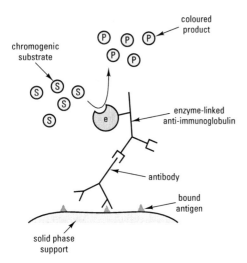

Fig. 38.9 Indirect ELISA.

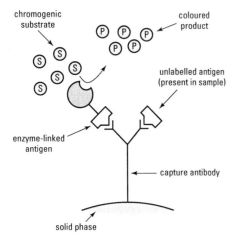

Fig. 38.10 Competitive ELISA

Indirect ELISA

This is used for antibody detection, with a specific antigen attached to a solid support. When the appropriate antibody is added, it binds to the antigen and will not be washed away during rinsing. Bound antibody is then detected using an enzyme-linked anti-immunoglobulin, e.g. a rabbit IgG antibody raised against human IgG (Fig. 38.9). One advantage of the indirect assay is that a single enzyme-linked anti-immunoglobulin can be used to detect several different antibodies, since the specificity is provided by the bound antigen.

Competitive ELISA

Here, any antigen present in a test sample competes with added enzyme-labelled antigen for a limited number of binding sites on the capture antibody (Fig. 38.10). Most commercial systems use 96-well microplates (12 columns by 8 rows), where each well is coated with the appropriate antibody. Following addition of known volumes of (i) sample and (ii) enzyme-labelled antigen, the plates are incubated (typically, up to 1 h), then washed thoroughly to remove all unbound material. Bound enzyme-labelled antigen is then detected using a suitable substrate. Quantitative results can be obtained by measuring the absorbance of each well using a spectrophotometric microplate reader: the absorbance at a particular wavelength is *inversely* proportional to the amount of antigen present in the test sample. Calibration standards (p. 252) are required to convert the readings to an amount or concentration. Alternatively, the test can be carried out in positive/negative format. Box 38.2 gives practical details for a sandwich-type ELISA.

Dip-stick (chromatographic) immunoassays

Improvements in technology have led to the development of a range of diagnostic kits that can be used either in a medical setting ('near-patient testing') or in the home, including immunoassays for cancer screening and drug testing. The first of these available to the general public was the home pregnancy test kit which measures the hormone human chorionic gonadotrophin (hCG) in urine. The level of this hormone is raised substantially during pregnancy. The principle of this test, which is similar to other home test kits, is that coloured plastic microbeads are conjugated to an anti-hCG monoclonal antibody and then coated onto an absorbent plastic or cellulose strip. When this strip is dipped into urine any hCG present will bind to the anti-hCG antibody and both hormone–antibody complexes and free antibody will then move up the strip by capillary action until they reach a second hCG-specific antibody that is coupled chemically to the strip so that it cannot move. Any hormone–antibody complex will bind to this 'fixed' antibody (through the hCG component); due to the presence of the coloured plastic microbeads, a distinct line will then appear at this location on the strip, signifying the presence of the hormone. If no hormone–antibody complex is present, the first antibody will simply continue up the strip until it reaches an antibody raised against the anti-hCG antibody itself, coupled to the strip at a higher point. This will bind any free anti-hCG antibody in the sample and a coloured line will appear at this point. Since there will always be an excess of the original anti-hCG antibody conjugated to the plastic beads, the appearance of

Box 38.2 How to perform an ELISA assay

While the following example is for a sandwich (capture) ELISA assay in microplate format, the same general principles apply to the other types of ELISA:

1. **Prepare the apparatus.** Switch on all equipment required:
   (i) The microplate reader – used to measure the absorbance of the solution in each well: set the reader to the required wavelength.
   (ii) The microplate washer (where used) – each well must be washed at various stages during the procedure. When using an automated washer, first check that the wash bottle contains sufficient diluent and then test using an old microplate, to check that all wells are being washed correctly. Where required, use a wire needle to clean any blocked wash delivery tubes and repeat. For manual washing, use a wash bottle or multichannel pipettor – make sure you fill each well and empty out all of the wash solution at the end of each wash stage.
   (iii) The computer – this will contain the software required to label the wells, draw the calibration curve and calculate the results for test samples: fill out the ELISA template with details of the assays to be carried out.

2. **Prepare the various solutions to be analysed.** These include:
   (i) Test samples – make sure that each sample is identified with a code that enables you to record what each test well contains.
   (ii) Calibrators/standards – including 'cutoff' calibrators and known positive standards.
   (iii) Controls – positive and negative controls and blanks.

3. **Coat the wells with the capture antibody.** Typically 100 µl of a solution of the appropriate monoclonal or polyclonal antibody is added to each well and microplates are then incubated overnight at 4°C to allow binding to the well.

4. **Wash the wells.** Transfer the microplate to the washer (or wash manually) – wash six times to remove excess coating (capture) antibody. The final rinse should be programmed so that the washer leaves the wells empty of diluent.

5. **Add blocking solution to each well.** Typically 100 µl of an inert protein solution (bovine serum albumin) is added to each well to block any free binding sites on the well. Microplates are incubated at room temperature for 30 min and then washed, as in step 4.

6. **Add test samples, calibrators and controls to wells.** Typically 100 µl of appropriately diluted test sample, control, etc. is added to each well. Microplates are incubated at room temperature for 90 min and then washed as step 4.

7. **Add the detection antibody to each well.** Add 100 µl of monoclonal or polyclonal detection antibody labelled with a suitable enzyme (e.g. horseradish peroxidise, HRP) to each well. The microplate is then incubated at room temperature for 30–60 min and then washed, as in step 4.

8. **Add chromogenic substrate to each well.** For example, with HRP-labelled antigen add 100 µl of a standard solution of tetramethylbenzidine (TMB) and hydrogen peroxide to each well and re-incubate in darkness for 30 min, to allow colour development. The TMB is oxidised in the presence of hydrogen peroxide to produce a blue colour.

    SAFETY NOTE TMB and hydrogen peroxide are harmful by inhalation – use a fume hood.

9. **Stop the reactions.** For example, by adding 100 µl of 2 mol l^{-1} sulphuric acid to denature the enzyme. The colour of the oxidised TMB will change from blue to yellow as a result of the pH shift. While the human eye can readily distinguish different shades of blue, it is more difficult to visually assess different shades of yellow, once the reaction has been stopped.

10. **Measure the absorbance of each sample/calibrator/control well.** Transfer the microplate to the reader and assay at an appropriate wavelength: for TMB, use 450 nm.

11. **Interpret the results.** Check that the absorbance values of calibrators and control are within the required range. Then, for each sample, either record the absolute value (convert to concentration or amount e.g. using a calibration curve, p. 252) or record as 'positive' or 'negative' (based on values for 'cutoff' calibrators), as appropriate.

---

two coloured lines means that the hormone is present and the test result is positive, while if only the second coloured line appears, it means that the test is negative (this also serves as a control to confirm that the test is working – lack of any coloured lines indicates that the test has failed). Fig. 38.11 shows the general operating principles.

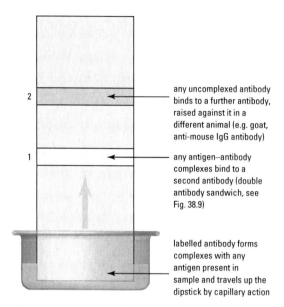

Fig. 38.11 Two-site dipstick immunoassay. For a positive test, coloured bands will appear at positions 1 and 2 while a negative test, as shown here, will give colour in band 2 only (any other colour options indicate a failed test).

Complement-based assays

Complement is a term applied to a group of around 20 different proteins which are involved in various aspects of immune function, including (i) control of inflammation, (ii) preparation of microbial cells for phagocytosis (opsonisation), (iii) lysis of target cells (e.g. antibody-coated cells) and (iv) activation of leucocytes (for details see Delves et al., 2006). At a practical level, the complement system can be used to detect (i) antibodies, (ii) antigens or (iii) antibody–antigen complexes, often *via* its lytic action on antibody-coated ('sensitised') cells. The operating principles of a complement fixation assay are illustrated in Fig. 38.12 for antibody detection, with the following steps:

1. Prepare serial doubling dilutions (pp. 137–8) of a test sample containing an unknown amount of antibody.

Understanding complement-based assays – *the principal benefit from using the complement system is that it operates as an amplifying cascade of reactions, providing higher sensitivity than assays based on agglutination or precipitin reactions. For example, antibodies can be detected at concentrations of less than $1\,\mu g\,ml^{-1}$.*

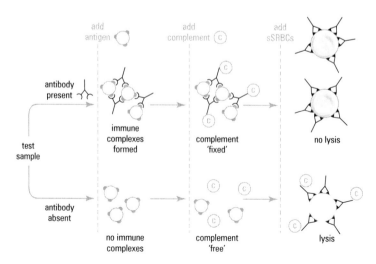

Fig. 38.12 Complement fixation test for antibody.

2. Add a standard amount of antigen: antigen–antibody complexes will be formed in direct proportion to the amount of antibody present in the sample.
3. Then, add a known amount of complement to the mixture: if immune complexes (Fig. 38.2) are present, they will 'fix' complement (i.e. the complement components will be consumed and no longer 'free' in solution), whereas if immune complexes are not present, or are at low concentration, complement will remain 'free' in solution, and available for subsequent reaction.
4. In the final step, add an indicator system for unbound complement – sensitised sheep red blood cells (sSRBCs) coated with rabbit anti-erythrocyte antibodies are often used. Unfixed complement is detected by lysis of the sensitised erythrocytes. The lowest dilution showing complement fixation (i.e. no erythrocyte lysis) gives the titre of antibody in the test sample.

Example Complement fixation tests are the basis of the Wasserman test to detect anti-treponemal antibodies in the serum of patients with syphilis.

The test can be modified to test for antigens in a sample, by titration against a fixed amount of added antibody. In all cases, suitable controls are required, to check for non-specific reactions and to confirm that immune complexes are not already present in the sample, since these will fix complement and invalidate the titration. Complement fixation tests are useful for many classes of IgG and for IgM, but not for most other antibody types.

Text references

Delves, P.J., Martin, S., Burton, D. and Roitt, I. (2006) *Roitt's Essential Immunology*, 10th edn. Blackwell, Oxford.

Hay, F.C., Westwood, O.M.R. and Nelson, P.N. (2002) *Practical Immunology*. Blackwell, Oxford.

Roitt, I.M., Broscoff, J. and Male, D.K. (eds) (2001) *Immunology*, 6th edn. Mosby, New York.

Sources for further study

Clausen, J. (1989) *Immunochemical Techniques for the Identification and Estimation of Macromolecules*, 3rd edn. Elsevier, Amsterdam.

Diamandis, E.P. and Christopoulos, T.K. (1997) *Immunoassay*. Academic Press, New York.

Gosling, J.R.G. (2000) *Immunoassays: a Practical Approach*. Oxford University Press, Oxford.

Lefkovits, I. (1997) *Immunology Methods Manual: The Comprehensive Sourcebook of Techniques*. Academic Press, New York.

Study exercises

38.1 Determine the amount of antigen in a test solution using Laurell rocket immunoelectrophoresis. The figure on the right represents the results from a rocket immunoelectrophoresis assay for a series of standards. Determine the amount of antigen in a test solution giving a rocket (precipitin line) height of 18.5 mm.

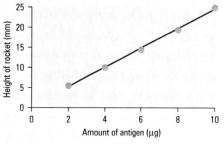

(continued)

Study exercises (continued)

38.2 Determine the amount of antigen in a test solution using single radial immunodiffusion (RID). Using the following data for a series of standards, determine the amount of antigen in a test solution giving a ring diameter of 6.5 mm.

Single radial immunodiffusion results for a series of standards containing 4–20 μg antigen

Amount of antigen (μg)	Ring diameter (mm)
4	3.5
8	5.5
12	7.0
16	8.0
20	9.0

38.3 Interpret data from a haemagglutination test. In some conditions, such as Hashimoto's disease, autoantibodies to the thyroid protein thyroglobulin are produced in large amounts. The level of these antibodies in serum can be measured by indirect haemagglutination: red blood cells coated with thyroglobulin are mixed with serial doubling dilutions of the test serum. The titre of antithyroglobulin autoantibodies is the lowest dilution at which haemagglutination occurs. The figure below represents the outcome of an antithyroglobulin haemagglutination test on a serum sample. The first well – numbered 1 – represents a 10-fold dilution of the serum sample, with serial doubling dilutions in subsequent wells.

(a) Write a brief explanation of the test results shown in the figure.
(b) Determine the autoantibody titre.
(c) Score the sample as 'positive' if the titre is more dilute than 1 in 5000 (i.e. 1 : >5000) and 'negative' if it is less dilute than 1 in 5000 (i.e. 1 : <5000).

```
 1  2  3  4  5  6  7  8  9  10 11 12
[⊙  ⊙  ○  ○  ○  ○  ○  ○  ○  ○  ⊙  ⊙]
```

39 Radioactive isotopes and their uses

Examples $^{12}_{6}C$, $^{13}_{6}C$ and $^{14}_{6}C$ are three of the isotopes of carbon. About 98.9% of naturally occurring carbon is in the stable $^{12}_{6}C$ form. $^{13}_{6}C$ is also a stable isotope but it only occurs at 1.1% natural abundance. Trace amounts of radioactive $^{14}_{6}C$ are found naturally; this is a negatron-emitting radioisotope (see Table 39.2).

The isotopes of a particular element have the same number of protons in the nucleus but different numbers of neutrons, giving them the same proton number (atomic number) but different nucleon numbers (mass number, i.e. number of protons + number of neutrons). Isotopes may be stable or radioactive. Radioactive isotopes (radioisotopes) disintegrate spontaneously at random to yield radiation and a decay product.

Radioactive decay

There are three forms of radioactivity (Table 39.1) arising from three main types of nuclear decay:

- Alpha decay involves the loss of a particle equivalent to a helium nucleus. Alpha (α) particles, being relatively large and positively charged, do not penetrate far in living tissue, but they do cause ionisation damage and this makes them generally unsuitable for tracer studies.
- Beta decay involves the loss or gain of an electron or its positive counterpart, the positron. There are three subtypes:
  (a) Negatron (β^-) emission: loss of an electron from the nucleus when a neutron transforms into a proton. This is the most important form of decay for radioactive tracers used in biology. Negatron-emitting isotopes of biological importance include ^3H, ^{14}C, ^{32}P and ^{35}S.
  (b) Positron (β^+) emission: loss of a positron when a proton transforms into a neutron. This only occurs when sufficient energy is available from the transition and may involve the production of gamma rays when the positron is later annihilated by collision with an electron.
  (c) Electron capture (EC): when a proton 'captures' an electron and transforms into a neutron. This may involve the production of X-rays as electrons 'shuffle' about in the atom (as with ^{125}I) and it frequently involves electron emission.
- Internal transition involves the emission of electromagnetic radiation in the form of gamma (γ) rays from a nucleus in a metastable state and always follows initial alpha or beta decay. Emission of gamma radiation leads to no further change in atomic number or mass.

Examples ^{226}Ra decays to ^{222}Rn by loss of an alpha particle, as follows:

$$^{226}_{88}Ra \rightarrow {}^{222}_{86}Rn + {}^{4}_{2}He^{2+}$$

^{14}C shows beta decay, as follows:

$$^{14}_{6}C \rightarrow {}^{14}_{7}N + \beta^-$$

^{22}Na decays by positron emission, as follows:

$$^{22}_{11}Na \rightarrow {}^{22}_{10}Ne + \beta^+$$

^{55}Fe decays by electron capture and the production of an X-ray, as follows:

$$^{55}_{26}Fe \rightarrow {}^{55}_{25}Mn + X$$

The decay of ^{22}Na by positron emission (β^+) leads to the production of a γ ray when the positron is annihilated on collision with an electron.

Note from the above that more than one type of radiation may be emitted when a radioisotope decays. The main radioisotopes used in biology and their properties are listed in Table 39.2.

Table 39.1 Types of radioactivity and their properties

Radiation	Range of maximum energies (MeV*)	Penetration range in air (m)	Suitable shielding material
Alpha (α)	4–8	0.025–0.080	Unnecessary
Beta (β)	0.01–3	0.150–16	Plastic (e.g. Perspex)
Gamma (γ)	0.03–3	1.3–13†	Lead

*Note that $1\,\text{MeV} = 1.6 \times 10^{-13}$ J
†Distance at which radiation intensity is reduced to half

Table 39.2 Properties of some isotopes used commonly in the life sciences. Physical data obtained from Lide (2006)

Isotope	Emission(s)	Maximum energy (MeV)	Half-life	Main uses	Advantages	Disadvantages
^{3}H	β^-	0.01861	12.3 years	Suitable for labelling organic molecules in wide range of positions at high specific activity	Relatively safe	Low efficiency of detection, high isotope effect, high rate of exchange with environment
^{14}C	β^-	0.15648	5715 years	Suitable for labelling organic molecules in a wide range of positions	Relatively safe, low rate of exchange with environment	Low specific activity
^{22}Na	β^+ (90%) + γ, EC	2.842 (β^+)	2.60 years	Transport studies	High specific activity	Hazardous
^{32}P	β^-	1.710	14.3 days	Labelling proteins and nucleotides (e.g. DNA)	High specific activity, ease of detection	Short half-life, hazardous
^{33}P	β^-	0.248	25 days	Labelling nucleotides and proteins	Safer than ^{32}P	Moderate half-life
^{35}S	β^-	0.167	87.2 days	Labelling proteins and nucleotides	Low isotope effect	Low specific activity
^{36}Cl	β^-, β^+, EC	0.709 (β^-) 1.142 (β^+, EC)	300 000 years	Transport studies	Low isotope effect	Low specific activity, hazardous
^{125}I	EC + γ	0.178 (EC)	59.9 days	Labelling proteins and nucleotides	High specific activity	Hazardous
^{131}I	$\beta^- + \gamma$	0.971 (β^-)	8.04 days	Labelling proteins and nucleotides	High specific activity	Hazardous

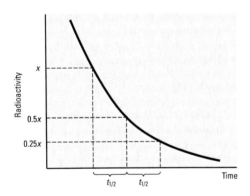

Fig. 39.1 Decay of a radioactive isotope with time. The time taken for the radioactivity to decline from x to $0.5x$ is the same as the time taken for the radioactivity to decline from $0.5x$ to $0.25x$, and so on. This time is the half-life ($t_{1/2}$) of the isotope.

Example For ^{35}S, with a half-life of 87.2 days (Table 39.2), the fraction of radioactivity remaining after 28 days would be worked out as follows: first, from Eqn [39.1] $x = (-0.693 \times 28) \div 87.2 = -0.222522936$, then using Eqn [39.1], $f = e^{-0.222522936} = 0.800496646$ (with appropriate rounding, 80.0% of original activity).

Each radioactive particle or ray carries energy, usually measured in electron volts (eV). The particles or rays emitted by a particular radioisotope exhibit a range of energies, termed an energy spectrum, characterised by the maximum energy of the radiation produced, E_{max} (Table 39.2). The energy spectrum of a particular radioisotope is relevant to the following:

- Safety: isotopes with the highest maximum energies will have the greatest penetrating power, requiring appropriate shielding (Table 39.1).
- Detection: different instruments vary in their ability to detect isotopes with different energies.
- Discrimination: some instruments can distinguish between isotopes, based on the energy spectrum of the radiation produced (p. 271).

The decay of an individual atom (a 'disintegration') occurs at random, but that of a population of atoms occurs in a predictable manner. The radioactivity decays exponentially, having a characteristic half-life ($t_{1/2}$). This is the time taken for the radioactivity to fall from a given value to half that value (Fig. 39.1). The $t_{1/2}$ values of different radioisotopes range from fractions of a second to more than 10^{19} years (see also Table 39.2). If $t_{1/2}$ is very short, as with ^{15}O ($t_{1/2} \approx 2$ min), then it is generally impractical to use the isotope in experiments because you would need to account for the decay during the experiment and counting period.

To calculate the fraction (f) of the original radioactivity left after a particular time (t), use the following relationship:

$$f = e^x, \text{ where } x = -0.693t/t_{\frac{1}{2}} \qquad [39.1]$$

Note that the same units must be used for t and $t_{1/2}$ in this equation.

Measuring radioactivity

The SI unit of radioactivity is the becquerel (Bq), equivalent to one disintegration per second (d.p.s.), but disintegrations per minute (d.p.m.) are also used. The curie (Ci) is a non-SI unit equivalent to the number of disintegrations produced by 1 g of radium (37 GBq). Table 39.3 shows the relationships between these units. In practice, most instruments are not able to detect all of the disintegrations from a particular sample, i.e. their efficiency is less than 100% and the rate of decay may be presented as counts $\min^{-1}$ (c.p.m.) or counts s^{-1} (c.p.s.). Most modern instruments correct for background radiation and inefficiencies in counting, converting count data to d.p.m. Alternatively, the results may be presented as the measured count rate, although this is only valid where the efficiency of counting does not vary greatly among samples.

Table 39.3 Relationships between units of radioactivity. For abbreviations, see text

1 Bq = 1 d.p.s.	1 Sv = 100 rem
1 Bq = 60 d.p.m.	1 Gy = 100 rad
1 Bq = 27 pCi	1 Gy ≈ 100 roentgen
1 d.p.s. = 1 Bq	1 rem = 0.01 Sv
1 d.p.m. = 0.0167 Bq	1 rad = 0.01 Gy
1 Ci = 37 GBq	1 roentgen ≈ 0.01 Gy
1 mCi = 37 MBq	
1 µCi = 37 kBq	

KEY POINT *The specific activity is a measure of the quantity of radioactivity present in a known amount of the substance:*

$$\text{specific activity} = \frac{\text{radioactivity (Bq, Ci, d.p.m., etc.)}}{\text{amount of substance (mol, g, etc.)}} \quad [39.2]$$

This is an important concept in practical work involving radioisotopes, since it allows interconversion of disintegrations (activity) and amount of substance (see Box 39.1).

Example If 0.4 ml of a ^{32}P-labelled DNA solution at a concentration of 50 µmol l^{-1} (amount = 0.4 × 50 ÷ 1000 = 0.02 µmol) gave a count of 2490 d.p.m. (= 41.5 Bq), using Eqn [39.2] this would correspond to a specific activity of 2490 ÷ 0.02 = 124500 dpm µmol^{-1} (or 2075 Bq µmol^{-1}).

Two SI units refer to doses of radioactivity and these are used when calculating exposure levels for a particular source. The sievert (Sv) is the amount of radioactivity giving a dose in humans equivalent to 1 gray (Gy) of X-rays: 1 Gy = an energy absorption of 1 J kg^{-1}. The dose received in most biological experiments is a negligible fraction of the maximum permitted exposure limit. Conversion factors from older units are given in Table 39.3.

The most important methods of measuring radioactivity for biological purposes are described below.

The Geiger–Müller (G–M) tube

This operates by detecting radiation when it ionises gas between a pair of electrodes across which a voltage has been applied. You should use a hand-held Geiger–Müller tube for routine checking for contamination (although it will not pick up ^{3}H activity).

The scintillation counter

This operates by detecting the scintillations (fluorescence 'flashes') produced when radiation interacts with certain chemicals called fluors (Fig. 39.2). In solid (or external) scintillation counters (often referred to as 'gamma counters') the radioactivity causes scintillations in a crystal of fluorescent material held close to the sample. This method is only suitable for radioisotopes producing penetrating radiation.

Liquid scintillation counters are mainly used for detecting beta decay and they are especially useful in biology. The sample is dispersed or dissolved in a suitable solvent containing the fluor(s) – the 'scintillation cocktail'. The radiation first interacts with the solvent, and the energy from this interaction is passed to the fluors which produce detectable light. The scintillations are measured by photomultiplier tubes which turn the light pulses into electronic pulses, the magnitude of which is directly related to

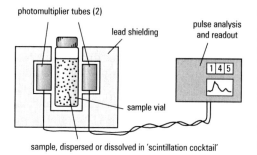

Fig. 39.2 Components of a scintillation counter. Note that in most modern instruments, all components are enclosed within a single cabinet.

> **Box 39.1 How to determine the specific activity of an experimental solution**
>
> Suppose you need to make up a certain volume of an experimental solution, to contain a particular amount of radioactivity. For example, 50 ml of a mannitol solution at a concentration of 25 mmol l^{-1}, to contain 5 Bq μl^{-1} – using a manufacturer's stock solution of ^{14}C-labelled mannitol (specific activity = 0.1 Ci mmol^{-1}).
>
> 1. **Calculate the total amount of radioactivity in the experimental solution**, in this example 5 × 1000 (to convert μl to ml) × 50 (50 ml required) = 2.5 × 10^5 Bq (i.e. 250 kBq).
>
> 2. **Establish the volume of stock radioisotope solution required**: for example, a manufacturer's stock solution of ^{14}C-labelled mannitol contains 50 μCi of radioisotope in 1 ml of 90% v/v ethanol:water. Using Table 39.3, this is equivalent to an activity of 50 × 37 = 1850 kBq. So, the volume of solution required is 250/1850 of the stock volume, i.e. 0.135 1 ml (135 μl).
>
> 3. **Calculate the amount of non-radioactive substance required** as for any calculation involving concentration (see pp. 136, 145), e.g. 50 ml (0.05 l) of a 25 mmol l^{-1} (0.025 mol l^{-1}) mannitol (relative molecular mass 182.17) will contain 0.05 × 0.025 × 182.17 = 0.2277 g.
>
> 4. **Check the amount of radioactive isotope to be added.** In most cases, this represents a negligible amount of substance, e.g. in this instance, 250 kBq of stock solution at a specific activity of 14.8 × 10^6 kBq mmol^{-1} (converted from 0.4 Ci mmol^{-1} using Table 39.3) is equal to 250/14 800 000 = 16.89 nmol, equivalent to approximately 3 μg mannitol. This can be ignored in calculating the mannitol concentration of the experimental solution.
>
> 5. **Make up the experimental solution** by adding the appropriate amount of non-radioactive substance and the correct volume of stock solution.
>
> 6. **Measure the radioactivity in a known volume of the experimental solution.** If you are using an instrument with automatic correction to Bq, your sample should contain the predicted amount of radioactivity, e.g. an accurately dispensed volume of 100 μl of the mannitol solution should give a corrected count of 100 × 5 = 500 Bq (or 500 × 60 = 30 000 d.p.m.).
>
> 7. **Note the specific activity of the experimental solution**: in this case, 100 μl (1 × 10^{-4} l) of the mannitol solution at a concentration of 0.025 mol l^{-1} will contain 25 × 10^{-7} mol (2.5 μmol) mannitol. Dividing the radioactivity in this volume (30 000 d.p.m.) by the amount of substance (Eqn [39.2]) gives a specific activity of 30 000/2.5 = 12 000 d.p.m. μmol^{-1}, or 12 d.p.m. nmol^{-1}. This value can be used:
>
>   (a) To assess the accuracy of your protocol for preparing the experimental solution: if the measured activity is substantially different from the predicted value, you may have made an error in making up the solution.
>
>   (b) To determine the counting efficiency of an instrument; by comparing the measured count rate with the value predicted by your calculations.
>
>   (c) To interconvert activity and amount of substance: the most important practical application of specific activity is the conversion of experimental data from counts (activity) into amounts of substance. This is only possible where the substance has not been metabolised or otherwise converted into another form; e.g. a tissue sample incubated in the experimental solution described above with a measured activity of 245 d.p.m. can be converted to nmol mannitol by dividing by the specific activity, expressed in the correct form. Thus 245/12 = 20.417 nmol mannitol.

Correcting for quenching – find out how your instrument corrects for quenching and check the quench indication parameter (QIP) on the printout, which measures the extent of quenching of each sample. Large differences in the QIP would indicate that quenching is variable among samples and might give you cause for concern.

the energy of the original radioactive event. The spectrum of electronic pulses is thus related to the energy spectrum of the radioisotope.

Modern liquid scintillation counters use a series of electronic 'windows' to split the pulse spectrum into two or three components. This may allow more than one isotope to be detected in a single sample, provided their energy spectra are sufficiently different (Fig. 39.3). A complication of this approach is that the energy spectrum can be altered by pigments and chemicals in the sample, which absorb scintillations or interfere with the transfer of energy to the fluor; this is known as quenching (Fig. 39.3).

Fig. 39.3 Energy spectra for three radioactive samples, detected using a scintillation counter. Sample *a* is a high energy β-emitter while *b* contains a low energy β-emitter, giving a lower spectral range. Sample *b'* contains the same amount of low energy β-emitter, but with quenching, shifting the spectral distribution to a lower energy band. The counter can be set up to record disintegrations within a selected range (a 'window'). Here, 'window a' could be used to count isotope *a* while 'window b' could give a value of isotope *b*, by applying a correction for the counts due to isotope *a*, based on the results from 'window a'. Dual counting allows experiments to be carried out using two isotopes (double labelling).

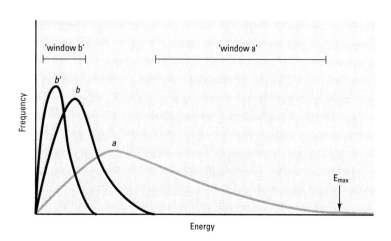

Liquid scintillation counting of high energy β-emitters – β-particles with energies greater than 1 MeV can be counted in water (Čerencov radiation), with no requirement for additional fluors (e.g. ^{32}P).

Most instruments have computer-operated quench correction facilities (based on measurements of standards of known activity and energy spectrum) which correct for such changes in counting efficiency.

Many liquid scintillation counters treat the first sample as a 'background', subtracting whatever value is obtained from the subsequent measurements as part of the procedure for converting to d.p.m. If not, you will need to subtract the background count from all other samples. Make sure that you use an appropriate background sample, identical in all respects to your radioactive sample but with no added radioisotope, in the correct position within the machine. Check that the background reading is reasonable (15–30 c.p.m. is a reasonable background for most radioisotopes). Tips for preparing samples for liquid scintillation counting are given in Box 39.2.

Gamma-ray (γ-ray) spectrometry

This is a method by which a mixture of γ-ray-emitting radionuclides can be resolved quantitatively by pulse-height analysis. It is based on the fact that pulse heights (voltages) produced by a photomultiplier tube are proportional to the amounts of γ-ray energy arriving at the scintillant or a lithium-drifted germanium detector. The lithium-drifted germanium detector, which is abbreviated to Ge(Li) – pronounced 'jelly' – provides high resolution (narrow peaks), essential in the analysis of complex mixtures, such as in biological material.

Autoradiography

This is a method where photographic film is exposed to the isotope. It is used mainly to locate radioactive tracers in thin sections of an organism or on chromatography papers and gels, but quantitative work is possible. The radiation interacts with the film in a similar way to light, silver grains being formed in the developed film where the particles or rays have passed through. The radiation must have enough energy to penetrate into the film, but if it has too much energy the grain formation may be too distant from the point where the isotope was located to identify precisely the point of origin (e.g. high energy β-emitters). Autoradiography is a relatively specialised method and individual lab protocols should be followed for particular isotopes/applications.

Box 39.2 Tips for preparing samples for liquid scintillation counting

Modern scintillation counters are very simple to operate; problems are more likely to be due to inadequate sample preparation than to incorrect operation of the machine. Common pitfalls are the following:

- **Incomplete dispersal of the radioactive compound in the scintillation cocktail.** This may lead to underestimation of the true amount of radioactivity present:

  (a) Water-based samples may not mix with the scintillation cocktail – change to an emulsifier-based cocktail. Take care to observe the recommended limits, upper and lower, for amounts of water to be added or the cocktail may not emulsify properly.

  (b) Solid specimens may absorb disintegrations or scintillations: extract radiochemicals using an intermediate solvent like ethanol (ideally within the scintillation vial) and then add the cocktail. Tissue-solubilising compounds such as Soluene are effective, particularly for animal material, but extremely toxic, so the manufacturer's instructions must be followed closely. Radioactive compounds on slices of agarose or polyacrylamide gels may be extracted using a product such as Protosol. Agarose gels can be dissolved in a small volume of boiling water.

  (c) Particulate samples may sediment to the bottom of the scintillation vial – suspend them by forming a gel. This can be done with certain emulsifier-based cocktails by adding a specific amount of water.

- **Chemiluminescence.** This is where a chemical reacts with the fluors in the scintillation cocktail causing spurious scintillations, a particular risk with solutions containing strong bases or oxidising agents. Symptoms include very high initial counts which decrease through time. Possible remedies are:

  (a) Leave the vials at room temperature for a time before counting. Check with a suitable blank that counts have dropped to an acceptable level.

  (b) Neutralise basic samples with acid (e.g. acetic acid or HCl).

  (c) Use a scintillation cocktail that resists chemiluminescence, such as Hiconicfluor.

  (d) Raise the energy of the lower counts detected to about 8 keV – most chemiluminescence pulses are weak (0–7 keV). This approach is not suitable for weak emitters, e.g. ^3H.

Biological applications for radioactive isotopes

The main advantages of using radioactive isotopes in biological experiments are:

- Radioactivity is readily detected. Methods of detection are sufficiently sensitive to measure extremely small amounts of radioactive substances.
- Studies can be carried out on intact, living organisms. If care is taken, minimal disruption of normal conditions will occur when radiolabelled compounds are introduced.
- Protocols are simple compared to equivalent methods for chemical analysis.

The main disadvantages are:

- The 'isotope effect'. Molecules containing different isotopes of the same atom may react at slightly different rates and behave in slightly different ways from the natural isotope. The isotope effect is more extreme the smaller the atom, and is most important for ^3H-labelled compounds of low molecular mass.
- The possibility of mistaken identity. The presence of radioactivity does not tell you anything about the compound in which the radioactivity is present: it could be different from the one in which it was applied, due to metabolism or spontaneous breakdown of a ^{14}C-containing organic compound.

Investigating the metabolic fate of radiolabelled compounds – you may need to separate individual metabolites before counting, e.g. using chromatography (Chapter 44), or electrophoresis (Chapter 46).

The main types of experiments are:

- Investigations of metabolic pathways: a radioactively labelled substrate is added (often to an *in vitro* experimental 'system' rather than a whole organism) and samples taken at different time intervals. By identifying the labelled compounds and plotting their appearance through time, an indication of the pathway of metabolism can be obtained.
- Translocation studies: radioisotopes are used to follow the fate of molecules within an organism. Uptake and translocation rates can be determined with relative ease.
- Ecological studies: radioisotopic tracers provide a convenient method for determining food web interrelationships and for investigating behaviour patterns, while environmental monitoring may involve following the 'spectral signature' of isotopes deliberately or accidentally released.
- Radio-dating: the age of plant or mineral samples can be determined by measuring the amount of a radioisotope in the sample. The age of the specimen can be found using the half-life by assuming how much was originally incorporated.
- Mutagenesis and sterilisation: radioactive sources can be used to induce mutations, particularly in micro-organisms. Gamma emitters of high energy will kill microbes and are used to sterilise equipment such as disposable Petri dishes.
- Assays: radioisotopes are used in several quantitative detection methods of value to biologists. Radioimmunoassay is described on p. 260. Isotope dilution analysis works on the assumption that introduced radiolabelled molecules will equilibrate with unlabelled molecules present in the specimen. The amount of substance initially present can be worked out from the change in specific activity of the radioisotope when it is diluted by the 'cold' material. A method is required whereby the substance can be purified from the sample and sufficient substance must be present for its mass to be measured accurately.

Billington *et al.* (1992) give further details and practical advice on using radioisotopes in biological experiments.

Example Carbon dating – living organisms have essentially the same ratio of ^{14}C to ^{12}C as the atmosphere; however, when an organism dies, its $^{14}C/^{12}C$ falls because the radioactive ^{14}C isotope decays. Since we know the half-life of ^{14}C (5715 years), a sample's $^{14}C/^{12}C$ ratio will allow us to estimate its age; e.g. if the ratio were exactly 1/8 of that in the atmosphere, the sample is three half-lives old and was formed 17 145 years before present. Such estimates carry an error of the order of 10% and are unreliable for samples older than 50 000 years, for which longer-lived isotopes can be used.

Working practices when using radioactive isotopes

By law, undergraduate work with radioactive isotopes must be very closely supervised. In practical classes, the protocols will be clearly outlined, but in project work you may have the opportunity to plan and carry out your own experiments, albeit under supervision. Some of the factors that you should take into account, based on the assumption that your department and laboratory are registered for radioisotope use, are discussed below:

1. Must you use radioactivity? If not, it may be legal requirement that you use an alternative method.
2. Have you registered for radioactive work? Normal practice is for all users to register with a local Radiation Protection Supervisor. Details of the project may have to be approved by the appropriate administrator(s). You may have to have a short medical examination before you can start work.

Registering for radioisotope work *– in the UK, institutions must be registered for work with specific radioisotopes under the* Radioactive Substances Act (1993).

Supervision of work with radioisotopes *– in the UK, the* Ionising Radiations Act (1985) *provides details of local arrangements for the supervision of radioisotope work.*

3. What labelled compound will you use? Radioactive isotopes must be ordered well in advance through your department's Radiation Protection Supervisor. Aspects that need to be considered include:
   (a) The radionuclide. With many organic compounds this will be confined to ^{3}H and ^{14}C (but see Table 39.2). The involvement of a significant 'isotope effect' may influence this decision (see p. 272).
   (b) The labelling position. This may be a crucial part of a metabolic study. Specifically labelled compounds are normally more expensive than those that are uniformly ('generally') labelled.
   (c) The specific activity. The upper limit for this is defined by the isotope's half-life, but below this, the higher the specific activity, the more expensive the compound.
4. Are suitable facilities available? You'll need a suitable work area, preferably out of the way of general lab traffic and within a fume cupboard for those cases where volatile radioactive substances are used or may be produced.

 SAFETY NOTE Planning radio-isotope work – each new experiment should be planned carefully and experimental protocols laid down in advance so you work as safely as possible and do not waste expensive radioactively labelled compounds.

In conjunction with your supervisor, decide whether your method of application will introduce enough radioactivity into the system, how you will account for any loss of radioactivity during recovery of the isotope and whether there will be enough activity to count at the end. You should be able to predict approximately the amount of radioactivity in your samples, based on the specific activity of the isotope used, the expected rate of uptake/exchange and the amount of sample to be counted. Use the isotope's specific activity to estimate whether the non-radioactive ('cold') compound introduced with the radiolabelled ('hot') compound may lead to excessive concentrations being administered. Advice for handling data is given in Box 39.1.

Carrying out a 'dry run' – consider doing this before working with radioactive compounds, perhaps using a dye to show the movement or dilution of introduced liquids, as this will lessen the risks of accident and improve your technique.

 Safety and procedural aspects

Make sure the bench surface is one that can be easily decontaminated by washing (e.g. Formica) and always use a disposable surfacing material such as Benchkote. It is good practice to carry out as many operations as possible within a Benchkote-lined plastic tray so that any spillages are contained. You will need a lab coat to be used exclusively for work with radioactivity, safety spectacles and a supply of thin latex or vinyl disposable gloves. Suitable vessels for liquid waste disposal will be required and special plastic bags for solids – make sure you know beforehand the disposal procedures for liquid and solid wastes. Wash your hands after handling a vessel containing a radioactive solution and again before removing your gloves. Gloves should be placed in the appropriate disposal bag as soon as your experimental procedures are complete.

 SAFETY NOTE Using Benchkote – the correct way to use Benchkote and similar products is with the waxed surface down (to protect the bench or tray surface) and the absorbent surface up (to absorb any spillage). Write the date in the corner when you put down a new piece. Monitor using a G–M tube and replace regularly under normal circumstances. If you are aware of spillage, replace immediately and dispose of correctly.

It is important to comply with the following guidelines:

- Read and obey the local rules for safe usage of radiochemicals.
- Maximise the distance between you and the source as much as possible.
- Minimise the duration of exposure.
- Wear protective clothing (properly fastened lab coat, safety glasses, gloves) at all times.
- Use appropriate shielding at all times (Table 39.1).
- Monitor your working area for contamination frequently.
- Mark all glassware, trays, bench work areas, etc., with tape incorporating the international symbol for radioactivity (Fig. 39.4).

Fig. 39.4 Tape showing the international symbol for radioactivity.

- Keep adequate records of what you have done with a radioisotope – the stock remaining and that disposed of in waste form must agree.
- Store radiolabelled compounds appropriately and return them to storage areas immediately after use.
- Dispose of waste promptly and with due regard for local rules.
- Make the necessary reports about waste disposal, etc., to your departmental Radiation Protection Supervisor.
- Clear up after you have finished each experiment.
- Wash thoroughly after using radioactivity.
- Monitor the work area and your body when finished.

Text references and sources for further study

Billington, D., Jayson, G.G. and Maltby, P.J. (1992) *Radioisotopes*. Bios, Oxford.

L'Annunziata, M.F. (ed.) (2003) *Handbook of Radioactivity Analysis*, 2nd edn. Academic Press, San Diego.

Lide, D.R. (ed.) (2006) *CRC Handbook of Chemistry and Physics*, 87th edn. CRC Press, Boca Raton.

Slater, J. (ed.) (2002) *Radioisotopes in Biology: a Practical Approach*, 2nd edn. Oxford University Press: Oxford.

Study exercises

39.1 Carry out a half-life calculation. A rat dropping found in a pyramid in Egypt had a $^{14}C:^{12}C$ ratio that was 57.25% of a modern-day standard. Use this value to estimate the approximate date when the rat visited the pyramid, to the nearest century.

39.2 Practise radioactivity interconversions. Express the following values in the alternative units indicated, with appropriate prefixes as necessary. Answer to three significant figures.
(a) 72 000 d.p.m. as Bq;
(b) 20 µCi as d.p.m.;
(c) 44 400 Bq as µCi;
(d) 6.3×10^5 d.p.m. mol^{-1} as Bq g^{-1}, for a compound with a relative molecular mass of 350;
(e) 3108 d.p.m. as pmol, for a sample of a standard where the specific activity is stated as 50 Ci mol^{-1}.

39.3 Use the concept of specific activity in calculations. A researcher wishes to estimate the rate of uptake of the sugar galactose by carrot cells in a suspension culture. She prepares 250 ml of the cell culture medium containing 10^7 cells per ml and unlabelled galactose at a concentration of 5 mmol l^{-1}. She then 'spikes' this with 5 µl (regard this as an insignificant volume) of radioactive standard containing 55 MBq of ^{14}C-labelled galactose (regard as an insignificant concentration). Answer to two significant figures.
(a) Calculate the specific activity of the galactose in the culture solution in Bq mol^{-1}.
(b) If the total cell sample takes up 79.2×10^5 Bq in a 2-hour period, calculate the galactose uptake rate in mol s^{-1} cell^{-1}.

40 Light measurement

The measurement of light is directly relevant to several aspects of biology, including: photosynthesis, photomorphogenesis and photoperiodism in plants; perception and thermoregulation in animals; and aspects of aquatic biology.

The nature of light

Light is most strictly defined as that part of the spectrum of electromagnetic radiation detected by the human eye. However, the term is also applied to radiation just outside that visible range (e.g. UV and infra-red 'light'). Electromagnetic radiation is emitted by the Sun and by other sources (e.g. an incandescent lamp) and the electromagnetic spectrum is a broad band of radiation, ranging from cosmic rays to radio waves (Fig. 40.1). Most biological experiments involve measurements within the UV, visible and infra-red (IR) regions (generally, within the wavelength range 200–1000 nm; see Table 40.1).

Radiation has the characteristics of a particle and of a vibrating wave, travelling in discrete particulate units, or 'packets', termed photons. A quantum is the amount of energy contained within a single photon (it is important not to confuse these two terms, although they are sometimes used interchangeably in the literature). In some circumstances, it is appropriate to measure light in terms of the number of photons, usually expressed directly in moles (6.02×10^{23} photons = 1 mol); older textbooks may use the redundant term Einstein as the unit of measurement, where 1 Einstein = 1 mol photons. Alternatively, the energy content (power) may be measured (e.g. in $W\,m^{-2}$). Radiation also behaves as a vibrating electrical and magnetic field moving in a particular direction, with the magnetic and electrical components vibrating perpendicular to one another and perpendicular to the direction of travel. The wave nature of radiation gives rise to the concepts of wavelength (λ, usually measured in nm), frequency (v, measured in s^{-1}), speed (c, the speed of electromagnetic radiation, which is $3 \times 10^8\,m\,s^{-1}$ in a vacuum), and direction. In other words, radiation is a vector quantity, where

$$c = \lambda v \qquad [40.1]$$

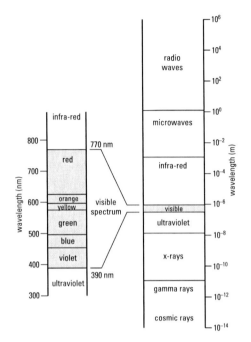

Fig. 40.1 The electromagnetic spectrum, Methods for Physical and Chemical Analysis of Fresh Waters, 2nd edn. International Biological Programme, No. 8, Blackwell Scientific (Golterman, H.L., Clymo, R.S. and Ohnstad, M.A.M., 1978).

Table 40.1 Wavelength ranges for UV, visible and IR radiation, compiled from several sources

Type	Wavelength range (nm)
UVC	100–290*
UVB	290–320
UVA	320–400
Violet	390–450
Blue	450–500
Green	500–560
Yellow	560–600
Orange	600–620
Red	620–770
IR	>770

* Note that UVC is absorbed by the Earth's ozone layer.

Photometric and radiometric measurements

Photometric measurements

These are based on the energy perceived by a 'standard' human eye, with maximum sensitivity in the yellow–green region, around 555 nm (Fig. 40.2). The unit of measurement is the candela, a base unit in the SI system, defined in terms of the visual appearance of a specific quantity of platinum at its freezing point. Derived units are used for the luminous flow (lumen) and luminous flow per unit area (lux). These units were once used in photobiology and you may come across them in older literature. However, it is now recognised that such measurements are of little direct relevance to biologists, including even those who may wish to study visual responses, because they are not based on fundamental physical principles.

KEY POINT *The human eye rapidly compensates for changes in light climate by varying the size of the pupil and is a very poor source of information on light quantity. It is important to make light measurements using reliable instruments and to express these measurements in appropriate units.*

Radiometric measurements

The radiometric system is based on physical properties of the electromagnetic radiation itself, expressed either as the number of photons, or their energy content. The following terms are used (units of measurement in parentheses):

- Photon flux (mol photons s^{-1}) is the number of photons arriving at an object within the specified time interval.
- Photon exposure (mol photons m^{-2}) is the total number of photons received by an object, usually expressed per unit surface area.
- Photon flux density (mol photons m^{-2} s^{-1}), or PFD, is the most commonly used term to describe the number of photons arriving at a particular surface, expressed per unit surface area and per unit time interval.
- Photosynthetically active radiation, or PAR, is radiation within the waveband 400–700 nm, since the photosynthetic pigments (chlorophylls, carotenoids, etc.) show maximum absorption within this band.
- Photosynthetic photon flux density (mol photons m^{-2} s^{-1}), or PPFD, is the number of photons within the waveband 400–700 nm arriving at a particular surface, expressed per unit surface area and per unit time interval. Often this term is used interchangeably with PFD.
- Irradiance (J m^{-2} s^{-1} = W m^{-2}) is the amount of energy arriving at a surface, expressed per unit surface area and per unit time interval.
- Photosynthetic irradiance (W m^{-2}), or PI, is the energy of radiation within the waveband 400–700 nm arriving at a surface, expressed per unit surface area and per unit time interval.

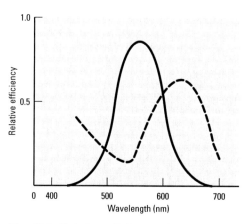

Fig. 40.2 Relative efficiency of vision (solid line) and photosynthesis (dotted line) as a function of wavelength.

Choice of measurement scale

Photon flux density

This is often the most appropriate unit of measurement for biological systems where individual photons are involved in the underlying process, e.g. in photosynthetic studies, where PPFD is measured, since each photochemical reaction involves the absorption of a single photon by a pigment molecule. Most modern light-measuring instruments (radiometers) can measure this quantity, giving a reading in μmol photons m^{-2} s^{-1}.

Irradiance

This is appropriate if you are interested in the energy content of the light, e.g. if you are studying energy balance or thermal effects. Many radiometers measure photosynthetic irradiance within the waveband 400–700 nm, giving a reading in W m^{-2}. It is possible to make an approximate conversion between PPFD and PI measurements, providing the spectral properties of the light source are known (see Table 40.2).

Table 40.2 Approximate conversion factors for a photosynthetic irradiance (PI) of 1 W m^{-2} to photosynthetic photon flux density (PPFD)

Source	PPFD (μmol photons m^{-2} s^{-1})
Sunlight	4.6
'Cool white' fluorescent tube	4.6
Osram 'daylight' fluorescent tube	4.6
Quartz-iodine lamp	5.0
Tungsten bulb	5.0

(*Source:* Lüning, 1981. Reproduced by kind permission of Blackwell Publishing Ltd.)

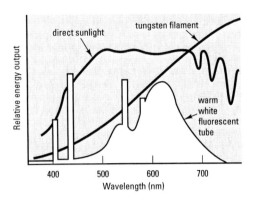

Fig. 40.3 Spectral distribution of energy output from various sources. (Adapted from Golterman *et al.*, 1978.

Spectral distribution

This can be determined using a spectroradiometer, e.g. to compare different light sources (Fig. 40.3). A spectroradiometer measures irradiance or photon flux density in specific wavebands. This instrument consists of a monochromator (p. 281) to allow separate narrow wavebands (5–25 nm bandwidth) to be measured by a detector; some instruments provide a plot of the spectral characteristics of the source.

Using a radiometer ('light meter')

The main components of a radiometer are:

- Receiver: either flat-plate, hemispherical or spherical, depending upon requirements. Most incorporate a protective diffuser, to reduce reflection.
- Detector: either thermoelectric or photoelectric. Some photoelectric detectors suffer from fatigue, with a decreasing response on prolonged exposure: check the manufacturer's handbook for exposure times.
- Processor and readout device to convert the output from the detector into a visible reading, in analogue or digital form.

Box 40.1 gives practical details of the steps involved in using a radiometer. In your write-up, give full details of how the measurement was made, e.g. the type of light source, instrument used, where the sensor was placed, whether an average was calculated, etc.

Box 40.1 Measuring photon flux density or irradiance using a battery-powered radiometer

1. **Check the battery.** Most instruments have a setting that gives a direct readout of battery voltage. Recharge if necessary before use.

2. **Select the appropriate type of measurement** (e.g. photon flux density or irradiance over the PAR waveband, or an alternative range): the simpler instruments have a selection dial for this purpose.

3. **Place the sensor in the correct location and position** for the measurement: it may be appropriate to make several measurements at different positions and take an average.

4. **Choose the most appropriate scale** for the readout device: for needle-type meters, the choice of maximum reading is usually selected by a dial, within the range 0.3 to 30 000. Start at a high range and work down until the reading is on the scale. Your final scale should be chosen to provide the most accurate reading, e.g. a reading of 15 μmol photons m^{-2} s^{-1} should be made using the 0–30 scale, rather than a higher range.

5. **Read the value from the meter.** For needle-type instruments there may be two scales, the upper one marked from 0 to 10 and the lower one from 0 to 3: make sure you use the correct one, e.g. a half-scale deflection on the 0–30 scale is 1.5.

6. **Check that the answer is realistic,** e.g. full sunlight has a PPFD of up to 2 000 μmol photons m^{-2} s^{-1} (PI $\equiv$ 400 W m^{-2}), though the value will be far lower on a dull or cloudy day, while the PPFD at a distance of 1 m from a mercury lamp is around 150 μmol photons m^{-2} s^{-1}, and 50 μmol photons m^{-2} s^{-1} at the same distance from a fluorescent lamp.

Text references

Golterman, H.L., Clymo, R.S. and Ohnstad, M.A.M. (1978) *Methods for Physical and Chemical Analysis of Fresh Waters*. Blackwell Scientific Publications, Oxford.

Lüning, K.J. (1981) 'Light', in C.S. Lobban and M.J. Wynne (eds), *The Biology of Seaweeds*, pp. 326–55, Blackwell Scientific, Oxford.

Sources for further study

Anon. *International Light Homepage*.
Available: http://www.intl-lighttech.com
Last accessed: 01/04/07.
[Information on products including several types of detector and meter. Also includes access to a light measurement handbook in .pdf format]

Boyd, R.W. (1983) *Radiometry and the Detection of Optical Radiation*. Wiley, New York.

National Physics Laboratory *Beginner's Guide to Measurement – Light*.
Available: http://www.npl.co.uk/publications/light/
Last accessed: 01/04/07.
[Gives some background on the development and use of the candela, the SI unit of luminous intensity, and its relationship with human vision]

Study exercises

40.1 Carry out a calculation involving interconversion of photosynthetic photon flux and photosynthetic irradiance. For sunlight with a photosynthetic photon flux density (PPFD) of $1610\,mol\,m^{-2}$, what is the total amount of photosynthetic energy (in joules) falling on a leaf of area $45\,cm^2$ over a 30-minute experimental period?

40.2 Compare the photosynthetic photon flux densities (PPFD) of different locations. The following values represent typical light levels for a range of different situations, expressed either as photosynthetic irradiance (PI) or PPFD:
(a) outside, on a sunny day: $PI = 300\,W\,m^{-2}$;
(b) in a room lit by 'cool white' fluorescent tubes: $PPFD = 6.50\,nmol\,photons\,cm^{-2}\,s^{-1}$;
(c) under the leaf canopy in a forest: $PPFD = 275\,\mu mol\,photons\,m^{-2}\,s^{-1}$;
(d) in a growth cabinet lit by a bank of 'daylight' fluorescent lights: $PI = 35.2\,W\,m^{-2}$;
(e) in a room lit by a tungsten bulb: $PI = 1.15\,mW\,cm^{-2}$.

Convert all to PPFD in the same units (expressed to three significant figures) and then rank the locations in order of decreasing PPFD.

40.3 Research the various types of commercial radiometers (light meters) available on the Web. Make a list of the features that you might want to find out about and see how much information you can find.

41 Basic spectroscopy

Using spectroscopy – this technique is valuable for:

- tentatively identifying compounds, by determining their absorption or emission spectra;
- quantifying substances, either singly or in the presence of other compounds, by measuring the signal strength at an appropriate wavelength;
- determining molecular structure;
- following reactions, by measuring the disappearance of a substance, or the appearance of a product as a function of time.

The absorption and emission of electromagnetic radiation of specific energy (wavelength) are characteristic features of many molecules, involving the movement of electrons between different energy states, in accordance with the laws of quantum mechanics and spectroscopic techniques are used to measure and interpret such interactions between molecules and radiation. Electrons in atoms or molecules are distributed at various energy levels, but are mainly at the lowest energy level, usually termed the ground state. When exposed to energy (e.g. from electromagnetic radiation), electrons may be excited to higher energy levels (excited states), with the associated absorption of energy at specific wavelengths giving rise to an absorption spectrum. One quantum of energy is absorbed for a single electron transition from the ground state to an excited state. On the other hand, when an electron returns to its ground state, one quantum of energy is released; this may be dissipated to the surrounding molecules (as heat) or may give rise to an emission spectrum. The energy change (ΔE) for an electron moving between two energy states, E_1 and E_2, is given by the equation:

$$\Delta E = E_1 - E_2 = h\nu \qquad [41.1]$$

where h is the Planck constant (p. 185) and ν is the frequency of the electromagnetic radiation expressed in Hz or s^{-1}). Frequency is related to wavelength (λ, usually expressed in nm) and the speed of electromagnetic radiation, c (p. 185) by the expression:

$$\nu = c \div \lambda \qquad [41.2]$$

UV/visible spectrophotometry

This is a widely used technique for measuring the absorption of radiation in the visible and UV regions of the spectrum. A spectrophotometer is an instrument designed to allow precise measurement at a particular wavelength, while a colorimeter is a simpler instrument, using filters to measure broader wavebands, e.g. light in the green, red or blue regions of the visible spectrum; Jones *et al.* (2007) gives details of how to use a colorimeter.

Principles of light absorption

Two fundamental principles govern the absorption of light passing through a solution:

- The absorption of light is exponentially related to the number of molecules of the absorbing solute that are encountered, i.e. the solute concentration [C].
- The absorption of light is exponentially related to the length of the light path through the absorbing solution, l.

These two principles are combined in the Beer–Lambert relationship (sometimes referred to simply as 'Beer's Law'), which is usually expressed in terms of absorbance (A) – the logarithm of the ratio of the incident light (I_0) to the emergent light (I):

$$A = \varepsilon l [C] \qquad [41.3]$$

Definition

Absorbance (A) – this is given by:

$A = \log_{10}(I_0/I)$.

Usually shown as A_x where 'x' is the wavelength, in nanometres. As an example, for incident light (I_0) = 1.00 and emergent light (I) = 0.16 (expressed in relative terms), $A = \log_{10}(1.00 \div 0.16) = \log_{10} 6.25 = 0.796$ (to three significant figures).

where A is absorbance, ε is a constant for the absorbing substance and the wavelength, termed the absorption coefficient or absorptivity, and $[C]$ is expressed either as mol l^{-1} or g l^{-1} (see p. 135) and l is given in cm.

 KEY POINT The Beer–Lambert relationship, expressed in mathematical form in Eqn [41.3], states that there is a direct linear relationship between the concentration of a substance in a solution, [C], and the absorbance of that solution, A.

This relationship is extremely useful, since most spectrophotometers are constructed to give a direct measurement of absorbance (A), sometimes also termed extinction (E), of a solution (older texts may use the outdated term optical density, OD). Note that for substances obeying the Beer–Lambert relationship, A is linearly related to $[C]$. Absorbance at a particular wavelength is often shown as a subscript, e.g. A_{550} represents the absorbance at 550 nm. The proportion of light passing through the solution is known as the transmittance (T), and is calculated as the ratio of the emergent and incident light intensities.

Some instruments have two scales:

- an exponential scale from zero to infinity, measuring absorbance;
- a linear scale from 0 to 100, measuring (per cent) transmittance.

For most practical purposes, the Beer–Lambert relationship applies and you should use the absorbance scale.

UV/visible spectrophotometer

The principal components of a UV/visible spectrophotometer are shown in Figure 41.1. High intensity tungsten bulbs are used as the light source in basic instruments, capable of operating in the visible region (i.e. 400–700 nm). Deuterium lamps are used for UV spectrophotometry (200–400 nm); these lamps are fitted with quartz envelopes, since glass does not transmit UV radiation.

A major improvement over the simple colorimeter is the use of a diffraction grating to produce a parallel beam of monochromatic light from the (polychromatic) light source. In practice the light emerging from such a monochromator is not of a single wavelength, but is a narrow band of wavelengths. This bandwidth is an important characteristic, since it determines the wavelengths used in absorption measurements – the bandwidth of basic spectrophotometers is around 5–10 nm while research instruments have bandwidths of less than 1 nm.

Bandwidth is affected by the width of the exit slit (the slit width), since the bandwidth will be reduced by decreasing the slit width. To obtain accurate data at a particular wavelength setting, the narrowest possible slit width should be used. However, decreasing the slit width also reduces the amount of light reaching the detector, decreasing the signal-to-noise ratio. The extent to which the slit width can be reduced depends upon the sensitivity and stability of the detection/amplification system and the presence of stray light.

Most UV/visible spectrophotometers are designed to take cuvettes with an optical path length of 10 mm. Disposable plastic cuvettes are suitable for routine work in the visible range using aqueous and alcohol-based solvents, while glass cuvettes are useful for other organic solvents. Glass cuvettes are manufactured to more exacting standards, so use optically matched glass cuvettes for accurate work, especially at low absorbances

Definition

Transmittance (T) – this is usually expressed as a percentage, at a particular wavelength, T_x, where

$T_x = (I/I_0) \times 100 \, (\%)$.

As an example, for incident light (I_0) = 1.00 and emergent light (I) = 0.275 (expressed in relative terms) then transmittance, $T = (0.275 \div 1.00) \times 100 = 27.5\%$.

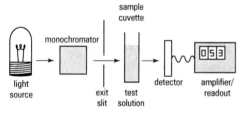

Fig. 41.1 Components of a UV/visible spectrophotometer.

 SAFETY NOTE Working with spectrophotometers – take care not to spill water into the inside of the instrument, due to the risk of electric shock during use (switch off at mains and seek assistance if this should happen).

Using plastic disposable cuvettes – these are adequate for work in the near-UV region, e.g. for enzyme studies using nicotinamide coenzymes, at 340 nm (p. 282), as well as the visible range.

Examples The molar absorptivity of NADH is 6.22×10^3 l mol^{-1} cm^{-1} at 340 nm. For a test solution giving an absorbance of 0.21 in a cuvette with a light path of 5 mm, using Eqn [41.3] this is equal to a concentration of:

$0.21 = 6.22 \times 10^3 \times 0.5 \times [C]$
$[C] = 0.0000675$ mol l^{-1}
(or 67.5 µmol l^{-1}).

The specific absorptivity (10 g l^{-1}) of double-stranded DNA is 200 at 260 nm, therefore a solution containing 1 g l^{-1} will have an absorbance of $200/10 = 20$. For a DNA solution, giving an absorbance of 0.35 in a cuvette with a light path of 1.0 cm, using Eqn [41.3] this is equal to a concentration of:

$0.35 = 20 \times 1.0 [C]$
$[C] = 0.0175$ g l^{-1}
(equivalent to 17.5 µg ml^{-1}).

Chlorophylls *a* and *b* of vascular plants and green algae can be extracted in 90% v/v acetone/water, assayed by measuring the absorbance of the mixed solution at two wavelengths, according to the formulae:

Chlorophyll *a* (mg l^{-1}) =
$11.93 A_{664} - 1.93 A_{647}$

Chlorophyll *b* (mg l^{-1}) =
$20.36 A_{647} - 5.5 A_{664}$

Note: different equations are required for other solvents.

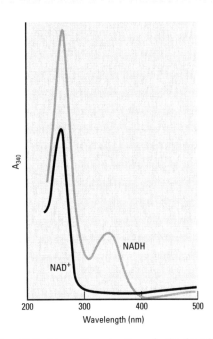

Fig. 41.2 Absorption spectra of nicotinamide adenine dinucleotide in oxidised (NAD$^+$) and reduced (NADH) form. Note the 340 nm absorption peak (A_{340}), used for quantitative work.

(< 0.1), where any differences in the optical properties of cuvettes for reference and test samples will be pronounced. Glass and plastic absorb UV light and quartz cuvettes must be used at wavelengths below 300 nm.

 KEY POINT Before taking a measurement, make sure that cuvettes are clean, unscratched, dry on the outside, filled to the correct level and located in the correct position in their sample holders.

Proteins and nucleic acids in biological samples can accumulate on the inside faces of glass/quartz cuvettes, so remove any deposits using acetone on a cotton bud, or soak overnight in 1 mol l^{-1} nitric acid. Corrosive and hazardous solutions must be used in cuvettes with tightly fitting lids, to prevent damage to the instrument and to reduce the risk of accidental spillage.

Basic instruments use photocells similar to those used in simple colorimeters or photodiode detectors. In many cases, a different photocell must be used at wavelengths above and below 550–600 nm, due to differences in the sensitivity of such detectors over the visible waveband. The detectors used in more sophisticated instruments give increased sensitivity and stability when compared to photocells.

Digital displays are increasingly used in preference to traditional needle-type meters, as they are not prone to parallax errors and misreading of the absorbance scale. Some digital instruments can be calibrated to give a direct readout of the concentration of the test substance.

 KEY POINT Basic spectrophotometers are most accurate within the absorbance range from 0.00 to 1.00 and your calibration standards and test solutions should be prepared to give readings within this range.

Types of UV/visible spectrophotometer

Basic instruments are single beam spectrophotometers in which there is only one light path. The instrument is set to zero absorbance using a blank solution, which is then replaced by the test solution, to obtain an absorbance reading. An alternative approach is used in double beam spectrophotometers, where the light beam from the monochromator is split into two separate beams, one beam passing through the test solution and the other through a reference blank. Absorbance is then measured by an electronic circuit which compares the output from the reference (blank) and sample cuvettes. Double beam spectrophotometry reduces measurement errors caused by fluctuations in output from the light source or changes in the sensitivity of the detection system, since reference and test solutions are measured at the same time (Box 41.1). Recording spectrophotometers are double beam instruments, designed for use with a chart recorder, either by recording the difference in absorbance between reference and test solutions across a predetermined waveband to give an absorption spectrum (Fig. 41.2), or by recording the change in absorbance at a particular wavelength as a function of time (e.g. in an enzyme assay, Chapter 55).

Quantitative spectrophotometric analysis

A single (purified) substance in solution can be quantified using the Beer–Lambert relationship (Eqn [41.3]), provided its absorptivity is known at a particular wavelength (usually at the absorption maximum for the

Box 41.1 How to use a spectrophotometer

1. **Switch on and select the correct lamp** for your measurements (e.g. deuterium for UV, tungsten for visible light).

2. **Allow up to 15 min for the lamp to warm up** and for the instrument to stabilise before use.

3. **Select the appropriate wavelength:** on older instruments a dial is used to adjust the monochromator, while newer machines have microprocessor-controlled wavelength selection.

4. **Select the appropriate detector:** some instruments choose the correct detector automatically (on the basis of the specified wavelength), while others have manual selection.

5. **Choose the correct slit width** (if available): this may be specified in the protocol you are following, or may be chosen on the manufacturer's recommendations.

6. **Insert appropriate reference blank(s):** single beam instruments use a single cuvette, while double beam instruments use two cuvettes (a matched pair for accurate work). The reference blanks should match the test solution in all respects apart from the substance under test, i.e. they should contain all reagents apart from this substance. *Make sure that the cuvettes are positioned correctly, with their polished (transparent) faces in the light path, and that they are accurately located in the cuvette holder(s).*

7. **Check/adjust the 0% transmittance:** most instruments have a control which allows you to zero the detector output in the absence of any light (termed 'dark current' correction). Some microprocessor-controlled instruments carry out this step automatically.

8. **Set the absorbance reading to zero:** usually *via* a dial, or digital readout.

9. **Analyse your samples:** replace the appropriate reference blank with a test sample, allow the absorbance reading to stabilise (5–10 s) and read the absorbance value from the meter/readout device. For absorbance readings greater than 1 (i.e. < 10% transmittance), the signal-to-noise ratio is too low for accurate results. Your analysis may require a calibration curve or you may be able to use the Beer–Lambert relationship (Eqn [41.3]) to determine the concentration of test substance in your samples.

10. **Check the scale zero at regular intervals** using a reference blank, e.g. after every 10 samples.

11. **Check the reproducibility of the instrument:** measure the absorbance of a single solution several times during your analysis. It should give the same value.

Problems (and solutions): inaccurate/unstable readings are most often due to incorrect use of cuvettes, e.g. dirt, fingerprints or test solution on outside of cuvette (wipe the clear faces using a soft tissue before insertion into the cuvette holder and handle only by the opaque faces), condensation (if cold solutions aren't allowed to reach room temperature before use), air bubbles (which scatter light and increase the absorbance; tap gently to remove), insufficient solution (causing refraction of light at the meniscus), particulate material in the solution (check for 'cloudiness' in the solution and centrifuge before use, where necessary) or incorrect positioning in light path (locate in correct position).

Measuring high absorbances in colorimetric analysis – if any final solution has an absorbance that is too high to be read with accuracy on your spectrophotometer (i.e. $A > 2$), it is bad practice to dilute the solution so that it can be measured. This dilutes both the sample molecules and the colour reagents to an equal extent. Instead, you should dilute the original sample and reassay.

substance, since this will give the greatest sensitivity). The molar absorptivity is the absorbance given by a solution with a concentration of 1 mol l^{-1} ($= 1 \text{ kmol m}^{-3}$) of the compound in a light path of 1 cm. The appropriate value may be available from tabulated spectral data (e.g. Anon., 1963), or it can be determined experimentally by measuring the absorbance of known concentrations of the substance (Box 41.1) and plotting a standard curve (see Chapter 37). This should confirm that the relationship is linear over the desired concentration range and the slope of the line will give the molar absorptivity.

The specific absorptivity is the absorbance given by a solution containing 10 g l^{-1} (i.e. 1% w/v) of the compound in a light path of 1 cm. This is useful for substances of unknown molecular weight, e.g. proteins or nucleic acids, where the amount of substance in solution is expressed in terms of its mass, rather than as a molar concentration. For

Measuring low absorbances in colorimetric analysis – *for accurate readings with highly dilute solutions, use a cuvette with a long optical path length (e.g. 5 cm, rather than 1 cm).*

use in Eqn [41.3], the specific absorptivity should be divided by 10 to give the solute concentration in $g\,l^{-1}$.

This simple approach cannot be used for mixed samples where several substances have a significant absorption at a particular wavelength. In such cases, it may be possible to estimate the amount of each substance by measuring the absorbance at several wavelengths, e.g. protein estimation in the presence of nucleic acids (pp. 370–1). Further details of spectrophotometric methods of determining the amount of protein in an aqueous sample are given in Chapter 50.

Fluorescence

With most molecules, after electrons are raised to a higher energy level by absorption of electromagnetic radiation, they soon fall back to the ground state by radiationless transfer of energy (heat) to the solvent. However, with some molecules, the events shown in Figure 41.3 may occur, i.e. electrons may lose only part of their energy by non-radiant routes and the rest may be emitted as electromagnetic radiation, a phenomenon known as fluorescence. Since not all of the energy that was absorbed is emitted (due to non-radiant loss), the wavelength of the fluorescent light is longer than the absorbed light (longer wavelength = lower energy). Thus, a fluorescent molecule has both an absorption spectrum and an emission spectrum. The difference between the excitation wavelength (λ_{ex}) and the emission wavelength (λ_{em}), measured in nm, is known as the Stokes shift, and is fundamental to the sensitivity of fluorescence techniques. The existence of a Stokes shift means that emitted light can be detected against a low background, independently of the excitation wavelength.

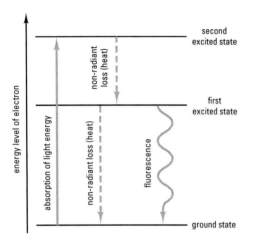

Fig. 41.3 Energy levels and energy transitions in fluorescence.

Most fluorescent molecules have the following features:

- A highly conjugated system (alternating double and single bonds), involving aromatic or heterocyclic rings, usually containing O or N (as heteroatoms).
- A condensed system of fused rings, with one or more heteroatoms.
- Electron-donating groups such as –OH, –OCH$_3$, –NH$_2$ and –NR$_2$, together with electron-attracting groups elsewhere in the molecule, in conjugation with the electron-donating groups.
- A rigid, planar structure.

Figure 41.4 illustrates many of these features for fluorescein, used in a range of biological applications including nucleic acid detection, fluorescent antibody tests, etc.

Fluorescence spectrophotometry

The principal components of a fluorescence spectrophotometer (fluorimeter) are shown in Figure 41.5. The instrument contains two monochromators, one to select the excitation wavelength and the other to monitor the light emitted, usually at 90° to the incident beam (though light is actually emitted in all directions). As an example, the wavelengths used to measure the highly fluorescent compound aminomethylcoumarin are 388 nm (excitation) and 440 nm (emission). Some examples of biomolecules with intrinsic fluorescence are given in Table 41.1 and the use of fluorogenic enzyme substrates is discussed further in Chapter 55, p. 382.

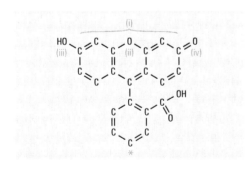

Fig. 41.4 Fluorescein – a widely used fluorescent label, showing (i) a planar conjugated system of fused rings; (ii) heteroatoms within the conjugated structures; (iii) an electron-donating group; and (iv) an electron-attracting group. In fluorogenic enzyme substrates (p. 382), linkage is usually *via* one of the two hydroxyl groups, while in fluoroscein-labelled proteins, linkage is *via* an isothiocyanate group (N═C═S) on the lowermost ring (*).

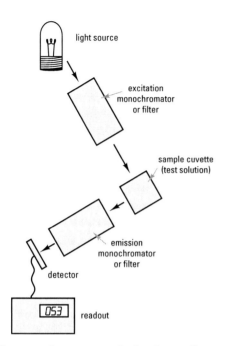

Fig. 41.5 Components of a fluorimeter (fluorescence spectrophotometer). Note that sample cuvettes for fluorimetry must have clear sides all round.

Compared with UV/visible spectrophotometry, fluorescence spectroscopy has certain advantages, including:

- Enhanced sensitivity (up to 1000-fold), since the emitted light is detected against a background of zero, in contrast to spectrophotometry where small changes in signal are measured against a large 'background' (see Eqn [41.3]).
- Increased specificity, because not one but two specific wavelengths are required for a particular compound.

However, there are also certain drawbacks:

- Not all compounds show intrinsic fluorescence, limiting its application. However, some non-fluorescent compounds may be coupled to fluorescent dyes, or fluorophores (e.g. proteins may be coupled to fluorescamine).
- The light emitted can be less than expected due to quenching, i.e. when substances in the sample (e.g. oxygen) either interfere with energy transfer, or absorb the emitted light (in some instances, the sample molecules may self-quench if they are present at high concentration).

The sensitivity of fluorescence has made it invaluable in techniques in which specific antibodies are linked to a fluorescent dye, including:

- fluorescence immunoassay (FIA);
- immunohistochemistry, which requires the use of a fluorescence microscope, e.g. using fluorescent antibodies, or fluorescent *in situ* hybridisation (FISH) for nucleic acid detection;
- flow cytometry – a fluorescence-activated cell sorter (FACS) uses cell surface protein-specific monoclonal antibodies labelled with fluorescent dyes to separate and enumerate cells such as lymphocytes.

Phosphorescence and luminescence

A phenomenon related to fluorescence is phosphorescence, which is the emission of light following intersystem crossing between electron orbitals (e.g. between excited singlet and triplet states). Light emission in phosphorescence usually continues after the exciting energy is no longer applied and, since more energy is lost in intersystem crossing, the emission wavelengths are generally longer than with fluorescence. Phosphorescence has limited applications in biomolecular sciences.

Luminescence (or chemiluminescence) is another phenomenon in which light is emitted, but here the energy for the initial excitation of electrons is provided by a chemical reaction rather than by electromagnetic radiation. An example is the action of the enzyme luciferase, extracted from fireflies, which catalyses the following reaction:

$$\text{luciferin} + \text{ATP} + \text{O}_2 \Rightarrow \text{oxyluciferin} + \text{AMP} + \text{PP}_i + \text{CO}_2 + \text{light} \quad [41.4]$$

The light produced is either yellow–green (560 nm) or red (620 nm). This system can be used in biomolecular analysis of ATP, either to determine ATP concentration in a biological sample, or to follow a coupled reaction (p. 382). Measurement can be performed using the

Table 41.1 Examples of compounds with intrinsic fluorescence

Drugs
Aspirin, morphine, barbiturates, propranolol, ampicillin, tetracyclines

Vitamins
Riboflavin, vitamins A, B6 and E, nicotinamide

Pollutants
Naphthalene, anthracene, benzopyrene

photomultiplier tubes of a scintillation counter (p. 269) to detect the emitted light, with calibration of the output using a series of standards of known ATP content.

Atomic spectroscopy

Atoms of certain metals will absorb and emit radiation of specific wavelengths when heated in a flame, in direct proportion to the number of atoms present. Atomic spectrophotometric techniques measure the absorption or emission of particular wavelengths of UV and visible light, to identify and quantify such metals.

Flame atomic emission spectrophotometry (or flame photometry)

The principal components of a flame photometer are shown in Figure 41.6. A liquid sample is converted into an aerosol in a nebuliser (atomiser) before being introduced into the flame, where a small proportion (typically fewer than 1 in 10 000) of the atoms will be raised to a higher energy level, releasing this energy as light of a specific wavelength, which is passed through a filter to a photocell detector. Flame photometry is used to measure the alkali metal ions K^+, Na^+, and Ca^{2+} in biological fluids; Box 41.2 gives details of the basic procedure.

Atomic absorption spectrophotometry (or flame absorption spectrophotometry)

This technique is applicable to a broad range of metal ions, including those of Pb, Cu, Zn, etc. It relies on the absorption of light of a specific wavelength by atoms dispersed in a flame. The appropriate wavelength is provided by a cathode lamp, coated with the element to be analysed,

SAFETY NOTE *working in atomic spectroscopy – the use of high pressure gas cylinders can be particularly hazardous. Always consult a member of staff before using such apparatus.*

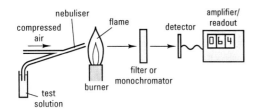

Fig. 41.6 Components of a flame photometer.

SAFETY NOTE *Carrying out acid digestion – always work within a fume hood and wear gloves and safety glasses throughout the procedure. Rinse any spillages with a large volume of water.*

Plotting calibration curves in quantitative analysis – *do not force your calibration line to pass through zero if it clearly does not. There is no reason to assume that the zero value is any more accurate than any other reading you have made.*

Box 41.2 How to use a flame photometer

When using a flame photometer:
- **Allow time for the instrument to stabilise.** Switch on the instrument, light the flame and wait at least 5 minutes before analysing your solutions.
- **Check for impurities in your reagents.** For example, if you are measuring K^+ in an acid digest of some biological material, check the K^+ content of a reagent blank, containing everything except the biological material, processed in exactly the same way as the samples. Subtract this value from your sample values to obtain the true K^+ content. If using acid digestion, always work within a fume hood and wear gloves and safety glasses at all times. Rinse any spillages with a large volume of water.
- **Quantify your samples using a calibration curve** (Chapter 37). Calibration standards should cover the expected concentration range for the test solutions – your calibration curve may be non-linear (especially at concentrations above $1\,\text{mmol}\,l^{-1}$, i.e. $1\,\text{mol}\,m^{-3}$ in SI units).
- **Analyse all solutions in duplicate,** so that reproducibility can be assessed.
- **Check your calibration.** Make repeated measurements of a standard solution of known concentration after every six or seven samples, to confirm that the instrument calibration is still valid.
- **Consider the possibility of interference.** Other metal atoms may emit light which is detected by the photocell, since the filters cover a wider waveband than the emission line of a particular element. This can be a serious problem if you are trying to measure low concentrations of a particular metal in the presence of high concentrations of other metals (e.g. Na^+ in sea water), or other substances which form complexes with the test metal, suppressing the signal (e.g. phosphate). Atomic absorption spectrophotometry is an alternative approach for such samples.

Practical advantages of atomic absorption spectrophotometry over flame photometry – these include:

- improved sensitivity;
- increased precision;
- decreased interference.

focused through the flame and onto the detector. When the sample is introduced into the flame, it will decrease the light detected in direct proportion to the amount of metal present. Newer variants of this method include flameless atomic absorption spectrophotometry and atomic fluorescence spectrophotometry, both of which are more sensitive than the flame absorption technique.

Text references

Anon. (1963) *Tables of Spectrophotometric Absorption Data for Compounds used for the Colorimetric Detection of Elements (International Union of Pure and Applied Chemistry)*. Butterworth-Heinemann, London.

Jones, A.M., Reed, R.H. and Weyers, J.D.B. (2007) *Practical Skills in Biology*, 4th edn. Prentice Hall, Harlow.

Sources for further study

Anon. *Spectrophotometry Protocols*. Available: http://www.protocol-online.org/prot/General_Laboratory_Techniques/Spectrophotometry/
Last accessed: 01/04/07.
[Gives examples of spectrophotometric analyses of selected biomolecules]

Burgess, C. and Jones, D.G. (1995) *Spectrophotometry, Luminescence and Colour*. Elsevier, Amsterdam.

Gore, M.G. (ed.) (2000) *Spectrophotometry and Spectrofluorimetry: A Practical Approach*, 2nd edn. Oxford University Press, Oxford.

Van Dyke, K., van Dyke, C. and Woodfork, K. (2002) *Luminescence Biotechnology: Instruments and Applications*. CRC Press, Boca Raton.

Wilson, K. and Walker, J. (eds.) (2005) *Principles and Techniques of Practical Biochemistry and Molecular Biology*, 6th edn. Cambridge University Press, Cambridge.
[Also covers a range of other topics, including chromatography and centrifugation]

Study exercises

41.1 Write a protocol for using a spectrophotometer. After reading this chapter, prepare a detailed stepwise protocol explaining how to use one of the spectrophotometers in your department. Ask another student or a tutor to evaluate your protocol and provide you with feedback.

41.2 Use the Beer–Lambert relationship in quantitative spectrophotometric analysis. Calculate the following (express your answer to three significant figures):

(a) The concentration of NADH ($\mu mol\,l^{-1}$) in a test solution giving an absorbance at 340 nm (A_{340}) of 0.53 in a cuvette with a path length of 1 cm, based on a molar absorptivity for NADH of $6220\,l\,mol^{-1}\,cm^{-1}$ at this wavelength.

(b) The amount of NADH (nmol) in 20 ml of a test solution where $A_{340} = 0.62$ in a cuvette with a path length of 5 cm, based on a molar absorptivity for NADH of $6220\,l\,mol^{-1}\,cm^{-1}$ at this wavelength.

(c) The mass concentration ($\mu g\,ml$) of double-stranded DNA in a test solution giving an absorbance at 260 nm (A_{260}) of 0.57 in a cuvette of path length 5 mm, based on an absorptivity of $20\,l\,g^{-1}\,cm^{-1}$.

(d) The amount (ng) of double stranded DNA in a 50 μl sub-sample from a test solution where $A_{260} = 0.31$ in a cuvette of path length 1 cm, based on an absorptivity of $20\,l\,g^{-1}\,cm^{-1}$.

(e) The chlorophyll *a* content, in μg (g fresh weight)$^{-1}$, of a plant leaf of fresh weight 1.56 g extracted in 50 ml of 90% acetone, where $A_{664} = 0.182$ at 664 nm and $A_{647} = 0.035$, based on the following equation:

Chlorophyll *a* (mg l^{-1}) = 11.93 A_{664} − 1.93 A_{647}

41.3 Determine the molar absorptivity of a substance in aqueous solution. A solution of *p*-nitrophenol containing $8.8\,\mu g\,ml^{-1}$ gave an absorbance of 0.535 at 404 nm in a cuvette of path length 1 cm. What is the molar absorptivity of *p*-nitrophenol at 404 nm, expressed to three significant figures? (Note: M_r of *p*-nitrophenol is 291.27.)

(continued)

Study exercises (continued)

41.4 Determine the concentration of metal ions based on atomic spectroscopy of test and standard solutions. The following data represent a set of calibration standards for K^+ in aqueous solution, measured by flame photometry:

Absorbance of standard solutions containing K^+ at up to 0.5 mmol l^{-1}

K^+ concentration (mmol l^{-1})	Absorbance
0	0.000
0.1	0.155
0.2	0.279
0.3	0.391
0.4	0.537
0.5	0.683

Draw a calibration curve using the above data and use this to estimate the amount of K^+ in a test sample prepared by digestion of 0.482 g of tissue in a final volume of 25 ml of solution, giving an absorbance of 0.429 when measured at the same time as the standards shown above. Express your answer in μmol K^+ (g tissue)$^{-1}$, to three significant figures. See also study exercise 37.1 for a similar exercise, based on atomic absorption spectrophotometry of Zn.

41.5 Explain the basis of fluorescence. Why is the wavelength of the emitted light longer than that used to excite a fluorescent molecule?

41.6 Account for the difference in design between spectrophotometer and fluorimeter cuvettes. Why does a spectrophotometer cuvette have only two clear sides while a fluorimeter cuvette has four clear sides?

42 Advanced spectroscopy and spectrometry

Identifying compounds – the techniques described in this chapter can often provide sufficient information to identify a compound with a low probability of error.

The following techniques are unlikely to be encountered in the early stages of your course, though they may be presented as a demonstration. Alternatively, you may be shown spectra for interpretation and analysis. The various types of spectroscopy, and their relationship with the electromagnetic spectrum, are shown in Table 42.1.

Definitions

Spectroscopy – any technique involving the production and subsequent recording of a spectrum of electromagnetic radiation, usually in terms of wavelength or energy.

Spectrometry – any technique involving the measurement of a spectrum, e.g. of electromagnetic radiation, molecular masses.

Wavenumber – the reciprocal of the wavelength (expressed as cm^{-1}): a term used widely in IR spectroscopy, but rarely in other types of analysis.

Infra-red (IR) and Raman spectroscopy

Both of these techniques involve the measurement of frequencies produced by the vibration of chemical bonds (bending and stretching). The IR/Raman region is generally considered to be from 800 to 2500 nm (for near-IR) and up to 16 000 nm (for mid-IR). Near-IR spectroscopy involves recording the spectrum in that region in a manner analogous to UV/visible spectroscopy, and quantitative analysis is possible. However, the most widely used technique is mid-IR spectroscopy, which allows identification of groups or atoms in a sample compound, but is inappropriate for quantitative measurement. A peak at a particular frequency can be identified by reference to libraries or computer databases of IR spectra, e.g. a peak at a wavenumber of 1730–1750 cm^{-1} corresponds to a carbonyl group – ($\sim$C$=$O), which is present in fatty acids and proteins. The 1400–600 cm^{-1} region is known as the 'fingerprint' region because no two compounds give identical spectra.

The value of IR spectroscopy is greatly enhanced by Fourier transformation (FT), named after the mathematician J.B. Fourier. FT is a procedure for interconverting frequency functions and time or distance functions. In FT-IR, information is obtained from an interferometer, which splits the incident beam so that it passes through both the sample and a reference. When the beam is recombined, interference patterns arise because the two path lengths are different. The interference pattern has the same relationship to a normal spectrum as a hologram has to a picture, and integral computers use FT to convert the pattern into a spectrum in under a minute. The overall result is a greatly enhanced signal-to-noise ratio.

Understanding the origins of IR and Raman spectra – the IR spectrum is due to changes in charge displacement in bonds. The Raman spectrum is due to changes in polarisability in bonds.

Table 42.1 The electromagnetic spectrum and types of spectroscopy

Type of radiation	Origin	Wavelength	Type of spectroscopy
γ-rays	Atomic nuclei	<0.1 nm	γ-ray spectroscopy
X-rays	Inner shell electrons	0.1–1.0 nm	X-ray fluorescence (XRF)
Ultraviolet (UV)	Ionisation	10–200 nm	UV spectroscopy
UV/visible	Valency electrons	200–800 nm	UV/visible spectroscopy
Infra-red	Molecular vibrations	0.8–25 μm	Infrared spectroscopy (IR) and Raman
Microwaves	Electron spin alignment	400 μm–30 cm	Electron spin resonance (ESR)
Radiowaves	Nuclear spin resonance	>100 cm	Nuclear magnetic resonance (NMR)

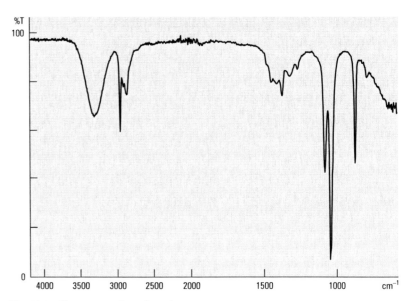

Fig. 42.1 IR spectrum for ethanol.

IR spectra

A typical IR spectrum is shown in Figure 42.1 and you should note the following points:

- The x-axis, the wavelength of the radiation, is given in wavenumbers ($\bar{v}$) and expressed in reciprocal centimetres (cm^{-1}). You may still see some spectra from old instruments using microns (μ, equivalent to the SI unit 'micrometres', μm, at 1×10^{-6} m) for wavelength.
- The y-axis, expressing the amount of radiation absorbed by the molecule, is usually shown as % transmittance. When no radiation is absorbed (all is transmitted through the sample) there is 100% transmittance while 0% transmittance implies all radiation is absorbed at a particular wavenumber. Since the y-axis scale goes from 0 to 100% transmittance, the absorption peaks are displayed *down* from the 100% line; this is *opposite* to most other common spectra.
- The cells holding the sample usually display imperfections and are not completely transparent to IR radiation, even when empty. Therefore the base line of the spectrum is rarely set on 100% transmittance and quantitative applications of IR spectroscopy are more complex than for UV/visible spectrophotometry (p. 282).

Applications of IR and Raman spectroscopy

The principal use of IR and Raman spectroscopy is in the identification of drugs (e.g. penicillin), small peptides, pollutants and food contaminants. When an IR spectrometer is coupled to a gas–liquid chromatograph, it can be used for the analysis of drug metabolites.

One possible future use of IR is for the non-invasive analysis of organic compounds that are of diagnostic significance (e.g. glucose in diabetics). This involves illuminating a small area of skin with IR radiation and using the resultant IR spectrum to quantify a specific compound: an example of such a system is the non-invasive blood glucose monitor described by Malchoff *et al.*, 2002.

The use of wavenumber – this is a well-established convention, since high wavenumber = high frequency = high energy = short wavelength. Expression of the IR range, 4000 cm^{-1} to 650 cm^{-1}, is in 'easy' numbers and the high energy is found on the left-hand side of the spectrum. Note that IR spectroscopists often refer to wavenumbers as 'frequencies', e.g. 'the peak of the C=O stretching 'frequency' is at 1720 cm^{-1}'.

Using IR – the rapid analysis possible with gaseous samples makes IR ideal for studying CO_2 metabolism in photosynthesis and respiration (p. 400).

Nuclear magnetic resonance (NMR)

Electromagnetic radiation (at radiofrequencies of 1–500 MHz) is used to identify and monitor compounds. This is possible because of differences in the magnetic states of atomic nuclei, involving very small transitions in energy levels. The atomic nuclei of isotopes of many elements are magnetic because they are charged and have spin. Typical magnetic nuclei are ^{1}H, ^{13}C, ^{14}N, ^{15}N, ^{19}F and ^{31}P. When these nuclei interact with a uniform external magnetic field, they behave like tiny compass needles and align themselves in a direction either parallel or antiparallel to the field. The two orientations have different energies, with the parallel direction having a lower energy than the antiparallel (Fig. 42.2). The energy difference between the two levels (ΔE) corresponds to a precise electromagnetic frequency (v), according to similar quantum principles to those for the excitation of electrons (p. 280). When a sample containing an isotope with a magnetic nucleus is placed in a magnetic field and exposed to an appropriate radiofrequency, transitions between the energy levels of magnetic nuclei will occur when the energy gap and the applied frequency are in *resonance* (i.e. when they are matched exactly). Differences in energy levels, and hence resonance frequencies (v_0), depend on the magnitude of the applied magnetic field (B_0) and the magnetogyric ratio (λ), according to the equation:

$$v_0 = \lambda B_0 / 2\pi \qquad [42.1]$$

The magnetogyric ratio varies from one isotope to another, so NMR is performed at different frequencies for different nuclei at any given value of B_0. The principal components of an NMR spectrometer are shown in Figure 42.3.

For magnetic nuclei in a given molecule, an NMR spectrum is generated because, in the presence of the applied field, different nuclei experience different local magnetic fields depending on the arrangement of electrons in their vicinity. The effective field (B) at the nucleus can be expressed as:

$$B = B_0(1 - \sigma) \qquad [42.2]$$

where σ (the shielding constant) expresses the contribution of the small secondary field generated by nearby electrons. The magnitude of σ is dependent on the electronic environment of a nucleus, so nuclei in different environments give rise to different resonance frequencies, according to the equation:

$$v_0 = \lambda B_0(1 - \sigma)/2\pi \qquad [42.3]$$

 SAFETY NOTE *Working with NMR spectrometers – these instruments can generate very high magnetic fields and can exert strong magnetic forces on materials such as steel items brought close to the magnet (<1m). NMR magnets can also destroy data stored in magnetic form – keep items such as memory sticks, laptops and credit cards well away.*

Example For an external magnetic field of 2.5 T (Tesla), ΔE for ^{1}H is 6.6×10^{-26} J, and, since $\Delta E = hv$, the corresponding frequency (v) is 100 MHz; for ^{13}C in the same field, ΔE is 1.7×10^{-26} J, and v is 25 MHz.

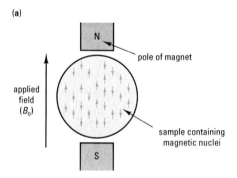

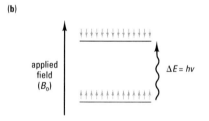

Fig. 42.2 Effect of an applied magnetic field, B_0, on magnetic nuclei. (a) Nuclei in magnetic field have one of two orientations – either with the field or against the field (in the absence of an applied field, the nuclei would have random orientation). (b) Energy diagram for magnetic nuclei in applied magnetic field.

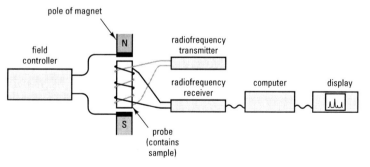

Fig. 42.3 Components of an NMR spectrometer.

Measuring chemical shifts – ppm is not a concentration term in NMR, but is used to reflect the very small frequency changes that occur relative to the reference standard, measured in proportional terms.

The separation of resonance frequencies from a reference value is termed the *chemical shift*, and is expressed in dimensionless terms, as parts per million (ppm). By convention, the chemical shift is positive if the sample nucleus is less shielded than the reference and negative if it is more shielded.

Understanding NMR spectra of biomolecules

Figure 42.4 illustrates the principle underlying this technique – the different chemical environments of the hydrogen atoms in a simple biomolecule such as acetic acid result in two different ^{1}H resonances or 'signals', one corresponding to those of protons in ~CH_3, and the other from that in ~COOH. Furthermore, the relative intensities of the NMR signals, as measured by their areas, are proportional to the number of contributing nuclei, so the relative areas of the peaks due to the protons in ~CH_3 and (undissociated) ~COOH would be 3:1. Similarly, with ^{31}P NMR of a biologically important molecule such as ATP, there are signals corresponding to the α, β and γ phosphates, the P nuclei of which are in different chemical environments (Fig. 42.5). Thus every molecule that contains one or more magnetic nuclei has its own characteristic NMR fingerprint that may be used for identification and analysis. Spectra such as those shown in Figs 42.4 and 42.5 can be obtained using FT of a large number of individual responses to radiowave pulses. Other factors such as the spin-lattice relaxation time (Gadian, 1996) can affect signal intensities (peak sizes), and resonances may be split into several lines due to spin–spin coupling (interactions between neighbouring nuclei).

In terms of resolution, narrow signals are obtained only from molecules that are fairly mobile, so most high resolution studies are carried out using solutions. However, since many metabolites in biological samples (cells, tissues, etc.) are in aqueous solution and freely mobile, they can give rise to high resolution spectra, as shown in Figure 42.5 for the ^{31}P-containing compounds, ATP, phosphocreatine and inorganic phosphate. On the other hand ^{31}P signals from immobile molecules such as DNA or phospholipids are very broad, and could be as wide as the whole scale shown in Fig. 42.5.

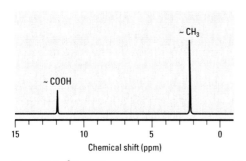

Fig. 42.4 ^{1}H NMR spectrum of acetic acid (CH_3COOH). The relative areas of the two signals are 1:3 and the frequencies (chemical shifts) are expressed in terms of ppm, relative to the reference signal (tetramethyl silane, TMS).

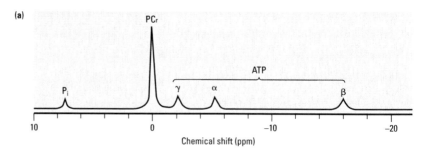

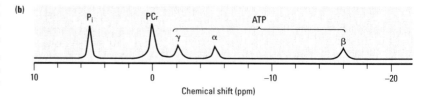

Fig. 42.5 Typical ^{31}P NMR spectrum from intact skeletal muscle (a) at rest and (b) after a fatiguing series of contractions. In (b) ATP levels are preserved at the expense of phosphocreatinine (PCr). Also, hydrolysis of ATP during contraction results in a large phosphate (P_i) peak which is shifted to the right compared to (a), reflecting the decrease in intracellular pH that accompanies glycolysis.

Table 42.2 The relative merits and disadvantages of various magnetic nuclei in biomolecular studies

Nucleus	Relative sensitivity	Natural abundance	Comments
^{1}H	100	99.98%	Multiple, but specific, spectral lines are obtained for individual biomolecules. For mixtures, the ubiquitous occurrence of ^{1}H gives complex, overlapping signals that are often difficult to interpret. Gives a large solvent peak with aqueous samples (can be avoided by using D_2O as solvent). Mainly used for structural studies of pure macromolecules. Essential for MRI.
^{31}P	6.6	100%	Very useful for studies on living systems, with narrow resonance peaks and a wide range of chemical shifts for different molecules. Spectra are simpler and easier to interpret than for ^{1}H, but are not as distinctive: different compounds may give similar ^{31}P spectra. Several important P-containing compounds (including ATP, ADP and inorganic phosphate) can be detected in intact cells – useful in bioenergetic studies.
^{13}C	0.016	1.1%	Gives narrow signals and a wide range of chemical shifts. Resolution is better than for ^{1}H, and a wide range of organic biomolecules can be detected. Low natural abundance gives low sensitivity, extending the time required to accumulate spectra. However, low natural abundance also means that specific metabolites can be selected for ^{13}C isotope enrichment, allowing particular metabolic pathways to be investigated, e.g. carbon assimilation.

Using NMR – *in contrast to most conventional metabolic studies, NMR is non-invasive and the time course of metabolic reactions can be followed using a single experimental subject or preparation, eliminating variation between samples.*

Understanding MRI – *body tissues have different water contents (typically 60–90% w/w). Signals arising from protons in water in different tissues can be used to differentiate between tissues (e.g. 'grey' and 'white' matter in the brain), and between normal tissue and tumours. Fat deposits can be detected, due to the difference between the ^{1}H signal from fatty acids and that of the water in the surrounding tissue.*

Biomolecular applications of NMR

The sensitivity of NMR has improved dramatically with the development of more powerful magnets. Details of the major uses of various magnetic nuclei are given in Table 42.2. The principal applications include:

- studies of the structure and function of macromolecules and biological systems, such as membranes;
- metabolic investigations on living organisms, including humans, since NMR can be used to obtain a 'fingerprint' of a particular molecule and changes in the intensity of spectra can be used for kinetic studies (e.g. Fig. 42.5). This involves use of 'surface coils' as sources of radiation, or placing the organism within the core of an electromagnet;
- measurement of intracellular pH by determination of the chemical shift of the phosphate peak, as this changes with pH in a predictable manner;
- magnetic resonance imaging (MRI), which is a form of proton NMR that uses a field gradient (as opposed to a uniform field) to produce signals that are translated by computers into anatomical images (Gadian, 1996).

Electron spin resonance (ESR)

This technique is based on energy transitions of spinning electrons in a magnetic field. As with NMR, the low energy state occurs when the electromagnetic field generated by the spinning electron is parallel to the externally generated field, while the high energy state occurs when the electron-generated field is antiparallel. ESR is very useful for studying metalloproteins and can be used to monitor the activity of such proteins (e.g. cytochrome oxidase) in intact mitochondria or chloroplasts. It can also be used to detect free radicals, for example in irradiated foodstuffs.

> **Understanding mass spectrometry** – since this technique does not involve the production and measurement of electromagnetic spectra and is not based on quantum principles, it should not be referred to as a spectroscopic technique.

Mass spectrometry (MS)

This technique involves the disintegration of organic compounds into fragment ions in a gas phase. These ions are accelerated to specific velocities using an electric field and then separated on the basis of their different masses. Each fragment of a particular mass is detected sequentially with time. The most widely used method in MS is electron impact ionisation (EI), where the electron source is a heated metal, such as a tungsten filament, subjected to an appropriate potential gradient. The stream of emitted electrons may then interact with biomolecules (M) in the sample by either:

1. electron removal – where an electron in a bond within the sample molecule is 'knocked out' by bombarding electrons, leaving the bond with only one unpaired electron and resulting in the production of a cationic free radical, i.e. $M + e^- \rightarrow M^{+\cdot} + 2e^-$;
2. electron capture – where addition of an extra electron results in the production of an anionic free radical, i.e. $M + e^- \rightarrow M^{-\cdot}$.

Where it is not known whether cations or anions are formed, the radical is given the symbol $\cdot$, i.e. as $M^{\cdot}$. The first ions formed (parent ions) are unstable, and rapidly undergo further disintegration to give smaller fragments and daughter ions, subsequently separated in the mass spectrometer (Fig. 42.6), to give a mass spectrum such as that shown in Figure 42.7.

For *in vivo* metabolic studies, the technique of isotope ratio mass spectroscopy (IRMS) is useful since it removes the need to use radioactive isotopes. This technique exploits the ability of MS to distinguish between isotopes such as ^{13}C and ^{12}C. For example, compounds containing ^{13}C have a greater mass than the same compound containing ^{12}C and can be differentiated in the mass spectrum. By selectively labelling a key metabolite with a non-radioactive isotope, the fate of this metabolite can be followed by MS analysis of sequential samples. Normally, the materials used are combusted in oxygen to give gases such as $^{13}CO_2$ and $^{15}NO_2$, followed by exposure of such gases to an EI source; useful metabolic data can be obtained by determining the isotope ratio, e.g. of $^{13}CO_2^{\cdot}$ and $^{12}CO_2^{\cdot}$.

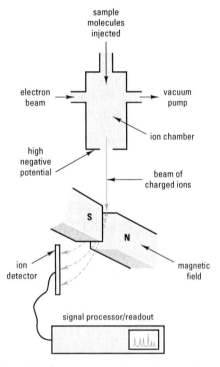

Fig. 42.6 Components of an electron-impact mass spectrometer.

> **Example** IRMS is used in the diagnosis of gastric ulcers caused by *Helicobacter pylori*. This bacterium produces urease, degrading urea to CO_2 and NH_3. A patient drinks a solution of ^{13}C-labelled urea and, if the organism is present in the stomach, $^{13}CO_2$ will appear at up to 5% v/v in the patient's breath, compared with the expected natural abundance value of 1.1% v/v.

Pyrolysis-mass spectrometry (PY-MS)

This is another variant of mass spectroscopy that is useful for biomolecular applications. The molecules in a sample are volatilised by heating to a precisely controlled high temperature for a specific time period. The volatile material is then removed by a vacuum, ionised by EI, and subjected to MS. Complex mixtures of pyrolysis products are produced, but interpretation is made easier by computer-based multivariate analysis. A useful application of PY-MS is in microbial identification, especially with small samples of organisms that are difficult to culture, e.g. mycobacteria.

Fast atom bombardment–mass spectrometry (FAB-MS)

This is particularly useful for biomolecular applications because it can be used with aqueous solutions. The solution containing the analyte is mixed with glycerol and applied to a probe which is inserted into a

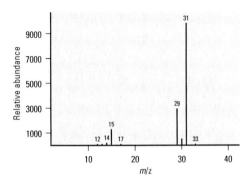

Fig. 42.7 Mass spectrum for methanol. m/z = charge ratio.

vacuum chamber. The mixture is bombarded with a high velocity stream of atoms (usually argon or xenon) rather than electrons, which induces fragmentation of the biomolecules and allows production of the mass spectrum. One problem with FAB-MS is the suppression effect phenomenon – with a mixed sample, not all components may be equally accessible to the atomic bombardment depending on the way they are distributed in the sample–glycerol mixture. This can be overcome by coupling FAB-MS to techniques that can give high resolution separation of components, e.g. gas–liquid chromatography (GC, p. 306), high performance liquid chromatography (HPLC, p. 305), or capillary electrophoresis (p. 331). Johnstone and Rose (1996) give further information on biomolecular aspects of MS.

Text references

Malchoff, C.D., Shoukri, K., Landau, J.I. and Buchert, J.M. (2002) Novel non-invasive blood glucose monitor. *Diabetes Care* **25**: 2268–2275.
Available: care.diabetesjournals.org/cgi/content/full/25/12/2268
Last accessed: 01/04/07.

Gadian, D.G. (1996) *NMR and its Applications to Living Systems*, 2nd edn. Oxford University Press, Oxford.

Johnstone, R.A.W. and Rose, M.E. (1996) *Mass Spectrometry for Chemists and Biochemists*, 2nd edn. Cambridge University Press, Cambridge.

Sources for further study

Guntzler, H. and Gremlich, H.U. (2002) *IR Spectroscopy: an Introduction*. Wiley-VCH, Mannheim.

Hore, P.J., Jones, J.A. and Wimperis, S. (2000) *NMR: the Toolkit*. Oxford University Press, Oxford.
[A 'primer' text – succinct, and includes relevant biological applications]

Keeler, J. (2005) *Understanding NMR Spectroscopy*. Wiley, New York.

Larkin, P.J. (2004) *An Introduction to Infrared and Raman Spectroscopy*. Jones and Bartlett, Sudbury.

Macomber, R.H. (2000) *A Complete Introduction to Modern NMR Spectrometry*. Wiley, New York.

[An accessible introductory text, with relevant biomedical information and examples]

Siudzak, G. (1996) *Mass Spectrometry for Biotechnology*. Academic Press, New York.
[Written for bioresearchers, it covers principles and applications without extensive use of maths]

Smith, R.M. and Butsch, K.L. (2004) *Understanding Mass Spectra: a Basic Approach*. Wiley, New York.

Westbrook, C. and Roth, C. (2005) *MRI in Practice*, 3rd edn. Blackwell, Oxford.
[Covers the basics of magnetic resonance imaging in medicine]

Study exercises

42.1 Interpret IR signals. The figure shows the IR spectra of acetic acid (ethanoic acid). Using the information provided below, identify the resonance corresponding to (a) the O–H and (b) C=O of the carboxylic acid group in the acetic acid spectrum.

Typical IR absorption ranges for different functional groups

Bond	Location	Wavenumber (cm^{-1})
C–O	esters, alcohols	1000–1300
C=O	ketones, aldehydes, carboxylic acids, esters, amides	1680–1750
O–H	carboxylic acids (H-bonded)	2500–3300
O–H	alcohols (H-bonded)	3230–3350
O–H	free	3580–3670
N–H	amines	3100–3500

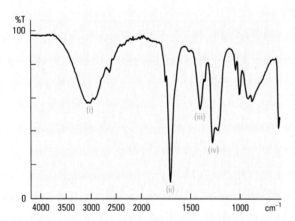

42.2 Investigate appropriate reference compounds for biological applications of NMR. Why are reference compounds important when measuring chemical shifts in NMR? Which compounds are suitable references for (a) ^{1}H, (b) ^{13}C and (c) ^{31}P NMR?

42.3 Test your understanding of the NMR terminology. Distinguish between the term ppm as used in NMR and as a concentration term (p. 146).

42.4 Interpret NMR signals from biomolecules in body fluids. A ^{1}H NMR spectrum (range 1–4.5 ppm) of a human urine sample obtained from a healthy adult is shown in the first figure below. The spectrum reveals a large number of components/peaks. The two most intense of these peaks (at approximately 3.05 and 4.05 ppm) correspond to creatinine (C), which is abundant in healthy urine. The molecular structure of creatinine is shown in the second figure below. Explain why the the ^{1}H NMR 'fingerprint' of creatinine in urine consists of just two discrete peaks.

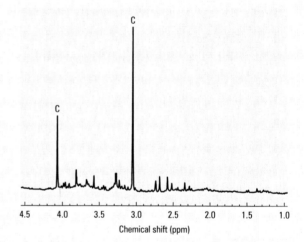

43 Centrifugation

 SAFETY NOTE *Working with centrifuges – these instruments generate very high centrifugal forces and should only be used after suitable instruction and training – if in doubt, ask.*

Particles suspended in a liquid will move at a rate that depends on:

- the applied force – particles in a liquid within a gravitational field, e.g. a stationary test tube, will move in response to the earth's gravity;
- the density difference between the particles and the liquid – particles less dense than the liquid will float upwards while particles denser than the liquid will sink;
- the size and shape of the particles;
- the viscosity of the medium.

For most biological particles (cells, organelles or molecules) the rate of flotation or sedimentation in response to the earth's gravity is too slow to be of practical use in separation.

 KEY POINT A centrifuge is an instrument designed to produce a centrifugal force far greater than the earth's gravity, by spinning the sample about a central axis (Fig. 43.1). Particles of different size, shape or density will thereby sediment at different rates, depending on the speed of rotation and their distance from the central axis.

Table 43.1 Relationship between speed (r.p.m.) and acceleration (relative centrifugal field, RCF) for a typical bench centrifuge with an average radius of rotation, $r_{av} = 115$ mm

r.p.m.	RCF*
500	30
1000	130
1500	290
2000	510
2500	800
3000	1160
3500	1570
4000	2060
4500	2600
5000	3210
5500	3890
6000	4630

*RCF values rounded to nearest 10.

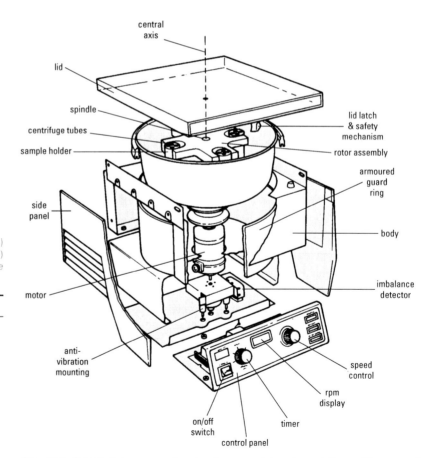

Fig. 43.1 Principal components of a low-speed bench centrifuge. Diagram of Fisher Scientific Model MSE Centaur 2, low-speed bench centrifuge, reproduced by kind permission of Fisher Scientific UK Ltd.

Working in SI units – to convert RCF to acceleration in SI units, multiply by $9.80\,m\,s^{-2}$, e.g. an RCF of 290 is equivalent to an acceleration of $290 \times 9.80 = 2842\,m\,s^{-2}$.

Examples Suppose you wanted to calculate the RCF of a bench centrifuge with a rotor of $r_{av} = 95\,mm$ running at a speed of 3000 r.p.m. Using Eqn [43.1] the RCF would be: $1.118 \times 95 \times (3)^2 = 956\,g$.

You might wish to calculate the speed (r.p.m.) required to produce a relative centrifugal field of $2000\,g$ using a rotor of $r_{av} = 85\,mm$. Using Eqn [43.2] the speed would be: $945.7\sqrt{(2000 \div 85)} = 4587$ r.p.m.

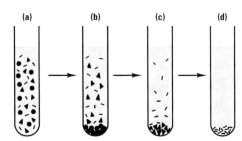

Fig. 43.2 Differential sedimentation. (a) Before centrifugation, the tube contains a mixed suspension of large, medium and small particles of similar density. (b) After low-speed centrifugation, the pellet is predominantly composed of the largest particles. (c) Further high-speed centrifugation of the supernatant will give a second pellet, predominantly composed of medium-sized particles. (d) A final ultracentrifugation step pellets the remaining small particles. Note that all of the pellets apart from the final one will have some degree of cross-contamination.

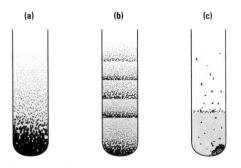

Fig. 43.3 Density gradients. (a) A continuous (linear) density gradient. (b) A discontinuous (stepwise) density gradient, formed by layering solutions of decreasing density on top of each other. (c) A single-step density barrier, designed to allow selective sedimentation of one type of particle.

How to calculate centrifugal acceleration

The acceleration of a centrifuge is usually expressed as a multiple of the acceleration due to gravity ($g = 9.80\,m\,s^{-2}$), termed the relative centrifugal field (RCF, or 'g value'). The RCF depends on the speed of the rotor (n, in revolutions per minute, r.p.m.) and the radius of rotation (r, in mm) where:

$$\mathrm{RCF} = 1.118\,r\left(\frac{n}{1000}\right)^2 \qquad [43.1]$$

This relationship can be rearranged, to calculate the speed (r.p.m.) for specific values of r and RCF:

$$n = 945.7\sqrt{\left(\frac{\mathrm{RCF}}{r}\right)} \qquad [43.2]$$

However, you should note that RCF is not uniform within a centrifuge tube: it is highest near the outside of the rotor (r_{max}) and lowest near the central axis (r_{min}). In practice, it is customary to report the RCF calculated from the average radius of rotation (r_{av}), as shown in Fig. 43.5. It is also worth noting that RCF varies relative to the *square* of the speed: thus the RCF will be doubled by an increase in speed of approximately 41% (Table 43.1).

Centrifugal separation methods

Differential sedimentation (pelleting)

By centrifuging a mixed suspension of particles at a specific RCF for a particular time, the mixture will be separated into a pellet and a supernatant (Fig. 43.2). The successive pelleting of a suspension by spinning for a fixed time at increasing RCF is widely used to separate organelles from cell homogenates. The same principle applies when cells are harvested from a liquid medium.

Density gradient centrifugation

The following techniques use a density gradient, a solution which increases in density from the top to the bottom of a centrifuge tube (Fig. 43.3).

- Rate-zonal centrifugation. By layering a sample onto a shallow pre-formed density gradient, followed by centrifugation, the larger particles will move faster through the gradient than the smaller ones, forming several distinct zones (bands). This method is time-dependent, and centrifugation *must* be stopped before any band reaches the bottom of the tube (Fig. 43.4).
- Isopycnic centrifugation. This technique separates particles on the basis of their buoyant density. Several substances form density gradients during centrifugation (e.g. sucrose, CsCl, Ficoll, Percoll, Nycodenz). The sample is mixed with the appropriate substance and then centrifuged – particles form bands where their density corresponds to that of the medium (Fig. 43.4). This method requires a steep gradient and sufficient time to allow gradient formation and particle redistribution, but is unaffected by further centrifugation.

Bands within a density gradient can be sampled using a fine Pasteur pipette, or a syringe with a long, fine needle. Alternatively, the tube may be punctured and the contents (fractions) collected dropwise in several tubes. For accurate work, an upward displacement technique can be used (see Ford and Graham, 1991, or Graham, 2001).

Working with silicone oil – the density of silicone oil is temperature-sensitive so work in a location with a known, stable temperature or the technique may fail.

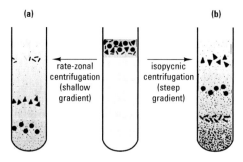

Fig. 43.4 Density gradient centrifugation. The central tube shows the position of the sample prior to centrifugation, as a layer on top of the density gradient medium. Note that particles sediment on the basis of size during rate-zonal centrifugation (a), but form bands in order of their densities during isopycnic centrifugation (b). ●, large particles, intermediate density; ▲, medium-sized particles, low density; ━, small particles, high density.

Recording usage of high-speed centrifuges and ultracentrifuges – most departments have a log book (for samples/speeds/times): make sure you record these details, as the information is important for servicing and replacement of rotors.

 SAFETY NOTE **Changing a rotor** – if you ever have to change a rotor, make sure that you carry it properly (don't knock/drop it), that you fit it correctly (don't cross-thread it, and tighten to the correct setting using a torque wrench) and that you store it correctly (clean it after use and don't leave it lying around).

Density barrier centrifugation

A single step density barrier (Fig. 43.3c) can be used to separate cells from their surrounding fluid, e.g. using a layer of silicone oil adjusted to the correct density using dinonyl phthalate. Blood cell types can be separated using a density barrier of e.g. Ficoll.

Types of centrifuge and their uses

Low-speed centrifuges

These are bench-top instruments for routine use, with a maximum speed of 3000–6000 r.p.m. and RCF up to $6000\,g$ (Fig. 43.1). They are used to harvest cells, larger organelles (e.g. nuclei, chloroplasts) and coarse precipitates (e.g. antibody–antigen complexes, p. 258). Most modern machines also have a sensor that detects any imbalance when the rotor is spinning and cuts off the power supply (Fig. 43.1). However, some of the older models do not, and must be switched off as soon as any vibration is noticed, to prevent damage to the rotor or harm to the operator. Box 43.1 gives details of operation for a low-speed centrifuge.

Microcentrifuges (microfuges)

These are bench-top machines, capable of rapid acceleration up to 12 000 r.p.m. and $10\,000\,g$. They are used to sediment small sample volumes (up to 1.5 ml) of larger particles (e.g. cells, precipitates) over short timescales (typically, 0.5–15 min). They are particularly useful for the rapid separation of cells from a liquid medium, e.g. silicone oil microcentrifugation.

Continuous flow centrifuges

Useful for harvesting large volumes of cells from their growth medium. During centrifugation, the particles are sedimented as the liquid flows through the rotor.

High-speed centrifuges

These are usually larger, free-standing instruments with a maximum speed of up to 25 000 r.p.m. and RCF up to $60\,000\,g$. They are used for microbial cells, many organelles (e.g. mitochondria, lysosomes) and protein precipitates. They often have a refrigeration system to keep the rotor cool at high speed. You would normally use such instruments only under direct supervision.

Ultracentrifuges

These are the most powerful machines, having maximum speeds in excess of 30 000 r.p.m. and RCF up to $600\,000\,g$, with sophisticated refrigeration and vacuum systems. They are used for smaller organelles (e.g. ribosomes, membrane vesicles) and biological macromolecules. You would not normally use an ultracentrifuge, though your samples may be run by a member of staff.

Rotors

Many centrifuges can be used with tubes of different size and capacity, either by changing the rotor, or by using a single rotor with different buckets/adaptors.

- Swing-out rotors: sample tubes are placed in buckets which pivot as the rotor accelerates (Fig. 43.5a). Swing-out rotors are used on many low-speed centrifuges: their major drawback is their extended path

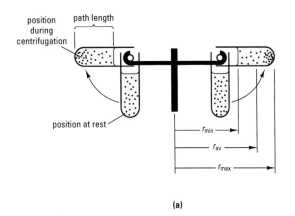 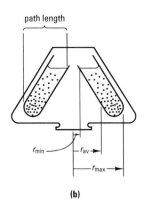 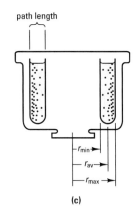

Fig. 43.5 Rotors: (a) swing-out rotor; (b) fixed angle rotor; (c) vertical tube rotor.

length and the resuspension of pellets due to currents created during deceleration.
- Fixed-angle rotors: used in many high-speed centrifuges and microcentrifuges (Fig. 43.5b). With their shorter path length, fixed rotors are more effective at pelleting particles than swing-out rotors.

Box 43.1 How to use a low-speed bench centrifuge

1. **Choose the appropriate tube size and material for your application**, with caps where necessary. Most low-speed machines have four-place or six-place rotors – use the correct number of samples to *fill* the rotor assembly whenever possible.

2. **Fill the containers to the appropriate level**: do not overfill, or the sample may spill during centrifugation.

3. **It is vital that the rotor is balanced during use**. Therefore, *identical* tubes must be prepared, to be placed opposite each other in the rotor assembly. This is particularly important for density gradient samples, or for samples containing materials of widely differing densities, e.g. soil samples, since the density profile of the tube will change during a run. However, for low-speed work using small amounts of particulate matter in aqueous solution, it is sufficient to counterbalance a sample with a second tube filled with water, or a saline solution of similar density to the sample.

4. **Balance each pair of sample tubes** (plus the corresponding caps, where necessary) to within 0.1 g using a top-pan balance; add liquid dropwise to the lighter tube, until the desired weight is reached. Alternatively, use a set of scales. For small sample volumes (up to 10 ml) added to disposable, lightweight plastic tubes, accurate pipetting of your solution may be sufficient for low-speed use.

5. **For centrifuges with swing-out rotors**, check that each holder/bucket is correctly positioned in its locating slots on the rotor and that it is able to swing freely. All buckets must be in position on a swing-out rotor, even if they do not contain sample tubes – buckets are an integral part of the rotor assembly.

6. **Load the sample tubes into the centrifuge**. Make sure that the outside of the centrifuge tubes, the sample holders and sample chambers are dry: any liquid present will cause an imbalance during centrifugation, in addition to the corrosive damage it may cause to the rotor. For sample holders where rubber cushions are provided, make sure that these are correctly located. Balanced tubes must be placed opposite each other – use a simple code if necessary, to prevent mix-ups.

7. **Bring the centrifuge up to operating speed** by gentle acceleration. Do not exceed the maximum speed for the rotor and tubes used.

8. **If the centrifuge vibrates at any time during use, switch off** and find the source of the problem.

9. **Once the rotor has stopped spinning, release the lid and remove all tubes**. If any sample has spilled, make sure you clean it up thoroughly using a non-corrosive disinfectant, e.g. Virkon, so that it is ready for the next user.

10. **Close the lid (to prevent the entry of dust) and return all controls to zero**.

- Vertical tube rotors: used for isopycnic density gradient centrifugation in high-speed centrifuges and ultracentrifuges (Fig. 43.5c). They cannot be used to harvest particles in suspension as a pellet is not formed.

Centrifuge tubes

These are manufactured in a range of sizes (from 1.5 ml up to 1000 ml) and materials. The following aspects may influence your choice:

- Capacity. This is obviously governed by the volume of your sample. Note that centrifuge tubes must be completely full for certain applications, e.g. for high-speed work.
- Shape. Conical-bottomed centrifuge tubes retain pellets more effectively than round-bottomed tubes, while the latter may be more useful for density gradient work.
- Maximum centrifugal force. Detailed information is supplied by the manufacturers. Standard Pyrex glass tubes can only be used at low centrifugal force (up to $2000\,g$).
- Caps. Most fixed-angle and vertical tube rotors require tubes to be capped, to prevent leakage during use and to provide support to the tube during centrifugation. For low speed centrifugation, caps must be used for any hazardous samples. Make sure you use the correct caps for your tubes.
- Solvent resistance. Glass tubes are inert, polycarbonate tubes are particularly sensitive to organic solvents (e.g. ethanol, acetone), while polypropylene tubes are more resistant. See manufacturer's guidelines for detailed information.
- Sterilisation. Disposable plastic centrifuge tubes are often supplied in sterile form. Glass and polypropylene tubes can be repeatedly sterilised. Cellulose ester tubes should *not* be autoclaved. Repeated autoclaving of polycarbonate tubes may lead to cracking/stress damage.
- Opacity. Glass and polycarbonate tubes are clear, while polypropylene tubes are more opaque.
- Ability to be pierced. If you intend to harvest your sample by puncturing the tube wall, cellulose acetate and polypropylene tubes are readily punctured using a syringe needle.

 SAFETY NOTE Working with centrifuge tubes – never be tempted to use a tube or bottle which was not designed to fit the machine you are using (e.g. a general-purpose glass test tube, or a screw-capped bottle), or you may damage the centrifuge and cause an accident.

Using microcentrifuge tubes – the integral push-on caps of microcentrifuge tubes must be correctly pushed home before use or they can come off during centrifugation.

 SAFETY NOTE Balancing tubes – never balance centrifuge tubes 'by eye' – use a balance. Note that a 35 ml tube full of liquid at an RCF of $3000\,g$ has an effective weight greater than a large adult man.

Balancing the rotor

For the safe use of centrifuges, the rotor must be balanced during use, or the spindle and rotor assembly may be damaged permanently; in severe cases, the rotor may fail and cause a serious accident.

KEY POINT It is vital that you balance your loaded centrifuge tubes before use. As a general rule, balance all sample tubes to within 1% or better, using a top-pan balance or scales. Place balanced tubes opposite each other.

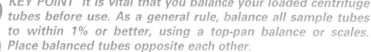 ## Safe practice

Given their speed of rotation and the extremely high forces generated, centrifuges have the potential to be extremely dangerous, if used incorrectly. You should only use a particular centrifuge if you fully understand the operating principles: if unsure, check with a member of staff. For safety reasons, all centrifuges are manufactured with an armoured casing that should contain any fragments in cases of rotor failure. Machines usually have a safety lock to prevent the motor from being switched on unless the lid is closed and to stop the lid from being opened while the rotor is moving. Don't be tempted to use older machines without a safety lock, or centrifuges where the locking mechanism is damaged/inoperative. Be particularly careful to make sure that hair and clothing are kept well away from moving parts.

Text references

Ford, T.C. and Graham, J.M. (1991) *An Introduction to Centrifugation.* Bios, Oxford.

Graham, J. (2001) *Biological Centrifugation.* Bios, Oxford.

Sources for further study

Boyer, R.F. (2005) *Biochemistry Laboratory: Modern Theory and Techniques.* Benjamin Cummings, San Francisco.

Rickwood, D., Ford, T. and Steensgard, J. (1994) *Centrifugation: Essential Data.* Wiley, New York.

Study exercises

43.1 Decide on the type of centrifuge required for a particular application. What centrifuge would you use for each of the following?
 (a) Separating mitochondria from a cell homogenate.
 (b) Separating yeast cells from their surrounding medium in an experiment to study the uptake of a radiolabelled amino acid as a function of time.
 (c) Harvesting cells from a bioreactor containing 25 litres of growth medium.

43.2 Determine centrifugal acceleration for a specific centrifuge. Calculate the centrifugal acceleration (RCF, i.e. the g value) of the following (express your answers to three significant figures):
 (a) A bench centrifuge with an average radius of rotation of 125 mm operating at 4000 r.p.m.
 (b) A bench-top microcentrifuge with an average radius of rotation of 60 mm, operating at 12 000 r.p.m.
 (c) An ultracentrifuge with an average radius of rotation of 186 mm, operating at 30 000 r.p.m.

43.3 Determine the speed required to give a particular centrifugal acceleration (g value). Calculate the speed required (r.p.m. value) for each of the following (express your answers to three significant figures):
 (a) RCF of 1500 g, using a bench centrifuge with an average radius of rotation of 95 mm.
 (b) RCF of 50 000 g, using a high-speed centrifuge with an average radius of rotation of 135 mm.
 (c) RCF of 13 000 g, using a bench-top microcentrifuge with an average radius of rotation of 5.5 cm.

43.4 Calculate the difference in centrifugal acceleration (g value) between the top, middle and bottom of a centrifuge tube. Assuming that the minimum, average and maximum radial distances of a centrifuge tube in a swing-out rotor of a bench centrifuge operating at 5000 r.p.m. are 45 mm, 70 mm and 95 mm respectively, what are the corresponding g values at the top, middle and bottom of the tube when the centrifuge is operating? (Express your answer to three significant figures.)

44 Chromatography – separation methods

Making compromises in chromatography – the process is often a three-way compromise between:

- separation of analytes
- time of analysis
- volume of eluent.

Thus, if you have a large sample volume and you want to achieve a good separation of a mixture of analytes, the time taken for chromatography will be lengthy.

 SAFETY NOTE The solvents used as the mobile phases of chromatographic systems are often toxic and may produce noxious fumes – where necessary, work in a fume hood.

Chromatography is used to separate the individual constituents within a sample on the basis of differences in their physical characteristics, e.g. molecular size, shape, charge, volatility, solubility and/or adsorptivity. The essential components of a chromatographic system are:

- A stationary phase, either a solid, a gel or an immobilised liquid, held by a support matrix.
- A chromatographic bed: the stationary phase may be packed into a glass or metal column, spread as a thin layer on a sheet of glass or plastic, or adsorbed on cellulose fibres (paper).
- A mobile phase, either a liquid or a gas which acts as a solvent, carrying the sample through the stationary phase and eluting from the chromatographic bed.
- A delivery system to pass the mobile phase through the chromatographic bed.
- A detection system to monitor the test substances (Chapter 45).

Individual substances interact with the stationary phase to different extents as they are carried through the system, enabling separation to be achieved.

 KEY POINT In a chromatographic system, those substances which interact strongly with the stationary phase will be retarded to the greatest extent while those which show little interaction will pass through with minimal delay, leading to differences in distances travelled or elution times.

Types of chromatographic system

Chromatographic systems can be categorised according to the form of the chromatographic bed, the nature of the mobile and stationary phases and the method of separation.

Thin-layer chromatography (TLC) and paper chromatography

Here, you apply the sample as a single spot near one end of the sheet, by microsyringe or microcapillary. This sheet is allowed to dry fully, then it is transferred to a glass tank containing a shallow layer of solvent (Fig. 44.1). Remove the sheet when the solvent front has travelled across 80–90% of its length.

You can express movement of an individual substance in terms of its relative frontal mobility, or R_F value, where:

$$R_F = \frac{\text{distance moved by substance}}{\text{distance moved by solvent}} \quad [44.1]$$

Alternatively, you may express movement with respect to a standard of known mobility, as R_X, where:

$$R_X = \frac{\text{distance moved by test substance}}{\text{distance moved by standard}} \quad [44.2]$$

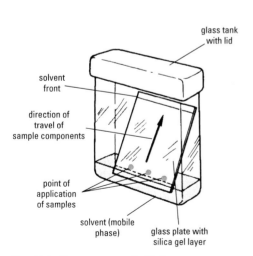

Fig. 44.1 Components of a TLC system.

Using a TLC system – it is essential that you allow the solvent to pre-equilibrate in the chromatography tank for at least 2 h before use, to saturate the atmosphere with vapour. Deliver drops of sample with a blunt-ended microsyringe. Make sure you know exactly where each sample is applied, so that R_F values can be calculated.

The R_F (or R_X) value is a constant for a particular substance and solvent system (under standard conditions) and closely reflects the partitioning of the substance between the stationary and mobile phases. Tabulated values are available for a range of biological molecules and solvents (e.g. Stahl, 1969, or Touchstone, 1980). However, you should analyse one or more reference compounds on the same sheet as your unknown sample, to check their R_F values.

Column chromatography

Here, you pack a glass column with the appropriate stationary phase and equilibrate the mobile phase by passage through the column, either by gravity (Fig. 44.2), or using a low pressure peristaltic pump. You can then introduce the sample to the top of the column, to form a discrete band of material. This is then flushed through the column by the mobile phase. If the individual substances have different rates of migration, they will separate within the column, eluting at different times as the mobile phase travels through the column.

You can detect eluted substances by collecting the mobile phase as it elutes from the column in a series of tubes (discontinuous monitoring), either manually or with an automatic fraction collector. Fractions of 2–5% of the bed volume are usually collected and analysed, e.g. by chemical assay. You can now construct an elution profile (or chromatogram) by plotting the amount of substance against either time, elution volume or fraction number, which should give a symmetrical peak for each substance (Fig. 44.3).

You can express the migration of a particular substance at a given flow rate in terms of its retention time (t), or elution volume (V_e). The separation efficiency of a column is measured by its ability to distinguish between two similar substances, assessed in terms of:

- selectivity (α), measured using the following equation, which takes into account the retention times of the two peaks (i.e. t_a and t_b), plus the column dead time (t_0):

$$\alpha = \frac{t_b - t_0}{t_a - t_0} \qquad [44.3]$$

The column dead time is the time it takes for an unretained compound to pass through the column without any interaction with the stationary phase.

- resolution (R), quantified in terms of the retention time and the base width (W) of each peak:

$$R = \frac{2(t_a - t_b)}{W_a + W_b} \qquad [44.4]$$

where the subscripts a and b refer to substances a and b respectively (Fig. 44.3). For most practical purposes, R values of 1 or more are satisfactory, corresponding to 98% peak separation for symmetrical peaks.

Non-symmetrical peaks may result from column overloading, co-elution of solutes, poor packing of the stationary phase, or interactions between the substances and the support material.

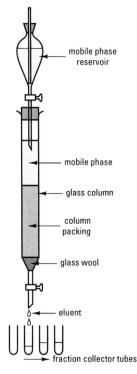

Fig. 44.2 Equipment for column chromatography (gravity feed system).

Chromatography – Separation Methods

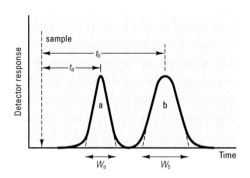

Fig. 44.3 Peak characteristics in a chromatographic separation, i.e. a chromatogram. For symbols, see Eqn [44.4].

Using HPLC – this is a versatile form of chromatography, used with a wide variety of stationary and mobile phases, to separate individual compounds of a particular class of molecules on the basis of size, polarity, solubility or adsorption characteristics.

High performance liquid chromatography (HPLC)

Column chromatography originally used large 'soft' stationary phases that required low pressure flow of the mobile phase to avoid compression; separations were usually time-consuming and of low resolution ('low performance'). Subsequently, the production of small, incompressible, homogeneous particulate support materials and high pressure pumps with reliable, steady flow rates have enabled high performance systems to be developed. These systems operate at pressures up to 10 MPa, forcing the mobile phase through the column at a high flow rate to give rapid separation with reduced band broadening, due to smaller particle size.

HPLC columns are usually made of stainless steel, and all components, valves, etc., are manufactured from materials which can withstand the high pressures involved. The two main systems are:

- Isocratic separation: a single solvent (or solvent mixture) is used throughout the analysis.
- Gradient elution separation: the composition of the mobile phase is altered using a microprocessor-controlled gradient programmer, which mixes appropriate amounts of two different substances to produce the required gradient.

Most HPLC systems are linked to a continuous monitoring detector of high sensitivity, e.g. proteins may be detected spectrophotometrically by monitoring the absorbance of the eluent at 280 nm as it passes through a flow cell (cuvette). Other detectors can be used to measure changes in fluorescence, current or potential (Chapter 45). Most detection systems are non-destructive, which means that you can collect eluent with an automatic fraction collector for further study (Fig. 44.4).

The speed and sensitivity of HPLC make this the method of choice for the separation of many small molecules of biological interest, normally using reverse phase partition chromatography (p. 306). Separation of macromolecules (especially proteins and nucleic acids) usually requires 'biocompatible' systems in which stainless steel components are replaced by titanium, glass or fluoroplastics, using lower pressures to avoid denaturation, e.g. the Pharmacia FPLC system. Such separations are carried out using ion-exchange, gel permeation and/or hydrophobic interaction chromatography (p. 308).

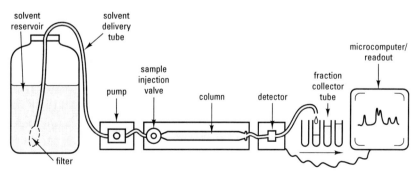

Fig. 44.4 Components of an HPLC system.

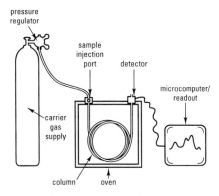

Fig. 44.5 Components of a GC system.

Applications of gas chromatography – GC is used to separate volatile, non-polar compounds: substances with polar groups must be converted to less polar derivatives prior to analysis, in order to prevent adsorption on the column, resulting in poor resolution and peak tailing.

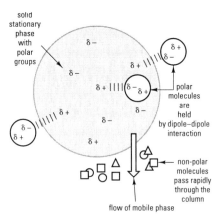

Fig. 44.6 Adsorption chromatography (polar stationary phase).

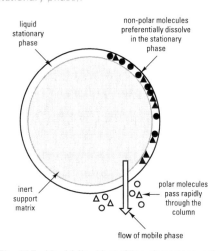

Fig. 44.7 Liquid–liquid partition chromatography, e.g. reverse-phase HPLC.

Gas chromatography (GC)

Modern GC uses capillary chromatography columns (internal diameter 0.1–0.5 mm) up to 50 m in length (Fig. 44.5). The stationary phase is generally a cross-linked silicone polymer, coated as a thin film on the inner wall of the capillary: at normal operating temperatures, this behaves in a similar manner to a liquid film but is far more robust. The mobile phase ('carrier gas') is usually nitrogen or helium. Selective separation is achieved as a result of the differential partitioning of individual compounds between the carrier gas and silicone polymer phases. The separation of most biomolecules is influenced by the temperature of the column, which may be constant during the analysis ('isothermal' – usually 50–250 °C) or, more commonly, may increase in a pre-programmed manner (e.g. from 50 °C to 250 °C at 10 °C per min). Samples are injected onto the 'top' of the column, through a sample injection port containing a gas-tight septum. The output from the column can be monitored by flame ionisation, electron capture or thermal conductivity (Chapter 45).

Spectrometric detection systems include mass spectrometry (GC-MS) and infra-red spectroscopy (GC-IR). GC can only be used with samples capable of volatilisation at the operating temperature of the column, e.g. short chain fatty acids. Other substances may need to be chemically modified to produce more volatile compounds, e.g. long chain saturated fatty acids (Chapter 51) are usually analysed as methyl esters while monosaccharides (Chapter 52) are converted to their trimethylsilyl derivatives.

Separation methods

Adsorption chromatography

This is a form of solid–liquid chromatography. The stationary phase is a porous, finely divided solid which adsorbs molecules of the test substance on its surface due to dipole–dipole interactions, hydrogen bonding and/or van der Waals interactions (Fig. 44.6). The range of adsorbents is limited, e.g. polystyrene-based resins (for non-polar molecules), silica, aluminium oxide and calcium phosphate (for polar molecules). Most adsorbents must be activated by heating to 110–120 °C before use, since their adsorptive capacity is significantly decreased in the presence of bound water. Adsorption chromatography can be carried out in column or thin-layer form, using a wide range of organic solvents.

Partition chromatography

This is based on the partitioning of a substance between two liquid phases, in this instance the stationary and mobile phases. Substances which are more soluble in the mobile phase will pass rapidly through the system while those which favour the stationary phase will be retarded (Fig. 44.7). In normal phase partition chromatography the stationary phase is a polar solvent, usually water, supported by a solid matrix (e.g. cellulose fibres in paper chromatography) and the mobile phase is an immiscible, non-polar organic solvent. For reverse-phase partition chromatography the stationary phase is a non-polar solvent (e.g. a C_{18} hydrocarbon, such as octadecylsilane) which is chemically bonded to a porous support matrix (e.g. silica), while the mobile phase can be chosen from a wide range of polar solvents, usually water or an aqueous buffered solution containing one or more organic solvents, e.g. acetonitrile. Solutes interact with the stationary phase through non-polar interactions and so the *least* polar solutes elute last from the column. Solute retention and separation are controlled by changing the composition of the mobile phase (e.g. % v/v acetonitrile). Reverse-phase high performance liquid chromatography (RP–HPLC) is used to separate a broad range of non-polar, polar and ionic biomolecules, including peptides, proteins, oligosaccharides and vitamins for ionic and ionisable solutes.

Ion-exchange chromatography

Here, separations are carried out using a column packed with a porous matrix which has a large number of ionised groups on its surfaces, i.e. the stationary phase is an ion-exchange resin. The groups may be cation or anion exchangers, depending upon their affinity for positive or negative ions. The net charge on a particular resin depends on the pK_a of the ionisable groups and the pH of the solution, in accordance with the Henderson–Hasselbalch equation (p. 156).

For most practical applications, you should select the ion-exchange resin and buffer pH so that the test substances are strongly bound by electrostatic attraction to the ion-exchange resin on passage through the system, while the other components of the sample are rapidly eluted (Fig. 44.8). You can then elute the bound components by raising the salt concentration of the mobile phase, either stepwise or as a continuous gradient, so that exchange of ions of the same charge occurs at oppositely charged sites on the stationary phase. Weakly bound sample molecules will elute first, while more strongly bound molecules will elute at a higher concentration. Computer-controlled gradient formers are available: if two or more components cannot be resolved using a linear salt gradient, an adapted gradient can be used in which the rate of change in salt concentration is decreased over the range where these components are expected to elute. To maximise resolution in IEC (and HIC) keep your columns as short as possible. Once the sample components have been separated, they should be eluted as quickly as possible from the column in order to avoid band broadening resulting from diffusion of sample ions in the mobile phase.

Ion-exchange chromatography can be used to separate mixtures of a wide range of ionic biomolecules, including amino acids, peptides, proteins and nucleotides. Electrophoresis (Chapters 46 and 47) is an alternative means of separating charged biomolecules.

Gel permeation chromatography (GPC) or gel filtration

Here, the stationary phase is in the form of beads of a cross-linked gel containing pores of a discrete size (Fig. 44.9). The size of the pores is controlled so that at the molecular level, the pores act as 'gates' that will exclude large molecules and admit smaller ones. However, this gating effect is not an all or nothing phenomenon: molecules of intermediate size partly enter the pores. A column packed with such beads will have within it two effective volumes that are potentially available to sample molecules in the mobile phase, i.e. V_i, the volume surrounding the beads, and V_{ii}, the volume within the pores. If a sample is placed at the top of such a column, the mobile phase will carry the sample components down the column, but at different rates according to their molecular size. A very large molecule will have access to all of V_i but to none of V_{ii}, and will therefore elute in the minimum possible volume (the 'void volume', or V_o, equivalent to V_i). A very small molecule will have access to all of V_i and all of V_{ii}, and therefore

Selecting a separation method – it is often best to select a technique that involves direct interaction between the substance(s) and the stationary phase, (e.g. ion exchange, affinity or hydrophobic interaction chromatography), due to their increased capacity and resolution compared to other methods, (e.g. partition or gel permeation chromatography).

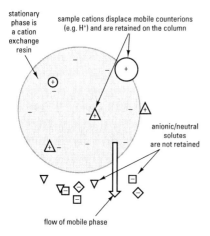

Fig. 44.8 Ion-exchange chromatography (cation exchanger).

Using a gel permeation system – keep your sample volume as small as possible, in order to minimise band broadening due to dilution of the sample during passage through the column.

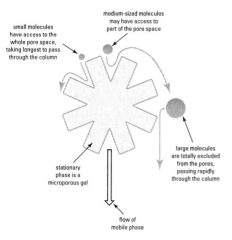

Fig. 44.9 Gel permeation chromatography.

Table 44.1 Fractionation ranges of selected gel permeation chromatography media

M_r	Medium
50–1000	Sephadex G15
	Biogel P-2
1000–5000	Sephadex G-25
1500–30 000	Sephadex G-50
	Biogel P-10
4000–150 000	Sephadex G-100
5000–250 000	Sephadex G-200
20 000–1 500 000	Sephacryl S 300
60 000–20 000 000	Sepharose 4B

it has to pass through the total liquid volume of the column (V_t, equivalent to $V_i + V_{ii}$) before it emerges. Molecules of intermediate size have access to all of V_i but only part of V_{ii}, and will elute at a volume between V_o and V_t, in order of decreasing size depending on their access to V_{ii}.

Cross-linked dextrans (e.g. Sephadex), agarose (e.g. Sepharose) and polyacrylamide (e.g. Bio-gel) can be used to separate mixtures of macromolecules, particularly enzymes, antibodies and other globular proteins (see Table 44.1 for details). Selectivity in GPC is solely dependent on the stationary phase, with the mobile phase being used solely to transport the sample components through the column. Thus, it is possible to estimate the molecular mass of a sample component by calibrating a given column using molecules of known molecular mass and similar shape. A plot of elution volume (V_e) against $\log_{10}$ molecular mass is approximately linear. A further application of GPC is the general separation of low molecular mass and high molecular mass components, e.g. 'desalting' a protein extract by passage through a Sephadex G-25 column is faster and more efficient than dialysis.

Hydrophobic interaction chromatography (HIC)

This technique is used to separate proteins, and exploits the fact that many proteins have hydrophobic sites, with hydrophobic amino acid residues (e.g. leucine, isoleucine, valine, phenylalanine) on their surfaces. Proteins differ in the nature and extent of these hydrophobic regions. The underlying principle is similar to that of RP-HPLC in that it involves hydrophobic interactions between the sample components and a non-polar stationary phase. However, RP-HPLC is only useful for analytical separations of proteins where retention of biological activity is not required, and is unsuitable for separation of native proteins for several reasons:

- the stationary phase in RP-HPLC columns tends to be densely packed with hydrophobic groups, leading to tight protein binding, possibly through multi-site attachment;
- the use of polar organic solvents may lead to protein denaturation;
- the high pressures used to obtain rapid flow rates in HPLC may also denature proteins.

The groups used on HIC stationary phases are both less densely packed and less hydrophobic than those used in RP-HPLC, and this results in milder adsorption of proteins (octyl or phenyl groups are commonly used in HIC, rather than octadecyl groups). Furthermore, retention and elution can be achieved using aqueous solutions so that an individual protein can be isolated with its 3-D conformation intact.

Separations are based on interactions between the three components of the system, i.e. the hydrophobic stationary phase, the hydrophobic sample molecules, and the aqueous stationary phase. In an aqueous environment, hydrophobic groups tend to associate, and this results in certain proteins binding to the stationary phase, where the strength of binding is related to the degree of hydrophobicity of the protein (Fig. 44.10). This tendency is promoted by the presence of certain salts, most commonly ammonium sulphate, that produce 'salting out' effects. Elution of components can be achieved by a variety of means:

- reducing the ammonium sulphate concentration of the mobile phase, to decrease the 'salting out' effect;

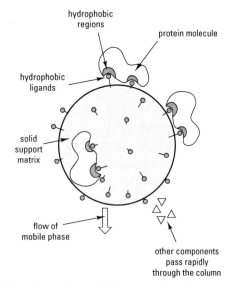

Fig. 44.10 Hydrophobic interaction chromatography. Hydrophobic interactions between ligands and hydrophobic amino acid residues release 'structured' water, making the interactions thermodynamically favourable.

Using HIC for purification – HIC can be used immediately after salt precipitation, when the salt concentration is high. If the desired biomolecule is then eluted using a gradient of reducing salt concentration, it may be possible to follow HIC directly by IEC (which requires an initial buffer of low ionic strength) without changing the buffer.

Avoiding protein precipitation in HIC – *certain protein components may precipitate in high ionic strength buffers. If this occurs, dilute your sample and inject larger volumes.*

- changing the salt in the mobile phase to one that does not promote salting out;
- including non-ionic detergents (e.g. Triton X-100) to reduce hydrophobic interactions;
- including aliphatic alcohols, reducing the polarity of the mobile phase;
- changing the pH;
- reducing the temperature.

Affinity chromatography (AC)

Affinity chromatography enables biomolecules to be purified on the basis of their biological specificity rather than by differences in physicochemical properties, and a high degree of purification (>1000-fold) can be expected. It is especially useful for isolating small quantities of material from large amounts of contaminating substances. The technique involves the immobilisation of a complementary binding substance (the ligand) onto a solid matrix in such a way that the specific binding affinity of the ligand is preserved. When a biological sample is applied to a column packed with this affinity support matrix, the molecule of interest will bind specifically to the ligand, while contaminating substances will be washed through with buffer (Fig. 44.11). Elution of the desired molecule can be achieved by changing the pH or ionic strength of the buffer, to weaken the non-covalent interactions between the molecule and the ligand, or by the addition of other substances that have greater affinity for the ligand.

In AC, the ligand must show specific, but reversible, interaction with the molecule to be purified. It must also contain a reactive functional group, independent of the biospecific site, that will allow covalent attachment to the matrix. The support matrix must be free of non-specific adsorption effects and have sufficient reactive functional groups for the attachment of ligands. Agarose is an ideal support matrix for use in AC. If the ligand is small and the molecule to be purified is large, binding may be restricted due to the proximity of the matrix surface. This problem may be overcome by the introduction of a 'spacer arm' (e.g. a hexane group) between the ligand and the matrix (Fig. 44.11).

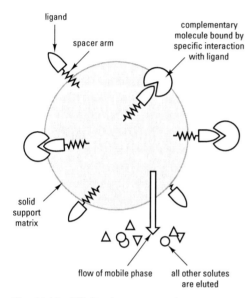

Fig. 44.11 Affinity chromatography.

A potential disadvantage of the specificity of AC is that a new ligand may have to be sought for each individual separation – a potentially time-consuming and expensive process. It may be more practical to use 'group-specific adsorbents' which contain ligands that have an affinity for a class of biochemically related substances. Examples of group-specific adsorbents include:

Examples Biospecific molecules used in affinity chromatography include:

- enzymes and inhibitors/cofactors/substrates;
- hormones and receptor proteins;
- antibodies and antigens;
- complementary base sequences in DNA and RNA (p. 369).

Elution of substances from an affinity system – *make sure that your elution conditions do not affect the interaction between the ligand and the stationary phase, or you may elute the ligand from the column.*

- Lectins, which are a group of proteins produced from moulds, plants and animals, that bind reversibly with specific sugar residues. They are very useful for the purification of sugar-containing macromolecules such as glycoproteins, serum lipoproteins, and membrane proteins such as receptors. Different lectins show different specificities, e.g. concanavalin A and lentil lectin bind to sugars having –OH groups at C-3, C-4 and C-5 (i.e. mannose and glucose), while wheatgerm lectin binds to N-acetyl glucosamine residues. Once bound, substances can be resolved by eluting the column with a salt gradient, or by including free sugars to act as competitive binding agents. Lectins are sometimes called agglutinins because of their ability to agglutinate different types of erythrocyte (e.g. A, B and O) by binding to the specific receptors on their surfaces.

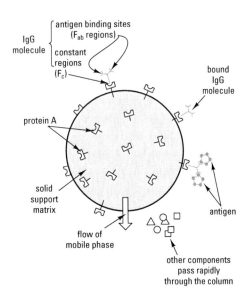

Fig. 44.12 Affinity chromatography using protein A as a ligand to bind IgG antibodies via the F_c region.

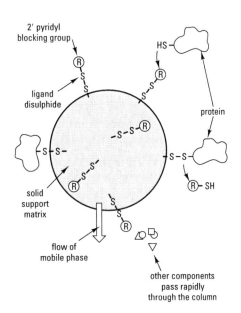

Fig. 44.13 Covalent chromatography. A protein with a free surface thiol group reduces the ligand disulphide, releasing the blocking group and forming a new disulphide.

- Protein A, which is a commercially available surface protein from the bacterium *Staphylococcus aureus* that has a specific binding capacity for the constant (F_c) region of IgG-type antibodies from most mammalian species (Fig. 44.12). As well as being useful in the purification of IgG, protein A can also be used for the isolation of other molecules, as long as an antibody can be raised against them. This is possible because interaction of protein A with IgG occurs via the F_c segment of immunoglobulins of this class, irrespective of their F_{ab} components (see Fig. 38.1). Immobilisation of a specific IgG antibody onto a protein A-agarose column produces an affinity support that can be used to purify the desired antigen.
- Immobilised dyes, which can be used to purify a wide range of enzymes and proteins. One of the most widely used dyes is Cibacron Blue F3G-A, available as an affinity support as Blue Sepharose. This dye has some structural similarities to nucleotide cofactors such as NAD^+ and $NADP^+$, and is useful for the purification of enzymes that require such cofactors. Chemical modification of such dyes increases their specificity, and it is likely that computer modelling will play a role in the future development of new dye-based affinity supports. Dyes are also used in the related non-chromatographic technique of affinity precipitation. This makes use of two dye molecules linked *via* a spacer molecule; when this 'bis-ligand' is added to a solution containing the molecule to be purified, the ligand specifically binds two desired molecules, forming an insoluble complex. Alternatively, a heterobifunctional ligand can be used – a combination of a specific ligand and a group that can be used to initiate precipitation once the desired molecule has bound to the ligand. Such interactions are equivalent to the precipitin reaction in immunology (p. 258).
- Poly(U)-agarose is an affinity support that can be used for the isolation of mRNA because of the biospecific hybridisation of poly(U) with the poly(A) 'tail' sequence characteristic of mRNAs. It can also be used to isolate proteins and enzymes that bind to RNA, such as reverse transcriptases.

Covalent chromatography

This is a variant of affinity chromatography which involves the formation of fairly strong, but reversible, covalent bonds between the affinity support and the molecule to be purified. One type of covalent chromatography is used for the purification of proteins containing thiol (–SH) groups (Fig. 44.13). If the sample is applied to a column containing matrix-attached disulphide 2′-pyridyl groups, thiol-containing proteins displace the 2′-pyridyl group on the support and become immobilised *via* disulphide bridges. Elution of bound components can be achieved by including thiol reagents such as cysteine, glutathione, 2-mercaptoethanol or dithiothreitol (DTT) in the mobile phase. Another form of covalent chromatography uses immobilised boronic acid, which binds certain carbohydrate groups (e.g. in glycoproteins).

Immobilised metal affinity chromatography (IMAC)

This exploits the ability of certain metal ions, especially Ni^{2+}, Cu^{2+}, Zn^{2+}, Hg^{2+} and Cd^{2+}, to bind to proteins by forming coordination complexes with imidazole groups of histidine, indole groups of tryptophan, or thiol groups of cysteine residues. The metal ion may be immobilised by chelation with iminodiacetic acid covalently bound to agarose. Proteins that bind to the metal ions can be eluted using free metal-binding ligands (e.g. amino acids) in the mobile phase.

A very useful application of IMAC is in the purification of 'histidine-tagged' recombinant proteins, e.g. using pET plasmid vectors (see p. 448). The technique exploits the affinity of histidine-containing proteins for a chelating gel, typically Ni^{2+} immobilised on agarose. The histidine tag present at the N- or C-terminus of a recombinant protein forms complexes with the transition metal and is therefore bound to the gel. The proteins can be desorbed (eluted) from the gel with increasing concentration of a substrate with an affinity for the chelated metal ion, e.g. imidazole, using either a gradient or a stepwise approach. Alternatively, chelating agents (e.g. EDTA) or low pH can be used for protein elution.

Optimising chromatographic separations

In any chromatographic technique, sample components leave the column at different elution volumes/different times, and are then monitored by a suitable detector (Chapter 45). The responses of the detector are recorded on a chart or a screen in the form of a *chromatogram*. Ideally the sample biomolecules will be completely separated and detection of components will result in a series of discrete individual peaks corresponding to each type of biomolecule (e.g. Fig. 44.3). However, to minimise the possibility of overlapping peaks, or of peaks composed of more than one component, it is important to maximise the separation efficiency of the technique, which depends on:

Assaying analytes – remember that you **cannot** quantify *a particular analyte* without first identifying it: the presence of a single peak on a chromatogram does not prove conclusively that a single analyte is present.

- the selectivity, as measured by the relative retention times of the two components, or by the volume of mobile phase between the peak maxima of the two components after they have passed through the column; this depends on the ability of the chromatographic method to separate two components with similar properties;
- the band-broadening properties of the chromatographic system, which influence the width of the peaks; these are mainly due to the effects of diffusion.

Separating small biomolecules – these will diffuse faster than large molecules, so they should be separated using faster flow rates.

The separation efficiency, or resolution of two adjacent components, can be defined in terms of selectivity and peak broadening, using Eqn [44.4]. In practical terms, good resolution is achieved when there is a large 'distance' (either time or volume) between peak maxima, and the peaks are as narrow as possible. The resolution of components is also affected by the relative amount of each substance: for systems showing low resolution, it can be difficult to resolve small amounts of a particular component in the presence of larger amounts of a second component. If you cannot obtain the desired results from a poorly resolved chromatogram, other chromatographic conditions, or even different methods, should be tried in an attempt to improve resolution. For liquid chromatography, changes in the following factors may improve resolution:

- Stationary phase particle size – the smaller the particle, the greater the area available for partitioning between the mobile phase and the stationary phase. This partly accounts for the high resolution observed with HPLC and FPLC compared with low pressure methods.

- The slope of the salt gradient in eluting IEC or HIC columns, e.g. using computer-controlled adapted gradients (p. 375).
- In low-pressure liquid chromatography, the flow rate of the mobile phase must be optimised because this influences two band-broadening effects which are dependent on diffusion of sample molecules, i.e. (i) the flow rate must be slow enough to allow effective partitioning between the mobile phase and the stationary phase, and (ii) it must be fast enough to ensure that there is minimal diffusion along the column once the molecules have been separated. To allow for these opposing influences, a compromise flow rate must be used.
- If you prepare your own columns, they must be packed correctly, with no channels present that might result in uneven flow and eddy diffusion.

Learning from experience – if you are unable to separate your biomolecule using a particular method, do not regard this as a failure, but instead, think about what this tells you about either the substance(s) or the sample.

Text references

Stahl, E. (1969) *Thin Layer Chromatography – a Laboratory Handbook*, 2nd edn. Springer-Verlag, Berlin.

Touchstone, J.C. (1980) *Thin Layer Chromatography: Quantitative Environmental and Clinical Applications*. Wiley, New York.

Sources for further study

Bliesner, D.M. (2006) *Validating Chromatographic Methods: a Practical Guide*. Wiley, New York.

Boyer, R.F. (2005) *Biochemistry Laboratory: Modern Theory and Techniques*. Benjamin Cummings, San Francisco.
[Also covers other topics, including centrifugation and spectroscopy]

Braithwaite, A. and Smith, F.J. (1996) *Chromatographic Methods*, 5th edn. Blackie Academic and Professional, London.

Cazes, J. (2005) *Encyclopedia of Chromatography*. CRC Press, Boca Raton.

Gooding, K.M. and Regnier, F.E. (2002) *HPLC of Biological Macromolecules Revised and Expanded*. Marcel Dekker, New York.

McMaster, M. (2005) *LC-MS – a Practical User's Guide*. Wiley, New York.
[Covers principles and practice of HPLC systems and MS detectors]

Miller, J.M. (2005) *Chromatography: Concepts and Contrasts*, 2nd edn. Wiley, New York.

Vijayalashmi, M.A. (2002) *Biochromatography*. Taylor & Francis, London.

Wellings, D.A. (2006) *A Practical Handbook of Preparative HPLC*. Elsevier, Amsterdam.
[Gives advice on column preparation and process optimisation]

Wilson, K. and Walker, J. (eds) (2005) *Principles and Techniques of Practical Biochemistry and Molecular Biology*, 6th edn. Cambridge University Press, Cambridge.
[Also covers a range of other topics, including spectroscopy and centrifugation]

Study exercises

44.1 Calculate R_F and R_X values from a chromatogram. The figure represents the separation of three pigments by thin-layer chromatography.

(a) What is the R_F value of each pigment?
(b) What is the mobility of pigments B and C, relative to pigment A (R_A)?

Express both answers to three significant figures.

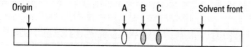

Thin-layer chromatographic separation of a mixture of three pigments (A, B and C).

44.2 Calculate the resolution and selectivity of two components from a chromatogram. Two compounds were separated by column chromatography, giving retention times of 4 min 30 s for A and 6 min 12 s for B, while a compound that was completely excluded from the stationary phase was eluted in 1 min 35 s. The base width of peak A was 40 s and the base width of peak B was 44 s. Calculate (a) the selectivity and (b) the resolution for these two compounds (express all answers to three significant figures).

44.3 Predict chromatographic behaviour on ion-exchange columns using protein titration curves. The variation of net charge with pH for two proteins, A and B, is shown in the figure.

(a) What is the isoelectric point (pI) of each protein?
(b) If a mixture of these proteins was passed down an anion exchange column (e.g. Q-Sepharose) equilibrated with buffer at pH 7.0, how would each protein behave in terms of elution and retention?
(c) If the proteins were bound to a cation exchanger (e.g. S-Sepharose) at pH 5.5, predict the order in which the proteins would elute using a salt gradient.

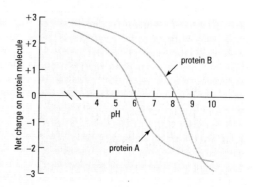

44.4 Test your understanding of the action of ammonium sulphate on proteins. Explain why mobile phases containing a concentration of ammonium sulphate promote the binding of many proteins to HIC columns.

45 Chromatography – detection and analysis

Once the individual components in a sample mixture have been separated by a given chromatographic technique (Chapter 44), a suitable detection system is required to monitor and record the elution of the components of the mixture as they pass through a 'flow cell'.

KEY POINT *The most appropriate detector depends on the type of chromatography and the application: ideally, the detector should show high sensitivity, a low detection limit and minimal noise or drift. These terms are defined in Chapter 49.*

Liquid chromatography detectors

UV/visible detectors

Understanding UV detection of proteins – note that the absorbance at 280 nm is mostly due to tryptophan and tyrosine residues; if a protein contains low amounts of these amino acids, its absorbance at this wavelength will be low.

These are widely used and have the advantages of versatility, sensitivity and stability. Such detectors are of two types: fixed wavelength and variable wavelength. Fixed-wavelength detectors are simple to use, with low operating costs. They usually contain a mercury lamp as a light source, emitting at several wavelengths between 254 and 578 nm; a particular wavelength is selected using suitable cut-off filters. The most frequently used wavelengths for analysis of biomolecules are 254 nm (for nucleic acids) and 280 nm (for proteins). Variable-wavelength detectors use a deuterium lamp and a continuously adjustable monochromator for wavelengths of 190–600 nm. For both types of detector, sensitivity is in the absorbance range 0.001–1.0 (down to $\cong$ 1 ng), with noise levels as low as 4×10^{-5}. Note that sensitivity is partly influenced by the path length of the flow cell (typically 10 mm). Monitoring at short wavelength UV (e.g. below 240 nm) may give increased sensitivity but decreased specificity, since many biological molecules absorb in this range. Additional problems with short wavelength UV detection include instrument instability, giving a variable baseline, and absorption by components of the eluting buffer (e.g. TRIS, which absorbs at 206 nm).

An important development in chromatographic monitoring is diode array detection (DAD). The incident light comprises the whole spectrum of light from the source, which is passed through a diffraction grating and the diffracted light detected by an array of photodiodes. A typical DAD can measure the absorbance of each sample component at 1–10 nm intervals over the range 190–600 nm. This gives an absorbance spectrum for each eluting substance which may be used to identify the compound and give some indication as to its purity. An example of a three-dimensional diode array spectrum is shown in Figure 45.1.

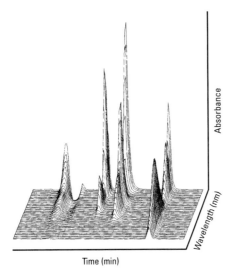

Fig. 45.1 Diode array detector absorption spectra of the eluent from an HPLC separation of a mixture of four steroids, taken every 15 seconds.

Fluorescence detectors

Many biomolecules, including some vitamins, nucleotides and porphyrins show natural fluorescence (Table 41.1), or can be made to fluoresce by pre-column or post-column derivatisation with a fluorophore. Fluorescence detection is more sensitive than UV/visible detection (p. 382), and may allow analysis in the pg range. A fluorescence detector consists of a light source (e.g. a xenon lamp), a diffraction grating to supply light at the excitation wavelength, and a photomultiplier to monitor the emitted light (usually arranged to be at right angles to the

Overcoming interference with fluorescence detectors – use a dual flow cell to offset background fluorescence due to components of the mobile phase.

Maximising sensitivity with fluorescence detectors – the concentration of other sample components, e.g. pigments, must not be so high that they cause quenching of fluorescence.

Optimising electrochemical detection – the mobile phase must be free of any compounds that might give a response; all constituents must be of the highest purity.

excitation beam). The use of instruments with a laser light source can give an extremely narrow excitation waveband, and increased sensitivity and specificity.

Electrochemical detectors

These offer very high sensitivity and specificity, with the possibility of detection of fg amounts of electroactive compounds such as catecholamines, vitamins, thiols, purines, ascorbate and uric acid. The two main types of detector, amperometric and coulometric, operate on similar principles, i.e. by measuring the change in current or potential as sample components pass between two electrodes within the flow cell. One of these electrodes acts as a reference (or counter) electrode (e.g. calomel electrode), while the other – the working electrode – is held at a voltage that is high enough to cause either oxidation or reduction of sample molecules. In the oxidative mode, the working electrode is usually glassy carbon, while in reductive mode a mercury electrode is used. In either case, a current flow between the electrodes is induced and detected.

Gas chromatography detectors

The most commonly used detectors for GC analysis of biomolecules are:

Flame ionisation detector (FID)

This is a widely used detector, being particularly useful for the analysis of a broad range of organic biomolecules. It involves passing the exit gas stream from the column through a hydrogen flame that has a potential of >100 V applied across it (Fig. 45.2). Most organic compounds, on passage through this flame, produce ions and electrons that create a small current across the electrodes, and this is amplified for measurement purposes. The FID is very sensitive (down to ≈0.1 pg), with a linear response over a wide concentration range. One drawback is that the sample is destroyed during analysis.

Thermal conductivity detector (TCD)

This simple detector is based on changes in the thermal conductivity of the gas stream brought about by the presence of separated sample molecules. The detector elements are two electrically heated platinum wires, one in a chamber through which only the carrier gas flows (the reference detector cell), and the other in a chamber that takes the gas flow from the column (the sample detector cell). In the presence of a constant gas flow, the temperature of the wires (and therefore their electrical resistance) is dependent on the thermal conductivity of the gas. Analytes in the gas stream are detected by temperature-dependent changes in resistance dependent on the thermal conductivity of each separated molecule; the size of the signal is directly related to concentration of the analyte.

The advantages of TCD include its applicability to a wide range of organic and inorganic biomolecules and its non-destructive nature, since the sample can be collected for further study. Its major limitation is its low sensitivity (down to ≈10 ng), compared with other systems.

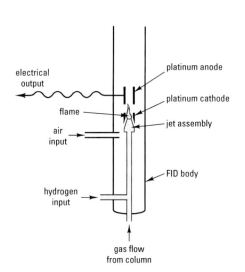

Fig. 45.2 Components of a flame ionisation detector (FID).

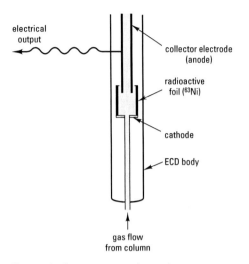

Fig. 45.3 Components of an electron capture detector (ECD).

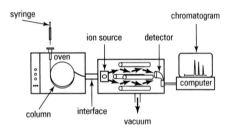

Fig. 45.4 Schematic diagram of a GCMS instrument.

Interpreting chromatograms – never assume that a single peak is a guarantee of purity: there may be more than one compound with the same chromatographic characteristics.

Electron capture detector (ECD)

This highly sensitive detector (Fig. 45.3) is useful for the detection of certain compounds with electronegative functional groups, e.g. halogens, peroxides and quinones. The gas stream from the column passes over a β-emitter (p. 267) such as ^{63}Ni, which provides electrons that cause ionisation of the carrier gas (e.g. nitrogen). When carrier gas alone is passing the β-emitter, its ionisation results in a constant current flowing between two electrodes placed in the gas flow. However, when electron-capturing sample molecules are present in the gas flow, a decrease in current is detected. An example of the application of the ECD is in detecting and quantifying chlorinated pesticides.

Interfacing GC or HPLC with mass spectrometry

Mass spectrometry (p. 294) used in conjunction with chromatographic methods can provide a powerful tool for identifying the components of complex mixtures, e.g. aqueous pollutants. The procedure requires computer control of instruments and data storage/analysis (Fig. 45.4). One drawback is the limited capacity of the mass spectrometer – due to its vacuum requirements – compared with the volume of material leaving the chromatography column. For capillary GC, the relatively small output can be fed directly into the ionisation chamber of the mass spectrometer. For packed column GC, a 'jet separator' is used to remove most of the carrier gas from the molecules to be analysed before they enter the spectrometer. Similarly, in HPLC, devices have been developed for solving the problem of large solvent volumes, e.g. by splitting the eluent from the column so only a small fraction reaches the mass spectrometer.

The computer-generated outputs from the mass spectrometer are similar to chromatograms obtained from other methods, and show peaks corresponding to the elution of particular components. However, it is then possible to select an individual peak and obtain a mass spectrum for the component in that peak to aid in its identification (p. 294). This has helped to identify hundreds of components present in biological systems, including flavour molecules in food, drug metabolites and water pollutants.

Coupling capillary GC columns with FT-IR spectrometers (pp. 289–90) provides another powerful means of separating and identifying compounds in complex biological mixtures.

Recording and interpreting chromatograms

Recording detector output

For analytical purposes, the detector output is usually connected to a computer-based data acquisition and analysis system. This consists of a personal computer (PC) with data acquisition hardware to convert the analogue detector signal to digital format, plus software to control the data acquisition process, store the signal information and display the resulting chromatogram. The software will also detect peaks and calculate their retention times and sizes (areas) for quantitative analysis. The software often incorporates functions to control the chromatographic equipment, enabling automatic operation. In sophisticated systems, the detector output may be compared with that from a 'library' of chromatograms for known compounds, to suggest possible identities of unknown sample peaks.

Chromatography – Detection and Analysis

Using a fraction collector – make sure you can relate individual fractions to the position of peaks on the chromatogram. Most fraction collectors send a signal to the recorder each time a fraction is changed.

In simpler chromatographic systems, you may need to use a chart recorder for detector output. Two important settings must be considered before using a chart recorder:

1. The baseline reading – this should be set only after a suitable quantity of mobile phase has passed through the column (prior to injection of the sample) and stability is established. The chart recorder is usually set a little above the edge of the chart paper grid, to allow for any baseline drift.
2. The detector range – this must be set to ensure that the largest peaks do not go off the top of the chart. Adjustment may be based on the expected quantity of analyte, or by trial-and-error process. Use the maximum sensitivity that gives intact peaks. If peaks are still too large, you may need to reduce the amount of sample used, or prepare and analyse a diluted sample.

Interpretation of chromatograms

Make sure you know the direction of the horizontal axis (usually, either volume or time) – it may run from right to left or *vice versa*, and make a note of the detector sensitivity on the vertical axis.

Ideally, the baseline should be 'flat' between peaks, but it may drift up or down due to a number of factors including:

Avoiding problems with air bubbles in liquid chromatography – always ensure that buffers are effectively degassed by vacuum treatment before use, and regularly clean the flow cell of the detector.

- air bubbles (in liquid chromatography); if the buffers used in the mobile phase are not effectively degassed, air bubbles may build up in the flow cell of the detector, leading to a gradual upward drift of the baseline, followed by a sharp fall when the accumulated air is released. Small air bubbles that do not become trapped may give spurious small peaks.
- changes in the composition of the mobile phase (e.g. in gradient elution, p. 307);
- tailing of material from previous peaks;
- carry-over of material from previous samples; this can be avoided by efficient cleaning of columns between runs – allow sufficient time for the previous sample to pass through the column before you introduce the next sample;
- loss of stationary phase from the column (column 'bleed'), caused by extreme elution conditions;

A peak close to the origin is likely to be due to non-retained sample molecules, flowing at the same rate as the mobile phase, or to artefacts, e.g. air (GC) or solvent (HPLC) in the sample. Whatever its origin, this peak can be used to measure the column dead time or column void volume (p. 304). No peaks from genuine sample components should appear before this peak.

Peaks can be denoted on the basis of their elution volume (used mainly in liquid chromatography) or their retention times, as in Fig. 45.5 (mainly in GC). If the peaks are not narrow and symmetrical, they may contain more than one component. Where peaks are more curved on the trailing side compared with the leading side, this may indicate too great an association between the component and the stationary phase, or may result from overloading of the column.

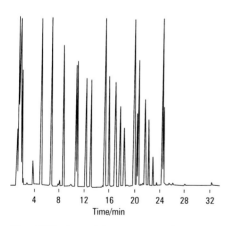

Fig. 45.5 A multicomponent chromatogram. Separation of many compounds, some that are well resolved, e.g. peaks at 12–13 min and others that are not, e.g. peaks at 24–25 min.

Quantifying biomolecules – note that quantitative analysis often requires assumptions about the identity of separated components (p. 349) and that further techniques may be required to provide information about the nature of the biomolecules present, e.g. mass spectrometry (see Chapter 42).

Using an external standard – samples and standards should be analysed more than once, to confirm the reproducibility of the technique.

Using an internal standard – you should add the internal standard to the sample at the first stage in the extraction procedure, so that any loss or degradation of test substance during purification is accompanied by an equivalent change in the internal standard, as long as the extraction characteristics of the internal standard and the test substance are very similar.

Quantitative analysis

Most detectors and chemical assay systems give a linear response with increasing amounts of the test substance over a given 'working range' of concentration. Alternative ways of converting the measured response to an amount of substance are:

- External standardisation: this is applicable where the sample volume is sufficiently precise to give reproducible results (e.g. HPLC, column chromatography). You measure the peak areas (or heights) of known amounts of the substance to give a calibration factor or calibration curve which can be used to calculate the amount of test substance in the sample.
- Internal standardisation: where you add a known amount of a reference substance (not originally present in the sample) to the sample, to give an additional peak in the elution profile. You determine the response of the detector to the test and reference substances by analysing a standard containing known amounts of both substances, to provide a response factor (r), where

$$r = \frac{\text{peak area (or height) of test substance}}{\text{peak area (or height) of reference substance}} \quad [45.1]$$

Use this response factor to quantify the amount of test substance (Q_t) in a sample containing a known amount of the reference substance (Q_r), from the relationship:

$$Q_t = \frac{\text{peak area (or height) of test substance}}{\text{peak area (or height) of reference substance}} \times (Q_r/r) \quad [45.2]$$

Internal standardisation should be the method of choice wherever possible, since it is unaffected by small variations in sample volume (e.g. for GC microsyringe injection). The internal standard should be chemically similar to the test substance(s) and must give a peak that is distinct from all other substances in the sample. An additional advantage of an internal standard which is chemically related to the test substance is that it may show up problems due to changes in detector response, incomplete derivatisation, etc. A disadvantage is that it may be difficult to fit an internal standard peak into a complex chromatogram.

Sources for further study

Baugh, P. (1994) *Gas Chromatography: a Practical Approach*. Oxford University Press, Oxford.

Braithwaite, A. and Smith, F.J. (1996) *Chromatographic Methods*, 5th edn. Blackie Academic and Professional, London.

Cazes, J. (2005) *Encyclopedia of Chromatography*. CRC Press, Boca Raton.

Gooding, K.M. and Regnier, F.E. (2002) *HPLC of Biological Macromolecules Revised and Expanded*. Marcel Dekker, New York.

Grob, R.L. and Barry, E.F. (2004) *Modern Practice of Gas Chromatography*, 4th edn. Wiley, New York.

McKay, P. *An Introduction to Chromatography*. Available: http://www.accessexcellence.org/LC/SS/chromatography_background.html
Last accessed: 01/04/07.

Miller, J.M. (2005) *Chromatography: Concepts and Contrasts*, 2nd edn. Wiley, New York.

Scott, R.P.W. (1996) *Chromatography Detectors: Design, Function and Operation*. Marcel Dekker, New York.

Study exercises

45.1 Test your knowledge of detector terminology. Explain what the following acronyms stand for: (a) FID; (b) TCD; (c) ECD; (d) DAD.

45.2 Check your knowledge of liquid chromatography detectors. Make a list of the various major types of liquid chromatography detectors in order, from highest to lowest sensitivity. Which of these methods is most versatile and why?

45.3 Consider how fluorescence detection can be applied to non-fluorescent molecules. Fluoresence detection of natural molecules is limited to molecules such as porphyrins, nucleotides and some vitamins, e.g. riboflavin. What procedures can convert a non-fluorescent molecule into a form suitable for sensitive fluorescence detection? How does this increase the selectivity of the analysis?

45.4 Calculate the amount of substance in a chromatographic separation using an internal standard. The chromatogram shown in the figure on the right represents the separation of three carbohydrates by gas chromatography. Peak A corresponds to mannitol, peak B to sucrose and peak C to the internal standard, trehalose, at 1.50 mg. The results shown in the table were obtained for the retention times and areas of the three peaks.

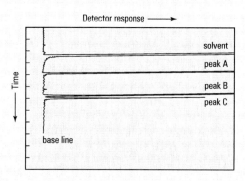

Gas-liquid chromatographic separation of three carbohydrates (A, B and C). Note that the plotter shows peaks A and B to be off-scale; however, the instrument still gives a valid area measurement for these two peaks (see table).

Retention times and peak areas for carbohydrates A, B and C

Peak	Retention time (min)	Area (relative)
A	3.92	2060
B	5.82	1898
C	6.03	604

Given a response factor (r) of 1.26 for mannitol and 0.92 for sucrose (relative to the internal standard), determine the amount of A and B in the sample (express your answer to three significant figures).

46 Principles and practice of electrophoresis

> **KEY POINT** Electrophoresis is a separation technique based on the movement of charged molecules in an electric field. Dissimilar molecules move at different rates and the components of a mixture will be separated when an electric field is applied. It is a widely used technique, particularly for the analysis of complex mixtures or for the verification of purity (homogeneity) of isolated biomolecules.

Understanding electrophoresis – this is, in essence, an incomplete form of electrolysis (p. 339), since the applied electrical field is switched off well before sample molecules reach the electrodes.

While electrophoresis is mostly used for the separation of charged macromolecules, techniques are available for high resolution separations of small molecules such as amino acids (for example, by capillary electrophoresis, p. 331). This chapter deals mainly with the electrophoretic separation of proteins, which has many applications including clinical diagnosis. However, the principles apply equally to other molecules (separation of nucleic acids is considered in Chapter 60).

The electrophoretic mobility of a charged molecule depends on:

- Net charge – negatively charged molecules (anions) migrate towards the anode (+), while positively charged molecules (cations) migrate towards the cathode (−); highly charged molecules move faster towards the electrode of opposite charge than those with lesser charge.
- Size – frictional resistance exerted on molecules moving in a solution means that smaller molecules migrate faster than large molecules.
- Shape – the effect of friction also means that the shape of the molecule will affect mobility, e.g. globular proteins compared with fibrous proteins, linear DNA compared with circular DNA.
- Electrical field strength – mobility increases with increasing electrical potential (voltage), but there are practical limitations to using high voltages, especially due to heating effects.

The combined influence of net charge and size means that mobility (μ) is determined by the charge : density ratio or the charge : mass ratio, according to the formula:

$$\mu = \frac{qE}{r} \qquad [46.1]$$

where q is the net charge on the molecule, r is the molecular radius and E is the field strength.

Definitions

Electrophoretic mobility – the rate of migration of a particular type of molecule in response to an applied electrical field.

Electrical field strength – a measure of the magnitude of an electrical field, usually expressed in terms of electrical potential difference per unit length, as $V\,m^{-1}$.

Electrophoresis and the separation of proteins

The net charge of a sample molecule determines its direction of movement and significantly affects its mobility. The net charge of a protein molecule is pH dependent, and is determined by the relative numbers of positively and negatively charged amino acid side chains at a given pH (Table 46.1). The degree of ionisation of each amino acid side chain is pH dependent, resulting in a variation of net charge on the protein at different pH values (Fig. 46.1). Since an individual protein will have a unique content of ionisable amino acids, each protein will have a characteristic 'titration curve' when net charge is plotted against pH. Thus, electrophoresis is always carried out at *constant* pH and a suitable buffer must be present along with the sample in order to maintain that pH (Chapter 25). If the

Table 46.1 pK_a values of ionisable groups in selected amino acid residues of proteins

Group/residue	pK_a^*
Terminal carboxyl	3.1
Aspartic acid	4.4
Glutamic acid	4.4
Histidine	6.5
Terminal amino	8.0
Cysteine	8.5
Tyrosine	10.0
Lysine	10.0
Arginine	12.0

*Note that these are typical values: the pK_a will change with temperature and ionic strength. Acidic residues will tend to be negatively charged at pH values above their pK_a, while basic residues will tend to be positively charged below their pK_a.

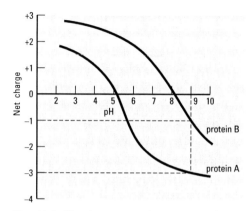

Fig. 46.1 Titration curves for two proteins, A and B containing different proportions of acidic and basic amino acid residues.

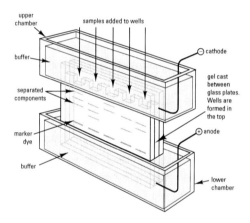

Fig. 46.2 Apparatus for vertical slab electrophoresis (components move downwards from wells, through the gel matrix).

Minimising diffusion – make the sample zone as narrow as possible, and fix and stain protein bands as soon as possible after the run.

> Definition
>
> **Ohm's law** $V = IR$, where V = voltage, I = current and R = resistance.

Optimising electrophoresis – attempting to minimise heat production using very low currents is not practical, since it leads to long separation times, and therefore to increased diffusion.

proteins shown in Figure 46.1 were subjected to electrophoresis at pH 9.0, and if the proteins were of similar size and shape, then the rate at which protein A (net charge, −3) migrates towards the anode would be faster than that for protein B (net charge, −1). Separation of proteins is usually carried out at alkaline pH, where most proteins carry a net negative charge.

Basic apparatus

Most types of electrophoresis using supporting media (described below) are simple to carry out and the apparatus can be easily constructed, although inexpensive equipment is commercially available. High resolution techniques such as 2D-electrophoresis and capillary electrophoresis require more sophisticated equipment, both for separation and analysis (p. 331).

Simple electrophoretic separations can be performed either vertically (Fig. 46.2) or horizontally (Fig. 60.2). The electrodes are normally made of platinum wire, each in its own buffer compartment. In vertical electrophoresis, the buffer solution forms the electrical contact between the electrodes and the supporting medium in which the sample separation takes place. In horizontal electrophoresis, electrical contact can be made by buffer-soaked paper 'wicks' dipping in the buffer reservoir and laid upon the supporting medium. The buffer reservoir normally contains a divider acting as a barrier to diffusion (but not to electrical current), so that localised pH changes which occur in the region of the electrodes (as a result of electrolysis, p. 339) are not transmitted to the supporting medium or the sample. Individual samples are spotted onto a solid supporting medium containing buffer or are applied to 'wells' formed in the supporting medium. The power pack used for most types of electrophoresis should be capable of delivering ≈500 V and ≈100 mA.

Using a supporting medium

The effects of convection currents (resulting from the heating effect of the applied field) and the diffusion of molecules within the buffer solution can be minimised by carrying out the electrophoresis in a porous supporting medium. This contains buffer electrolytes and the sample is added in a discrete location or zone. When the electrical field is applied, individual sample molecules remain in sharp zones as they migrate at different rates. After separation, post-electrophoretic diffusion of selected biomolecules (e.g. proteins) can be avoided by 'fixing' them in position on the supporting medium, e.g. using trichloracetic acid (TCA).

The heat generated during electrophoresis is proportional to the square of the applied current and to the electrical resistance of the medium: even when a supporting medium is used, heat production will lead to zone broadening by increasing the rate of diffusion of sample components and buffer ions. Heat denaturation of sample proteins may also occur, resulting in loss of biological activity e.g. with enzymes (p. 385). Another problem is that heat will reduce buffer viscosity, leading to a decrease in resistance. If the electrophoresis is run at constant voltage, Ohm's law dictates that as resistance falls, the current will increase, leading to further heat production. This can be avoided by using a power pack that provides constant power. In practice, most electrophoresis equipment incorporates a cooling device;

even so, distortions of an electrophoretic zone from the ideal 'sharp, linear band' can often be explained by inefficient heat dissipation.

Types of supporting media

These can be subdivided into:

- Inert media – these provide physical support and minimise convection: separation is based on charge density only (e.g. cellulose acetate).
- Porous media – these introduce molecular sieving as an additional effect: their pore size is of the same order as the size of molecules being separated, restricting the movement of larger molecules relative to smaller ones. Thus, separation depends on both the charge density and the size of the molecule.

With some supporting media, e.g. cellulose acetate, a phenomenon called electro-endosmosis or electro-osmotic flow (EOF) occurs. This is due to the presence of negatively charged groups on the surface of the supporting medium, attracting cations in the electrophoresis buffer solution and creating an electrical double layer. The cations are hydrated (surrounded by water molecules) and when the electric field is applied, they are attracted towards the cathode, creating a flow of solvent that opposes the direction of migration of anionic biomolecules towards the anode. The EOF can be so great that weakly anionic biomolecules (e.g. antibodies) may be carried towards the cathode.

Where necessary, EOF can be avoided by using supporting media such as agarose or polyacrylamide (p. 323), but it is not always a hindrance to electrophoretic separation. Indeed, the phenomenon of EOF is used in the high resolution technique of capillary electrophoresis (p. 331).

Cellulose acetate

Acetylation of the hydroxyl groups of cellulose produces a less hydrophilic structure than cellulose in the form of paper: as a result it holds less water and diffusion is reduced, with a corresponding increase in resolution. Cellulose acetate is often used in the electrophoretic separation of plasma proteins in clinical diagnosis – it can be carried out quickly (≈ 45 min) and its resolution is adequate to detect gross differences in various types of protein (e.g. paraproteins in myeloma). Cellulose acetate has a fairly uniform pore structure and the pores are large enough to allow unrestricted passage of all but the largest of molecules as they migrate through the medium.

Agarose

Agarose is the neutral, linear polysaccharide component of agar (from seaweed), consisting of repeating galactose and 3,6-anhydrogalactose subunits (Fig. 46.3). Powdered agarose is mixed with electrophoresis buffer at concentrations of 0.5–3.0% w/v, boiled until the mixture becomes clear, poured onto a glass plate, then allowed to cool until it forms a gel. Gelation is due to the formation of hydrogen bonds both between and within the agarose polymers, resulting in the formation of pores. The pore size depends on the agarose concentration. Low concentrations produce gels with large pores relative to the size of proteins, allowing them to migrate relatively unhindered through the gel,

Definition

Electro-osmotic flow – the osmotically driven mass flow of water resulting from the movement of ions in an electrophoretic system.

Handling cellulose acetate – the fragile strips must be carefully handled, avoiding touching the flat surfaces with your fingers.

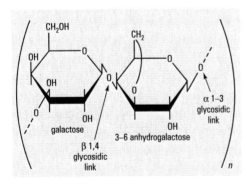

Fig. 46.3 Structure of agarose. Additional sulphate and pyruvyl groups are attached at selected hydroxyls in the polymer.

as determined by their individual charge densities. Low concentrations of agarose gel are suitable for techniques such as immunoelectrophoresis (p. 260) and isoelectric focusing (p. 330), where charge is the main basis of separation. The smaller pores produced by higher concentrations of agarose may result in molecular sieving.

When agarose gels are used for the separation of DNA, the large fragment size means that molecular sieving is observed, even with low concentration gels. This is the basis of the electrophoretic separation of nucleic acids (see Chapter 60).

Polyacrylamide

Polyacrylamide gel electrophoresis (PAGE) has a major role in protein analysis, both for one-dimensional and two-dimensional separations. The gel is formed by polymerising acrylamide monomer into long chains and cross-linking these chains using N, N'-methylene bisacrylamide (often abbreviated to 'bis'). The process is shown in Figure 46.4. In most protocols, polymerisation is initiated by free radicals produced by ammonium persulphate in the presence of N, N, N', N'-tetramethylethylenediamine (TEMED). The photodecomposition of riboflavin can also be used as a source of free radicals.

Advantages of polyacrylamide gels – in addition to their versatility in terms of pore size, these gels are chemically inert, stable over a wide range of pH, ionic strength and temperature, and transparent.

The formation of polyacrylamide from its acrylamide monomers is extremely reproducible under standard conditions, and electrophoretic separations are correspondingly precise. The pore size, and hence the extent of molecular sieving, depends on the total concentration of monomer (% T), i.e. acrylamide plus bisacrylamide in a fixed ratio. This means that pores in the gel can be 'tailored' to suit the size of biomolecule to be separated: gels containing 3% acrylamide have large pores and are used in methods where molecular sieving should be avoided (e.g. in isoelectric focusing, p. 330), while higher concentrations of acrylamide (5–30% T) introduce molecular sieving to various degrees depending on the size of the sample components, i.e. with 30% acrylamide gels, molecules as small as M_r 2000 may be subject to molecular sieving. Gels of $<2.5\%$ are necessary for molecular sieving of molecules of $M_r > 10^6$, but such gels are almost fluid and require 0.5% agarose to make them solid. Note that a gel of 3% will separate DNA by molecular sieving, due to the large size of the nucleic acid molecules (p. 431).

 SAFETY NOTE Preparing gels for PAGE – both acrylamide and bisacrylamide are extremely potent neurotoxins, so you must wear plastic gloves when handling solutions containing these reagents. Although the polymerised gel is non-toxic, it is still advisable to wear gloves when handling the gel, because some monomer may still be present.

Before you embark on a particular PAGE protein separation, you will need to think about the general strategy for that separation and make certain choices, including whether to use:

Preparing polyacrylamide gels – most solutions used for gel preparation can be made in advance, but the ammonium persulphate solution must be prepared immediately before use.

- Rod or slab gels – flat slab gels are formed between glass plates, using plastic spacers 0.75–1.5 mm thick: rod gels are made in narrow bore tubes. For most separations using several samples, a slab gel saves time because up to 25 samples can be separated under identical conditions in a single gel, while rod gels can only be used for individual samples. Rectangular slab gels are also easier to read, by densitometry, and photograph. However, rod gels are useful in preliminary separations, for determining a suitable pH and gel concentration, and for applications where the gel is sliced in order to extract and assay proteins of interest.

Selecting and using gels in 2-D electrophoresis – a rod gel may be used for the first dimension and a slab gel for the second dimension.

- Dissociating or non-dissociating conditions – the most widely used PAGE protein separation technique uses an ionic detergent, usually sodium dodecyl sulphate (SDS), which dissociates proteins into their

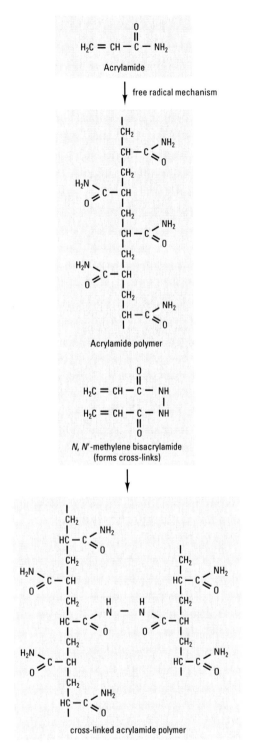

Fig. 46.4 Reactions involved in the formation of polyacrylamide gels.

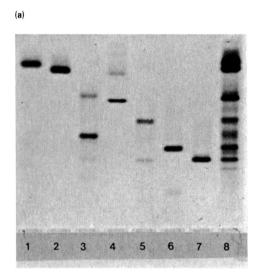

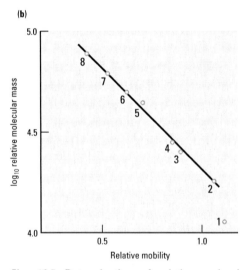

Fig. 46.5 Determination of relative molecular mass (M_r) of proteins by SDS-PAGE: (a) gel samples: 1, cytochrome c; 2, myoglobin; 3, γ-globulin; 4, carbonic anhydrase; 5, ovalbumin; 6, albumin; 7, transferrin; 8, mixture of samples 1–7 (photo courtesy of Pharmacia Biotech). (b) plot of log M_r against distance travelled through gel.

Table 46.2 Molecular masses of standard proteins used in electrophoresis

Protein	M_r	$\log_{10} M_r$
Cytochrome c	11 700	4.068
Myoglobin	17 200	4.236
γ-globulin (light chain)	23 500	4.371
Carbonic anhydrase	29 000	4.462
Ovalbumin	43 000	4.634
γ-globulin (heavy chain)	50 000	4.699
Human albumin	68 000	4.832
Transferrin	77 000	4.886
Myosin (heavy chain)	212 000	5.326

Following the progress of PAGE – add bromophenol blue solution (0.002% w/v) to the sample in the ratio 1:25 (dye: sample). This highly ionic, small M_r dye migrates with the electrophoretic front.

Choosing a buffer system for PAGE – discontinuous systems are more time-consuming to prepare, but have the advantage over continuous systems in that relatively large volumes of dilute sample can be used and good resolution is still obtained.

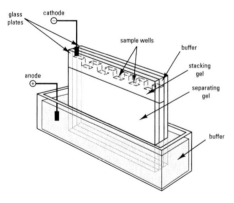

Fig. 46.6 Apparatus for discontinuous electrophoresis.

Choosing a pH for protein electrophoresis – many proteins have isoelectric points in the range pH 4–7 and in response to electrophoresis with buffers in the region pH 8.0–9.5, most proteins will migrate towards the anode.

individual polypeptide subunits and gives a uniform net charge along each denatured polypeptide. This technique, known as SDS-PAGE, requires only μg amounts of sample and is quick and easy to carry out. On the other hand, if it is necessary to preserve the native protein conformation and biological activity, non-dissociating conditions are used, i.e. no SDS is added. In SDS-PAGE the sample protein is normally heated to 100 °C for 2 min, in buffer containing 1% (w/v) SDS and 1% (w/v) 2-mercaptoethanol, the latter to cleave any disulphide bonds. The resultant polypeptides bind to SDS in a constant weight ratio, with 1.4 g of SDS per g of protein. As a result, the intrinsic net charge of each polypeptide is 'swamped' by the negative charge imposed by SDS, and there is a uniform negative charge per unit length of polypeptide. Since the polypeptides now have identical charge densities, when they are subject to PAGE (with SDS present) using a gel of appropriate pore size, molecular sieving will occur and they will migrate strictly according to polypeptide size. This not only gives effective separation, but the molecular mass of a given polypeptide can be determined by comparing its mobility to polypeptides of known molecular mass run under the same conditions (Fig. 46.5). Several manufacturers (e.g. Pharmacia, Sigma) supply molecular mass standard kits which may include polypeptides of M_r 11 700 to 212 000 (Table 46.2), together with details of their preparation and use. Where necessary, the treated sample can be concentrated by ultrafiltration and the buffer composition can be altered by diafiltration (p. 379).

- Continuous or discontinuous buffer systems – a continuous system is where the same buffer ions are present in the sample, gel and buffer reservoirs, all at the same pH. The sample is loaded directly onto a gel (the 'separating gel' or 'resolving gel') that has pores small enough to introduce molecular sieving. In contrast, discontinuous systems have different buffers in the gel compared to the reservoirs, both in terms of buffer ions and pH. The sample is loaded onto a large-pore 'stacking gel', previously polymerised on top of a small-pore separating gel (Fig. 46.6). The individual proteins in the sample concentrate into very narrow zones during their migration through the large-pore gel and stack up according to their charge densities, prior to separation in the small-pore gel, giving enhanced results compared with continuous systems.
- pH and buffer for the separation – PAGE can be carried out between pH 3 and 10. The pH is not critical for continuous SDS-PAGE, since SDS-treated polypeptides are negatively charged over a wide pH range. However, when using non-dissociating systems, the pH can be critical, particularly if the biological activity of the molecules is to be retained.
- Gel concentration – at one extreme, a gel with a very high percentage T might totally exclude the sample components, while a gel with a very low percentage T might lead to all SDS-treated proteins migrating at the same rate, i.e. with the electrophoretic front. A sensible approach is to set up a series of rod gels in the range 5–15% T and observe the separation and resolution obtained. Alternatively, a *gradient gel* can be used, in which the percentage T increases, and hence pore size decreases, in the direction of protein migration. A useful gradient for a preliminary experiment would be 5–20% T. Such

Table 46.3 Preparation of gels for PAGE and SDS-PAGE. The gel solutions are made by mixing the components in the proportions and in the order shown. Figures are ml of each solution required to give the stated % gel strength

Solution (added in order shown)	PAGE			SDS-PAGE		
	3.5% gel (T = 3.6%)	5% gel (T = 5.1%)	7.5% gel (T = 7.7%)	5% gel (T = 5.1%)	7.5% gel (T = 7.7%)	10% gel (T = 10.2%)
1. Distilled water	19.3	14.9	7.5	14.9	7.5	—
2. TRIS-glycine buffer, pH 8.9, 0.1 mol l^{-1}	33.0	33.0	33.0	—	—	—
3. Imidazole buffer, pH 7.0, 0.1 mol l^{-1} plus 0.2% w/v SDS	—	—	—	33.0	33.0	33.0
4. Acrylamide solution 22.2% w/v and 0.6% w/v bis	10.4	14.8	22.2	14.8	22.2	29.7
5. Ammonium persulphate solution, 0.15% w/v	3.2	3.2	3.2	3.2	3.2	3.2
6. TEMED	0.1	0.1	0.1	0.1	0.1	0.1
Final volume (ml)	66.0	66.0	66.0	66.0	66.0	66.0

Separating protein mixtures – for high resolution, a combination of dissociating and discontinuous PAGE in slab gels is the system of choice.

What to do if your polyacrylamide gels fail to set – polymerisation is inhibited by oxygen, so solutions should be degassed, and the surfaces of the polymerisation mixture exposed to air should be overlaid with water; if your gels still do not polymerise, the most common cause is the use of 'old' ammonium persulphate stock solution. If low pH buffers are used, polymerisation may be delayed because TEMED is required in the free base form.

gels are able to resolve protein mixtures with a wide range of molecular masses. Furthermore, as proteins migrate into regions of ever-decreasing pore size, the movement of the leading edge of a zone will become increasingly restricted. This allows the trailing edge to catch up, resulting in considerable zone sharpening. Gradient formers are fairly simple to make, and are commercially available.

Practical details of the preparation of PAGE and SDS-PAGE gels are given in Table 46.3 (see Westermeier, 2004, or Gersten, 1996 for further details).

Post-electrophoretic procedures – handling of the supporting medium, staining and analysis

For protein electrophoresis, the following stages are appropriate: details for nucleic acids are given on pp. 431–2.

Handling
All types of supporting medium should be handled carefully: wearing gloves is advisable, for safety and to avoid transfer of proteins from the skin. Cellulose acetate strips should be immediately transferred to a fixing and staining solution. Agarose gels should be dried quickly before staining. Polyacrylamide gels in vertical slabs must be freed carefully from one of the glass plates in which they are formed, taking care to lever the glass at a point well away from the part of the gel containing the wells: once free, the gels should be immediately transferred to fixing or staining solution.

Rod gels are recovered by a process called 'rimming', a technique that takes a little practice to master. To remove the gel from the tube, hold the tube in one hand and, using a syringe with a long blunt needle, squirt water between the gel and the inner wall of the tube: while squirting, move the needle gently up and down and rotate the tube. If the gel does not become free, perform the same procedure at the other end of the tube, until the gel is released. If necessary, a Pasteur pipette bulb applied to one end can be used to push the gel out of the tube.

Fixing, staining and destaining

To prevent the separated proteins from diffusing, they are usually fixed in position. Proteins on cellulose acetate strips can be fixed and stained using Ponceau S (0.2% w/v in a solution of 3% v/v aqueous trichloracetic acid, TCA).

Fixing is also required for most types of gel electrophoresis: 3% v/v TCA is often used. The most widely used stain for protein separations in gels is Coomassie Blue R-250 (where R = reddish hue): the detection limit is $\approx 0.2\,\mu g$ and staining is quantitative up to $20\,\mu g$ for some proteins. It is normal for background staining of the gel to occur, and removal of background colour ('destaining') can be achieved either by diffusion or electrophoresis. To destain by diffusion, transfer the gel to isopropanol:acetic acid:water (12.5:10:77.5 v/v/v) and allow to stand for 48 h, or change the solution several times to speed up the staining process. Electrophoretic destaining can remove Coomassie Blue, which is anionic: stained gels are placed between porous plastic sheets with electrodes on each side, and the tank is filled with 7% acetic acid. Passing a current of up to 1.0 A through the gel for around 30–40 min. should result in substantial destaining.

If you need greater sensitivity (e.g. for ng to fg amounts), or when using high resolution techniques such as 2D-electrophoresis (p. 330), silver staining can be used. Depending on the protocol chosen and the proteins being stained, the silver technique can be 5 to 200-fold more sensitive than Coomassie Blue. The method involves a fixation step (e.g. with TCA), followed by exposure to silver nitrate solution and development of the stain. The silver ions are thought to react with basic and thiol groups in proteins, and subsequent reduction (e.g. by formaldehyde at alkaline pH, or by photodevelopment) leads to deposits of silver in the protein bands. Most proteins stain brown or black, but lipoproteins may stain blue, and some glycoproteins stain yellow or red. Some proteins lacking in amino acids with reducing groups (e.g. those lacking cysteine residues) may stain negatively, i.e. the bands are more transparent than the background staining of the gel. Although many protocols have been published, silver stain kits are commercially available, e.g. from Bio-Rad.

Although silver staining has clear advantages in terms of sensitivity, for routine work it is more laborious and expensive to carry out than the Coomassie Blue method. It also requires high purity water, otherwise significant background staining occurs. Another feature is that the staining can be non-specific, since DNA and polysaccharides may stain on the same gel as proteins.

Other methods for the detection of separated components include:

- autoradiography for proteins labelled with ^{32}P or ^{125}I (p. 271);
- fluorescence for proteins pre-labelled with fluorescent dyes (p. 353);
- periodic acid-Schiff (PAS) stain using dansyl hydrazine for glycoproteins.

Handling gels – *avoid touching gels with paper as it sticks readily and is difficult to remove without tearing.*

Avoiding overloaded gels and band distortion – *determine the protein concentration of the sample beforehand. Around $100\,\mu g$ of a complex mixture, or $1\text{--}10\,\mu g$ of an individual component, will be sufficient, but bear in mind that underloading may result in bands being too faint to be detected.*

Optimising resolution – *keep the sample volume as small as possible; methods for concentrating protein solutions are considered on pp. 376–9. For vertical slab gels and for rod gels, include 10% w/v sucrose or glycerol to increase density and allow buffer solution to be overlaid on the sample without dilution.*

Understanding the terminology of high purity water – *this is usually produced by reverse osmosis and its purity is often expressed with respect to its resistivity: ultrapure water has a resistivity of $18\ M\Omega\ cm^{-1}$ (informally referred to as 'eighteen megaohm water').*

Origins of 'Western' blotting – following the description of 'Southern' blotting of DNA by Dr. Ed Southern (p. 433), other points of the compass have been used to describe other forms of blotting, with 'Western' blotting for proteins.

Example For detection of lactate dehydrogenase (LDH): when a solution containing lactate, NAD$^+$, phenazine methasulphate (PMS) and methyl thiazolyl tetrazolium (MTT) is added to a gel containing LDH, a series of redox reactions occurs in the enzyme-containing regions, starting with the oxidation of lactate to pyruvate, and proceeding via NAD$^+$, PMS and MTT to the eventual reduction of MTT to formazan dye. After incubation at 37 °C in the dark (since MTT is light-sensitive), LDH is detected by the appearance of blue–black bands on the gel.

Detecting enzymes in gels – minimise zone spreading by incorporating substrates, etc. in a thin agarose indicator gel poured over the separating gel.

Measuring peak areas – a valid 'low-tech' alternative to computer-based systems is to cut out and weigh peaks from a recorder chart.

Blotting

The term 'blotting' refers to the transfer of separated proteins from the gel matrix to a thin sheet such as nitrocellulose membrane (commercially available from e.g. Millipore, Amersham). The proteins bind to this membrane, and are immobilised. Blotting of proteins is usually achieved by electrophoretic transfer, and this process is normally referred to as Western blotting (see also Southern blotting and Northern blotting for DNA and RNA respectively, pp. 433–4). Its major advantage is that the immobilised proteins on the surface of the membrane are readily accessible to detection reagents, and staining and destaining can be achieved in less than 5 min. Use of labelled antibodies to detect specific proteins (immunoblotting) can take less than 6 h. In addition, it is easy to dry and store Western blots for long periods, for further analysis.

Detection of enzymes

If you need to detect enzyme activity you should use a non-denaturing gel. The gel matrix will hinder diffusion of the enzyme, but will allow access to the small molecular weight substrates, co-factors and dyes necessary to localise enzyme activity *in situ*. Most methods for detecting enzymes on gels are modifications of protocols originally developed by histochemists, e.g.:

- NAD$^+$-requiring oxidoreductases can be detected by incubating the gels with substrate, NAD$^+$ and a solution of a tetrazolium salt which, when reduced, forms an insoluble coloured formazan dye.
- Transferases and isomerases can be detected by coupling their reactions to an oxidoreductase-requiring reaction, visualised as described above.
- Hydrolases can be detected using appropriate chromogenic or fluorogenic substrates.
- Phosphatases can be detected by precipitating any phosphate released from the substrate with Ca^{2+}.

For more details, see Manchenko (2002).

Recording and quantification of results

A number of expensive, dedicated instruments are available for the analysis of gels, e.g. laser densitometers. Alternatively, gel scanning attachments can be purchased for standard spectrophotometers, allowing measurement and recording of the absorbance of the Coomassie Blue stained bands at 560–575 nm: for instruments connected to a computer, quantification of individual components can be achieved by integrating the areas under the peaks.

You can photograph gels using a conventional camera (fine grain film), digital camera, or using a photocopier. Alternatively, a dedicated image capture and analysis system may be used, e.g. GelDoc. The gel should be placed on a white glass transilluminator. A red filter will increase contrast with bands stained with Coomassie Blue. If the gels themselves need to be retained, they can be preserved in 7% acetic acid. Alternatively, they can be dried using a commercially available gel dryer.

Text references

Gersten, D. (1996) *Gel Electrophoresis: Proteins* (Essential Techniques Series). Wiley, New York.

Manchenko, G.P. (2002) *Handbook of Detection of Enzymes on Electrophoretic Gels,* 2nd edn. CRC Press, Boca Raton.

Westermeier, R. (2004) *Electrophoresis in Practice: A Guide to Methods and Applications of DNA and Protein Separations,* 4th edn. Wiley VCH, Berlin.

Sources for further study

Anon. *The American Electrophoresis Society Homepage.*
Available: http://www.aesociety.org
Last accessed: 01/04/07.
[Includes details on a variety of electrophoretic techniques]

Anon. *British Society for Proteomics Research Homepage.*
Available: http://www.bspr.org/
Last accessed: 01/04/07.
[Formerly the British Electrophoresis Society, includes links to other relevant websites and databases]

Hames, B.D. and Rickwood, D. (1998) *Gel Electrophoresis of Proteins: A Practical Approach,* 3rd edn. Oxford University Press, Oxford.

Martin, R.M. (1996) *Gel Electrophoresis: Nucleic Acids* (Introduction to Biotechniques Series). Bios, Oxford. [An introductory text, starting from first principles, with practical examples]

Rabilloud, T. (2004) *Proteome Research: Two-dimensional Gel Electrophoresis and Identification Methods.* Springer, Heidelberg.

Tietz, D. (1998) *Nucleic Acid Electrophoresis.* Springer-Verlag, Berlin.

Study exercises

46.1 Find out why the net charge on a protein molecule varies with pH. Identify the amino acids primarily responsible for determining the net charge on a protein molecule and draw simple diagrams to represent the ionisation of their side chains, indicating how you would expect these side chains to be charged at acid, neutral and alkaline pH values.

46.2 Explain the function of the various reagents used in SDS-PAGE analysis of proteins. What is the function of each of the following SDS-PAGE reagents? (a) acrylamide; (b) N,N'-methylene bisacrylamide; (c) ammonium persulphate; (d) N,N,N',N'-tetramethylethylenediamine; (e) sodium dodecyl sulphate; (f) 2-mercaptoethanol; (g) buffer; (h) Coomassie Blue R-250; (i) bromophenol blue.

46.3 Test your knowledge of 'blotting' terminology. What is Western blotting, and how does it differ from Northern and Southern blotting?

46.4 Determine the M_r of subunits of a protein by SDS-PAGE. A number of proteins of known molecular weight were analysed by SDS-PAGE and the results are shown in the table. In the same experiment, a sample of purified lactate dehydrogenase (LDH), was treated in identical fashion and run, producing a band with a relative mobility of 0.77. (a) What is the M_r of the polypeptide produced from the LDH sample?

Relative mobilities of molecular weight standards by SDS-PAGE

Protein	M_r	Relative mobility
Myoglobin	17 200	1.00
Carbonic anhydrase	29 000	0.84
Ovalbumin	43 000	0.67
Human albumin	68 000	0.51
Transferrin	77 000	0.46

(b) If the M_r of native LDH is 136 000, what do you conclude about the subunit structure of this enzyme?

46.5 Find out about 'troubleshooting' problems with gels. Investigate the pitfalls of working with and polymerising gels.

47 Advanced electrophoretic techniques

Using electrophoresis in proteomics – note that many of the techniques described in this chapter are employed to generate the information used in bioinformatic analysis (Chapter 11).

Although the resolution obtained using the basic electrophoretic techniques described in Chapter 46 is adequate for many biomolecular applications, a number of advanced techniques are available that give very high resolution and which can be used with very small amounts of sample material.

Isoelectric focusing (IEF)

In contrast to electrophoresis, which is carried out at constant pH, IEF is carried out using a pH gradient. The gradient is formed using small molecular mass ampholytes, which are analogues and homologues of polyamino-, polycarboxylic acids that collectively have a range of isoelectric points (pI values) between pH 3 and 10. The mixture of ampholytes (p. 152), either in a gel or in free solution, is placed between the anode in acid solution (e.g. H_3PO_4), and the cathode in alkaline solution (e.g. NaOH). When an electric field is applied, each ampholyte migrates to its own pI and forms a stable pH gradient which will persist for as long as the field is applied. When a protein sample is applied to this gradient separation is achieved, since individual proteins will migrate to their isoelectric points. The net charge on the protein when first applied will depend on the specific 'titration curve' for that protein (Fig. 47.1). As an example, consider two proteins, X and Y, having pI values of pH 5 and pH 8 respectively, which are placed together on the gradient at pH 6 (Fig. 47.2). At that pH, protein X will have a net negative charge, and will migrate towards the anode, progressively losing charge until it reaches its pI (pH 5) and stops migrating. Protein Y will have a net positive charge at pH 6, and so will migrate towards the cathode until it reaches its pI (pH 8).

Using a polyacrylamide gel as a supporting medium and a narrow pH gradient, proteins differing in pI by 0.01 units can be separated. Even greater resolution is possible in free solution (e.g. in capillary electrophoresis, p. 331). Such resolution is possible because protein molecules that diffuse away from the pI will acquire a net charge (negative at increased pH, positive at decreased pH) and immediately be focused back to their pI. This focusing effect will continue for as long as the electric field is applied.

A useful variant of IEF is in obtaining *titration curves* for proteins. A pH gradient is set up, and the sample applied in a line at a right angle to the gradient. The net charge on a given protein will vary according to its position on the gradient – when electrophoresis is carried out at right angles to the pH gradient, the protein will migrate at a velocity and direction governed by that charge. When stained, each protein will appear as a continuous curved line, corresponding to its titration curve (Fig. 47.3). This technique can be usefully performed during protein purification, prior to ion-exchange chromatography (p. 307): by obtaining the titration curve for a protein of interest and those of major contaminants, the mobile phase pH that gives optimal separation can be selected.

In IEF, it is important that electro-osmotic flow (EOF, p. 322) is avoided, as this would affect the ability of the proteins to remain stationary at their pIs. For gel IEF, polyacrylamide minimises EOF, while capillary IEF uses narrow bore tubing with an internal polymer coating (p. 334).

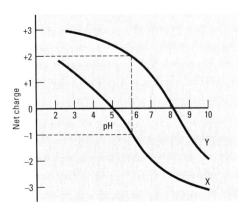

Fig. 47.1 Titration curves for two proteins, X and Y.

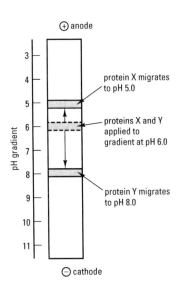

Fig. 47.2 The migration of two proteins, X and Y, in response to a pH gradient.

Two-dimensional electrophoresis

The most commonly used version of this high resolution technique involves separating proteins by charge in one dimension using IEF in polyacrylamide gel, followed by separation by molecular mass in the second dimension using denaturing SDS-PAGE (p. 325). The technique allows up to 1000 proteins to be separated from a single sample. Typically, the first dimension IEF run (pH 3–10) is carried out on gel strips of length 7–24 cm. Strips are run at a voltage of 500–3500 V for 1.5 h, then at 3500 V for a further 4 h. Gel strips can then be used immediately, or frozen until required.

It is common for the second-dimension SDS-PAGE separation to be carried out on a discontinuous slab gel 0.5–1.5 mm thick, which includes a low percentage T stacking gel and a separating gel with an exponential gradient of 10–16% T. The separating gel can be prepared in advance, but the stacking gel must be formed shortly before addition of the rod gel from the one dimensional run.

After equilibration with the buffer used in SDS-PAGE, the IEF gel strip is loaded onto the 2D gel (still between the glass plates in which it was formed) and sealed in position using acrylamide or agarose. Before the sealing gel sets, a well should be formed in it at one end to allow addition of molecular mass markers. The second-dimension is run at 100–200 V until the dye front is ≈1 cm from the bottom edge of the slab. After running, the gel is processed for the detection of polypeptides, e.g. using Coomassie Blue or silver stain. Analysis of the complex patterns that result from 2D electrophoresis requires computer-aided gel scanners to acquire, store and process data from a gel, such as that shown in Figure 47.4. These systems can compare, adjust and match up patterns from several gels, allowing both accurate identification of spots and quantification of individual proteins. Allowance is made for the slight variations in patterns found in different runs, using internal references ('landmarks'), which are either added standard proteins or particular spots known to be present in all samples.

Capillary electrophoresis

The technique of capillary electrophoresis (CE) combines the high resolving power of electrophoresis with the speed and versatility of HPLC (p. 305). The technique largely overcomes the major problem of carrying out electrophoresis without a supporting medium, i.e. poor resolution due to convection currents and diffusion. A capillary tube has a high surface area:volume ratio, and consequently the heat generated as a result of the applied electric current is rapidly dissipated. A further advantage is that very small samples (5–10 nl) can be analysed. The versatility of CE is demonstrated by its use in the separation of a range of biomolecules, e.g. amino acids, proteins, nucleic acids, drugs, vitamins, organic acids and inorganic ions; CE can even separate neutral species, e.g. steroids, aromatic hydrocarbons (see Weinberger, 2000).

The components of a typical CE apparatus are shown in Figure 47.5. The capillary is made of fused silica and externally coated with a polymer for mechanical strength. The internal diameter is usually 25–50 µm, a compromise between efficient heat dissipation and the need for a

Avoiding streaking in 2D electrophoresis – ensure that the sample contains no particulate material (e.g. from protein aggregation); filter or centrifuge before use.

Maximising resolution in 2D electrophoresis of proteins – try to minimise nucleic acid contamination of your sample (see Chapter 54), as they may interact with polypeptides/proteins, affecting their movement in the gel.

Freezing gel strips – be sure to mark the identity and orientation of each gel strip before freezing, e.g. by inserting a fine wire into one end of the strip. Note that if urea is used in the gel strip, it will form crystals on freezing.

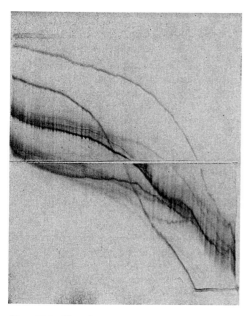

Fig. 47.3 Titration curves of bovine muscle proteins produced by electrofocusing-electrophoresis (photo courtesy of Pharmacia Biotech).

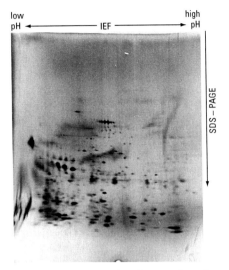

Fig. 47.4 Two-dimensional separation of proteins from 100x concentrated urine (2.5 μg total protein; silver stain. Courtesy of T. Marshall and K.M. Williams).

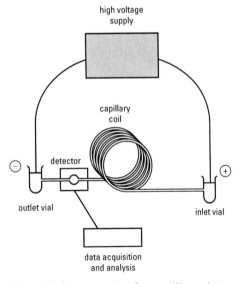

Fig. 47.5 Components of a capillary electrophoresis system.

light path that is not too short for detection using UV/visible spectrophotometry. A gap in the polymer coating provides a window for detection purposes. Samples are injected into the capillary by a variety of means, e.g. electrophoretic loading or displacement. In the former, the inlet end of the capillary is immersed in the sample and a pulse of high voltage is applied. The displacement method involves forcing the sample into the capillary, either by applying pressure in the sample vial using an inert gas, or by introducing a vacuum at the outlet. The detectors used in CE are similar to those used in chromatography (p. 314), e.g. UV/visible spectrophotometric systems. Fluorescence detection is more sensitive, but this may require sample derivitisation. Electrochemical and conductivity detection is also used in some applications, e.g. conductivity detection of inorganic cations such as Na^+ and K^+.

Electro-osmotic flow (EOF), described on page 322, is essential to the most commonly used types of CE. The existence of EOF in the capillary is the result of the net negative charge on the fused silica surface at pH values over 3.0. The resulting solvent flow towards the cathode is greater than the attraction of anions towards the anode, so they will flow towards the cathode (note that the detector is situated at the cathodic end of the capillary). The greater the net negative charge on an anion, the greater is its resistance to the EOF and the lower its mobility. Separated components migrate towards the cathode in the order: (1) cations, (2) neutral species, (3) anions.

Capillary zone electrophoresis (CZE)

This is the most widely used form of CE, and is based on electrophoresis in free solution and EOF, as discussed above. Separations are due to the charge : mass ratio of the sample components, and the technique can be used for almost any type of charged molecule, and is especially useful for peptide separation and confirmation of purity.

Micellar electrokinetic chromatography (MEKC)

This technique involves the principles of both electrophoresis and chromatography. Its main strength is that it can be used for the separation of neutral molecules as well as charged ones. This is achieved by including surfactants (e.g. SDS, Triton X-100) in the electrophoresis buffer at concentrations that promote the formation of spherical micelles, with a hydrophobic interior and a charged, hydrophilic surface. When an electric field is applied, these micelles will tend to migrate with or against the EOF depending on their surface charge. Anionic surfactants like SDS are attracted by the anode, but if the pH of the buffer is high enough to ensure that the EOF is faster than the migration velocity of the micelles, the net migration is in the direction of the EOF, i.e. towards the cathode. During this migration, sample components partition between the buffer and the micelles (acting as a pseudo-stationary phase); this may involve both hydrophobic and electrostatic interactions. For neutral species it is only the partitioning effect that is involved in separation; the more hydrophobic a sample molecule, the more it will interact with the micelle, and the longer will be its migration time, since the micelle resists the EOF. The versatility of MEKC enables it to be used for separations of biomolecules as diverse as amino acids and polycyclic hydrocarbons. MEKC is also known as micellar electrokinetic capillary chromatography (MECC).

Chiral capillary electrophoresis (CCE)

Resolution of a pair of chiral enantiomers (optical isomers) represents one of the biggest challenges for separation science, because each member of the pair will have identical physicochemical properties. CE offers an effective method of separating enantiomers by inducing a 'chiral selector' in the electrophoresis medium. The most commonly used chiral selectors are cyclodextrins such as the highly sulphated cyclodextrins (HSCDs) (Fig. 47.6). As the enantiomers migrate along the capillary, one will tend to interact more strongly than the other and its mobility will be reduced relative to the other. Figure 47.7 shows separation of the R and S forms of amphetamine using 5% HSCD in the electrophoresis buffer. The R-form has greater affinity for the HSCD used, so its retention time on the capillary is longer than that of the S-form.

Note that HSDCs can also be used in HPLC, but CCE is more effective, with shorter development times and lower reagent costs.

Capillary gel electrophoresis (CGE)

The underlying principle of this technique is directly comparable with that of conventional PAGE, i.e. the capillary contains a polymer that acts as a molecular sieve. As charged sample molecules migrate through the polymer network, larger molecules are hindered to a greater extent than smaller ones and will tend to move more slowly. CGE differs from CZE

Distinguishing between stereoisomers – the R and S convention involves prioritising atoms or groups bonded to an asymmetric carbon atom in order of their atomic number. With the smallest atom or group pointing away from you, note the size of the remaining three groups or atoms. If the configuration in order of increasing size is clockwise, this is termed the R-configuration (L. rectus); if the order is anticlockwise, this is the S-configuration (L. sinister). Note that the older, alternative terminology of 'D' (L. dextro) and 'L' (L. laevo) isomers is widely used in the life sciences (p. 352), but the two terminologies do not always coincide (i.e. R is not always D, and S is not always L).

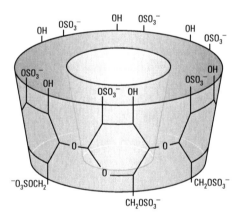

Fig. 47.6 General structure of highly sulphated cyclodextrins (HSCDs). The central cavity of the cyclodextrin (shaded blue) can interact differently with the R and S isomers of a biomolecule, enabling separation in CE. Copyright © 1999–2002 Beckman Coulter, Inc. http://www.beckmancoulter.com/products/splashpage/chiral38/default.asp/

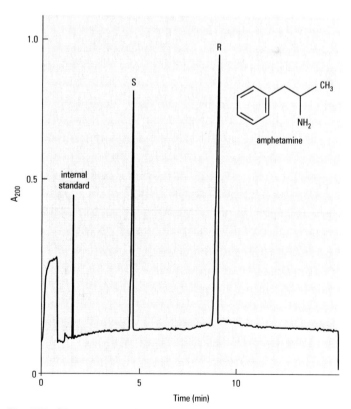

Fig. 47.7 CE separation of R and S enantiomers of amphetamine using Beckman Coulter highly sulphated gamma cyclodextrin. Copyright © 1999–2002 Beckman Coulter, Inc. http://www.beckmancoulter.com/products/splashpage/chiral38/default.asp/

and MEKC in that the inner surface of the capillary is polymer-coated in order to prevent EOF; this means that for most applications (e.g. polypeptide or oligonucleotide separations) sample components will migrate towards the anode at a rate determined by their size. The technique also differs from conventional PAGE in that a 'polymer network' is used rather than a gel: the polymer network may be polyacrylamide or agarose.

CGE offers the following advantages over conventional electrophoresis:

- efficient heat dissipation means that a high electrical field can be applied, giving shorter separation times;
- detection of the separated components as they move towards the anodic end of the capillary (e.g. using a UV/visible detector) means that staining is unnecessary;
- automation is feasible.

Choosing a detector for capillary electrophoresis – most types of HPLC detector are suitable for CE and related applications (see Chapter 45).

Capillary isoelectric focusing (CIEF)

This is used mainly for protein separation. Here, the principles of IEF are valid as long as EOF is prevented by using capillaries that are polymer-coated on their inner surface. Sample components migrate to their isoelectric points and become stationary. Once separated (<10 min), the components must be mobilised so that they flow past the detector. This is achieved by changing the NaOH solution in the cathodic reservoir with an NaOH/NaCl solution. When the electric field is reapplied, Cl^- enters the capillary, causing a decrease in pH at the cathodic end and the subsequent migration of sample components.

Text reference

Weinberger, R. (2000) *Practical Capillary Electrophoresis*. Academic Press, New York.

Sources for further study

Anon. *The American Electrophoresis Society Homepage*. Available: http://www.aesociety.org
Last accessed: 01/04/07.
[Includes details on electrophoretic techniques including IEF and CE]

Anon. *Cyclodextrin Resource*. Available: http://www.cyclodex.com/index.html
Last accessed: 01/04/07.
[Information on cyclodextrin structure, applications, etc.]

Cunico, R.L., Gooding, K.M. and Wehr, T. (1998) *Basic HPLC and CE of Biomolecules*. Bay Bioanalytical, Hercules, CA.

Khaledi, M.G. (1998) *High Performance Capillary Electrophoresis: Theory, Techniques and Applications*. Wiley, New York.
[Detailed manual of CE techniques and procedures]

Palfrey, S.M. (1999) *Clinical Applications of Capillary Electrophoresis*. Humana Press, New Jersey.

Strenge, M.A. and Lagu, A.L. (2004) *Capillary Electrophoresis of Proteins and Peptides*. Humana Press, New Jersey.

Westermeier, R. (2004) *Electrophoresis in Practice: A Guide to Methods and Applications of DNA and Protein Separations*, 3rd edn. Wiley-VCH, Berlin.
[Covers the principles underlying IEF]

Weston, A. and Brown, P.R. (1997) *HPLC and CE: Principles and Practice*. Academic Press, New York.
[Covers practical aspects including applications, optimisation and troubleshooting]

Study exercises

47.1 Consider the requirements for sample application in PAGE and IEF. Explain why in PAGE the sample is applied in a discrete narrow band, usually at the cathodic end of the gel, while in IEF the sample can be applied at any point along the length of the gel without concern about location or narrowness of the sample zone.

47.2 Choose a suitable CE technique for a range of applications. Which variant of CE would you consider most suitable for the following applications? (a) Separation of small peptides of similar size; (b) separation of oligonucleotides of different lengths; (c) separation of proteins with different isoelectric points; (d) a sample of urine tested for several anabolic steroids.

47.3 Check the similarities and differences between electrophoresis and isoelectric focusing. Complete the following table, indicating (i) whether each of the factors listed is relevant to each technique and (ii) briefly explain why, in each case.

Potential factors influencing electrophoresis (E) and isoelectric focusing (IEF)

Factor	E	IEF
Dependence upon ionisable groups in proteins		
Requirement for the application of an electric field		
Dependence upon differences in net charge between sample molecules		
Dependence upon differences in pI values between sample molecules		
Requirement for buffer at constant pH		
Requirement for a pH gradient		
Requirement for ampholytes to be included in solution		
Use of supporting medium with molecular sieving		
Treatment of samples with SDS is possible		
Diffusion of separated proteins is significant		

47.4 Identify proteins by 2D electrophoresis. The figure represents the separation of a mixture of 10 proteins by 2D electrophoresis. Using the information in the table, identify the individual proteins A–J.

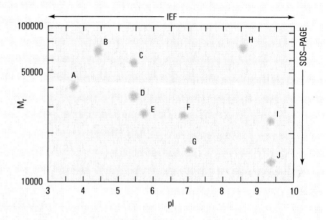

M_r and pI values of selected protein subunits

Protein	Subunit M_r	Subunit pI
Nerve growth factor	13 300	9.3
Ribonuclease	13 700	7.8
Haemoglobin	16 000	7.0
β-lactoglobulin	17 500	5.2
Ceramide trihexosidase	22 000	3.0
Trypsinogen	24 500	9.3
Triose phosphate isomerase	27 000	5.7
Galactokinase	27 000	5.7
Arginase	30 000	9.2
Glycerol-3-phosphate dehydrogenase	34 000	6.4
Alcohol dehydrogenase	35 000	5.4
Deoxyribonuclease II	38 000	10.2
Aldolase	40 000	5.2
Pepsinogen	41 000	3.7
Hexokinase	51 000	5.3
Lipoxidase	54 000	5.7
Catalase	57 500	5.4
Alkaline phosphatase	69 000	4.4
Acetylcholine esterase	70 000	4.5
Glyceraldehyde-3-phosphate dehydrogenase	72 000	8.5

48 Electroanalytical techniques

Electrochemical methods are used to quantify a broad range of different biomolecules, including ions, gases, metabolites, drugs and hormones.

KEY POINT *The basis of all electrochemical analysis is the transfer of electrons from one atom or molecule to another atom or molecule in an obligately coupled oxidation–reduction reaction (a redox reaction).*

It is convenient to separate such redox reactions into two half-reactions and, by convention, each is written as:

$$\text{oxidised form} + \text{electron(s)}\ (n e^-) \underset{\text{oxidation}}{\overset{\text{reduction}}{\rightleftharpoons}} \text{reduced form} \quad [48.1]$$

You should note that the half-reaction is reversible: by applying suitable conditions, reduction *or* oxidation can take place. As an example, a simple redox reaction occurs when metallic zinc (Zn) is placed in a solution containing copper ions (Cu^{2+}), as follows:

$$Cu^{2+} + Zn \rightarrow Cu + Zn^{2+} \quad [48.2]$$

The half-reactions are (i) $Cu^{2+} + 2e^- \rightarrow Cu$ and (ii) $Zn^{2+} + 2e^- \rightarrow Zn$. The oxidising power of (i) is greater than that of (ii), so in a coupled system the latter half-reaction proceeds in the opposite direction to that shown above, i.e. as $Zn - 2e^- \rightarrow Zn^{2+}$. When Zn and Cu electrodes are placed in separate solutions containing their ions, and connected electrically (Fig. 48.1), electrons will flow from the Zn electrode to Cu^{2+} via the Cu electrode due to the difference in oxidising power of the two half-reactions.

By convention, the electrode potential of any half-reaction is expressed relative to that of a standard hydrogen electrode (half-reaction $2H^+ + 2e^- \rightarrow H_2$) and is called the standard electrode potential, E°. Table 48.1 shows the values of E° for selected half-reactions. With any pair of half-reactions from this series, electrons will flow from that having the lowest electrode potential to that of the highest. E° is determined at pH = 0. It is often more appropriate to express standard electrode potentials at pH 7 for biological systems, and the symbol $E^{\circ'}$ is used: in all circumstances, it is important that the pH is clearly stated.

The arrangement shown in Figure 48.1 represents a simple galvanic cell where two electrodes serve as the interfaces between a chemical system and an electrical system. For analytical purposes, the magnitude of the potential (voltage) or the current produced by an electrochemical cell is related to the concentration (strictly the activity, a, p. 146) of a particular chemical species. Electrochemical methods offer the following advantages:

- excellent detection limits, and wide operating range (10^{-1}–10^{-8} mol l^{-1});
- measurements may be made on very small volumes (μl) allowing small amounts (pmol) of sample to be measured in some cases;
- miniature electrochemical sensors can be used for certain *in vivo* measurements, e.g. pH, glucose, oxygen content.

Definitions

Oxidation – loss of electrons by an atom or molecule (or gain of O atoms, loss of H atoms, increase in positive charge, or decrease in negative charge).

Reduction – gain of electrons by an atom or molecule (or loss of O atoms, gain of H atoms, decrease in positive charge or increase in negative charge).

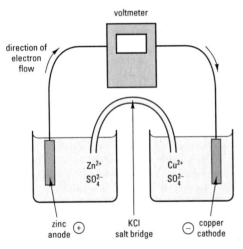

Fig. 48.1 A simple galvanic electrochemical cell. The KCl salt bridge allows migration of ions between the two compartments but prevents mixing of the two solutions.

Definition

Galvanic cell – an electrochemical cell in which reactions occur spontaneously at the electrodes when they are connected externally by a conductor, producing electrical energy.

Table 48.1 Standard electrode potentials* (E^0) for selected half-reactions

Half-reaction	E^0 at 25 °C (V)
$Cl_2 + 2e^- \rightleftharpoons 2Cl^-$	+1.36
$O_2 + 4H^+ + 4e^- \rightleftharpoons 2H_2O$	+1.23
$Br_2 + 2e^- \rightleftharpoons 2Br^-$	+1.09
$Ag^+ + e^- \rightleftharpoons Ag$	+0.80
$Fe^{3+} + e^- \rightleftharpoons Fe^{2+}$	+0.77
$I_3^- + 2e^- \rightleftharpoons 3I^-$	+0.54
$Cu^{2+} + 2e^- \rightleftharpoons Cu$	+0.34
$Hg_2Cl_2 + 2e^- \rightleftharpoons 2Hg + 2Cl^-$	+0.27
$AgCl + e^- \rightleftharpoons Ag + Cl^-$	+0.22
$Ag(S_2O_3)_2^{3-} + e^- \rightleftharpoons Ag + 2S_2O_3^{2-}$	+0.01
$2H^+ + 2e^- \rightleftharpoons H_2$	+0.00
$AgI + e^- \rightleftharpoons Ag + I$	−0.15
$PbSO_4 + 2e^- \rightleftharpoons Pb + SO_4^{2-}$	−0.35
$Cd^{2+} + 2e^- \rightleftharpoons Cd$	−0.40
$Zn^{2+} + 2e^- \rightleftharpoons Zn$	−0.76

*From Milazzo *et al.* (1978).

Using a calomel electrode – always ensure that the KCl solution is saturated by checking that KCl crystals are present.

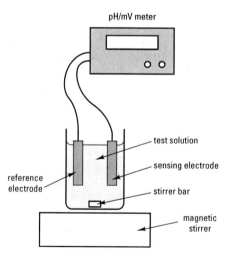

Fig. 48.2 Components of a potentiometric cell.

Potentiometry and ion-selective electrodes

Operating principles

These systems involve galvanic cells (p. 336) and are based on measurement of the potential (voltage) difference between two electrodes in solution when no net current flows between them: no net electrochemical reaction occurs and measurements are made under equilibrium conditions. These systems include methods for measuring pH, ions and gases such as CO_2 and NH_3. A typical potentiometric cell is shown in Figure 48.2. It contains two electrodes:

1. A 'sensing' electrode, the half-cell potential of which responds to changes in the activity (concentration) of the substance to be measured; the most common type of indicator electrodes are ion-selective electrodes (ISE).
2. A 'reference' electrode, the potential of which does not change, forming the second half of the cell.

To assay a particular analyte, the potential difference between these electrodes is measured by a millivolt meter (e.g. a standard pH meter).

Reference electrodes for potentiometry are of three main types:

1. The standard hydrogen electrode, which is the reference half-cell electrode, defined as 0.0 V at all temperatures, against which values of E° are expressed. H_2 gas at 1 atmosphere pressure is bubbled over a platinum electrode immersed in an acid solution with an activity of unity. This electrode is rarely used for analytical work, since it is unstable and other reference electrodes are easier to construct and use.
2. The calomel electrode (Fig. 48.3), which consists of a paste of mercury covered by a coat of calomel (Hg_2Cl_2), immersed in a saturated solution of KCl. The half-reaction: $Hg_2Cl_2 + 2e^- \rightarrow 2Hg + 2Cl^-$ gives a stable standard electrode potential of +0.24 V.
3. The silver/silver chloride electrode. This is a silver wire coated with AgCl and immersed in a solution of constant chloride concentration. The half-reaction: $AgCl + e^- \rightarrow Ag + Cl^-$ gives a stable, standard electrode potential of +0.20 V.

 KEY POINT *Ion-selective electrodes (ISEs) are based on measurement of a potential across a membrane which is selective for a particular analyte.*

An ISE consists of a membrane, an internal reference electrode, and an internal reference electrolyte of fixed activity. The ISE is immersed in a sample solution that contains the analyte of interest, along with a reference electrode. The membrane is chosen to have a specific affinity for a particular ion, and if the activity of this ion in the sample differs from that in the reference electrolyte, a potential develops across the membrane that is dependent on the ratio of these activities. Since the potentials of the two reference electrodes (internal and external) are fixed, and the internal electrolyte is of constant activity, the measured potential, E, is dependent on the membrane potential and is given by the Nernst equation (Robinson and Stokes, 2002) as:

$$E = K + 2.303\frac{RT}{zF} \log [a] \qquad [48.3]$$

Using ISEs, including pH electrodes – standards and samples must be measured at the same temperature, as the Nernst equation shows that the measured potential is temperature-dependent.

where K represents a constant potential which is dependent on the reference electrode, z represents the net charge on the analyte, $[a]$ represents the activity of analyte in the sample and all other symbols and constants have their usual meaning (p. 185). For a series of standards of known activity, a plot of E against $\log[a]$ should be linear over the working range of the electrode, with a slope of $2.303\,RT/zF$ (0.059 V at 25 °C). Although ISEs strictly measure *activity*, the potential differences can be approximated to concentration as long as (i) the analyte is in dilute solution (p. 144), (ii) the ionic strength of the calibration standards matches that of the sample, e.g. by adding appropriate amounts of a high ionic strength solution to the standards, and (iii) the effect of binding to sample macromolecules (e.g. proteins, nucleic acids) is minimal. Potentiometric measurements on undiluted biological fluids, e.g. K^+ and Na^+ levels in plasma, tissue fluids or urine, are likely to give lower values than flame emission spectrophotometry, since the latter procedure measures total ion levels, rather than just those in aqueous solution.

All of the various types of membrane used in ISEs operate by incorporating the ion to be analysed into the membrane, with the accompanying establishment of a membrane potential. The scope of electrochemical analysis has been extended to measuring gases and non-ionic compounds by combining ISEs with gas-permeable membranes, enzymes, and even immobilised bacteria or tissues.

Glass membrane electrodes

The most widely used ISE is the glass membrane electrode for pH measurement (p. 154). The membrane is thin glass (50 μm wall thickness) made of silica which contains some Na^+. When the membrane is soaked in water, a thin hydrated layer is formed on the surface in which negative oxide groups (Si-O$^-$) in the glass act as ion exchange sites. If the electrode is placed in an acid solution, H^+ exchanges with Na^+ in the hydrated layer, producing an external surface potential: in alkaline solution, H^+ moves out of the membrane in exchange for Na^+. Since the inner surface potential is kept constant by exposure to a fixed activity of H^+, a consistent, accurate potentiometric response is observed over a wide pH range. Glass electrodes for other cations (e.g. Na^+, NH_4^+) have been developed by changing the composition of the glass, so that it is predominantly sensitive to the particular analyte, though the specificity of such electrodes is not absolute. The operating principles and maintenance of such electrodes are broadly similar to those for pH electrodes (p. 154).

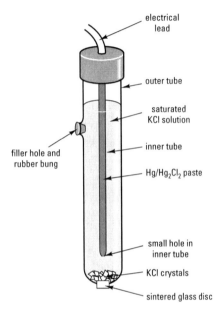

Fig. 48.3 A calomel reference electrode.

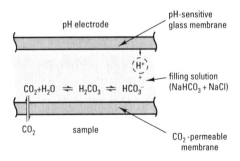

Fig. 48.4 Underlying principles of a gas-sensing electrode.

Using CO_2 electrodes – applications include measurement of blood pCO_2 and in enzyme studies where CO_2 is utilised or released: calibration of the electrode is accomplished using 5% v/v and 10% v/v mixtures of CO_2 in an inert gas equilibrated against the measuring solution.

Gas-sensing glass electrodes

Here, an ISE in contact with a thin external layer of aqueous electrolyte (the 'filling solution') is kept close to the glass membrane by an additional, outer membrane that is selectively permeable to the gas of interest. The arrangement for a CO_2 electrode is shown in Figure 48.4: in this case the outer membrane is made of CO_2-permeable silicone rubber. When CO_2 gas in the sample selectively diffuses across the membrane and dissolves in the filling solution (in this case an aqueous $NaHCO_3/NaCl$ mixture), a change in pH occurs due to the shift in the equilibrium:

$$CO_2 + H_2O \rightleftharpoons H_2CO_3 \rightleftharpoons H^+ + HCO_3^- \qquad [48.4]$$

The pH change is 'sensed' by the internal ion-selective pH electrode, and its response is proportional to the partial pressure of CO_2 of the solution

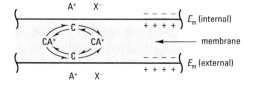

Fig. 48.5 Underlying principles of a liquid membrane ion-selective electrode. A^+ – analyte; C – neutral carrier ionophore; E_m – surface potential; membrane potential = E_m(internal) – E_m(external).

> **Definition**
>
> **Ionophore** – a compound that enhances membrane permeability to a specific ion: an ionophore may be incorporated into an ISE to detect that ion.

> **Definitions**
>
> **Electrolytic cell** – an electrochemical cell in which reactions are driven by the application of an external voltage greater than the spontaneous potential of the cell.
>
> **Electrolysis** – a non-spontaneous chemical reaction resulting from the application of a potential to an electrode.

(pCO_2). A similar principle operates in the NH_3 electrode, where a Teflon membrane is used, and the filling solution is NH_4Cl.

Liquid and polymer membrane electrodes

In this type of ISE, the liquid is a water-insoluble viscous solvent containing a soluble ionophore, i.e. an organic ion-exchanger, or a neutral carrier molecule, that is specific for the analyte of interest. When this liquid is soaked into a thin membrane such as cellulose acetate, it becomes effectively immobilised. The arrangement of analyte (A^+) and ionophore in relation to this membrane is shown in Figure 48.5. The potential on the inner surface of the membrane is kept constant by maintaining a constant activity of A^+ in the internal solution, so the potential change measured is that which results from A^+ in the sample interacting with the ionophore in the outer surface of the membrane.

A relevant example of a suitable ionophore is the antibiotic valinomycin (p. 395), which specifically binds K^+. Other ionophores have been developed for measurement of e.g. NH_4^+, Ca^{2+}, Cl^-. In addition, electrodes have been developed for organic species by using specific ion-pairing reagents in the membrane that interact with ionic forms of the organic compound, e.g. with drugs such as 5,5-diphenylhydantoin.

Solid-state membrane electrodes

These contain membranes made from single crystals or pressed pellets of salts of the analyte. The membrane material must show some permeability to ions and must be virtually insoluble in water. Examples include:

- the fluoride electrode, which uses LaF_3 impregnated with Eu^{2+} (the latter to increase permeability to F^-). A membrane potential is set up when F^- in the sample solution enters spaces in the crystal lattice;
- the chloride electrode, which uses a pressed-pellet membrane of Ag_2S and $AgCl$.

Voltammetric methods

Voltammetric methods are based on measurements made using an electrochemical cell in which electrolysis is occurring. Voltammetry, sometimes also called amperometry, involves the use of a potential applied between two electrodes (the working electrode and the reference electrode) to cause oxidation or reduction of an electroactive analyte. The loss or gain of electrons at an electrode surface causes current to flow, and the size of the current (usually measured in mA or μA) is directly proportional to the concentration of the electroactive analyte. The materials used for the working electrode must be good conductors and electrochemically inert, so that they simply transfer electrons to and from species in solution. Suitable materials include Pt, Au, Hg and glassy carbon.

Two widely used devices that operate on the voltammetric principle are the oxygen electrode and the glucose electrode. These are sometimes referred to as amperometric sensors.

Oxygen electrodes

The Clark (Rank) oxygen electrode

These instruments measure oxygen in solution using the polarographic principle, i.e. by monitoring the current flowing between two electrodes

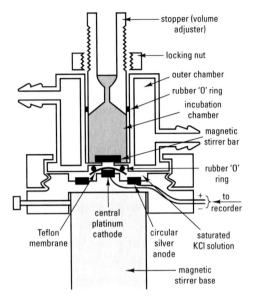

Fig. 48.6 Transverse section through a Clark (Rank) oxygen electrode.

when a voltage is applied. The most widespread electrode is the Clark type (Fig. 48.6), manufactured by Rank Bros, Cambridge, UK, which is suitable for measuring O_2 concentrations in cell, organelle and enzyme suspensions. Pt and Au electrodes are in contact with a solution of electrolyte (normally saturated KCl). The electrodes are separated from the medium by a Teflon membrane, permeable to O_2. When a potential is applied across the electrodes, this generates a current proportional to the O_2 concentration (Wilson and Walker, 2005). The reactions can be summed up as:

$4Ag \rightarrow 4Ag^+ + 4e^-$ (at silver anode)

$O_2 + 2e^- + 2H^+ \rightleftharpoons H_2O_2$ (in electrolyte solution; O_2 replenished by diffusion from test solution)

$H_2O_2 + 2e^- + 2H^+ \rightleftharpoons 2H_2O$ (at platinum cathode)

Setting up and using a Clark (Rank) oxygen electrode

Box 48.1 describes the steps involved: if you are setting up from scratch, perform steps 1–13; if a satisfactory membrane is already in place, start at step 7. Box 48.2 gives advice on how to use a chart recorder with an oxygen electrode, including how to calculate rates of oxygen consumption/production, using data for oxygen saturation of air-equilibrated water shown in Table 48.2.

The temperature of the incubation vessel should be controlled by passing water (e.g. from a water bath) through the outer chamber (Fig. 48.6). Cells or organelles may be present in the solution added to the incubation chamber or can be added via the hole in the stopper using a syringe, as can chemicals such as inhibitors (Box 48.2). Take care not to introduce air bubbles, and remove any that appear by gently raising and lowering the stopper. The electrode can be used repeatedly, providing the membrane is

Table 48.2 Oxygen saturation values for distilled water and sea water at standard atmospheric pressure and a range of temperatures (derived from Green and Carritt, 1967)

Temperature (°C)	Distilled water			Sea water		
	mol m^{-3} (mmol l^{-1})	mg l^{-1} (p.p.m.)	ml l^{-1}	mol m^{-3} (mmol l^{-1})	mg l^{-1} (p.p.m.)	ml l^{-1}
0	0.460	14.7	10.3	0.359	11.5	8.04
2	0.435	13.9	9.75	0.342	10.9	7.65
4	0.413	13.2	9.24	0.326	10.4	7.30
6	0.392	12.5	8.78	0.311	9.95	6.97
8	0.373	11.9	8.35	0.298	9.54	6.66
10	0.355	11.4	7.95	0.285	9.12	6.37
12	0.339	10.8	7.59	0.273	8.74	6.11
14	0.324	10.4	7.25	0.261	8.35	5.85
16	0.310	9.92	6.94	0.251	8.03	5.62
18	0.297	9.50	6.65	0.241	7.71	5.40
20	0.285	9.12	6.38	0.232	7.42	5.19
22	0.274	8.77	6.12	0.224	7.17	5.00
24	0.263	8.42	5.89	0.215	6.88	4.82
26	0.253	8.1	5.67	0.208	6.66	4.64
28	0.244	7.81	5.46	0.200	6.40	4.48
30	0.235	7.52	5.27	0.193	6.18	4.32
37	0.211	6.75	4.71	0.174	5.57	3.88

Notes: mol m^{-3} ≡ mmol l^{-1} ≡ µmol ml^{-1} ≡ nmol µl^{-1}. Tabulated values assume atmospheric pressure = 101.3 kPa (≈760 mm Hg); for more accurate work, a correction for any deviation can be made by multiplying the appropriate figures from the table by the ratio of the real pressure to the assumed pressure.

satisfactory: remove solutions carefully (e.g. using a pipette, or vacuum line and trap). Keep water in the chamber when not in use. Replace the membrane if:

- the reading becomes noisy;
- the electrode will not zero after adding sodium dithionite;
- the response becomes too slow (check by switching off stirrer – oxygen concentration should drop rapidly as the available oxygen is consumed).

 KEY POINT Successful operation of a Clark (Rank) oxygen electrode requires an intact Teflon membrane and clean electrodes. When replacing the Teflon membrane, you should always check that the electrode surfaces are clean (shiny); if necessary, use a mild abrasive paste to remove any oxidised material.

Box 48.1 How to set up a Clark (Rank) oxygen electrode

1. **Detach the base of the incubation vessel** (see Fig. 48.6) by unscrewing the locking ring.
2. **Add enough saturated KCl to cover the electrodes.**
3. **Cut a 1 mm square hole in the centre of a 10 × 10 mm square of lens tissue** and place this on the KCl solution so that the hole is over the central platinum cathode.
4. **Cut a 10 × 10 mm square of Teflon membrane and place over the lens tissue;** seal by gently lowering the incubation vessel and tightening the locking ring, making sure that the rubber O-ring is correctly positioned over the membrane.
   (a) Do not overtighten the locking ring.
   (b) Take care not to trap air bubbles beneath the membrane.
   (c) Make sure that the membrane does not become twisted.
5. **Clamp the electrode over the magnetic stirrer base using the clamping screw.**
6. **Connect the electrode leads to the polarising unit/recording device** (silver anode to positive, platinum cathode to negative). You can either use a digital readout system on the control unit or the output can be passed to a chart recorder, giving a readout of oxygen status as a function of time (see Box 48.2). Check that the polarising voltage is set to 0.60 V and adjust, if necessary, using the 'polarising voltage' control (typically, this requires a small screwdriver to adjust).
7. **Add air-saturated experimental solution and a small Teflon-coated magnetic stirrer bar to the chamber.** The volume of the incubation chamber can be adjusted by moving the locking nut on the stopper. To adjust, add the appropriate amount of liquid to the chamber using a pipette, insert the stopper and screw the locking nut until the solution just fills the incubation chamber.
8. **Gently push the stopper (volume adjuster) into position,** making sure that no air bubbles are trapped in the chamber, and switch on the stirrer. Adjust the rate of stirring using the 'stirrer' control, if required.
9. **Set the zero:** remove the stopper and add a few crystals of sodium dithionite – the dithionite ions ($S_2O_4^{2-}$) will be oxidised to sulphite (SO_3^{2-}) and sulphate ions (SO_4^{2-}), thereby consuming all of the oxygen in the solution within 3–5 min (an alternative is to bubble N_2 through the solution for > 10 min). Once the reading has stabilised, if this is not at zero, then adjust using the 'set zero' control (in practice, the zero is usually quite stable and often needs no adjustment).
10. **Adjust the sensitivity** – rinse out the dithionite solution thoroughly/repeatedly, then replace with air-equilibrated water and allow the reading to stabilise (this may take 5–10 min.). Then adjust the 'sensitivity' control of the electrode system to set the oxygen saturated value at an appropriate point (for a controller with a digital readout, this is best set to 100, and then the values represent percentage oxygen saturation). To check that the instrument is working correctly, switch off the stirrer for a few moments – the reading should drop quickly as oxygen is consumed at the electrode surface and should then rise rapidly when the stirrer is switched back on.
11. **Rinse the incubation chamber thoroughly and add fresh experimental solution.** Make sure that all traces of sodium dithionite are removed.
12. **Carry out your experiment.**
13. **Remove the solution and check the calibration.** If the reading for air – equilibrated water is different, the electrode's sensitivity or the temperature may have changed and you may need to recalibrate and repeat the measurement.

Box 48.2 How to convert a chart recorder trace to a rate of O_2 consumption or production

Having set up your oxygen electrode (Box 48.1) or probe, the most common experiments are those in which you measure the rate of oxygen consumption (respiration) or production (photosynthesis) in a fixed volume of solution (e.g. in the chamber of a Rank (Clark) oxygen electrode. Typically, this involves attaching the electrode to a chart recorder and following the change in oxygen concentration with time. The principal steps are as follows:

1. **Calculate the amount of oxygen in the electrode chamber** – multiply the appropriate oxygen-saturation concentration, from Table 48.2, by the volume of the chamber. For example, if you are working in mmol l^{-1} ($= \mu$mol ml^{-1}) then at 20°C, air-equilibrated (oxygen-saturated) distilled water contains 0.285 μmol ml^{-1}. For an electrode chamber of volume 5 ml, there will be $0.285 \times 5 = 1.425$ μmol oxygen in the water within the chamber at saturation.

2. **Set the zero on the recorder** – following the addition of a small amount of sodium dithionite to the chamber, to remove oxygen (Box 48.1), the trace is monitored until it has stabilised. Then adjust the 'zero' or 'back off' control of the recorder until the chart trace is set to zero on the chart paper: this is marked as 'A' on Fig. 48.8.

3. **Set the oxygen-saturated reading on the recorder** – rinse out the dithionite solution thoroughly/repeatedly, replace with air-equilibrated water and allow the reading to stabilise (this may take 3–5 min). Then adjust the 'sensitivity' control of the chart recorder to set the oxygen-saturated value at an appropriate point on the chart – for respiration measurements, this can be close to the full width of the chart paper, whereas for photosynthesis measurements it is more usual to set it closer to the mid-point of the chart paper, to allow measurements over 100% to be recorded. In the example shown in Fig. 48.8, where the chart paper is calibrated in mm divisions, the oxygen saturated reading is set at 80 divisions (80 mm), marked as 'B'. Note that the recorder pen can be lifted between readings, and during rinsing of the chamber, to reduce the amount of 'noise' on the chart.

4. **Calculate the amount of oxygen equivalent to a single division of the chart paper** – divide the amount of oxygen in the chamber by the number of divisions (mm) to give the amount per division (e.g. $1.425 \div 80 = 0.0178125$ μmol ($= 17.8125$ nmol per division).

5. **Carry out your experiment** – the example shown in Fig. 48.8 represents the consumption of oxygen during respiration of a suspension of yeast cells, marked as 'C'. Continue the recording until you are satisfied that the rate has remained stable for at least 5 min. In many experiments, you will be adding substances that change the rate (e.g. metabolic inhibitors, or substrates) – this can be done by injecting a solution containing the substance through the small hole in the stopper of the electrode (keep the volume of this solution to < 1% of the total volume of the chamber, to minimise the effect of the added solution).

6. **Convert the gradient of the trace into a rate of oxygen consumed or produced** – using a transparent ruler, draw the line of best fit through that part of the trace showing a stable relationship (see Fig. 48.8). Use this line of best fit to work out the gradient of the trace, by converting from linear dimensions to an amount per unit time: e.g. the gradient shown in Fig. 48.8 is equivalent to a decrease of 24 chart divisions (mm) in 3 cm. Using the value calculated in step 4, the amount of oxygen consumed is thus $17.8125 \times 24 = 427.5$ nmol. For a chart speed of 0.2 cm min^{-1}, this represents the change occurring over a time interval of $3 \div 0.2 = 15$ min. Thus the rate of oxygen consumed is $427.5 \div 15 = 28.5$ nmol min^{-1}.

7. **Express the rate on an appropriate basis** – in most instances, this will be either per cell (e.g. determined using a haemocytometer, p. 236) or per unit mass of protein (p. 353), chlorophyll (p. 282), or an equivalent marker. For the example shown in Fig. 48.8, if there were 2.4×10^7 yeast cells per ml of suspension, then in the 5 ml chamber of the electrode there would be a total of $2.4 \times 10^7 \times 5 = 1.2 \times 10^8$ yeast cells, giving a rate of $28.5 \div (1.2 \times 10^8) = 2.375 \times 10^{-7}$ nmol min^{-1} $cell^{-1}$, probably better expressed as 237.5 amol min^{-1} $cell^{-1}$. Note that when inhibitors/substrates are used, you should always wait until a new stable rate has been achieved (for 5 min) before using the trace to determine the new rate of oxygen consumption/production.

Definition

Biosensor – a device for measuring a substance, combining the selectivity of a biological reaction with that of a sensing electrode.

Oxygen probes

Clark-type oxygen electrodes are also available in probe form for immersion in the test solution (Fig. 48.7), e.g. for field studies, allowing direct measurement of oxygen status *in situ*, in contrast to chemical assays. The main point to note is that the solution must be stirred during

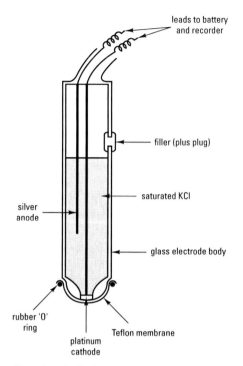

Fig. 48.7 A Clark-type oxygen probe.

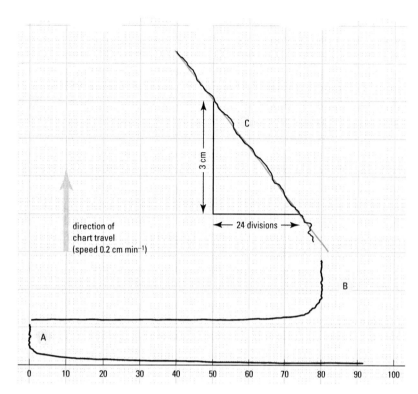

Fig. 48.8 Chart recorder output for a Clark (Rank) oxygen electrode. A = zero oxygen; B = oxygen saturation; C = respiration of yeast.

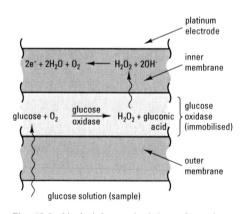

Fig. 48.9 Underlying principles of a glucose electrode.

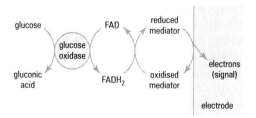

Fig. 48.10 Operating principles of a glucose biosensor. In the ExacTech system, the mediator is a ferrocene derivative.

measurement, to replenish the oxygen consumed by the electrode ('boundary layer' effect).

Glucose electrodes

These are simple types of biosensor; a simple design is shown in Figure 48.9. It consists of a Pt electrode, overlaid by two membranes. Sandwiched between these membranes is a layer of the immobilised enzyme glucose oxidase. The outer membrane is glucose-permeable and allows glucose in the sample to diffuse through to the glucose oxidase layer, where it is converted to gluconic acid and H_2O_2. The inner membrane is selectively permeable to H_2O_2, which is oxidised to O_2 at the surface of the Pt electrode. The current arising from this release of electrons is proportional to the glucose concentration in the sample within the range 10^{-7}–10^{-3} mol l^{-1}. The efficiency of a biosensor can be increased by including an electrode surface that is capable of capturing electrons directly. This is the basis of recently developed blood glucose meters for diabetics such as the ExacTech meter (Medisense-Abbott Laboratories). Here a pinprick blood sample is applied to a test strip containing glucose oxidase: if glucose is present a series of reactions occurs as shown in Figure 48.10.

The electrons produced are donated to the electrode surface, producing an initial current that is proportional to the initial rate at which glucose is oxidised which, in turn, is proportional to the glucose concentration in the blood sample. The future design of biosensors is likely to involve microchip technology in which the biocatalyst and the transducer are even more intimately linked (for more details, see Eggins, 2002).

Electrochemical detectors used in chromatography operate by voltammetric principles and currents are produced as the mobile phase flows over electrodes set at a fixed potential: to achieve maximum sensitivity, this potential must be set at a level that allows electrochemical reactions to occur in all analytes of interest.

Coulometric methods

Here, the charge required to completely electrolyse a sample is measured: the time required to titrate an analyte is measured at constant current and related to the amount of analyte using Faraday's law. There are few biological applications of this technique, though it is sometimes used for determination of Cl^- in serum and body fluids.

Text references

Eggins, B.R. (2002) *Chemical Sensors and Biosensors*. Wiley, New York.

Green, E.J. and Carritt, D.E. (1967) New Tables for Oxygen Saturation of Seawater, *Journal of Marine Research*, **25**, 140–7.

Milazzo, G., Caroli, S. and Sharma, V.K. (1978) *Tables of Standard Electrode Potentials*. Wiley, London.

Robinson, R.A. and Stokes, R.H. (2002) *Electrolyte Solutions*. Dover Publications, New York.

Wilson, K. and Walker, J. (eds) (2005) *Principles and Techniques of Practical Biochemistry and Molecular Biology*, 6th edn. Cambridge University Press, Cambridge.
[Also covers a range of other topics, including chromatography, spectroscopy and centrifugation]

Sources for further study

Anon. *Rank Dissolved Oxygen Electrode*. Available: http://www.rankbrothers.co.uk/prod1.htm Last accessed: 01/04/07.
[Homepage of the manufacturing company – contains technical details and contact information]

Fisher, A.C. (1996) *Electrode Dynamics*. Oxford University Press, Oxford.

Golterman, H.L., Clymo, R.S. and Ohnstad, M.A.M. (1978) *Methods for Physical and Chemical Analysis of Fresh Waters*. Blackwell Scientific Publications, Oxford.

Knopf, G.K. and Bassi, A.S. (2006) *Smart Biosensor Technology*. Taylor & Francis, New York.

Linek, V. (1988) *Measurement of Oxygen by Membrane-covered Probes*. Ellis-Horwood, Chichester.

Sanders, G.H.W., Compton, R.G. and Compton, S. (1996) *Electrode Potentials*. Oxford University Press, Oxford.
[Covers the underlying theory, plus applications relating to ion-selective electrodes]

Study exercises

48.1 Test your understanding of the circumstances under which ion-selective electrodes can be used. List the three main assumptions underlying the measurement of ion concentration using an ion-selective electrode.

48.2 Test your knowledge of biosensor design. List the two main functional components of biosensor such as the glucose electrode. Research a specific application – note that it should be different from the ones used in this chapter (e.g. using the Web).

(continued)

Study exercises (continued)

48.3 Calculate the oxygen content of specified volumes of water at defined temperatures. Using the information in Table 48.2 and assuming air equilibration at 101.3 kPa, what is the amount of oxygen in each of the following (give all answers to three significant figures):

(a) 4 ml of distilled water at 20 °C (express your answer in µmol);

(b) 20 ml of sea water at 12 °C (express your answer in µmol);

(c) 10 ml of distilled water at 15 °C (express your answer in µmol);

(d) 250 ml of distilled water at 37 °C (express your answer in mg);

(e) 200 ml of sea water at 25 °C (express your answer in ml).

48.4 Calculate respiration using data from an oxygen electrode readout. An oxygen electrode was set up to contain 5.0 ml of air-equilibrated distilled water at 30 °C, giving a full-scale reading on a chart recorder (200 divisions on the chart recorder paper), while a zero reading gave no deflection (0 divisions on the chart recorder paper). A yeast cell suspension (containing 4.1×10^9 cells in total) was added to the same electrode system along with a suitable growth medium, giving an approximately linear trace on the chart recorder with a slope of -25 units per cm, at a chart speed of 0.5 cm min^{-1}. What is the respiratory oxygen consumption in nmol min^{-1} (10^9 cells)$^{-1}$? (Give your answer to three significant figures.)

48.5 Determine net photosynthetic rate from oxygen electrode readings. The data below represent oxygen electrode readings taken every minute from 0 min for an illuminated suspension of a cyanobacterium in a 5.0 ml chamber of an oxygen electrode at 20 °C: 70.4, 73.6, 75.3, 78.5, 83.6, 85.9, 88.2, 91.4, 94.1. The system was calibrated to read 100.0 for air-equilibrated distilled water and 0.0 for anaerobic water, in effect making the scale read in terms of percentage oxygen saturation. The chlorophyll a content of an equivalent amount of cyanobacterial suspension to that used in the electrode chamber was measured at 2.13 µg ml^{-1}. What is the net photosynthetic rate of the cyanobacterial suspension, in µmol min^{-1} (mg chlorophyll a)$^{-1}$? (Express your answer to three significant figures.)

Assaying Biomolecules and Studying Metabolism

49 Analysis of biomolecules: fundamental principles

Definitions

Accuracy – the closeness of an individual measurement, or a mean value based on a number of measurements, to the true value.

Concentration range – the range of values from the detection limit to the upper concentration at which the technique becomes inaccurate or imprecise.

Detection limit – the minimum concentration of an analyte that can be detected at a particular confidence level.

Drift – 'baseline' movement in a particular direction: drift can be a problem between analyses (e.g. using a spectrophotometer for colorimetric analysis), or for a single analysis (e.g. when separating biomolecules by chromatography).

Noise – random fluctuations in a continuously monitored signal.

Precision – the extent of mutual agreement between replicate data values for an individual sample.

Quality assurance – procedures to monitor, document and audit a process against a set of criteria, usually defined externally (e.g. ISO 9000), to confirm that the process meets the specified requirements.

Quality control – procedures designed to ensure that the performance of a process meets a particular standard, e.g. analysis of a test sample containing a known amount of a substance, to confirm that the measurement falls within an acceptable range of the true value.

Replicate – repeated measurement.

Selectivity – the extent to which a method is free from interference due to other substances in the sample.

Sensitivity – the ability to discriminate between small differences in analyte concentration.

Validation – the process whereby the accuracy and precision of a particular analytical method are checked in relation to specific standards, using an appropriate reference material containing a known amount of analyte.

Biochemical analysis involves the characterisation of biological components within a sample using appropriate laboratory techniques. Most analytical methods rely on one or more chemical or physical properties of the test substance (the analyte) for detection and/or measurement. There are two principal approaches:

1. Qualitative analysis – where a sample is assayed to determine whether a biomolecule is present or absent. As an example, a blood sample might be analysed for a particular drug or a specific antibody (p. 257), or a bacterial cell might be probed for a nucleic acid sequence (p. 434).
2. Quantitative analysis – where the quantity of a particular biomolecule in a sample is determined, either as an amount (e.g. as g, or mol) or in terms of its concentration in the sample (e.g. as $g l^{-1}$, or $mol\, m^{-3}$). For example, a blood sample might be analysed to determine its pH ($-\log_{10} [H^+]$), alcohol concentration in $mg\, ml^{-1}$, or glucose concentration in $mmol\, l^{-1}$. Skoog et al. (2006) give details of methods.

Your choice of approach will be determined by the purpose of the investigation and by the level of accuracy and precision required (p. 175). Many of the basic quantitative methods described in Chapters 50–53 rely on chemical reactions of the analyte and involve assumptions about the nature of the test substance and the lack of interfering compounds in the sample: such assumptions are unlikely to be wholly valid at all times. If you need to make more exacting measurements of a particular analyte, it may be necessary to separate it from the other components in the sample, e.g. using chromatography (Chapter 44), ultracentrifugation (Chapter 43), or electrophoresis (Chapter 46), and then identify the separated components, e.g. using spectroscopic methods (Chapter 41). However, each stage in the separation and purification procedure may introduce further errors and/or loss of sample, as described in detail for protein purification in Chapter 54.

 KEY POINT In general, you should aim to use the simplest procedure that satisfies the purpose of your investigation – there is little value in using a complex, time-consuming or costly analytical procedure to answer a simple problem where a high degree of accuracy is not required.

Most of the routine methods based on chemical analysis are destructive, since the analyte is usually converted to another substance which is then assayed, e.g. in colorimetric assays of the major types of biomolecules (Table 49.1). In contrast, many of the analytical methods based on physical properties are non-destructive (e.g. the intrinsic absorption and emission of electromagnetic radiation in spectrophotometry, p. 280, or nuclear magnetic resonance techniques, p. 291). Non-destructive methods are often preferred, as they allow the further characterisation of a particular sample. Most biological methods are destructive: bioassays are often sensitive to interference and require validation (p. 238).

Evaluating a new method – results from a novel technique can be compared with an established 'standardised' technique by measuring the same set of samples by each method and analysing the results by correlation (p. 500).

Table 49.1 Some examples of colorimetric assays

Analyte	Reagent/wavelength
Amino acids	ninhydrin/540 nm (p. 352)
Proteins	biuret/520 nm (p. 353)
	Folin–Ciocalteau/600 nm (p. 354)
Carbohydrates	anthrone/625 nm (p. 364)
Reducing sugars	dinitrosalicylate/540 nm (p. 365)
DNA	diphenylamine/600 nm (p. 371)
RNA	orcinol/660 nm (p. 372)

Interpreting results from 'spiked' samples – remember that such procedures tell you nothing about the extraction efficiency of biomolecules from a particular sample, e.g. during homogenisation (Chapter 36).

Criteria for the selection of a particular analytical method:

- the required level of accuracy and precision;
- the number of samples to be analysed;
- the amount of each sample available for analysis;
- the physical form of the samples;
- the expected concentration range of the analyte in the samples;
- the sensitivity and detection limit of the technique;
- the likelihood of interfering substances;
- the speed of analysis;
- the ease and convenience of the procedure;
- the skill required by the operator;
- the cost and availability of the equipment.

Validity and quantitative analysis

Before using a particular procedure, you should consider its possible limitations in terms of:

- measurement errors, and their likely magnitude: these might include processing errors (e.g. in preparing solutions and making dilutions), instrumental errors (e.g. a pH electrode that has not been set up correctly), calibration errors (e.g. converting a digital readout analyte concentration) and errors due to the presence of interfering substances;
- sampling errors: these may occur if the material used for analysis is not representative, e.g. due to biological differences between the individual organisms used in the sampling procedure (p. 480).

Replication will allow you to make quantitative estimates of several potential sources of error: for example, repeated measurements of the same sample can provide information on the precision of the analytical method, e.g. by calculating the coefficient of variation (p. 486), while measurements of several different samples can provide information on biological and sample variability, e.g. by calculating the standard deviation (Chapter 66). Analyses of different sets of samples at different times (e.g. on different days) can provide information on 'between batch' variability, as opposed to 'within batch' variability (based on a single set of analytical data).

The reliability of a particular method can be assessed by measuring 'standards' (sometimes termed 'controls'). These are often prepared in the laboratory by adding a known amount of analyte to a real sample (this is often termed 'spiking' a sample), or by preparing an artificial sample containing a known amount of analyte along with other relevant components (e.g. the major sample constituents and possible interfering substances). In many instances, several standards (including a 'blank' or 'zero') are assayed to construct a 'standard curve', which is then used to convert sample measurements to amounts of analyte (see Chapter 37 for details). Such standard curves form the basis of many routine laboratory assays: while hand-drawn linear calibration curves are sufficient for basic assays, more complex curves are often fitted to a particular mathematical function using a computer program (e.g. bioassays, p. 238). Standards can also be used to check the calibration of a particular method: a mean value based on repeated measurements of an individual 'standard' can be compared with the true value using a modified t test (p. 498, Eqn 67.1), in which there is only one standard error term, i.e. that associated with the measured values.

Validation of a particular method can be important in certain circumstances, e.g. in a forensic science or a clinical biochemistry laboratory, where particular results can have important implications. Such laboratories operate strict validation procedures, including: (i) adherence to standard operating procedures for each analytical method; (ii) calibration of assays using certified reference materials containing known amount of analyte and traceable to a national reference laboratory; (iii) effective systems for internal quality control and external quality assurance (p. 349); (iv) detailed record-keeping, covering all aspects of the analysis and recording of results. Although such rigour is not required for routine analysis, the general principles of standardisation, calibration, assessment of performance and record-keeping are equally valid for all analytical work.

Text reference

Skoog, D.A., Holler, F.J. and Crouch, S.R. (2006) *Instrumental Analysis Principles*. Brooks-Cole, Pacific Grove.

Sources for further study

Center for Drug Evaluation and Research *Guidance for Industry Bioanalytical Method Validation*. Available: http://www.fda.gov/CDER/GUIDANCE/4252fnl.htm
Last accessed: 01/04/07.
[US FDA guidance/details of procedures]

Evans, G. (2003) *Handbook of Bioanalysis and Drug Metabolism*. CRC Press, Boca Raton.

International Union of Biochemistry and Molecular Biology *Homepage*. Available: http://www.iubmb.unibe.ch/
Last accessed: 01/04/07.
[Promotes international standardisation of methods, nomenclature and symbols in biomolecular analysis]

Laboratory of the Government Chemist. *Valid Analytical Measurement Homepage*. Available: http://www.vam.org.uk
Last accessed: 01/04/07.

Venn, R.F. (2000) *Principles and Practice of Bioanalysis*. Taylor and Francis, London.

Wilson, K. and Walker, J. (eds) (2000) *Principles and Techniques of Practical Biochemistry*, 5th edn. Cambridge University Press, Cambridge.

Study exercises

49.1 Check your understanding of the fundamental principles of biomolecular analysis. Distinguish between each of the following pairs of terms:

(a) qualitative analysis and quantitative analysis;
(b) sensitivity and selectivity;
(c) accuracy and precision;
(d) validation and replication;
(e) noise and drift.

49.2 Research sources of calibration standards and reference materials. Using the Web, find a supplier of a validated reference standard for the following:

(a) cortisol in serum;
(b) cotinine in urine.

49.3 Compare equipment specifications. The table (alongside) gives operational details for two glucose meters, A and B.

Which meter has:

(a) the lower detection limit?
(b) the wider linear dynamic range?
(c) the greater accuracy?
(d) the better level of precision?
(e) the smaller sample volume?

Performance of two glucose meters

Aspect	Meter A	Meter B
Units of measurement	mmol l^{-1}	mg dl^{-1}
Detection limit	0.10	1.00
Linearity	to 20.0	to 250.0
Accuracy	±0.050	±1.50
Relative standard deviation (%)	3.5	2.5
Sample volume	40 µl	0.05 ml

50 Assaying amino acids, peptides and proteins

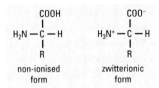

Fig. 50.1 Structure of α-amino acids. R – side chain (see Table 50.1 for examples).

Table 50.1 The 20 amino acids incorporated into protein, grouped according to their side chains

Name	Three letter code	Capital letter code
Aliphatic side chains		
Glycine	gly	G
Alanine	ala	A
Valine	val	V
Leucine	leu	L
Isoleucine	ile	I
Aromatic side chains		
Phenylalanine	phe	F
Tyrosine	tyr	Y
Tryptophan	trp	W
S-containing side chains		
Cysteine	cys	C
Methionine	met	M
Side chains with –OH groups*		
Serine	ser	S
Threonine	thr	T
Basic side chains		
Histidine	his	H
Lysine	lys	K
Arginine	arg	R
Acidic side chains		
Aspartate	asp	D
Glutamate	glu	E
Amide side chains		
Asparagine	asn	N
Glutamine	gln	Q
Cyclic structure (imino acid)		
Proline	pro	P

*tyr also has an OH group.

Definitions

Simple protein – a protein composed entirely of amino acid residues.
Conjugated protein – a protein containing a non-amino acid component (the prosthetic group), in addition to a polypeptide component (the apoprotein).

Proteins are linear polymers, formed by the linkage of α-amino acids (Fig. 50.1) *via* peptide bonds to create polypeptide chains. The 20 amino acids most commonly incorporated into proteins are listed in Table 50.1. Others may be found when a protein is hydrolysed, due to post-translational modification (e.g. hydroxyproline).

 KEY POINT In all amino acids apart from glycine, the α-carbon atom is asymmetrical, resulting in stereoisomers. Most life scientists use the D- and L- nomenclature (Fischer convention) when referring to stereoisomers of amino acids, although the R- and S- convention (p. 333) can be a useful alternative when considering enzymatic reaction mechanisms.

While most of the chemical properties of stereoisomers are identical, enzymes (including those involved in protein synthesis) can distinguish between them and, with the exception of a few bacterial components, only the L-isomers of amino acids are normally used in protein synthesis.

Amino acids and their properties

There is no agreed formal classification of amino acids but, in terms of protein structure and function, the side chains of amino acid residues can be grouped into:

- those that carry a charge (e.g. asp, glu, lys, arg, his) – they may participate in electrostatic interactions and charge–dipole (hydrophilic) interactions;
- those that contain O, N or S (e.g. ser, met, thr, asn, gln, cys, tyr) – they can participate in hydrogen bonding and hydrophilic interactions;
- those that are hydrocarbon in nature (e.g. ala, leu, val, ile, phe, trp) – they may participate in hydrophobic interactions;
- cysteine – this amino acid can form intra- and inter-polypeptide disulphide bridges.

Detection and quantification of amino acids

The primary amino group of amino acids will react with ninhydrin to give a purple-coloured product – this reaction can be used for qualitative assay, e.g. to detect individual amino acids in chromatography (p. 303), or for quantitative colorimetric assay by measuring the absorbance at 570 nm using a spectrophotometer (p. 283). Note that different amino acids give different amounts of coloured product on reaction with ninhydrin, so careful standardisation is needed. The secondary amino groups of the cyclic imino acids hydroxyproline and proline give yellow products with ninhydrin and are assayed at 440 nm.

Proteins and their properties

The diversity of protein structure must be taken into account when considering a suitable analytical method. Globular proteins (e.g. albumin, antibodies) are relatively soluble in dilute salt solutions while fibrous proteins (e.g. collagens, keratin) are typically insoluble, unless hydrolysed (e.g. using 6 mol l^{-1} HCl). Conjugated proteins (e.g. metalloproteins, glycoproteins, nucleoproteins) also include non-amino acid components, which may influence the choice of analytical method.

Definitions

Peptide – a general term for a chain of amino acids linked by peptide bonds, often applied to shorter polymers of up to 50 amino acid residues.

Polypeptide – a single chain of amino acid residues, typically applied to longer polymers of over 50 amino acid residues.

Oligopeptide – a very short chain, typically containing 6–10 amino acid residues.

Chemical properties of proteins and peptides

The features most commonly exploited for quantitative analysis are:

- the peptide bond, e.g. in the biuret reaction (Box 50.1);
- the phenolic group of tyr and the indole group of trp, which react with the oxidising agents phosphotungstic and phosphomolybdic acids in the Folin–Ciocalteau reagent to produce a blue colour. This is combined with the biuret reaction in the Lowry method (Box 50.1);
- dye binding to hydrophobic regions, e.g. Bradford assay (Box 50.1);
- the primary amino groups of lys residues, the guanidino group of arg residues and the N-terminal amino acid residue will react with ninhydrin, allowing colorimetric assay similar to that described above for isolated amino acids. However, a more sensitive method uses the reaction of these groups with fluorescamine to give an intensely fluorescent product, which you can measure by spectrofluorimetry (p. 284).

Box 50.1 Methods of determining the amount of protein/peptide in an aqueous solution

For all of the following methods, the amounts are appropriate for semi-micro cuvettes (1.5 ml volume, path length 1 cm). Appropriate controls (blanks) must be analysed, to assess possible interference (e.g. due to buffers, etc.).

Biuret method

This is based on the specific reaction between cupric ions (Cu^{2+}) in alkaline solution and two adjacent peptide bonds, as found in proteins and peptides. As such, it is not significantly affected by differences in amino acid composition.

1. **Prepare protein standards over an appropriate range** (typically, between 1 and 10 mg ml^{-1}).
2. **Add 1 ml of each standard solution to separate test tubes. Prepare a reagent blank, using 1 ml of distilled water, or an appropriate solution.**
3. **Add 1 ml of each unknown solution to separate test tubes.**
4. **Add 1 ml of biuret reagent (1.5 g $CuSO_4 \cdot 5H_2O$, 6.0 g sodium potassium tartrate in 300 ml of 10 w/v NaOH) to all standard and unknown tubes and to the reagent blank.**
5. **Incubate at 37 °C for 15 min.**
6. **Read the absorbance of each solution at 520 nm against the reagent blank.** The violet colour is stable for several hours.

The main limitation of the biuret method is its lack of sensitivity – it is unsuitable for solutions with a protein content of less than 1 mg ml^{-1}.

Direct measurement of UV absorbance (Warburg–Christian method)

Proteins and peptides absorb EMR maximally at 280 nm (due to the presence of aromatic amino acids) and this forms the basis of the method. The principal advantages of this approach are its simplicity and the fact that the assay is non-destructive. The most common interfering substances are nucleic acids, which can be assessed by measuring the absorbance at 260 nm: a pure solution of protein will have a ratio of absorption (A_{280}/A_{260}) of approximately 1.8, decreasing with increasing nucleic acid contamination. Note also that any free aromatic amino acids in your solution will absorb at 280 nm, leading to an overestimation of protein content. The simplest procedure, which includes a correction for small amounts of nucleic acid, is as follows (use quartz cuvettes throughout):

1. **Measure the absorbance of your solution at 280 nm (A_{280}):** if A_{280} is greater than 1, dilute by an appropriate amount and remeasure (see p. 283).
2. **Repeat at 260 nm (A_{260}).**
3. **Estimate the approximate protein content using the following relationship:**

$$[\text{protein}]\,\text{mg ml}^{-1} = 1.45\,A_{280} - 0.74\,A_{260} \quad [50.1]$$

This equation is based on the work of Warburg and Christian (1942) for enolase. For other proteins, it should not be used for quantitative work, since it gives only a rough approximation of the amount present, due to variations in aromatic amino acid composition.

(continued)

Box 50.1 (continued)

Lowry (Folin–Ciocalteau) method

This is a colorimetric assay, based on a combination of the biuret method, described above, and the oxidation of tyrosine and tryptophan residues with Folin–Ciocalteau reagent to give a blue–purple colour. The method is extremely sensitive (down to a protein/peptide content of $20\,\mu g\,ml^{-1}$), but is subject to interference from a wide range of non-protein substances, including many organic buffers (e.g. TRIS, HEPES), EDTA, urea and certain sugars. The choice of an appropriate standard is important, as the intensity of colour produced for a particular protein/peptide is dependent on the amount of aromatic amino acids present.

1. **Prepare protein standards within an appropriate range for your samples** (the method can be used from 0.02 to $1.00\,mg\,ml^{-1}$).
2. **Add 1 ml of each standard solution to separate test tubes. Prepare a reagent blank, using 1 ml of distilled water, or an appropriate solution.**
3. **Add 1 ml of each of your unknown solutions to separate test tubes.**
4. **Then, add 5 ml of 'alkaline solution' (prepared by mixing 2% w/v Na_2CO_3 in $0.1\,mol\,l^{-1}$ NaOH, 1% w/v aqueous $CuSO_4$ and 2% w/v aqueous NaK tartrate in the ratio 100:1:1.** Mix thoroughly and allow to stand for at least 10 min.
5. **Add 0.5 ml of Folin–Ciocalteau reagent (commercial reagent, diluted 1:1 with distilled water on the day of use).** Mix rapidly and thoroughly and then allow to stand for 30 min.
6. **Read the absorbance of each sample at 600 nm.**

Dye-binding (Bradford) method

Coomassie brilliant blue combines with proteins and peptides to give a dye–protein complex with an absorption maximum of 595 nm. This provides a simple and sensitive means of measuring protein content, with few interferences. However, the formation of dye–protein complex is affected by the number of basic amino acids within a protein, so the choice of an appropriate standard is important. The method is sensitive down to a protein content of approximately $5\,\mu g\,ml^{-1}$ but the relationship between absorbance and concentration is often non-linear, particularly at high protein content.

1. **Prepare protein standards over an appropriate range (between 5 and $100\,\mu g\,ml^{-1}$).**
2. **Add $100\,\mu l$ of each standard solution to separate test tubes. Prepare a reagent blank, using $100\,\mu l$ of distilled water, or an appropriate solution** (note that these small volumes must be accurately dispensed, e.g. using a calibrated pipettor, p. 130).
3. **Add $100\,\mu l$ of your unknown solutions to separate test tubes.**
4. **Add 5.0 ml of Coomassie brilliant blue G250 solution ($0.1\,g\,l^{-1}$).**
5. **Mix and incubate for at least 5 min: read the absorbance of each solution at 595 nm.**

Other methods are less widely used. They include determination of the total amount of nitrogen in solution (e.g. using the Kjeldahl technique) and calculating the protein content, assuming a nitrogen content of 16%. An alternative approach is to precipitate the protein (e.g. using trichloracetic acid, tannic acid or salicylic acid) and then measure the turbidity of the resulting precipitate (using a nephelometer, or a spectrophotometer, p. 283).

Advice on preparing standard (calibration) curves – see p. 252 for practical guidance.

Exploiting the biological properties of proteins and peptides – relevant analytical methods include immunoassay (p. 257), enzymatic analysis (p. 381) and affinity chromatography (p. 309), based on specific biological interactions.

Most assays for proteins and peptides do not give absolute values, but require standard solutions, containing appropriate amounts of a particular protein, to be analysed at the same time, so that a standard curve can be constructed. Bovine serum albumin (BSA) is commonly used as a protein standard. However, you may need an alternative standard if the protein you are assaying has an amino acid composition which is markedly different from that of BSA, depending on your chosen method.

Physical properties of proteins and peptides

The characteristics that can be used for separation and analysis include:

- absorbance of aromatic amino acid residues (phe, tyr and trp) at 280 nm (Warburg–Christian method, Box 50.1);
- prosthetic groups of conjugated proteins – these often have characteristic absorption maxima, e.g. the haem group of haemoglobin absorbs strongly at 415 nm, allowing quantitative assay;

Definitions

Lipoproteins – globular, micelle-like particles consisting of a non-polar core of triacylglyceryl and cholesteryl esters surrounded by a coating of proteins, phospholipids and cholesterol: involved in the transport of lipids in blood.

Glycoproteins – compounds consisting of proteins covalently attached to carbohydrate residues in post-translational modifications of the protein constituents.

Purifying a protein – the practical procedures involved in the purification of a particular protein from a biological sample are considered in detail in Chapter 54.

SAFETY NOTE Hydrolysis of proteins – this is usually achieved using 6 mol l^{-1} HCl at 110°C for 24–72 h: care is required when working with hot acids.

Investigating the secondary and tertiary structure of a protein – computer-based molecular modelling programs can be used to predict the 3D conformation of a protein from information based on the primary structure (Chapter 11).

- density – most proteins have a density of 1.33 kg l^{-1}, but lipoproteins have a lower density. This can be used to separate lipoproteins from other classes of protein, or to subdivide the various classes of lipoproteins;
- net charge – proteins and peptides differ in the types and number of amino acids with ionisable groups in their side chains. The ionisation of these side chains is pH dependent, resulting in variation of net charge with pH (p. 320). This property is exploited in the techniques of electrophoresis (Chapter 46), isoelectric focusing (Chapter 47) and ion-exchange chromatography (Chapter 44);
- water solubility and surface hydrophobicity – these are exploited in salt fractionation (p. 376) and hydrophobic interaction chromatography (p. 308).

Determining the primary structure of a protein

The term 'primary structure' refers to the linear sequence of amino acid residues along the polypeptide (p. 353): ultimately this determines the 3D shape of the molecule and hence its biological properties.

Amino acid analysis

Once a particular protein has been purified (Chapter 54), the first step in determining its primary structure is to hydrolyse the polypeptide and then determine the constituent amino acids. This can be achieved by several techniques, including thin layer chromatography and high performance liquid chromatography (Chapter 44). Often a dedicated amino acid analyser is used for quantitative assays: a polystyrene resin-based cation exchange column is used to separate the amino acids on the basis of ion-exchange and hydrophobic interactions.

Sequence analysis

Having obtained the amino acid composition of the protein, the next step is to determine the order of the amino acid residues along the polypeptide chain. This may be achieved by:

1. Cleavage of the polypeptide at specific peptide bonds resulting in smaller fragments (peptides); e.g. using cyanogen bromide (CNBr), which cleaves peptide bonds formed by the carboxyl group of met, or by proteolytic enzymes, e.g. trypsin (cleaving after lys and arg residues) and chymotrypsin (cleaving after phe, tyr and trp).
2. Separation of the peptides, e.g. by column chromatography.
3. Determination of the sequence of each peptide by a process called Edman degradation. This involves the selective removal of the N-terminal amino acid by treating the fragments with phenyl isothiocyanate ($C_6H_5N{=}C{=}S$) followed by acid hydrolysis; the substituted phenylthiohydantoin formed can then be identified by chromatography. The process can be repeated as many as 50 times, releasing the N-terminal residue on each occasion, and allowing the sequence of each fragment to be established.
4. Matching of the sequenced, overlapping peptide fragments, to determine the overall sequence of amino acid residues in the original protein.

Text reference

Warburg, O. and Christian, W. (1942) *Isolierung und Kristallisation des Garungferments Enolase. Biochemische Zeitschrift*, **310**, pp. 384–421.

Sources for further study

Dey, P.M., Harbourne, J.B. and Rogers, L.E. (1997) *Methods in Plant Biochemistry: Amino Acids, Proteins and Nucleic Acids*. Academic Press, New York.

White, J.S. and White, D.C. (2002) *Proteins, Peptides and Amino Acids Sourcebook*. Humana Press, Totowa, NJ.

Whitford, D. (2005) *Proteins: Structure and Function*. Wiley, Chichester.

Study exercises

50.1 Test your knowledge of the basis of protein assays. Match up each of the following methods with the corresponding underlying reaction:
Method: (a) Warburg–Christian; (b) Bradford; (c) Lowry (Folin–Ciocalteau); (d) Biuret.
Reaction: (i) Coomassie Blue binding; (ii) Copper(II) ion–peptide bond interaction; (iii) Absorption of UV radiation by aromatic amino acids; (iv) A combination of the copper(II) ion–peptide bond reaction and the oxidation of tyrosine and tryptophan residues.

50.2 Determine the amount of amino acid in samples, using a 'standard curve' (see also study exercise 37.2 for an equivalent exercise for protein assay). The following data are for a set of calibration standards, measured using the ninhydrin method, which is non-linear at high amino acid concentrations.

Absorbance of calibration standards

Amino acid concentration in assay tube ($\mu g\,ml^{-1}$) (total assay volume 5 ml)	Absorbance of standard (A_{570})
0.0	0.000
2.0	0.151
4.0	0.305
6.0	0.481
8.0	0.623
10.0	0.751
12.0	0.866
14.0	0.935
16.0	0.977
18.0	1.012
20.0	1.028

(a) Draw a calibration curve by hand using graph paper and estimate the concentration of amino acid in three different test samples giving the following A_{570} values: (i) 0.341 (ii) 0.686 (iii) 0.969

(b) Assuming that these A_{570} values were obtained using the following dilutions and sample volumes in a total assay volume of 5 ml for all samples and standards alike, calculate the amino acid concentration of each of the three original samples, consisting of: (i) 100 μl of undiluted sample; (ii) 50 μl of a 10-fold dilution of sample; (iii) 80 μl of a 25-fold dilution of sample. Give all answers in $mg\,ml^{-1}$, to one decimal place.

50.3 Determine the amino acid sequence of a polypeptide from cleavage fragment data. The following data represent the results for a polypeptide of unknown amino acid sequence, when a subsample of the polypeptide was digested to completion with trypsin while another subsample was cleaved with cyanogen bromide (CNBr). In both cases, four peptide fragments were produced. These peptide fragments produced were then subjected to Edman degradation, and the following results were obtained for each peptide, in order of successive rounds of Edman degradation:

Tryptic digestion products: fragment (i): val asn lys; fragment (ii): ala gly met ser arg; fragment (iii): trp phe met ala ala; fragment (iv): his gly met ala glu lys

CNBr cleavage products: fragment (i): ala ala; fragment (ii): his gly met; fragment (iii): ser arg trp phe met; fragment (iv): ala glu lys val asn lys ala gly met

What is the sequence of the intact polypeptide?

51 Assaying lipids

Table 51.1 Some examples of fatty acids

No. of carbon atoms	Systematic name	Trivial name
Saturated fatty acids (no C=C bonds)		
12	n-Dodecanoic	Lauric
14	n-Tetradecanoic	Mystiric
16	n-Hexadecanoic	Palmitic
18	n-Octadecanoic	Stearic
20	n-Eicosanoic	Arachidic
22	n-Docosanoic	Behenic
Mono-unsaturated fatty acids (one C=C bond)		
12	cis-9-Dodecenoic	Lauroleic
14	cis-9-Tetradecenoic	Myristoleic
16	cis-9-Hexadecenoic	Palmitoleic
18	cis-9-Octadecenoic	Oleic
20	cis-9-Eicosenoic	Gadoleic
22	cis-9-Docosenoic	Erucic

Note: palmitic, stearic, palmitoleic and oleic acids are quantitatively the most common fatty acids in the majority of organisms. Major polyunsaturated fatty acids include linoleic acid (C_{18}, two double bonds), linolenic acid (C_{18}, three double bonds) and arachidonic acid (C_{20}, four double bonds).

The term 'lipid' is used to describe a broad group of compounds with a wide variety of chemical structures, physical properties and biological functions. Lipids in biological systems are often combined with either proteins (e.g. in lipoproteins) or polysaccharides (e.g. in lipopolysaccharides). This chapter provides only a brief outline of the techniques used in lipid analysis (for further details, see Gunstone et al., 1994).

KEY POINT The defining feature of lipids is their relative insolubility in water: consequently, they are extracted from biological material using organic solvents, e.g. acetone, ether and chloroform. Because of their diversity and complexity, they are often referred to by more than one name, and the various types of non-systematic names can be confusing.

Biological lipids are often subdivided into two main types, each of which contains fatty acids as a major structural component:

Simple or neutral lipids

These are esters of fatty acids and an alcohol (e.g. Fig. 51.1). Fatty acids are straight-chain carboxylic acids, typically with an even number of carbon atoms and chain lengths of C_{12}–C_{22}, which may be saturated or unsaturated (Table 51.1). The greater the chain length and the fewer the number of double bonds, the higher the melting point, making most long chain saturated fatty acids solids at room temperature. Glycerol is the most common alcohol found in simple lipids, though higher M_r alcohols occur in waxes, and cyclic alcohols (sterols, e.g. cholesterol) occur in bile acids, steroid hormones and vitamins (e.g. vitamin K). While glycerol is a liquid at room temperature, cholesterol remains solid up to 150 °C.

Neutral fats (triglycerides or triacylglycerol) are esters of fatty acids and glycerol, as shown in Fig. 51.1. Many animal triglycerides ('fats') contain mainly saturated fatty acids and are solids at room temperature, while plant triglycerides ('oils') often have shorter chain lengths and a greater degree of unsaturation and are liquids at room temperature. Waxes are esters of fatty acids with alcohols of higher M_r than glycerol. The major biological functions of simple lipids include (i) energy storage, e.g. oils in plant seeds, (ii) insulation, e.g. subcutaneous fat deposits in whales, and (iii) waterproofing, e.g. waxes in the cuticles of plant leaves.

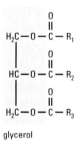

Fig. 51.1 General structure of a neutral lipid: the structure of a triacylglycerol. R = remainder of fatty acid.

Complex, compound or conjugated lipids

These are acyl esters of glycerol, or the amino alcohol sphingosine, that also include a hydrophilic group (e.g. a phosphoryl or carbohydrate group). They are often described in terms of this hydrophilic group, e.g. phospholipids contain a phosphoryl group while glycolipids contain a carbohydrate group.

KEY POINT An important feature of complex lipids is their amphipathic nature, i.e. each molecule has a polar (hydrophilic) and a non-polar (hydrophobic) region.

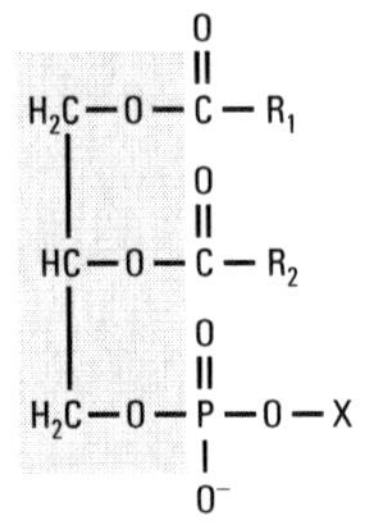

Fig. 51.2 General structure of a phospholipid. R — remainder of fatty acid, X — hydrophilic group, e.g. $-CH_2CH_2NH_3^+$ in phosphatidyl ethanolamine and $-CH_2CH_2N(CH_3)_3^+$ in phosp3hatidyl choline.

SAFETY NOTE Working with solvents – chloroform and benzene must be used with care, due to their high toxicity: they are often replaced by dichloromethane and toluene respectively. All mixing and pouring steps should be carried out in a spark-free fume cupboard. Note that ethers may form explosive peroxides on prolonged storage. Lipid extracts extracted in flammable solvents must be stored in a spark-proof refrigerator, not in routine lab fridges.

Definition

Emulsion – a colloidal mixture where one liquid is dispersed (but not dissolved) in another liquid, e.g. lipids often aggregate to form micelles in aqueous solutions.

The two major types of complex lipid are:

1. Phospholipids – the most common types (phosphoglycerides) are based on phosphatidic acid, with two fatty acids esterified to glycerol. Most phospholipids also contain a hydrophilic amino alcohol or a similar group, attached to the phosphoryl group (Fig. 51.2). The principal phospholipid classes are: (i) phosphatidyl cholines (or lecithins), which form stable emulsions with water and dissolve completely on addition of bile salts – these are insoluble in acetone, a feature that enables lecithins to be separated from most other lipids; (ii) phosphatidyl ethanolamines (or cephalins) – unlike lecithins, they are insoluble in ethanol and methanol; (iii) phosphatidyl serine; (iv) phosphatidyl inositol; and (v) plasmalogens.

2. Sphingolipids – these incorporate the amino dialcohol, sphingosine, rather than glycerol. Fatty acids are linked to sphingosine *via* an amide bond to form ceramides, which include: (i) cerebrosides (glycosphingolipids); (ii) sulphatides (sulphated cerebrosides); (iii) gangliosides (glycosphingolipids containing sialic acid residues); and (iv) phosphosphingolipids, including sphingomyelins, which are esters of a ceramide and phosphoryl choline.

Complex lipids have important roles in biological membranes: phospholipids are major structural components while sphingolipids are involved in cell–cell recognition and similar membrane features, e.g. glycosphingolipids are determinants of human ABO blood groups.

Extraction and analysis of lipids

Solvent extraction

Lipid extraction is carried out in a suitable organic solvent, using a homogeniser to disrupt cell and tissue structure (Chapter 36). There are two possible approaches:

1. Total extraction of all lipids, followed by separation of the different lipid classes. A commonly used solvent mixture is methanol: chloroform: water $(2:1:0.5, v/v/v)$. Adding an equal volume of aqueous 1% w/v NaCl solution to the extract results in the formation of two layers: the lipids are present in the lower layer, while the upper aqueous layer contains other biomolecules (e.g. proteins). Hexane: isopropanol $(3:2, v/v)$ can be used instead of chloroform: methanol, since it is less toxic and it dissolves very little non-lipid material.

2. Selective lipid extraction. Neutral lipids, e.g. within storage tissue, can be extracted by relatively non-polar solvents including hexane, diethyl ether and chloroform. Extraction of membrane lipids (e.g. phospholipids) requires disruption of the membrane using more polar solvents (e.g. methanol or ethanol), with selective precipitation by adding cold, dry acetone. Glycolipids can be extracted using acetone.

The principal disadvantages of solvent extraction methods include: (i) the requirement for large volumes of potentially hazardous solvents; and (ii) the possible formation of an emulsion, with incomplete extraction of component lipids.

Adsorption chromatography

Silica gel, octadecylsilane-bonded silica or ion-exchange resins can be used to bind solvent-extracted lipids by a combination of polar, ionic and

Preventing oxidative rancidity – all lipids will oxidise (become rancid) when exposed to air in daylight, due to hydrolysis and/or photo-oxidation. Keep extracted lipids in the dark, and add an antioxidant such as butylated hydroxytoluene for longer-term storage. Minimise oxidation by flushing vessels with nitrogen gas and bubbling solvents with (oxygen-free) nitrogen during analysis.

van der Waals forces. In practice, a glass column is packed with a slurry of adsorbent in an appropriate organic solvent, and the lipid extract (dissolved in the same solvent) is applied to the top of the column. Lipids can then be selectively eluted; a mixture can be broadly separated into neutral lipids, glycolipids and phospholipids using solvents of increasing polarity, e.g. chloroform → acetone → methanol. It is possible to further separate the lipid subfractions on silica gel columns, as follows:

- Neutral lipids can be separated and eluted using hexane containing increasing proportions of diethyl ether (0 → 100% v/v) in the order: hydrocarbons, cholesterol esters, triacylglycerols, free fatty acids, cholesterol, diacylglycerols, then monoacylglycerols.
- Glycolipids and sulpholipids can be separated by first eluting the glycolipids with chloroform:acetone (1:1 v/v), then using acetone to elute the sulpholipids.
- Phospholipids can be eluted using chloroform containing increasing proportions of methanol (5% → 50% v/v) in the order: phosphatidic acid, phosphatidyl ethanolamine, phosphatidyl serine, phosphatidyl choline, phosphatidyl inositol, sphingomyelin.

Thin layer chromatography (TLC) of lipids

This can be used to separate lipid mixtures into their constituents, or to quantify particular lipids, as part of an analytical procedure. The principle of the technique is described in Chapter 44.

Understanding silica gel codes – silica gel is manufactured as particles of 10–15 μm diameter, containing pores of diameter 40, 60, 80, 100 or 150 Å (Ångstrom, p. 178), where Å = 10^{-10} m (thus, silica gel 60 has a pore size of 60 Å). Silica gel G contains calcium sulphate to assist binding to the glass support, while silica gel H has no binder.

Silica gel G60 is the most frequently used stationary phase, acting as a polar absorbent. When the separation is carried out with a non-polar mobile phase, non-polar lipids will migrate more rapidly (i.e. they will have high R_F values, p. 303), while polar lipids will migrate more slowly. By increasing the polarity of the solvent, the R_F values of the polar lipids can be increased. Your choice of solvent will depend on the lipids in the extract:

- For a broad range of neutral lipids, a typical solvent system is hexane:diethyl ether:glacial acetic acid at 80:20:2 (v/v/v), separating in the following order of decreasing R_F: steryl esters, wax esters, fatty acids, methyl esters, triacylglycerols, fatty acids, fatty alcohols, sterols, 1,2-diacylglycerols, monoacylglycerols.
- For polar lipids, silica gel H is preferred, as silica gel G prevents the separation of acidic phospholipids. Most solvent systems for polar lipids are based on chloroform:methanol:water, e.g. at 65:25:4 (v/v/v). In this system, the relative order of migration is: monogalactosyldiacylglycerol, cerebrosides, phosphatidic acid, cardiolipin, lysophosphatidyl ethanol-amine, phosphatidyl ethanolamine and digalactosyldiacylglycerol, sulphatides, phosphatidyl choline, phosphatidyl inositol, sphingomyelin and phosphatidyl serine.

 SAFETY NOTE Working with lipid stains – these often contain corrosive or toxic reagents, so all manipulations should be performed in a fume cupboard.

 KEY POINT No single solvent system will completely separate all lipid components in a single TLC procedure. However, 2D TLC may separate up to 200 individual constituents from a sample.

SAFETY NOTE *Spraying TLC plates with sulphuric acid solutions – this must be carried out on a suitable support (e.g. disposable paper or card) within a fume hood or a commercial spray cabinet. Wear safety glasses and gloves throughout the procedure.*

Lipids can also be separated by reversed-phase TLC (RP-TLC), where the silica gel is made non-polar (e.g. by silanisation), and highly polar solvents are used as the mobile phase: here, the polar lipids will have the highest R_F values and the non-polar lipids the lowest R_F values. Lipids separated by TLC can be located by staining. Several stains are non-specific, enabling almost all types of lipid to be visualised, while others will locate particular lipid classes. Staining can be carried out by immersion of the TLC plate in the stain, or by spraying. Non-specific staining methods include:

- Spraying with iodine solution (e.g. 1–3% w/v in chloroform) – most lipids appear as brown spots on a yellow background, though glycolipids stain weakly by this method.
- Treatment with a strong oxidising agent followed by charring (e.g. spray with 5% v/v sulphuric acid in ethanol, followed by heating in an oven at 180 °C for 30–60 min) – lipids appear as black deposits; the detection limit is 1–2 µg. Scanning densitometry (Chapter 45) can provide quantitative information.
- Fluorescent stains (e.g. the widely used 2′,7′-dichlorofluorescein, DCF, at 0.1–0.2% w/v in ethanol). Under UV, lipids appear as bright yellow spots on a yellow–green background; the limit of detection is 5 µg. Other fluorescent stains such as 1-anilo-8-naphthalene sulphonate (ANS, 0.1% w/v in water) can detect ng quantities of lipid.

Quantitative assay of lipids and their components

While TLC can be used to quantify particular lipids, it is more common to assay the compounds released on hydrolysis of simple or complex lipids, namely the alcohols, fatty acids or other components, rather than the native lipid. Alkali hydrolysis of lipids containing fatty acids results in the formation of a soap (i.e. saponification), e.g. the incubation of tripalmitin with KOH yields potassium palmitate and glycerol. Acid hydrolysis of triacylglycerols releases 'free' fatty acids and glycerol.

Basic information about the relative size of the fatty acid component of oils and fats is given by the saponification value, determined by titration against 0.8 mol l^{-1} KOH; the lower the saponification value, the higher the M_r of the fatty acids. The degree of unsaturation of the fatty acids is given by the iodine number; the higher the iodine number, the greater the content of unsaturated fatty acids.

To obtain the iodine number for a particular fat, the free iodine remaining after reaction is titrated against 0.1 mol l^{-1} sodium thiosulphate (Na$_2$S$_2$O$_3$) using a trace amount of starch as an indicator, giving a titration volume for the test solution, V_t. A blank containing no fat is also titrated, to establish the volume of sodium thiosulphate required to titrate the initial free iodine, V_o. This allows the amount of iodine that reacts with the fat to be calculated according to the formula:

$$\text{iodine number (in g)} = \frac{1.27(V_0 - V_t)}{m} \quad [51.1]$$

where m is the mass of test fat (in g), and V_o and V_t are expressed in ml.

Definitions

Saponification – the hydrolysis of an ester under alkaline conditions, to form an alcohol and the salt of the acid.

Saponification value – the amount of KOH (in mg) required to completely saponify 1 g of fat. Typically within the range 150–300.

Iodine number – the amount of iodine (g) absorbed by 100 g of fat, due to the reaction of iodine with C=C bonds within the fat. Typically within the range 30–200.

Example A 0.15 g sample of cod liver oil was titrated against 0.1 mol l^{-1} Na$_2$S$_2$O$_3$, giving a titration volume of 31.7 ml (V_t), compared with a blank of 49.8 ml (V_o). Substituting into Eqn [51.1] gives an iodine number of [1.27 (49.8 − 31.7)] ÷ 0.15 = 153 (to three significant figures).

Measurement of glycerol content

Glycerolipids can be quantified by measuring the glycerol released on hydrolysis. A widely used method involves a coupled enzyme assay (p. 382), with the following reactions:

$$\text{triacylglycerol} \xrightarrow{\textit{lipase}} \text{glycerol} + \text{fatty acids} \quad [51.2]$$

$$\text{glycerol} + \text{ATP} \xrightleftharpoons{\textit{glycerol kinase, } Mg^{2+}} \text{glycerol-3-phosphate} + \text{ADP} \quad [51.3]$$

$$\text{ADP} + \text{phosphoenolpyruvate} \xrightleftharpoons{\textit{pyruvate kinase}} \text{ATP} + \text{pyruvate} \quad [51.4]$$

$$\text{pyruvate} + \text{NADH} + \text{H}^+ \xrightleftharpoons{\textit{lactate dehydrogenase}} \text{lactate} + \text{NAD}^+ \quad [51.5]$$

The glycerol concentration is determined by measuring the decrease in absorbance of NADH at 340 nm (A_{340}), compared to that of a blank with no added lipase.

Understanding coupled enzyme assays based on NADH/NAD$^+$ interconversion – these are explained in more detail on p. 382, while the procedure required to convert changes in A_{340} to [NADH] is given on p. 282.

Measurement of cholesterol content

Cholesterol in cholesterol esters can be estimated after hydrolysis to free cholesterol using cholesterol esterase. The subsequent assay is as follows:

$$\text{cholesterol} + \text{O}_2 \xrightarrow{\textit{cholesterol oxidase}} \text{cholest-4-en-3-one} + \text{H}_2\text{O}_2 \quad [51.6]$$

The hydrogen peroxide produced as a result of the action of cholesterol oxidase can be measured amperometrically or colorimetrically, via the peroxidase-catalysed reaction:

$$2\text{H}_2\text{O}_2 + \text{phenol} + \text{4-aminoantipyrene} \xrightarrow{\textit{peroxidase}} \text{quinoneimine} + 4\text{H}_2\text{O} \quad [51.7]$$

This is the basis of many commercially available cholesterol testing kits. An alternative approach is to measure the cholest-4-en-3-one directly, *via* its absorbance maximum at 240 nm.

Alternative approaches to measuring H_2O_2 – the hydrogen peroxide produced from the oxidation of cholesterol in [Eqn 51.6] can be converted to water and oxygen by the enzyme catalase ($2H_2O_2 \rightarrow 2H_2O + O_2$) and the oxygen produced can be quantified, e.g. using an oxygen electrode (p. 340).

Gas chromatography (GC) of lipids and lipid components

This technique is used for the quantitative and qualitative analysis of a broad range of lipids. Volatile lipids may be analysed without modification, while non-volatile lipids must first be converted to a more volatile form, either by degradation (e.g. phospholipids), or derivatisation (p. 306). The most effective GC columns are support-coated open tubular (SCOT) capillary columns, with a thin film (0.1–10 μm) of the stationary phase coated onto the internal wall (p. 305). The choice of stationary phase depends on the components to be separated, e.g. some non-polar stationary phases cannot resolve methyl esters of saturated and mono-unsaturated fatty acids. Many stationary phases are based on silicone greases or polysiloxane, ranging from the non-polar dimethyl polysiloxanes (e.g. OV-101) to the polar trifluoropropyl methyl polysiloxanes (e.g. OV-210). Other polar stationary phases are based on polyethylene glycol (e.g. Carbowax 20M).

Example Fatty acids are derivatised to their methyl esters by boiling for 2–3 min with boron trifluoride (BF$_3$) solution (14% w/v in methanol), prior to GC analysis, to increase their volatility and stability in the gas phase. The separated components are normally detected by flame ionisation (p. 315).

Text reference

Gunstone, F.D. and Harwood, J.L. (2007) *The Lipid Handbook*, 3rd edn. CRC Press, Boca Raton.

Sources for further study

Anon. *Cyberlipid Center Homepage.* Available: http://www.cyberlipid.org Last accessed: 01/04/07.

Byrdwell, W.C. (2006) *Modern Methods for Lipid Analysis by Liquid Chromatography/Mass Spectrometry and Related Techniques.* American Oil Chemists Society, Boulder, CO.

Gunstone, F.D. and Padley, F.B. (1997) *Lipid Technologies and Applications.* Marcel Dekker, New York.

Gurr, M.I., Harwood, J.L. and Frayn, K.N. (2002) *Lipid Biochemistry.* Blackwell, Oxford.

Hemming, F.W., Hawthorne, J.N. and White, D.A. (1996) *Lipid Analysis.* Bios, Cambridge.

Study exercises

51.1 Consider how to avoid oxidative rancidity. What practical steps could you take to avoid lipid oxidation during extraction and analysis?

51.2 Use the Web to discover more about lipid structure and analysis. Visit the Cyberlipid Center website (at http://www.cyberlipid.org/index.htm), a resource giving details of a wide range of lipid structures, including triglycerides, phospholipids and glycolipids: (a) identify the main lipid components of beeswax and (b) suggest which separation methods might be appropriate for the analysis of these components.

51.3 Derive the equation for calculation of the iodine number of a lipid. Given that the following reactions (i)–(iii) are involved in the volumetric analysis for calculating the iodine number of a lipid, show how Eqn 51.1 is derived:

(i) $-CH=CH- + ICl \rightarrow -(I)CH-CH(Cl)-$

(ii) $ICl + KI \rightarrow HCl + I_2$

(iii) $I_2 + 2Na_2S_2O_3 \rightarrow 2NaI + Na_2S_4O_6$

51.4 Interpret a GC chromatogram for a lipid mixture. The chromatogram shown in the figure below represents a GC analysis of triacylglycerols in butter fat. There are 15 peaks, corresponding to lipids with even numbers of acyl carbon atoms between 26 and 54. How would you expect the lipids of different hydrocarbon lengths to be associated with the retention times of the different peaks? (Give a brief explanation of the reasoning underlying your expectation.)

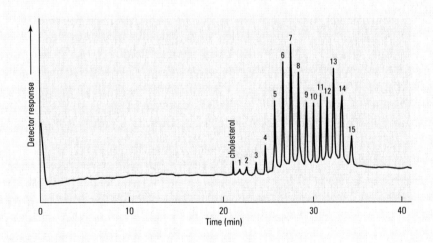

52 Assaying carbohydrates

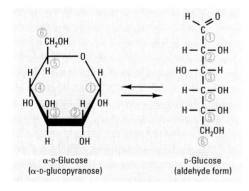

Fig. 52.1 Ring and straight chain forms of glucose. Note that β-D-glucose has H and OH groups reversed at C-1 in the pyranose form.

These are compounds with a formula based on $C_x(H_2O)_x$. They play a key role in energy metabolism, and are essential constituents of cell walls and membranes. They may exist individually, or as heteropolymers, e.g. linked to protein in glycoproteins (where carbohydrates form the minor component), or proteoglycans (where they form the major component).

> **KEY POINT** The identification and quantitative analysis of carbohydrates in biological samples can be difficult, due to structural and physicochemical similarities between related compounds. Several routine analytical methods cannot distinguish between isomeric forms of a particular carbohydrate.

Monosaccharides

The simplest carbohydrates are the monosaccharide sugars, which are polyhydroxy aldehydes and polyhydroxy ketones (so-called aldoses and ketoses), typically with three to seven carbon atoms per molecule. The common names end with the suffix '-ose', e.g. glucose (Fig. 52.1), which contains six carbon atoms and is a hexose. The simplest aldose is the three-carbon compound (triose) glyceraldehyde (Fig. 52.2a), and the other aldoses can be considered to be derived from glyceraldehyde by the addition of successive secondary alcohol groups (H–C–OH). In a similar manner, all ketoses can be considered to be structurally related to the triose dihydroxyacetone (Fig. 52.2b).

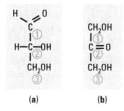

Fig. 52.2 Structure of (a) glyceraldehyde and (b) dihydroxyacetone.

Monosaccharides are assigned as D or L isomers according to a convention based on D-glyceraldehyde as the reference compound, with the carbon atoms numbered from the end of the chain containing the reactive group. Most of the carbohydrates found in biological systems are D isomers. While trioses and tetroses exist in linear form, pentoses and larger monosaccharides can be represented either as a linear structure (Fischer form) or as a ring structure (Haworth form), as in Fig. 52.1. The cyclic structure results from the reaction of the carbonyl group (aldehyde or ketone) at one end of the molecule with a hydroxyl group at the other end of the chain, forming a hemiacetal or hemiketal, as shown for glucose (Fig. 52.1) and fructose (Fig. 52.3).

The formation of a hemiacetal or hemiketal creates another asymmetric carbon atom, so that two ring forms exist – one with the –OH group on this asymmetric carbon positioned below the plane of the ring (α) and another with the –OH group above the plane of the ring (β). These different isomeric forms are called anomers and are particularly difficult to separate by chromatographic methods. The six-membered ring structure shown for glucose has a structure similar to that of pyran and is termed glucopyranose. A few monosaccharides have a five-membered ring structure similar to that of furan, e.g. fructofuranose.

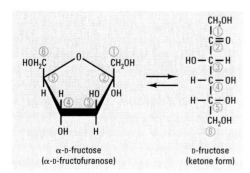

Fig. 52.3 Ring and straight chain forms of fructose. Note that β-D-fructose has OH and CH₂OH groups reversed at C-2 in the furanose form.

While the open chain form is present in very small amounts in aqueous solution it has a very reactive carbonyl group which is responsible for the reducing properties of sugars: several analytical methods are based on this feature (see p. 364).

Studying the biological roles of glycosides – these compounds are often formed in plants during the detoxification of certain compounds, or to control plant hormone activity.

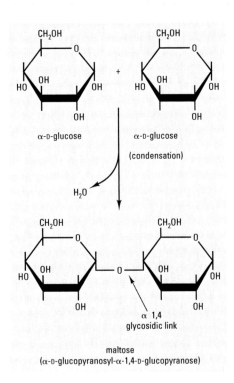

Fig. 52.4 Formation of a glycosidic link ($\alpha 1 \rightarrow 4$) in the disaccharide maltose.

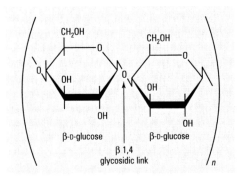

Fig. 52.5 The repeating structure of cellulose ($\beta 1 \rightarrow 4$ glycosidic links) $n \approx 5000$.

 SAFETY NOTE Using the anthrone method – note that this involves hot H_2SO_4 – wear gloves, carry out the heating step in a fume hood and rinse all spillages thoroughly with excess water.

Glycosides

These are formed when a covalent bond, or glycosidic link, is created between the hemiacetal or hemiketal group of a carbohydrate (e.g. a monosaccharide) and the hydroxyl group of a second compound (e.g. a polyhydroxy alcohol, or another monosaccharide). If the sugars are not joined *via* their reactive groups, they will still show reducing properties, as in the disaccharide maltose (Fig. 52.4). In contrast, sucrose is a non-reducing disaccharide, since the hemiacetal and hemiketal groups of glucose and fructose are involved in the formation of the glycosidic link.

Polymeric carbohydrates have important biological roles – oligosaccharides contain up to 10 sugars, while polysaccharides are larger polymers with M_r of up to several million, e.g. glycogen, amylopectin, cellulose. The structure of polysaccharides is not fully defined, in contrast to proteins (Chapter 50) and nucleic acids (Chapter 53). Polysaccharide structure is the result of the separate actions of a number of biosynthetic enzymes, producing a range of molecules that may vary in the number of monosaccharide residues and the types of glycosidic bond. The terminology used to describe the glycosidic links in such compounds denotes the anomer involved (α or β) and the C atoms involved in the link, e.g. $\alpha 1 \rightarrow 4$ in maltose (Fig. 52.4), $\beta 1 \rightarrow 4$ in cellulose (Fig. 52.5).

Extraction and analysis of carbohydrates

While most low M_r carbohydrates are soluble in water, ethanol:water (80% v/v) is more often used, since polysaccharides and other biological macromolecules are insoluble in aqueous ethanol. In contrast, polysaccharides are more diverse, and the isolation procedures vary greatly, e.g. boiling water, mild acid or mild alkali can be used to solubilise storage polysaccharides, e.g. starch, while more vigorous treatment is required for structural polysaccharides, e.g. 24% w/v KOH for cellulose. Among the techniques used to purify extracted polysaccharides are gel permeation chromatography (p. 307) and ultracentrifugation (p. 299).

Identification and quantification of carbohydrates

This can be achieved by a variety of procedures, including:

Chemical methods Several monosaccharide assay methods are based on the reductive capacity of the aldehyde or ketone groups (Table 52.1). A widely used method for quantitative analysis is that based on reduction of 3,5-dinitrosalicylate. This method is also suitable for glycosides, provided the reducing carbonyl groups are not involved in the glycosidic links. Certain polysaccharides react with iodine in acid solution to form coloured complexes: starch gives a blue colour, while glycogen gives a red–brown colour. The anthrone method (typically using $0.1\,g\,l^{-1}$ anthrone in H_2SO_4, at $100\,°C$ for $10\,min$, then assayed at $630\,nm$) is an alternative assay for estimating total carbohydrate content. Careful choice of calibration standards is required for quantitative work – try to use a standard that matches the likely composition of the samples. While such chemical methods can give a general indication of the relative amount of carbohydrate in a sample, they can provide little useful information on the types of carbohydrate present.

Enzymatic methods These offer a higher degree of specificity in monosaccharide assay, and may allow differentiation between the

Table 52.1 Methods for carbohydrate analysis

Method	Principle	Comments
Chemical assay		
Benedict's test (and Fehling's test)	Reduction of Cu^{2+} to Cu^{+} in presence of reducing sugar; alkaline solution plus heat results in formation of Cu_2O; solution turns from blue, through yellow, to red.	Usually presence/absence test; quantitative assay involves measurement of Cu_2O formed.
Dinitrosalicylate (DNS)	Reduction of DNS (yellow) to orange-red derivative; alkaline solution plus heat (100 °C, 10 min).	Quantitative: read at 540 nm.
Enzymatic assay		
Glucose oxidase (coupled reaction)	β-D-glucose + O_2 $\xrightarrow{\text{glucose oxidase}}$ gluconic acid + H_2O_2 H_2O_2 + reduced dye $\xrightarrow{\text{peroxidase}}$ H_2O + oxidised dye	Mutarotation allows reaction to reach completion; hydrogen peroxide formed may be measured using peroxidase; ABTS* is a suitable dye. Assay at 437 nm.

*ABTS = 2,2'-azino-di-[3-ethylbenzthiazoline]-6-sulphonate.

Other applications of glycosidase assays – some microbial identification schemes are based on the detection of specific glycosidase enzymes, e.g. β-glucuronidase for *E. coli* (p. 383).

Measuring carbohydrate migration in TLC systems – the distance migrated by glucose is taken as a reference ($R_F = 100$), and the migration of other carbohydrates is given as the R_G value, where R_G is:

$$\frac{\text{distance moved by carbohydrate}}{\text{distance moved by glucose}} \times 100$$

various stereoisomeric and anomeric forms. The glucose-specific method based on glucose oxidase shown in Table 52.1 also forms the basis of the glucose electrode (see Chapter 48). Hydrogen peroxide may be assayed using peroxidase and a suitable chromogenic substrate, or by electrochemical methods, e.g. using an amperometric sensor. Alternatively, the consumption of oxygen in the initial reaction can be measured using an oxygen electrode (p. 339). Glycosidases can be used to hydrolyse specific disaccharides or polysaccharides into their constituent monosaccharides, which can then be identified and quantified, e.g. α-glucosidase, which hydrolyses the $\alpha(1 \rightarrow 4)$ linkage between the glucose residues in maltose (Fig. 52.4). However, you should note that it is rare for such enzymes to show absolute specificity for a particular substrate, so you should be alert to the possibility of interference due to related compounds in the sample, or to impurities in the enzyme preparation.

Chromatographic methods Traditional methods include paper and thin layer chromatographic procedures. Suitable supports for TLC include microcrystalline cellulose (in which the sugars partition between the mobile phase and the cellulose-bound water complex) and silica gel. A wide range of traditional solvent systems can be used (Stahl, 1965, gives details) and your choice of mobile phase will depend upon the expected composition of the mixture, e.g. cellulose with ethyl acetate : pyridine : water (100 : 35 : 25 v/v/v) gives a good separation of pentoses and hexoses, and will resolve glucose and galactose, as well as some disaccharides. Staining methods usually exploit the reducing properties of carbohydrates. Of the high resolution techniques, HPLC (p. 305) is the preferred method for the analysis of simple monosaccharide mixtures, and for oligosaccharide analysis and purification. Ion-exchange columns are often used for these purposes, with refractive index, or electrochemical, detection of separated components (Chapter 44). GC is more suitable for complex monosaccharide mixtures, and can analyse subnanomolar amounts of carbohydrates and their derivatives, e.g. polyols, including glycerol (p. 357). However, a

SAFETY NOTE Working with trimethylsilylating reagents – these compounds are extremely reactive and must be handled with care, using gloves and a fume hood. Since these reagents react violently with water, samples are dried and then redissolved in an organic solvent before trimethylsilylation.

SAFETY NOTE Working with acids – HCl gas is produced when concentrated HCl is heated: always wear gloves and work inside a fume hood.

Complex heteropolymers that contain carbohydrates – these include:

- **lignins** – structural polymers in plants: composed of amino acids (phenylalanine and tyrosine) together with carbohydrates, and extremely resistant to hydrolysis;
- **peptidoglycan** – found in the bacterial cell wall: composed of amino acids and carbohydrates, providing strength and rigidity;
- **proteoglycans** – found in animal/human connective tissues: composed of a core protein with many carbohydrate side chains attached, responsible for hydration, lubrication, resistance to compressive forces and mediation of cellular interactions.

preliminary step is necessary to produce volatile derivatives of the carbohydrates in the mixture, e.g. methylation or, more often, trimethylsilation (adding hexamethyldisilazane and trimethylchlorosilane at 2 : 1 v/v at room temperature rapidly produces trimethylsilyl ethers – TMS derivatives). Efficient separations can be obtained with a non-polar stationary phase, e.g. methylpolysiloxy gum (OV-1): the use of SCOT columns (p. 361) can enable over 30 components to be resolved. A disadvantage of GC methods is that the carbohydrates are first converted to a derivative form, then quantified by destructive means, preventing further analysis.

Capillary electrophoresis (Chapter 47) is a powerful technique for the separation of carbohydrates, though it is usual to modify uncharged carbohydrates by reductive amination to form primary amines to allow their separation within the capillary.

Characterisation of polysaccharides

Polysaccharides are characterised according to the relative proportions of the constituent sugar residues, the various types of glycosidic links and the M_r. To investigate the composition of a given type of polysaccharide, its glycosidic links must be hydrolysed (e.g. by heating with concentrated HCl at 60 °C for 30 min), followed by separation, identification and quantification of the individual components. The position of glycosidic linkages can be determined by methylation of all free hydroxyl groups of the polysaccharide followed by complete hydrolysis to give a mixture of partially methylated monosaccharides. These are then reduced to alditols, and acetylated. The partially methylated alditol acetates can be identified by GC-mass spectrometry (p. 295), allowing the types of glycosidic links to be deduced from the positions of the acetylated groups. The α and β configuration of the glycosidic linkages can be determined by enzymic assay or by NMR spectroscopy (p. 291).

Text references and sources for further study

Chaplin, M.F. and Kennedy, J.F. (1994) *Carbohydrate Analysis: A Practical Approach*. Oxford University Press, Oxford.

Davis, B.G. and Fairbanks, A.J. (2002) *Carbohydrate Chemistry*. Oxford University Press, Oxford.
[A primer on carbohydrate chemistry]

Lindhorst, T.K. (2000) *Essentials of Carbohydrate Chemistry and Biochemistry*, 2nd edn. Wiley-VCH, Berlin.

Rassi, Z.E. (2002) *Carbohydrate Analysis by Modern Chromatography and Electrophoresis*. Elsevier, Amsterdam.

Robyt, J.F. (1998) *Essentials of Carbohydrate Chemistry*. Springer-Verlag, Berlin.

Stahl, E. (1965) *Thin Layer Chromatography – a Laboratory Handbook*. Springer, Berlin.

Study exercises

52.1 Check your understanding of the terminology of carbohydrate structure. For each of the following terms (i) provide a definition, and (ii) give an example of a representative compound:
   (a) aldose;
   (b) pentose;
   (c) hemiacetal;
   (d) pyranoside;
   (e) β anomer.

52.2 Identify sugars by thin layer chromatography (TLC). A typical TLC separation of sugars on silica gel is shown in the figure opposite.
   (a) Identify the sugars present in the hydrolysed raffinose sample. Check the structure of raffinose, to see whether your findings are consistent with the monosaccharide constituents.
   (b) If the relative frontal mobility of unhydrolysed raffinose relative to glucose (R_G, Chapter 44) is known to be 0.34, and the distance migrated by glucose is 35 mm, how far (in mm) would you expect the raffinose to have migrated?

52.3 Consider the operating principles underlying chemical methods for quantitative assay of carbohydrates. Why is the dinitrosalicylate (DNS) method unsuitable for the assay of the disaccharide sucrose, yet it is suitable for measuring its constituent monosaccharides, i.e. glucose and fructose? Is DNS unsuitable for all disaccharides?

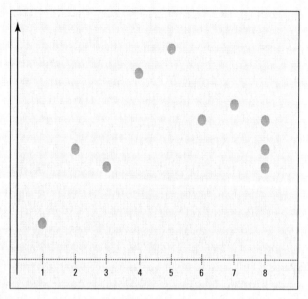

Thin-layer chromatographic separation of sugars: lane 1, lactose; lane 2, glucose; lane 3, galactose; lane 4, xylose; lane 5, ribose; lane 6, fructose; lane 7, arabinose; lane 8, hydrolyzed raffinose (dotted line indicates origin and arrow indicates direction of solvent flow).

53 Assaying nucleic acids and nucleotides

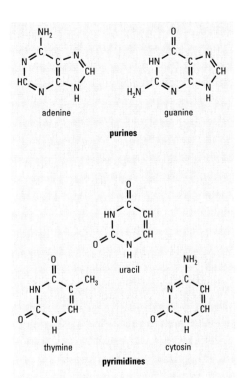

Fig. 53.1 Nitrogenous bases in nucleic acids.

Nucleic acids are nitrogen-containing compounds of high M_r, often found within nucleic acid–protein (nucleoprotein) complexes in cells. The two main groups of nucleic acids are:

1. Deoxyribonucleic acid (DNA) – found in chromosomes and the principal molecule responsible for the storage and transfer of genetic information (Chapter 58).
2. Ribonucleic acid (RNA) – involved with the DNA-directed synthesis of proteins in cells. Three principal types of RNA exist: messenger RNA (mRNA), ribosomal RNA (rRNA) and transfer RNA (tRNA). In some viruses, RNA acts as the genetic material. In eukaryotes, mRNA molecules initially synthesised in the nucleus from genomic DNA (nascent mRNA) will contain several sequences – called introns – that are not transcribed into protein. These are successively excised in the nucleus, leaving only coding sequences (exons) in the mRNA that migrates to the cytoplasm, to be translated at the ribosome.

 KEY POINT *Nucleic acids are important in the transmission of information within cells, and one of the most important aspects of nucleic acid analysis is to decipher the coded information within these molecules (see also Chapters 11 and 60).*

The structure of nucleic acids

Nucleic acids are polymers of nucleotides (polynucleotides), where each nucleotide consists of:

- a nitrogenous base, of which there are five main types. Two have a purine ring structure, i.e. adenine (A) and guanine (G), and three have a pyrimidine ring, i.e. thymine (T), uracil (U) and cytosine (C), as shown in Figure 53.1. Their carbon atoms are numbered C-1, C-2, etc;
- a pentose sugar, which is ribose in RNA and deoxyribose in DNA. The carbon atoms are denoted as C-1′, C-2′, etc. and deoxyribose has no hydroxyl group on C-2′ (Fig. 53.2). The C-1′ of the sugar is linked either to the N-9 of a purine or the N-1 of a pyrimidine;
- a phosphate group, which links with the sugars to form the sugar–phosphate backbone of the polynucleotide chain.

A compound with sugar and base only is called a nucleoside (Fig. 53.2) and the specific names given to the various nucleosides and nucleotides are listed in Table 53.1. The individual nucleotides within nucleic acids are linked by phosphodiester bonds between the 3′ and 5′ positions of the sugars (Fig. 53.3).

 KEY POINT *RNA and DNA differ both in the nature of the pentose sugar residue, and in their base composition: both types contain adenine, guanine and cytosine, but RNA contains uracil while DNA contains thymine.*

Fig. 53.2 A nucleoside triphosphate – deoxyadenosine 5′ triphosphate (dATP).

Table 53.1 Nomenclature of nucleosides and nucleotides

Base	Nucleoside	Nucleotide
Adenine	Adenosine	Adenylic acid
Guanine	Guanosine	Guanylic acid
Uracil*	Uridine	Uridylic acid
Cytosine	Cytidine	Cytidylic acid
Thymine†	Thymidine	Thymidylic acid

*In DNA.
†In RNA.

Minimising damage to chromosomes – *chromosomes vary in size from 0.3 to 200 megabase pairs (Mb), so some breaks in DNA inevitably occur during manipulation. Shear effects can be minimised by using wide-mouthed pipettes, gentle mixing, by avoiding rotamixing, and by precipitating DNA with ethanol at −20 °C.*

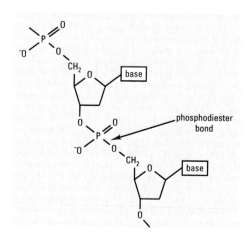

Fig. 53.3 Linkage of nucleotides in nucleic acids.

Working with precipitated DNA – *this can be spooled from solution by winding it around a glass rod.*

Differences also exist in the conformation of the two types of nucleic acid. DNA typically exists as two interwoven helical polynucleotide chains, with their structure stabilised by hydrogen bonds between matching base pairs on the adjacent strands: A always pairs with T (two hydrogen bonds), and G with C (three hydrogen bonds). This complementarity is important since it stabilises the DNA duplex (double helix) and provides the basis for replication and transcription (Chapter 60). While most double stranded (ds) DNA molecules are in this form, i.e. as a double helix, some viral DNA is single stranded (ss). Intact DNA molecules are very large indeed, with high M_r values (e.g. 10^9). In the main, RNA is single stranded and in the form of a gentle right-handed helix stabilised by base-stacking interactions, although some sections of RNA (i.e. tRNA) have regions of self-complementarity, leading to base pairing. Typical values for M_r of RNA range from 10^4 for tRNA to 10^6 for other types.

Extraction and purification of nucleic acids

The types of nucleic acid most commonly isolated are chromosomal DNA, plasmid DNA and mRNA. Irrespective of the source, extraction and purification involve the following stages, in sequence:

- disruption of cells to release their contents;
- removal of non-nucleic acid components (e.g. protein), leaving DNA and/or RNA;
- concentration of the remaining nucleic acids.

DNA isolation procedures

Specific details for the preparation of plasmid DNA from bacterial cells are given in Chapter 62. For other sources of DNA, such as mammalian tissue or plant material, the following steps are required:

1. Homogenisation – tissues can be disrupted by the methods described in detail in Chapter 36, e.g. by lysis in a buffered solution containing the detergent sodium dodecyl sulphate (SDS) or Triton X-100.
2. Enzymic removal of protein and RNA – using proteinase (e.g. proteinase K at $0.1\,\text{mg ml}^{-1}$) and ribonuclease (typically at $0.1\,\mu\text{g ml}^{-1}$) for 1–2 h.
3. Phenol–chloroform extraction – to remove any remaining traces of contaminating protein.
4. Precipitation of nucleic acids – usually by adding twice the volume of ethanol.
5. Solubilisation in an appropriate volume of buffer (pH 7.5) – ribo-nuclease is often added to remove any traces of contaminating RNA.

Density gradient centrifugation (p. 298) is an alternative approach to the separation of DNA from contaminating RNA, as described below.

RNA isolation procedures

Each mammalian cell contains about 10 pg of RNA, made up of rRNA (80–85%), tRNA (10–15%) and mRNA (1–5%). While rRNA and tRNA components are of discrete sizes, mRNA is heterogeneous and varies in length from several hundred to several thousand nucleotides.

Avoiding contamination of glassware by RNases – autoclave all glassware before use, to denature RNases. These enzymes are present in skin secretions: use gloves at all times and use plasticware wherever possible.

Avoiding RNase degradation of RNA – endogenous RNases can be inhibited by including diethyl pyrocarbonate (DEP) (at 0.1% v/v g ml^{-1}) in solutions used for RNA extraction.

Purification of mRNA – following wholecell extraction of RNA, mRNA can be separated from other types by affinity chromatography (p. 309) using poly (U)-Sepharose which binds to the poly (A) 'tail' sequence at the 3' end of mRNA molecules.

Measuring nucleic acids by spectrophotometry – ideally, the nucleic acid extracts should be prepared to give A_{260} values of between 0.10 and 0.50, for maximum accuracy and precision.

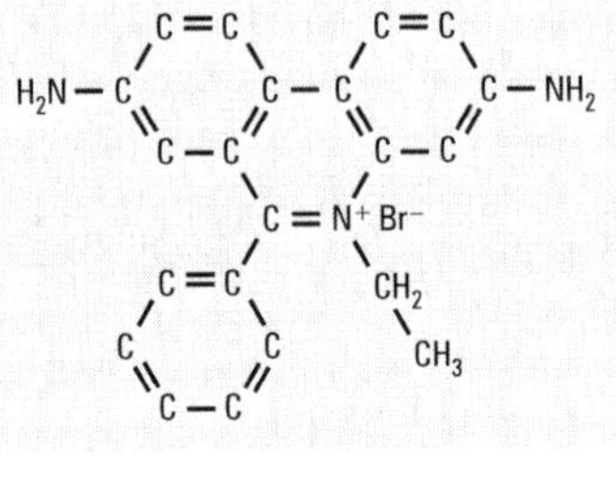

Fig. 53.4 Ethidium bromide (EtBr), a fluorescent molecule used for the detection and assay of DNA.

> **KEY POINT** *RNA is more difficult to purify than DNA, partly because of degradation during the extraction process due to the action of contaminating ribonucleases, and partly because the rigorous treatment required to dissociate the RNA from protein in ribosomes may fragment the polyribonucleotide strands.*

RNA can be prepared either from the cytoplasm of cells (to give mainly rRNA, tRNA and fully processed mRNA), or from whole cells, in which case nascent mRNA from the nucleus will also be present (p. 373):

- Preparation of cytoplasmic RNA involves lysis of cells or protoplasts with a hypotonic buffer, leaving the nuclei intact. Cell debris and nuclei are then removed by centrifugation (Chapter 43), and sodium dodecyl sulphate (SDS) is added to the supernatant (the cytoplasmic fraction) to inhibit ribonuclease. Proteinase K can be added to release rRNA from ribosomes. Phenol–chloroform extraction removes contaminating proteins, as for DNA preparation (p. 446), and the RNA present in the aqueous phase can then be precipitated by addition of twice the volume of ethanol.

- Preparation of whole-cell RNA requires more vigorous cell lysis, e.g. with a solution containing 6 mol l^{-1} guanidinium chloride and 2-mercaptoethanol: this effectively denatures any ribonuclease present. Caesium chloride is added to the extract to give a final concentration of 2.4 mol l^{-1}, and the solution is processed by density gradient centrifugation (Chapter 43) at 100 000 g for 18 h, using a cushion of 5.7 mol l^{-1} CsCl. DNA and protein remain in the upper layers of CsCl, while RNA forms a pellet at the base of the tube. The RNA pellet is redissolved in buffer, and precipitated in cold ethanol.

Separating nucleic acids

Electrophoresis is the principal method used for separating nucleic acids. At alkaline pH values, linear DNA and RNA molecules have a uniform net negative charge per unit length due to the charge on the phosphoryl group of the backbone. Electrophoresis using a supporting medium that acts as a molecular sieve (e.g. agarose or polyacrylamide, p. 322) enables DNA fragments or RNA molecules to be separated on the basis of their relative sizes (see Chapter 46 for further details).

Quantitative analysis of nucleic acids

Measuring nucleic acid content

The concentration of reasonably pure samples of DNA or RNA can be measured by spectrophotometry (p. 282). In contrast, measurement of the nucleic acid content of whole cell or tissue homogenates requires chemical methods, since the homogenates will contain many interfering substances. The principles involved in each technique are as follows:

- Spectrophotometry – DNA and RNA both show absorption maxima at ≈260 nm, due to the conjugated double bonds present in their constituent bases. At 260 nm, an A_{260} value of 1.0 is given by a 50 μg ml^{-1} solution of dsDNA, or a 40 μg ml^{-1} solution of ssRNA. If the absorbance at 280 nm is also measured, protein contamination can be quantified. Pure nucleic acids give A_{260}/A_{280} ratios of 1.8–2.0, and a value below 1.6 indicates significant protein contamination. Further purification steps are required for contaminated samples, e.g. by

SAFETY NOTE Working with ethidium bromide – this compound is highly toxic and mutagenic. Avoid skin contact (wear gloves) and avoid ingestion. Use a safe method of disposal (e.g. adsorb from solution using an appropriate adsorbant, e.g. activated charcoal).

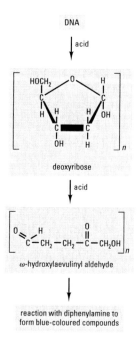

Fig. 53.5 The diphenylamine reaction for assay of DNA.

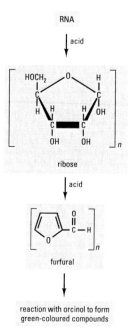

Fig. 53.6 The orcinol reaction for assay of RNA.

repeating the phenol–chloroform extraction step. RNA contamination of a DNA preparation is indicated if A_{260} decreases when the sample is treated with 2.5 μl of RNase at $20\,\mu g\,\mu l^{-1}$. DNA contamination of an RNA preparation might be suspected if the sample is very viscous, and this can be confirmed by electrophoresis.

- Spectrofluorimetry – this is the best approach for samples where the DNA concentration is too low to allow direct assay by the spectrophotometric method described above. The method uses the fluorescent dye ethidium bromide (Fig. 53.4), which binds to dsDNA by insertion between stacked base pairs, a phenomenon termed 'intercalation'. The fluorescence of ethidium bromide is enhanced 25-fold when it interacts with dsDNA. Since ssDNA gives no significant enhancement of fluorescence, dsDNA can be quantified in the presence of denatured DNA. The concentration of dsDNA in solution, $[dsDNA]_x$, can be calculated by comparing its fluorescence (excitation, 525 nm; emission, 590 nm) with that of a standard of known concentration, $[dsDNA]_{std}$, using the relationship:

$$[dsDNA]_x = \frac{[dsDNA]_{std} \times \text{fluorescence of unknown}}{\text{fluorescence of standard}} \quad [53.1]$$

- Chemical methods – these are mostly based on colorimetric reactions with the pentose groups of nucleic acids. The total DNA concentration can be measured by the diphenylamine reaction (Fig. 53.5), which is specific for 2-deoxypentoses. The diphenylamine reaction involves heating 2 ml of DNA solution with 4 ml of freshly prepared diphenylamine reagent (diphenylamine, at $10\,g\,l^{-1}$ in glacial acetic acid, plus 25 ml concentrated sulphuric acid) for 10 min in a boiling water bath. The acids cleave some of the phosphodiester bonds, and hydrolyse the glycosidic links between the deoxyribose and purines. Deoxyribose residues are converted to ω-hydroxylaevulinyl aldehyde (Fig. 53.5), which reacts with the diphenylamine to produce a blue pigment, assayed at 600 nm. By constructing a DNA standard curve ranging from 0 to $400\,\mu g\,ml^{-1}$, the DNA concentration of the unknown can be determined.

RNA concentration can be measured by the orcinol reaction, which is a general assay for pentoses. The orcinol reaction involves heating 2 ml of RNA solution with 3 ml of orcinol reagent (prepared by dissolving 1 g of $FeCl_3 \cdot 6H_2O$ in 1 litre of concentrated HCl, and adding 35 ml of 6% w/v orcinol in ethanol) in a boiling water bath for 20 min. The acid cleaves some phosphodiester bonds and hydrolyses the glycosidic links between the ribose and purines. The hot acid also converts the ribose to furfural (Fig. 53.6), which reacts with orcinol in the presence of ferric ions to produce green-coloured compounds, assayed at 660 nm. A standard curve for RNA ranging from 0 to $400\,\mu g\,ml^{-1}$ is used to determine the RNA concentration of the unknown. The orcinol reaction is less specific than the diphenylamine reaction, as deoxyribose reacts to some extent and DNA gives about 10% of the colour given by the same concentration of RNA. If the DNA concentration of the extract is known (e.g. from a diphenylamine assay), the contribution of DNA to A_{660} can be measured for a standard prepared to have the same DNA concentration as the sample: the A_{660} due to DNA can then be subtracted from the result of the orcinol reaction, and the remaining A_{660} value is then used to determine RNA concentration from the RNA standard curve.

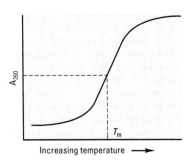

Fig. 53.7 Effect of temperature on the absorbance (A_{260}) of DNA in solution. T_m is the temperature for the mid-point of the absorbance change.

Assaying the relative proportions of base pairs in DNA

Double stranded DNA has a lower molar absorptivity at 260 nm than single stranded DNA, due to electron interactions between the stacked base pairs and hydrogen bonding between the complementary bases of dsDNA. When a dsDNA solution is heated slowly, there is little change in absorbance until a temperature is reached where the hydrogen bonds are broken and the DNA strands separate. At this so-called 'melting temperature' (T_m), the A_{260} value increases sharply (Fig. 53.7). This temperature-dependent increase in absorbance is often referred to as the 'hyperchromic effect'. The actual value of the T_m (and hence the stability of dsDNA) is dependent on the base-pair composition of the DNA being studied: GC base pairs have three double bonds while AT base pairs have two, and the higher the % GC content, the higher the value of the T_m. The length of the molecule also affects the T_m, since short dsDNA molecules do not show a hyperchromic effect. Studies on a number of DNA samples have shown the following relationship, for GC contents between 30 and 70%:

$$\% \text{ GC} = 2.44(T_m - 69.3) \qquad [53.2]$$

The relative proportion of AT base pairs can then be calculated by subtracting the % GC value from 100, giving % AT.

Sources for further study

(See also Chapter 11 for nucleic acid databases and related information.)

Blackburn, G.M., and Gait, M.G., Loakes, D. and Williams, D.M. (2005) *Nucleic Acids in Chemistry and Biology*, 3rd edn. Royal Society of Chemistry London.

Neidle, S. (2002) *Nucleic Acid Structure and Recognition*. Oxford University Press, Oxford.

Walker, J. and Rapley, R. (2000) *The Nucleic Acids Protocols Handbook*. Humana Press, New Jersey.

Study exercises

53.1 Calculate the concentration of DNA in a sample using a fluorescence assay (see also exercises 41.2c and 41.2d, for equivalent exercises using UV spectrophotometry). A volume of 10 μl of a DNA standard solution (of concentration 50 μg ml^{-1}) was added to 3 ml of an ethidium bromide solution (containing 0.5 mg l^{-1} ethidium bromide, buffered at pH 7.5) in a fluorimeter cuvette, mixed, then read in a spectrofluorimeter using an excitation wavelength of 525 nm and an emission wavelength of 590 nm, giving a relative fluorescence intensity (RFI) of 58. This was repeated using 10 μl of a DNA solution of unknown concentration instead of the DNA standard solution, giving an RFI of 32. What is the DNA concentration of the unknown? (Express your answer to three significant figures.)

53.2 Investigate the binding properties of ethidium bromide. Find out (i) why ethidium bromide fluorescence is enhanced by DNA, and (ii) under what circumstances ethidium bromide can also bind to RNA.

53.3 Consider how to avoid problems in DNA spectrofluorometric analysis caused by the presence of RNA in a sample. What steps could you take to ensure that all the fluorescence obtained from a cell extract was due to the reaction of ethidium bromide with DNA rather than from DNA plus contaminating RNA?

54 Protein purification

Much of the current knowledge about metabolic and physiological events has been gained from *in vitro* studies of purified proteins. Such studies range from investigations into the kinetics and regulation of enzymes (Chapter 55) to the determination of the structure of a protein and its relationship to function. In addition, certain purified proteins have a role as therapeutic agents, e.g. Factor VIII in the treatment of haemophilia.

Protein purification and proteomics – an important aspect of proteomics (p. 68) is the characterisation of individual proteins, using some of the techniques described in this chapter.

KEY POINT The purification of most proteins involves a series of procedures, based on differential solubility and/or chromatography, that selectively separate the protein of interest from contaminating proteins and other material. It is rare for purification to be a one-step process.

Deciding on your objectives

Purity, yield and cost are the major considerations, and the relative importance of these factors will depend on the purpose of the purification. For studies on the biological activity of a protein (e.g. an enzyme), only μg amounts may be required and, as long as there are no interfering substances present, 100% purity may not be necessary. For structural studies, mg amounts may be needed and the protein must be pure. A protein produced for industrial applications (e.g. an enzyme for use in starch degradation) needs to be produced in large amounts (g or kg), but purity usually is not essential. In contrast, commercially produced proteins for therapeutic use need to be free of any significant contamination.

Preventing loss of protein during purification – silicone-coated glass containers prevent the adsorption of proteins to the surface of a glass vessel.

The cost of the purification will depend on the nature of the source material, the number of purification steps required, and the price of materials for the separation techniques employed. Inevitably, some protein will be lost at each stage of the purification procedure.

KEY POINT You should aim to use the smallest number of purification steps that will give you the yield and purity required for your application.

Preliminary considerations

Finding out about the protein

Rather than approaching every purification step on a 'trial and error' basis, try to find out as much as you can about the physical and biological properties of the desired protein before you begin your practical work. Even if the protein to be purified is novel, it is likely that similar proteins will be described in the literature. Information on the isoelectric point (pI) and M_r, of both the protein and of the likely major contaminants, is particularly useful for ion-exchange and gel permeation chromatography (p. 307). Knowledge of other factors such as metal ion and co-enzyme requirements, presence of thiol groups and known inhibitors can indicate useful chromatographic steps, and may allow you to take steps to preserve the tertiary structure and biological activity of the desired protein during the purification process.

Definition

Isoelectric point (pI) – the pH value at which a protein has a net charge of zero, i.e. where positive and negative charges on amino acid side chains are balanced.

Source material for protein purification – an important advance has been the use of genetically engineered organisms, designed to produce large amounts of a particular protein (p. 446).

Choosing the source material

If you have a choice of starting material, use a convenient source in which the protein of interest is abundant. This can vary from animal tissues (e.g. heart, kidney or liver) obtained from an abattoir or from laboratory animals, to plant material, or to microbial cells from a laboratory culture.

Homogenisation and solubilisation

Unless the protein is extracellular, the cells in the source material need to be disrupted and homogenised by one of the methods described in Chapter 36. Following homogenisation, proteins from the cytosol or extracellular fluid will normally be present in a soluble form, but membrane-bound proteins and those within organelles will require further treatment. Isolation of organelles by differential centrifugation (p. 298) will provide a degree of purification, while membrane-bound proteins will need organic solvents or detergents to render them soluble. Once the protein of interest is in a soluble form, any particulate material in the extract should be removed, e.g. by centrifugation or filtration.

Storing protein solutions during a purification procedure – overnight storage at 4°C is acceptable, but it is advisable to include bacteriostatic agents (e.g. azide at 0.5% w/v) and protease inhibitors: freezing may be required for longer-term storage – use liquid nitrogen or dry ice/methanol for rapid freezing. If freezing and thawing might denature the sample, include 20% v/v glycerol in buffers, to allow storage at −20°C in liquid form.

Preserving enzyme activity during purification

Once an enzyme has been released from its intracellular environment, it will encounter potentially adverse conditions that may result in denaturation and permanent inactivation, e.g. a lower pH, an oxidising environment or exposure to lysosomal proteases. During homogenisation, and in some or all of the purification steps, buffers should contain reagents that can counteract these potentially damaging effects (see Chapter 36). Fewer precautions may be needed with extracellular enzymes. For most proteins, the initial procedures should be carried out at 4°C to minimise the risk of proteolysis.

Devising a strategy for protein purification

Any successful protein purification scheme will exploit the unique properties of the desired protein in terms of its size, net charge, hydrophobic nature, biological activity, etc. The chromatographic techniques that separate on the basis of these properties are detailed in Chapter 44. However, the *order* in which the various purification steps are carried out needs some thought, and each stage needs to be considered in relation to the following factors:

Working with recombinant proteins – these are often produced with either N-terminal or C-terminal hexahistidine 'tags', enabling them to be purified readily using immobilised metal affinity chromatography (p. 310).

- Capacity – the amount of material (volume or concentration) that the technique can handle. High capacity techniques such as ammonium sulphate precipitation (see later) and ion-exchange chromatography (p. 307) should be used at an early stage, to reduce the sample volume.
- Resolution – the efficiency of separating one component from another. At later stages of the purification, the sample volume will be considerably reduced, but any contaminating proteins may well have very similar properties to those of the protein of interest. Therefore, a high resolution (but low capacity) technique will be required, such as co-valent chromatography or immobilised metal affinity chromatography (IMAC).

Measuring yield – note that yields of > 100% may be obtained if an inhibitor is lost during purification.

- Yield – the amount of protein recovered at each step, expressed as a percentage of the initial amount (p. 376). For example, yields of >80% can be obtained with ammonium sulphate precipitation. With certain

Table 54.1 Principal causes of decreased yield in protein purification

Cause	Possible solution
Denaturation	Include EDTA or reducing agents in buffers; avoid extreme temperatures
Inhibition	Check buffer composition for possible inhibitors
Proteolysis	Include protease inhibitors in buffers
Non-elution	Alter salt concentration or pH of the eluting buffer
Co-factor loss	Recombine fractions on a trial-and-error basis

Calculating total protein content – multiply the protein concentration by the total volume, making sure that the units are consistent, e.g. mg ml^{-1} × ml. A similar procedure is required to determine the total amount of enzyme.

Using immunoassays – note that inactive or denatured forms of the protein may be detected along with the biologically active form.

Example For an enzyme extract having a specific activity of 250.2 U mg^{-1} compared to an initial specific activity of 45.5 U mg^{-1}, using Eqn [54.1] gives a purification factor of 250.2 ÷ 45.5 = 5.5-fold (to one decimal place).

types of affinity chromatography, where harsh elution conditions need to be employed, yields may be quite low (e.g. <20%). Some of the other possible reasons for a decreased yield are considered in Table 54.1.

An ideal purification procedure will have the following features:

1. an initial step that has a high capacity, but not necessarily a high resolution;
2. a series of chromatographic steps, each of which exploits a different property of the protein to increase purity;
3. a minimum number of manipulations and changes in conditions (e.g. in buffer composition) between steps.

The following series of steps would meet the above conditions:

Step 1 Ammonium sulphate precipitation – a high capacity, low resolution technique that yields a protein solution with a greatly reduced volume and a high concentration of ammonium sulphate.

Step 2 Hydrophobic interaction chromatography (HIC) – an absorption technique (p. 308) using a starting buffer with a high ammonium sulphate concentration. During development of the column, the ionic strength of the mobile phase is reduced in a gradient elution.

Step 3 Ion-exchange chromatography (IEC) – another absorption technique (p. 307). Here the starting buffer has a low ionic strength, and a gradient of increasing ionic strength is used to separate components.

Step 4 Gel permeation chromatography (GPC) – a low capacity method (p. 307) where separation is independent of the composition of the mobile phase.

The sample from step 1 may be able to be applied directly to the HIC column, and that from step 2 may be used for IEC without changing the buffer. A concentration step (pp. 376–9) would be required before step 4. The above scheme is very much an idealised approach: in practice, buffers may need to be changed by dialysis or ultrafiltration, as described later. However, the principle of using the minimum number of manipulations to obtain the desired purification remains valid.

Monitoring purification

At each step, the separated material is usually collected as a series of 'fractions'. Each fraction must be assayed for the protein of interest, and fractions containing that protein are pooled prior to the next step. The assay performed will depend on the properties of the protein of interest but specific enzyme assays or immunoassays are most commonly used. The protein concentration (p. 353) and the volume of the pooled fraction must be determined, and these values, together with those obtained for the amount of the protein of interest, are used to determine the purity and yield after each step. For enzymes, the biological activity is measured and used to calculate the specific activity (the enzyme activity per unit mass of protein) as described on page 381. By determining the specific activity of the enzyme at each step, the degree of purification, or purification factor (*n*-fold purification), can be obtained from this relationship:

$$\text{Purification factor} = \frac{\text{specific activity after a particular step}}{\text{specific activity of initial sample}} \quad [54.1]$$

Table 54.2 Example of a record of purification for an enzyme

Step	Procedure	Total protein (mg)	Total enzyme (U)*	Specific activity (U mg^{-1})	Purification factor (n-fold)	Enzyme yield (%)
Initial sample	—	210	1984	9.45	—	100.0
Step 1	'Salting out'	112	1740	15.54	1.6	87.7
Step 2	HIC	30	1701	56.70	6.0	85.7
Step 3	IEC	6.0	1604	267.33	28.3	80.8
Step 4	GPC	3.0	1550	516.66	54.7	78.1

*U = unit (p. 381).

Increased purification usually represents a decrease in total protein relative to the biological activity of the protein of interest, though in some instances it may reflect the loss of an inhibitor during a purification step.

Calculation of the yield of enzyme at each step is straightforward, since:

$$\text{Yield} = \frac{\text{total enzyme activity after a particular step}}{\text{total enzyme activity in initial sample}} \times 100 \ (\%) \quad [54.2]$$

Note that the yield equation uses the total amount of enzyme and is therefore unaffected by the volume of the solutions involved. You should make a record of the progress of your purification procedure at each step, as in Table 54.2.

Example For an enzyme extract having a total activity of 8521 U compared to an initial total activity of 9580 U, using Eqn [54.2] gives a yield of [8521 ÷ 9580] × 100 = 88.9% (to one decimal place).

Assaying enzymes – if possible, use a preliminary screening technique to detect active fractions prior to quantitative assay, e.g. using tetrazolium dyes in microtitre plates to detect oxidoreductases (p. 328).

Measuring the mass of proteins – this is sometimes expressed in daltons (Da), or kilodaltons (kDa), where 1 dalton = 1 atomic mass unit. M_r is an alternative, numerically equivalent expression (p. 145).

Monitoring progress using electrophoresis

Polyacrylamide gel electrophoresis (PAGE) of subsamples from each step will give an indication of the number of proteins still present: ideally, a pure protein will give only one band after silver staining (p. 332); however, this stain is so sensitive that even trace impurities can be detected in what you expect to be a pure sample, so don't be dismayed by the appearance of the gel after staining.

Carrying out SDS-PAGE and isoelectric focusing (or, ideally, 2D-PAGE, p. 331) on subsamples also gives information on the M_r and pI of any remaining contaminants, and this can be taken into account when planning the next chromatographic step.

Differential solubility separation techniques

Often the volume and protein concentration of the initial soluble extract will be quite high. Application of a differential solubility technique at this stage results in the precipitation of selected proteins. These can be recovered by filtration or centrifugation, washed, and then resuspended in an appropriate buffer. This will reduce the sample volume and may give a small degree of purification, making the sample more suitable for subsequent chromatographic steps.

Ammonium sulphate precipitation ('salting out')

This is the most widely used differential solubility technique, having the advantage that most precipitated enzymes are not permanently denatured, and can be redissolved with restoration of activity. Precipitation depends on the existence of hydrophobic 'patches' on the surface of proteins, inducing a reorganisation of water molecules in their vicinity. When

Table 54.3 Amount of $(NH_4)_2SO_4$ $(g\,l^{-1})$ required for a particular percentage saturation

Final concentration (%) →	20	30	40	50	60	70	80	90	100
Initial concentration (%) ↓	Ammonium sulphate added $(g\,l^{-1})$								
0	107	166	229	295	366	442	523	611	707
10	54	111	171	236	305	379	458	545	636
20	—	56	115	177	244	316	392	476	565
30		—	57	119	184	253	328	408	495
40			—	59	122	190	262	340	424
50				—	61	127	197	272	353
60					—	63	131	204	283
70						—	66	136	212
80							—	68	141
90								—	71

Using ammonium sulphate – although the phenomenon of salting out is seen with several salts, $(NH_4)_2SO_4$ is the most widely used because it is highly soluble (saturating at $\approx 4\,mol\,l^{-1}$), inexpensive, and can be obtained in very pure form.

ammonium sulphate is added to the extract, it dissolves to give ions that become hydrated, leaving fewer water molecules in association with the protein. As a result, the hydrophobic patches become 'exposed', and hydrophobic interactions between different protein molecules lead to their aggregation and precipitation. The basis of fractionation in this method is that, as the salt concentration of the extract is increased, proteins with larger or more abundant hydrophobic patches will precipitate before those with smaller or fewer patches.

Unless you can obtain information from the literature about the $(NH_4)_2SO_4$ concentration that will precipitate your target protein, fractionation is done on a trial-and-error basis. The $(NH_4)_2SO_4$ concentration is expressed in terms of the percentage saturation value: Table 54.3 shows the amount of $(NH_4)_2SO_4$ required to give 20–100% saturation. For each percentage saturation chosen, the $(NH_4)_2SO_4$ salt should be added slowly while stirring, and the mixture left at 4 °C for 1 h before centrifuging at 3000 g for 40 min. For an effective separation, start with the maximum percentage saturation that does not precipitate the protein of interest, then increase the percentage saturation by the minimum amount that will then precipitate it. The proteins precipitated between any two values of percentage saturation (say between 30 and 50%) are referred to as a 'cut'. An alternative approach is to add saturated $(NH_4)_2SO_4$ solution – the volume (V_a) to be added to an initial volume of solution (V_i) with an initial saturation S_i, to give a final saturation S_f, is given by the equation:

Example To prepare a solution of 40% $(NH_4)_2SO_4$ saturation $(S_f = 0.4)$ from 100 ml (V_i) containing no added $(NH_4)_2SO_4$ using Eqn 54.3:
$V_a = 100(0.4 - 0) \div (1 - 0.4) = 67\,ml$ of saturated $(NH_4)_2SO_4$ solution, giving a volume of 167 ml.
For a 40–60% 'cut' of this sample $(V_i = 167\,ml)$, using Eqn [54.3]:
$V_a = 167(0.6 - 0.4) \div (1 - 0.6) = 83.5\,ml$ of saturated $(NH_4)_2SO_4$ solution, to give a final volume of 250.5 ml.

$$V_a = \frac{V_i(S_f - S_i)}{1 - S_f} \qquad [54.3]$$

where S_i and S_f are expressed as fractional saturation, e.g. $S_f = 0.5 \equiv 50\%$ saturation, and both volumes are expressed in the same terms, e.g. ml.

Precipitation by changing pH

Purifying bacterial proteins – adjusting the extract to pH 5 can be useful, since many bacterial proteins have pI values in this region.

Proteins are least soluble at their isoelectric points because, at that pH, there is no longer the repulsion that occurs between positively or negatively charged protein molecules at physiological pH values. If the precipitated proteins are required for further purification, it is essential that the protein of interest is not irreversibly denatured. The method is probably best employed by precipitating contaminating proteins, leaving the desired protein in solution. Use citric acid for <pH 3, acetic acid for <pH 4, and sodium carbonate or ethanolamine for >pH 8.

Heat denaturation

Exposure of most proteins to high temperatures disrupts their conformation through effects on non-covalent interactions such as hydrogen bonds and van der Waals forces. However, different proteins are denatured, and hence precipitated, at different temperatures, and this can provide a basis for the separation of some heat-stable proteins. By incubating small aliquots ($\approx$1 ml) of extract for 1 min at a range of temperatures between 45 and 65 °C, it is possible to determine the temperature that gives maximum precipitation of contaminating protein with minimal inactivation of the desired protein.

> **Example** The HPII catalase of stationary phase *E. coli* remains intact when heated to 55°C for 15 min, aiding its separation from other proteins.

Solvent and polymer precipitation methods

Organic solvents (e.g. acetone, ethanol) cause precipitation of proteins by lowering the dielectric constant of the solution. Performing the precipitation at 0 °C minimises permanent denaturation. Stepwise concentration (% v/v) increments are used, giving 'cuts' of precipitated proteins, as with ammonium sulphate precipitation.

Organic polymers, particularly polyethylene glycol (PEG), also lower the dielectric constant, but at lower concentrations than with acetone or ethanol. The most commonly used PEG preparations have M_r values of 6000 or 20 000, and these can be removed from the sample by ultrafiltration. PEG precipitation does not involve salts, so it may be a useful preliminary step prior to ion-exchange chromatography which starts with low salt buffer. Also, since the techniques of PEG and ammonium sulphate precipitation involve different principles, they can be used sequentially.

> **Definition**
>
> **Dielectric constant** ($\mathcal{E}$) – a dimensionless measure of the screening effect on the force (F) between two charges (q_1 and q_2) due to the presence of solvent, from the equation $F = (kq_1q_2) \div (\mathcal{E}r^2)$, where r is the distance between the two particles and k is a constant. The dielectric constant for water is high ($\approx$80), while most organic solvents have lower values, in the range 1–10.

Concentration by ultrafiltration

This involves forcing water and small molecules through a semi-permeable membrane using high pressure or centrifugation. A range of membranes with 'nominal' molecular weight cut-offs between 500 and 300 000 are commercially available (e.g. Amicon, Millipore), with pore sizes of 0.1–10 µm. Concentration of small samples (<5 ml) can be achieved using either a membrane backed by an absorbent pad (e.g. Minicon, Fig. 54.1, available with M_r cut-off from 5000 to 30 000), or by using a centrifugal concentrator (e.g. Vivaspin, Centricon). Larger volumes (up to 400 ml) can be concentrated using a stirred ultrafiltration chamber (an ultrafiltration 'cell') where the liquid is forced through the membrane using nitrogen or an inert gas.

Ultrafiltration not only concentrates the sample, but also may give a degree of purification. It can also be used to change the buffer composition by diafiltration (see below). Note that molecular weight cut-off values are quoted for globular proteins – fibrous proteins of higher M_r may pass through the ultrafiltration membrane.

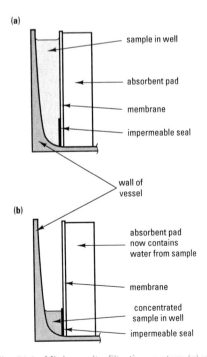

Fig. 54.1 Minicon ultrafiltration system (a) with sample added, (b) after ultrafiltration.

Removing salts and changing the buffer

Although you should aim to use the minimum number of manipulations to obtain the desired purification, it may be necessary to remove salts or to change the buffer before the next step will work effectively (e.g. when carrying out IEC, the ionic strength or the pH of the sample may need to be changed before the target protein will bind to the column). Several methods are available, including:

- Dialysis: the sample is placed in a bag consisting of semipermeable membrane (e.g. Visking tubing) and placed in at least 20 volumes of the required buffer. The membrane allows molecules of $M_r < 20\,000$ to pass freely, while retaining larger molecules (note that dialysis is not suitable for small proteins). The small molecules diffuse through the membrane until the osmotic pressure between the sample and the dialysis buffer is equalised. Several changes of dialysis buffer may be required to obtain the desired buffer conditions within the sample. Normally dialysis is carried out overnight at 4 °C. Efficient stirring is required – use a magnetic stirrer and stirrer bar.
- Diafiltration: quicker than dialysis and more suitable for larger volumes. It involves addition of the desired buffer to the sample solution, followed by ultrafiltration. Several buffer addition and ultrafiltration steps may be necessary to obtain the desired sample conditions.
- Gel permeation chromatography (gel filtration): a gel filtration medium of small pore size (e.g. Sephadex G-25) is used to prepare a column of ≈ 5 times the volume of the sample. When the sample passes through the column, the large protein molecules will elute with the void volume, while salt ions are retained. This method is only suitable for small volumes, and it results in dilution of the sample.

Using Visking tubing – this must be boiled before use to ensure a uniform pore size and to remove heavy metal contaminants.

Avoiding protein precipitation during dialysis or gel filtration – make sure your buffer pH is either above or below the pI of the proteins.

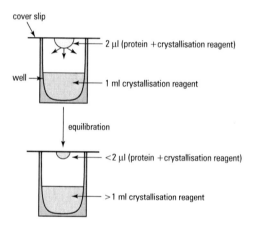

Fig. 54.2 The hanging drop technique for protein crystallisation.

Protein crystallisation

Much of the information on protein structure that is now available on protein sequence databases (see Table 11.1) has been obtained by X-ray crystallographic studies on protein crystals. To obtain protein crystals, the protein should be purified to at least 90–95% and its concentration in solution should be 10–20 mg ml^{-1}. The method most widely used for crystallisation is the hanging drop vapour diffusion technique (Fig. 54.2). Here, a small droplet of the protein sample (usually about 1–2 μl) is mixed with a crystallisation reagent on a siliconised glass coverslip and inverted over a microtitre plate well containing 0.5–1.0 ml of the same crystallisation reagent. The presence of protein in the droplet means that the initial concentration of crystallisation reagent in the droplet is less than that in the well. As a result, water will diffuse from the droplet into the vapour phase until an equilibrium is achieved between the droplet, the vapour phase and the solution in the well. During this equilibration, the protein is concentrated within the droplet, enhancing the formation of protein crystals. The essential components of a crystallisation reagent are (i) protein precipitant and (ii) buffer. The specific reagents differ, depending on the type of protein to be crystallised. Commercially available reagents include the PEG/Ion Screen reagents (Hampton Research, USA) and the Clear Strategy Screen reagents (Molecular Dimensions, UK). An image of protein crystals produced by this method is shown in Figure 54.3.

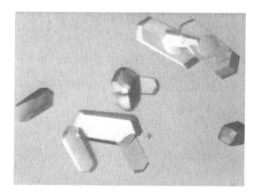

Fig. 54.3 Image of protein crystals produced by the hanging drop method (image courtesy of S. Charnock).

Sources for further study

Coligan, J.E., Dunn, B.M., Speicher, D.W. and Wingfield, P.T. (2003) *Short Protocols in Protein Science*. Wiley, New York.
[Provides details of over 500 specific methods]

Cutler, P. (2003) *Protein Purification Protocols*, 2nd edn. Humana Press, New Jersey.

Howard, G.C. and Brown, W.E. (2002) *Practical Methods in Advanced Protein Chemistry*. CRC Press, Boca Raton.

Hunte, C., von Jagow, G. and Schägger, H. (2003) *Membrane Protein Purification and Crystallisation: a Practical Guide*, 2nd edn. Academic Press, San Diego.
[Gives specific procedures for isolation of membrane proteins]

Rabilloud, T. (1999) *Proteome Research: Two-Dimensional Gel Electrophoresis and Identification Methods*. Springer, New York.

Roe, S. (2001) *Protein Purification Applications: A Practical Approach*, 2nd edn. Oxford University Press, Oxford.

Rosenberg, I.M. (2005) *Protein Analysis and Purification: Benchtop Techniques*, 2nd edn. Springer-Verlag, Berlin.

Scopes, R.K. (1994) *Protein Purification: Principles and Practice*, 3rd edn. Springer-Verlag, Berlin.

Study exercises

54.1 Practise ammonium sulphate fractionation calculations. For a mixture of three proteins, A, B and C, in 100 ml of aqueous solution, you are provided with the information that protein A precipitates at a $(NH_4)_2SO_4$ concentration of 40%, and that protein B precipitates in the 40–60% 'cut', while protein C is not precipitated by 60% $(NH_4)_2SO_4$. Determine how much solid $(NH_4)_2SO_4$ (in grams) would be needed to achieve 40% saturation, and once protein A is removed, how much $(NH_4)_2SO_4$ will then need to be added to the solution to reach 60% saturation. What will be the fate of protein C in this procedure?

54.2 Select a chromatographic technique for use with a particular protein. Using the information given in Chapter 44, which affinity techniques would be most suitable for the following proteins: (a) an enzyme that uses $NAD^+/NADH$ as a co-factor, (b) a genetically engineered protein with a polyhistidine tag at its N-terminal end (c) a glycoprotein?

54.3 Calculate yield and relative purification of an enzyme. The table shows representative data for an enzyme purification involving five stages. Calculate the yield and *n*-fold purification (relative to the initial values) at each stage. Give all answers to three significant figures.

Volume, protein concentration and enzyme activity at several stages during a protein purification procedure

Stage	Total volume of extract (ml)	Enzyme activity (U ml^{-1})	Protein concentration (mg ml^{-1})
1	78	1.2	4.8
2	70	1.2	1.7
3	10	6.4	2.5
4	5	10.8	1.4
5	0.6	65.7	3.4

55 Enzyme studies

Enzymes are globular proteins that increase the rate of specific biochemical reactions. Each enzyme operates on a limited number of substrates of similar structure to generate products under well-defined conditions of concentration, pH, temperature, etc. In metabolism, groups of enzymes work together in sequential pathways to carry out complex molecular transformations, e.g. the multi-reaction conversion of glucose to lactate (glycolysis).

> **KEY POINT** *Enzymes are categorised according to the chemical reactions they catalyse, leading to a four-figure Enzyme Commission (EC) code number and a systematic name for each enzyme. Most enzymes also have a recommended trivial name, often denoted by the suffix 'ase'.*

Example Enzyme EC 1.1.1.1 is usually known by its trivial name, alcohol dehydrogenase.

Measuring enzyme reactions

Activity

This is measured in terms of the rate of enzyme reaction. Activity may be expressed directly as amount of substrate utilised per unit time (e.g. nmol min^{-1}, etc.), or in terms of the non-SI international unit (U, or sometimes IU), defined as the amount of enzyme which will convert 1 µmol of substrate to product(s) in 1 min under specified conditions. However, the recommended (SI) unit of enzyme activity is the katal (kat), which is the amount of enzyme which will convert 1 mol of substrate to product(s) in 1 s under optimal conditions, determined from the following equation:

$$\text{enzyme activity (kat)} = \frac{\text{substrate converted (mol)}}{\text{time (s)}} \quad [55.1]$$

Example An enzyme preparation that converted 295 µmol of substrate to product in 15 min would have an activity, expressed in terms of non-SI units (U) of: 285 ÷ 15 = 19.7 U (to three significant figures). Using Eqn [55.1], the same preparation would have an activity, expressed in katals, of: 295 × 10^{-6} ÷ 900 = 3.27 × 10^{-7} = 329 nkat (to three significant figures).

This unit is relatively large (1 kat = 6 × 10^7 U) so SI prefixes are often used, e.g. nkat or pkat (p. 183). Note that the units involve amount of substrate (mol), not concentration (mol l^{-1} or mol m^{-3}).

For enzymes with macromolecular substrates of unknown molecular weight (e.g. deoxyribonuclease, amylase), activity can be expressed as the mass of substrate consumed (e.g. ng DNA min^{-1}), or amount of product formed (e.g. nmol glucose min^{-1}). You must ensure that your units clearly specify the substrate or product used, especially when the enzyme transformations involve different numbers of substrate or product molecules. Specific activity, expressed in terms of the amount or mass of substrate or product, is useful for comparing the purity of different enzyme preparations (e.g. p. 375).

Example The hydrolysis of one molecule of maltose to give two glucose molecules by α-glucosidase means that enzyme activity specified in terms of substrate consumption (nmol maltose) would be half the value expressed with respect to product formation (nmol glucose).

The turnover number of an enzyme is the amount of substrate (mol) converted to product in 1 s by 1 mol of enzyme operating under optimum conditions. In practice, this requires information on the molecular weight of the enzyme, the amount of enzyme present and its maximum activity.

Definition

Specific activity – enzyme activity (e.g. kat, U, ng min^{-1}) expressed per unit mass of protein present (e.g. mg, µg).

Types of assay

The rate of substrate utilisation or product formation must be measured under controlled conditions, using some characteristic which changes in direct proportion to the concentration of the test substance.

Spectrophotometric assays

Many substrates and products absorb visible or UV light and the change in absorbance at a particular wavelength provides a convenient assay method (p. 282). In other cases, a product may be measured by a colorimetric chemical reaction.

Several assays are based on interconversion of the nicotinamide adenine dinucleotide coenzymes NAD^+ or $NADP^+$ which are reversibly reduced in many enzymic reactions. The reduced form (either NADH or NADPH) can be detected at 340 nm, where the oxidised form has negligible absorbance (p. 282). An alternative approach is to use a coupled enzyme assay, where a product of the test enzyme is used as a substrate for a second enzyme reaction which involves oxidation/reduction of nucleotide coenzymes. Such assays are particularly useful for continuous monitoring of enzyme activity and for reactions where the product from the test substance is too low to detect by other methods, since coupled assays are more sensitive. Note that the reaction of interest (test enzyme) must be the rate-limiting process, not the indicator reaction (second enzyme).

Example Phosphoenolpyruvate carboxylase (PEP carboxylase) can be assayed by coupling to malate dehydrogenase (MDH): the PEP carboxylase reaction converts phosphoenolpyruvate to oxaloacetate, which is then oxidised to malate by MDH (present in excess), with a stoichiometric (1:1) reduction of NAD^+. The coupled assay is monitored spectophotometrically as an increase in A_{340} against time (p. 282).

Radioisotopic assays

These are useful where the substrate and product can be easily separated, e.g. in decarboxylase assays using a ^{14}C-labelled substrate, where gaseous $^{14}CO_2$ is produced.

Working with radioisotopes – this is covered in Chapter 39.

Electrochemical assays

Enzyme reactions involving acids and bases can be monitored using a pH electrode (p. 154), though the change in pH will also affect the activity of the test enzyme. An alternative approach is to measure the amount of acid or alkali required to maintain a constant preselected pH in a pH-stat.

An oxygen electrode can be used if O_2 is a substrate or a product (p. 340). Other ion-specific electrodes can monitor ammonia, nitrate, etc.

Chromogenic and fluorogenic enzyme substrates

Artificial substrates that generate either (i) a coloured product (i.e. they are chromogens) or (ii) a fluorescent product (fluorogens) are used widely for the detection and quantification of enzymes, especially when no suitable spectrophotometric assay is available for the natural product or the natural substrate. In most cases, these artificial substrates are analogues of the natural substrates, composed of a 'core molecule' that is either fluorescent (p. 284) or coloured, coupled to another group by a covalent bond. The addition of this group to the core molecule usually either causes a decrease in fluorescence and a shift in the excitation and emission signal to longer wavelengths, or converts the coloured compound to a non-coloured form. The covalent bond between core molecule and added group is recognised and cleaved by the target enzyme to liberate the core molecule, thereby generating either a fluorescent or a coloured product that can be measured fluorimetrically (p. 284) or spectrophotometrically (p. 281). The principle is best illustrated using a specific example of each type of substrate (see Fig. 55.1). Such substrates serve as a sensitive and specific means of detecting individual enzymes in the presence of a range of other biomolecules. Applications include:

Understanding fluorescence – the basis of fluorescence is explained on p. 284.

- biochemical identification of bacteria, on the basis of enzyme profiles (p. 218, *see also* Fig. 55.1);

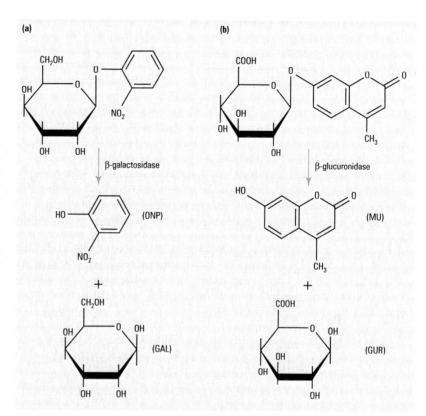

Fig. 55.1 (a) The chromogenic substrate *ortho*-nitrophenyl-β-D-galactoside (ONP-β-GAL); (b) the fluorogenic substrate 4-methylumbelliferyl-β-D-glucuronide (MU-β-GUR). Both substrates show minimal colour and/or fluorescence due to the coupling of the core molecule to a carbohydrate group. ONP-β-GAL is cleaved by the enzyme β-galactosidase to liberate ONP, which is yellow (usually assayed spectrophotometrically at 420 nm). In contrast, the cleavage of MU-β-GUR by the enzyme β-glucuronidase liberates MU, which is strongly fluorescent under UV light (e.g. excitation wavelength 360 nm, emisson wavelength 440 nm). These two substrates are used to detect and differentiate between coliform bacteria (which contain β-galactosidase) and *Escherichia coli* (which contains both β-galactosidase and β-glucuronidase), e.g. in the IDEXX Colilert system for water analysis (see: http://www.idexx.com/water/colilert/).

Example Detection of enzymes that degrade β-lactam antibiotics in resistant bacteria can be carried out using the chromogenic substrate nitrocefin, which changes from pale yellow to red when hydrolysed by β-lactamases.

Table 55.1 Core molecules used in chromogenic and fluorogenic enzyme substrates

Core molecule	Reaction
Alizarin	Red colour
Aminomethyl coumarin	Blue fluorescence
Fluorescein	Green fluorescence
Indoxyl (and derivatives)	Blue colour (and others)
Methylumbelliferone	Blue fluorescence
Nitroaniline	Yellow colour
Ortho-nitrophenol	Yellow colour
Tetramethylbenzidine	Blue (yellow at low pH)
Rhodamine	Red fluorescence

- enzyme-linked immunoassays, e.g. ELISA (p. 261);
- nucleic acid hybridisation and blotting (p. 433);
- detection of specific enzymes in living cells;
- quantitative assay for enzyme activity – e.g. fluorogenic substrates for HIV protease.

Note that care is required for quantitative fluorimetric assay, since impurities in the enzyme preparation can lead to background fluoresence or to quenching (reduction) of the signal: appropriate controls must be run at the same time as test samples. A wide range of chromogenic core molecules (chromophores) are used in artificial substrates (Table 55.1), including: nitrophenols (e.g. ONP, Fig. 55.1); nitroanilines (e.g. Z-arginine-*para*-nitroanilide derivatives for assaying trypsin activity); indoxyl substrates (e.g. 5-bromo-4-chloro-3-indolyl-β-D-galactoside, also known as X-GAL, which is used widely to detect the activity of the *lacZ* gene in molecular biology, p. 448), along with derivatives of alizarin and 3,3′,5,5′-tetramethylbenzidine (Fig. 55.2). Similarly, several different fluorophores are commercially available, including coumarin derivatives (e.g. 4-methylumbelliferone, MU, Fig. 55.1, and 7-amino-4-methylcoumarin, AMC, which fluoresce blue), fluorescein (Fig. 41.4) and resorufin (Table 55.1). For further details see James (1994).

Methods of monitoring substrate utilisation/product formation

Continuous assays (kinetic assays)

The change in substrate or product is monitored as a function of time, to provide a progress curve for the reaction (Fig. 55.3). These curves start off

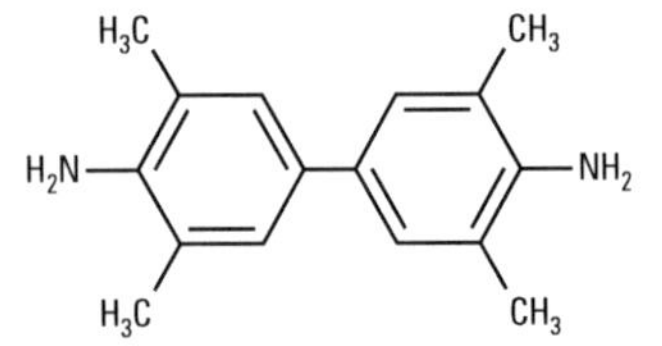

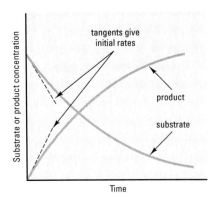

Fig. 55.2 The chromophore 3,3',5,5'-tetramethylbenzidine (TMB), used in peroxide assays, e.g. in ELISA (p. 263).

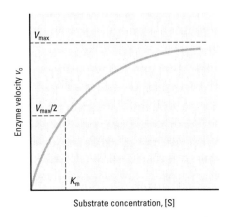

Fig. 55.3 Enzyme reaction progress curve: substrate utilisation/product formation as a function of time.

Fig. 55.4 Effect of substrate concentration on enzyme activity.

in a near-linear manner, decreasing in slope as the reaction proceeds and substrate is used up. The initial velocity of the reaction (v_0) is obtained by drawing a tangent to the curve at zero time and measuring its slope. Continuous monitoring can be used when the test substance can be assayed rapidly (and non-destructively), e.g. using a chromogenic substrate. Reaction rate analysers allow simultaneous addition of reactant(s) or enzyme, mixing and measurement of absorbance: this enables the initial rate to be determined accurately.

Discontinuous assays (fixed time assays)

It is sometimes necessary to measure the amount of substrate consumed or product formed after a fixed time period, e.g. where the test substance is assayed by a (destructive) colorimetric chemical method. It is vital that the time period is kept as short as possible, with the change in substrate concentration limited to around 10%, so that the assay is within the linear part of the progress curve (Fig. 55.3). A continuous assay may be carried out as a preliminary step, in order to determine whether the reaction is approximately linear over the time period to be used in the fixed time assay.

Enzyme kinetics

For most enzymes, when the initial reaction rate (v_0) using a fixed amount of an enzyme is plotted as a function of the concentration of a single substrate [S] with all other substrates present in excess, a rectangular hyperbola is obtained (Fig. 55.4). At low substrate concentrations v_0 is directly proportional to [S], with a decreasing response as substrate concentration is increased until saturation is achieved. The shape of this plot can be described by a mathematical relationship, known as the Michaelis–Menten equation:

$$v_0 = \frac{V_{max}[S]}{K_m + [S]} \quad [55.2]$$

This equation makes use of two kinetic constants:

- V_{max}, the maximum velocity of the reaction (at infinite substrate concentration).
- K_m, the Michaelis constant, is the substrate concentration where $v_0 = \frac{1}{2}V_{max}$.

V_{max} is a function of the amount of enzyme and is the appropriate rate to use when determining the specific activity of a purified enzyme. The Michaelis constant is expressed in terms of substrate concentration (mol l^{-1}) and is independent of enzyme concentration. K_m is derived from the individual rate constants of the reaction – for example, with a single substrate (S) and single product (P) enzymic reaction, the process can be described as follows:

$$E + S \underset{k_2}{\overset{k_1}{\rightleftharpoons}} ES \overset{k_3}{\longrightarrow} E + P \quad [55.3]$$

where E = enzyme, ES = enzyme–substrate complex, and k_1, k_2 and k_3 are rate constants. The Michaelis constant can be expressed as:

$$K_m = \frac{k_2 + k_3}{k_1} \quad [55.4]$$

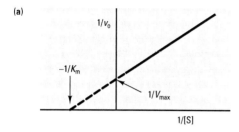

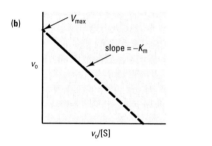

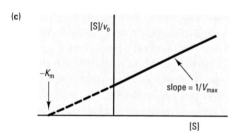

Fig. 55.5 Graphical transformations for determining the kinetic constants of an enzyme. (a) Lineweaver–Burk plot. (b) Eadie–Hofstee plot. (c) Hanes–Woolf plot.

Example The EMBOSS free, open-source software package (at http://emboss.sourceforge.net/) can be used to find the K_m and V_{max} of an enzyme, using Michaelis–Menten and Hanes–Woolf plots.

For many enzymes, $k_3 \ll k_2$, and Eqn [55.4] simplifies to:

$$K_m = \frac{k_2}{k_1} = \frac{[E][S]}{[ES]} \qquad [55.5]$$

When this applies, K_m provides a measure of the affinity of an enzyme for the substrate, and this is an important characteristic of each particular enzyme. Values for mammalian enzymes usually fall within the range 10^{-2}–10^{-5} mol l^{-1}. Thus an enzyme with a large K_m usually has a low affinity for its substrate, while an enzyme with a small K_m usually has a high affinity. K_m values can be used to select a substrate concentration that will give maximum reaction velocity (p. 386) or to compare the affinities of different substrates for a given enzyme, or the same substrate with different enzymes.

Your first step in determining the kinetic constants for a particular enzyme is to measure the rate of reaction at several substrate concentrations, as in Fig. 55.4. There are various ways to obtain K_m and V_{max} from such data, mostly involving drawing a graph representing a linear transformation of Eqn [55.2]:

- The Lineweaver–Burk plot: a graph of the reciprocal of the reaction rate ($1/v_0$) against the reciprocal of the substrate concentration ($1/[S]$) gives $-1/K_m$ as the intercept of the x-axis and $1/V_{max}$ as the intercept of the y-axis (Fig. 55.5a). The slope of the plot is most affected by the least accurate values, i.e. those measured at low substrate concentration.
- The Eadie–Hofstee plot: v_0 against $v_0/[S]$, where the intercept on the y-axis gives V_{max} and the slope equals $-K_m$ (Fig. 55.5b).
- The Hanes–Woolf plot: $[S]/v_0$ against $[S]$, giving $-K_m$ as the intercept of the x-axis and $1/V_{max}$ from the slope (Fig. 55.5c).
- The direct linear plot (Isenbhal and Cornish) is another option that may give more accurate estimates of kinetic constants than the other plots.

There are several computer packages that will plot the above relationships and calculate the kinetic constants from a given set of data using linear regression analysis (p. 501). While the Eadie–Hofstee and Hanes–Woolf plots distribute the data points more evenly than the Lineweaver–Burk plot, the best approach to such data is to use non-linear regression on untransformed data. This is usually outside the scope of the simpler computer programs, though tailor-made commercial packages can carry out such analyses. Note also that some enzymes, particularly those involved in the control of metabolism, do not show Michaelis–Menten kinetics.

Factors affecting enzyme activity

If you want to measure the maximum rate of a particular enzyme reaction, you will need to optimise the following:

Cofactors

Many enzymes require appropriate concentrations of specific cofactors for maximum activity. These are subdivided into coenzymes (soluble, low molecular weight organic compounds which are actively involved in catalysis by accepting or donating specific chemical groups, i.e. they are cosubstrates of the enzyme; examples include NAD$^+$ and ADP); and activators (inorganic metal ions, required for maximal activity, e.g. Mg^{2+}, K$^+$).

Removing a bottle of freeze-dried enzyme from a freezer or fridge – do not open it until it has been warmed to room temperature or water may condense on the contents – this will make weighing inaccurate and may lead to loss of enzyme activity.

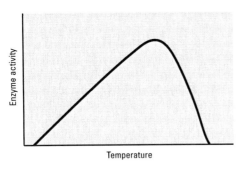

Fig. 55.6 Effect of temperature on enzyme activity.

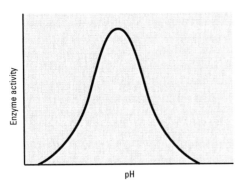

Fig. 55.7 Effect of pH on enzyme activity.

Definitions

Inhibition – the reduction in enzyme activity due to the presence of another compound (an inhibitor).

Negative feedback – a process within a metabolic pathway where the product of a reaction inhibits an enzyme earlier in the pathway, resulting in a reduction in product formation.

Temperature

Enzyme activity increases with temperature, until an optimum is reached. Above this point, activity decreases as a result of protein denaturation (Fig. 55.6). Note that the optimum temperature for enzyme *activity* may not be the same as that for maximum *stability* (enzymes are usually stored at temperatures near to or below 0 °C, to maximise stability).

pH

Enzymes work best at a particular pH, due to changes in ionisation of the substrates or of the amino acid residues within the enzyme (Fig. 55.7). Most enzyme assays are performed in buffer solutions (p. 156), to prevent changes in pH during the assay. Note that some enzymes have different pH optima for different substrates.

Substrate concentration

A substrate must be present in excess to ensure maximum reaction velocity. The K_m of an enzyme can be used to predict the substrate concentration required for the enzyme to operate at or near its maximum rate – this occurs when all active sites of the enzyme are filled. Using the Michaelis–Menten equation, the fraction of active sites filled (f_{ES}) for a given reaction velocity, v_o, is given by:

$$f_{ES} = \frac{v_o}{V_{max}} = \frac{[S_o]}{[S_o] + K_m} \qquad [55.6]$$

When $[S_o] = K_m$, 50% of active sites are filled, and the reaction proceeds at $\frac{1}{2}V_{max}$. When $[S_o]$ is 10-fold greater than K_m, 91% of active sites are filled. When $[S_o]$ is 100-fold greater than K_m, 99% of active sites are filled and the reaction proceeds at 99% of V_{max}.

> **KEY POINT** Note that temperature and pH optima are dependent upon reaction conditions, including cofactor and substrate concentrations – you should therefore specify the experimental conditions under which such optima are determined.

Enzyme inhibition

It is important to investigate the inhibition of enzyme activity by specific molecules and ions because:

- enzyme inhibition is an important control mechanism in biological systems (e.g. negative feedback by a product);
- many drugs act by inhibiting enzymes;
- the action of many toxins can be explained by enzyme inhibition.

In terms of their interaction with enzymes, inhibitors can be involved in either reversible or irreversible reactions, and the inhibition can be competitive, non-competitive or uncompetitive, depending on whether inhibition can be reduced by an increase in the concentration of the natural substrate.

Irreversible inhibitors

These are substances, usually not of biological origin, which react covalently with an enzyme (E), preventing substrate binding or catalysis,

Determining the effects of an inhibitor on enzyme kinetics – measure enzyme activity at ⩾ 6 different substrate concentrations, both in the presence and absence of inhibitor (see Fig. 55.9).

e.g. iodoacetamide, which binds to thiol groups in enzymes as follows:

$$E\text{—}CH_2\text{—}SH + ICH_2CONH_2 \longrightarrow E\text{—}CH_2\text{—}S\text{—}CH_2CONH_2 + HI \quad [55.7]$$

The thiol group may be within the active site, forming part of the binding and/or the catalytic sites, or it may be further from the active site and affect the 3D conformation of the enzyme.

The toxicity of heavy metal ions (e.g. Hg^{2+}) is largely due to their irreversible effects on enzyme activity. Other examples include the inhibition of acetylcholinesterase by organophosphate pesticides and nerve agents. In such cases, it is often an active site serine residue that is affected, e.g. with diisopropylfluorophosphate (DFP):

$$E\text{—}CH_2OH + F\text{—}\overset{\overset{R}{|}}{\underset{\underset{R}{|}}{P}}\text{=}O \rightarrow E\text{—}CH_2O\text{—}\overset{\overset{R}{|}}{\underset{\underset{R}{|}}{P}}\text{=}O + HF \quad [55.8]$$

where R = isopropyl residue.

Reversible inhibitors

These inhibitors do not react covalently with an enzyme, but show rapid reversible binding and dissociation. The velocity of the enzyme-catalysed reaction is reduced by the formation of enzyme–inhibitor (EI) or enzyme–substrate–inhibitor (ESI) complexes. Reversible inhibitors are subdivided according to their effects on K_m and V_{max}:

- Competitive inhibitors bind reversibly to groups in the active site. Often the inhibitor resembles the substrate, and the occupation of the active site by an inhibitor molecule prevents a substrate molecule from binding to the same active site (Fig. 55.8a). The enzyme (E) can bind to the substrate (S), to form an ES complex, or to the inhibitor (I), to form EI, but cannot bind to both (ESI). A competitive inhibitor lowers the rate of catalysis by reducing the proportion of enzyme molecules bound to the substrate. It is possible to reverse competitive inhibition by increasing the substrate concentration; V_{max} is unaffected, but K_m increases (see p. 388). An example of competitive inhibition is the inhibition of succinate dehydrogenase by malonate, an analogue of the natural substrate, succinate.

- Non-competitive reversible inhibition occurs when an inhibitor binds to an enzyme whether or not the active site is occupied by the substrate. This type of inhibition often involves natural inhibitors of enzymes, and is important in the control of metabolism. The inhibitor binds to a site other than the active site (Fig. 55.8b). K_m is unchanged because the affinity of substrate molecules that bind to any uninhibited enzyme is unaffected, but V_{max} decreases due to the concentration of active enzyme molecules being effectively reduced by the presence of inhibitor.

- Uncompetitive inhibition, where the inhibitor binds only after the substrate has bound to the enzyme (Fig. 55.8c). It is most common in reactions involving more than one substrate.

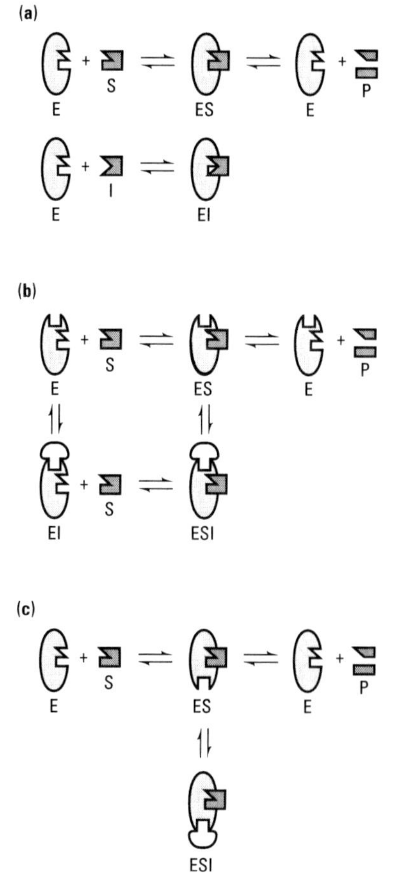

Fig. 55.8 Representation of (a) competitive inhibition, where the inhibitor (I) binds to the same site on the enzyme (E) as the substrate (S); (b) simple linear non-competitive inhibition; (c) uncompetitive inhibition (P — products).

Using enzyme inhibition kinetics to identify the type of inhibitor

Enzyme kinetics can be used to distinguish between the different forms of inhibition and to provide quantitative information on the effectiveness

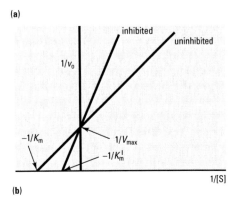

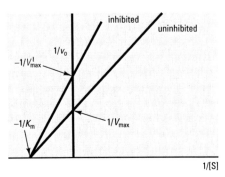

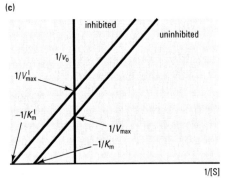

Fig. 55.9 Lineweaver–Burk plots showing the effect of (a) competitive inhibition, (b) simple linear non-competitive inhibition, (c) uncompetitive inhibition.

of various inhibitors, by allowing the dissociation constant of the enzyme–inhibitor complex, K_i, to be determined. K_i is an expression that relates to the strength of binding of the inhibitor to the enzyme.

Competitive inhibition

With inhibitor present, the enzyme can react with the substrate (Eqn [55.3]), but can also react reversibly with the inhibitor (I), to give an inactive enzyme–inhibitor complex (EI), as follows:

$$E + I \underset{k_2'}{\overset{k_1'}{\rightleftharpoons}} EI \qquad [55.9]$$

As a result, the presence of I decreases the amount of free enzyme [E] available for interaction with S, i.e. [E] = [E_T] − [ES] − [EI], where E_T is the total amount of enzyme present. The concentration of the EI complex depends on the concentration of free inhibitor and on the dissociation constant, K_i, where:

$$K_i = \frac{[E][I]}{[EI]} = \frac{k_2'}{k_1'} \qquad [55.10]$$

The Michaelis–Menten equation (Eqn [55.2]) must be modified to account for the presence of inhibitor and for a competitive inhibitor, and results in the Lineweaver–Burk plot shown in Figure 55.9a. The lines in the presence and absence of inhibitor intersect on the ordinate (y-axis), indicating that with competitive inhibition, the inhibitory effect disappears at high substrate concentration (when 1/[S] = 0, [S] = ∞). The value of K_i can be calculated from knowledge of the K_m values obtained in the presence and absence of inhibitor (intercepts on the x-axis), and the following relationship applies:

$$\text{slope}_{\text{inhibited}} = \text{slope}_{\text{uninhibited}} \frac{(1 + [I])}{K_i} \qquad [55.11]$$

Non-competitive inhibition

In this case, not only can I bind E, but I can also bind to the ES complex to give an inactive ESI complex, i.e. E + I ⇌ EI and ES + I ⇌ ESI. In the simplest case, the binding of S has no effect on the binding of I, and *vice versa*, so the dissociation constant of ESI is the same as that of ES, i.e. the K_m. A typical Lineweaver–Burk plot for a non-competitive inhibitor is shown in Fig. 55.9b. This indicates that this type of inhibitor decreases V_{max} but does not affect K_m. Effectively, this means that the inhibitor removes a certain fraction of active enzyme from operation, no matter what the concentration of substrate. The V_{max} changes by a factor of $(1 + [I])/K_i$ and K_i can be obtained by comparison of the slopes obtained with the uninhibited enzyme and the inhibited enzyme, as described above for a competitive inhibitor.

Uncompetitive inhibition

An uncompetitive inhibitor leads to a Lineweaver–Burk plot as shown in Fig. 55.9c. Parallel lines are obtained, and both K_m and V_{max} are affected.

Here K_i pertains to the dissociation of I from ESI. Uncompetitive inhibitors are fairly rare.

Regulatory enzymes

In cell metabolism, groups of enzymes work together in sequential pathways to carry out a given metabolic process. In such enzyme systems, the reaction product of the first enzyme becomes the substrate for the next. Most of the enzymes in each system obey Michaelis–Menten kinetics. However in each system, there is at least one enzyme – often the first in the sequence – that sets the rate of the overall sequence (the flux through the pathway) because it catalyses the slowest or rate-limiting step. These regulatory enzymes exhibit increased or decreased catalytic activity in response to certain signals. By the action of such regulatory enzymes, the rate of each metabolic sequence is constantly adjusted to meet changes in the cell's demands for energy and biosynthesis.

The activities of regulatory enzymes are altered by non-covalent binding of various types of signal molecules, which are generally small metabolites or cofactors termed 'modulators'. Such enzymes are generally described as 'allosteric'.

Properties of allosteric enzymes

The following general characteristics enable allosteric enzymes to be identified:

- Feedback inhibition: in some multi-enzyme systems the regulatory enzyme is specifically inhibited by the end-product of the pathway whenever the end-product increases in excess of the cell's needs. The end-product does not bind to the active site but binds to another, specific, site termed the regulatory site. This binding is non-covalent and readily reversible. Thus if the concentration of the end-product decreases, the rate of enzyme activity increases.
- Modulators for allosteric enzymes may be either inhibitory or stimulatory. An activator is often the substrate itself, and regulatory enzymes for which substrate and modulator are identical are called homotropic. When a modulator is a molecule other than the substrate the enzyme is heterotropic. Some enzymes have two or more modulators.
- Each enzyme molecule has one or more regulatory or allosteric sites for binding the modulator. Just as the enzyme's active site is specific for its substrate, the allosteric site is specific for its modulator. Enzymes with several modulators generally have different specific binding sites for each. In homotropic enzymes the active site and regulatory site are the same.
- These enzymes are generally larger and more complex than simple enzymes and normally consist of two or more subunits. A useful technique for preliminary determination of subunit structure is SDS-PAGE (Chapter 46).
- Michaelis–Menten kinetics are not followed. When v_o is plotted against [S] a sigmoid saturation curve normally results (Fig. 55.10), rather than the hyperbolic curve shown by non-regulatory enzymes (Fig. 55.4). With sigmoidal kinetics, although a value of [S] can be determined at which v_o is half maximal, this value is not equivalent to K_m. Instead the

> **Definition**
>
> **Allosteric enzyme** – an enzyme having more than one shape or conformation induced by the binding of modulators (Greek allos, 'other'; stereos, 'solid' or 'shape').

> **Example** In the multi-stage conversion of L-threonine into L-isoleucine in bacteria, the first enzyme in the sequence (threonine dehydratase) is inhibited by isoleucine, the final product.

> **Example** Aspartate transcarbomylase, the principal regulatory enzyme in the synthetic pathway for cytidine triphosphate, has six catalytic subunits and six regulatory subunits.

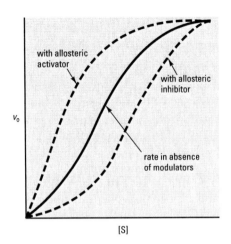

Fig. 55.10 Allosteric effects in enzyme kinetics. The graph shows the rate of enzyme reaction (v_o) against substrate concentration [S].

Example The binding of adrenaline to skeletal muscle receptors triggers reactions that culminate in phosphorylation of a serine residue in glycogen phosphorylase: this activates the enzyme, increasing glycogen breakdown.

symbol $[S]_{0.5}$, or $K_{0.5}$, is often used to represent the substrate concentration giving half maximum velocity of the reaction catalysed by an allosteric enzyme. Sigmoid kinetic behaviour generally reflects cooperative interactions between multiple protein subunits, i.e. changes in the structure of one subunit result in structural changes in adjacent subunits that affect substrate binding. In general, allosteric activators cause the curve to become more nearly hyperbolic (Fig. 55.10), with a decrease in $K_{0.5}$, but no change in V_{max}, and therefore v_o is higher for any value of substrate concentration. However, some allosteric enzymes respond to an activator by an increase in V_{max} with little change in $K_{0.5}$. In contrast, an allosteric inhibitor may produce a more sigmoidal curve with an increase in $K_{0.5}$ (Fig. 55.10).

Covalently modified regulatory enzymes

Some enzymes are regulated by covalent modification, e.g. by phosphorylation of the hydroxyl groups of specific serine, threonine or tyrosine residues. Depending on the enzyme, this may result in activation or inactivation: in most cases, the non-phosphorylated form of the enzyme is relatively inactive. Often these modifications are under hormonal control. Some regulatory enzymes are influenced both by covalent modification and by allosteric effects.

Text reference

James, A.L. (1994) Enzymes in Taxonomy and Diagnostic Bacteriology. In M. Goodfellow and A.G. McDonnell (eds) *Chemical Methods in Bacterial Systematics*, pp. 471–92. Wiley, London.

Sources for further study

Anon. *International Union of Biochemistry and Molecular Biology*. Available: http://www.iubmb.unibe.ch/ Last accessed: 01/04/07.
[Links to enzyme nomenclature database and other relevant websites]

Bisswanger, H. (2002) *Enzyme Kinetics: Principles and Methods*. Wiley–VCH, Weinheim.

Cook, P.F. and Cleland, W.W. (2007) *Enzyme Kinetics and Mechanism*. Garland Science, New York.

Copeland, R.A. (2000) *Enzymes: A Practical Introduction to Structure, Mechanism, and Data Analysis*, 2nd edn. Wiley, New York.
[Covers basic principles, kinetics and experimental systems]

Cornish-Bowden, A. (2004) *Fundamentals of Enzyme Kinetics*. Portland Press, London.

Eisethal, R. and Danson, M. (2002) *Enzyme Assays: A Practical Approach*, 2nd edn. Oxford University Press, Oxford.

Marangoni, A.G. (2002) *Enzyme Kinetics: a Modern Approach*. Wiley, New York.

Purich, D.L. and Allison, R.D. (2000) *Handbook of Biochemical Kinetics*. Academic Press, New York.

Segel, I.H. (1993) *Enzyme Kinetics: Behavior and Analysis of Rapid Equilibrium and Steady-State Enzyme Systems*. Wiley, New York.
[In-depth coverage of theoretical and practical aspects]

Taylor, K.B. (2002) *Enzyme Kinetics and Mechanisms*. Kluwer, Dordrecht.
[Deals with steady-state enzyme kinetics]

Study exercises

55.1 Determine enzyme activity and purity (specific activity). An extract of α-glucosidase containing 2.5 mg of protein was incubated with p-nitrophenol-α-D-glucose at 1.0 mmol l^{-1} and assayed by following the liberation of p-nitrophenol at 404 nm. This gave a progress curve with an initial velocity equal to an absorbance change of 0.25 min^{-1}. Given that the conversion of 10 nmol of substrate to product results in an absorbance change of 0.122:

(a) Determine the amount of product formed in nmol min^{-1};
(b) Calculate the enzyme activity in kat;
(c) Determine the specific activity, in terms of (i) nmol min^{-1} (mg protein)$^{-1}$ and (ii) kat (mg protein)$^{-1}$.

(Quote all answers to three significant figures.)

55.2 Estimate K_m and V_{max} for an enzyme. The table below shows representative data for enzyme activity as a function of substrate concentration. Use a suitable graphical method to determine the kinetic constants K_m and V_{max} of the enzyme (give your answers to three significant figures).

Enzyme activity at various substrate concentrations

Substrate concentration (μmol l^{-1})	Enzyme activity (U)
0	0
5	16.2
7	20.5
10	28.1
15	32.7
25	39.2
50	56.7
100	68.2

55.3 Identify the type of enzyme inhibition shown by therapeutic drugs. In the treatment of acquired immunodeficiency syndrome (AIDS), a possible mode of therapy is to inhibit the reverse transcriptase (RT) of the human immunodeficiency virus (HIV), which is required for the retrovirus to be propagated by RNA-directed DNA synthesis. In the figure, one of the substrates for RT is thymidine (a); two drugs, AZT (b) and HBY097 (c) are known to inhibit HIV RT.

Look at the structures and predict the type of inhibition (i.e. competitive or non-competitive) likely to be shown by each drug. Outline an experiment that would enable you to confirm the type of inhibition by investigating enzyme kinetics and explain how you would interpret the results.

56 Membrane transport processes

Choosing an experimental system – transport studies can be carried out using whole multicellular organisms (e.g. fish), tissue slices/discs from animals or plants, bacterial cell suspensions or subcellular membrane vesicles, prepared by homogenisation and fractionation (Chapter 36).

All cells operate a variety of processes whereby solutes move across membranes, including those processes involved in the uptake of nutrients and the loss of metabolic waste products at the plasma membrane, or the intracellular transport of substances across internal membranes, e.g. mitochondria. Membrane transport processes serve to regulate the intracellular environment, creating suitable conditions for internal metabolism. In microbes, the plasma membrane plays the principal role in the selective movement of biomolecules into, or out of, the cell. In multicellular animals and plants, membrane transport processes in the outermost epithelial layers are especially important in determining the solute composition of tissues and organs. Individual organs often have specialised transport functions, e.g. in animals, nutrient absorption by intestinal epithelial cells and urea excretion *via* the proximal tubular epithelial cells of the kidney or, in plants, the uptake of nutrients by root hair cells.

Measuring solute transport

Usually, the rate of movement of a solute in a particular biological system is determined using a radioisotope-labelled (p. 267) or a stable isotope-labelled solute (p. 294) as a 'tracer'. Typically, in solute uptake (influx), the labelled solute is added to the external medium and its accumulation is then monitored by assaying samples of cells or tissue after known time intervals, enabling you to plot the accumulation of tracer as a function of time. Loss of solute (efflux) is studied using cells or tissues pre-loaded with labelled solute and then transferred to a tracer-free incubation medium: samples of the medium can be assayed after known time intervals, allowing the time course to be followed without destruction of the biological material. For microbial cell suspensions, filtration or silicone oil microcentrifugation (p. 299) may be used to separate the cells from their incubation medium.

KEY POINT Determination of the relationship between the amount of tracer and the amount of solute (the 'specific activity', p. 270) enables the uptake or loss of solute to be expressed in terms of the net amount of solute transported per unit biomass (e.g. per mg protein, or per cell) per unit time. If the relationship between biomass and membrane area is known, the net uptake or loss of solute can be expressed per m² surface area per unit time.

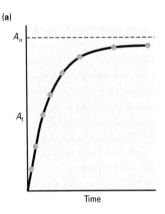

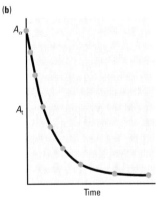

Fig. 56.1 Time course of (a) tracer influx and (b) tracer efflux (A_t). A_α, tracer activity at equilibrium.

Short-term tracer studies will give approximate values for the unidirectional flux rate from the labelled side of the membrane: longer-term studies may show more complex kinetics, due to two-way tracer movement or to secondary transport within the cells.

For non-metabolised solutes that are accumulated to particular levels within individual cells, e.g. inorganic ions such as K^+, Na^+, Cl^-, the time course for tracer accumulation or loss will often show exponential kinetics, as the transport process allows the tracer to equilibrate across the cell membrane (Fig. 56.1). Such exponential plots can be transformed to give linear plots using the relationship $\ln(1 - A_t/A_\alpha)$ for influx, where A_α is the activity at equilibrium and A_t is the activity after time = t, while a

Example Suppose 5 ml of a red cell suspension containing 2 mg protein ml⁻¹ accumulated 1050 Bq of radiolabelled solute of specific activity 210 Bq nmol⁻¹ in 8 min. The net rate of solute accumulation over this time period, expressed in nmol mg protein⁻¹ min⁻¹, would be 1050 ÷ 210 ÷ 10 ÷ 8 = 0.0625 (≡ 62.5 pmol mg protein⁻¹ min⁻¹).

Membrane Transport Processes

plot of $\ln A_t/A_\alpha$ should be linear when the cell behaves as a single compartment with respect to tracer exchange. The slope of such a plot gives a rate constant for exchange (k) that can be used to quantify the unidirectional solute movement, either influx or efflux (ϕ) from the following equation:

$$\phi = \frac{k[C_i]}{M_a} \qquad [56.1]$$

where $[C_i]$ is the internal solute concentration, determined either from tracer equilibration or by chemical assay (p. 349), and M_a is the membrane surface area, expressed per unit volume. For some cells and tissues, an initial, short-term component is observed, due to tracer exchange between the extracellular space and the bulk medium – this can be subtracted from the longer-term component, allowing correction for extracellular tracer content.

Metabolised solutes pose additional problems, since they do not simply equilibrate across membranes – for example, glucose may be used as a source of carbon or energy after being transported into a cell. Transport studies of metabolised solutes may need to account for such processes, e.g. by separating and assaying the various labelled metabolites. A simpler alternative is to use a non-metabolised analogue of the solute, enabling transmembrane movement to be studied in the absence of metabolism and simplifying the interpretation of tracer studies.

Identifying the transport mechanism for a particular solute

Many studies are carried out to provide information on the nature of the transport process. Solute movement is often subdivided into a number of categories:

Simple diffusion – passive transport

In accordance with Fick's first law of diffusion, an uncharged molecule can move through the lipid bilayer of a membrane with a net flux, J, determined by its lipid solubility and by the difference in concentration (strictly, the difference in activity) across the membrane, according to the relationship:

$$J = P([C_o] - [C_i]) \qquad [56.2]$$

where P is the permeability coefficient for the substance while $[C_o]$ and $[C_i]$ are external and internal solute concentrations respectively. The permeability coefficients of various solutes correlate with their partition coefficients between non-polar organic solvents and water, reflecting their relative solubility in the lipid phase. At a practical level, the flux rate calculated from Eqn [56.2] will apply only to the initial state, since diffusion will reduce the concentration gradient and hence the flux rate.

At a practical level, a number of characteristics can be used to recognise simple diffusion, including:

- The measured rates of transport are consistent with the lipid permeabilities of artificial lipid bilayers (e.g. values in Stein, 1997).
- The rate of transport is directly proportional to the concentration gradient (Eqn [56.2]), giving a straight line relationship with no evidence of saturation, in contrast to *all* other transport systems.
- Movement is insensitive to inhibitors, including structural analogues of the solute, metabolic inhibitors and chemical inactivators.

Example For a tissue with a surface area of 24 000 m² per m³, an internal K⁺ concentration of 230 mol m⁻³, and an influx rate constant of 0.000036 s⁻¹, substitution into Eqn [56.1] gives an influx of: (0.000036 × 230) ÷ 24 000 = 3.45 × 10⁻⁷ mol m⁻² s⁻¹ (345 nmol m⁻² s⁻¹).

Example 3-O-methyl glucose and 2-deoxyglucose are non-metabolised analogues of glucose, while methyl-β-D-thiogalactoside is a non-metabolised analogue of lactose.

Definition

Fick's first law of diffusion – each solute diffuses in a direction that eliminates its concentration gradient and at a rate proportional to the size of the gradient.

Example For a solute with a flux of 9 × 10⁻⁶ mol m⁻² s⁻¹, an internal concentration of 250 mol m⁻³ and an external concentration of 15 mol m⁻³, rearranging Eqn [56.2] gives P = (9 × 10⁻⁶) ÷ (250 − 15) = 3.83 × 10⁻⁸ m s⁻¹ (to three significant figures).

- Passive diffusion lacks specificity – chemically similar solutes with comparable lipid solubilities will show similar rates of transmembrane movement.

Simple diffusion can account for the movement of only a limited number of low M_r, uncharged substances, including O_2, CO_2, NH_3 and non-polar hydrocarbons. There is negligible transbilayer diffusion of all other polar and ionic solutes, due to their hydrophilic, lipophobic nature and to the presence of a surrounding shell of water molecules, preventing movement through the hydrophobic interior of the membrane.

KEY POINT *All solutes not transported across membranes by simple diffusion are moved by protein-mediated transporters, often termed 'permeases', 'porters' or 'translocators'.*

Passive diffusion of water (osmosis) – this is a special case, due to the high permeability of biological membranes to water and the high concentration of water on each side of the membrane.

Facilitated diffusion – passive transport

Here, the transmembrane movement is the result of specific membrane proteins (often termed 'uniporters'). While net movement is energetically 'downhill', as in simple diffusion, the rate of movement is more rapid than would be predicted from Eqn [56.2], since the individual uniporter molecules act as solute-specific channels across the membrane. Examples include glucose transport across the erythrocyte membrane, maltodextrin transport across the outer membrane of Gram-negative bacteria and the stretch-activated ion channels in stomatal membranes. The diagnostic features of facilitated diffusion include:

- The measured rates of transport are far greater than can be accounted for by simple diffusion and Fick's first law: for example, the erythrocyte glucose system operates at more than 10 000 times the rate predicted from permeability coefficients of synthetic lipid bilayers.
- The system exhibits specificity, transporting the solute but not chemically related compounds with a different 3D shape.
- The rate of transport should show saturation kinetics similar to those observed for enzymes (Fig. 56.2), with a hyperbolic relationship between the rate of transport (v) and solute concentration [C] according to the relationship:

$$v = \frac{V_{max}[C]}{K_s + [C]} \qquad [56.3]$$

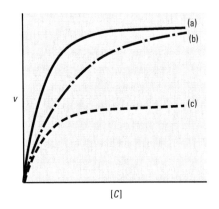

Fig. 56.2 Relationship between the rate of transport (v) and solute concentration [C] for (a) uninhibited 'control', (b) with a competitive inhibitor, (c) with a non-competitive inhibitor.

where V_{max} is the maximum rate of transport and K_s, the half saturation constant, is a measure of the affinity of the transport system for the solute – a low K_s generally indicates a high-affinity transport system while a high K_s shows that the transport system has a low affinity for the solute.

- Movement is sensitive to competitive inhibitors – the presence of structurally similar compounds will reduce the rate of solute transport. Competitive inhibitors decrease the affinity of the permease (increased K_s), but V_{max} is not affected (Fig. 56.2).
- Facilitated diffusion systems are susceptible to inactivation by non-competitive inhibitors that act as protein denaturants, e.g. heavy metal salts. Such irreversible inactivation confirms that the transport process is mediated by a protein. At low concentrations of inhibitor, only some of the permease molecules will be inactivated, giving a decreased V_{max} but unchanged K_s (Fig. 56.2).

Determining kinetic parameters in transport studies – since Eqn [56.3] is equivalent to the Michaelis–Menten equation, the same approaches can be used to determine V_{max} and K_s, based on measurements of the rate of transport at several different solute concentrations (see p. 384).

Example For K^+ (a monovalent cation) in a cell with an internal concentration of 450 mol m^{-3} in an experimental solution at an external concentration of 10 mol m^{-3}, substitution into Eqn [56.4] with a value for the gas constant, $R = 8.31443$ J K^{-1} mol^{-1} and Faraday's constant, $F = 9.648675 \times 10^4$ J V^{-1} mol^{-1} at 20°C (T = 293.15 K) gives a transmembrane equilibrium potential, E_n = 2.303 [(8.31443 × 293.15) ÷ (+1 × 9.648675 × 10^4)] log (10 ÷ 450) = -0.0961782 V (-96.2 mV to three significant figures).

Example A membrane potential of -60 mV (interior negative) can account for a ten-fold passive accumulation of a monovalent cation and a ten-fold passive exclusion of a monovalent anion.

Measuring membrane potentials – depending on the cell type, this can be carried out using microelectrodes, lipophilic cations or fluorescent dyes (see p. 407).

If facilitated diffusion of an uncharged solute occurs, then net solute movement will occur down the transmembrane concentration gradient. However, for charged solutes (ions), net movement will depend on the transmembrane electrical potential (the membrane potential), as well as the concentration gradient. The transmembrane equilibrium potential for a particular ion, E_n, can be calculated from the Nernst equation, as:

$$E_n = 2.303 \frac{RT}{zF} \log \frac{[C_o]}{[C_i]} \qquad [56.4]$$

where all symbols have their usual meanings (p. 185, see also Chapter 48). A comparison of the equilibrium potential and the membrane potential will show whether the ion is in electrochemical equilibrium across the membrane, or whether it has been transported against the electrochemical potential gradient.

 KEY POINT *The transmembrane electrical potential, responsible for the facilitated diffusion of ions, is usually established as a direct result of the action of one or more ion-pumping, active transport systems, e.g. H^+-ATPases.*

Ionophores (Table 56.1) can be viewed as simple models of facilitated diffusion systems: they can be used to study the effects of an increase in the permeability of membranes to a single solute or group of solutes, to create transmembrane movements of ions, or to monitor specific ions in solution.

Table 56.1 Ionophores and inhibitors of transmembrane solute movement*

Ionophores	Effect
valinomycin	electrogenic carrier: increased permeability to K^+ and Rb^+
gramicidin A	electrogenic channel: uniport of monovalent ions, particularly H^+
monensin	electroneutral carrier: exchange of Na^+ or Li^+ for H^+
nigericin	electroneutral carrier: exchange of K^+ or Rb^+ for H^+
A-23187	electroneutral carrier: exchange of divalent cations (e.g. Ca^{++}, Mg^{++}) for $2H^+$
2,4-dinitrophenol (DNP)	electrogenic carrier: increased H^+ permeability (protonophore)
carbonylcyanide m-chlorophenyl-hydrazone (CCCP)	electrogenic carrier: increased H^+ permeability (protonophore)
carbonylcyanide p-trifluoromethoxy-phenylhydrazone (FCCP)	electrogenic carrier: increased H^+ permeability (protonophore)
nonactin	increased permeability of monovalent cations
amphotericin B	anion channel (weak selectivity)
nystatin	anion channel (weak selectivity)
Transport inhibitors[†]	**Effect**
oligomycin	inhibition of F_0F_1-type H^+-ATPases (mitochondria, bacteria, chloroplasts)
ouabain	inhibition of Na^+-K^+-ATPases (mammalian cell plasma membrane)
digitoxigenin	inhibition of Na^+-K^+-ATPases (mammalian cell plasma membrane)
vanadate	inhibition of E_1E_2-type ATPases (plant plasma membrane)
N,N'-dicyclohexylcarbodiimide (DCCD)	inhibition of F_0F_1 and V-type ATPases (mitochondria and plant tonoplast)
N-ethylmaleimide (NEM)	inhibition of plant tonoplast V-type ATPases
phlorizin	inhibition of Na^+-glucose symporters (intestinal epithelial cells)
furosemide	inhibition of cotransporters (mammalian cell membrane)
cytochalasin B	inhibition of glucose transporters (mammalian cell membrane)
amiloride	inhibition of Na^+/H^+ antiporters (mammalian and plant cells)
4,4'-diisothiocyano-2,2'-stilbene-disulphonate (DIDS)	inhibition of mammalian anion antiporters

*Safety note: all of these inhibitors are highly toxic and must be handled carefully, observing the appropriate safety precautions.
[†]Details of electron transport inhibitors are given in Chapter 57.

Active transport

The principal criterion to establish active transport is that movement of the solute occurs against its electrochemical potential gradient. The energy for such 'uphill' transport can be provided by one of two mechanisms:

1. Primary active transport. The movement of solute molecules is coupled to the hydrolysis of ATP, *via* a membrane-bound solute-specific ATPase, or to a redox chain. For example, the Na$^+$-K$^+$ ATPase of mammalian cells is a primary active transporter, moving 3 Na$^+$ ions outwards and 2 K$^+$ ions inwards for every molecule of ATP hydrolysed, i.e. it is *electrogenic*, with a net charge separation across the membrane. Some of the most important primary active transport processes are those which create a transmembrane H$^+$ gradient, or proton-motive force (Δp), either *via* a H$^+$-translocating ATPase or an electron transport chain (p. 407). In many membranes, Δp provides the major driving force for the movement of other solutes (see Nicholls and Ferguson, 1997, for details). Some Gram-negative bacteria use a variation of phosphoryl-driven active transport to accumulate certain sugars – this is known as 'group translocation' since it involves the simultaneous phosphorylation of each sugar molecule as it is transported. Because the cell membrane is impermeable to sugar phosphates, the bacterial 'phosphotransferase' system can concentrate these solutes within cells.

2. Secondary active transport, or secondary transport. Here, the movement of solute is coupled to the 'downhill' movement of another solute, either *via* a 'symporter' (co-transporter) that simultaneously transports the two solutes in the same direction, or an 'antiporter' (counter-transporter) that simultaneously moves the two solutes in opposite directions across the membrane (Fig. 56.3). These systems are 'active' only in the sense that the 'downhill' movement of the other solute can only occur after primary active transport has established a transmembrane solute gradient (e.g. for H$^+$ or Na$^+$). Thus, lactose uptake in bacterial cells is coupled to the transmembrane proton gradient, *via* the lactose–H$^+$ symporter, while Na$^+$ efflux from plant cells is mediated by an Na$^+$–H$^+$ antiporter. In both instances, the H$^+$ gradient is created by a primary active transport system that pumps protons out of the cell, generating a proton-motive force that can be utilised by secondary transport systems. Similarly, the Na$^+$-K$^+$ primary active transport system creates an Na$^+$ gradient that can be utilised for nutrient uptake in mammalian cells, e.g. *via* the Na$^+$-glucose symporter (a secondary transporter). Primary and secondary transport systems can be envisaged as operating as a 'circuit' for the movement of ions across membranes (Fig. 56.3).

In addition to the characteristics of specificity, saturation and inhibition shared by facilitated diffusion, active transport systems can be identified by the following features:

- The transmembrane distribution of the solute is not consistent with Eqn [56.2] (where $J = 0$ at equilibrium, i.e. $[C_o] = [C_i]$) for an uncharged solute, or Eqn [56.4] for an ion.
- The active component will be sensitive to metabolic inhibitors, e.g. ATPase inhibitors, for primary active transport, or specific inhibitors of secondary transporters (see Table 56.1).

Definition

Proton-motive force – an electrochemical gradient of protons (H$^+$) generated by active transport of H$^+$ across a membrane: the gradient can then be used to carry out work, e.g. solute transport or ATP synthesis.

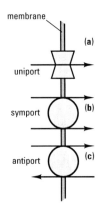

Fig. 56.3 Diagrammatic representation of various membrane transport systems, indicating the direction of movement of transported solute molecules.

Interpreting inhibitor experiments – while short-term effects may be unique to a particular inhibitor, in the longer term more general, non-specific effects are likely. For example, inhibitors of primary active transport will eventually affect secondary transport systems, due to their effects on ion gradients and membrane potentials.

- For secondary transport systems, dissipation of the primary ion gradient using a suitable ionophore will inhibit transport of the solute. Conversely, artificially generated ion gradients can be used to drive solute movement, either in whole cells or in membrane vesicle preparations.

Studying membrane transport using patch-clamp techniques

In this technique, part of a membrane (a 'patch') is sealed to the end of a heat-polished micropipette of diameter $\approx 1\,\mu m$ and the flow of current across the membrane patch is then measured for different bathing solutions and/or transmembrane voltages, set at particular values ('clamped') by a feedback amplifier. The size and orientation of the membrane patch can be controlled, as shown in Figure 56.4. Since the membrane patch can be relatively small, the operation of ion-transporting proteins can be seen as changes in current flow, due to the opening or closing of individual membrane 'channels' (voltage-sensitive transporters). While the physiological factors controlling the *in vivo* operation of such channels remain unclear, patch clamping provides a powerful means of studying channel-mediated ion movements *in vitro*, enabling investigation of:

- the voltage required to produce 'channel' opening;
- the membrane conductance under conditions where the ion concentrations on both sides of the membrane are known;
- temporal effects, including opening and closing times, frequency of opening, etc.;
- the ion selectivity of particular 'channels';
- the effects of inhibitors, ionophores, etc.

When whole cells are used, the flow of ions through primary active ion transport systems (ion 'pumps') can be studied, since changes in current flow represent the sum of many transporters, acting together. Patch-clamp systems have provided a considerable amount of information on the transport properties of individual membranes, e.g. the plasma membrane and tonoplast of plant cells (see Yeo and Flowers, 2007, for details).

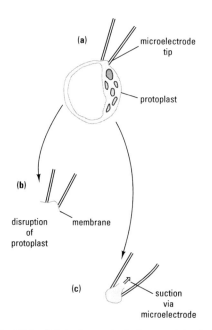

Fig. 56.4 Patch-clamp techniques: (a) whole plant cell, (b) inside-out membrane patch, (c) outside-out membrane patch.

Text references and sources for further study

Baldwin, S.A. (2000) *Membrane Transport: A Practical Approach*. Oxford University Press, Oxford.

Bernhardt, I. and Ellory, J.C. (2003) *Red Cell Membrane Transport in Health and Disease*. Springer-Verlag, Heidelberg.

Blatt, M. (2003) *Membrane Transport in Plants. Annual Plant Reviews*, Vol. 15. Blackwell, Oxford.

Nicholls, D.G. and Ferguson, S.J. (2002) *Bioenergetics*, 2nd edn. Academic Press, London.

Stein, W.D. (1997) *Channels, Carriers and Pumps: An Introduction to Membrane Transport*. Academic Press, New York.

Yan, Q. (2003) *Membrane Transporters: Methods and Protocols*. Humana Press, New York.

Yeo, A.R. and Flowers, T.J. (2007) *Plant Solute Transport*. Blackwell, Oxford.

Study exercises

56.1 Calculate flux rates of various solutes based on data from tracer studies. Using the following data for rate constant (k), internal concentration ($[C_i]$) and membrane surface area (M_a), calculate the unidirectional flux of each solute (express your answers to three significant figures, in all of the study exercises in this chapter):

(a) Solute uptake for a cell where $k = 0.0001\,s^{-1}$, $[C_i] = 150\,mol\,m^{-3}$ and $M_a = 50\,000\,m^2\,m^{-3}$.

(b) K^+ efflux from a protozoan, where $k = -1.65 \times 10^{-4}\,s^{-1}$, $[C_i] = 106\,mol\,m^{-3}$ and $M_a = 8.25 \times 10^4\,m^2\,m^{-3}$.

(c) Na^+ uptake into a bacterial cell, where $k = 4.75 \times 10^{-5}\,s^{-1}$, $[C_i] = 10.5\,mmol\,dm^{-3}$ and $M_a = 1.66 \times 10^{-9}\,m^2\,dm^{-3}$.

56.2 Determine passive transmembrane fluxes of uncharged solutes. What is the direction and magnitude of the net passive transmembrane flux for each of the following cases, all involving uncharged solutes:

(a) Movement of glycerol in a halotolerant bacterial cell with an internal concentration of $10\,mol\,m^{-3}$, an external concentration of $1.5\,mol\,m^{-3}$ and a permeability coefficient of $2 \times 10^{-7}\,m\,s^{-1}$.

(b) Glucose movement in an animal cell with an internal concentration of $8.65\,mol\,m^{-3}$, an external concentration of $0.03\,mol\,m^{-3}$ and a permeability coefficient of $1.6 \times 10^{-9}\,m\,s^{-1}$.

(c) Methanol flow in a fungal cell with an internal concentration of $0.25\,mmol\,dm^{-3}$, an external concentration of $3.45\,mmol\,dm^{-3}$ and a permeability coefficient of $2.6 \times 10^{-3}\,cm\,s^{-1}$.

56.3 Interpret kinetic parameters of transport systems. Details of the kinetic parameters for the transport of an amino acid into cells of three different types are given in the table at the top of the next column.

Kinetic parameters for amino acid uptake into three different cell types

Cell type	V_{max}	K_s
A	$15.2\,pmol\,dm^{-3}\,s^{-1}$	$1.24 \times 10^{-6}\,pmol\,dm^{-3}$
B	$9.6\,nmol\,m^{-3}\,s^{-1}$	$6.54 \times 10^5\,nmol\,m^{-3}$
C	$84 \times 10^{-11}\,mol\,m^{-3}\,s^{-1}$	$2.54 \times 10^{-4}\,mol\,m^{-3}$

(a) Which cell type has the highest maximum rate of transport?
(b) Which cell type has the highest affinity for the amino acid?
(c) Which cell type gives the fastest rate of uptake at an amino acid concentration of $1 \times 10^{-4}\,mol\,m^{-3}$?

56.4 Compare calculated transmembrane equilibrium potentials with the membrane potential to decide whether an ion has been actively transported. At $25\,°C$, a marine algal cell has a measured plasma membrane potential of $-56\,mV$ and the following internal (cytoplasmic) and external ion concentrations:

Internal and external ion concentrations for a marine alga

Ion	Internal concentration ($mol\,m^{-3}$)	External concentration ($mol\,m^{-3}$)
K^+	197	10
Na^+	8.4	225
Cl^-	27.3	235

(a) What is the transmembrane equilibrium potential for each of these ions?
(b) Which ions appear to be in electrochemical equilibrium and which appear to be actively transported (and in what direction)?

57 Photosynthesis and respiration

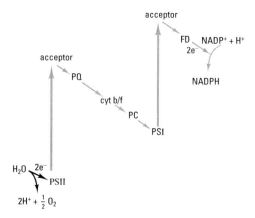

Fig. 57.1 Non-cyclic electron transport from H_2O to $NADP^+$ in plant photosynthesis (Z scheme) showing the principal components: PS II, photosystem II; PQ, plastoquinone; cyt, cytochrome; PC, plastocyanin; PS I, photosystem I; FD, ferredoxin; $NADP^+$, nicotinamide adenine dinucleotide phosphate.

Measuring photosynthetic pigments – equations for spectrophotometric analysis of chlorophylls are given on p. 282.

Definitions

C_3 plants – those in which phosphoglyceric acid is the first stable product.

C_4 plants – those in which malate or aspartate are the principal products, via oxaloacetic acid, in a reaction catalysed by phosphoenolpyruvate (PEP) carboxylase.

CAM plants – those in which malic acid is produced, via oxaloacetic acid, by PEP carboxylase activity at night, with little or no net assimilation of CO_2 during the day (crassulacean acid metabolism).

Techniques for investigating photosynthesis and respiration are considered together, since:

- Gas exchange processes in these two processes may be studied using similar techniques, and interpreted in similar stoichiometric terms, e.g. by relating changes in O_2 or CO_2 to those of other reactants or products.
- Membrane-bound electron transport systems are involved in the production of a transmembrane H^+ gradient that drives ATP synthesis via an ATPase embedded in the same membrane. Measurement of electron transport in photosynthesis and respiration often makes use of redox dyes as artificial electron donors or acceptors, with spectrophotometric analysis of the reactions.
- Some electron transport inhibitors can affect both photosynthesis and respiration, while others are specific to one or the other process.

 KEY POINT Photosynthesis and respiration are complex, multi-stage metabolic pathways involving a large number of cellular components: they may be studied as a whole, or as isolated components and individual metabolic reactions, to gain a deeper insight into the underlying processes.

Photosynthesis

This is the conversion of light energy to chemical energy by photoautotrophic plants and bacteria. The overall process is often summarised in terms of the synthesis of carbohydrate, $(CH_2O)_n$, driven by the energy within photons of light ($h\nu$), as:

$$nCO_2 + 2nH_2O^* + 4nh\nu \rightarrow (CH_2O)_n + nO_2^* + nH_2O \qquad [57.1]$$

where the asterisks show that all of the O_2 is derived from the photolysis of water. Carbohydrate synthesis occurs as a result of two distinct processes, namely (i) light reactions and (ii) dark reactions. In the former, light energy is absorbed by pigments within membrane-bound reaction centres (photosystems). The water-splitting light reactions generate electrons, protons and molecular oxygen. Electron and proton movement via a series of membrane-bound redox carriers leads to the reduction of $NADP^+$ and the generation of ATP (photophosphorylation), driven by a transmembrane H^+ gradient, or proton-motive force (p. 407). The light reactions of plant-type photosynthesis are often represented by the 'Z scheme' (Fig. 57.1).

The dark reactions use the products of the light reactions to 'fix' CO_2, via a series of soluble enzymes known as the Calvin cycle, or reductive pentose phosphate cycle. The primary carboxylating enzyme in temperate green plants (C_3 plants) and cyanobacteria is ribulose bisphosphate carboxylase ('Rubisco'). This enzyme catalyses the addition of CO_2 to ribulose bisphosphate (C_5), producing two molecules of phosphoglyceric acid (C_3) which are then reduced to triose phosphate (C_3) by enzymes that utilise the NADPH and ATP generated by the light reactions. Thus both light and dark reactions operate in a coupled manner when photosynthetically active cells are illuminated with visible light.

> **Definitions**
>
> **Net photosynthesis** – the net rate of CO_2 uptake or O_2 production, including respiratory and photorespiratory gas exchange.
>
> **Gross photosynthesis** – the rate of CO_2 uptake or O_2 production, allowing for gas exchange due to respiratory and photorespiratory activity.

Correcting for photorespiration and respiration – measuring the rate of O_2 consumption or CO_2 production in the first few minutes after the plant material has been transferred from the light to darkness will give you the best estimate of photorespiratory and respiratory gas exchange. Use this rate to convert net photosynthesis to gross photosynthesis.

Preventing oxygen supersaturation in oxygen electrode studies – lower the O_2 content of your experimental solutions by bubbling them with N_2 before use, so that the O_2 evolved during the experiment remains in solution.

Fig. 57.2 A portable infra-red gas analyser, used to measure photosynthetic CO_2 uptake (leaf cuvette/probe shown in left hand, with analysis unit and user interface below). Courtesy of PP Systems, Amesbury, MA, USA (http://www.ppsystems.com).

Measuring photosynthetic activity

Eqn [57.1] shows that the rate of photosynthesis can be measured in terms of the amount of O_2 evolved or CO_2 fixed. However, several other metabolic processes may lead to concurrent changes in O_2 and/or CO_2 status, particularly:

- Photorespiration – light-dependent O_2 uptake, due to the oxygenase activity of Rubisco, leading to the production of glycollic acid (C_2) and the subsequent release of CO_2 due to the operation of a 'scavenging' pathway where two molecules of glycollate are converted to one of phosphoglyceric acid (C_3). The relative rates of photosynthesis and photorespiration depend upon the concentration of O_2 and CO_2 at the active site of Rubisco – in C_3 plants, photorespiration is highest under conditions of high temperature, high light and low water availability, while C_4 plants show low levels of photorespiration.

- Respiration – generation of ATP *via* carbohydrate oxidation, as in non-photosynthetic cells (p. 405). In most photoautotrophs, respiratory activity is likely to be lower during the day than at night. However, plant tissues often respond to wounding or environmental stress by *increasing* their rate of respiration – this may be significant if you are measuring photosynthetic activity in tissue fragments, isolated cells or protoplasts. If you wish to calculate gross rates of photosynthesis, you must make appropriate allowances for both respiration and photorespiration.

Measurement of oxygen production This is most often used with aquatic systems, including algae and photosynthetic bacteria, higher plant cells, protoplasts, chloroplast suspensions (p. 247) or isolated thylakoids. Oxygen in solution can be determined by end-point chemical analysis (e.g. the Winkler method, often used in ecological and field studies) or, more conveniently, by continuous monitoring using an oxygen electrode and chart recorder, to give the rate of net O_2 production (Chapter 48) under various conditions, e.g. light, temperature, etc. For photosynthesis–irradiance ($P–I$) curves using an O_2 electrode, you should note that the electrode assembly can act as a lens, so the light within the chamber can be higher than that measured at the outside surface. The rate of gross photosynthesis is obtained by correcting for O_2 uptake when the electrode assembly is transferred to darkness (e.g. using a thick black cloth).

Measurement of carbon dioxide uptake For higher plant studies, this is most easily achieved using an infra-red gas analyser, or IRGA (Fig. 57.2). Most modern instruments are portable and generally incorporate: (i) a cuvette, with a transparent window that attaches to a whole leaf or a known area of leaf providing a gas-tight seal, and within which the air is stirred by a fan; (ii) a gas supply system that allows control over input gases; (iii) miniaturised infra-red analysers (p. 289) to detect differences in CO_2 and H_2O content of input and output gas streams; (iv) systems for measuring (and possibly controlling) other environmental variables, e.g. light and temperature; (v) an on-board microprocessor for calculating results as rates of photosynthesis and transpiration and estimating the leaf's internal pCO_2 (C_i) and for exporting these and other data (e.g. time, leaf temperature, photosynthetic photon flux density, etc.) to PCs. Portable IRGAs are particularly useful for the rapid construction of photosynthesis–irradiance ($P–I$) and photosynthesis–pCO_2 ($P–C_i$) curves used to estimate photosynthetic efficiency, photosynthetic capacity and compensation points.

Definitions

Photosynthetic efficiency – the rate of photosynthesis when limited either (i) by supply of photons or (ii) by supply of CO_2, determined from the initial gradient of a P–I or P–C_i curve respectively.

Photosynthetic capacity – the rate of photosynthesis under saturating photon flux density (PFD) or CO_2 supply, determined from the upper asymptote of a P–I or P–C_i curve respectively.

Compensation point – PFD or pCO_2 at which net photosynthesis = 0, i.e. where gross photosynthesis, respiration and photorespiration are balanced: from the intercept of a P–I or P–C_i curve (x-axis).

Interpreting ^{14}C fixation data – short-term studies of a few minutes may estimate gross photosynthesis, while longer-term studies will give a rate closer to net photosynthesis, since some of the fixed ^{14}C will be respired.

Studying the release of photoassimilated ^{14}C in aquatic photoautotrophs – the loss of glycollate and other metabolites can be quantified by acidifying a known amount of medium, to drive off unfixed ^{14}C, then counting directly, or after an appropriate separation procedure.

Example For air, with $[^{13}CO_2]$ at 0.144 μmol l^{-1} and $[^{12}CO_2]$ at 12.915 μmol l^{-1}, using Eqn [57.2] gives a molar abundance ratio, $R = 0.144 \div 12.95 = 0.011\ 15$ (to four significant figures).

Measurement of radiocarbon (^{14}C) fixation The tracer may be supplied as $^{14}CO_2$ for gas exchange studies or, more readily, as $H^{14}CO_3^-$ for studies in aqueous systems, since there will be an interconversion of soluble CO_2, HCO_3^- and $H_2CO_3^{2-}$ in accordance with Eqn [48.4]. In aqueous systems, the plant material is incubated in medium containing ^{14}C-labelled bicarbonate for a known time period, then removed and prepared for liquid scintillation counting (p. 269). Microalgal cells and photosynthetic bacteria can be separated from the experimental medium by filtration, or silicone oil microcentrifugation (p. 299) – mild acid treatment (e.g. 50 mmol l^{-1} HCl) will ensure that any unfixed ^{14}C is released as $^{14}CO_2$. To express the results in terms of the amount of C assimilated, you will need to calculate (i) the total inorganic C content of the medium, i.e. $CO_2 + HCO_3^- + H_2CO_3^{2-}$, obtained from pH and alkalinity measurements or by IR spectroscopy of the CO_2 produced when a known amount of the medium is acidified and (ii) the relationship between the total inorganic C content and the amount of ^{14}C tracer added (i.e. the specific activity of the experimental solution, p. 270).

One of the advantages of studying the photoassimilation of ^{14}C is that the radiotracer can be used to follow the fate of fixed carbon, by separating and fractionating the various cellular components prior to counting, e.g. by sequential solvent extraction of the plant material in 80% ethanol (low molecular weight solutes) then boiling water (polysaccharides), leaving a residual fraction (structural polysaccharides, proteins and nucleic acids). More sophisticated separation techniques must be used to quantify the amount of radioactivity within individual biomolecules, e.g. using column chromatography (p. 304) and autoradiography (p. 271), or 2-D thin layer chromatography (p. 303) and autoradiography. High specific activity $H^{14}CO_3^-$ and short incubation times of a few seconds are required to study the early stages in carbon photoassimilation.

Using stable carbon isotopes Measurements of stable carbon isotopes in organic material and in the air passing over photosynthetically active plant material can be used as a probe of photosynthetic physiology. 'Light' and 'heavy' isotopes of carbon (^{12}C and ^{13}C respectively) have different diffusion rates within the leaf and Rubisco also discriminates against ^{13}C, leading to changes in the isotope composition of photoassimilated carbon compared to that of air. By measuring this discrimination, the physiological processes associated with carbon fixation and metabolism can be investigated in a time-integrated manner. Analysis of stable isotope composition makes use of mass spectrometric measurements of the ratio of heavy-to-light isotopes in the sample under investigation and in a predefined standard.

The mass spectrometer (p. 294) compares the mass-to-charge ratio (m/z) of the 44 and 45 masses of $^{12}CO_2$ and $^{13}CO_2$ respectively, to give the isotopic composition of (i) a sample and (ii) a standard, in terms of the ^{13}C molar abundance ratio (R), where:

$$R = \frac{[^{13}CO_2]}{[^{12}CO_2]} \qquad [57.2]$$

These values are then used to calculate the isotopic composition of the sample relative to the standard as a $\delta^{13}C$ value, usually expressed in parts per thousand (‰, see p. 146 – note that this is a proportion, and *not* a unit – spoken as 'per mil'). The original standard was CO_2 derived from a

Example For air, where $R = 0.01115$, $\delta^{13}C = (0.01115 - 0.01124) \div 0.01124 \times 1000 = -8.0‰$ (to one decimal place). $\delta^{13}C$ for plant tissue where Rubisco is the main carboxylating enzyme is *more negative* than $\delta^{13}C$ for C_4 plant tissue, since Rubisco discriminates strongly against ^{13}C (e.g. $R = 0.01095$, $\delta^{13}C = -25.8‰$). $\delta^{13}C$ for C_4 plant tissue where PEP carboxylase is the main carboxylating enzyme is *less negative* than $\delta^{13}C$ for C_3 plant tissue, since this enzyme shows little discrimination against ^{13}C, and the main effect is due to diffusion within the leaf (e.g. $R = 0.01111$, $\delta^{13}C = -11.6‰$).

fossil Belemnite from the Pee Dee deposits in South Carolina (PDB standard: $R = 0.01124$). The $\delta^{13}C$ value of a sample in parts per thousand is then calculated as:

$$\delta^{13}C = \frac{R_{\text{sample}} - R_{\text{standard}}}{R_{\text{standard}}} \times 1000 \qquad [57.3]$$

Isotopic composition values may also be used to calculate discrimination (Δ), which takes into account the isotopic composition of the standard, source and sample, according to the following equation:

$$\Delta = \frac{\delta_a - \delta_p}{1 + \delta_p} \qquad [57.4]$$

where δ_a and δ_p are the carbon isotope discrimination values for air (source) and plant (sample) respectively, expressed in fractional terms (see worked examples in margin). In contrast to $\delta^{13}C$ values which are negative when PDB is used as the standard, values for Δ are positive for plant material, since ^{12}C is favoured over ^{13}C. The advantage in using Δ notation is that it is more straightforward to make comparisons, avoiding the possible confusion involved in discussing more and less negative $\delta^{13}C$ values. Like $\delta^{13}C$ values, Δ values are usually expressed in parts per thousand (‰). Values for Δ values in C_3 plants, where Rubisco is the main carboxylating enxyme, usually fall within the range 16–25‰, since Rubisco shows substantial discrimination against ^{13}C. In contrast, Δ values for C_4 plants, where PEP is the main carboxylating enzyme, are lower – typically in the range 3–7‰ – since PEP carboxylase shows minimal discrimination against ^{13}C.

Example Using Eqn [57.4], Δ for C_3 plant tissue with a $\delta^{13}C$ of $-25.8‰$ is $[-0.008 - (-0.0258)] \div [1 + (-0.0258)] = 0.0182714 = 18.2‰$.
Using Eqn [57.4], Δ for C_4 plant tissue with a $\delta^{13}C$ of $-11.5‰$ is $[-0.008 - (-0.0115)] \div [1 + (-0.0115)] = 0.0035407 = 3.5‰$.

Advances in the study of hydrogen, oxygen and nitrogen isotope ratios have further extended the range of tools available to physiologists to investigate areas including plant water relations and nitrogen economy at the individual plant and ecosystem levels. Since the stable carbon isotope ratio of a plant is a broad measure of the overall balance between carbon uptake (photosynthesis) and water loss (transpiration), Δ is found to vary with changes in transpiration rate, with a lower value corresponding to a higher water use efficiency. It has been proposed that concurrent analysis of $^{18}O/^{16}O$ ratios can enable researchers to interpret variations in Δ in terms of water use efficiency: this can be useful in plant breeding studies, where improved water use efficiency results in closely coupled ^{13}C and ^{18}O signals.

Chlorophyll fluorescence This can be used to provide a non-destructive indication of photosynthetic function. When light energy is absorbed by the light-harvesting apparatus of green plants, several reactions compete for the deactivation of excited chlorophyll molecules. Principally, there are three possible outcomes: (i) the light energy may be trapped and used to reduce photosystem II (PS II) and drive photochemical reactions (Fig. 57.1); (ii) the energy may be lost as heat; or (iii) the energy may be re-emitted as fluorescence, measured at wavelengths around 685 nm using a purpose-built chlorophyll fluorometer. Since these three processes are mutually exclusive, measurements of chlorophyll fluorescence can provide information about changes in photochemistry and heat dissipation, since the maximum potential fluorescence yield is 'quenched' (reduced) by (i) photochemical and (ii) non-photochemical processes. Most measurements are made using pulse-amplified modulated (PAM) fluorometers (Fig. 57.3), where the light source used to generate the chlorophyll fluorescence can be

Understanding fluorescence – the underlying basis of this phenomenon is explained on p. 284.

turned on and off at high frequency (pulse modulated), enabling the chlorophyll fluorescence signal to be separated from all other light, and therefore allowing measurements to be made in light-adapted leaves under 'natural' conditions.

To interpret fluorescence changes, the relative contribution of the two quenching processes must be established. The starting point is to consider what happens when a strong light is shone on a dark-adapted leaf: there is a rapid rise in chlorophyll fluorescence to a maximum value, as photosynthetic electron transport is rapidly saturated and no further electrons can be accepted by PS II. If the light remains on, the chlorophyll fluorescence falls progressively, due to photochemical processes (e.g. activation of carbon-fixing enzymes) and non-photochemical processes (dissipation of chlorophyll energy as heat). By measuring changes in chlorophyll fluorescence, the relative contributions of these two processes can be quantified. At a practical level, this is achieved by first shining a single flash of high-intensity light (the saturating pulse) onto a leaf in the dark-adapted state and the resulting maximum fluorescence signal, F_m (see Fig. 57.4 for detail) is then used to determine the potential maximum quantum yield of PS II as:

$$\text{Maximum quantum yield PS II} = (F_m - F_0) \div F_m \qquad [57.5]$$

The leaf is then allowed to adapt to ambient light, and the saturation pulse is then reapplied: the reduction in maximum fluorescence (to $F_{m'}$) is the result of non-photochemical quenching (NPQ), i.e. dissipation of energy as heat (Fig. 57.4). This can be determined from the relationship:

$$\text{NPQ} = (F_m - F_{m'}) \div F_{m'} \qquad [57.6]$$

Non-photochemical quenching processes protect the protein components of the photosynthetic apparatus from oxidative damage due to excess

Fig. 57.3 A pulse-amplified modulated fluorometer (PAM), used to measure chlorophyll fluorescence (probe shown in left hand, attached to leaf, with analysis/recording unit on shoulder strap). Courtesy of Heinz Walz GmBH, Effeltrich, Germany (http://www.walz.com).

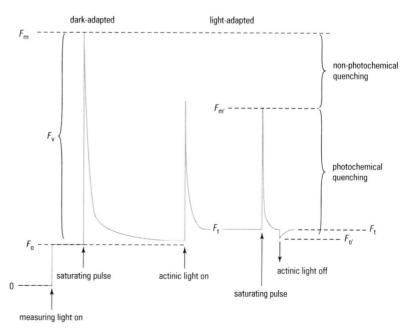

Fig. 57.4 Representation of chlorophyll fluorescence measurement using the saturation pulse method (for details, see text).

light. The remaining fraction is equivalent to photochemical quenching (qP), a measure of the proportion of PS II centres that are 'open', calculated from the relationship:

$$qP = (F_{m'} - F_t) \div (F_{m'} - F_{0'}) \qquad [57.7]$$

The quantum yield of PS II ($\Phi_{PS\,II}$), which is a measure of the efficiency of PS II photochemistry, can be determined as:

$$\Phi_{PS\,II} = (F_{m'} - F_t) \div F_{m'} \qquad [57.8]$$

Using chlorophyll fluorescence measurements – these can give insights into the stress responses of plants, e.g. the extent of photoinhibition under field conditions – further details are provided by Maxwell and Johnson (2000).

The light and dark reactions of photosynthesis are coordinated, since electron transport generates the ATP and NADPH required for CO_2 fixation. For example, a reduction in the rate of CO_2 assimilation due to a metabolic inhibitor will be mirrored by a decrease in $\Phi_{PS\,II}$ and, in many cases, by an increase in NPQ as the light intensity is increased to levels in excess of those required for photosynthesis.

Studying photosynthetic electron transport

Robert Hill first showed that isolated chloroplasts can evolve O_2 in the absence of CO_2 fixation, as long as they are provided with an artificial electron acceptor that intercepts electrons from the photosynthetic electron transport chain. The net result of this 'Hill reaction' is the photolysis of water and the reduction of the Hill acceptor (A), as follows:

$$2nH_2O + 4nA \rightarrow 4nAH + nO_2 \qquad [57.9]$$

Measuring photosynthetic quotients – in studies where both O_2 and CO_2 are measured, the photosynthetic quotient (PQ) can be determined from the relationship: $PQ = O_2$ evolved $\div CO_2$ consumed. In the simplest case, where fixed carbon accumulates as carbohydrate, a value of 1 should be obtained. However, the PQ may vary, depending on the amount of carbon incorporated into fats, proteins, etc. and the utilisation of photosynthetic energy for other metabolic processes and growth, e.g. NO_3^- assimilation, carbon storage.

The practical significance of the Hill reaction is that it allows the photochemical reactions of the electron transport system to be studied independently of the dark reactions of photosynthesis. Many of the Hill acceptors are redox dyes that show changes in absorbance as they are reduced, e.g. ferricyanide ($Fe(CN)_6^{3-}$), which accepts electrons from the PS I complex, or 2,6-dichlorophenolindophenol (DCPIP), which intercepts electrons from the electron transport chain between PS II and PS I (Fig. 57.1). The reduction of these artificial electron acceptors can be followed spectrophotometrically, allowing the Hill reaction to be quantified. Alternatively, an O_2 electrode can be used to follow O_2 evolution from PS II in the presence of various artificial electron acceptors. Redox mediators can be used to investigate the following aspects of photosynthetic electron flow:

Measuring the Hill reaction using DCPIP – since the dye will revert to its oxidised (blue) form as soon as the chloroplasts are removed from the light, you must measure the absorbance as quickly as possible.

- the activity of PS II and PS I operating in series can be studied in whole chloroplasts using a suitable terminal acceptor, e.g. ferricyanide;
- the activity of PS II can be studied in whole chloroplasts using DCPIP (membrane-permeable), or in fragmented chloroplasts using ferricyanide;
- the activity of PS I can be measured if PS II is blocked by the herbicide diuron (DCMU) and if electron flow is maintained by the addition of an artificial electron donor, e.g. ascorbate plus DCPIP. Note that this cannot be measured by O_2 evolution, since PS II is inoperative – methyl viologen can be used as an electron acceptor from PS I and is rapidly reoxidised, consuming O_2 in a reaction which can be measured using an O_2 electrode (as O_2 *uptake*, rather than O_2 production);

Table 57.1 Some inhibitors of photosynthetic electron transport*

Inhibitor	Target site
Hydroxylamine (NH$_2$OH)	photolysis of H$_2$O
DCMU (3(3,4-dichlorophenyl)-1,1-dimethylurea)	electron transport from PS II to plastoquinone
DBMIB (2,5-dibromo-3-methyl-6-isopropyl-p-benzoquinone)	electron flow from plastoquinone to cytochrome f
Methyl viologen (Paraquat)	electron flow from PS I to NADP$^+$
Atrazine (2-chloro-4-(2-propylamino)-6-ethylamine-5-triazine)	electron flow to plastoquinone
HOQNO (2-heptyl-4-hydroxyquinoline-N-oxide)	electron flow between quinones and cytochromes
DSPD (disalicylidinepropanediamine)	electron flow via ferredoxin

*See also Table 56.1 for uncouplers and ATPase inhibitors.

 SAFETY NOTE Working with inhibitors – make sure you treat all inhibitors as potentially toxic, observing appropriate safety precautions, and that you know the procedure to be followed in case of accident or spillage.

- the sites of action of inhibitors (Table 57.1) can be determined by measuring their effects on the various components of the electron transport system, measured using redox dyes or the methyl viologen/O$_2$ electrode system.

Respiration

At the cellular level, respiration can be considered as the oxidation of organic compounds coupled to the production of so-called high-energy intermediates, such as ATP. The overall process is often represented in terms of the oxidation of glucose, as:

$$C_6H_{12}O_6 + 12O_2^* + 6H_2O + nADP + nP_i \rightarrow$$
$$6CO_2 + 12H_2O^* + nATP + nH_2O \quad [57.10]$$

the asterisks showing that O$_2$ is converted to water, and where n has a *maximum* value of 38. The process can be divided into three principal stages, namely (i) glycolysis, (ii) the tricarboxylic acid (TCA) cycle and (iii) oxidative phosphorylation. The glycolytic pathway operates in the cytosol and results in the partial breakdown of glucose (C$_6$) to two molecules of pyruvate (C$_3$) plus two molecules of NADH and two molecules of ATP. The mitochondrial TCA cycle involves the sequential dismantling of pyruvate to CO$_2$ (*via* an intermediate step involving decarboxylation to acetyl-coA), producing one molecule of FADH$_2$, four molecules of NADH and one of GTP for each pyruvate. The final stage occurs at the inner mitochondrial membrane – the movement of electrons and protons along the respiratory transport chain from reductant to O$_2$ as the terminal electron acceptor creates a transmembrane H$^+$ gradient that leads to the net synthesis of ATP, as summarised in Figure 57.5. Thus, as in photosynthesis, respiration involves enzyme-catalysed interconversion of organic compounds plus membrane-bound electron transport reactions that are coupled to ATP synthesis.

The enzymic reactions of glycolysis and the TCA cycle can be studied using ^{14}C-labelled intermediates and pulse-chase experiments, using techniques similar to those used to investigate the dark reactions of photosynthesis (p. 399), or by purification and characterisation of the individual enzymes (p. 381). Oxidative phosphorylation can be studied using techniques appropriate for electron transport reactions.

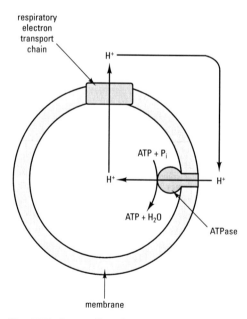

Fig. 57.5 Proton flow in mitochondrial oxidative phosphorylation: the creation of a transmembrane H$^+$ gradient due to the respiratory electron transport chain drives ATP synthesis via a membrane-bound ATPase.

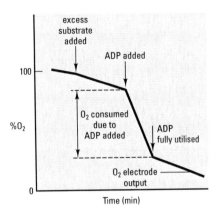

Fig. 57.6 Representative data for O_2 uptake by a suspension of mitochondria in response to additions of substrate and ADP.

Interpreting respiratory quotients – the complete oxidation of carbohydrates should give values close to 1.0, in agreement with Eqn [57.11], while the oxidation of fats will give values close to 0.7 and protein oxidation will produce values of about 0.8.

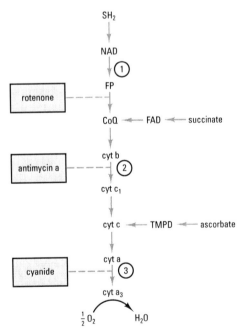

Fig. 57.7 The mitochondrial electron transport system, showing the sites of action of various inhibitors (dotted lines) and the three sites of ATP synthesis (numbered 1–3). S, substrate; FP, flavoprotein; CoQ, coenzyme Q; cyt, cytochrome; TMPD, tetramethylphenylenediamine.

Measuring the respiratory activity of cells and tissues

The overall rate of cellular respiration is usually measured in terms of the amount of O_2 consumed by a known amount of material in a given time, e.g. as μmol O_2 mg protein^{-1} min^{-1}.

The principal techniques include:

- Manometry – this traditional approach involves measuring either the pressure change (Warburg manometer) or volume change (Gilson manometer) as gases are produced or consumed during respiration. Manometry is relatively insensitive and may be subject to large measurement errors: the principal application is in studying the relationship between the amounts of O_2 consumed and CO_2 evolved in terms of the respiratory quotient (RQ) for a particular substrate, from the relationship:

$$RQ = \frac{CO_2 \text{ evolved}}{O_2 \text{ consumed}} \quad [57.11]$$

The calculations involved in determining CO_2 production or O_2 consumption are complex and are specific to individual instruments – manufacturer's guidelines should be followed carefully.

- Oxygen electrode studies – this has largely replaced manometry, as the apparatus is more versatile and less complex to set up and interpret, giving a continuous readout of O_2 status via a chart recorder (p. 341). Net O_2 evolution can be measured using whole cells, mitochondrial suspensions (p. 243) or sub-mitochondrial preparations.

Using an oxygen electrode to study respiratory electron transport

Intact mitochondria suspended in an isotonic medium will show little respiratory activity unless supplied with (i) a suitable electron donor or substrate (e.g. NADH), (ii) ADP and (iii) P_i: this is termed respiratory control and the mitochondria are said to be tightly coupled, since there is a close link between electron transport, O_2 consumption and phosphorylation. The extent of this coupling can be determined using an O_2 electrode (p. 339) to measure O_2 consumption in the presence of substrate, P_i and ADP, with that in the absence of ADP (Fig. 57.6), as:

$$\text{respiratory control ratio} = \frac{\text{rate of } O_2 \text{ consumption with ADP}}{\text{rate of } O_2 \text{ consumption without ADP}} \quad [57.12]$$

Freshly prepared, tightly coupled mitochondria should have a respiratory control ratio of ≥ 4.

To investigate the relationship between the number of ATP molecules produced per substrate molecule, substrate and P_i are supplied in excess and the O_2 uptake produced by a known amount of ADP is measured using an O_2 electrode (Fig. 57.7), allowing the P/O ($\equiv$ADP/O) ratio for a particular substrate to be calculated as:

$$\text{P/O ratio} = \frac{\mu\text{mol ADP added}}{2 \times \mu\text{mol } O_2 \text{ consumed}} \quad [57.13]$$

Note that the amount of O_2 is multiplied by 2, since O_2 is a diatomic molecule. The P/O ratio is determined by the site at which electrons are

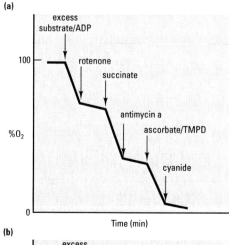

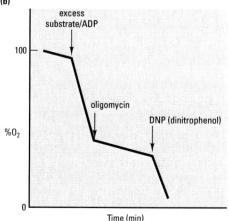

Fig. 57.8 Representative data for O_2 uptake by mitochondria in response to the addition of (a) electron transport inhibitors and substrates, or (b) inhibitors of ATP synthesis and uncouplers. Steeper slopes indicate faster rates of O_2 uptake.

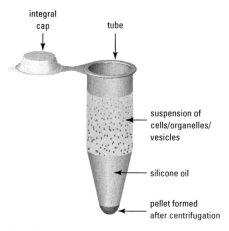

Fig. 57.9 Silicone oil microcentrifugation system: cells, organelles or membrane vesicles are pelleted below the silicone oil layer as a result of their higher density, while the less dense suspension medium remains about the silicone oil layer on centrifugation.

transferred to the respiratory chain (Fig. 57.8). Thus NADH should give a P/O ratio of 3, while succinate and $FADH_2$ should have values of 2. Artificial electron donors to cytochrome *c* have P/O ratios of 1, e.g. ascorbate/tetramethylphenylenediamine (TPMD). However, measured P/O ratios are always less than these values, due to partial uncoupling and the action of ATPases.

The effects of inhibitors of respiratory electron transport on the rate of oxygen consumption of mitochondria can be used to determine their site of action (Fig. 57.7). Thus, an inhibitor of electron flow will prevent O_2 uptake while an artificial substrate that donates electrons at a location beyond the site of inhibition will restore O_2 uptake, as shown in Fig. 57.8.

'Uncouplers' act by increasing the permeability of membranes to protons, causing the dissipation of the transmembrane H^+ gradient. Addition of an uncoupler increases electron flow along the respiratory chain in the absence of phosphorylation, i.e. there is loss of respiratory control. Uncouplers can reverse the effects of inhibitors such as oligomycin, whose target is the mitochondrial ATPase (Table 56.1). Oligomycin prevents the return of H^+ to the interior of the mitochondrion, thereby inhibiting respiratory electron transport and O_2 uptake – the addition of an uncoupler (e.g. CCCP) reverses this inhibition.

Measuring the components of the proton-motive force in chloroplasts and mitochondria

According to chemiosmotic principles, the transmembrane electrochemical potential gradient of protons drives ATP synthesis in photophosphorylation and oxidative phosphorylation via a membrane-bound ATPase (Fig. 57.5). This gradient is often expressed as the proton-motive force, or pmf (Δp), expressed in mV as:

$$\Delta p = E_m - 59(pH_i - pH_o) \qquad [57.14]$$

at 25 °C, where E_m is the transmembrane electrical potential (mV) and pH_i and pH_o are internal and external pH values. The individual components of the proton-motive force can be determined by a variety of methods:

- Transmembrane electrical potential, by the distribution of a suitable radiolabelled lipophilic cation, e.g. tetraphenylphosphonium (TPP^+), in accordance with the Nernst equation (see p. 395); by the measurement of K^+ or Rb^+ in the presence of the ionophore valinomycin (p. 395); or by the quenching of a fluorescent dye (e.g. cyanine or oxanol dyes).
- Transmembrane pH gradient from the equilibrium distribution of a radiolabelled weak permeant acid, e.g. 5,5-dimethyl-2,4-oxazolidinedione (DMO); by quenching of a fluorescent pH probe, e.g. 9-aminoacridine; or by ^{31}P-NMR (p. 292).

For radiolabelled probes, organelles or membrane vesicles can be separated from the bathing medium by silicone oil microcentrifugation (Fig. 57.9), with suitable correction for carry-over of external medium, e.g. using a membrane-impermeant solute, labelled with a second radioisotope. Care is required in such double-labelled experiments, as carry-over can represent a significant component, leading to substantial measurement errors if uncorrected.

Typical measurements for Δp across energy-transducing membranes are 180–200 mV (inside-negative): in mitochondria, E_m is the principal component while ΔpH represents the largest component of Δp in thylakoid membranes.

Text reference

Maxwell, K. and Johnson, G.N. (2000) Chlorophyll fluorescence – a practical guide. *Journal of Experimental Botany*, **51**, pp. 659–68.

Sources for further study

DeEll, J.R. and Tolvonen, P.M.A. (2003) *Practical Applications of Chlorophyll Fluorescence in Plant Biology*. Kluwer, Dordrecht.

James, D.C. and Matthews, G.S. (1991) *Understanding the Biochemistry of Respiration*. Cambridge University Press, Cambridge.

Jones, R.L., Buchanan, B.B. and Guissem, W. (2000) *Biochemistry and Molecular Biology of Plants*. Wiley, Chichester.
[Covers photosynthesis, respiration and photorespiration, along with other topics]

McDonald, M.S. (2003) *Photobiology of Higher Plants*. Wiley, Chichester.

Nobel, P.S. (2005) *Physicochemical and Environmental Plant Physiology*. Academic Press, Oxford.

Ridge, I. (ed.) (2002) *Plants*. Oxford University Press, Oxford.
[Chapter 2 is a good general introduction to the physiology of photosynthesis]

Taiz, L. and Zeiger, E. (2006) *Plant Physiology*, 4th edn. Sinauer, Sunderland, USA.
[Provides a detailed treatment of photosynthesis]

West, J.B. (2004) *Respiratory Physiology: the Essentials*, 7th edn. Lippincott, Williams and Wilkins, Philadelphia.
[Provides a comprehensive guide to the human respiratory system]

Wilson, K. and Walker, J. (eds) (2005) *Principles and Techniques of Practical Biochemistry and Molecular Biology*, 6th edn. Cambridge University Press, Cambridge.
[Also covers other topics, including spectroscopy and centrifugation]

Study exercises

(See also study exercises 48.4 and 48.5 for calculations of respiratory oxygen consumption and photosynthetic oxygen evolution based on oxygen electrode measurements.)

57.1 Check your understanding of photosynthesis and respiration. Having read through this chapter, distinguish between each of the following pairs of terms:

(a) net photosynthesis and gross photosynthesis;
(b) C_3 photosynthesis and C_4 photosynthesis;
(c) the oxygenase and carboxylase functions of Rubisco;
(d) coupling and uncoupling;
(e) respiratory quotient and respiratory control ratio.

57.2 Calculate rates of photosynthetic carbon fixation from ^{14}C data. A sample of the photosynthetic cyanobacterium *Spirulina platensis* containing 0.29 mg chlorophyll a (chl a) was incubated in the light for 3 min in 10 ml of an aqueous solution containing inorganic carbon (HCO_3^-) at a concentration of 12 mmol l^{-1} and at a specific activity of 12 Bq pmol^{-1} ($\equiv 0.72$ d.p.m. nmol^{-1}) At the end of this period, the cyanobacterial cells were separated from the medium and assayed by liquid scintillation counting, giving a quench-corrected value of 2808 d.p.m. What is the photosynthetic rate, expressed in terms of μmol C fixed per mg chl a per minute? (Express your answer to three significant figures.)

Study exercises (continued)

57.3 Interpret results from an experiment using DCPIP to study photosynthetic electron transport. The table below shows data for the change in A_{600} of a chloroplast suspension incubated with the Hill acceptor DCPIP, plus different levels of the herbicide simazine.

A_{600} measurements for a chloroplast suspension incubated with DCPIP and various levels of simazine

Time (s)	Control (no simazine)	Simazine at:		
		1.0 nmol l^{-1}	10.0 nmol l^{-1}	100.0 nmol l^{-1}
0	0.462	0.461	0.464	0.462
30	0.372	0.394	0.41	0.436
60	0.320	0.345	0.371	0.413
90	0.284	0.315	0.343	0.397
120	0.253	0.284	0.317	0.385

(a) What is the reason for the rapid decrease in A_{600} in the control sample, with no added simazine?

(b) How do you account for the fact that increasing amounts of simazine lead to a reduction in the extent of change in A_{600} with time?

(c) Plot a graph of A_{600} against time to show the composite data from the table. Use this to determine the initial rate of change in A_{600}. Then, calculate the percentage inhibition at each simazine concentration. Finally, plot percentage inhibition against simazine concentration, to determine the concentration causing 50% inhibition (I_{50}); give your answer to three significant figures.

57.4 Check your understanding of respiratory quotients. Which of the following values would you expect for the respiratory quotient of aerobic bacteria growing on the following substrates as their principal source of carbon and energy: (a) glucose; (b) triglycerides; (c) casein:

(i) 0.7
(ii) 0.8
(iii) 1.0

Give brief explanations for your choices.

57.5 Calculate the P/O ratio for a respiring mitochondrial suspension. An oxygen electrode chamber containing a mitochondrial suspension (0.5 ml), NADH (0.5 ml) and buffer (4 ml) showed a fall in oxygen concentration from 218 to 174 µmol l^{-1} on addition of 0.05 ml of a 25 mmol l^{-1} ADP solution. What was the P/O ratio over this period?

57.6 Calculate mitochondrial proton-motive force. A suspension of mitochondria in buffer at an external pH of 7.2 had a measured internal (matrix) pH of 8.6 and a measured transmembrane potential (inner membrane) of −85.6 mV. Calculate the proton-motive force across the inner membrane (express your answer to three significant figures). What is your interpretation of the calculated value?

Genetics

58 Mendelian genetics

Gregor Mendel, an Austrian monk, made pioneering studies of the genetics of eukaryotic organisms in the middle of the nineteenth century. He made crosses between different forms of flowering plants. Through careful examination and numerical analysis of the observable characteristics, or phenotype, of the parents and their progeny, Mendel was able to deduce much about their genetic characteristics, or genotype. The principles derived from these experiments explain the basis of heredity, and hence underpin our understanding of sexual reproduction, biodiversity and evolution. Mendelian genetics is concerned primarily with the transmission of genetic information, as opposed to molecular genetics which deals with the molecular details of the genome and techniques for altering genes (see Chapters 60–62) and genome analysis (genomics) in Chapter 11.

 KEY POINT *A common initial stumbling block in genetics is terminology. In many cases the definitions are interdependent, so your success in this subject depends on your grasp of all of the definitions and underlying ideas explained below.*

Important terms and concepts

Each character in the phenotype is controlled by the organism's genes, the basic units of inheritance. Each gene includes the 'genetic blueprint' (DNA) which usually defines the amino acid sequence for a specific polypeptide or protein – often an enzyme or a structural protein. The protein gives rise to the phenotype through its activity in metabolism or its contribution to the organism's structure. The full complement of genes in an individual is known as its genome. Individual genes can exist in different forms, each of which generally leads to a different form of the protein it codes for. These different gene forms are known as alleles.

In eukaryotes, the genes are located in a particular sequence on chromosomes within the nucleus. The number of chromosomes per cell is characteristic for each organism (its chromosome number, n). For example, the chromosome number for man is 23. In cells of most 'higher' organisms, there are two of each of the chromosomes ($2n$). This is known as the diploid state. As a result of the process of meiosis which precedes reproduction, special haploid cells are formed (gametes) which contain only one of each chromosome ($1n$). In sexual reproduction, haploid gametes from two individuals fuse to form a zygote, a diploid cell with a new genome, which gives rise to a new individual through the process of mitosis. Cell numbers are increased by this process, producing genetically identical cells.

Organisms vary in the span of the diploid and haploid phases. In some 'lower' organisms the haploid phase is the longer lasting form; in most 'higher' organisms the diploid phase is dominant. Since each diploid individual carries two of each chromosome, it has two copies of each gene in every cell. The number of alleles of each character present depends on whether the relevant genes are the same – the homozygous state – or whether they are different – the heterozygous state. Hence, while there

---

Definitions

Phenotype – the observable characteristics of an individual organism; the consequence of the underlying genotype and its interaction with the environment.

Genotype – an individual's genetic make-up, i.e. the organism's genes.

Essential vocabulary for Mendelian genetics – make sure you know what the following terms and symbols mean:
- chromosome, sex chromosome, autosome;
- gene, allele, locus;
- dominant, recessive, lethal;
- haploid, diploid, gamete, zygote;
- heterozygous, homozygous;
- P, F1, F2, ♀, ♂.

Genomes and chromosome numbers – remember that mitochondria and chloroplasts contain DNA molecules, but these are not included when calculating the chromosome number.

Definitions

Meiosis – division of a diploid cell which results in haploid daughter cells carrying half the original number of chromosomes. Occurs during gamete (sperm and egg) formation.

Mitosis – division of a cell into two new cells, each with the same chromosome number. Occurs in somatic cells, e.g. during growth, development, repair, replacement.

Example In the garden pea, *Pisum sativum*, studied by Mendel, the yellow seed allele Y was found to be dominant over the green seed allele y. A cross of YY × yy genotypes would give rise to Yy in all of the F_1 generation, all of which would thus have the yellow seed phenotype. If the F_1 generation were interbred, this Yy × Yy cross would lead to progeny in the next, F_2, generation with the genotypes YY : Yy : yy in the expected ratio 1 : 2 : 1. The expected phenotype ratio would be 3 : 1 for yellow : green seed.

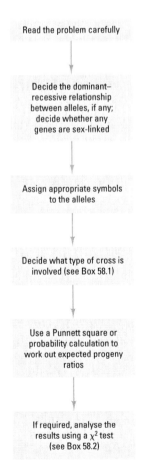

Fig. 58.1 Flowchart for tackling problems in Mendelian genetics.

may be many alleles for any given gene, an individual could have two at most, and might only have one if it were homozygous. This will depend on the alleles present in the parental gametes that fused when the zygote was formed. Offspring that inherit a different combination of alleles at the two loci compared with their parents are known as recombinants.

The basis of Mendel's experiments and of many exercises in genetics are crosses, where individuals showing particular phenotypes are mated and the phenotypes of the offspring, or F_1 generation, are studied (see Box 58.1). If homozygous individuals carrying alternative alleles for a character are crossed, one of the alleles may be dominant, and all the F_1 generation will show that character in their phenotype. The character not evident is said to be recessive.

In describing crosses, geneticists denote each character with a letter of the alphabet, a capital letter being used for the dominant allele and lower case for the recessive. Taking, for example, a gene with dominant and recessive forms 'A' and 'a' respectively, there are three possibilities for each individual: it can be (a) homozygous recessive aa; (b) homozygous dominant AA; or (c) heterozygous Aa.

The reasons for dominance might relate to the activity of an enzyme coded by the relevant gene; for example, Mendel's yellow pea allele is dominant because the gene involved codes for the breakdown of chlorophyll. In the homozygous recessive case, none of the functional enzyme will be present, chlorophyll breakdown cannot occur and the seeds remain green. Not all alleles exhibit dominance in this form. In some cases, the heterozygous state results in a third phenotype (incomplete or partial dominance); in others, the heterozygous individual expresses both genotypes (codominance). Another possible situation is epistasis, in which one gene affects the expression of another gene that is independently inherited. Note also that many genes, such as those coding for human blood groups, have multiple alleles.

 KEY POINT Genetics problems may well involve one of the 'standard' crosses shown in Box 58.1. Before tackling the problem, try to analyse the information provided to see if it fits one of these types of cross. Figure 58.1 is a flowchart detailing the steps you should take in answering genetics problems.

Unless otherwise stated or obvious from the evidence, you should assume that the genes being considered in any given case are on separate chromosomes. This is important because it means that they will assort independently during meiosis. Thus, the fact that an allele of gene A is present in any individual will not influence the possibility of an allele of gene B being present. This allows you to apply simple probability in predicting the genetic make-up of the offspring of any cross (see p. 417).

Where genes are present on the same chromosome, they are said to be linked genes, and thus it would appear that they would not be able to assort independently during metaphase 1 of meiosis, when homologous chromosomes are independently orientated. However, although physically attached to each other, they may become separated when crossing over occurs between the homologous chromosomes at an early stage of meiosis. Exchange of genetic information between homologous chromosomes is called recombination. Linkage can be detected from a cross between

Pedigree notation:

○ normal female

□ normal male

● ■ female or male with an inherited condition

○—□ mating

offspring

Example of simple family pedigree:

This diagram shows the offspring of a normal male and female. The two daughters are normal, but the son has the inherited condition.

Fig. 58.2 Pedigree notation and family trees.

Denoting linked genes – these are often shown diagrammatically, with a double line indicating the chromosome pair. For example, the two possible linkages for the genotype AaBb would be shown as:

Using probability calculations – this can be simpler and faster than Punnett squares when two or more genes are considered.

individuals heterozygous and homozygous recessive for the relevant genes, e.g. AaBb × aabb. If the genes A and B are on different chromosomes, we expect the ratio of AaBb, aabb, Aabb and aaBb to be 1:1:1:1 in the F_1. However, if the dominant alleles of both genes occur on the same chromosome, the last two combinations will occur, but rarely. Just how rarely depends on how far apart they lie on the relevant chromosome – the further apart, the more likely it is that crossing over will occur. This is the basis of chromosome mapping (see Box 58.1).

Another complication you will come across is sex-linked genes. These occur on one of the X or Y chromosomes that control sex. Because one or other of the sexes – depending on the organism – is determined as XX and the other as XY (see Box 58.1), this means that rare recessive genes carried on the X chromosome may be expressed in XY individuals. Sex-linked genes are sometimes obvious from differences in the frequencies of phenotypes in male and female offspring. Pedigree charts (Fig. 58.2) are codified family trees that are often used to show the inheritance and expression of sex-linked characteristics through various generations.

Analysis of crosses

There are two basic ways of working out the results of crosses from known or assumed genotypes:

- The Punnett square method provides a good visual indication of potential combinations of gametes for a given cross. Lay out your Punnett squares consistently as shown in Figure 58.3. Then group together the like genotypes to work out the genotype ratio and proceed to work out the corresponding phenotype ratio if required.
- Probability calculations are based on the fact that the chance of a number of independent events occurring is equal to the probabilities of each event occurring, multiplied together. Thus if the probability P of a child being a boy is 0.5 and the probability of the child of particular parents being blue-eyed is 0.5, then the probability of that couple having a blue-eyed son is $0.5 \times 0.5 = 0.25$, and that of having two blue-eyed boys is:

$$P = (0.5 \times 0.5) \times (0.5 \times 0.5) = 0.0625$$

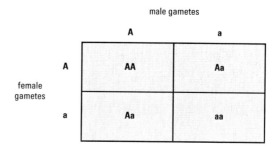

Fig. 58.3 Layout for a simple Punnett square for the cross Aa × Aa. The genotypic ratios for this cross are 1:2:1 for AA:Aa:aa, and the phenotypic ratio would be 3:1 for characteristic A to characteristic a. In this simple Punnett square, the allele frequencies are treated as equal ($f = 0.5$); if different from this, the probability of genotypes in each combination will be the relevant frequencies multiplied together.

Box 58.1 Types of cross and what you can (and can't) learn from them

Monohybrid cross – the simplest form of cross, considering two alleles of a single gene.

Example: $AA \times aa$

If only the parental *phenotypes* are known, you can't always deduce the parental genotypes from the phenotype ratio in the F_1. An individual of dominant phenotype in the F_1 could arise from a homozygous dominant or heterozygous genotype. However, crossing the F_1 generation with themselves to give an F_2 generation may provide useful information from the phenotype ratios that are found.

Dihybrid cross – a cross involving two genes, each with two alleles.

Example: $AaBB \times AaBb$

As with a monohybrid cross, you can't always deduce the parental genotype from the phenotype ratio in the F_1 generation alone.

Test cross – a cross of an unknown genotype with a homozygous recessive.

Example: $AABb \times aabb$

A test cross is one between an individual dominant for A and B with one recessive for both genes. The progeny will all be dominant for A, revealing the homozygous nature of the parent for this gene, but the progeny phenotypes will be split approximately 1:1 dominant to recessive for gene B, revealing the heterozygous nature of the parent for this gene. This type of cross reveals the unknown parental genotype in the proportions of phenotypes in the F_1.

Sex-linked cross – a cross involving a gene carried on the X chromosome; this can be designated as dominant or recessive using appropriate superscripts (e.g. X^A and X^a).

Example: $X^A X^a \times X^A Y$

In sex-linked crosses you need to know the basis of sex determination in the species concerned – which of XX or XY is male (for example, the former in birds, butterflies and moths, the latter in mammals and *Drosophila*). The expected ratios of phenotypes in the offspring will depend on this. Note that the recessive genotype a will be expressed in the $X^a Y$ case.

Crosses with linked genes – genes are linked if they are on the same chromosome. This is revealed from a cross between individuals heterozygous and homozygous recessive for the relevant genes.

Example 1 (genes on separate chromosomes): expected offspring frequency from the cross $AaBb \times aabb$ is AaBb, aabb, Aabb and aaBb in the ratio 1:1:1:1.

Example 2 (linked genes): the frequencies of the last two combinations in example 1 might be skewed according to the direction of parental linkage.

Chromosome mapping uses the frequency of crossing-over of linked genes to estimate their distance apart on the chromosome on the basis that crossing over is more likely, the further apart are the genes. So-called 'map units' are calculated on the following basis:

$$\frac{\text{No. of recombinant progeny}}{\text{Total no. of progeny}} \times 100 = \% \text{ crossing over}$$

[58.1]

By convention, 1% crossing over = 1 map unit (centimorgan, cM). The order of a number of genes can be worked out from their relative distances from each other. Thus, if genes A and B are 12 map units apart, while A and B are respectively 5 and 7 map units from C, the assumed order on the chromosome is ACB (Fig. 58.4).

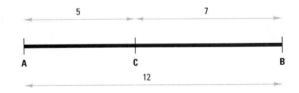

Fig. 58.4 Genetic map showing relative positions of genes A, B and C.

Carrying out genetic crosses – typically, these are performed using organisms that have a large number of offspring, to even out random variation. Other useful attributes include: short generation time; ease of maintenance; readily observed mutant phenotypes.

How do you decide whether the results of an experimental cross fit your expectation from theory as calculated above? This isn't easy, because of the element of chance in fertilisation. Thus, while you might expect to see a 3 : 1 phenotype ratio of progeny for a given cross, in 500 offspring you might actually observe a ratio of 379 : 121, which is a ratio of just over 3.13 : 1. Can you conclude that this is significantly different from 3 : 1 in the context of random error? The answer to this problem comes from statistics. However, the answer isn't certain, and your conclusion will be based on a balance of probabilities (see Box 58.2 and Table 58.1).

Population genetics

Population genetics is largely concerned with the frequencies of alleles in a population and how these may change in time. The Hardy–Weinberg Principle states that the frequency of alleles f remains the same between generations, unless influenced by some outside factor(s).

Box 58.2 Example of a chi² (χ^2) test

This test allows you to assess the difference between observed (O) and expected (E) values and is extremely useful in biology, particularly in determining whether progeny phenotype ratios fit your assumptions about their genotypes. The test is best illustrated by the use of an example. Assume that your null hypothesis (see p. 492) is that the phenotypic ratio is 3 : 1 and you observe that in 500 offspring the phenotype ratio is 379 : 121 whereas the expected ratio is 375 : 125.

Start the test by calculating the test statistic χ^2. The general formula for calculating χ^2 is:

$$\chi^2 = \sum \frac{(O - E)^2}{E} \qquad [58.2]$$

Limitation of the chi² test – note that the formula cited in Box 58.2 is valid only if expected numbers are greater than 5.

In this example, this works out as:

$$\chi^2 = \frac{(379 - 375)^2}{375} + \frac{(121 - 125)^2}{125} = \frac{16}{375} + \frac{16}{125} = 0.171$$

The probability associated with this value can be obtained from χ^2 tables for $(n - 1)$ degrees of freedom (d.f.), where n = the number of categories = number of phenotypes considered. Here the d.f. value is $2 - 1 = 1$. Since the χ^2 value of 0.171 is lower than the tabulated value for 1 d.f. (3.84, Table 58.1), we therefore accept the null hypothesis and conclude that the difference between observed and expected results is not significant (since $P > 0.05$). Had χ^2 been ≥ 3.84, then $P \leq 0.05$ and we would have rejected the null hypothesis and concluded that the difference was significant, i.e. that the progeny phenotype did not fit the expected ratio.

You can carry out χ^2 calculations using the CHITEST, CHIINV and CHIDIST functions in Microsoft Excel: the *Help* function within that program contains useful guidance.

Table 58.1 Values of chi² (χ^2) for which P = 0.05. The value for $(n - 1)$ degrees of freedom (d.f.) should be used, where n = the number of categories (= phenotypes) considered (normally fewer than 4 in genetics problems). If χ^2 is less than or equal to this value, accept the null hypothesis that the observed values arose by chance; if χ^2 is greater than this value, reject the null hypothesis and conclude that the difference between the observed and expected values is statistically significant

Degrees of freedom	χ^2 value for which $P = 0.05$
1	3.84
2	5.99
3	7.82
4	9.49

> **Definition**
>
> **Hardy–Weinberg Principle** – for a population in genetic equilibrium, the genotype frequencies at an autosomal locus will remain at particular equilibrium values, which can be expressed in terms of the allele frequencies at that locus.

To understand why this is the case, consider alleles H and h for a particular gene, which exist in the breeding population at frequencies p and q respectively. If the individuals carrying these alleles interbreed randomly, then the expected genotype and allele ratios in the F_1 generation can be calculated simply as:

$$f(\text{HH}) = p^2;$$
$$f(\text{Hh}) = 2pq; \quad \text{and}$$
$$f(\text{hh}) = q^2$$

If you wish to confirm this, lay out a Punnett square with appropriate frequencies for each allele. Now, by summation, the frequency of H in the $F_1 = p^2 + pq$ (a similar calculation can be made for allele h); and since in this example there are only two alleles, $p + q = 1$ and so $q = (1 - p)$. Substituting $(1 - p)$ for q, the frequency of H in the F_1 is thus:

$$p^2 + p(1 - p) = p^2 + p - p^2 = p$$

i.e. the frequency of the allele is unchanged between generations. A similar relationship exists for the other alleles.

The Hardy–Weinberg Principle was named after its first, independent, protagonists. It holds so long as the following criteria are satisfied:

1. random mating – so that no factors influence each individual's choice of a mate;
2. large population size – so that the laws of probability will apply;
3. no mutation – so that no new alleles are formed;
4. no emigration, immigration or isolation – so that there is no interchange of genes with other populations nor isolation of genes within the population;
5. no natural selection – so that no alleles have a reproductive advantage over others.

> **Example** Cystic fibrosis occurs in 0.04% of Caucasian babies. If this condition results from a double recessive allele aa, then following the Hardy–Weinberg Principle, $q^2 = 0.0004$ (0.04% expressed as a fraction of 1) and so $q = \sqrt{0.0004} = 0.02$, or 2%. Since $p = (1 - q)$, $p = 0.98$, or 98%. The frequency of carriers of cystic fibrosis in the Caucasian population (people having the alleles Aa) is given by $2pq$. From the above, $2pq = 2(0.02 \times 0.98) = 0.0392$. Hence 3.92% of the Caucasian population are carriers (roughly one in 25).

Population geneticists use the Hardy–Weinberg Principle to gain an idea of the rate of evolution and the influences on evolution. By ensuring that criteria 1–4 hold, if there are any changes in allele frequency between generations, then the rate of change of allele frequencies indicates the rate of evolutionary change (natural selection).

Sources for further study

Anon. *Online Mendelian Inheritance in Man*. Available: http://www.ncbi.nlm.nih.gov/entrez/query.fcgi?db=OMIM Last accessed: 01/04/07.

Blumberg, R.B. *MendelWeb*. Available: http://www.mendelweb.org/ Last accessed: 01/04/07. [A resource for students interested in the origins of classical genetics]

Falconer, D.S. and MacKay, T.F.C. (1996) *Introduction to Quantitative Genetics*, 4th edn. Longman, Harlow.

Klug, W.S. and Cummings, M.R. (2005) *Essentials of Genetics*, 5th edn. Prentice Hall, Upper Saddle River.

Mange, E.J. and Mange, A.P. (1998) *Basic Human Genetics*, 2nd edn. Sinauer, Sunderland.

Roberts, D.B. (1998) *Drosophila: a Practical Approach*, 2nd edn. Oxford University Press, Oxford.

Thomas, A. (2002) *Introducing Genetics: from Mendel to Molecule*. Taylor & Francis, Abingdon.

Winter, P.C., Hickey, G.I. and Fletcher, H.L. (1998) *Instant Notes in Genetics*. Bios, Oxford.

Study exercises

58.1 Use a Punnett square to predict the outcome of a cross. Lay out a Punnett square for a cross between genotypes RrOO × RrOo, where R is a semi-dominant gene for flower colour such that RR = red, Rr = pink and rr = white; and O is a dominant gene for corolla shape such that OO = closed corolla, Oo = closed corolla and oo = open corolla. From the Punnett square, derive both the genotypic and phenotypic ratios for the cross.

58.2 Use probability to predict the outcome of a cross. Two hazel-eyed parents are heterozygous for the eye-colour gene B. When expressed as bb, the individual is blue-eyed. Mum's hair is (genuine) blonde but Dad's is mousy-brown. In this case she is double recessive (mm) for the hair-colour gene M and he is heterozygous (Mm). What is the probability that they will have a blue-eyed, blonde daughter?

58.3 Work out a likely genetic scenario for a given set of results. Five tail-less female mice were crossed with normal males (with tails). There were 31 normal mice and 28 tail-less mice in the F_1 progeny. When pairs of tail-less mice from the F_1 generation were crossed, their (F_2) progeny were as follows: normal, 27; tail-less, 55; dead on birth, but tail-less, 30. In each case the ratio of males to females was roughly 50:50. Provide a logical explanation for these results.

58.4 Predict parental genotypes from the results of crosses involving sex-linked genes. A red-eye gene is known to be sex-linked in *Drosophila*; that is, the alleles R (red-eyed) or r (white-eyed) are carried on the X chromosome, while the Y chromosome does not carry these eye-colour alleles. In the fruit fly, XX = female and XY = male. Predict the possible parental genotypes from the following F_1 progeny ratios:

(a) 35 red-eyed female, 17 red-eyed male, 19 white-eyed male;

(b) 27 red-eyed male, 29 red-eyed female;

(c) 19 red-eyed female, 18 white-eyed female, 22 red-eyed male, 21 white-eyed male.

58.5 Carry out a χ^2 (chi^2) test. A geneticist expects the results of a test cross to be in the phenotype ratio 1:2:1. He observes 548 progeny from his cross in the ratio 125:303:120. What should he conclude?

59 Bacterial and phage genetics

> **Definitions**
>
> **Phage** – a bacterial virus (bacteriophage).
>
> **Prophage** – a bacterial virus genome integrated within the genome of a host bacterial cell.
>
> **Copy number** – the average number of copies of a particular cellular molecule.
>
> **Transposon** – a section of DNA coding for its own movement from one genomic location to another and carrying other genes in addition to those coding for transposition.
>
> **Merozygote** – a cell containing two copies of a part of its genome, i.e. a partial diploid (sometimes also termed a merodiploid).

In eukaryotic organisms, genetic reassortment usually involves the fusion of two haploid gametes to form a zygote and a new generation (p. 413).

KEY POINT Bacterial genetics is very different from eukaryotic genetics (Chapter 58), due to the nature of the bacterial genome, which typically consists of a single chromosome plus none, one or several types of plasmid and/or phage, depending on the particular strain of bacterium.

In most bacteria, the chromosome is usually a covalently closed circular DNA molecule, carrying genes for essential metabolic functions and structural components. As a consequence, bacteria can be regarded as haploid organisms. Plasmids are additional 'mini-chromosomes', typically coding for non-essential features, e.g. antibiotic resistance, heavy metal tolerance. They are often present at a higher copy number than the chromosome and may also carry genes within mobile transposable elements (transposons). A single bacterium may contain more than one type of plasmid (though not if they are closely related plasmids, i.e. from the same incompatibility group). Plasmids can be introduced into a bacterial cell by conjugation (p. 424), or by transformation (p. 450). Bacteria can be 'cured' of their plasmids by chemical treatment, e.g. using acridine dyes that interfere with replication, or by growth under particular conditions, e.g. at high temperatures, where plasmid replication is unable to keep up with cell division.

A phage may replicate inside a bacterial cell or, in selected instances (temperate phages, p. 208), may exist within the cell in a non-replicating (latent) state, termed a prophage. As such, phages represent additional genetic elements that may be present within a bacterial cell, forming an important component of several aspects of bacterial genetics at the practical level.

The principal characteristics of experimental bacterial genetics ('crosses') are:

- the processes are completely distinct from sexual reproduction in eukaryotes;
- the processes are directional, from a donor cell (exogenote) to a recipient cell (endogenote);
- usually, only part of the donor cell's genome is transferred;
- in several instances, the recipient cell becomes a merozygote, with more than one copy of a gene, or genes. The merozygote may be a transient or a stable state, depending on circumstances;
- recombination (synapsis and 'crossing over') may or may not be involved, depending on the process and strains involved.

Using and interpreting standard nomenclature in bacterial genetics – different forms of three letter abbreviations are used for:

- **phenotypic features** – non-italic text, with superscripts where appropriate, e.g. Lac$^+$;
- **genotypic features** – lower case italic or underlined text, with individual letters to denote individual genes, e.g. lacZ;
- **gene products** (polypeptides and proteins) – the non-italicised equivalent of the abbreviation for the gene and with a capitalised first letter, e.g. the LacZ protein, which is a β-galactosidase and is the product of the lacZ gene in Lac$^+$ cells.
- **transposon** – inactivated genes show the transposon after the individual gene, separated by two colons, e.g. lacZ::Tn6.

Several abbreviations may be combined, e.g. a single strain might have the phenotype Amp^r Lac$^+$ Trp$^-$.

KEY POINT Bacterial crosses are best described as gene transfer rather than gene exchange, since the latter term suggests reciprocal DNA movement.

Working with bacterial mutants

To study bacterial genetics, it is necessary to use mutant strains, which have phenotypic characteristics that allow them to be distinguished from wild-type strains. The principal types of bacterial mutant include:

- Morphological mutants, with different structural characteristics from the wild type, e.g. so-called 'rough' mutants of selected bacteria, such as *Streptococcus* and *Klebsiella*, are defective in their synthesis of capsular polysaccharides, giving small, dull colonies on agar-based media. This is in contrast to 'smooth' wild-type strains, where colonies are large and glistening due to the hydrophilic polysaccharide capsule. At a practical level, it is relatively straightforward to work with such mutants, since wild-type and mutant bacteria will grow on the same medium, the mutants having a feature that visibly distinguishes them from wild types. However, there are few examples in general use.

- Resistant mutants, which grow in the presence of an inhibitory substance such as an antibiotic, a toxic compound or a particular phage. The isolation of an ampicillin-resistant mutant (Amp^r phenotype) is made possible by including ampicillin in the growth medium, since the growth of sensitive wild-type cells (Amp^s phenotype) will be inhibited.

- Carbon source mutants, which are unable to use a particular substance as a source of carbon or energy, e.g. *E. coli* mutants unable to use lactose (Lac^- phenotype), in contrast to Lac^+ wild types. A Lac^- mutant would be unable to grow on a minimal medium containing lactose as the principal carbon source. In order to identify and characterise such mutants, a differential medium must be used, e.g. MacConkey agar (a complex medium containing lactose as a *supplementary* carbon source, plus a pH indicator dye – colonies of Lac^+ and Lac^- strains are distinguished on the basis of size and pigmentation, p. 213).

- Nutritional mutants, which have an additional requirement for a particular nutrient, compared to the wild type. For example, a strain auxotrophic for the amino acid tryptophan (Trp^-) would grow only if the medium contained tryptophan (e.g. in a minimal medium plus tryptophan). Since it is not possible to devise a single medium which will allow an auxotroph to be distinguished from the corresponding prototrophic wild type, the selection and identification of such mutants require a different approach, involving replica plating from (a) nutrient-rich medium onto (b) minimal medium and (c) minimal medium supplemented with the particular nutrient (Fig. 59.1).

- Mutants created by transposon mutagenesis, where a gene has been inactivated by the insertion of a transposon, e.g. *lacZ* inactivated by Tn10 (p. 420).

- Strains with different forms of a particular gene on the chromosome and also on a plasmid (partial diploids): the genes carried on the plasmid can be shown as follows: *lacZ*::Tn10/F'*lacZ*$^+$ with the slash separating the transposon-inactivated chromosomal gene from the F' plasmid with its functional *lacZ* gene.

Definitions

Minimal medium – a chemically defined medium, containing only sufficient nutrients to meet the requirements of wild-type cells, i.e. inorganic salts plus a particular carbon source.

Auxotrophs – mutants requiring an additional organic compound (e.g. an amino acid or growth factor) in order to grow in minimal medium.

Prototrophs – wild types (from which auxotrophs are derived) and all other strains capable of growth in minimal medium.

Differential medium – usually a complex medium with additional compounds that distinguish between two types of bacteria, e.g. wild-type and mutant strains, often using pH indicator dyes or chromogenic/fluorogenic substrates.

Visualising carbon source mutants – chromogenic and fluorogenic enzyme substrates (p. 382) can be used to detect particular phenotypes; for example, E. coli mutant strains without a functional LacZ gene would give non-coloured colonies on a medium containing X-galactoside (p. 383), in contrast to wild type strains, which would give blue-green colonies.

Take care when distinguishing between carbon source and nutritional mutants – note that a Lac$^-$ mutant is unable to grow if provided with lactose as the sole carbon source, while a Trp$^-$ mutant is unable to grow unless it is provided with tryptophan as a specific nutrient.

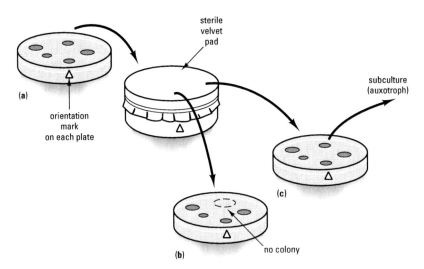

Fig. 59.1 Identification of auxotrophs using replica plating: a mixed suspension of wild-type and mutant cells is first cultured on the surface of a master plate containing nutrient-rich medium (a), then transferred to a sterile pad which is then used to inoculate minimal medium (b) and minimal medium plus a particular nutrient (c). Failure to grow on (b), coupled with growth on (c), implies auxotrophy for that particular nutrient and allows the auxotroph to be further subcultured and studied.

- Conditional lethal mutants, which have a defect that causes death under a specific set of circumstances (the 'restrictive condition'), e.g. in *E. coli* temperature-sensitive mutants that grow at 30 °C, but not 40 °C, often as a result of the temperature-dependent inactivation of mutant enzymes.

DNA transfer in bacteria

> **KEY POINT** *The traditional approach to genetic analysis in prokaryotes involves mapping the position of individual genes using information provided from 'crosses', based on gene transfer. Typically, the recipient cell will be a mutant and the donor DNA will carry wild-type genes, enabling the transfer of wild-type characteristics to be studied in the laboratory.*

The principal bacterial gene transfer processes are:

Natural transformation

In contrast to genetic engineering techniques, where DNA uptake is induced under specific laboratory conditions (p. 450), natural transformation involves the release of DNA to the external medium (e.g. death and lysis of the donor cell) and its subsequent uptake and incorporation into the genome of the recipient cell. Natural transformation is restricted to a limited number of bacterial groups and only occurs if the recipient cells are in the correct physiological state, termed *competence*, often in the early exponential (log) growth phase (p. 233). Transformation is a relatively rare event, occurring at frequencies of $\leqslant 1$ transformant per 10^3 cells. In a competent cell, DNA will be taken up and, if homologous (i.e. from another strain of the same species), may then be incorporated into the genome of the recipient cell by homologous recombination *via* a double 'crossover' (two recombination events, as shown in Figure 59.2).

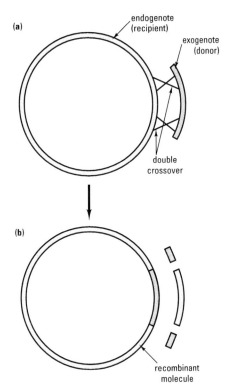

Fig. 59.2 Homologous recombination between donor DNA (from transformation, transduction or conjugation) and the recipient genome *via* a double crossover.

Performing co-transformation tests using auxotrophic recipients – controls must be set up without added donor DNA to determine the number of revertants (spontaneous mutants), giving the actual rate of transformation by subtraction.

The principal application of transformation has been in mapping the position of genes in those bacteria showing natural competence (e.g. *Bacillus subtilis*). The experiments are often easiest to perform with auxotrophic recipients and prototrophic (wild-type) donor DNA, since transformants can be selectively grown on media lacking one or more individual nutrients. The frequency of co-transformation of two genes is a measure of how close together they are likely to be on the donor DNA strand – a high co-transformation frequency implies that they are close together on the chromosome, reaching a recipient cell on the same fragment of DNA.

 KEY POINT *While co-transformation frequencies are inversely related to map distances, they are not directly equivalent to the recombination frequencies used in mapping eukaryotic genomes (p. 416), because they are also influenced by the size distribution of the fragments of donor DNA and by the likelihood of homologous recombination.*

Transformation mapping has several limitations, since it requires a fairly large number of complex replica plating experiments to produce a chromosomal map, and the relative position of genes that are very far apart cannot be determined directly – the 'jigsaw' requires a large number of available pieces, before the underlying structure can be seen. It is also insensitive for small map distances – two genes that are adjacent, or very close together, will give similar high co-transformation values.

Transduction

Here, the DNA exchange is mediated by a phage (p. 208), in one of two processes:

Performing generalised transduction in E. coli – P1 phage is often used, since it gives a high proportion of transducing particles (≈0.3% of the total), with a large fragment size (≈2.5% of the E. coli genome), making it easier to locate individual transductants.

1. Generalised transduction: occasionally, a fragment of chromosomal or plasmid DNA within an infected bacterial cell may be packaged within the protein coat of a phage, in place of the phage genome. This fragment might be derived from any part of the host cell genome (exogenote). After release, on lysis of the donor cell, the transducing particle may introduce the DNA fragment into a new recipient cell. The introduced DNA may then be incorporated into the host genome by a double crossover, in a homologous recombination event similar to that shown in Figure 59.2 for transformation. Generalised transduction can be used to establish gene order and for mapping purposes, using broadly similar principles to transformation, as only closely spaced genes will show co-transduction, with a co-transduction frequency that is inversely related to the distance between the two genes. Since a generalised transducing particle can carry a fairly small amount of DNA, the relative frequency of co-transduction can be used to provide finer detail of gene order over shorter distances than for transformation.

2. Specialised transduction: this is mediated only by a temperate phage, e.g. λ phage, which integrates at a specific site on the chromosome by a single crossover event. It involves a restricted number of genes – typically, a pair of genes on either side of the integration site for a particular temperate phage (Fig. 59.3). Specialised transduction will only occur if, on entry into the lytic cycle (phage replication, p. 208), there is an incorrect (abnormal) excision of the prophage and a part of the bacterial genome is incorporated, giving a specialised transducing

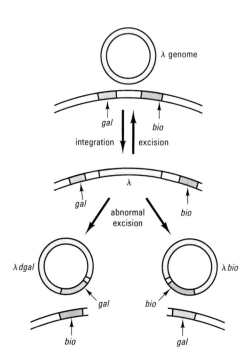

Fig. 59.3 Integration, excision and abnormal excision of λ phage, creating a modified phage genome (either λdgal or λbio), prior to specialised transduction.

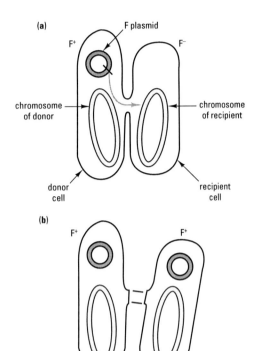

Fig. 59.4 Conjugation between F+ donor and F− recipient. A single strand of F plasmid DNA is cut and transferred in linear form to the recipient (a), followed by recircularisation and complementary strand synthesis, converting the recipient to F+ (b).

Example The *E. coli* strain HfrH has the F proplasmid integrated at a site close to genes for the synthesis of threonine, so these genes would be transferred early during HfrH × F− conjugation.

particle (a modified phage, carrying a bacterial gene). In the case of λ phage (Fig. 59.3), the transducing particle may carry either the biotin gene (λ*bio*), or the galactose gene (λ*gal*, also termed λ*dgal*, since the modified phage is defective due to the loss of an essential part of the phage genome, and therefore is incapable of replicating without the aid of a co-infecting, non-defective 'helper' phage). On entry into the recipient cell, the modified phage may become latent, creating a partial diploid. The practical applications of specialised transduction are limited to those genes flanking the integration sites of temperate phages. The most common investigations are of complementation in merozygotes containing two copies of a particular gene. Where donor and recipients of the same mutant phenotype have mutations in different genes, the creation of a partial diploid can produce a wild-type phenotype.

The integration and excision of phages is likely to be an important means of gene transfer in natural environments, playing a key role in the evolution and pathogenesis of bacteria, as has been demonstrated by recent genome sequencing projects that have demonstrated the role of prophages in toxigenic *Vibrio cholerae* and *E. coli*.

Conjugation

Here, the transfer occurs as a result of cell-to-cell contact, with direct transmission of DNA from donor to recipient cell. In *E. coli*, the donor cell carries specific surface pili (protein microtubules), allowing a donor cell to attach to receptors on the surface of a recipient cell and bringing the paired cells into close contact (Fig. 59.4). In the simplest instance, the donor cell carries an additional plasmid, the F plasmid (originally termed 'F factor'), that encodes the genes responsible for conjugation, including those for the protein subunits of the specialised pili. In conventional notation, the donor is termed F^+ and the recipient F^-. During conjugation between F^+ donor and F^- recipient, a single strand of the circular F plasmid is cut and transferred (in linear form) to the recipient cell, which becomes F^+ once the entire plasmid has been transferred and a complementary strand has been synthesised – this process takes a few minutes.

> KEY POINT It is important to understand that the process of bacterial conjugation is completely different from sexual reproduction in eukaryotic organisms, since there are no gametes and no zygote or offspring is formed. Consequently, it is more appropriate to describe the participants as donor and recipient (of genetic information) rather than as male or female.

At a practical level, crosses involving F^+ donors give little useful information, apart from mapping the position of genes on the F plasmid. However, two other types of donor are more useful:

1. Hfr strains, where the F plasmid DNA has become integrated into the chromosome, as a proplasmid, in an analogous manner to a temperate phage such as λ (Fig. 59.3). Such strains show a *high frequency of recombination* (hence Hfr), since chromosomal genes are transferred to the recipient cell at a far greater frequency than in crosses using F^+ donors. The Hfr × F^- cross is illustrated in Figure 59.5: a part of the F proplasmid is first to be transferred, followed by chromosomal DNA and, finally, the remaining fragment of the F proplasmid. After transfer, donor chromosomal DNA can be integrated into the

Example F'*lac* is an F' plasmid that incorporates wild-type chromosomal DNA coding for lactose utilisation.

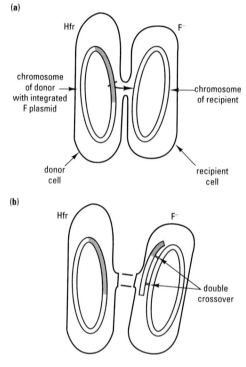

Fig. 59.5 Conjugation between Hfr donor and F⁻ recipient. The F proplasmid is cut and leads the chromosome (in single stranded form) from the donor (a), followed by homologous recombination in the recipient (b).

recipient's genome by homologous recombination (Fig. 59.2). The conjugating pair usually breaks apart before the process is complete and the recipient cell remains F⁻, since only the leading fragment of the F plasmid reaches the recipient cell. A number of different Hfr strains of *E. coli* are available, with the F proplasmid inserted at different chromosomal locations.

2. F' ('F prime') donors, with a modified F plasmid incorporating one or more chromosomal genes, formed as a result of defective excision of the F proplasmid from the chromosome in a similar manner to λ*bio* (Fig. 59.3). The F' plasmid will transfer its chromosomal genes at very high frequency to a recipient, which will also become F' once transfer of the F' plasmid DNA is complete. This process is sometimes referred to as F-duction (or, less appropriately, as 'sexduction').

Mapping with Hfr donors using interrupted conjugation

Mating an antibiotic-sensitive wild-type Hfr donor with an antibiotic-resistant mutant F⁻ recipient provides a simple means of detecting recombinants, since the donor cells will be unable to grow on a medium containing the antibiotic while unmodified recipient cells will have a different phenotype from the recombinant recipient cells. The method is even simpler for the most common crosses, using nutritional mutants, where a prototrophic Hfr donor is crossed with an auxotrophic F⁻ recipient and only the recombinants are able to grow on minimal medium with added antibiotic; this approach can be extended to multiple genes, using a multiple auxotroph F⁻ recipient. By carrying out a series of experiments where the conjugation process is terminated at different times (e.g. using vigorous mixing in a vortex mixer, p. 139), the time required to transfer a particular characteristic can be determined as the earliest time at which this interruption no longer prevents recombinants from appearing. Typical results from a cross involving several genetic markers are shown in Figure 59.6a: by plotting the result graphically and extrapolating the curves to their intersect with the *x*-axis, the *time of entry* of a particular marker can be determined accurately.

The different times of entry of each characteristic reflect their relative positions on the chromosome, with genes near to the origin of transfer of

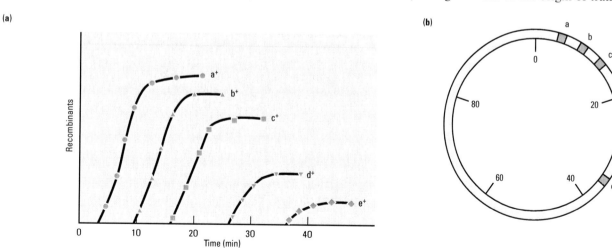

Fig. 59.6 Recombinants obtained from interrupted conjugation for a number of characters (a to e), plotted to show the time of entry of each character (a); the same data have been used to map the position of each character on the bacterial chromosome (b).

Genome mapping using interrupted conjugation – while this technique works well for widely spaced markers, it is less useful than transformation and transduction for closely linked genes.

the F proplasmid having short times of entry and those further along the chromosome having later times of entry. The times of entry can be used to locate the genes on a chromosome map (Fig. 59.6b), representing map distances relative to the origin of transfer. This map differs from those produced from transformation and transduction analysis, since it is a transfer order map, rather than a linkage map. However, as transfer of the entire chromosome requires ≈100 minutes, it is difficult to map genes far from the origin of transfer, due to the decreasing number of recombinants with increasing time (Fig. 59.6a): this is overcome by combining mapping data from several Hfr strains, each with the F plasmid integrated at a different position. In fact, this approach was originally used to demonstrate the circularity of the *E. coli* chromosome and, subsequently, to determine the location of over 1500 individual genes in this organism.

Genetic analysis using F' plasmids

Since cells containing an F' plasmid are stable merozygotes, their principal use is in studying the behaviour of genes under diploid conditions, for example:

- to determine whether a particular mutation (on the F' plasmid) is dominant or recessive over another (on the chromosome);
- to test whether a particular gene complements another: if two mutations are in different genes, then they can complement each other in a stable merozygote, while two mutations in the same gene cannot complement each other in a stable merozygote;
- to examine whether a particular regulatory gene located on an F' plasmid influences the expression of one or more chromosomal genes, i.e. whether it can operate in the *trans* position. The alternative test is to construct a merozygote with two genes in a chromosomal location (i.e. the *cis* position), for example, using specialised transduction (p. 423). This type of analysis has been used to investigate the molecular basis of gene expression and the function of regulatory regions of the chromosome.

Definitions

Trans (Latin 'across') – on separate DNA molecules.

Cis (Latin 'here') – on the same DNA molecule.

 KEY POINT Note that, while many of the above procedures have proved to be useful in the early stages of bacterial genome analysis, the techniques and methods of molecular biology (Chapters 60–62) can be used to provide more detailed genetic information, without the need to obtain certain types of mutant, or carry out specific crosses between particular donors and recipients.

Phage crosses

Performing phage crosses – large numbers of both types of phage must be used to ensure that every bacterial cell is infected with at least one of each type, i.e. there is a high multiplicity of infection.

In practical classes, you may carry out experiments using phages, often using T-even phages of *E. coli* (e.g. T4 coliphage). The methods used to map the phage genome superficially resemble those used for eukaryotic organisms (Chapter 58) since the 'progeny' of two 'parental' phages with contrasting genotypes are analysed for recombinants and the frequency of recombination is used to measure the distance between linked genes.

KEY POINT *A phage cross (sometimes termed a 'mating') does not involve meiosis, gamete production or zygote formation, but is a result of a mixed infection, where a single bacterial cell is infected with phages of the two original genotypes.*

Following a single round of replication, the 'progeny' from this mixed infection are screened for the various possible genotypes, as plaques on a 'lawn' of a susceptible strain of *E. coli*, using conventional phage culture techniques (p. 208).

Phage mutants

The phage genes most widely used in crosses are those affecting:

- plaque morphology – for example, mutants may form large plaques as opposed to small plaques, or plaques with a different margin, e.g. a light turbid halo as opposed to a dark halo;
- host range – e.g. T-even mutants may infect different strains of *E. coli* from the parental types;
- conditional lethality, especially T4 phage mutants of the *rII* locus, unable to grow on *E. coli* K12(λ) in contrast to wild-type T4.

Understanding conditional lethal mutants – an important group of phages are the so-called 'amber' mutants, with polypeptides containing a premature termination codon, UAG (p. 429); such mutant strains are propagated in 'amber suppressor' mutant strains of E. coli, which contain a mutant tRNA that recognises UAG as coding for an amino acid.

Phage mapping

By crossing a mutant for plaque morphology with a mutant for host range, the relative frequency of recombinants (i.e. progeny with double mutant and wild-type phenotypes) can be expressed in terms of the proportion of the total number of plaques, as a recombination frequency (*R*), where:

$$R = \frac{\text{number of double mutants and wild-type plaques}}{\text{total number of plaques}} \quad [59.1]$$

This value is sometimes multiplied by 100 and expressed as a percentage, or as map units, equivalent to 'centimorgans' (1 cM = 1% crossing over) in conventional Mendelian crosses (p. 416).

Learning from Benzer's work – detailed fine structure mapping of the rII region of T4 coliphage established that recombination can occur within a single gene, and even between adjacent nucleotides. Before these experiments, a gene was regarded as an indivisible unit – nowadays we appreciate that the smallest unit in genetics is the base pair.

The pioneering work of Seymour Benzer established that the *rII* region of the T4 genome consisted of two distinct genes (*rIIA* and *rIIB*), each of which can give rise to the same mutant phenotype. T4 *rII* mutants are most often used for demonstrating *trans* complementation between these two genes and for demonstrating the principles of fine structure mapping of this region of the genome, since the conditional lethality of the rII phenotype simplifies the search for rare recombinant wild-type phages, as only these recombinants are able to form plaques on *E. coli* K12(λ).

Sources for further study

Anon. *Information about bacteriophage lambda.*
Available: http://www.asm.org/division/M/fax/LamFax.html
Last accessed: 01/04/07.
[Provides an overview of phage λ]

Birge, E.A. (2005) *Bacterial and Bacteriophage Genetics*, 4th edn. Springer, New York.

Dale, J.W. and Park, S.F. (2004) *Molecular Genetics of Bacteria*, 4th edn. Wiley, Chichester.

Eynard, N., and Teisse, J. (2000) *Electrotransformation of Bacteria.* Springer-Verlag, Berlin.

Hardy, K.G. (1994) *Plasmids: A Practical Approach.* Oxford University Press, Oxford.

Hartl, D.L. and Jones, E.W. *GeNETics on the Web*. Available: http://www.jbpub.com/genetics/essentials3e/geNETics_on_the_web_Links.cfm Last accessed: 01/04/07.

Snyder, L. and Champness, W. (2002) *Molecular Genetics of Bacteria*, 2nd edn. American Society for Microbiology, Washington, DC.

Trun, N. and Trempy, J. (2003) *Fundamental Bacterial Genetics*. Blackwell, Oxford.

Study exercises

59.1 Test your understanding of bacterial and phage genetics. Distinguish between each of the following pairs of terms:
(a) zygote and merozygote;
(b) transformation and transposition;
(c) generalised and specialised transduction;
(d) prototroph and auxotroph;
(e) HFr and F′ strains.

59.2 Interpret the codes commonly used to describe bacterial mutant strains. Describe the *E. coli* phenotypes represented by the following:
(a) Ampr Lac⁺ Trp⁻
(b) His⁻ Lac⁻
(c) *glnA*::Tn10
(d) *trpA*::Tn 6/F′*trpA*⁺
(e) a Gal⁺ strain lysogenised with λ*dgal*

59.3 Interpret results from an HFr mapping experiment. Three different HFr strains were used to map the genome of a bacterium in a series of interrupted conjugation experiments. The strains transferred genes to the F′ recipient cell in the following orders:
Strain (a): X W C Z A
Strain (b): Y D F X W
Strain (c): Z A E B Y
Assuming that all of these genes are present only once on a single circular chromosome, what is their order?

59.4 Construct a chromosomal map. The table below represents data for an interrupted conjugation study using an HFr donor strain containing four 'marker' genes. These were distinguished by plating onto four different media, with the following major components:
Medium A: glucose, histidine (trace amount), tryptophan (trace amount), ampicillin.
Medium B: glucose, histidine (trace amount).
Medium C: glucose, tryptophan (trace amount).
Medium D: lactose, histidine (trace amount), tryptophan (trace amount).
(a) What is tested for by each of these four different media?
(b) Based on the composition of these media and the data below, what is the phenotype of (i) the HFr donor and (ii) the F⁻ recipient?
(c) Given a chromosome map size of 60 min, where would each of these genes be located on the bacterial chromosome?

Colony numbers following interrupted mating, with plating on medium A, B, C or D

Time (min)	Medium A	Medium B	Medium C	Medium D
0	0	0	0	0
10	0	32	0	0
20	0	287	38	0
30	34	339	182	0
40	156	341	226	28
50	179	338	229	89
60	180	340	227	95

60 Molecular genetics I – fundamental principles

Deoxyribonucleic acid (DNA) is the genetic material of all cellular organisms. Its structure is outlined in Chapter 53.

 KEY POINT *The sequence of the bases A, G, T and C carries the genetic information of the organism. A section of DNA that encodes the information for a single polypeptide or protein is referred to as a gene, while the entire genetic information of an organism is called the genome.*

The amount of DNA in the genome is usually expressed in terms of base pairs (bp), rather than M_r, and its size depends on the complexity of the organism: for example, the human papilloma virus has a genome of 8×10^3 base pairs (8 kbp), that of *Escherichia coli* is 4×10^6 base pairs (4 Mbp), while the human haploid genome is very large, comprising 3×10^9 base pairs (3000 Mbp). The human genome contains about 23 000 genes (distributed on 23 pairs of chromosomes), which represent only about 1 per cent of the total amount of genomic DNA (i.e. 99 per cent of human DNA is non-coding). Organisms with smaller genomes have smaller amounts of non-coding DNA: some viral genomes have 'overlapping' genes, where the same base sequences carry information for more than one protein.

The size of each individual gene varies considerably: the largest ones may exceed 10 Mbp. Chromosomes represent the largest organisational units of DNA: in eukaryotes, they are usually linear molecules, complexed with protein and RNA, varying in length from tens to hundreds of Mbp. The unit used to denote physical distance between genes (base pairs) differs from that used to describe genetic distance (centimorgan, cM), which is based on recombination frequency (p. 416). In humans, $1\,cM \approx 1\,Mbp$, though this relationship varies widely depending on recombination frequency within particular regions of a chromosome.

Each template for the synthesis of RNA (transcription) begins at a promoter site upstream of the coding sequence and terminates at a specific site at the end of the gene. The base sequence of this RNA is complementary to the 'template strand' and equivalent to the 'coding strand' of the DNA. In eukaryotic cells, transcription occurs in the nucleus, where the newly synthesised RNA, or primary transcript, is also subject to processing, or 'splicing', in which non-coding regions within the gene (introns) are excised, joining the coding regions (exons) together into a continuous sequence. Further processing results in the addition of a polyadenyl 'tail' at the 3' end and a 7-methylguanosine 'cap' at the 5' end of what is now mature eukaryotic messenger RNA (mRNA). The mRNA then migrates from the nucleus to the cytoplasm, where it acts as a template for protein synthesis (translation) at the ribosome: the translated portion of mRNA is read in coding units, termed codons, consisting of three bases.

Each codon corresponds to a specific amino acid, including a codon for the initiation of protein synthesis (Table 60.1). Individual amino acids are brought to the ribosome by specific transfer RNA (tRNA) molecules that recognise particular codons. The amino acids are incorporated into the growing polypeptide chain in the order dictated by the sequence of codons until a termination codon is recognised, after which the protein is released from the ribosome.

Definition

Units of nucleic acid size (length) –
Kilobase pair (kbp) = 10^3 base pairs
Megabase pair (Mbp) = 10^6 base pairs

Table 60.1 The Genetic Code – combinations of nucleotide bases coding for individual amino acids

1st Base	2nd Base				3rd Base
	U	C	A	G	
U	F	S	Y	C	U
	F	S	Y	C	C
	L	S	*	*	A
	L	S	*	W	G
C	L	P	H	R	U
	L	P	H	R	C
	L	P	Q	R	A
	L	P	Q	R	G
A	I	T	N	S	U
	I	T	N	S	C
	I	T	K	R	A
	M	T	K	R	G
G	V	A	D	G	U
	V	A	D	G	C
	V	A	E	G	A
	V	A	E	G	G

* = termination codons.
Standard abbreviations for the above amino acids are as given in Table 50.1.
Note that AUG (=M) is the initiation codon. The above codons are given for mRNA – the coding strand of DNA would have T in place of U, while the template strand would have complementary bases to those given above.

Good practice in molecular genetics –
this includes:

- accurate pipetting down to 1 µl, or less;
- steadiness of hand in sample loading;
- keeping enzyme solutions cold during use and frozen during storage;
- using sterile plasticware;
- using double-distilled ultrapure water;
- wearing disposable gloves to avoid contamination.

Using restriction digests of viral genomes – the controlled fragmentation pattern obtained when a phage or plasmid is cut with a specific restriction enzyme can be used to create a 'DNA ladder' – a set of DNA fragments of known length that can be used as reference standards in gel electrophoresis (p. 431).

Example ClustalW software can be used for multiple sequence alignment – it can be downloaded from the European Bioinformatics Institute website at: http://www.ebi.ac.uk/clustalw/

Using restriction enzymes in genetic engineering – those enzymes that cleave DNA to give single stranded 'sticky ends' are particularly useful in gene cloning (Chapter 62). Two different molecules of DNA cut with the same restriction enzyme will have complementary single stranded regions, allowing them to anneal as a result of hydrogen bonding between individual bases within these regions.

The techniques described in this and subsequent chapters are used widely in many aspects of molecular biology, including the identification and characterisation of genes, gene cloning and genetic engineering, medical genetics and genetic fingerprinting.

 KEY POINT An important characteristic of nucleic acids is their ability to hybridise: two single strands with complementary base pairs will hydrogen bond (anneal) to produce a duplex, as in conventional double stranded (ds)DNA. This duplex can be converted to single stranded (ss)DNA (i.e. 'melted') by conditions that disrupt hydrogen bonding, e.g. raising the temperature or addition of salt, and then reannealed by lowering the temperature or by removal of salt.

DNA can be easily purified and assayed (Chapter 53), and it is relatively stable in its pure form. Most of the problems of working with such a large biomolecule have been overcome by the following:

- using restriction enzymes (see below) to cut DNA precisely and reproducibly: mammalian DNA may yield millions of fragments, with sizes ranging from a few hundred base pairs to tens of kbp, while small viral genomes may give only a few fragments;
- electrophoretic methods for the separation of DNA fragments on the basis of their sizes (p. 431);
- segments of DNA can be detected by 'probes' that specifically hybridise with the DNA of interest (p. 433);
- once separated, specific segments of DNA can be obtained in almost unlimited quantities by insertion into vectors and multiplication within suitable host cells (DNA cloning, Chapter 62);
- methods are now available for rapidly determining the base sequence of DNA segments (p. 436), together with strategies for combining information from several adjacent segments to give contiguous sequences ('contigs') representing entire genomes in well-studied organisms (Chapter 11);
- specific target sequences of DNA within a genome can be amplified by more than a billion-fold by the polymerase chain reaction (PCR, see Chapter 61).

Producing DNA fragments

Fragments of DNA can be produced by mechanical shearing or ultrasonication (p. 245), producing a random array of fragment sizes. Reproducible cleavage can be achieved using type II restriction endonucleases (commonly called restriction enzymes) which recognise and cleave at a specific palindromic sequence of double stranded DNA (usually four or six base pairs), known as the restriction site. Each enzyme is given a code name derived from the name of the organism from which it is isolated, e.g. *Hin* dIII was the third restriction enzyme to be obtained from *Haemophilus influenzae* strain Rd (Fig. 60.1). Most restriction enzymes will cut each DNA strand at a slightly different position within the restriction site to produce short, single stranded regions known as cohesive ends, or 'sticky ends', as shown in Figure 60.1. A few restriction enzymes cleave DNA to give blunt-ended fragments (e.g. *Hin* dII or *Sma* I).

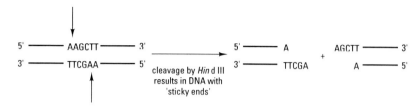

Fig. 60.1 Recognition base sequence and cleavage site for the restriction enzyme Hin dIII. This is the conventional representation of double-stranded DNA, showing the individual bases, where A is adenine, C cytosine, G guanine and T thymine. The cleavage site on each strand is shown by an arrow.

Maximising recovery of DNA – these large molecules are easily damaged by mechanical forces, e.g. vigorous shaking or stirring during extraction. In addition, all glassware must be scrupulously cleaned and gloves must be worn, to prevent deoxyribonuclease contamination of solutions.

Working with small volumes in molecular biology – use a pipettor of appropriate size (e.g. P2 or P20 Gilson Pipetman) with a fine tip. For very small volumes, pre-wet the tip before delivering the required volume.

 SAFETY NOTE Using an electrophoresis tank – always ensure that the top cover is in position, to prevent evaporation of buffer solution and to reduce the possibility of electric shock.

Separation of nucleic acids using gel electrophoresis

Separation of DNA by agarose and polyacrylamide gel electrophoresis
Electrophoresis is the term used to describe the movement of ions in an applied electrical field. DNA molecules are negatively charged, migrating through an agarose gel towards the anode at a rate which is dependent upon molecular size – smaller, compact DNA molecules can pass through the sieve-like agarose matrix more easily than large, extended fragments. Electrophoresis of plasmid DNA is usually carried out using a submerged agarose gel (Fig. 60.2). The amount of agarose is adjusted, depending on the size of the DNA molecules to be separated, e.g. 0.3% w/v agarose is used for large fragments (>20 000 bp) while 0.8% is used for smaller fragments. Very small fragments are best separated using a polyacrylamide gel (Table 60.2). Note the following:

- Individual samples are added to pre-formed wells using a pipettor. The volume of sample added to each well is usually less than 25 µl so a steady hand and careful dispensing are needed to pipette each sample.
- The density of the samples is usually increased by adding a small amount of glycerol or sucrose, so that each sample is retained within the appropriate well.
- A water-soluble anionic tracking dye (e.g. bromophenol blue or xylene cyanol) is also added to each sample, so that migration can be followed visually.

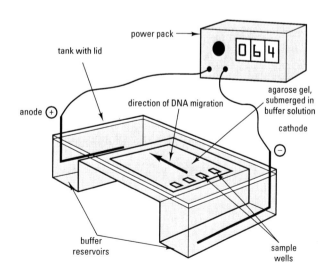

Fig. 60.2 Agarose gel electrophoresis of DNA.

Table 60.2 Gel concentrations for the separation of DNA of various sizes

Type of gel	% (w/v)	Range of resolution of DNA (bp)
Polyacrylamide	20.0	5–100
	15.0	20–150
	5.0	75–500
	3.5	100–1000
Agarose	2.0	100–5000
	1.2	200–8000
	0.8	400–20 000
	0.3	1000–70 000

Using DNA fragment size markers – for accurate determination of fragment size (length) your standards must have the same conformation as the DNA in your sample, i.e. linear DNA standards for linear (restriction) fragments and closed circular standards for plasmid DNA.

- DNA fragment size markers ('molecular weight' standards) are added to one or more wells. After electrophoresis, the relative positions of bands of known molecular weight can be used to prepare a calibration curve (usually by plotting $\log_{10}$ of the size (length) of each band against the distance travelled).
- The gel should be run until the bromophenol blue tracking dye has migrated across 80% of the gel (see manufacturer's instructions for appropriate voltages/times).
- After electrophoresis, the bands of DNA can be visualised by soaking the gel for around 5 min in $0.1–0.5\,mg\,l^{-1}$ ethidium bromide (Fig. 53.4), which binds to DNA by intercalation between the paired nucleotides of the double helix.

SAFETY NOTE Handling ethidium bromide – ethidium bromide is carcinogenic so always use gloves when handling stained gels and make sure you do not spill any staining solution, or use a safer alternative, such as SYBR Safe.

- Under UV light, bands of DNA are visible due to the intense fluorescence of the ethidium bromide. The limit of detection using this method is around 10 ng DNA per band. The migration of each band from the well can be measured using a ruler. Alternatively, a photograph can be taken, using a digital camera and adaptor, or a dedicated system, e.g. GelDoc from BioRad.
- If a particular band is required for further study (e.g. a plasmid), the piece of gel containing that band is cut from the gel using a clean scalpel. The DNA can be separated from the agarose by solubilising the gel slice and then binding the DNA to an anion exchange resin, followed by elution in water or buffer (e.g. TE buffer, containing $10\,mmol\,l^{-1}$ TRIS/HCl at pH 8.0 plus $1\,mmol\,l^{-1}$ EDTA). Various commercial kits based on this approach are also available, e.g. from Qiagen.

Electrophoretic separation of RNA

Total cellular RNA or purified mRNA can be separated on the basis of size by electrophoretic separations similar to those used for DNA fragments. However, under the conditions used to separate dsDNA, RNA molecules tend to develop a secondary structure, and this leads to anomalous mobilities. To eliminate RNA secondary structure, samples are pretreated by heating in dilute formamide or glyoxal, and electrophoresis is carried out in 'denaturing gels' which include buffers containing formaldehyde.

SAFETY NOTE Working with a UV radiation source – always wear suitable UV-filtering plastic safety glasses or goggles to protect your eyes.

Assaying nucleic acids in solution – double stranded DNA at $50\,\mu g\,ml^{-1}$ has an A_{260} of 1, and the same absorbance is obtained for single stranded DNA at $33\,\mu g\,ml^{-1}$ and (single stranded) RNA at $40\,\mu g\,ml^{-1}$. These values can be used to convert the absorbance of a test solution to a concentration of nucleic acid.

Pulsed field gel electrophoresis (PFGE)

If structural information is to be gained about large stretches of genomic DNA, then the order of the relatively short DNA segments (generated by

Definition

Chromosome walking – a method for analysing areas of interest in DNA, in which the end of a segment of DNA is used as a probe to locate other segments that overlap the first segment: long stretches of DNA can be analysed by subsequent use of probes made from the ends of successive overlapping segments.

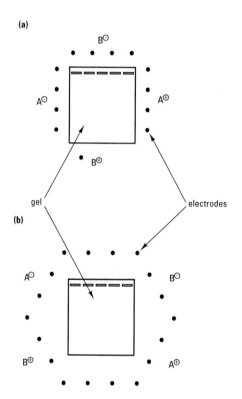

Fig. 60.3 The configuration of electrodes for conventional PFGE (a) and CHEF/PACE (b).

the restriction enzymes described above) needs to be established. This is technically possible by techniques such as 'chromosome walking', but is time-consuming and potentially difficult, especially when dealing with large chromosomes such as those from yeast (a few Mbp) and humans (50–100 Mb), in contrast to the smaller genomes of bacteria and viruses. The technique of PFGE allows separation of DNA fragments of up to ≈12 Mb.

Very large DNA fragments (>100 kbp) can be generated from chromosomal DNA by the use of certain restriction enzymes that recognise base sequences that are present at relatively low frequency, e.g. the enzyme *Not* I, which recognises a sequence of 8 bp rather than 4–6 bp. These enzymes are sometimes called 'rare cutters'. Genomic DNA prepared in the normal way is not suitable for digestion by these enzymes, as shearing during extraction fragments the DNA. Therefore, genomic DNA for analysis by PFGE is prepared as follows:

- cells are embedded in an agarose block;
- the block is incubated in solutions containing detergent, RNase and proteinase K, lysing the cells and hydrolysing RNA and proteins. The products of RNase and proteinase digestion diffuse away, leaving behind genomic DNA molecules exceeding several thousand kbp;
- the block is incubated *in situ* in a buffered solution containing an appropriate 'rare cutter': restriction fragments are produced, of up to ≈800 kbp.

PFGE differs from conventional electrophoresis in that it uses two or more alternating electric fields. An explanation for the effectivess of the technique is that large DNA fragments will be distorted by the voltage gradient, tending to elongate in the direction of the electric field and 'snaking' through pores in the gel. If the original electric field (Fig. 60.3a), is removed, and a second is applied at an angle to the first (Fig. 60.3b), the DNA must reorientate before it can migrate in the new direction. Larger (longer) DNA molecules will take more time to reorientate than smaller molecules, resulting in size-dependent separations.

The original configuration of electrodes used in PFGE is shown in Fig. 60.3a; this tends to produce 'bent' lanes that make lane-to-lane comparisons difficult. This can be overcome by using one of the many variants of the technique, one of which employs contour clamped homogeneous electric fields (CHEF). Here, multiple electrodes are arranged in a hexagonal array around the gel (Fig. 60.3b) and these are used to generate homogeneous electric fields with reorientation angles of up to 120°. A further development of CHEF involves programmable, autonomously controlled electrodes (PACE), which allows virtually unlimited variation of field and pulsing configurations, and can fractionate DNA molecules from 100 bp to 6 Mbp.

Identification of specific nucleic acid molecules using blotting and hybridisation techniques

Southern blotting

After separation by conventional agarose gel electrophoresis (p. 431), the fragments of DNA can be denatured and immobilised on a filter membrane using a technique named after its inventor, E.M. Southern.

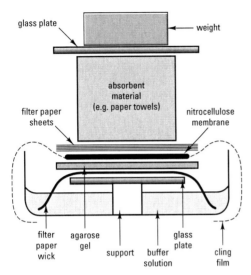

Fig. 60.4 Components of Southern blotting apparatus.

Biomedical applications of dot blotting – these include the detection of specific pathogenic microbes in samples and the identification of particular genetic markers, e.g. in inherited genetic disorders and forensic science.

Definition

Probe – a labelled DNA or RNA sequence used to detect the presence of a complementary sequence (by molecular hybridisation) in a mixture of DNA or RNA fragments.

Example For a simple tripeptide containing methionine, aspartate and phenylalanine, the synthetic oligonucleotide probes would include combinations of the following codons:
1st codon (met): ATG
2nd codon (asp): GAT, GAC
3rd codon (phe): TTT, TTC

The main features of the conventional apparatus for Southern blotting are shown in Figure 60.4. The principal stages in the procedure are:

1. To ensure efficient transfer, the gel is first soaked in HCl, leading to random cleavage of the DNA into smaller fragments.
2. The gel is subsequently soaked in alkali to denature the dsDNA to ssDNA, then neutralised: this is necessary to allow hybridisation with probe DNA after blotting.
3. A nitrocellulose or nylon membrane is then placed directly on the gel, followed by several layers of absorbent paper. The DNA is 'blotted' onto the filter as the buffer solution soaks into the paper by capillary action.
4. The filter is baked in a vacuum oven at 80 °C for 3–5 h or exposed to UV light, in order to 'fix' the DNA.
5. Specific DNA fragments are identified by incubation (6–24 h) with complementary labelled probes of ssDNA, which will hybridise with a particular sequence, followed by visualisation, often using an enzyme-based system. If radiolabelled probes are used (see below), the desired fragments are located by autoradiography (p. 271).

Modifications of the method use either a vacuum apparatus, an electric field ('electroblotting') or positive pressure to transfer the DNA fragments from the gel to the membrane. These modifications reduce the time for transfer by around 10-fold.

Northern blotting
This process is virtually identical to Southern blotting, but RNA is the molecule that is separated and probed.

Dot blotting or slot blotting
Here, samples containing denatured DNA or RNA samples are applied directly to the nitrocellulose membrane (*via* small individual round or slot-like templated holes) without prior digestion with restriction enzymes or electrophoretic separation. The 'blot' is then probed in a similar manner to that described for Southern blotting. This allows detection of a particular nucleic acid sequence in a sample.

Types of probe
The probes used in blotting and DNA hybridisation can be obtained from a variety of sources including:

- cDNA (complementary or copy DNA) which is produced from isolated mRNA using reverse transcriptase. This retroviral enzyme catalyses RNA-directed DNA synthesis (rather than the normal transcription of DNA to RNA). After the mRNA has been reverse transcribed, it is degraded by the addition of alkali or ribonuclease, leaving the ssDNA copy. This is then used as a template for a DNA polymerase, which directs the synthesis of the second complementary DNA strand to form dsDNA. This is denatured to ssDNA before use.
- Oligonucleotide probes (15–30 nucleotides) can be produced if the amino acid sequence of the gene product is known. Since the genetic code is degenerate, i.e. some amino acids are coded for by more than one codon (see Table 60.1) it may be necessary to synthesise a mixture of oligonucleotides to detect a particular DNA sequence: this mixture of oligonucleotides is termed a 'degenerate probe'.

- Specific genomic DNA sequences, where the gene has been characterised.
- PCR-generated fragments (Chapter 61).
- Heterologous probes, i.e. sequences for the same gene, or its equivalent, in another organism.

Labelling of probes

Detection of very low concentrations of target DNA sequences requires probes that can be detected with high sensitivity. This is achieved by radiolabelling, e.g. using ^{33}P, or by using enzyme-linked methods, which are often available in kit form:

- Radiolabelled probes can be made by several methods, including the nick-translation technique. This uses DNA polymerase I from *E. coli*, which has (i) an exonuclease activity that 'nicks' dsDNA and removes a nucleotide, and (ii) a polymerase activity that can replace this with a labelled deoxyribonucleotide. After several cycles of nick-translation, the labelled DNA is denatured to ssDNA for use as probes. Newer approaches include random priming of single stranded template DNA, followed by the synthesis of radiolabelled DNA fragments complementary to the template using a DNA polymerase and a radiolabelled deoxynucleoside triphosphate (dNTP). Probe hybridisation to a target sequence is detected by autoradiography. Oligonucleotide probes cannot be labelled using the above methods and require 'end labelling', where a kinase is used to replace the 5′ terminal phosphate group with a radiolabelled group.
- Enzyme-linked methods involve incorporating a modified nucleotide precursor, such as biotinylated dTTP, into the DNA by nick-translation or random priming. When the probe hybridises with the target sequence, it can be detected by addition of an enzyme (e.g. horse radish peroxidase) coupled to streptavidin. The streptavidin binds specifically to the biotin attached to the probe and the addition of a suitable fluorogenic or chromogenic substrate for the enzyme allows the probe to be located.

Using commercial kits – do not blindly follow the protocol given by the manufacturer without making sure you understand the principles of the method and the reasons for the procedure. This will help you to recognise when things go wrong and what you might be able to do about it.

Using labelled probes – an alternative approach is the chemiluminescent system based on dioxygenin-labelled nucleotides.

Hybridisation of probes

The stability of the duplex formed between the probe and its target is directly proportional to the number and type of complementary base pairs that can be formed between them: stability increases with the amount of G + C, since these bases form three hydrogen bonds per base pair, rather than two (p. 369). Duplex stability is also influenced by temperature, ionic strength and pH of the hybridisation buffer, and these can be varied to suit the stringency of hybridisation required:

- In 'low stringency' hybridisation, duplex formation with less than perfect complementarity is promoted, either by lowering the temperature, or increasing ionic strength.
- 'Stringent' hybridisation conditions usually involve high temperatures, or decreased ionic strength, and will sustain only perfectly matched duplexes.

Example Low stringency hybridisation can be useful when using a heterologous probe, i.e. from another species.

Example High stringency conditions can be used for detecting a single base change in a mutant gene using a dot-blot hybridisation procedure with a specific oligonucleotide probe.

DNA sequencing

By fragmenting target DNA with several restriction enzymes and then sequencing the overlapping fragments, it is possible to determine the nucleotide sequences of very large stretches of DNA, including entire

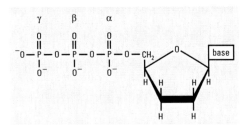

Fig. 60.5 A 2′,3′-dideoxynucleoside triphosphate.

genomes. Sequencing methods rely on polyacrylamide gel electrophoresis (p. 323). The Sanger, or chain termination, method is the most widely used in DNA sequencing. This makes use of dideoxynucleotides (Fig. 60.5), which have no –OH group at either the C-2 or C-3 of ribose. A dideoxynucleoside triphosphate (ddNTP) can be added to a growing DNA chain, but since it lacks an –OH group at the C-3 position it cannot form a phosphodiester bond with the incoming dNTP of the growing chain. Therefore a dideoxynucleotide acts as a terminator at the site it occupies. Details of theSanger sequencing method are given in Box 60.1 and Figure 60.6.

Box 60.1 DNA sequencing using the chain termination (Sanger) method

The DNA to be sequenced must first be obtained as single-stranded fragments, typically around 200 bp, e.g. using a denatured plasmid. Sequencing is often performed using a commercial kit and the principal stages are as follows:

1. **Set up the strand synthesis reaction:** four separate tubes are required, each containing a small amount of one of the dideoxynucleoside triphosphates. Each tube contains all of the other components required for DNA synthesis, i.e. (i) the DNA selected for analysis (template DNA), e.g. denatured plasmid, (ii) a DNA polymerase (e.g. thermostable *Taq* polymerase), (iii) a [^{35}S] oligonucleotide primer of known sequence, to allow synthesis of the complementary DNA strand, (iv) dGTP, dATP, dCTP, dTTP (in excess) together with (v) one type of ddNTP (i.e. ddGTP, ddATP, ddCTP or ddTTP) in limited concentration. New DNA strands are synthesised by addition of dNTPs to the primer, guided by the template DNA until a ddNTP is added. As an example of the principle of the method, consider the ddTTP tube. When T is required to pair with A on the template DNA strand, the dTTP will be competing with the ddTTP, but because the dTTP is in excess this will normally be added to the chain at the appropriate position. However on occasions a ddTTP will be inserted at a given site and this will terminate DNA synthesis on the template strand. Thus, synthesis will be halted at all possible sites where ddT has substituted for dT and several strands of different length will be formed in the reaction mixture, each ending with ddT.

2. **Terminate the strand synthesis reaction:** after 5–10 min at 37 °C add a 'stop' solution containing formamide to disrupt hydrogen bonding between complementary bases and incubate at 80 °C for 15 min, to produce ssDNA.

3. **Separate the fragments using polyacrylamide gel electrophoresis:** 4–6% (w/v) denaturing acrylamide gels are used, containing urea (46% w/v) as the denaturing agent. Thin gels (0.35 mm thickness) are used, typically 20 cm wide by 50 cm long. The products in each of the four tubes are placed in four lanes side by side on the same gel and are separated by electrophoresis at 35–40 W (up to 32 mA, 1.5 kV) for 2.5 h. The high voltage raises the gel temperature to ≈50 °C, helping denaturation. A single gel can separate DNA fragments that differ in length by a single nucleotide.

4. **Locate the positions of individual bands by autoradiography:** the gel is fixed for 15 min with 10% (v/v) acetic acid, covered with Whatman 3MM paper and Saran wrap, then vacuum-dried, to avoid quenching the radioisotope signal (p. 270) and to prevent it from sticking to the X-ray film. The gel is unwrapped and placed next to an X-ray film for 24 h.

5. **Read the gel:** the nucleotide sequence can be read directly from the band positions that represent the newly made DNA segments of varying lengths (Fig. 60.6). The smallest segment is represented by the band at the bottom, since it travels furthest in the gel. The nucleotide sequence of the template strand can be deduced directly from that of the new strand, since the base on the template strand will be represented by an incomplete chain that terminates with the complementary dd nucleotide, i.e. template A with ddT, G with ddC, T with ddA, and C with ddG (Fig. 60.6).

Troubleshooting and other points to note: streaking can be due to damage to wells, air bubbles in wells/gel, or contamination by dirt/dust; faint and fuzzy gels may be due to insufficient template DNA, primer and/or dNTPs, due to poor annealing of primer and template, or to errors in preparing or running the gel; bands in more than one lane can indicate contamination of the template DNA, more than one primer site on the template, or secondary structure of DNA, giving 'ghost' banding.

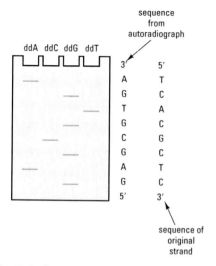

Fig. 60.6 Sanger sequencing gel, showing how the banding pattern is converted into a sequence of nucleotide bases.

Using PCR (p. 439) for sequencing – with the four ddNTPs tagged using different fluorophores, the products can be run in the same lane and detected by laser excitation at the end of the gel, with the sequence of fluorophores giving the base sequence.

DNA microarrays and bioinformatics – interpretation of DNA microarray data is a growing area of bioinformatics, requiring dedicated software, e.g. the TM$_4$ microarray analysis freeware programs available from the Institute of Genomic Research, at: http://www.tm4.org/

The alternative approach to sequencing is based on chemical degradation (the Maxam and Gilbert method), using different reagents to break the target DNA sequence into fragments which are then separated by polyacrylamide gel electrophoresis. The Maxam and Gilbert method is more involved than the Sanger technique, and is reserved for specialised applications, e.g. for studying the interaction between DNA and proteins.

Automated DNA sequencing

The development of automated DNA sequencing machines (so-called 'DNA sequenators') has enabled sequencing to be performed several orders of magnitude faster than with manual methods. Base-specificity is achieved by using primers labelled with fluorophores with different fluorescence characteristics in each of the four reaction tubes (Box 60.1). After the reactions are separately completed, the four sets of products can be pooled and fractionated by electrophoresis in a single lane. The fluorophore-labelled fragments are detected as they pass a scanning laser detector and the DNA sequence is determined by using both the specific wavelength emitted by the fragment (indicating the base) and by the migration time (indicating the fragment size, which corresponds to the base location in the DNA sequence). Nowadays, automated DNA sequencing tends to be performed by capillary electrophoresis (p. 331) rather than polyacrylamide gel electrophoresis.

DNA microarrays

This area has seen rapid, recent development. DNA microarrays ('gene chips') support the quantitative analysis of many different genes/sequences at the same time. This involves synthesising and fixing single-stranded probes for a large number (>1000) of different nucleotide sequences onto specific sites on a glass, quartz or silicon wafer. When single-stranded labelled nucleic acid is added, it will hybridise only with probes having a complementary sequence and these can be determined by the presence of an attached label, e.g. a fluorescent dye, detected using a scanning laser. One application involves the detection of labelled cDNA produced from the mRNA in an organism by using an 'expression' microchip, which contains probes for all of the expressed genes in an organism, enabling the relative expression of each gene to be determined from the fluorescence signal at each location on the microarray. Other applications include the use of single nucleotide polymorphism (SNP) microarrays to identify mutations and genetic variation, e.g. in screening for genetic diseases and in forensic DNA analysis. A commercial example is the Affymetrix GeneChip system.

Sources for further study

Anon. *NCBI Microarray Factsheet*. Available: http://www.ncbi.nlm.nih.gov/About/primer/microarrays.html Last accessed: 01/04/07.
[A basic guide to the principles of DNA microarrays]

Brown, T.A. (2002) *Essential Molecular Biology: A Practical Approach*, Vols 1 and 2, 2nd edn. Oxford University Press, Oxford.

Clark, M. (2003) *In Situ Hybridization: Laboratory Companion*. Wiley-VCH, Mannheim.

Darby, I.A. and Hewitson, T.D. (2003) *In Situ Hybridization Protocols*. Humana Press, New Jersey.

Hames, B.D. and Higgins, S.J. (1995) *Gene Probes: A Practical Approach*. Oxford University Press, Oxford.

Monaco, A.P. (1995) *Pulsed Field Gel Electrophoresis: A Practical Approach*. IRL Press, Oxford.

Nuber, U. (2005) *DNA Microarrays*. Bios (Advanced Methods Series). Taylor & Francis, Abingdon.

Sambrook, J. and Russell, D.W. (2001) *Molecular Cloning: A Laboratory Manual*, 3rd edn. Cold Spring Harbor Lab, Cold Spring Harbor.

Study exercises

60.1 Practise the calculations involved in working with nucleic acids. Imagine that you have been supplied with a freeze-dried sample of a single stranded DNA probe 24 bases in length, which you reconstitute in 1 ml of buffer (reconstituted probe). You take 5 μl, add it to 495 μl of water and determine the A_{260} as 0.14.

(a) What is the mass concentration of the probe in this solution, in μg ml^{-1}?

(b) Assuming the average M_r of a nucleotide is 325, what is the concentration of the probe in the solution, expressed in micromolar terms (i.e. as μmol l^{-1})?

(c) The concentration of probe required in a blotting procedure is 10 pmol ml^{-1} and a total of 10 ml of this solution is required. You have been asked to prepare 5 ml of a stock solution of the DNA probe at 100× the concentration required in the blotting procedure, so that 0.100 ml can be diluted to 10 ml for the final blotting procedure, while the rest will be stored at −20 °C, for subsequent experiments. How would you prepare the 100× stock solution?

Express all answers to three significant figures.

60.2 Select an electrophoretic technique for separating DNA fragments of different sizes. Which technique would you use to separate DNA fragments in the following ranges: (a) 1–70 kbp; (b) 75–500 bp; (c) 100–6 Mbp.

60.3 Test your understanding of the term 'stringency' in relation to nucleic acid probes. What does the term 'stringency' mean in the context of molecular biology? What factors cause increased stability between a nucleic acid probe and its target sequence? Give examples of when you might use 'low stringency' and 'high stringency' hybridisation.

60.4 Interpret the results of a Sanger sequencing gel. The figure represents a typical gel from a dideoxynucleotide sequencing procedure. Determine the sequence of the original DNA strand.

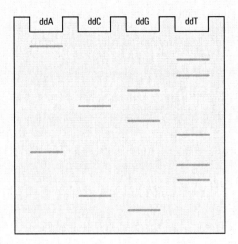

61 Molecular genetics II – PCR and related applications

Applications of PCR – these include:
- diagnosis and screening of genetic diseases and cancer;
- rapid detection of slowly growing microorganisms (e.g. mycobacteria) and viruses (e.g. HIV);
- HLA typing in transplantation;
- analysis of DNA in archival material;
- DNA fingerprinting in forensic science;
- preparation of nucleic acid probes;
- clone screening, mapping and sub-cloning (p. 446).

Understanding the terminology of PCR – cloned (amplified) DNA sequences produced using PCR techniques are often termed amplicons.

The polymerase chain reaction (PCR) is a rapid, inexpensive and simple means of producing µg amounts of DNA from minute quantities of template.

 KEY POINT *PCR offers an alternative approach to gene cloning (Chapter 62) for the production of many copies of an identical sequence of DNA. The starting material may be genomic DNA (e.g. from a single cell), RNA, DNA from archival specimens, cloned DNA or forensic samples.*

The technique uses *in vitro* enzyme-catalysed DNA synthesis to create millions of identical copies of DNA. If the base sequence of the adjacent regions of the DNA to be amplified is known, this enables synthetic oligonucleotide primers to be constructed that are complementary to these so-called 'flanking regions'. Initiation of the PCR occurs when these primers are allowed to hybridise (anneal) to the component single strands of the target DNA, followed by enzymatic extension of the primers (from their 3′ ends) using a thermostable DNA polymerase. A single PCR cycle consists of three distinct steps, carried out at different temperatures (Fig. 61.1), as follows:

1. Denaturation of dsDNA by heating to 94–98 °C separating the individual strands of the target DNA.
2. Annealing of the primers, which occurs when the temperature is reduced to 37–65 °C.
3. Extension of the primers by a thermostable DNA polymerase (e.g. *Taq* polymerase, isolated from *Thermus aquaticus*) at 72 °C; this step should last long enough to generate the PCR product: approximately 1 min of reaction time is required per kbp of sequence.

In the first cycle, the product from one primer is extended beyond the region of complementarity of the other primer, so each newly synthesised strand can be used as a template for the primers in the second cycle (Fig. 61.1). Successive cycles will thus generate an exponentially increasing number of DNA fragments, the termini of which are bounded by the 5′ ends of the primers (length of each fragment = length of primers + length of target sequence). Since the amount of DNA produced doubles in each cycle, the amount of DNA produced $= 2^n$, where n is the number of cycles. Up to 1 µg of amplified target DNA can be produced in 25–30 cycles from a single-copy sequence within 50 ng of genomic DNA, assuming close to 100% efficiency during the cycling process. After electrophoresis, the PCR product is normally present in sufficient quantity to be visualised directly, for example with ethidium bromide or SYBR Safe.

The temperature changes in PCR are normally achieved using a thermal cycler, which is simply a purpose-built incubator block that can be programmed to vary temperatures, incubation times and cycle numbers.

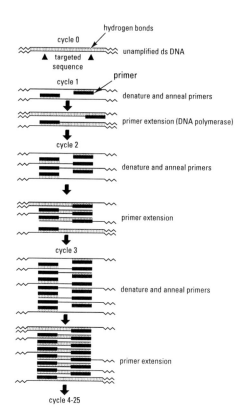

Fig. 61.1 The polymerase chain reaction (PCR).

 KEY POINT *PCR is so sensitive that one of the main problems associated with the technique is contamination with 'foreign' DNA. Great care is required to avoid sample contamination during in vitro amplification (Box 61.1).*

PCR components and conditions

These are readily available in kit form from commercial suppliers; they include a thermostable DNA polymerase, such as *Taq* polymerase, dNTPs, buffer, a detergent (e.g. Triton), KCl, $MgCl_2$ or $MgSO_4$, and primers. Protocols for PCR vary considerably for particular applications – a typical procedure is given in Box 61.1. However, if you are trying to develop your own PCR procedure, the following information may be useful:

- Primers need to be at least 18 nucleotides long (primers longer than 18 nucleotides should be unique, even in a large eukaryotic genome); both primers should have similar annealing temperatures, T_m, p. 372), with a minimal degree of self-complementarity (to avoid formation of secondary structures), and no complementarity to each other (so that primer dimers are not formed).
- For most applications, the final concentration of each primer should be $0.1–0.5\,\mu mol\,l^{-1}$, which gives an excess of primers of about 10^7 with respect to the template, e.g. target genomic DNA at a concentration of 50 ng per 10 μl reaction mixture.
- Annealing temperatures are based around T_m, the temperature at which 50% of the primers are annealed to their target sequence (Fig. 53.7). For primers of <20 bases, T_m can be roughly calculated in °C from the equation:

$$T_m = 4(G + C) + 2(A + T) \qquad [61.1]$$

where G, C, A and T are the number of bases in the primer. Using this as a starting point, the optimum annealing temperature can be determined by trial and error. The T_m values of the two primers should be within 5 °C of each other (and, therefore, be ideally identical in G + C content). The annealing temperature is then set 5 °C below the lowest T_m of the primer pair.
- The last two bases at the 3' end (where elongation is initiated) should be either G or C (3 hydrogen bonds, giving strong annealing), rather than either A or T (only 2 hydrogen bonds).
- dNTPs should be used at equal concentrations of $200\,\mu mol\,l^{-1}$, which should provide the initial excess required for incorporation into DNA.
- One of the key variables in PCR is the Mg^{2+} concentration; Mg^{2+} is required as a cofactor for the thermostable DNA polymerase (Anon., 2002). Excess Mg^{2+} stabilises dsDNA and may prevent complete denaturation of product at each cycle; it also promotes spurious annealing of primers, leading to the formation of undesired products. However, very low Mg^{2+} concentration impairs polymerisation. The purpose of gelatin and Triton X-100 is to stabilise the DNA polymerase during thermal cycling.
- The most frequently used thermostable DNA polymerase is *Taq* polymerase, which extends primers at a rate of 2–4 kbp per min at 72 °C. It should be used at a concentration of $\geqslant 1\,nmol\,l^{-1}$ ($\geqslant 0.1$ U per 5 μl reaction mixture). A disadvantage of *Taq* polymerase is that it has a relatively high rate of misincorporation of bases (one aberrant nucleotide per 100 000 nucleotides per cycle). Other polymerases are available (e.g. *Pfu* polymerase from *Pyrococcus furiosus*, KOD polymerase from *Thermococcus kodakarensis* or genetically modified forms of *Taq* polymerase), which have lower misincorporation rates.

Storing primers for PCR – primers are best stored in ammonia solution, which remains liquid at −20°C, avoiding the need for repeated freezing and thawing when dispensed. Before use, aliquots of stock solution should be heated in a fume cupboard, to drive off the ammonia.

Using dNTP solutions – make up stock solutions, pool in small volumes (50–100 μl of each dNTP) and store separately at −20°C.

Establishing the optimum concentration of Mg^{2+} – this can be determined by trial and error, and can be a useful starting point for any new PCR; try a range of different Mg^{2+} concentrations and select the one giving the strongest intensity of the target band following gel electrophoresis.

Understanding codes for primers and probes – oligomeric nucleotides are often referred to by the number of bases, e.g. a 20-mer primer will contain 20 nucleotide bases.

Box 61.1 How to carry out the polymerase chain reaction (PCR)

The protocol given below is typical for a standard PCR. Note that temperatures, incubation times and the number of cycles will vary with the particular application, as discussed in the text.

1. **Make sure you have the required apparatus and reagents to hand**, including: (i) a thermal cycler; (ii) template thermostable DNA ($\geqslant 50$ ng/μl); (iii) stock solution of all dNTPs (5 mmol l^{-1} for each dNTP); (iv) a DNA polymerase (at 5 U/μl); (v) primers at e.g. 30 μmol l^{-1}; (vi) stock buffer solution, e.g. containing 100 mmol l^{-1} TRIS (pH 8.4), 500 mmol l^{-1} KCl, 15 mmol l^{-1} MgCl$_2$, 1% (w/v) gelatin, 1% (v/v) Triton X-100 (this stock is often termed '10x PCR buffer stock').

2. **Prepare a reaction mixture:** for example, a mixture containing: 1.0 μl target DNA; 2.5 μl stock buffer solution; 1.0 μl primer 1; 1.0 μl primer 2; 1.0 μl of each of the stock solutions of dNTPs, 0.1 μl *Taq* polymerase; 15.4 μl distilled deionised water, to give a total volume of 25.0 μl.

3. **Use appropriate positive and negative controls:** a positive control is a PCR template that is known to work under the conditions used in the laboratory, e.g. a plasmid, with appropriate primers, known to amplify at the annealing temperature to be used. A commonly used negative control is the PCR mixture minus the template DNA, though negative controls can be set up lacking any one of the reaction components.

4. **Cycle in the thermal cycler:** for example, an initial period of 5 min at 94 °C, followed by 30 cycles of 94 °C for 1 min (denaturation), then 50 °C for 1 min (primer annealing; temperature depends upon G + C and A + T content, Eqn [61.1] p. 440), then 72 °C for 1 min (chain extension).

5. **Assess the effectiveness of the PCR:** for example, by gel electrophoresis and ethidium bromide staining.

Troubleshooting

- If no PCR product is detected, repeat the procedure, checking carefully that all components are added to the reaction mixture. If there is still no product, check that the annealing temperature is not too high, or the denaturing temperature is not too low.

- If too many bands are present, this may indicate that (i) the primers may not be specific, (ii) the annealing temperature is too low, or (iii) there is an excess of Mg^{2+}, dNTPs, primers or enzyme.

- Bands corresponding to primer-dimers indicate that (i) the 3' ends of the primers show partial complementarity, (ii) the annealing temperature is not high enough, or (iii) the concentration of primers is too high.

Avoiding contamination in diagnostic PCR

The sensitivity of PCR is also the major drawback, since the technique is susceptible to contamination, particularly from DNA from the skin and hair of the operator, from previous PCR products, from airborne microbes and from positive control plasmids. A number of routine precautions can be taken to avoid such contamination:

- use a laminar flow cabinet (p. 205) dedicated to PCR use and located in a separate lab from that used to store PCR products or prepare clones;

- keep separate supplies of pipettors, tips, microfuge tubes and reagents – these should be exclusive to the PCR, with separate sets for sample preparation, reagents and product analysis;

- autoclave all buffers, distilled deionised water, pipette tips and tubes;

- wear disposable gloves at all times and change them frequently: protective coverings for the face and hair are also advisable;

- avoid contamination due to carry-over by including dUTP in the PCR mixture instead of dTTP. Thus copies will contain U rather than T. Before the template denaturing step, treat the mixture with uracil-N-glycosylase (UNG, available commercially as AmpEase): this will destroy any strands containing U, i.e. any strands carried over from a previous reaction, or any contaminating material from another PCR. The target DNA will contain T, rather than U and will not be degraded by UNG. At the first heating step, the UNG will be denatured, so any newly synthesised U-containing copies will remain intact.

- for very sensitive work, use a strong UVC light (p. 276) inside the PCR workstation for 20–30 minutes before starting your work, to degrade any contaminating DNA.

 SAFETY NOTE Note that you must not expose your skin or eyes to such a UVC source.

PCR variations

Nested PCR

This can be used when the target sequence is known, but the number of DNA copies is very small (e.g. a single DNA molecule from a microbial genome), or if the sample is degraded (e.g. a forensic sample). The process involves two consecutive 'rounds' of PCR. The first PCR uses so-called 'external' primers, and the second PCR uses two 'internal' (or 'nested') primers that anneal to sequences within the product of the first PCR. This increases the likelihood of amplification of the target sequences by selecting for it using different primers during each round. Thus, nested PCR also increases the specificity of the reaction, since a single set of primers used in isolation may give a reasonable yield but several bands, while the use of a second set of primers ensures that a unique sequence is amplified, e.g. in microbial diagnostics.

Inverse PCR

This is a useful technique for amplifying a DNA sequence flanking a region of known base sequence (Fig. 61.2), e.g. to provide material for characterising an unknown region of DNA. The DNA is cut with a restriction enzyme so that both the region of known sequence and the flanking regions are included. This restriction fragment is then circularised and cut with a second restriction enzyme with specificity for a region in the known sequence. The now linear DNA will have part of the known sequence at each terminus, and by using primers that anneal to these parts of the known sequence, the unknown region can be amplified by conventional PCR. The product can then be sequenced and characterised (Chapter 60).

Reverse transcriptase-PCR (RT-PCR)

This technique is useful for detecting cell-specific gene expression (as evident by the presence of specific mRNA) when the amount of biological material is limited. Using either an oligo-dT primer to anneal to the 3′ polyadenyl 'tail' of the mRNA, or random hexamer primers, together with reverse transcriptase, cDNA is produced which is then amplified by PCR. RT-PCR is often a useful method of generating a probe, the identity of which can be confirmed by sequencing (p. 436).

Amplification fragment length polymorphism (AFLP)

This term refers to several closely related techniques in which a single oligonucleotide primer of arbitrary sequence is used in a PCR reaction under conditions of low stringency, so that the primer is able to anneal to a large number of different sites within the target DNA. Some of the multiple amplification products will be polymorphic (e.g. the presence or absence of a particular annealing site will result in presence or absence of a particular band on the gel, after PCR and electrophoresis). Such polymorphisms can be used to detect differences between dissimilar DNA sequences.

PCR and DNA fingerprinting techniques

Genetic mapping and sequencing studies have led to the discovery of highly variable regions in the non-coding regions of DNA between different individuals. These hypervariable regions, often termed 'minisatellite DNA', are found at many sites throughout the genome.

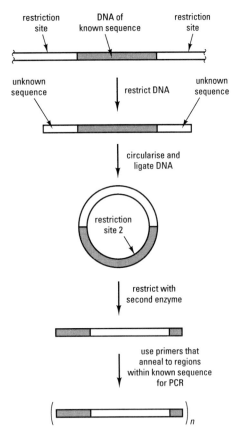

Fig. 61.2 Inverse PCR – basic principles.

Example Random amplification of polymorphic DNA (RAPD-PCR) can be used as a form of 'molecular typing' (p. 219), to identify a particular strain of an organism, e.g. in tracing the route of transmission of a pathogenic microbe.

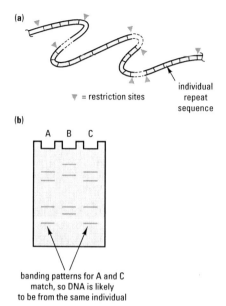

Fig. 61.3 Simplified representation of DNA fingerprinting: (a) tandem repeat sequences within DNA are cut using a restriction enzyme to yield fragments of different size; (b) these fragments are then separated by agarose gel electrophoresis on the basis of their size (M_r), giving a banding pattern that is a characteristic of the DNA used.

Each minisatellite contains a defined sequence of nucleotides which is repeated a number of times in a tandem fashion (Fig. 61.3); the greater the number of repeats, the longer the minisatellite. The number of tandem repeats in any particular minisatellite varies from one person to another (i.e. they have a variable number of tandem repeats or VNTRs). This is exploited in identifying individuals on the basis of their DNA profile, by carrying out the following steps:

1. Extraction of DNA from cells of the individual (e.g. leukocytes, buccal cells, spermatozoa).
2. Digestion of the DNA with a restriction enzyme that cuts at sites other than those within the minisatellite, to produce a series of fragments of different M_r.
3. Electrophoresis and Southern blotting of the restriction fragments using probes that are specific to the particular minisatellite.

The size of the fragments identified will depend on the number of minisatellites that each fragment contains, and the pattern obtained in the Southern blot is characteristic of the individual being profiled (Fig. 61.3).

PCR is widely used in DNA profiling in circumstances where there is a limited amount of starting material. By selecting suitable primers, highly variable regions can be amplified from very small amounts of DNA, and the information from several such regions is used to decide, with a very low chance of error, whether any two samples of DNA are from the same individual or not, e.g. in forensic science (see Kirby, 2002).

Real-time PCR

Conventional PCR techniques rely on end-point detection of amplified product, e.g. by electrophoretic separation and staining (p. 431). However, such methods are time-consuming and are only semi-quantitative, since they are based predominantly on the detection of an amplified fragment (band) in a sample, rather than being designed to give exact information on its abundance (copy number). Quantitative analysis is only feasible during the early stages of PCR, where reagents are in excess and where the amount of amplified product is small, thereby avoiding the problems of product hybridisation, which would compete with primer binding. Real-time PCR enables the simultaneous amplification and quantification of template DNA in a sample by establishing the number of copies present by working in the exponential (early) phase of amplification. Currently, while several different formats are available for real-time PCR, all rely on the generation of a fluorescent signal.

Dye fluorescence

This is the simplest and least expensive approach. A fluorescent dye such as SYBR Green is included in the PCR mixture and the level of fluorescence is monitored as the reaction proceeds; since SYBR Green binds strongly to dsDNA, showing an enhancement in fluorescence of over 100-fold, any increase in fluorescence is directly proportional to the amount of dsDNA produced. Calibration is achieved by running a series of standards containing known amounts of dsDNA. This approach works well for optimised single PCR product reactions where non-specific reactions are minimised.

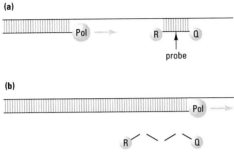

Fig. 61.4 Real-time PCR using a fluorescent reporter probe. In (a) the probe is bound to the target sequence while in (b) the 5' nuclease activity of the DNA polymerase (Pol) has cleaved the reporter dye (R) and quencher (Q), resulting in fluorescence.

Fluorescent reporter probes

Here a fluorescent reporter dye is covalently bound to the 5' end of an oligonucleotide probe while a 'quencher' group is attached to the 3' end (Fig. 61.4). The probe is designed to hybridise to an internal region of the target sequence. During PCR, when the DNA polymerase molecule reaches the hybridised probe, the 5' nuclease activity of the polymerase will cleave the reporter dye from the rest of the probe, causing an increase in fluorescence with each cycle that is in direct proportion to the amount of PCR product being formed, which is itself directly related to the original number of copies of the target DNA sequence. A commercial example is the TaqMan series of probes – while this approach is more accurate and reliable than dye fluorescence, it is also far more expensive, since a specific reporter probe must be synthesised for each target sequence. Other variants rely on changes in 3D conformation of the probe when it binds to the target sequence, causing an increase in fluorescence, e.g. Molecular Beacons and Scorpion probes.

 KEY POINT You should note that the PCR technique is continually being updated, with subtle new variations being produced on a regular basis – novel approaches and novel acronyms are very likely to be reported within the lifetime of this book.

Text references

Anon. *PCR Optimization.* Available:
http://www.promega.com/paguide/chap1.htm
Last accessed: 01/04/07.
[Details of how to optimise Mg^{2+} and polymerase concentration, etc.]

Kirby, L.T. (2002) *DNA Fingerprinting: An Introduction. Breakthroughs in Molecular Biology.* Oxford University Press, Oxford.

Sources for further study

Altshuler, M.L. (2006) *PCR Troubleshooting: the Essential Guide.* Caister Academic Press, Wymondham.

Diffenback, C.W. and Dveksler, G.S. (2003) *PCR Primer: a Laboratory Manual,* 2nd edn. Cold Spring Harbor Laboratory, Cold Spring Harbor.

Dorak, M.T. (2006) *Real-Time PCR.* Taylor & Francis, Abingdon.

McPherson, M.J. and Moller, S.G. (2006) *PCR Basics.* Taylor & Francis, Abingdon.

Study exercises

61.1 Investigate the role of magnesium ions in PCR. Find out why optimising the Mg^{2+} concentration is important for successful amplification of target DNA in PCR.

61.2 Check your knowledge of primer design. List the principal factors that should be taken into account when designing primers for PCR.

61.3 Carry out a T_m calculation for a primer. For the following primer, what would be the temperature at which 50% of the molecules are annealed to their target sequence (i.e. T_m)?

5'-TGCATGGCTGGATTAGCG-3'

Study exercises (continued)

61.4 Consider how the risk of contamination can be reduced in PCR. List the most important practical steps that can be taken to reduce the possibility of contamination.

61.5 Select appropriate primers for a specific PCR amplification. The sequence in the box below represents a region of DNA that is to be amplified by PCR.

Which of the following three pairs of primers is most suited to amplify a PCR product from the above sequence?
(a) 5'-TCGCTGAAGGACATGTCG-3' and
    5'-CGATCTAGGCGACATGTCC-3'
(b) 5'-TCGCTGAAGGACATGTCG-3' and
    5'-TCCTTAGGTCTCGACCGTA-3'
(c) 5'-TCGCTGAAGGACATGTCG-3' and
    5'-GTCTCGACCGTAGCTAGCATC-3'

61.6 Find out what types of DNA analysis might be suitable for forensic examination of degraded DNA samples. In 1918 Tsar Nicholas of Russia, his wife Tsarina Alexandra, and their five children were executed and buried in a shallow grave in Yekaterinberg, Russia. Samples from the bones of the family were analysed by the UK Forensic Science Service (FSS). What types of DNA-based tests would you expect to have been carried out to confirm that a family group was present in the grave?

```
5'-TCGCTGAAGGACATGTCGATGCTAGCTACGGTCGAGACGTAAGGACATGTCGGCTAGATCGC-3'
3'-AGCGACTTCCTGTACAGCTACGATCGATGCCAGCTCTGCATTCCTGTACAGCGGATCTAGCG-5'
```

62 Molecular genetics III – genetic engineering techniques

Advances in the procedures used to manipulate nucleic acids *in vitro* have increased our understanding of the structure and function of genes at the molecular level. Additionally, these techniques can be used to alter the genome of an organism (genetic engineering or gene cloning), e.g. to create a bacterium capable of synthesising a foreign protein such as a potentially useful hormone or vaccine component, or a novel protein. The procedures are often termed 'recombinant DNA technology', since they involve the creation of novel combinations of DNA (i.e. recombinant DNA) under controlled laboratory conditions.

In the UK, the Genetically Modified Organisms (Contained Use) Regulations (2000) *provide the regulatory framework for all research procedures involving the genetic modification of organisms.*

 KEY POINT *While genetic manipulation must be carried out under strict containment, in accordance with appropriate legislation, the procedures involved in the isolation, amplification, recombination and cloning of DNA are often used at undergraduate level, to illustrate the general features of the techniques.*

Basic principles

Gene cloning involves several steps:

1. Isolation of the DNA sequence (gene) of interest from the genome of an organism, or from a gene library. This usually involves DNA purification followed by enzymic digestion or mechanical fragmentation, to liberate the target DNA sequence, or PCR amplification (p. 439) of the target DNA sequence.
2. Creation of an artificial recombinant DNA molecule (rDNA), by inserting the target sequence into a DNA molecule capable of replicating in a host cell, i.e. a 'cloning vector'. Suitable cloning vectors for bacterial cells include plasmids (p. 448) bacteriophages (p. 449) and cosmids (p. 449).
3. Introduction of the recombinant DNA molecule into a suitable host, e.g. *E. coli*. The process is termed transformation when a plasmid is used, or transfection for viral nucleic acid.
4. Selection and growth of the transformed (or transfected) cell, using the techniques of cell culture (p. 203). Since a single transformed host cell can be grown to give a clone of genetically identical cells, each carrying the target DNA of interest, the technique is often referred to as 'gene cloning', or molecular cloning.

Definitions

Bacteriophage (phage) – a bacterial virus. Useful as they possess a means of penetrating the bacterial cell.

Cosmids – hybrid plasmid vectors containing the *cos* (cohesive) sites from phage λ, enabling *in vitro* packaging into phage capsids. Useful for cloning large segments of DNA, typically up to 50 kb.

Plasmids – circular molecules of DNA capable of autonomous replication within a bacterial cell. Can be isolated, manipulated and reintroduced into bacterial cells.

Subcloning – procedures that isolate and characterise smaller portions of a particular DNA sequence, e.g. during the search for a specific gene.

Transfection – in bacteria: uptake of viral nucleic acid; in eukaryotes: uptake of any naked, foreign DNA.

Transformation – in bacteria: stable incorporation of external DNA, e.g. a plasmid; in eukaryotes: the conversion of a cell culture of finite life to a continuous (immortal) cell line (also occurs in cancer).

Extraction and purification of plasmid DNA

Specific details of the steps involved in the isolation of DNA vary, depending upon the source material. However, the following protocol shows the principal stages in the purification of plasmid DNA from bacterial cells:

1. Cell wall digestion: incubation of bacteria in a lysozyme solution will hydrolyse the peptidoglycan cell wall. This is often carried out under isotonic conditions, to stop the cells from bursting open and releasing chromosomal DNA. Note that Gram-negative bacteria are relatively insensitive to lysozyme, requiring additional treatment to allow the

Preparing glassware – all glassware for DNA purification must be siliconised, to prevent adsorption of DNA. All glass and plastic items must be sterilised before use.

SAFETY NOTE Working with solvents in molecular biology – note that phenol is toxic and corrosive while chloroform is potentially carcinogenic (p. 134). Take appropriate safety precautions (e.g. wear gloves, extinguish all naked flames, use a fume hood, where available).

Using ultrapure sterile water in molecular biology – to avoid contamination, prepare all solutions using pre-sterilised (autoclaved) deionised purified water (e.g. MilliQ water).

Extracting DNA from organisms other than bacteria – different homogenisation techniques will be required (see Chapter 36). DNA can be extracted from the resulting homogenate by (i) complexation with the detergent cetrylmethylammonium bromide (CTAB), e.g. for plant DNA, or (ii) anion-exchange column chromatography (p. 307). Many laboratories now use small-scale chromatographic columns for routine extraction, e.g. Qiagen miniprep columns.

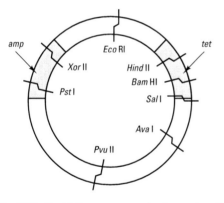

Fig. 62.1 Restriction map of the plasmid pBR322. The position of individual restriction sites is shown together with the genes for ampicillin resistance (*amp*) and tetracycline resistance (*tet*).

Definition

Type II restriction endonucleases – intracellular enzymes, produced by certain strains of bacteria: their function is to restrict the growth of phages within the cell by cutting phage DNA at particular sites (host DNA is protected by methylation at these sites).

enzyme to reach the peptidoglycan layer (p. 243), e.g. osmotic shock, or incubation with a chelating agent, e.g. ethylenediaminetetra-acetate (EDTA). The latter treatment will also inactivate any bacterial deoxyribonucleases (DNases) in the solution, preventing enzymic degradation of plasmid DNA during extraction.

2. Lysis using strong alkali (NaOH) and a detergent, e.g. sodium dodecyl sulphate, to solubilise the cellular membranes and partially denature the proteins. Neutralisation of this solution (e.g. using potassium acetate) causes the chromosomal DNA to aggregate as an insoluble mass, leaving the plasmid DNA in solution.

3. Removal of other macromolecules, particularly RNA and proteins, by, for example, enzymic digestion using ribonuclease and proteinase. Additional steps give further increases in purity: most laboratories now use small-scale chromatographic columns containing an anion exchange resin for routine extraction of nucleic acids. These are available in kit form, e.g. Wizard miniprep columns. This approach gives more reliable purification than traditional methods, such as phenol/chloroform extraction.

4. Precipitation of DNA by adding two volumes of absolute ethanol to one volume of aqueous extract, followed by centrifugation, to recover the DNA as a pellet. Further washing with 70% v/v ethanol will remove any residual salt from the previous stages. After removal of residual ethanol by evaporation, the purified DNA is dissolved in buffer and can be stored for months at $-20\,°C$.

The simplest approach to quantifying the amount of nucleic acid in an aqueous solution is to measure the absorbance of the solution at 260 nm (A_{260}) using a spectrophotometer, as detailed on page 283. Any contaminating protein would also contribute to A_{260}; protein can be detected by measuring the absorbance of the solution at 280 nm. Purified nucleic acids have a value for A_{260}/A_{280} of 1.8–2.0 and contaminating protein will give a lower ratio. If your solution gives a ratio substantially lower than this, you should carry out further purification steps.

Enzymatic manipulation of DNA

Restriction enzymes

Type II restriction endonucleases can be used to produce linear fragments of DNA with either single stranded 'sticky' ends or 'blunt' ends (Fig. 60.1): most commercial preparations of restriction enzymes will completely cleave a sample of DNA, producing a 'restriction digest', within 1 h at $37\,°C$. The position of individual restriction sites can be used to create a diagnostic restriction map for a particular molecule, e.g. a plasmid cloning vector (Fig. 62.1). An important additional feature is that, under appropriate conditions, the sticky ends of any two restriction fragments cut with the same enzyme (Fig. 62.2) can anneal (base pair), due to the formation of hydrogen bonds between individual bases within this region, allowing them to be joined together (ligated), irrespective of the sources of the two restriction fragments.

DNA ligase

To construct a recombinant DNA molecule, a suitable vector must be ligated to the DNA fragment to be cloned: this is performed using another microbial enzyme, DNA ligase (usually obtained from T4-infected

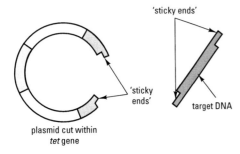

Fig. 62.2 Restriction of plasmid pBR322 and foreign DNA with Hin dIII.

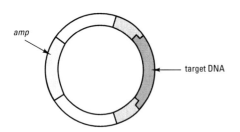

Fig. 62.3 Annealing and ligation of plasmid pBR 322 and target DNA to give a recombinant plasmid which confers resistance to ampicillin only. As target DNA has been inserted within the tet gene (Fig. 62.1), the gene is now discontinuous and inactive.

Understanding the nomenclature for plasmid vectors – pBR322 is named according to standard rules: 'p' indicates the vector is a plasmid, 'BR' identifies the researchers Bolivar and Rodriguez and '322' is the specific code given to this plasmid, to distinguish it from others developed by the same workers.

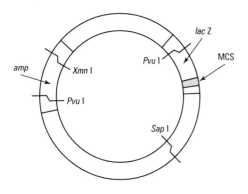

Fig. 62.4 Restriction map of the plasmid pUC19. The position of some individual restriction sites is shown together with the genes for ampicillin resistance (amp) and β-galactosidase (lacZ). MCS = multiple cloning site for 40 restriction enzymes within the lacZ gene.

E. coli). This ATP-dependent enzyme is capable of forming covalent phosphodiester bonds between annealed DNA molecules, thereby creating recombinant DNA (Fig. 62.3). When the two molecules involved are the cloning vector and the target DNA, the size of the recombinant molecule can be predicted (e.g. a plasmid of 4.5 kbp, plus a target DNA fragment of 2.5 kbp will give a recombinant molecule of 7 kbp), allowing separation and recovery by agarose gel electrophoresis (p. 431). Ligation is usually carried out at lower temperatures (to encourage annealing), over an extended time period of several hours (to allow the enzyme to operate), e.g. an overnight incubation at $\simeq 16\,°C$. The volume of the ligation mixture is kept as low as possible (typically, $<10\,\mu l$), with approximately a 2:1 ratio of target DNA to vector DNA, to encourage annealing between the two different types of DNA molecule and to reduce the chance of circularisation of vector or target DNA fragments.

Choosing a suitable vector

The features to be considered when selecting a suitable vector are:

- the ease of purification – e.g. reliable procedures have been developed to allow plasmid DNA to be extracted and purified from bacterial cells;
- the efficiency of insertion of the recombinant vector into a new host cell;
- the presence of single copies of suitable restriction sites;
- the presence of selectable markers (e.g. 'reporter genes', p. 450);
- the size of the DNA insert to be cloned;
- the copy number of the vector in the host cell.

Plasmids

The simplest bacterial cloning vectors are those based on small plasmids – circular DNA molecules, capable of autonomous replication within a bacterial cell. The general characteristics of plasmids are described in Chapter 59. One of the first cloning vectors to be developed for E. coli was pBR322, a genetically engineered plasmid of 4.4 kbp containing an origin of replication, two antibiotic resistance genes and single sites for a range of restriction enzymes (Fig. 62.1). Subsequently, other plasmid vectors have been developed, with additional features: for example, plasmids of the pUC series (Fig. 62.4) are now widely used for transformation of E. coli, having the following advantages over earlier types:

- high copy number: several hundred identical copies of the plasmid may be present in each bacterial cell, giving improved yield of plasmid DNA;
- single-step selection of recombinants using the $lacZ'$ gene (p. 450);
- clustering of restriction sites within a short region of the $lacZ'$ gene – restriction using two enzymes (a 'double digest') cuts a small fragment from this 'multiple cloning site' or 'polylinker', producing a cleaved plasmid with two different 'sticky ends' (e.g. Eco RI at one end and Hin dIII at the other), allowing complementary fragments of target DNA to be ligated in a particular orientation ('directional cloning').

The pET series of plasmid vectors (Novagen; Fig. 62.5) are now widely used, since they offer several additional advantages, including:

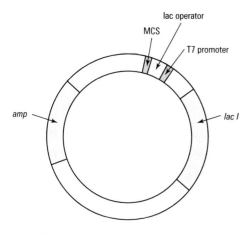

Fig. 62.5 Map of a pET plasmid vector (Novagen) showing the location of the genes for ampicillin resistance (*amp*) and lactose repressor protein (*lacI*) which controls expression of the T7 promoter via the lac operator (binding of the Lac repressor protein is countered by addition of isothiopropylgalactoside, IPTG). Expression of a gene cloned into the multiple cloning site (MCS) can thus be controlled, since DNA transcription only occurs when IPTG is added to the growth medium.

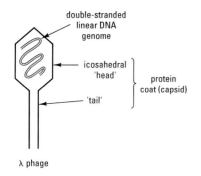

Fig. 62.6 Bacteriophage λ.

Definition

Phagemid – a recombinant entity created by fusion of a plasmid with the origin of replication of a phage: can be propagated and packaged into phage particles with the aid of a 'helper' phage.

- high levels of protein expression, as the cloned DNA is under the control of the T7 promoter and the lac operon system;
- specialised vectors and hosts for the production of soluble proteins, disulphide bond formation, protein export, etc;
- the attachment of a short histidine 'tag' on the end of an expressed protein, to enable the protein to be purified by immobilised metal affinity chromatography (IMAC), as described on page 310.

Phages

Bacteriophage vectors can offer several advantages over plasmid vectors:

- some phages have been engineered to carry larger fragments of foreign DNA – plasmid vectors work best with relatively short inserts of 1–2 kbp, while some phage vectors can accept 10 kbp or more of foreign DNA;
- a recombinant phage has a ready-made mechanism for transferring rDNA to a new host cell: the process is broadly similar to phage transduction, described on page 423.

A number of vectors have been developed from *E. coli* phage λ (genome = 49 kbp of dsDNA in linear form, Fig. 62.6). For example, λZAPII, an insertion vector capable of accepting up to 10 kbp of additional (foreign) DNA and with the *lacZ′* gene and multiple cloning site for selection, or λEMBL4, a replacement vector, where part of the vector DNA is removed and replaced by foreign DNA (up to 23 kbp) during the cloning procedure.

Most vectors based on λ are virulent phages, released by lysis of the infected cell at the end of the replication phase: such vectors are cloned using standard techniques for phages, e.g. as plaques of lysis on a 'lawn' of bacteria growing on an agar-based medium (p. 207). Note that suitable strains of *E. coli* must be used for each of the various cloning vectors, e.g. vectors containing *lacZ′* can only be used with those *E. coli* strains that are deficient in this part of their genome (*lacZ′* codes for the α-component of the enzyme β-galactosidase, present in wild-type *E. coli*). In some instances, a second phage (a 'helper phage') must be used along with the vector, to carry out some of the functions removed from the vector genome, e.g. phage packaging and assembly functions. In some instances, more specialised phage vectors may be required, e.g. M13mp19, a single stranded phage vector, sometimes used for sequencing studies.

Cosmids and other vectors

Cosmid vectors contain the single stranded cohesive terminal sequences from λ phage (so-called 'cos' sites) inserted into a plasmid cloning vector. They can be used to clone up to 40 kbp of foreign DNA, using the 'cos' sites to allow *in vitro* packaging into the λ phage coat: such large fragments are required for the production of gene libraries, as described below. Other hybrid vectors have been produced by the fusion of a phage and a plasmid vector – these phagemids can be manipulated in the laboratory as plasmids for ease of DNA uptake, or converted to phage-like particulate form for ease of storage.

Creating a gene library

For viruses with small genomes, an individual gene may be identified by hybridisation of a single fragment with a suitable probe, following

Screening clones – *a particular gene can be detected using Southern blotting (p. 433), or the expressed gene product (protein) may be detected by a suitable method, e.g. an immunoassay (p. 258). While it is relatively simple to describe the screening process, finding a specific gene (= clone) in a clone library is likely to be a time-consuming process.*

Using cloned cDNA – *information from the sequence of a particular cDNA clone may be useful in designing a suitable oligonucleotide probe to locate a particular gene in a genomic library.*

digestion with a particular restriction enzyme (Chapter 60). However, for most genomes, such a digest would contain a very complex mixture of fragments and a 'clone library' must be created. The isolation and identification of a particular gene are then carried out by screening individual clones from this library. Two types of clone library can be used:

1. A genomic library – prepared from the entire genome of the organism under study. The genome is fragmented to give overlapping fragments, e.g. by partial restriction (incomplete digestion, e.g. at low temperature), or by mechanical shearing. Individual fragments are then incorporated into a suitable vector, e.g. a cosmid, to create a vector library (e.g. a cosmid library). Each recombinant vector is then transferred to a separate host cell, which is cultured, giving a collection of transformants that represents the clone library. This approach is sometimes termed 'shotgun cloning'.
2. A cDNA library (cDNA = complementary, or copy DNA) – prepared by converting mRNA into DNA using the retroviral enzyme reverse transcriptase. Such a library consists only of genes expressed in those cells used to create the library, making individual genes easier to locate and identify, since the number of clones represented is usually smaller than for a genomic library. Another feature of cDNA is that it is complementary to the processed transcript (mRNA), so it contains no information on non-coding regions (introns) or transcriptional sequences (promoter regions, etc.).

Transferring rDNA to a suitable host cell

- Once a recombinant vector has been produced *in vitro*, it must be introduced into a suitable host cell. Packaged phage vectors have a built-in transfer mechanism, while naked phage DNA and plasmids must be introduced by treatments that cause a temporary increase in membrane permeability resulting in transformation. These treatments include:
- physicochemical shock treatment – Box 62.1 gives details of a typical procedure using $CaCl_2$ and heat shock treatment to transform *E. coli* with the plasmid pUC19;
- electroporation – cells or protoplasts are subjected to electric shock treatment (typically, $>10\,kV\,cm^{-1}$) for very short periods ($<10\,ms$);
- a range of techniques can be used for animal and plant cells, e.g. electroporation of protoplasts, or various micro-injection treatments, either using a microsyringe, DNA-coated microprojectiles ('biolistics') or cationic lipids (Fig. 62.7) which form lipid–DNA complexes that fuse with cell membranes and allow DNA entry. *Agrobacterium* can be used for rDNA transfer in certain plants.

Selection and detection of transformants

As the efficiency of transformation is very low, many of the plasmid vectors used in genetic engineering carry genes coding for antibiotic resistance, e.g. pBR322 carries separate genes for ampicillin resistance, *amp*, and tetracycline resistance, *tet* (Fig. 62.1). These genes act as 'markers' for the vector. One gene (e.g. *amp*) can be used to select for transformants, which would form colonies on an agar-based medium

Table 62.1 Examples of genes used to detect transformants

Gene	Product/assay
lacZ / *lacZ'*	β-galactosidase/chromogenic substrate (e.g. X-GAL) or fluorogenic substrate (e.g. MU-GAL)
uidA	β-glucuronidase/chromogenic substrate (e.g. X-GLUC) or fluorogenic substrate (e.g. MU-GLAC, p. 383)
lux	luciferase/bioluminescence in the presence of luciferin (p. 285)
bla / *amp*	β-lactamase/resistance to ampicillin
cat	chloramphenicol acetyltransferase/resistance to chloramphenicol
gfp	green fluorescent protein from *Aequorea victoria*, visualised under UV light

Recognising transformants – *after plating bacteria onto medium containing ampicillin, you may notice a few small 'satellite' colonies surrounding a single larger (transformant) colony. These satellite colonies are derived from non-transformed cells which survive due to the breakdown of antibiotic in the medium around the transformant colony, and should not be selected for subculture.*

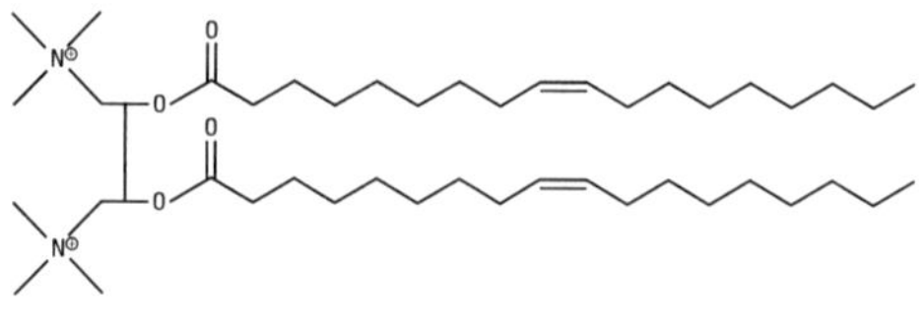

Fig. 62.7 General structure of a synthetic cationic lipid, e.g. Tfx.

containing the antibiotic, while non-transformed (ampicillin-sensitive) cells would be killed. The other gene (e.g. *tet*) can be used as a marker for the recombinant plasmid vector, since ligation of the target sequence into this gene causes insertional inactivation (Fig. 62.3). Thus, cells transformed with the recombinant plasmid will be resistant to ampicillin only, while cells transformed with the recircularised ('native') plasmid will be resistant to both antibiotics. These two types of transformant can be distinguished using replica plating (see p. 422). For those genes where the product is an enzyme, the presence or absence of the functional gene can be assessed using a suitable substrate. For example, the insertional inactivation of *lacZ'* can be detected by including a suitable inducer of β-galactosidase (e.g. isopropylthiogalactoside, IPTG) and the chromogenic substrate 5-bromo-4-chloro-3-indolyl-β-D-galactoside (X-GAL) within the agar medium: a transformant colony derived from the native plasmid will be blue-green while a transformant containing the recombinant molecule (inactive *lacZ'*) will grow to produce a white colony. A number of other relevant examples are given in Table 62.1.

Box 62.1 Transformation of *E. coli* and selection of transformants

The following procedure illustrates the principal stages of the process: the efficiency of transformation will depend on the choice of *E. coli* strain and plasmid, and on the handling procedures prior to and during the process – for example, some strains do not require all of the stages listed below (for further details, see Hanahan *et al.*, 1995). Sterile equipment and appropriate technique (Chapter 32) are required at all times:

1. **Grow the cells under appropriate conditions:** the best results are often obtained using actively growing mid-log phase cells, rather than cells that have been grown to stationary phase (p. 233). Actively growing cells can be stored under appropriate conditions, e.g. as a 'frozen stock' – in suspension in concentrated glycerol at $-85\,°C$.

2. **Induce cell competence by transfer to a suitable sterile transformation buffer:** this would normally contain divalent and monovalent salts, e.g. a salt solution containing $CaCl_2$ at $50–100\,mmol\,l^{-1}$, plus KCl and $MnCl_2$ at $10–20\,mmol\,l^{-1}$. Cells are usually kept on ice in this solution for 10–15 min, to encourage the binding of plasmid DNA, added at a later stage.

3. **Add reagents to increase the permeability of cellular membranes:** typically dimethyl sulphoxide (DMSO, at up to 7% v/v) and dithiothreitol (DTT at up to $0.2\,mmol\,l^{-1}$), incubated on ice for a few minutes. DMSO is readily oxidised, reducing the effectiveness of the transformation procedure, and should be stored at $-80\,°C$ when not in use, to minimise oxidation.

4. **Add plasmid DNA:** typically at 10–1000 ng per transformation. Maintain on ice for at least 10 min in a minimal volume of solution, to allow the plasmid to become associated with the cell surface.

5. **Heat shock the cells:** briefly raise the temperature to 42–45 °C for 60–120 s, then return to ice for a few minutes. It is important that this treatment gives a *rapid* change in temperature – use a small, thin-walled (disposable) sterile plastic tube containing the minimum volume of solution to maximise the rate of temperature change.

6. **Allow the cells to recover from heat shock and to express any new genes** (e.g. to synthesise enzymes conferring antibiotic resistance in transformed cells): add sterile nutrient broth (e.g. 1 ml) and then incubate at 37 °C for up to 60 min.

7. **Plate onto an appropriate medium, to allow selection and detection of transformants:** for example, using pUC19, surface spread (p. 206) the suspension onto a medium containing ampicillin (at $25\,mg\,l^{-1}$), plus IPTG (at $15\,mg\,l^{-1}$) and X-GAL (at $25\,mg\,l^{-1}$), to detect the *lacZ'* gene product, as described in the text.

Text reference

Hanahan, D., Jessee, J. and Bloom, F.R. (1995) 'Techniques for transformation of *E. coli*' in Glover, D.M. and Hames, B.D. *DNA Cloning I*, Practical Approach Series. IRL Press, Oxford.

Sources for further study

Brown, T.A. (2002). *Essential Molecular Biology: A Practical Approach*, Vols 1 and 2, 2nd edn. Oxford University Press, Oxford.
[A practical manual on the basic principles involved in gene cloning and recombinant DNA technology]

Brown, T.A. (2006) *Gene Cloning and DNA Analysis*, 5th edn. Blackwell, Oxford.

Dale, J.W. and Schantz, M.V. (2002) *From Genes to Genomes: Concepts and Applications of DNA Technology*. Wiley, Chichester.

Hardin, C., Edwards, J., Riell, A., Presutti, D., Miller, W. and Robertson, D. (2001) *Cloning, Gene Expression, and Protein Purification: Experimental Procedures and Process Rationale*. Oxford University Press, Oxford.

[A practical handbook and laboratory manual, with basic experimental protocols together with underlying theoretical principles]

Jones, P. (1997) *Vectors: Expression Systems: Essential Techniques*. Wiley, New York.

Lodge, J., Lund, P. and Michin, S. (2006) *Gene Cloning: Principles and Applications*. Taylor & Francis, Abingdon.

Primrose, S.B. and Twyman, R.M. (2004) *Principles of Gene Manipulation and Genomics*, 7th edn. Blackwell, Oxford.

Russell, P. (2005) *iGenetics: a Molecular Approach*. Addison-Wesley, Harlow.

Study exercises

62.1 Explain the terminology of restriction enzymes. Why were *Hin* dIII and *Eco* RI given these names?

62.2 Explain the purpose of the various stages involved in the extraction and purification of DNA. Having read through this chapter, briefly describe the main purpose of each of the following reagents within a DNA extraction protocol, and rearrange them into their order of use within the procedure:

(a) anion exchange resin column chromatography;
(b) incubation in 70% v/v ethanol : water;
(c) ribonuclease and proteinase digestion;
(d) incubation with lysozyme/EDTA;
(e) sodium dodecyl sulphate incubation;
(f) NaOH treatment, followed by neutralisation with potassium acetate.

62.3 Construct a restriction map. The figure alongside represents an electrophoretic separation of plasmid pRD, digested with different restriction enzymes, both individually and in combination. All of the enzymes cut the plasmid only once. Use this information to construct a simple restriction map of the plasmid. Key to lanes: 1 = DNA standards (0.5–8 kbp); 2 = *Eco* RI; 3 = *Hin* dIII; 4 = *Pvu* II; 5 = *Eco* RI and *Hin* dIII; 6 = *Eco* RI and *Pvu* II; 7 = *Hin* dIII and *Pvu* II; 8 = *Eco* RI, *Hin* dIII and *Pvu* II; 9 = DNA standards (0.5–8 kbp).

62.4 Calculate transformation efficiency. A volume of 100 µl of *E. coli* cells was transformed using 20 µl of plasmid DNA (prepared at 0.1 ng µl^{-1}) and then made up to a final volume of 500 µl in buffer. A 10-fold dilution of this suspension was prepared and 200 µl of this dilution was surface spread (p. 206) onto a suitable medium containing an antibiotic to select for transformants, giving 180 colonies on the spread plate after overnight incubation. What is the transformation efficiency, expressed as the number of transformants (colony-forming units, CFU) per µg DNA? (Give your answer in exponential notation, to three significant figures.)

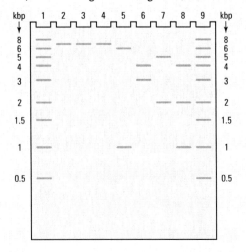

Analysis and Presentation of Data

61 Manipulating and transforming raw data

The process of discovering the meaning within your results can be fascinating. There are two main elements to this process:

- Exploratory data manipulation – this is used to investigate the nature of your results and suggest possible patterns and relationships within the data set. The aim is to generate hypotheses for further investigation. Exploratory techniques allow you to visualise the form of your data. They are ideal for examining results from pilot 'studies', but should be used throughout your investigations.
- Confirmatory analysis – this is used to test the hypotheses generated during the exploratory phase. The techniques required are generally statistical in nature and are dealt with in Chapter 66.

KEY POINT *Spreadsheets (Chapter 11) are invaluable tools for data manipulation and transformation: complex mathematical procedures can be carried out rapidly and the results visualised almost immediately using the inbuilt graphing functions. Spreadsheets also facilitate the statistical analysis of data (see Chapters 65 and 66).*

Summarising your results – original, unsummarised data belong only in your primary record, either in laboratory books or as computer records. You should produce summary tables to organise and condense original data.

Organising numbers

In order to organise, manipulate and summarise data, you should:

- Simplify the numbers, e.g. by rounding or taking means. This avoids the detail becoming overwhelming.
- Rearrange your data in as many ways as possible for comparison.
- Display in graphical form; this provides an immediate visual summary that is relatively easy to interpret.
- Look for an overall pattern in the data – avoid getting lost in the details at this stage.
- Look for any striking exceptions to that pattern (outliers) – they often point to special cases of particular interest or to errors in the data produced through mistakes during the acquisition, recording or copying of data.
- Move from graphical interpretations to well-chosen numerical summaries and/or verbal descriptions, including where applicable an explanatory hypothesis.

Colour	Tally	Total
Green	III	3
Blue	IIII III	8
Red	IIII	4
White	IIII IIII II	12
Black	I	1
Maroon	III	3
Yellow	II	2
		33

Fig. 61.1 An example of a tally chart

After collecting data, the first step is often to count how frequently each value occurs and to produce a frequency table. The frequency is simply the number of times a value occurs in the data set, and is, therefore, a count. The raw data could be acquired using a tally chart system to provide a simple frequency table. To construct a tally chart (e.g. Fig. 61.1):

Producing a histogram – a neatly constructed tally chart will double as a rough histogram.

- enter only one tally at a time;
- if working from a data list, cross out each item on the list as you enter it on to the tally chart, to prevent double entries;
- check that all values are crossed out at the end and that the totals agree.

Table 61.1 An example of a frequency table

Size class	Frequency	Relative frequency (%)
0–4.9	7	2.6
5–9.9	23	8.6
10–14.9	56	20.9
15–19.9	98	36.7
20–24.9	50	18.7
25–29.9	30	11.2
30–34.9	3	1.1
Total	267	99.8*

* ≠ 100 due to rounding error.

Stem	Leaves
7	23
7	55
7	6
7	9
8	000
8	233
8	45555
8	77
8	888899
9	0000111111
9	2333333
9	44555555555
9	66777777
9	88888999
10	00

Fig. 61.2 A simple 'stem and leaf' plot of a data set. The 'stem' shows the common component of each number, while the 'leaves' show the individual components, e.g. the top line in this example represents the numbers 72 and 73.

Example Allometry involves a logarithmic transformation of data that reveals aspects of the relationship between the dimensions of organisms (see p. 294).

Convert the data to a formal table when complete (e.g. Table 61.1). Because proportions are easier to compare than class totals, the table may contain a column to show the relative frequency of each class. Relative frequency can be expressed in decimal form (as a proportion of 1) or as a percentage (as a proportion of 100).

Graphing data

Graphs are an effective way to investigate trends in data and can reveal features that are difficult to detect from a table, e.g. skewness of a frequency distribution. The construction and use of graphs is described in detail in Chapter 62. When investigating the nature of your data, the main points are as follows:

- Make the values stand out clearly; attention should focus on the actual data, not the labels, scale markings, etc. (contrast with the requirements for constructing a graph for data presentation, see p. 387).
- Avoid clutter in the graph; leave out grid lines and try to use the simplest graph possible for your purpose.

 KEY POINT Use a computer spreadsheet with graphics options whenever possible: the speed and flexibility of these powerful tools should allow you to explore every aspect of your data rapidly and with relatively little effort (see Chapters 11 and 62).

Displaying distributions

A visual display of a distribution of values is often useful for variables measured on an interval or ratio scale (p. 156). The distribution of a variable can be displayed by a frequency table for each value or, if the possible values are numerous, groups (classes) of values of the variable. Graphically, there are two main ways of viewing such data:

- histograms, (see pp. 389, 390), generally used for large samples;
- stem and leaf plots (e.g. Fig. 61.2), often used for samples of less than 100: these retain the actual values and are faster to draw by hand. The main drawback is the limitation imposed by the choice of stem values since these class boundaries may obscure some features of the distribution.

These displays allow you to look at the overall shape of a distribution and to observe any significant deviations from the idealised theoretical ones. Where necessary, you can use data transformations to investigate any departures from standard distribution patterns such as the normal distribution or the Poisson distribution.

Transforming data

Transformations are mathematical functions applied to data values. They are particularly valuable where your results are related to areas and volumes (e.g. leaf area, body mass).

The most common use of transformations is to prepare data sets so that specific statistical tests may be applied. For instance, if you find that your data distribution is unimodal but not symmetrical (p. 415), it is often useful to apply a transformation that will redistribute the data values to

Using transformations – note that if you wish to conserve the order of your data, you will need to take negative values when using a reciprocal function (i.e. $-1/n^n$). This is essential when using a box plot to compare graphically the effects of transformations on the five-number summary of a data set.

form a symmetrical distribution. The object of this exercise is often to find the function that most closely changes the data into a standard normal distribution, allowing you to apply a wide range of parametric hypothesis-testing statistics (see Chapter 66). A frequently used transformation is to take logarithms of one or more sets of values: if the data then approximate to a normal distribution, the relationship is termed 'log-normal'.

Some general points about transformations are:

- They should be made on the raw data, not on derived data values: this is simpler, mathematically valid, and more easily interpreted.
- The transformed data can be analysed like any other numbers.
- Transformed data can be examined for outliers, which may be more important if they remain after transformation.

Figure 61.3 presents a ladder of transformations that will help you decide which transformations to try (see also Table 66.1). Note that percentage and proportion data are usually arc-sine transformed, which is a more complex procedure; consult Sokal and Rohlf (1994) for details.

Figure 61.4 illustrates the following 'quick-and-easy' way to choose a transformation:

1. Calculate the 'five-number summary' for the untransformed data (p. 419).
2. Present the summary graphically as a 'box-and-whisker' plot (p. 419).
3. Decide whether you need to correct for positive or negative skew (p. 421).
4. Apply one of the 'mild' transformations in Fig. 61.3 *on the five-number summary values only.*
5. Draw a new box-and-whisker plot and see whether the skewness has been corrected.
6. If the skewness has been undercorrected, try again with a stronger transformation. If it has been overcorrected, try a milder one.
7. When the distribution appears to be acceptable, transform the full data set and recalculate the summary statistics. If necessary use a statistical test to confirm that the transformed data are normally distributed (p. 430).
8. If no simple transformation works well, you may need to use non-parametric statistics when comparing data sets.

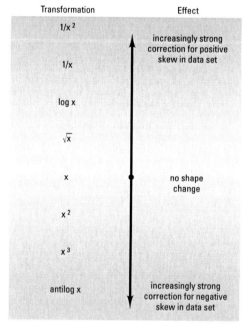

Fig. 61.3 Ladder of transformations (after J.W. Tukey)

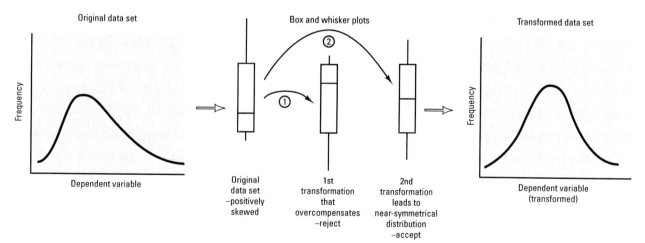

Fig. 61.4 Illustration of the processes of transforming a data set

Text references and sources for further study

Heath, D. (1995) *An Introduction to Experimental Design and Statistics for Biology*. UCL Press, London.

Quinn, G.P. and Keough, M.J. (2002) *Experimental Design and Data Analysis for Biologists*. Cambridge University Press, Cambridge.

Sokal, R.R. and Rohlf, F.J. (1994) *Biometry*, 3rd edn. W.H. Freeman, San Francisco.

Study exercises

The following study exercises can be most easily carried out using a spreadsheet such as Microsoft Excel.

61.1 Compare frequency distributions for samples with different sample sizes. Compute relative frequencies for each class in the table below. Graph the data as a frequency polygon.

Frequency distribution data

Hyphal length (μm)	Treatment B	Treatment A	Hyphal length (μm)	Treatment B	Treatment A
0	0	0	13	6	30
1	1	0	14	3	26
2	2	0	15	1	22
3	4	1	16	1	18
4	8	1	17	0	14
5	13	2	18	0	12
6	22	5	19	0	8
7	36	9	20	0	5
8	48	14	21	0	4
9	46	21	22	0	2
10	32	26	23	0	2
11	18	31	24	0	1
12	11	33	25	0	1

61.2 Use a stem and leaf plot. Work out the mean and standard deviation of the data contained in the following stem and leaf plot:

6	1334
6	56788
7	013344
7	556667899
8	11224
8	569
9	03
9	6

Stem and leaf plot

61.3 Use transformations to correct skew in a data set. Find a transformation that will make the data of treatment B in Study exercise 61.1 approximately symmetrical about the mean. Demonstrate graphically that you have accomplished this.

62 Using graphs

Fig. 62.1 Effect of antibiotic on yield of two bacterial isolates: ○, sensitive isolate; □, resistant isolate. Vertical bars show standard errors ($n = 6$).

Graphs can be used to show detailed results in an abbreviated form, displaying the maximum amount of information in the minimum space. Graphs and tables present findings in different ways. A graph (figure) gives a visual impression of the content and meaning of your results, while a table provides an accurate numerical record of data values. You must decide whether a graph should be used, e.g. to illustrate a pronounced trend or relationship, or whether a table (Chapter 63) is more appropriate.

A well-constructed graph will combine simplicity, accuracy and clarity. Planning of graphs is needed at the earliest stage in any write-up as your accompanying text will need to be structured so that each graph delivers the appropriate message. Therefore, it is best to decide on the final form for each of your graphs before you write your text. The text, diagrams, graphs and tables in a laboratory write-up or project report should be complementary, each contributing to the overall message. In a formal scientific communication it is rarely necessary to repeat the same data in more than one place (e.g. as a table and as a graph). However, graphical representation of data collected earlier in tabular format may be applicable in laboratory practical reports.

Practical aspects of graph drawing

The following comments apply to graphs drawn for laboratory reports. Figures for publication, or similar formal presentation, are usually prepared according to specific guidelines provided by the publisher/organiser.

> **KEY POINT** Graphs should be self-contained – they should include all material necessary to convey the appropriate message without reference to the text. Every graph must have a concise explanatory title to establish the content. If several graphs are used, they should be numbered, so they can be quoted in the text.

Selecting a title – it is a common fault to use titles that are grammatically incorrect: a widely applicable format is to state the relationship between the dependent and independent variables within the title, e.g. 'The relationship between enzyme activity and external pH'.

Remembering which axis is which – a way of remembering the orientation of the x axis is that x is a 'cross', and it runs 'across' the page (horizontal axis) while y is the first letter of yacht, with a large vertical mast (vertical axis).

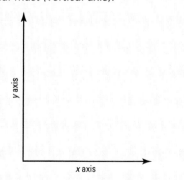

- Consider the layout and scale of the axes carefully. Most graphs are used to illustrate the relationship between two variables (x and y) and have two axes at right angles (e.g. Fig. 62.1). The horizontal axis is known as the abscissa (x axis) and the vertical axis as the ordinate (y axis).
- The axis assigned to each variable must be chosen carefully. Usually the x axis is used for the independent variable (e.g. treatment) while the dependent variable (e.g. biological response) is plotted on the y axis (p. 185). When neither variable is determined by the other, or where the variables are interdependent, the axes may be plotted either way round.
- Each axis must have a descriptive label showing what is represented, together with the appropriate units of measurement, separated from the descriptive label by a solidus or 'slash' (/), as in Fig. 62.1, or by brackets, as in Fig. 62.2.
- Each axis must have a scale with reference marks ('tics') on the axis to show clearly the location of all numbers used.
- A figure legend should be used to provide explanatory detail, including a key to the symbols used for each data set.

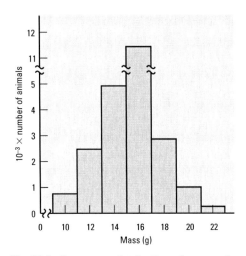

Fig. 62.2 Frequency distribution of masses for a sample of animals (sample size 24 085); the size class interval is 2 g.

Example For a data set where the smallest number on the log axis is 12 and the largest number is 9000, three-cycle log–linear paper would be used, covering the range 10–10 000 (Fig 62.3).

Handling very large or very small numbers

To simplify presentation when your experimental data consist of either very large or very small numbers, the plotted values may be the measured numbers multiplied by a power of 10: this multiplying power should be written immediately before the descriptive label on the appropriate axis (as in Fig. 62.2). However, it is often better to modify the primary unit with an appropriate prefix (p. 159) to avoid any confusion regarding negative powers of 10.

Size

Remember that the purpose of your graph is to communicate information. It must not be too small, so use at least half an A4 page and design your axes and labels to fill the available space without overcrowding any adjacent text. If using graph paper, remember that the white space around the grid is usually too small for effective labelling. The shape of a graph is determined by your choice of scale for the x and y axes which, in turn, is governed by your experimental data. It may be inappropriate to start the axes at zero (e.g. Fig. 62.1). In such instances, it is particularly important to show the scale clearly, with scale breaks where necessary, so the graph does not mislead. Note that Fig. 62.1 is drawn with 'floating axes' (i.e. the x and y axes do not meet in the lower left-hand corner), while Fig. 62.2 has clear scale breaks on both x and y axes.

Graph paper

In addition to conventional linear (squared) graph paper, you may need the following:

- Probability graph paper. This is useful when one axis is a probability scale (e.g. p. 430)
- Log–linear graph paper. This is appropriate when one of the scales shows a logarithmic progression, e.g. the exponential growth of cells in liquid culture (p. 272). Log–linear paper is defined by the number of logarithmic divisions (usually termed 'cycles') covered (e.g. Fig. 62.3), so make sure you use a paper with the appropriate number of cycles for your data. An alternative approach is to plot the log-transformed values on 'normal' graph paper.

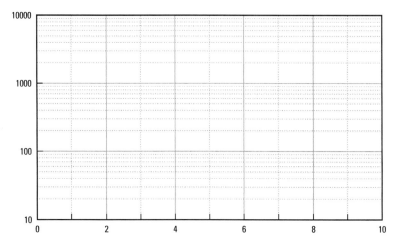

Fig. 62.3 Representation of three-cycle log–linear graph paper, marked up to show a y axis (log) scale from 10 to 10 000 and an x axis (linear) scale from 0 to 10.

- Log–log graph paper. This is appropriate when both scales show a logarithmic progression, e.g. in allometry (p. 294).

Types of graph

Different graphical forms may be used for different purposes, including:

- Plotted curves – used for data where the relationship between two variables can be represented as a continuum (e.g. Fig. 62.4).
- Scatter diagrams – used to visualise the relationship between individual data values for two interdependent variables (e.g. Fig. 62.5) often as a preliminary part of a correlation analysis (p. 434).
- Three-dimensional graphs show the interrelationships of three variables, often one dependent and two independent (e.g. Fig. 62.6). A contour diagram is an alternative method of representing such data.
- Histograms represent frequency distributions of continuous variables (e.g. Fig. 62.7). An alternative is the tally chart (p. 383).
- Frequency polygons emphasise the form of a frequency distribution by joining the coordinates with straight lines, in contrast to a histogram. This is particularly useful when plotting two or more frequency distributions on the same graph (e.g. Fig. 62.8).
- Bar charts represent frequency distributions of a discrete qualitative or quantitative variable (e.g. Fig. 62.9). An alternative representation is the line chart (Fig. 66.3, p. 429).
- Pie charts illustrate portions of a whole (e.g. Fig. 62.10).
- Pictographs give a pictorial representation of data (e.g. Fig. 62.11).

The plotted curve

This is the commonest form of graphical representation used in biology. The key features are outlined below and given in checklist form in Box 62.1, while Box 62.2 gives advice on using Microsoft Excel.

Choosing between a histogram and a bar chart – use a histogram for continuous quantitative *variables and a* bar chart for discrete *variables (see Chapter 25 for details of these types of measurement scale).*

Using computers to produce graphs – never allow a computer program to dictate size, shape and other aspects of a graph: find out how to alter scales, labels, axes, etc., and make appropriate selections (see Box 62.2 for Microsoft Excel). Draw curves freehand if the program only has the capacity to join the individual points by straight lines.

Box 62.1 Checklist for the stages in drawing a graph

The following sequence can be used whenever you need to construct a plotted curve: it will need to be modified for other types of graph.

1. **Collect all of the data values and statistical values** (in tabular form, where appropriate).
2. **Decide on the most suitable form of presentation**: this may include transformation (p. 384) to convert data to linear form.
3. **Choose a concise descriptive title**, together with a reference (figure) number and date, where necessary.
4. **Determine which variable is to be plotted on the *x* axis and which on the *y* axis.**
5. **Select appropriate scales for both axes** and make sure that the numbers and their location (scale marks) are clearly shown, together with any scale breaks.
6. **Decide on appropriate descriptive labels for both axes**, with SI units of measurement, where appropriate.
7. **Choose the symbols for each set of data points** and decide on the best means of representation for statistical values.
8. **Plot the points** to show the coordinates of each value with appropriate symbols.
9. **Draw a trend line for each set of points.** Use a see-through ruler, so you can draw the line to have an equal number of points on either side of it.
10. **Write a figure legend**, to include a key that identifies all symbols and statistical values and any descriptive footnotes.

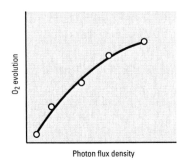

Fig. 62.4 Plotted curve: the rate of photosynthetic O_2 evolution as a function of photon flux density.

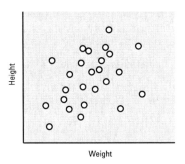

Fig. 62.5 Scatter diagram: height and weight of individual animals in a sample.

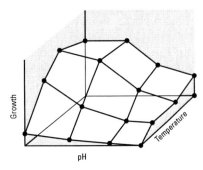

Fig. 62.6 Three-dimensional graph: growth of an organism as a function of temperature and pH.

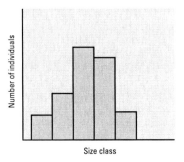

Fig. 62.7 Histogram: the number of plants within different size classes.

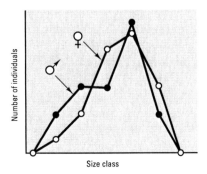

Fig. 62.8 Frequency polygon: frequency distributions of male and female animals according to size.

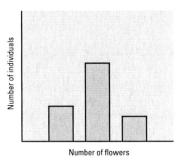

Fig. 62.9 Bar chart: number of flowers per plant.

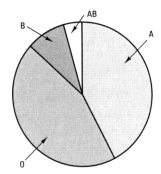

Fig. 62.10 Pie chart: relative abundance of human blood groups.

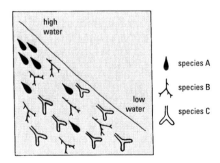

Fig. 62.11 Pictograph: distribution of plants on a rocky shore.

Box 62.2 How to create and amend graphs within a spreadsheet (Microsoft Excel) for use in coursework reports and dissertations

Microsoft Excel can used to create plotted curves, bar charts and histograms of reasonable quality, but only if you know how to amend the default settings to improve the overall effect, so that they meet the standard required for practical and project reports. As with a hand-drawn graph, the basic stages in graph drawing (Box 62.1) still apply.

Producing a plotted curve or bar chart

1. **Create the appropriate type of graph (chart) for your data.** Enter your data into the spreadsheet (e.g. in two columns), select the data array (highlight the appropriate cells by clicking and holding down the left mouse button and dragging) then use the 'Chart Wizard' icon (or the *Insert > Chart* option) to select a particular *Chart type*. For a plotted curve, use the *XY (Scatter)* option with unconnected points as the *Chart sub-type* – the line will be added at a later stage (note: never use the *Line Plot* because this will create a graph with evenly spaced x axis entries, regardless of their actual values). For a bar chart (p. 389), select the *Column* option. Work through each step of the Chart Wizard, selecting appropriate options for *Chart source*, *Chart options* (including labels for x and y axes) and *Location*. The graphs shown in Fig. 62.12(a) and Fig. 62.13(a) were producing using the default settings in Excel 2003.

2. **Change the default settings to improve the overall appearance.** Consider each element of the image in turn, including the overall size, height and width of your graph (click on the chart using the left-hand mouse button to show the 'sizing handles' and drag these to resize). The graphs shown in Fig. 62.12(b) and Fig. 62.13(b) were produced by altering the settings, typically by moving the cursor over the feature and clicking the right mouse button to reveal an additional menu of editing/formatting options.

The examples given below are for illustrative purposes only, and should not necessarily be regarded as a prescriptive list:

Example for a plotted curve (compare Fig. 62.12(a) with Fig. 62.12(b)):

- Grey background removed using *Clear* function, and border line around graph removed using *Format Chart Area > Patterns > Border > None*.
- Horizontal gridlines can be either removed using *Clear* function or, if desired (Fig. 62.12(b)), changed using *Chart Options > Gridlines* to show major and minor gridlines, then selecting *Format Gridlines* option to change the *Colour*, *Scale* and *Weight* of the gridlines, to make them more like those of conventional graph paper.
- Unnecessary legend box on right-hand side removed using the *Clear* option.
- Axes reformatted using the *Format Axis* menu (*Patterns, Scale* and *Number* options).

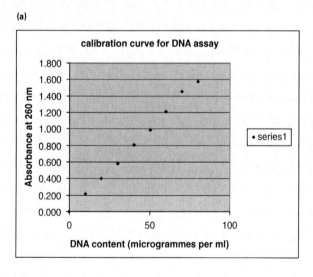

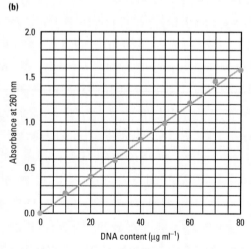

Fig. 1 Calibration curve for DNA assay. Performed using an A100X spectrophotometer (on 01.04.07). Values shown are averages of triplicate measurements.

Fig. 62.12 Examples of a plotted curve produced in Microsoft Excel using (a) default settings and (b) modified (improved) settings

(continued)

Box 62.2 (continued)

- Axis label changes achieved using the *Format Axis Title > Font* menu options (including superscript for selected parts – alternatively, cut and paste symbols/text from Word).
- Data point style changed using the *Format Data Series > Patterns > Marker > Custom* options for *Style, Foreground* (edge of symbol), *Background* (centre of symbol) and *Size* (note: *Line* function not used, since it joins individual data points, rather than giving a linear plot).
- Straight line of best fit added using the *Format Data Series > Add Trendline > Type > Linear* option, with the line thickness then adjusted using the *Format Trendline > Patterns > Custom > Weight* option (see also Box 50.2, p. 306).

Example for a bar chart (compare Fig. 62.13(a) with Fig. 62.13(b)):
- Grey background removed using *Clear* function, and border line around graph removed using *Format Chart Area > Patterns > Border > None*; colour of horizontal gridlines changed to light grey using *Format Gridlines > Patterns > Custom > Colour* option.
- Unnecessary legend box on right-hand side removed using the *Clear* option.
- Axes reformatted (tic marks, scales and font) using the *Format Axis* menu (*Patterns, Scale* and *Font* options).
- Axis labels changed to Times New Roman regular 12-point font size using *Format Axis Title > Font, > Font style* and *> Size* options.
- Bar colour removed using *Format Data Series > Patterns > Area*, selecting the white colour (not option *None*).

- Individual numbers shown using the *Format Data Series > Labels > Value* option; bar width altered using *Format Data Series > Options > Gap Width*.

Note that for both types of graph, it is better not to use the default *Chart Title* option within Excel, which places the title at the top of the chart (as in Fig. 62.13(a)), but to cut and paste your untitled graph into a word processor such as Microsoft Word and then type a formal figure legend below the graph itself (as in Fig. 62.12(b) and Fig. 62.13(b)). However, once your graph is embedded into Word, it is generally best not to make further amendments – go back to the original Excel file, make the required change(s) and reinsert the graph into Word.

Producing a histogram

The histogram function in Excel requires a little more effort to master, compared with the 'Chart Wizard' graph-drawing function. Essentially, a histogram is a graphical display of frequencies (counts) for a continuous quantitative variable (pp. 389, 395), where the data values are grouped into discrete classes. It is possible to select the upper limit for each class into which you want your data to be grouped (termed 'bin range values' in Excel).

An alternative approach is to let the software select the group ranges for you: Excel selects evenly distributed bins between the minimum and maximum values (this is often less effective than selecting your own class boundaries).

The following steps outline the procedure used to create the histograms shown in Fig. 62.14 for the table of data opposite (length, in mm, of 24 leaf petioles from a single plant).

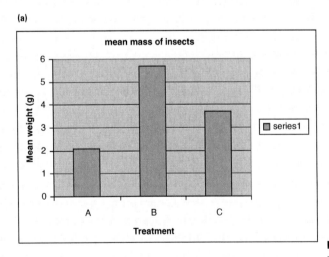

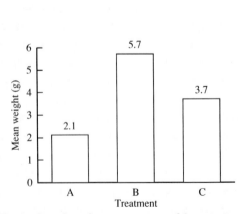

Fig. 2 Bar chart for mean mass of insects treated using three different methods. Key: A = chloroform, B = saline, C = formalin.

Fig. 62.13 Examples of a bar chart produced in Microsoft Excel using (a) default settings and (b) modified (improved) settings

(continued)

Box 62.2 (continued)

7.2	6.5	7.1	8.5	6.6	7.2
7.0	7.3	8.6	9.1	7.5	8.3
7.1	5.7	7.3	7.6	6.9	7.1
8.3	7.6	5.4	8.6	7.9	8.0

1. **Enter the raw data values,** e.g. as a single column of numbers, or as an array, as above.
2. **Decide the class intervals to be used.** Base your choice on the number of data points and the maximum and minimum values (for a small data set, you may be able to do this by quickly examining the data set, whereas for a large data set, use the Excel – *COUNT, MAX* and *MIN* functions) – a typical histogram would have 4–10 classes depending on the level of discrimination required. Enter the upper limit for each class (bin range values) in a separate array of cells close to the data, in ascending order (in the above example, 6, 7, 8 and 9 were chosen – the few data values above the final bin value will be shown on the histogram as a group labelled 'more').
3. **Select the histogram function, then input your data and class-interval values.** From the dropdown menu under *Tools* select *Data Analysis > Histogram* (if the *Data Analysis* option is not already activated on the PC you are using, select using the *Add Ins* function within this menu, but note that this addition may require the Microsoft Office software CD) then *Histogram*. Input your data into the *Input Range* box (highlight the appropriate cells by clicking and dragging while holding down the left mouse button) and similarly input the *Bin Range* values into the appropriate box (if this is left empty,

Excel will select default *Bin Range* values) – most of the remaining boxes can be left empty, though you must click the last box to get a *Chart Output*, otherwise the software will give the numerical counts for each group, without a histogram. Click OK and entries in a new worksheet will be created showing the upper limits of each group (column labelled *Bin*) along with the counts for each group (column labelled *Frequency*), plus a poorly constructed chart based on default settings, as shown in Fig. 62.14(a) (note that the default output is a bar chart rather than a histogram, since there are gaps between the groups).

Example for a histogram Fig. 62.14(b) shows the following modifications compared with Fig. 62.14(a):
- Chart resized to increase height (use sizing handles); grey background removed using *Clear* function and border line around graph removed using *Format Chart Area > Patterns > Border > None*.
- Title and unnecessary legend box removed using *Clear* option; y scale amended using *Format Axis > Scale* command; x labels (classes) amended by typing class limits directly into the cells containing the bin values.
- Axes reformatted using the *Format Axis* menu, and new text typed directly into the axis label box (double-click on the chart to access).
- Bar colour changed to light grey using *Format Data Series > Patterns > Area* option.
- Bar chart converted to true histogram (no gaps between bars) using the *Format Data Series > Options menu,* setting *Gap Width* = 0.
- Figure legend added below figure in Microsoft Word, following copying and pasting of the Excel histogram into a Word file.

(a)

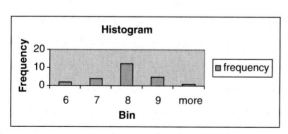

(b)

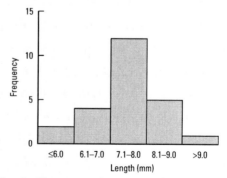

Fig. 3 **Histogram of petiole lengths (mm) in a single plant** (n = 24).

Fig. 62.14 Examples of histogram output from Microsoft Excel using (a) default settings and (b) modified (improved) settings

Choosing graphical symbols – plotted curves are usually drawn using a standard set of symbols: ●, ○, ■, □, ▲, △, ◆, ◇. By convention, paired symbols ('closed' and 'open') are often used to represent 'plus' (treatment) and 'minus' (control) treatments.

Adding error bars to Microsoft Excel graphs – you can do this as follows (for Excel 2003):
1. On your graph, right-click on one of the data points of the series to which you want to add error bars.
2. Select Format Data Series and use either the Y Error Bars tab (vertical error bars) or the X Error Bars tab (horizontal error bars):
   - select the type of error bar that you want under Display (e.g. Both);
   - select the method to be used to input the error values under Error amount; typically, you will want to select Custom, to input pre-calculated numerical values. Then, in the Plus and Minus boxes, either specify the range of cells in your worksheet where the error values are located, or enter the relevant numbers directly, separated by commas.

Conveying the correct message – the golden rule is: 'always draw the simplest line that fits the data reasonably well and is biologically reasonable'.

Extrapolating plotted curves – try to avoid the need to extrapolate by better experimental design.

Data points

Each data point must be shown accurately, so that any reader can determine the exact values of x and y. In addition, the results of each treatment must be readily identifiable. A useful technique is to use a dot for each data point, surrounded by a hollow symbol for each treatment (see Fig. 62.1). An alternative is to use symbols only (Fig. 62.4), though the coordinates of each point are defined less accurately. Use the same symbol for the same entity if it occurs in several graphs and provide a key to all symbols.

Statistical measures

If you are plotting average values for several replicates and if you have the necessary statistical knowledge, you can calculate the standard error (p. 420, or the 95 per cent confidence limits (p. 434) for each mean value and show these on your graph as a series of vertical bars (see Fig. 62.1). Make it clear in the legend whether the bars refer to standard errors or 95 per cent confidence limits and quote the value of n (the number of replicates per data point). Another approach is to add a least significant difference bar (p. 431) to the graph.

Interpolation

Once you have plotted each point, you must decide whether to link them by straight lines or a smoothed curve. Each of these techniques conveys a different message to your reader. Joining the points by straight lines may seem the simplest option, but may give the impression that errors are very low or non-existent and that the relationship between the variables is complex. Joining points by straight lines is appropriate in certain graphs involving time sequences (e.g. the number of animals at a particular site each year), or for repeat measurements where measurement error can be assumed to be minimal (e.g. recording a patient's temperature in a hospital, to emphasise any variation from one time point to the next). However, in most plotted curves the best straight line or curved line should be drawn (according to appropriate mathematical or statistical models, or by eye), to highlight the relationship between the variables – after all, your choice of a plotted curve implies that such a relationship exists. Don't worry if some of your points do not lie on the line: this is caused by errors of measurement and by biological variation (p. 157). Most curves drawn by eye should have an equal number of points lying on either side of the line. You may be guided by 95 per cent confidence limits, in which case your curve should pass within these limits wherever possible.

Curved lines can be drawn using a flexible curve, a set of French curves, or freehand. In the latter case, turn your paper so that you can draw the curve in a single, sweeping stroke by a pivoting movement at the elbow (for larger curves) or wrist (for smaller ones). Do not try to force your hand to make complex, unnatural movements, as the resulting line will not be smooth.

Extrapolation

Be wary of extrapolation beyond the upper or lower limit of your measured values. This is rarely justifiable and may lead to serious errors. Whenever extrapolation is used, a dotted line ensures that the reader is aware of the uncertainty involved. Any assumptions behind an extrapolated curve should also be stated clearly in your text.

Drawing a histogram – *each datum is represented by a column with an area proportional to the magnitude of y: in most cases, you should use columns of equal width, so that the height of each column is then directly proportional to y. Shading or stippling may be used to identify individual columns, according to your needs.*

The histogram

Whereas a plotted curve assumes a continuous relationship between the variables by interpolating between individual data points, a histogram involves no such assumptions. Histograms are also used to represent frequency distributions (p. 383), where the y axis shows the number of times a particular value of x was obtained (e.g. Fig. 62.2). As in a plotted curve, the x axis represents a continuous quantitative variable that can take any value within a given range (e.g. plant height), so the scale must be broken down into discrete classes and the scale marks on the x axis should show either the mid-points (mid-values) of each class, or the boundaries between the classes.

The columns are contiguous (adjacent to each other) in a histogram, in contrast to a bar chart, where the columns are separate because the x axis of a bar chart represents discrete values.

Interpreting graphs

The process of analysing a graph can be split into five phases:

1. Consider the context. Look at the graph in relation to the aims of the study in which it was reported. Why were the observations made? What hypothesis was the experiment set up to test? This information can usually be found in the Introduction or Results section of a report. Also relevant are the general methods used to obtain the results. This might be obvious from the figure title and legend, or from the Materials and Methods section.
2. Recognise the graph form and examine the axes. First, what kind of graph is presented (e.g. histogram, plotted curve)? You should be able to recognise the main types summarised on pp. 389–90 and their uses. Next, what do the axes measure? You should check what quantity has been measured in each case and what units are used.
3. Look closely at the scale of each axis. What is the starting point and what is the highest value measured? For the x-axis, this will let you know the scope of the treatments or observations (e.g. whether they lasted for 5 min or 20 years; whether a concentration span was two-fold or fifty-fold). For each axis, it is especially important to note whether the values start at zero; if not, then the differences between any treatments shown may be magnified by the scale chosen (see Box 62.3).

Examining graphs – *don't be tempted to look at the data displayed within a graph before you have considered its context, read the legend and decided the scale of each axis.*

4. Examine the symbols and curves. Information will be provided in the key or legend to allow you to determine what these refer to. If you have made your own photocopy of the figure, it may be appropriate to note this directly on it. You can now assess what appears to have happened. If, say, two conditions have been observed while a variable is altered, when exactly do they differ from each other; by how much; and for how long?
5. Evaluate errors and statistics. It is important to take account of variability in the data. For example, if mean values are presented, the underlying errors may be large, meaning that any difference between two treatments or observations at a given x-value could simply have arisen by chance. Thinking about the descriptive statistics used (Chapter 65) will allow you to determine whether apparent differences could be significant in both statistical and biological senses.

Box 62.3 How graphs can misrepresent and mislead

1. **The 'volume' or 'area' deception** – this is mainly found in histogram or bar chart presentations where the size of a symbol is used to represent the measured variable. For example, the amount of hazardous waste produced in different years might be represented on a chart by different sizes of a chemical drum, with the y axis (height of drum) representing the amount of waste. However, if the symbol retains its *shape* for all heights as in Fig. 62.15(a), its *volume* will increase as a cubic function of the height, rather than in direct proportion. To the casual observer, a two-fold increase may look like an eight-fold one, and so on. Strictly, the *height* of the symbol should be the measure used to represent the variable, with no change in symbol width, as in Fig. 62.15(b).

2. **Effects of a non-zero axis** – A non-zero axis acts to emphasise the differences between measures by reducing the range of values covered by the axis. For example, in Fig. 62.16(a), it looks as if there are large differences in mass between males and females; however, if the scale is adjusted to run from zero, then it can be seen that the differences are not large as a proportion of the overall mass. Always scrutinise the scale values carefully when interpreting any graph.

3. **Use of a relative rather than absolute scale** – this is similar to the above, in that data compared using relative scales (e.g. percentage or ratio) can give the wrong impression if the denominator is not the same in all cases. In Fig. 62.17(a), two treatments are shown as equal in *relative* effect, both resulting in 50 per cent relative response compared (say) to the respective controls. However, if treatment A is 50 per cent of a control value of 200 and treatment B is 50 per cent of a control value of 500, then the actual difference in *absolute* response would have been masked, as shown by Fig. 62.17(b).

4. **Effects of a non-linear scale** – when interpreting graphs with non-linear (e.g. logarithmic) scales, you may interpret any changes on an imagined linear scale. For example, the pH scale is logarithmic, and linear changes on this scale mean less in terms of absolute H^+ concentration at high (alkaline) pH than they do at low (acidic) pH. In Fig. 62.18(a), the cell density in two media is compared on a logarithmic scale, while in Fig. 62.18(b), the same data are graphed on a linear scale. Note, also, that the log y-axis scale in Fig. 62.18(a) cannot be shown to zero, because there is no logarithm for 0.

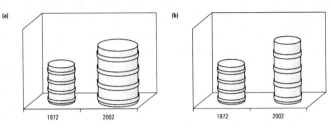

Fig. 62.15 Increase in pesticide use over a 30 year period

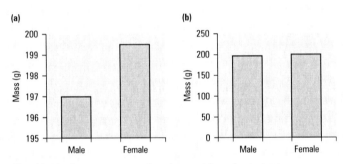

Fig. 62.16 Average mass of males and females in test group

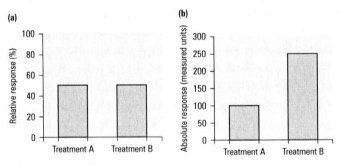

Fig. 62.17 Responses to treatments A and B

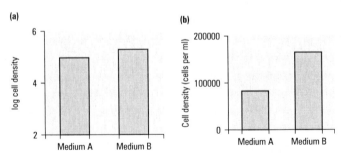

Fig. 62.18 Effect of different media on cell density

(continued)

Box 62.3 (continued)

5. **Unwarranted extrapolation** – a graph may be extrapolated to indicate what would happen if a trend continued, as in Fig. 62.19(a). However, this can only be done under certain assumptions (e.g. that certain factors will remain constant or that relationships will hold under new conditions). There may be no guarantee that this will actually be the case. Figure 62.19(b) illustrates other possible outcomes if the experiment were to be repeated with higher values for the *x*-axis.

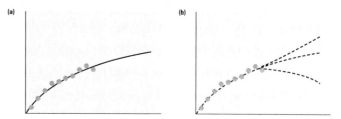

Fig. 62.19 Extrapolation of data under different assumptions

6. **Failure to account for data point error** – this misrepresentation involves curves that are overly complex in relation to the scatter in the underlying data. When interpreting graphs with complex curves, consider the errors involved in the data values. It is probably unlikely that the curve would pass through all the data points unless the errors were very small. Figure 62.20(a) illustrates a curve that appears to assume zero error and is thus overly complex, while Fig. 62.20(b) shows a curve that takes possible errors of the points into account.

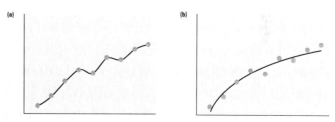

Fig. 62.20 Fitted curves under different assumptions of data error

7. **Failure to reject outlying points** – this is a special case of the previous example. There may be many reasons for outlying data, from genuine mistakes to statistical 'freaks'. If a curve is drawn through such points on a graph, it indicates that the point carries equal weight with the other points, when in fact, it should probably be ignored. To assess this, consider the accuracy of the measurement, the number and position of adjacent points, and any special factors that might be involved on a one-off basis. Figure 62.21(a) shows a curve where an outlier (arrowed) has perhaps been given undue weight when showing the presumed relationship. If there is good reason to think that the point should be ignored, then the curve shown in Fig. 62.21(b) would probably be more valid.

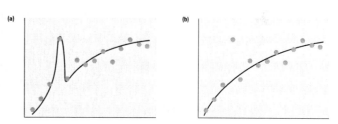

Fig. 62.21 Curves with and without outlier taken into account

8. **Inappropriate fitted line** – here, the mathematical function chosen to represent a trend in the data might be inappropriate. A straight line might be fitted to the data, when a curve would be more correct, or vice versa. These cases can be difficult to assess. You need to consider the theoretical validity of the model used to generate the curve (this is not always stated clearly). For example, if a straight line is fitted to the points, the implicit underlying model states that one factor varies in direct relation to another, when the true situation may be more complex. In Fig. 62.22(a), the relationship has been shown as a linear relationship, whereas an exponential relationship, as shown in Fig. 62.22(b), could be more correct.

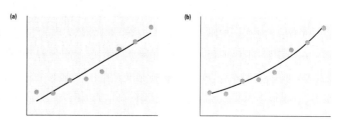

Fig. 62.22 Different mathematical model used to represent trends in data

Understanding graphs within scientific papers – the legend should be a succinct summary of the key information required to interpret the figure without further reference to the main text. This is a useful approach when 'skimming' a paper for relevant information (pp. 18–19).

Sometimes graphs are used to mislead. This may be unwitting, as in an unconscious favouring of a 'pet' hypothesis of the author. Graphs may be used to 'sell' a product in the field of advertising or to favouring of a viewpoint as, perhaps, in politics. Experience in drawing and interpreting graphs will help you spot these flawed presentations, and understanding how graphs can be erroneously presented (Box 62.3) will help you avoid the same pitfalls.

Sources for further study

Briscoe, M.H. (1996) *Preparing Scientific Illustrations: a Guide to Better Posters, Presentations and Publications.* Springer-Verlag, New York.

Carter, M., et al. *Graphing with Excel.* Available: http://www.ncsu.edu/labwrite/res/gt/gt-menu.html Last accessed 09/04/07.
[Online tutorial from the US-NSF LabWrite 2000 Project.]

Institute of Biology (2000) *Biological Nomenclature: Recommendations on Terms, Units and Symbols.* Institute of Biology, London.
[Includes a section on presentation of data.]

Robbins, N.B. (2005) *Creating More Effective Graphs.* Wiley, New York.

Study exercises

62.1 Select appropriate graphical presentations. Choose an appropriate graphical form for each of the following examples:
(a) Interaction between pH and cation concentration on enzyme activity.
(b) Proportion of different petal shapes in a countywide survey of *Primula vulgaris* flowers.
(c) Relationship between pulse rate and age in humans.
(d) Number of bacteria per field of view.
(e) Effect of copper concentration on the activity of an enzyme.

62.2 Create a pie chart. Display the following information in the form of a pie chart. Do *not* use a spreadsheet for this exercise.

Behaviour of male brown bear under zoo conditions

Behaviour category	Occasions observed
Eating	37
Scratching	8
Sexual display	13
Mating	2
Sleeping	25

62.3 Create a frequency distribution histogram. The following table gives data for the haemoglobin levels of 100 people. Plot a histogram showing the frequency distribution of the data. Write a brief description of the important features of the distribution.

Haemoglobin content ($g\,l^{-1}$) in blood

11.1	14.2	13.5	9.8	12.0	13.9	14.1	14.6	11.0	12.3
13.4	12.9	12.9	10.0	13.1	11.8	12.6	10.7	8.1	11.2
13.8	12.4	12.9	11.3	12.7	12.4	14.6	15.1	11.2	9.7
11.3	14.7	10.8	13.3	11.9	11.4	12.5	13.0	11.6	13.1
9.3	13.5	14.6	11.2	11.7	10.9	12.4	12.0	12.1	12.6
10.9	12.1	13.4	9.5	12.5	11.6	12.2	8.8	10.7	11.1
10.2	11.7	10.4	14.0	14.9	11.5	12.0	13.2	12.1	13.3
12.4	9.4	13.2	12.5	10.8	11.7	12.7	14.1	10.4	10.5
13.3	10.6	10.5	13.7	11.8	14.1	10.3	13.6	10.4	13.9
11.7	12.8	10.4	11.9	11.4	10.6	12.7	11.4	12.9	12.1

62.4 Find examples of misleading graphs. Create a portfolio of examples of misleading graphs taken from newspapers. For each graph, state what aspect is misleading (see Box 62.3) and, where possible, attempt to show the data correctly in a new graph.

63 Presenting data in tables

A table is often the most appropriate way to present numerical data in a concise, accurate and structured form. Assignments and project reports should contain tables that have been designed to condense and display results in a meaningful way and to aid numerical comparison. The preparation of tables for recording primary data is discussed on p. 192.

Decide whether you need a table, or whether a graph is more appropriate. Histograms and plotted curves can be used to give a visual impression of the relationships within your data (p. 389). On the other hand, a table gives you the opportunity to make detailed numerical comparisons.

KEY POINT Always remember that the primary purpose of your table is to communicate information and allow appropriate comparison, not simply to put down the results on paper.

Preparation of tables

Title

Every table must have a brief descriptive title. If several tables are used, number them consecutively so they can be quoted in your text. The titles within a report should be compared with one another, making sure they are logical and consistent and that they describe accurately the numerical data contained within them.

Constructing titles – take care over titles as it is a common mistake in student practical reports to present tables without titles, or to misconstruct the title.

Structure

Display the components of each table in a way that will help the reader understand your data and grasp the significance of your results. Organise the columns so that each category of like numbers or attributes is listed vertically, while each horizontal row shows a different experimental treatment, organism, sampling site, etc. (as in Table 63.1). Where appropriate, put control values near the beginning of the table. Columns that need to be compared should be set out alongside each other. Use rulings to subdivide your table appropriately, but avoid cluttering it up with too many lines.

Saving space in tables – you may be able to omit a column of control data if your results can be expressed as percentages of the corresponding control values.

Table 63.1 Characteristics of selected photoautotrophic microbes

Division	Species	Optimum [NaCl]* (mol m^{-3})	Intracellular carbohydrate	
			Identity	Quantity† (nmol (g dry wt)$^{-1}$)
Chlorophyta	Scenedesmus quadruplicatum	340	Sucrose	49.7
	Chlorella emersonii	780	Sucrose	102.3
	Dunaliella salina	4700	Glycerol	910.7
Cyanobacteria	Microcystis aeruginosa	<20‡	None	0.0
	Anabaena variabilis	320	Sucrose	64.2
	Rivularia atra	380	Trehalose	ND

* Determined after 28-day incubation in modified Von Stosch medium.
† Individual samples, analysed by gas–liquid chromatography.
‡ Poor growth in all media with added NaCl (minimum NaCl concentration 5 mol m^{-3}).
ND Sample lost: no quantitative data.

Headings and sub-headings

These should identify each set of data and show the units of measurement, where necessary. Make sure that each column is wide enough for the headings and for the longest data value.

Numerical data

Within the table, do not quote values to more significant figures than necessary, as this will imply spurious accuracy (pp. 157, 408). By careful choice of appropriate units for each column you should aim to present numerical data within the range 0 to 1 000. As with graphs, it is less ambiguous to use derived SI units, with the appropriate prefixes, in the headings of columns and rows, rather than quoting multiplying factors as powers of 10. Alternatively, include exponents in the main body of the table (see Table 24.1), to avoid any possible confusion regarding the use of negative powers of 10.

Other notations

Avoid using dashes in numerical tables, as their meaning is unclear; enter a zero reading as '0' and use 'NT' not tested or 'ND' if no data value was obtained, with a footnote to explain each abbreviation. Other footnotes, identified by asterisks, superscripts or other symbols in the table, may be used to provide relevant experimental detail (if not given in the text) and an explanation of column headings and individual results, where appropriate. Footnotes should be as condensed as possible. Table 63.1 provides examples.

Statistics

In tables where the dispersion of each data set is shown by an appropriate statistical parameter, you must state whether this is the (sample) standard deviation, the standard error (of the mean) or the 95 per cent confidence limits and you must give the value of n (the number of replicates). Other descriptive statistics should be quoted with similar detail, and hypothesis-testing statistics should be quoted along with the value of P (the probability). Details of any test used should be given in the legend, or in a footnote.

Text

Sometimes a table can be a useful way of presenting textual information in a condensed form (see examples on pp. 347 and 371).

When you have finished compiling your tabulated data, carefully double-check each numerical entry against the original information, to ensure that the final version of your table is free from transcriptional errors. Box 63.1 gives a checklist for the major elements of constructing a table.

Examples If you measured the width of a fungal hypha to the nearest one-tenth of a micrometre, quote the value in the form '52.6 µm'.
Quote the width of a fungal hypha as 52.6 µm, rather than 0.000 052 6 m or 52.6 10^{-6} m.

Saving further space in tables – in some instances a footnote can be used to replace a whole column of repetitive data.

Using spreadsheets and word-processing packages – these can be used to prepare high-quality versions of tables for project work (Box 63.2).

Box 63.1 Checklist for preparing a table

Every table should have the following components:

1. **A title**, plus a reference number and date where necessary.
2. **Headings for each column and row**, with appropriate units of measurement.
3. **Data values**, quoted to the nearest significant figure and with statistical parameters, according to your requirements.
4. **Footnotes** to explain abbreviations, modifications and individual details.
5. **Rulings to emphasise groupings** and distinguish items from each other.

Box 63.2 How to use a word processor (Microsoft Word) or a spreadsheet (Microsoft Excel) to create a table for use in coursework reports and dissertations

Creating tables with Microsoft Word: Word-processed tables are suitable for text-intensive or number-intensive tables, although in the second case entering data can be laborious. When working in this way, the natural way to proceed is to create the 'shell' of the table, add the data, then carry out final formatting on the table.

1. **Move the cursor to the desired position in your document.** This is where you expect the top left corner of your table to appear. Go to the *Table* command, then choose *Insert > Table*.
2. **Select the appropriate number of columns and rows.** Don't forget to add rows and columns for headings. As default, a full-width table will appear, with single rulings for all cell boundaries, with all columns of equal width and all rows of equal height.

Example of a 4 x 3 Table:

3. **Customise the columns.** By placing the cursor over the vertical rulings then 'dragging', you can adjust their width to suit your heading text entries, which should now be added.

Heading 1	Heading 2	Heading 3	Heading 4

4. **Work through the table adding the data.** Entries can be numbers or text.

Heading 1	Heading 2	Heading 3	Heading 4
xx	xx	xx	xx
xx	xx	xx	

5. **Make further adjustments to column and row widths to suit.** For example, if text fills several rows within a cell, consider increasing the column width, and if a column contains only single or double digit numbers, consider shrinking its width. To combine cells, first highlight them, then use *Table > Merge Cells*. You may wish to reposition text within a cell using *Format > Paragraph > Spacing*.

Heading 1	Heading 2	Heading 3	Heading 4
xx	xx	xx	xx
	xx	xx	xx

6. **Finally, remove selected borders to cells.** One way to do this is using the table borders options accessed from the *Table borders* button on the toolbar, so that your table looks like the examples shown in this chapter.
7. **Add a table title.** This should be positioned *above* the table (*c.f.* a figure title and legend p. 390), legend and footnotes.

Final version of the Table:

Table xx. A table of some data

Heading 1	Heading 2[a]	Heading 3	Heading 4
Aaa	xx	xx	xx
	xx	xx	xx

[a]An example of a footnote

Creating tables with Microsoft Excel:
Tables derived from spreadsheets are effective when you have lots of numerical data, especially when these are stored or created using the spreadsheet itself. When working in this way, you can design the table as part of an output or summary section of the spreadsheet, add explanatory headings, format, then possibly export to a word processor when complete.

1. **Design the output or summary section.** Plan this as if it were a table, including adding text headings within cells.

(continued)

Box 63.2 (continued)

2. **Insert appropriate formulae within cells to produce data.** If necessary, formulae should draw on the other parts of the spreadsheet.

17				
18	Heading 1	Heading 2	Heading 3	Heading 4
19	Aaa	=A1	=C3*5	=SDEV(A1:A12)
20	Bbb	=A2	=F45/G12	=SDEV(B1:B12)
21				
22				

3. **Format the cells.** This is important to control the number of decimal places presented (*Format > Cells > Number*).
4. **Adjust column width to suit.** You can do this via the column headings, by placing the cursor over the rulings between columns then 'dragging'.

17				
18	Heading 1	Heading 2	Heading 3	Heading 4
19	Aaa	=A1	=C3*5	=SDEV(A1:A12)
20	Bbb	=A2	=F45/G12	=SDEV(B1:B12)
21				
22				

5. **Add rulings as appropriate.** Use the borders menu on the toolbar as described above.

17				
18	Heading 1	Heading 2	Heading 3	Heading 4
19	Aaa	=A1	=C3*5	=SDEV(A1:A12)
20	Bbb	=A2	=F45/G12	=SDEV(B1:B12)
21				

6. **Add 'real' data values to the spreadsheet.** This should result in the summary values within the table being filled. Check that these are presented with the appropriate number of significant figures (p. 408).
7. **The table can now be copied and pasted to a Word document.** If you wish to link the spreadsheet and the word-processed document so that the latter is updated whenever changes are made to the spreadsheet values, then click *Paste* on the *Formatting* toolbar. Click *Paste Options* next to the data, and then select *Match Destination Table Style and Link to Excel* or *Keep Source Formatting and Link to Excel*.

Sources for further study

Kirkup, L. (1994) *Experimental Methods: An Introduction to the Analysis and Presentation of Data.* Wiley, New York.

Simmonds, D. and Reynolds, L. (1994) *Data Presentation and Visual Literacy in Medicine and Science.* Butterworth-Heinemann, London.

Study exercises

63.1 Redesign a table of data. Using the following example, redraft the table to improve layout and correct inconsistencies.

Concentrations of low molecular weight solutes in bacteria

Bacterium	Constituent	Concentration
Escherichia coli	Proline	21.0 mmol l^{-1}
Escherichia coli	Trehalose	1.547×10^2 kmol m^{-3}
Bacillus subtilis	Proline	39.7 mmol l^{-1}
Bacillus subtilis	Glutamate	0.0521 mmol cm^{-3}
*Staphylococcus aureus**	Glutamate	15 260 mmol m^{-3}
Escherichia coli	Glutamate	0.50% w/v
Bacillus subtilis	Trehalose	<0.001% w/v

*Proline and trehalose were not measured.

(continued)

Study exercises (continued)

63.2 Devise a text-based table. After reading through this chapter and working from memory, draw up a table listing the principal components of a typical table in the first column, and brief comments on the major features of each component in the second column.

63.3 Interpret data from a table. The table below shows representative data for energy, protein and niacin (vitamin B3) as a function of age for human males and females. The typical average mass ('weight') of each class is also shown.

(a) In what age class does the mass of females exceed that of males?
(b) In what age class is the energy requirement of males greatest in proportion to the energy requirement of females?
(c) Over what age range do males and females have the same protein requirement?
(d) In what age class do males have their highest niacin requirement, and how does this compare with females?

Selected age-related characteristics of human males and females

Age (years)	Average mass (kg)		Energy requirement (MJ)		Protein requirement (g)		Niacin requirement (mg)	
	Male	Female	Male	Female	Male	Female	Male	Female
<1	8	8	3.4	3.2	14	14	6	6
1–3	13	13	5.2	4.9	16	16	9	9
4–6	20	20	7.2	6.5	24	24	12	12
7–10	28	28	8.2	7.3	28	28	13	13
11–14	45	46	9.3	7.7	45	46	17	15
15–18	66	55	11.5	8.8	59	44	20	15
19–50	79	63	10.6	8.1	63	50	19	15
>50	77	65	9.5	7.7	63	50	15	13

64 Hints for solving numerical problems

The following boxes give advice on dealing with numerical procedures:
- *Box 22.2: preparing solutions (p. 130)*
- *Box 23.1: molar concentrations (p. 139)*
- *Box 46.1: cell-counting chambers (p. 275)*
- *Box 46.2: plate (colony) counts (p. 276)*
- *Box 50.1: calibration curves (p. 304)*
- *Box 56.1: radioactivity (p. 349)*

Biology often requires a numerical or statistical approach. Not only is mathematical modelling an important aid to understanding, but computations are often needed to turn raw data into meaningful information or to compare them with other data sets. Moreover, calculations are part of laboratory routine, perhaps required for making up solutions of known concentration (see p. 130 and below) or for the calibration of a microscope (see p. 258). In research, 'trial' calculations can reveal what input data are required and where errors in their measurement might be amplified in the final result (see p. 187).

 KEY POINT *If you find numerical work difficult, practice at problem-solving is especially important.*

Practising at problem-solving:

- demystifies the procedures involved, which are normally just the elementary mathematical operations of addition, subtraction, multiplication and division (Table 64.1);
- allows you to gain confidence so that you don't become confused when confronted with an unfamiliar or apparently complex form of problem;
- helps you recognise the various forms a problem can take as, for instance, in crossing experiments in classical genetics (Box 53.1).

Table 64.1 Sets of numbers and operations

Sets of numbers	
Whole numbers:	0, 1, 2, 3, …
Natural numbers:	1, 2, 3, …
Integers:	… −3, −2, −1, 0, 1, 2, 3, …
Real numbers:	integers and anything between (e.g. −5, 4.376, 3/16, π, $\sqrt{5}$)
Prime numbers:	subset of natural numbers divisible by 1 and themselves only (i.e. 2, 3, 5, 7, 11, 13, …)
Rational numbers:	p/q where p (integer) and q (natural) have no common factor (e.g. 3/4)
Fractions:	p/q where p is an integer and q is natural (e.g. −6/8)
Irrational numbers:	real numbers with no exact value (e.g. π)
Infinity:	(symbol ∞) is larger than any number (technically not a number as it does not obey the laws of algebra)
Operations and symbols	
Basic operators:	+, −, × and ÷ will not need explanation; however, / may substitute for ÷, ∗ may substitute for × or this operator may be omitted
Powers:	a^n, i.e. 'a to the power n', means a multiplied by itself n times (e.g. $a^2 = a \times a =$ 'a squared', $a^3 = a \times a \times a =$ 'a cubed'). n is said to be the index or exponent. Note $a^0 = 1$ and $a^1 = a$
Logarithms:	the common logarithm (log) of any number x is the power to which 10 would have to be raised to give x (i.e. the log of 100 is 2; $10^2 = 100$); the antilog of x is 10^x. Note that there is no log for 0, so take this into account when drawing log axes by breaking the axis. Natural or Napierian logarithms (ln) use the base e ($= 2.71828\ldots$) instead of 10
Reciprocals:	the reciprocal of a real number a is $1/a$ ($a \neq 0$)
Relational operators:	$a > b$ means 'a is greater (more positive) than b', $<$ means less than, $\leqslant$ means less-than-or-equal-to and $\geqslant$ means greater-than-or-equal-to
Proportionality:	$a \propto b$ means 'a is proportional to b' (i.e. $a = kb$, where k is a constant). If $a \propto 1/b$, a is inversely proportional to b ($a = k/b$)
Sums:	Σx_i is shorthand for the sum of all x values from $i = 0$ to $i = n$ (more correctly the range of the sum is specified under the symbol)
Moduli:	$|x|$ signifies modulus of x, i.e. its absolute value (e.g. $|4| = |-4| = 4$)
Factorials:	$x!$ signifies factorial x, the product of all integers from 1 to x (e.g. $3! = 6$). Note $0! = 1! = 1$

Steps in tackling a numerical problem

The step-by-step approach outlined below may not be the fastest method of arriving at an answer, but most mistakes occur where steps are missing, combined or not made obvious, so a logical approach is often better.

Have the right tools ready

Scientific calculators (p. 117) greatly simplify the numerical part of problem-solving. However, the seeming infallibility of the calculator may lead you to accept an absurd result that could have arisen because of faulty key-pressing or faulty logic. Make sure you know how to use all the features on your calculator, especially how the memory works; how to introduce a constant multiplier or divider; and how to obtain an exponent (note that the 'exp' button on most calculators gives you 10^x, not 1^x or y^x; so 1×10^6 would be entered as $\boxed{1}\boxed{\text{exp}}\boxed{6}$, *not* $\boxed{10}\boxed{\text{exp}}\boxed{6}$).

Approach the problem thoughtfully

If the individual steps have been laid out on a worksheet, the 'tactics' will already have been decided. It is more difficult when you have to adopt a strategy on your own, especially if the problem is presented as a story and it isn't obvious which equations or rules need to be applied.

- Read the problem carefully as the text may give clues as to how it should be tackled. Be certain of what is required as an answer before starting.
- Analyse what kind of problem it is, which effectively means deciding which equation(s) or approach will be applicable. If this is not obvious, consider the dimensions/units of the information available and think how they could be fitted to a relevant formula. In examinations, a favourite ploy of examiners is to present a problem such that the familiar form of an equation must be rearranged (see Table 64.2 and Box 64.1). Another is to make you use two or more equations in series (see Box 64.2). If you are unsure whether a recalled formula is correct, a dimensional analysis can help: write in all the units for the variables and make sure that they cancel out to give the expected answer.
- Check that you have, or can derive, all of the information required to use your chosen equation(s). It is unusual but not unknown for examiners to supply redundant information. So, if you decide not to use some of the information given, be sure why you do not require it.
- Decide in what format and units the answer should be presented. This is sometimes suggested to you. If the problem requires many changes in the prefixes to units, it is a good idea to convert all data to base SI units (multiplied by a power of 10) at the outset.
- If a problem appears complex, break it down into component parts.

Present your answer clearly

The way you present your answer obviously needs to fit the individual problem. The example shown in Box 64.2 has been chosen to illustrate several important points, but this format would not fit all situations. Guidelines for presenting an answer include:

(a) Make your assumptions explicit. Most mathematical models of biological phenomena require that certain criteria are met before they can be

---

Tracing errors in mathematical problems – this is always easier when all the stages in a calculation are laid out clearly.

Using a spreadsheet for numerical problems – this may be very useful in repetitive work or for 'what if?' case studies (see Chapter 11).

Table 64.2 Simple algebra – rules for manipulating

If $a = b + c$, then $b = a - c$ and $c = a - b$
If $a = b \times c$, then $b = a \div c$ and $c = a \div b$
If $a = b^c$, then $b = a^{1/c}$ and $c = \log a \div \log b$
$a^{1/n} = \sqrt[n]{a}$
$a^{-n} = 1 \div a^n$
$a^b \times a^c = a^{(b+c)}$ and $a^b \div a^c = a^{(b-c)}$
$(a^b)^c = a^{(b \times c)}$
$a \times b = \text{antilog}(\log a + \log b)$

Presenting calculations in assessed work – always show the steps in your calculations; most markers will only penalise a mistake once and part marks will be given if the remaining operations are performed correctly. This can only be done if those operations are visible.

> **Box 64.1 Example of using the algebraic rules of Table 64.2**
>
> **Problem: if $a = (b - c) \div (d + e^n)$, find e**
>
> 1. Multiply both sides by $(d + e^n)$; formula becomes: $a(d + e^n) = (b - c)$
>
> 2. Divide both sides by a; formula becomes:
> $$d + e^n = \frac{b - c}{a}$$
>
> 3. Subtract d from both sides; formula becomes:
> $$e^n = \frac{b - c}{a} - d$$
>
> 4. Raise each side to the power $1/n$; formula becomes:
> $$e = \left\{\frac{b - c}{a} - d\right\}^{1/n}$$

legitimately applied (e.g. 'assuming the tissue is homogeneous ...'), while some approaches involve approximations that should be clearly stated (e.g. 'to estimate the mouse's skin area, its body was approximated to a cylinder with radius x and height y ...').

(b) Explain your strategy for answering, perhaps giving the applicable formula or definitions that suit the approach to be taken. Give details of what the symbols mean (and their units) at this point.

(c) Rearrange the formula to the required form with the desired unknown on the left-hand side (see Table 64.2).

(d) Substitute the relevant values into the right-hand side of the formula, using the units and prefixes as given (it may be convenient to convert values to SI beforehand). Convert prefixes to appropriate powers of 10 as soon as possible.

(e) Convert to the desired units step by step, i.e. taking each variable in turn.

(f) When you have the answer in the desired units, rewrite the left-hand side and underline the answer for emphasis. Make sure that the result is presented to an appropriate number of significant figures (see below).

Units – never write any answer without its unit(s) unless it is truly dimensionless.

Check your answer

Having written out your answer, you should check it methodically, answering the following questions:

- Is the answer realistic? You should be alerted to an error if a number is absurdly large or small. In repeated calculations, a result standing out from others in the same series should be double-checked.
- Do the units make sense and match up with the answer required? Don't, for example, present a volume in units of m^2.
- Do you get the same answer if you recalculate in a different way? If you have time, recalculate the answer using a different 'route', entering the numbers into your calculator in a different form and/or carrying out the operations in a different order.

Rounding: decimal places and significant figures

Rounding off to a specific number of significant figures – do not round off numbers until you arrive at the final answer or you will introduce errors into the calculation.

In many instances, the answer you produce as a result of a calculation will include more figures than is justified by the accuracy and precision of the original data. Sometimes you will be asked to produce an answer to a specified number of decimal places or significant figures and other times you will be expected to decide for yourself what would be appropriate.

Box 64.2 Model answer to a typical biological problem

Problem

Estimate the total length and surface area of the fibrous roots on a maize seedling from measurements of their total fresh weight and mean diameter. Give your answers in m and cm² respectively to four significant figures.

Measurements

Fresh weight[a] = 5.00 g, mean diameter[b] = 0.5 mm.

Answer

Assumptions: (1) the roots are cylinders with constant radius[c] and the 'ends' have negligible area; (2) the root system has a density of 1 000 kg m⁻³ (i.e. that of water[d]).

Strategy: from assumption (1), the applicable equations are those concerned with the volume and surface area of a cylinder (Table 64.3), namely:

$$V = \pi r^2 h \qquad [64.1]$$
$$A = 2\pi r h \text{ (ignoring ends)} \qquad [64.2]$$

where V is volume (m³), A is surface area (m²), $\pi \approx 3.14159$, h is height (m) and r is radius (m). The total length of the root system is given by h and its surface area by A. We can find h by rearranging eqn [64.1] and then substitute its value in eqn [64.2] to get A.

To calculate total root length: rearranging eqn [64.1], we have $h = V/\pi r^2$. From measurements[e], $r = 0.25$ mm $= 0.25 \times 10^{-3}$ m.

From density = weight/volume,

V = fresh weight/density
 = 5 g/1000 kg m⁻³
 = 0.005 kg/1000 kg m⁻³
 = 5×10^{-6} m³

Total root length,

$h = V/\pi r^2$
5×10^{-6} m³/3.14159 × (0.25 × 10⁻³ m)²

∴ Total root length = 25.46 m

To calculate surface area of roots: substituting value for h obtained above into eqn [64.2], we have:

Root surface area
 = 2 × 3.14159 × 0.25 × 10⁻³ m × 25.46 m
 = 0.04 m²
 = 0.04 × 10⁴ cm²
 (there being 100 × 100 = 10⁴ cm² per m²)

∴ Root surface area = 400.0 cm²

Notes

(a) The fresh weight of roots would normally be obtained by washing the roots free of soil, blotting them dry and weighing.

(b) In a real answer you might show the replicate measurements giving rise to the mean diameter.

(c) In reality, the roots will differ considerably in diameter and each root will not have a constant diameter throughout its length.

(d) This will not be wildly inaccurate as about 95 per cent of the fresh weight will be water, but the volume could also be estimated from water displacement measurements.

(e) Note conversion of measurements into base SI units at this stage and on line 3 of the root volume calculation. Forgetting to halve diameter measurements where radii are required is a common error.

 KEY POINT Do not simply accept the numerical answer from a calculator or spreadsheet, without considering whether you need to modify this to give an appropriate number of significant figures or decimal places.

Rounding to n decimal places

This is relatively easy to do.

1. Look at the number to the right of the nth decimal place.
2. If this is less than five, simply 'cut off' all numbers to the right of the nth decimal place to produce the answer (i.e. round down).
3. If the number is greater than five, 'cut off' all numbers to the right of the nth decimal place and add one to the nth decimal place to produce the answer (i.e. round up).

Examples
The number 4.123 correct to two decimal places is 4.12
The number 4.126 correct to two decimal places is 4.13
The number 4.1251 correct to two decimal places is 4.13
The number 4.1250 correct to two decimal places is 4.12
The number 4.1350 correct to two decimal places is 4.14
The number 99.99 correct to one decimal place is 100.0.

4. If the number is 5, then look at further numbers to the right to determine whether to round up or not.
5. If the number is *exactly* 5 and there are no further numbers to the right, then round to the nearest even number. *Note:* When considering a large number of calculations, this procedure will not affect the overall mean value. Some rounding systems do the opposite to this (i.e. round to the nearest odd number), while others always round up where the number is exactly 5 (which *will* affect the mean). Take advice from your tutor and stick to one system throughout a series of calculations.

Whenever you see any number quoted, you should assume that the last digit has been rounded. For example, in the number 22.4, the '.4' is assumed to be rounded and the calculated value may have been between 22.35 and 22.45.

Quoting to n significant figures

The number of significant figures indicates the degree of approximation in the number. For most cases, it is given by counting all the figures except zeros that occur at the beginning or end of the number. Zeros *within* the number are always counted as significant. The number of significant figures in a number like 200 is ambiguous and could be one, two or three; if you wish to specify clearly, then quote as e.g. 2×10^2 (one significant figure), 2.0×10^2 (two significant figures), etc. to avoid spurious accuracy (pp. 157, 400). When quoting a number to a specified number of significant figures, use the same rules as for rounding to a specified number of decimal places, but do not forget to keep zeros before or after the decimal point. The same principle is used if you are asked to quote a number to the 'nearest 10', 'nearest 100', etc.

Examples
The number of significant figures in 194 is 3
The number of significant figures in 2305 is 4
The number of significant figures in 0.003482 is 4
The number of significant figures in 210×10^8 is 3 (21×10^9 would be 2).

When deciding for yourself how many significant figures to use, adopt the following rules of thumb:

- Always round *after* you have done a calculation. Use *all* significant figures available in the measured data during a calculation.
- If adding or subtracting with measured data, then quote the answer to the number of decimal places in the data value with the least number of decimal places (e.g. $32.1 - 45.67 + 35.6201 = 22.1$, because 32.1 has one decimal place).
- If multiplying or dividing with measured data, keep as many significant figures as are in the number with the least number of significant places (e.g. $34901 \div 3445 \times 1.3410344 = 13.59$, because 3445 has four significant figures).
- For the purposes of significant figures, assume 'constants' (e.g. number of mm in a metre) have an infinite number of significant figures.

Examples
The number of significant figures in 3051.93 is 6
To five significant figures, this number is 3051.9
To four significant figures, this number is 3052
To three significant figures, this number is 3050
To two significant figures, this number is 3100
To one significant figure, this number is 3000
3051.93 to the nearest 10 is 3050
3051.93 to the nearest 100 is 3100
Note that in this last case you must include the zeros before the decimal point to indicate the scale of the number (even if the decimal point is not shown). For a number less than 1, the same would apply to the zeros before the decimal point. For example, 0.00305193 to three significant figures is 0.00305. Alternatively, use scientific notation (in this case, 3.05×10^{-3}).

Some reminders of basic mathematics

Errors in calculations sometimes appear because of faults in mathematics rather than computational errors. For reference purposes, Tables 64.1–64.3 give some basic mathematical principles that may be useful. Eason et al. (1992) or Stephenson (2003) should be consulted for more advanced/specific needs.

Table 64.3 Geometry and trigonometry – analysing shapes

Shape/object	Diagram	Perimeter	Area
Two-dimensional shapes			
Square		$4x$	x^2
Rectangle		$2(x+y)$	xy
Circle		$2\pi r$	πr^2
Ellipse		$\pi[1.5(a+b) - \sqrt{a*b}]$ (approx.)	πab
Triangle (general)		$x+y+z$	$0.5zh$
(right-angled)		$x+y+r$ $\sin\theta = y/r$, $\cos\theta = x/r$, $\tan\theta = y/x$; $r^2 = x^2 + y^2$	$0.5xy$

Shape/object	Diagram	Surface area	Volume
Three-dimensional shapes			
Cube		$6x^2$	x^3
Cuboid		$2xy + 2xz + 2yz$	xyz
Sphere		$4\pi r^2$	$4\pi r^3/3$
Ellipsoid		no simple formula	πrab
Cylinder		$2\pi rh + 2\pi r^2$	$\pi r^2 h$
Cone and pyramid		$0.5PL + B$	$BL/3$

Key: x, y, z = sides a, b = half minimum and maximum axes; r = radius or hypotenuse; h = height; B = base area; L = perpendicular height; P = perimeter of base.

Percentages and proportions

A percentage is just a fraction expressed in terms of hundredths, indicated by putting the percentage sign (%) after the number of hundredths. So 35% simply means 35 hundredths. To convert a fraction to a percentage, just multiply the fraction by 100. When the fraction is in decimal form, multiplying by 100 to obtain a percentage is easily achieved just by moving the decimal point two places to the right.

> **Examples**
> 1/8 as a percentage is $1 \div 8 \times 100 = 100 \div 8 = 12.5\%$
> 0.602 as a percentage is $0.602 \times 100 = 60.2\%$.

To convert a percentage to a fraction, just remember that, since a percentage is a fraction multiplied by 100, the fraction is the percentage divided by 100. For example: $42\% = 42/100 = 0.42$. In this example, since we are dealing with a decimal fraction, the division by 100 is just a matter of moving the decimal point two places to the left (42% could be written as 42.0%). Percentages greater than 100% represent fractions greater than 1. Percentages less than 1 may cause confusion. For example, 0.5% means half of one per cent (0.005) and must not be confused with 50% (which is the decimal fraction 0.5).

> **Examples**
> 190% as a decimal fraction is $190 \div 100 = 1.9$
> 5/2 as a percentage is $5 \div 2 \times 100 = 250\%$.

To find a percentage of a given number, just express the percentage as a decimal fraction and multiply the given number. For example: 35% of 500 is given by $0.35 \times 500 = 175$. To find the percentage change in a quantity, work out the difference (= value 'after' − value 'before'), and divide this difference by the original value to give the fractional change, then multiply by 100.

> **Example** A population falls from 4 million to 3.85 million. What is the percentage change? The decrease in numbers is $4 - 3.85 = -0.15$ million. The fractional decrease is $-0.15 \div 4 = -0.0375$ and we multiply by 100 to get the percentage change = minus 3.75%.

Exponents

Exponential notation is an alternative way of expressing numbers in the form a^n ('a to the power n'), where a is multiplied by itself n times. The number a is called the base and the number n the exponent (or power or index). The exponent need not be a whole number, and it can be negative if the number being expressed is less than 1. See Table 64.2 for other mathematical relationships involving exponents.

> **Example** $2^3 = 2 \times 2 \times 2 = 8$.

Scientific notation

In scientific notation, also known as 'standard form', the base is 10 and the exponent a whole number. To express numbers that are not whole powers of 10, the form $c \times 10^n$ is used, where the coefficient c is normally between 1 and 10. Scientific notation is valuable when you are using very large numbers and wish to avoid suggesting spurious accuracy. Thus if you write 123 000, this may suggest that you know the number to ±0.5, whereas 1.23×10^5 might give a truer indication of measurement accuracy (i.e. implied to be ±500 in this case). Engineering notation is similar, but treats numbers as powers of 10 in groups of 3, i.e. $c \times 10^0$, 10^3, 10^6, 10^9, etc. This corresponds to the SI system of prefixes (p. 159).

> **Example** Avogadro's number, ≈ 602 352 000 000 000 000 000 000, is more conveniently expressed as $6.023\,52 \times 10^{23}$.

A useful property of powers when expressed to the same base is that when multiplying two numbers together, you simply add the powers, while if dividing, you subtract the powers. Thus, suppose you counted 8 bacteria in a known volume of a 10^{-7} dilution (see p. 132), there would be 8×10^7 in the same volume of undiluted solution; if you now dilute this 500-fold (5×10^2), then the number present in the same volume would be $8/5 \times 10^{(7-2)} = 1.6 \times 10^5 = 160\,000$.

Logarithms

When a number is expressed as a logarithm, this refers to the power n that the base number a must be raised to give that number. Any base could be used, but the two most common are 10, when the power is referred to as $\log_{10}$ or simply log, and the constant e (2.718282), used for mathematical convenience in certain situations, when the power is referred to as $\log_e$ or ln (natural logarithm). Note that (a) logs need not be whole numbers; (b) there is no log value for the number zero; and (c) that $\log_{10} = 0$ for the number 1.

To obtain logs, you will need to use the log key on your calculator, or special log tables (now largely redundant). To convert back (antilog), use

- the $\boxed{10^x}$ key, with $x = $ log value;
- the $\boxed{\text{inverse}}$ then the $\boxed{\log}$ key; or
- the $\boxed{y^x}$ key, with $y = 10$ and $x = $ log value.

If you have used log tables, you will find complementary antilogarithm tables to do this.

There are many uses of logarithms in biology, including pH ($= -\log[H^+]$), where $[H^+]$ is expressed in $mol\,l^{-1}$ (see p. 146); the exponential growth of micro-organisms, where if log(cell number) is plotted against time, a straight-line relationship is obtained (see p. 272); and allometric studies of growth and development, where if data are plotted on log axes, a series of straight-line relationships may be found (see p. 294).

Examples The logarithm to the base 10 ($\log_{10}$) of 1000 is 3, since $10^3 = 1000$. The logarithm to the base e ($\log_e$ or ln) of 1000 is 6.907755 (to six decimal places).

Examples (use to check the correct use of your own calculator)
102963 as a $\log_{10} = 5.012681$ (to six decimal places)
$10^{5.012681} = 102962.96$
(Note loss of accuracy due to loss of decimal places.)
102963 as a natural logarithm (ln) $= 11.542125$ (to six decimal places), thus $2.718282^{11.542125} = 102963$.

Linear functions and straight lines

One of the most straightforward and widely used relationships between two variables x and y is that represented by a straight-line graph, where the corresponding mathematical function is known as the equation of a straight line, where:

$$y = a + bx \quad \quad [64.3]$$

In this relationship, a represents the intercept of the line on the y (vertical) axis, i.e. where $x = 0$, and therefore $bx = 0$, while b is equivalent to the slope (gradient) of the line, i.e. the change in y for a change in x of 1. The constants a and b are sometimes given alternative symbols, but the mathematics remains unchanged, e.g. in the equivalent expression for the slope of a straight line, $y = mx + c$. Figure 64.1 shows what happens when these two constants are changed, in terms of the resultant straight lines.

The two main applications of the straight-line relationship are:

1. Function fitting. Here, you determine the mathematical form of the function, i.e. you estimate the constants a and b from a data set for x and y, either by drawing a straight line by eye (p. 394) and then working out the slope and y intercept, or by using linear regression (p. 435) to obtain the most probable values for both constants. When putting a straight line of best fit by eye on a hand-drawn graph, note the following:

    - Always use a *transparent* ruler, so you can see data points on either side of the line.
    - For a data series where the points do not fit a perfect straight line, try to have an equal number of points on either side of the line, as in

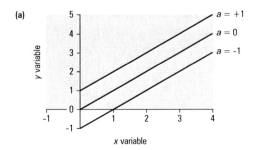
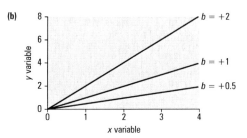

Fig. 64.1. Straight-line relationships ($y = a + bx$), showing the effects of (a) changing the intercept at constant slope, and (b) changing the slope at constant intercept.

Fig. 64.2(a), and try to minimise the average distance of these points from the line.

- Once you have drawn the line of best fit use this line, rather than your data values, in all subsequent procedures (e.g. in a calibration curve, Chapter 50).
- Tangents drawn to a curve give the slope (gradient) at a particular point, e.g. in an enzyme reaction progress curve, Fig. 52.2. These are best drawn by bringing your ruler up to the curve at the exact point where you wish to estimate the slope and then trying to make the two angles immediately on either side of this point approximately the same, by eye (Fig. 64.2(b)).
- Once you have drawn the straight line or tangent, choose two points reasonably far apart at either end of your line and then draw construction lines to represent the change in y and the change in x between these two points: make sure that your construction lines are perpendicular to each other. Determine the slope as the change in y divided by the change in x (Fig. 64.2).

2. Prediction. Where a and b are known, or have been estimated, you can use eqn [64.3] to predict any value of y for a specified value of x, e.g. during exponential growth of a cell culture (p. 272), where $\log_{10}$ cell number (y) increases as a linear function of time (x): note that in this example the dependent variable has been transformed to give a linear relationship. You will need to rearrange eqn [64.3] in cases where a prediction of x is required for a particular value of y (e.g. in calibration curves, p. 303, or bioassays, p. 295), as follows:

$$x = (y - a) \div b \qquad [64.4]$$

This equation can also be used to determine the intercept on the x (horizontal) axis, i.e. where $y = 0$.

Hints for some typical problems

Calculations involving proportions or ratios

The 'unitary method' is a useful way of approaching calculations involving proportions or ratios, such as those required when making up solutions from stocks (see also Chapter 23) or as a subsidiary part of longer calculations.

1. If given a value for a multiple, work out the corresponding value for a single item or unit.
2. Use this 'unitary value' to calculate the required new value.

Calculations involving series

Series (used in e.g. dilutions, see also p. 131) can be of three main forms:

1. Arithmetic, where the *difference* between two successive numbers in the series is a constant, e.g. 2, 4, 6, 8, 10, ...
2. Geometric, where the *ratio* between two successive numbers in the series is a constant, e.g. 1, 10, 100, 1000, 10000, ...
3. Harmonic, where the values are reciprocals of successive whole numbers, e.g. $1, \frac{1}{2}, \frac{1}{3}, \frac{1}{4}, \ldots$

Note that the logs of the numbers in a geometric series will form an arithmetic series (e.g. 0, 1, 2, 3, 4, ... in the above case). Thus, if a

---

Examples
Using eqn [64.3], the predicted value for y for a linear function where $a = 2$ and $b = 0.5$, where $x = 8$ is: $y = 2 + (0.5 \times 8) = 6$.
Using eqn [64.4], the predicted value for x for a linear function where $a = 1.5$ and $b = 2.5$, where $y = -8.5$ is:
$x = (-8.5 - 1.5) \div 2.5 = -4$.
Using eqn [64.4] the predicted x intercept for a linear function where $a = 0.8$ and $b = 3.2$ is: $x = (0 - 0.8) \div 3.2 = -0.25$.

Example A lab schedule states that 5 g of a compound with a relative molecular mass of 220 are dissolved in 400 ml of solvent. For writing up your Materials and Methods, you wish to express this as $mol\,l^{-1}$.
1. If there are 5 g in 400 ml, then there are $5 \div 400$ g in 1 ml.
2. Hence, 1000 ml will contain $5 \div 400 \times 1000$ g = 12.5 g.
3. $12.5\,g = 12.5 \div 220\,mol = 0.0568\,mol$, so [solution] = $56.8\,mmol\,l^{-1}$ ($= 56.8\,mol\,m^{-3}$).

Examples For a geometric dilution series involving ten-fold dilution steps, calculation of concentrations is straightforward, e.g. two serial decimal dilutions (= 100-fold dilution) of a solution of NaCl of $250\,mmol\,l^{-1}$ will produce a dilute solution of $250 \div 100 = 2.5\,mmol\,l^{-1}$. Similarly, for an arithmetic dilution series, divide by the overall dilution to give the final concentration, e.g. a sixteen-fold dilution of a solution of NaCl of $200\,mg\,ml^{-1}$ will produce a dilute solution of $200 \div 16 = 12.5\,mg\,ml^{-1}$.

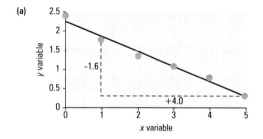
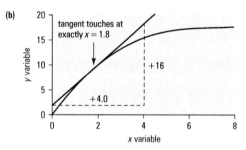

Fig. 64.2. Drawing straight lines. (a) Simple linear relationship, giving a straight line with an intersect of 2.3 and a slope of $-1.6 \div 4.0 = 0.4$. (b) Tangent drawn to a curve at $x = 1.8$, giving a slope of $16 \div 4 = 4$.

quantity y varies with a quantity x such that the rate of change in y is proportional to the value of y (i.e. it varies in an exponential manner), a semi-log plot of such data will form a straight line. This form of relationship is relevant for exponentially growing cell cultures (p. 272) and radioactive decay (p. 373).

Statistical calculations

The need for long, complex calculations in statistics has largely been removed because of the widespread use of spreadsheets with statistical functions (Chapter 11) and specialised programs such as SPSS and Minitab. It is, however, important to understand the principles behind what you are trying to do (see Chapters 65 and 66) and interpret the program's output correctly, either using the 'help' function or a reference manual.

Problems in Mendelian genetics

These cause difficulties for many students. The key is to recognise the different types of problems and to practise so you are familiar with the techniques for solving them. Chapter 53 deals with the different types of crosses you will come across and methods of analysing them, including the use of the χ^2 (Chi^2) test.

Text references

Eason, G., Coles, C.W. and Gettinby, G. (1992) *Mathematics and Statistics for the Bio-Sciences*. Ellis Horwood, Chichester.

Stephenson, F.H. (2003) *Calculations in Molecular Biology and Biotechnology*. Academic Press, London.

Sources for further study

Anon. *S.O.S. Mathematics*. Available: http://www.sosmath.com/
Last accessed 09/04/07.
[A basic Web-based guide with very wide coverage.]

Britton, N.F. (2003) *Essential Mathematical Biology*. Springer-Verlag. New York.
[Describes a range of mathematical applications in population dynamics, epidemiology, genetics, biochemistry and medicine.]

Cann, A.J. (2002) *Maths from Scratch for Biologists*. Wiley, Chichester.
[Deals with basic manipulations, formulae, units, molarity, logs and exponents as well as basic statistical procedures.]

Causton, D.R. (1992) *A Biologist's Basic Mathematics*. Cambridge University Press, Cambridge.

Forster, P.C. (1999) *Easy Mathematics for Biologists*. Harwood, Amsterdam.
[Covers basic principles, units, logarithms, ratios and proportions, concentrations, equations, rates and graphs.]

Harris, M., Taylor, G. and Taylor, J. (2005) *Catch Up Maths and Stats for the Life and Medical Sciences*. Scion, Bloxham.
[Covers a range of basic mathematical operations and statistical procedures.]

Koehler, K.R. *College Physics for Students of Biology and Chemistry*. Available: http://www.rwc.uc.edu/koehler/biophys/text.html.
Last accessed 09/04/07.
[A 'hypertextbook' written for first-year undergraduates. Assumes that you have a working knowledge of algebra.]

Lawler, G. (2006) *Understanding Maths. Basic Mathematics Explained*. Studymates, Abergele.

Study exercises

64.1 Rearrange the following formulae.
(a) If $y = ax + b$, find b
(b) If $y = ax + b$, find x
(c) If $x = y^3$, find y
(d) If $x = 3^y$, find y
(e) If $x = (1-y)(z^p + 3)$, find z
(f) If $x = (y-z)^{1/n}/pq$, find n

64.2 Work with decimal places or significant figures. Give the following numbers to the accuracy indicated:

(a) 214.51 to three significant figures
(b) 107 029 to three significant figures
(c) 0.0450 to one significant figure
(d) 99.817 to two decimal places
(e) 99.897 to two decimal places
(f) 99.997 to two decimal places
(g) 6255 to the nearest 10
(h) 134 903 to the nearest ten thousand

State the following:
(i) the number of significant figures in 3400
(j) the number of significant figures in 3400.3
(k) the number of significant figures in 0.00167
(l) the number of significant figures in 1.00167
(m) the number of decimal places in 34.46
(n) the number of decimal places in 0.00167

64.3 Carry out calculations involving percentages. Answer the following questions, giving your answers to two decimal places:

(a) What is 6/35ths expressed as a percentage?
(b) What is 0.0755 expressed as a percentage?
(c) What is 4.35% of 322?
(d) A rat's weight increased from 55.23 g to 75.02 g. What is the percentage increase in its weight?

64.4 Practise using exponents and logarithms.
(a) Write out (i) the charge on an electron and (ii) the speed of light *in vacuo* in longhand (i.e. without using powers of 10). See Table 26.4 for values given in scientific notation.
(b) Compute the following values: $10^{3.624}$; $\log(6.37)$; $e^{-2.32}$; $\ln(1123)$; $6^{3.2}$. A calculator should be used, but round the output to give five significant figures.

64.5 Practise using geometric formulae. Select appropriate formulae from Table 64.3, and use them to compare the surface-area-to-volume ratios of the following:

(a) a spherical unicellular alga of diameter 30 µm;
(b) a cylindrical root of radius 0.5 mm (ignoring ends).

64.6 Practise working with linear functions (note also that Chapter 50 includes study exercises based on linear functions and plotting straight lines). Assuming a linear relationship between x and y, calculate the following (give your answers to three significant figures):

(a) x, where $y = 7.0$, $a = 4.5$ and $b = 0.02$;
(b) x, where $y = 15.2$, $a = -2.6$ and $b = -4.46$;
(c) y, where $x = 10.5$, $a = 0.2$ and $b = -0.63$;
(d) y, where $x = 4.5$, $a = -1.8$ and $b = 4.1$.

65 Descriptive statistics

Whether obtained from observation or experimentation, most data in biology exhibit variability. This can be displayed as a frequency distribution (e.g. Fig. 62.7). Descriptive (or summary) statistics quantify aspects of the frequency distribution of a sample. You can use them to:

- condense a large data set for presentation in figures or tables;
- provide estimates of parameters of the frequency distribution of the population being sampled (p. 427).

 KEY POINT *The appropriate descriptive statistics to choose depend on both the type of data, i.e. quantitative, ranked, or qualitative, and the nature of the underlying population frequency distribution.*

If you have no clear theoretical grounds for assuming what the underlying frequency distribution is like, graph one or more sample frequency distributions, ideally with a sample size >100. The tally system for recording data (see p. 383) can give an immediate visual indication of the frequency distribution as data are collected.

The methods used to calculate descriptive statistics depend on whether data have been grouped into classes. You should use the original data set if it is still available, because grouping into classes loses information and accuracy. However, large data sets may make calculations unwieldy, and are best handled using computer programs.

Three important features of a frequency distribution that can be summarised by descriptive statistics are:

- the sample's location, i.e. its position along a given dimension representing the dependent (measured) variable (Fig. 65.1);
- the dispersion of the data, i.e. how spread out the values are (Fig. 65.2);
- the shape of the distribution, i.e. whether symmetrical, skewed, U-shaped, etc. (Fig. 65.3).

Measuring location

Here, the objective is to pinpoint the 'centre' of the frequency distribution, i.e. the value about which most of the data are grouped. The chief measures of location are the mean, median and mode. Figure 65.4 shows how to choose among these for a given data set.

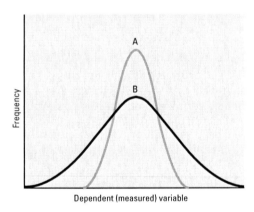

Fig. 65.1 Two distributions with different locations but the same dispersion. The data set labelled B could have been obtained by adding a constant to each datum in the data set labelled A.

Fig. 65.2 Two distributions with different dispersions but the same location. The data set labelled A covers a relatively narrow range of values of the dependent (measured) variable, while that labelled B covers a wider range.

Example Box 65.1 shows a set of data and the calculated values of the measures of location, dispersion and shape for which methods of calculation are outlined here. Check your understanding by calculating the statistics yourself and confirming that you arrive at the same answers.

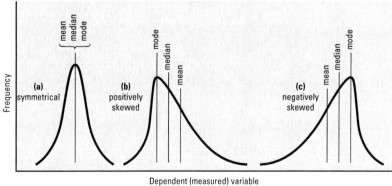

Fig. 65.3 Symmetrical and skewed frequency distributions, showing relative positions of mean, median and mode.

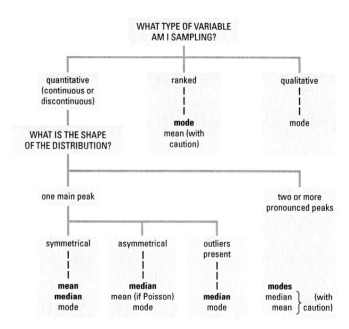

Fig. 65.4 Choosing a statistic for characterising a distribution's location. Statistics written in bold are the preferred option(s).

Use of symbols – Y is used in Chapters 65 and 66 to signify the dependent variable in statistical calculations (following the example of Sokal and Rohlf, 1994, Heath, 1995 and Wardlaw, 2000). Note, however, that some authors use X or x in analogous formulae and many calculators refer to e.g. $\bar{x}$, Σx^2, etc., for their statistical functions.

Definition

Rank – the position of a data value when all the data are placed in order of ascending magnitude. If ties occur, an average rank of the tied variates is used. Thus, the rank of the datum 6 in the sequence 1,3,5,6,8,8,10 is 4; the rank of each datum with value 8 is 5.5.

Definition

An outlier – any datum that has a value much smaller or bigger than most of the data.

Mean

The mean (denoted $\bar{Y}$ and also referred to as the arithmetic mean) is the average value of the data. It is obtained from the sum of all the data values divided by the number of observations (in symbolic terms, $\Sigma Y/n$). The mean is a good measure of the centre of symmetrical frequency distributions. It uses all of the numerical values of the sample and therefore incorporates all of the information content of the data. However, the value of a mean is greatly affected by the presence of outliers (extreme values). The arithmetic mean is a widely used statistic in biology, but there are situations when you should be careful about using it (see Box 65.2 for examples).

Median

The median is the mid-point of the observations when ranked in increasing order. For odd-sized samples, the median is the middle observation; for even-sized samples it is the mean of the middle pair of observations. Where data are grouped into classes, the median can only be estimated. This is most simply done from a graph of the cumulative frequency distribution, but can also be worked out by assuming the data to be evenly spread within the class. The median may represent the location of the main body of data better than the mean when the distribution is asymmetric or when there are outliers in the sample.

Mode

The mode is the most common value in the sample. The mode is easily found from a tabulated frequency distribution as the most frequent value. If data have been grouped into classes then the term modal class is used for the class containing most values. The mode provides a rapidly and easily found estimate of sample location and is unaffected by outliers. However,

Box 65.1 Descriptive statistics for an illustrative sample of data

Value (Y)	Frequency (f)	Cumulative frequency	fY	fY²
1	0	0	0	0
2	1	1	2	4
3	2	3	6	18
4	3	6	12	48
5	8	14	40	200
6	5	19	30	180
7	2	21	14	98
8	0	21	0	0
Totals	$21 = \Sigma f (=n)$		$104 = \Sigma fY$	$548 = \Sigma fY^2$

In this example, for simplicity and ease of calculation, integer values of Y are used. In many practical exercises, where continuous variables are measured to several significant figures and where the number of data values is small, giving frequencies of 1 for most of the values of Y, it may be simpler to omit the column dealing with frequency and list all the individual values of Y and Y² in the appropriate columns. To gauge the underlying frequency distribution of such data sets, you would need to group individual data into broader classes (e.g. all values between 1.0 and 1.9, all values between 2.0 and 2.9, etc.) and then draw a histogram (p. 395). Calculation of certain statistics for data sets that have been grouped in this way (e.g. median, quartiles, extremes) can be tricky and a statistical text should be consulted.

Statistic	Value*	How calculated
Mean	4.95	$\Sigma fY / n$, i.e. 104/21
Median	5	Value of the $(n+1)/2$ variate, i.e. the value ranked $(21+1)/2 = 11$th (obtained from the cumulative frequency column)
Mode	5	The most common value (Y value with highest frequency)
Upper quartile	6	The upper quartile is between the 16th and 17th values, i.e. the value exceeded by 25% of the data values
Lower quartile	4	The lower quartile is between the 5th and 6th values, i.e. the value exceeded by 75% of the data values
Semi-interquartile range	1.0	Half the difference between the upper and lower quartiles, i.e. $(6-4)/2$
Upper extreme	7	Highest Y value in data set
Lower extreme	2	Lowest Y value in data set
Range	5	Difference between upper and lower extremes
Variance (s^2)	1.65	$s^2 = \dfrac{\Sigma fY^2 - (\Sigma fY)^2 / n}{n-1}$ $= \dfrac{548 - (104)^2/21}{20}$
Standard deviation (s)	1.28	$\sqrt{s^2}$
Standard error (SE)	0.280	$s/\sqrt{n}$
95% confidence limits	4.36 – 5.54	$\bar{Y} \mp t_{0.05}[20] \times SE$, (where $t_{0.05}[20] = 2.09$, Table 66.2)
Coefficient of variation (CoV)	25.9%	$100 s / \bar{Y}$

*Rounded to three significant figures (see p. 408), except when it is an exact number.

Describing the location of qualitative data – the mode is the only statistic that is suitable for this task. For example, 'the modal (most frequent) eye colour was crimson'.

the mode is affected by chance variation in the shape of a sample's distribution and it may lie distant from the obvious centre of the distribution.

The mean, median and mode have the same units as the variable under discussion. However, whether these statistics of location have the same or similar values for a given frequency distribution depends on the symmetry and shape of the distribution. If it is near-symmetrical with a single peak, all three will be very similar; if it is skewed or has more than one peak, their values will differ to a greater degree (see Fig. 65.3).

Measuring dispersion

Here, the objective is to quantify the spread of the data about the centre of the distribution. Figure 65.5 indicates how to decide which measure of dispersion to use.

Range

The range is the difference between the largest and smallest data values in the sample (the extremes) and has the same units as the measured variable. The range is easy to determine, but is greatly affected by outliers. Its value may also depend on sample size: in general, the larger this is, the greater will be the range. These features make the range a poor measure of dispersion for many practical purposes.

Example In a sample of data with values 3, 7, 15, 8, 5, 10 and 4, the range is 12 (i.e. the difference between the highest value, 15, and the lowest value, 3).

Semi-interquartile range

The semi-interquartile range is an appropriate measure of dispersion when a median is the appropriate statistic to describe location. For this, you need to determine the first and third quartiles, i.e. the medians for those

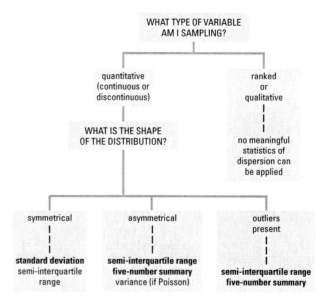

Fig. 65.5 Choosing a statistic for characterising a distribution's dispersion. Statistics written in bold are the preferred option(s). Note that you should match statistics describing dispersion with those you have used to describe location, i.e. standard deviation with mean, semi-interquartile range with median.

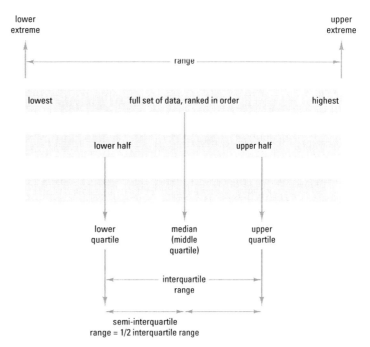

Fig. 65.6 Illustration of median, quartiles, range and semi-interquartile range

data values ranked below and above the median of the whole data set (see Fig. 65.6). To calculate a semi-interquartile range for a data set:

1. Rank the observations in ascending order.
2. Find the values of the first and third quartiles.
3. Subtract the value of the first quartile from the value of the third.
4. Halve this number.

For data grouped in classes, the semi-interquartile range can only be estimated. Another disadvantage is that it takes no account of the shape of the distribution at its edges. This objection can be countered by using the so-called 'five-number summary' of a data set, which consists of the three quartiles and the two extreme values; this can be presented on graphs as a box and whisker plot (see Fig. 65.7) and is particularly useful for summarising skewed frequency distributions. The corresponding 'six-number summary' includes the sample's size.

Variance and standard deviation

For symmetrical frequency distributions, an ideal measure of dispersion would take into account each value's deviation from the mean and provide a measure of the average deviation from the mean. Two such statistics are the sample variance, which is the sum of squares ($\Sigma(Y - \bar{Y})^2$) divided by $n - 1$ (where n is the sample size), and the sample standard deviation, which is the positive square root of the sample variance.

The variance (s^2) has units that are the square of the original units, while the standard deviation (s, or SD) is expressed in the original units, one reason s is often preferred as a measure of dispersion. Calculating s or s^2 longhand is a tedious job and is best done with the help of a calculator or

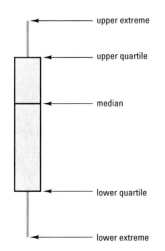

Fig. 65.7 A box and whisker plot, showing the 'five-number summary' of a sample as it might be used on a graph.

computer. If you don't have a calculator that calculates s for you, an alternative formula that simplifies calculations is:

$$s = +\sqrt{\frac{\Sigma Y^2 - (\Sigma Y)^2/n}{n-1}} \quad [65.1]$$

To calculate s using a calculator:

1. Obtain ΣY, square it, divide by n and store in memory.
2. Square Y values, obtain ΣY^2, subtract memory value from this.
3. Divide this answer by $n - 1$.
4. Take the positive square root of this value.

Take care to retain significant figures, or errors in the final value of s will result. If continuous data have been grouped into classes, the class mid-values or their squares must be multiplied by the appropriate frequencies before summation (see example in Box 65.1). When data values are large, longhand calculations can be simplified by coding the data, e.g. by subtracting a constant from each datum, and decoding when the simplified calculations are complete (see Sokal and Rohlf, 1994).

Coefficient of variation

The coefficient of variation (CoV) is a dimensionless measure of variability relative to location that expresses the sample standard deviation, usually as a percentage of the sample mean, i.e.

$$\text{CoV} = 100s/\bar{Y} \, (\%) \quad [65.2]$$

This statistic is useful when comparing the relative dispersion of data sets with widely differing means or where different units have been used for the same or similar quantities.

A useful application of the CoV is to compare different analytical methods or procedures, so that you can decide which involves the least proportional error – create a standard stock solution, then base your comparison on the results from several subsamples analysed by each method. You may find it useful to use the CoV to compare the precision of your own results with those of a manufacturer, e.g. for an autopippettor (p. 123). The smaller the CoV, the more precise (repeatable) is the apparatus or technique (note: this does not mean that it is necessarily more *accurate*, see p. 157).

Measuring the precision of the sample mean as an estimate of the true value using the standard error

Most practical exercises are based on a limited number of individual data values (a sample) that are used to make inferences about the population from which they were drawn. For example, the haemoglobin content might be measured in blood samples from 100 adult females and used as an estimate of the adult female haemoglobin content, with the sample mean ($\bar{Y}$) and sample standard deviation (s) providing estimates of the true values of the underlying population mean (μ) and the population standard deviation (σ). The reliability of the sample mean as an estimate of the true (population) mean can be assessed by calculating a statistic termed the standard error of the sample mean (often abbreviated to standard error or SE), from:

$$\text{SE} = s/\sqrt{n} \quad [65.3]$$

Using a calculator for statistics – make sure you understand how to enter individual data values and which keys will give the sample mean (usually shown as $\bar{X}$ or $\bar{x}$) and sample standard deviation (often shown as σ_{n-1}). In general, you should not use the population standard deviation (usually shown as σ_n).

Example Consider two methods of bioassay for a toxin in fresh water. For a given standard, Method A gives a mean result of = 50 'response units' with $s = 8$, while Method B gives a mean result of = 160 'response units' with $s = 18$. Which bioassay gives the more reproducible results? The answer can be found by calculating the CoV values, which are 16 per cent and 11.25 per cent respectively. Hence, Method B is the more precise, even though the absolute value of s is larger.

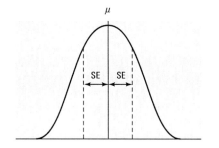

Fig. 65.8 Frequency distribution of sample means around the population mean (μ). Note that SE is equivalent to the standard deviation of the sample means, for sample size n.

Strictly, the standard error is an estimate of the dispersion of repeated sample means around the true (population) value: if several samples were taken, each with the same number of data values (n), then their means would cluster around the population mean (μ) with a standard deviation equal to SE, as shown in Fig. 65.8. Therefore, the *smaller* the SE, the more reliable the sample mean is likely to be as an estimate of the true value, since the underlying frequency distribution would be more tightly clustered around μ. At a practical level, eqn [65.3] shows that SE is directly affected by the dispersion of individual data values within the sample, as represented by the sample standard deviation (s). Perhaps more importantly, SE is inversely related to the *square root* of the number of data values (n). Therefore, if you wanted to increase the precision of a sample mean by a factor of 2 (i.e. to reduce SE by half), you would have to increase n by a factor of 2^2 (i.e. four-fold).

Summary descriptive statistics for the sample mean are often quoted as $\bar{Y} \pm \text{SE}(n)$, with the SE being given to one significant figure more than the mean. For example, summary statistics for the sample mean and standard error for the data shown in Box 65.1 would be quoted as 4.95 ± 0.280 ($n = 21$). You can use such information to carry out a t-test between two sample means (Box 66.1); the SE is also useful because it allows calculation of confidence limits for the sample mean (p. 434).

Describing the 'shape' of frequency distributions

Frequency distributions may differ in the following characteristics:

- number of peaks;
- skewness or asymmetry;
- kurtosis or pointedness.

The shape of a frequency distribution of a small sample is affected by chance variation and may not be a fair reflection of the underlying population frequency distribution: check this by comparing repeated samples from the same population or by increasing the sample size. If the original shape were due to random events, it should not appear consistently in repeated samples and should become less obvious as sample size increases.

Genuinely bimodal or polymodal distributions may result from the combination of two or more unimodal distributions, indicating that more than one underlying population is being sampled (Fig. 65.9). An example of a bimodal distribution is the height of adult humans (females and males combined).

A distribution is skewed if it is not symmetrical, a symptom being that the mean, median and mode are not equal (Fig. 65.3). Positive skewness is

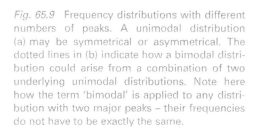

Fig. 65.9 Frequency distributions with different numbers of peaks. A unimodal distribution (a) may be symmetrical or asymmetrical. The dotted lines in (b) indicate how a bimodal distribution could arise from a combination of two underlying unimodal distributions. Note here how the term 'bimodal' is applied to any distribution with two major peaks – their frequencies do not have to be exactly the same.

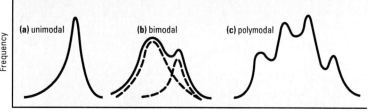

Box 65.2 Three examples where simple arithmetic means are inappropriate

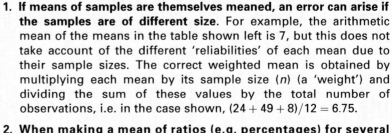

Mean	n
6	4
7	7
8	1

1. **If means of samples are themselves meaned, an error can arise if the samples are of different size.** For example, the arithmetic mean of the means in the table shown left is 7, but this does not take account of the different 'reliabilities' of each mean due to their sample sizes. The correct weighted mean is obtained by multiplying each mean by its sample size (n) (a 'weight') and dividing the sum of these values by the total number of observations, i.e. in the case shown, $(24 + 49 + 8)/12 = 6.75$.

2. **When making a mean of ratios (e.g. percentages) for several groups of different sizes, the ratio for the combined total of all the groups is not the mean of the proportions for the individual groups.** For example, if 20 rats from a batch of 50 are male, this implies 40% are male. If 60 rats from a batch of 120 are male, this implies 50% are male. The mean percentage of males $(50 + 40)/2 = 45\%$ is *not* the percentage of males in the two groups combined, because there are $20 + 60 = 80$ males in a total of 170 rats = 47.1% approx.

3. **If the measurement scale is not linear, arithmetic means may give a false value.** For example, if three media had pH values 6, 7 and 8, the appropriate mean pH is not 7 because the pH scale is logarithmic. The definition of pH is $-\log_{10}[H^+]$, where $[H^+]$ is expressed in mol l^{-1} ('molar'); therefore, to obtain the true mean, convert data into $[H^+]$ values (i.e. put them on a linear scale) by calculating $10^{(-pH\,value)}$ as shown. Now calculate the mean of these values and convert the answer back into pH units. Thus, the appropriate answer is pH 6.43 rather than 7. Note that a similar procedure is necessary when calculating statistics of dispersion in such cases, so you will find these almost certainly asymmetric about the mean.

pH value	$[H^+]$ (mol l^{-1})
6	1×10^{-6}
7	1×10^{-7}
8	1×10^{-8}
mean	3.7×10^{-7}
$-\log_{10}$ mean	6.43

Mean values of log-transformed data are often termed geometric means – they are sometimes used in microbiology and in cell culture studies, where log-transformed values for cell density counts are averaged and plotted (p. 272), rather than using the raw data values. The use of geometric means in such circumstances serves to reduce the effects of outliers on the mean.

where the longer 'tail' of the distribution occurs for higher values of the measured variable; negative skewness where the longer tail occurs for lower values. Some biological examples of characteristics distributed in a skewed fashion are volumes of plant protoplasts, insulin levels in human plasma and bacterial colony counts.

Kurtosis is the name given to the 'pointedness' of a frequency distribution. A platykurtic frequency distribution is one with a flattened peak, while a leptokurtic frequency distribution is one with a pointed peak (Fig. 65.10). While descriptive terms can be used, based on visual observation of the shape and direction of skew, the degree of skewness and kurtosis can be quantified and statistical tests exist to test the 'significance' of observed values (see Sokal and Rohlf, 1994), but the calculations required are complex and best done with the aid of a computer.

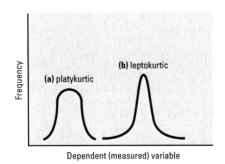

Fig. 65.10 Examples of the two types of kurtosis

Using computers to calculate descriptive statistics

There are many specialist statistical packages (e.g. SPSS) that can be used to simplify the process of calculation of statistics. Note that correct interpretation of the output requires an understanding of the terminology used and the underlying process of calculation, and this may best be obtained by working through one or more examples by hand before using these tools. Spreadsheets offer increasingly sophisticated statistical analysis functions, some examples of which are provided in Box 65.3 for Microsoft Excel.

Box 65.3 How to use a spreadsheet (Microsoft Excel) to calculate descriptive statistics

Method 1: Using spreadsheet functions to generate the required statistics.

Suppose you had obtained the following set of data, stored within an array (block of columns and rows) of cells (A2:L6) within a spreadsheet:

	A	B	C	D	E	F	G	H	I	J	K	L
1	My data set											
2	4	4	3	3	5	4	3	7	7	3	5	3
3	6	2	9	7	3	4	5	6	6	9	4	8
4	5	3	2	5	4	5	7	2	8	3	6	3
5	11	3	5	2	4	3	7	8	4	4	4	3
6	3	6	8	5	6	4	3	4	3	6	10	5

The following functions could be used to extract descriptive statistics from this data set:

Descriptive statistic	Example of use of function[a,b]	Result for the above data set
Sample size n	=COUNT((A2:L6)	60
Mean	=AVERAGE(A2:L6)[c]	4.9
Median	=MEDIAN(A2:L6)	4.0
Mode	=MODE(A2:L6)	3
Upper quartile	=QUARTILE(A2:L6,3)[d]	6.0
Lower quartile	=QUARTILE(A2:L6,1)	3.0
Semi-interquartile range	=QUARTILE(A2:L6,3)-QUARTILE(A2:L6,1)	3.0
Upper extreme	=QUARTILE(A2:L6,4) or =MAX(A2:L6)	11
Lower extreme	=QUARTILE(A2:L6,0) or =MIN(A2:L6)	2
Range	=MAX(A2:L6)- MIN(A2:L6)[e]	9.0
Variance	=VAR(A2:L6)	4.464
Standard deviation	=STDEV(A2:L6)	2.113
Standard error	=STDEV(A2:L6)/(SQRT(COUNT(A2:L6)))[f]	0.273
Coefficient of variation	=100*STDEV(A2:L6)/AVERAGE(A2:L6)	43.12%

Notes:
[a] Typically, in an appropriate cell, you would *Insert > Function > COUNT*, then select the input range and press return.
[b] Other descriptive statistics can be calculated – these mirror those shown in Box 65.1, but for this specific data set.
[c] There is no 'MEAN' function in Microsoft Excel.
[d] The first argument within the brackets relates to the array of data, the second relates to the quartile required (consult the *Help* feature for further information).
[e] There is no direct 'RANGE' function in Microsoft Excel.
[f] There is no direct 'STANDARD ERROR' function in Microsoft Excel. The SQRT function returns a square root and the COUNT function determines the number of filled data cells in the array.

(continued)

Box 65.3 (continued)

Method 2: Using the *Tools > Data Analysis* option – This can automatically generate a table of descriptive statistics for the data array selected, although the data must be presented as a single row or column. This option might need to be installed for your network or personal computer before it is available to you (in the latter case use the *Add Ins > Analysis ToolPak* option from the *Tools* menu – consult the *Help* feature for details). Having entered or rearranged your data into a row or column, the steps involved are as follows:

1. Select *Tools > Data Analysis*.
2. From the *Data Analysis box*, select *Descriptive Statistics*.
3. Input your data location into the *Input Range* (left-click and hold down to highlight the column of data).
4. From the menu options, select *Summary Statistics* and *Confidence Level for Mean: 95%*.
5. When you click *OK* you should get a new worksheet, with descriptive statistics and confidence limits shown. Alternatively, at step 3, you can select an area of your current worksheet as a data output range (select an area away from any existing content as these cells would otherwise be overwritten by the descriptive statistics output table).
6. Change the format of the cells to show each number to an appropriate number of decimal places. You may also wish to make the columns wider so you can read their content.
7. For the data set shown above, the final output table should look as shown in Table 65.1.

Table 65.1 Descriptive statistics for a data set.

Column1[a,b]	
Mean	4.9
Standard error	0.27
Median	4.0
Mode	3
Standard deviation	2.113
Sample variance	4.464
Kurtosis	0.22
Skewness	0.86
Range	9.00
Minimum	2.0
Maximum	11.0
Sum	294
Count	60
Confidence level (95.0%)	0.55

Notes:
[a] These descriptive statistics are specified (and are automatically presented in this order) – any others required can be generated using Method 1.
[b] A more descriptive heading can be added if desired - this is the default.

Text references and sources for further study

Heath, D. (1995) *An Introduction to Experimental Design and Statistics for Biology*. UCL Press, London.

Schmuller, J. (2005) *Statistical Analysis with Excel for Dummies*. Wiley, Hoboken, New Jersey.

Sokal, R.R. and Rohlf, F.J. (1994) *Biometry*, 3rd edn. W.H. Freeman, San Francisco.

Wardlaw, A.C. (2000) *Practical Statistics for Experimental Biologists*, 2nd edn. Wiley, New York.

Study exercises

65.1 Choose appropriate descriptive statistics. State for the following data and distribution types what statistics of location and dispersion (Figs. 65.4 and 65.5) you should choose.

(a) Colour of each egg in nests as judged against a colour chart.
(b) Clutch size (number of eggs) in bird nests visited.
(c) Beak lengths for chicks hatched in nests, where there is a pronounced difference in beak length between the sexes.

65.2 Practise calculating descriptive statistics. Using the data set given below, calculate the following statistics:

(a) range
(b) variance
(c) standard deviation
(d) coefficient of variation
(e) standard error.

Answers (b) to (e) should be given to four significant figures.

65.3 Calculate and interpret standard errors. Two samples, A and B, gave the following descriptive statistics (measured in the same units): Sample A, mean = 16.2, standard deviation = 12.7, number of data values = 12; Sample B, mean = 13.2, standard deviation 14.4, number of data values = 20. Which has the lower standard error in absolute terms and in proportion to the sample mean? (Express answers to three significant figures.)

65.4 Compute a mean value correctly. A researcher finds that the mean diameter of limpets on three seashore stones designated A, B and C is 3.0, 2.5 and 2.0 mm respectively. He computes the mean limpet diameter as 2.5 mm, but forgets that the sample sizes were 24, 37 and 6 respectively. What is the true mean diameter of the limpets in this sample? (Answer to three significant figures.)

Set of data

9	6	7	5	7	7	8	6	5	5	7	8
5	8	7	7	6	7	8	6	5	7	7	6
3	6	8	9	9	6	7	8	5	6	5	5
8	8	7	5	6	5	8	6	7	5	7	6
5	6	7	8	7	6	7	7	8	8	9	4

66 Choosing and using statistical tests

This chapter outlines the philosophy of hypothesis-testing statistics, indicates the steps to be taken when choosing a test, and discusses features and assumptions of some important tests. For details of the mechanics of tests, consult appropriate texts (e.g. Sokal and Rohlf, 1994; Heath, 1995; Wardlaw, 2000). Most tests are now available in statistical packages for computers (see p. 78) and many in spreadsheets (p. 70).

To carry out a statistical test:

1. Decide what it is you wish to test (create a null hypothesis and its alternative).
2. Determine whether your data fit a standard distribution pattern.
3. Select a test and apply it to your data.

Setting up a null hypothesis

Hypothesis-testing statistics are used to compare the properties of samples either with other samples or with some theory about them. For instance, you may be interested in whether two samples can be regarded as having different means, whether the counts of an organism in different quadrats can be regarded as randomly distributed, or whether property A of an organism is linearly related to property B.

> **KEY POINT** You can't use statistics to prove any hypothesis, but they can be used to assess how likely it is to be wrong.

Statistical testing operates in what at first seems a rather perverse manner. Suppose you think a treatment has an effect. The theory you actually test is that it has no effect; the test tells you how improbable your data would be if this theory were true. This 'no effect' theory is the null hypothesis (NH). If your data are very improbable under the NH, then you may suppose it to be wrong, and this would support your original idea (the 'alternative hypothesis'). The concept can be illustrated by an example. Suppose two groups of subjects were treated in different ways, and you observed a difference in the mean value of the measured variable for the two groups. Can this be regarded as a 'true' difference? As Fig. 66.1 shows, it could have arisen in two ways:

- Because of the way the subjects were allocated to treatments, i.e. all the subjects liable to have high values might, by chance, have been assigned to one group and those with low values to the other (Fig. 66.1(a)).
- Because of a genuine effect of the treatments, i.e. each group came from a distinct frequency distribution (Fig. 66.1(b)).

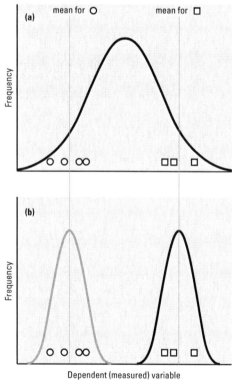

Fig. 66.1 Two explanations for the difference between two means. In case (a) the two samples happen by chance to have come from opposite ends of the same frequency distribution, i.e. there is no true difference between the samples. In case (b) the two samples come from different frequency distributions, i.e. there is a true difference between the samples. In both cases, the means of the two samples are the same.

A statistical test will indicate the probabilities of these options. The NH states that the two groups come from the same population (i.e. the treatment effects are negligible in the context of random variation). To test this, you calculate a test statistic from the data, and compare it with tabulated critical values giving the probability of obtaining the observed or a more extreme result by chance (see Boxes 66.1 and 66.2). This probability is sometimes called the significance of the test.

Note that you must take into account the degrees of freedom (d.f.) when looking up critical values of most test statistics. The d.f. is related to the size(s) of the samples studied; formulae for calculating it depend on the test being used. Biologists normally use two-tailed tests, i.e. we have no expectation beforehand that the treatment will have a positive or negative effect compared with the control (in a one-tailed test we expect one particular treatment to be bigger than the other). Be sure to use critical values for the correct type of test.

By convention, the critical probability for rejecting the NH is 5 per cent (i.e. $P = 0.05$). This means we reject the NH if the observed result would have come up by chance a maximum of one time in twenty. If the modulus of the test statistic is less than or equal to the tabulated critical value for $P = 0.05$, then we accept the NH and the result is said to be 'not significant' (NS for short). If the modulus of the test statistic is greater than the tabulated value for $P = 0.05$, then we reject the NH in favour of the alternative hypothesis that the treatments had different effects and the result is 'statistically significant'.

Two types of error are possible when making a conclusion on the basis of a statistical test. The first occurs if you reject the NH when it is true and the second if you accept the NH when it is false. To limit the chance of the first type of error, choose a lower probability, e.g. $P = 0.01$, but note that the critical value of the test statistic increases when you do this and results in the probability of the second error increasing. The conventional significance levels given in statistical tables (usually 0.05, 0.01, 0.001) are arbitrary. Increasing use of statistical computer programs now allows the actual probability of obtaining the calculated value of the test statistic to be quoted (e.g. $P = 0.037$).

Note that if the NH is rejected, this does not tell you which of many alternative explanations are true. Also, it is important to distinguish between statistical significance and biological relevance: identifying a statistically significant difference between two samples doesn't mean that this will carry any biological importance.

Comparing data with parametric distributions

A parametric test is one that makes particular assumptions about the mathematical nature of the population distribution from which the samples were taken. If these assumptions are not true, then the test is obviously invalid, even though it might give the answer we expect. A non-parametric test does not assume that the data fit a particular pattern, but it may assume some things about the distributions. Used in appropriate circumstances, parametric tests are better able to distinguish between true but marginal differences between samples than their non-parametric equivalents (i.e. they have greater 'power').

The distribution pattern of a set of data values may be biologically relevant, but it is also of practical importance because it defines the type of statistical tests that can be used. The properties of the main distribution types found in biology are given below with both rules of thumb and more rigorous tests for deciding whether data fit these distributions.

Binomial distributions

These apply to samples of any size from populations when data values occur independently in only two mutually exclusive classes (e.g. type A or type B). They describe the probability of finding the different possible combinations of the attribute for a specified sample size k

---

Definition

Modulus – the absolute value of a number, e.g. modulus $-3.385 = 3.385$.

Quoting significance – the convention for quoting significance levels in text, tables and figures is as follows:
$P > 0.05$ = 'not significant' (or NS)
$P \leqslant 0.05$ = 'significant' (or *)
$P \leqslant 0.01$ = 'highly significant' (or **)
$P \leqslant 0.001$ = 'very highly significant' (or ***)

Thus, you might refer to a difference in means as being 'highly significant ($P \leqslant 0.01$)'. For this reason, the word 'significant' in its everyday meaning of 'important' or 'notable' should be used with care in scientific writing.

Choosing between parametric and non-parametric tests – always plot your data graphically when determining whether they are suitable for parametric tests as this may save a lot of unnecessary effort later.

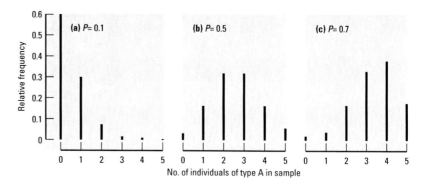

Fig. 66.2 Examples of binomial frequency distributions with different probabilities. The distributions show the expected frequency of obtaining n individuals of type A in a sample of 5. Here P is the probability of an individual being type A rather than type B.

(e.g. out of 10 specimens, what is the chance of 8 being type A?). If p is the probability of the attribute being of type A and q the probability of it being type B, then the expected mean sample number of type A is kp and the standard deviation is $\sqrt{kpq}$. Expected frequencies can be calculated using mathematical expressions (see Sokal and Rohlf, 1994). Examples of the shapes of some binomial distributions are shown in Fig. 66.2. Note that they are symmetrical in shape for the special case $p = q = 0.5$ and the greater the disparity between p and q, the more skewed the distribution.

Some biological examples of data likely to be distributed in binomial fashion are: possession of two alleles for seed-coat morphology (e.g. smooth and wrinkly); whether an organism is infected with a microbe or not; whether an animal is male or female. Binomial distributions are particularly useful for predicting gene segregation in Mendelian genetics and can be used for testing whether combinations of events have occurred more frequently than predicted (e.g. more siblings being of the same sex than expected). To establish whether a set of data is distributed in binomial fashion: calculate expected frequencies from probability values obtained from theory or observation, then test against observed frequencies using a χ^2 test (p. 330) or a G test (see Wardlaw, 2000).

Poisson distributions

These apply to discrete characteristics that can assume low whole-number values, such as counts of events occurring in area, volume or time. The events should be 'rare' in that the mean number observed should be a small proportion of the total that could possibly be found. Also, finding one count should not influence the probability of finding another. The shape of Poisson distributions is described by only one parameter, the mean number of events observed, and has the special characteristic that the variance is equal to the mean. The shape has a pronounced positive skewness at low mean counts, but becomes more and more symmetrical as the mean number of counts increases (Fig. 66.3).

Some examples of characteristics distributed in a Poisson fashion are: number of plants in a quadrat; number of microbes per unit volume of medium; number of animals parasitised per unit time; number of radioactive disintegrations per unit time. One of the main uses for the

Tendency towards the normal distribution – under certain conditions, binomial and Poisson distributions can be treated as normally distributed:

- where samples from a binomial distribution are large (i.e. > 15) and p and q are close to 0.5;
- for Poisson distributions, if the number of counts recorded in each outcome is greater than about 15.

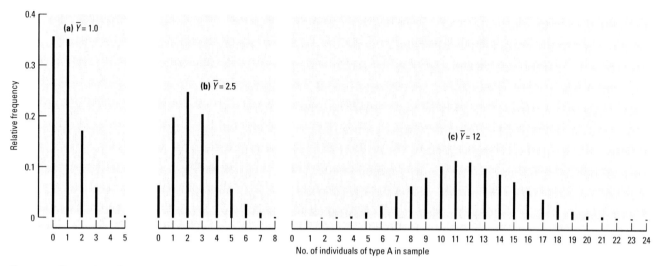

Fig. 66.3 Examples of Poisson frequency distributions differing in mean. The distributions are shown as line charts because the independent variable (events per sample) is discrete.

Poisson distribution is to quantify errors in count data such as estimates of cell densities in dilute suspensions (see p. 275). To decide whether data are Poisson distributed:

- Use the rule of thumb that if the coefficient of dispersion ≈ 1, the distribution is likely to be Poisson.
- Calculate 'expected' frequencies from the equation for the Poisson distribution and compare with actual values using a χ^2 test or a G-test.

It is sometimes of interest to show that data are *not* distributed in a Poisson fashion, e.g. the distribution of parasite larvae in hosts. If $s^2/\bar{Y} > 1$, the data are 'clumped' and occur together more than would be expected by chance; if $s^2/\bar{Y} < 1$, the data are 'repulsed' and occur together less frequently than would be expected by chance.

Definition

Coefficient of dispersion $= s^2/\bar{Y}$. This is an alternative measure of dispersion to the coefficient of variation (p. 420).

Normal distributions (Gaussian distributions)

These occur when random events act to produce variability in a continuous characteristic (quantitative variable). This situation occurs frequently in biology, so normal distributions are very useful and much used. The bell-like shape of normal distributions is specified by the population mean and standard deviation (Fig. 66.4): it is symmetrical and configured such that 68.27 per cent of the data will lie within ±1 standard deviation of the mean, 95.45 per cent within ±2 standard deviations of the mean, and 99.73 per cent within ±3 standard deviations of the mean.

Some biological examples of data likely to be distributed in a normal fashion are: fresh weight of plants of the same age; linear dimensions of bacterial cells; height of either adult female or male humans. To check whether data come from a normal distribution, you can:

- Use the rule of thumb that the distribution should be symmetrical and that nearly all the data should fall within ±3s of the mean and about two-thirds within ±1s of the mean.
- Plot the distribution on normal probability graph paper. If the distribution is normal, the data will tend to follow a straight line (see Fig. 66.5). Deviations from linearity reveal skewness and/or kurtosis

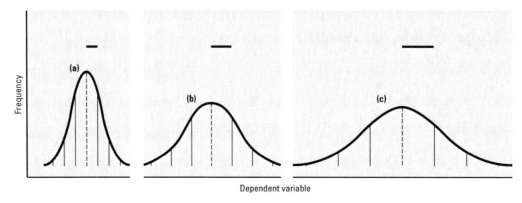

Fig. 66.4 Examples of normal frequency distributions differing in mean and standard deviation. The horizontal bars represent population standard deviations for the curves, increasing from (a) to (c). Vertical dashed lines are population means, while vertical solid lines show positions of values ±1, 2 and 3 standard deviations from the means.

(see p. 422), the significance of which can be tested statistically (see Sokal and Rohlf, 1994).

- Use a suitable statistical computer program to generate predicted normal curves from the $\bar{Y}$ and s values of your sample(s). These can be compared visually with the actual distribution of data and can be used to give 'expected' values for a χ^2 test or a G-test.

The wide availability of tests based on the normal distribution and their relative simplicity means you may wish to transform your data to make them more like a normal distribution. Table 66.1 provides transformations that can be applied (see also Fig. 61.3). The transformed data should be tested for normality as described above before proceeding – don't forget that you may need to check that transformed variances are homogeneous for certain tests (see below).

A very important theorem in statistics, the Central Limit Theorem, states that as sample size increases, the distribution of a series of means from any frequency distribution will become normally distributed. This fact can be used to devise an experimental or sampling strategy that ensures that data are normally distributed, i.e. using means of samples as if they were primary data.

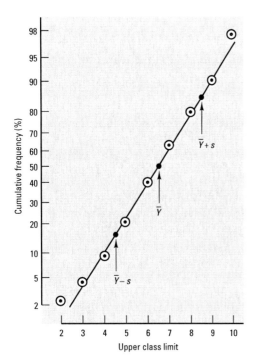

Fig. 66.5 Example of a normal probability plot. The plotted points are from a small data set where the mean $Y = 6.93$ and the standard deviation $s = 1.895$. Note that values corresponding to 0% and 100% cumulative frequency cannot be used. The straight line is that predicted for a normal distribution with $Y = 6.93$ and $s = 1.895$. This is plotted by calculating the expected positions of points for $Y \pm s$. Since 68.3% of the distribution falls within these bounds, the relevant points on the cumulative frequency scale are $50 \pm 34.15\%$; thus this line was drawn using the points (4.495, 15.85) and (8.285, 84.15) as indicated on the plot.

Table 66.1 Suggested transformations altering different types of frequency distributions to the normal type. To use, modify data by the formula shown; then examine effects with the tests described on pp. 427–30.

Type of data; distribution suspected	Suggested transformation(s)
Proportions (including percentages); binomial	arcsine $\sqrt{x}$ (also called the angular transformation)
Scores; Poisson	$\sqrt{x}$ or $\sqrt{(x + 1/2)}$ if zero values present
Measurements; negatively skewed	x^2, x^3, x^4, etc. (in order of increasing strength)
Measurements; positively skewed	$1/\sqrt{x}, \sqrt{x}, \ln x, 1/x$ (in order of increasing strength)

> **Definition**
>
> **Homogeneous variance** – uniform (but not necessarily identical) variance of the dependent variable across the range of the independent variable. The term homoscedastic is also used in this sense. The opposite of homogeneous is heterogeneous (= hereoscedastic).

> **Understanding 'degrees of freedom'** – this depends on the number of values in the data set analysed, and the method of calculation depends on the statistical test being used. It relates to the number of observations that are free to vary before the remaining quantities for a data set can be determined.

> **Checking the assumptions of a test** – always acquaint yourself with the assumptions of a test. If necessary, test them before using the test.

Choosing a suitable statistical test

Comparing location (e.g. means)

If you can assume that your data are normally distributed, the main test for comparing two means from independent samples is Student's t-test (see Boxes 66.1 and 66.2, and Table 66.2). This assumes that the variances of the data sets are homogeneous. Tests based on the t-distribution are also available for comparing means of paired data or for comparing a sample mean with a chosen value.

When comparing means of two or more samples, analysis of variance (ANOVA) is a very useful technique. This method also assumes data are normally distributed and that the variances of the samples are homogeneous. The samples must also be independent (e.g. not subsamples). The test statistic calculated is denoted F and it has two different degrees of freedom related to the number of means tested and the pooled number of replicates per mean. The nested types of ANOVA are useful for letting you know the relative importance of different sources of variability in your data. Two-way and multi-way ANOVAs are useful for studying interactions between treatments.

For data satisfying the ANOVA requirements, the least significant difference (LSD) is useful for making planned comparisons among several means (see Sokal and Rohlf, 1994). Any two means that differ by more than the LSD will be significantly different. The LSD is useful for showing on graphs.

The chief non-parametric tests for comparing the locations of two samples are the Mann–Whitney U-test and the Kolmogorov–Smirnov test. The former assumes that the frequency distributions of the samples are similar, whereas the latter makes no such assumption. In both cases the sample's size must be $\geqslant 4$ and for the Kolmogorov–Smirnov test the samples must have equal sizes. In the Kolmogorov–Smirnov test, significant differences found with the test could be due to differences in location or shape of the distribution, or both.

Table 66.2 Critical values of Student's t statistic (for two-tailed tests). Reject the null hypothesis at probability P if your calculated t value equals or exceeds the value shown for the appropriate degrees of freedom = $(n_1 - 1) + (n_2 - 1)$.

Degrees of freedom	Critical values for $P = 0.05$	Critical values for $P = 0.01$	Critical values for $P = 0.001$
1	12.71	63.66	636.62
2	4.30	9.92	31.60
3	3.18	5.84	12.94
4	2.78	4.60	8.61
5	2.57	4.03	6.86
6	2.45	3.71	5.96
7	2.36	3.50	5.40
8	2.31	3.36	5.04
9	2.26	3.25	4.78
10	2.23	3.17	4.59
12	2.18	3.06	4.32
14	2.14	2.98	4.14
16	2.12	2.92	4.02
20	2.09	2.85	3.85
25	2.06	2.79	3.72
30	2.04	2.75	3.65
40	2.02	2.70	3.55
60	2.00	2.66	3.46
120	1.98	2.62	3.37
∞	1.96	2.58	3.29

Box 66.1 How to carry out a t-test

The t-test was devised by a statistician who used the pen name 'Student', so you may see it referred to as Student's t-test. It is used when you wish to decide whether two samples come from the same population or from different ones (Fig. 66.1). The samples might have been obtained by selective observation (Chapter 27) or by applying two different treatments to an originally homogeneous population (Chapter 31).

The null hypothesis (NH) is that the two groups can be represented as samples from the same overlying population (Fig. 66.1(a)). If, as a result of the test, you accept this hypothesis, you can say that there is no significant difference between the group means.

The alternative hypothesis is that the two groups come from different populations (Fig. 66.1(b)). By rejecting the NH as a result of the test, you can accept the alternative hypothesis and say that there is a significant difference between the sample means, or, if an experiment were carried out, that the two treatments affected the samples differently.

How can you decide between these two hypotheses? On the basis of certain assumptions (see below), and some relatively simple calculations, you can work out the probability that the samples came from the same population. If this probability is very low, then you can reasonably reject the NH in favour of the alternative hypothesis, and if it is high, you will accept the NH.

To find out the probability that the observed difference between sample means arose by chance, you must first calculate a 't value' for the two samples in question. Some computer programs (e.g. Minitab) provide this probability as part of the output, otherwise you can look up statistical tables (e.g. Table 66.2). These tables show 'critical values' – the borders between probability levels. If your value of t equals or exceeds the critical value for probability P, you can reject the NH at this probability ('level of significance'). Note that:

- for a given difference in the means of the two samples, the value of t will get larger the smaller the scatter within each data set; and
- for a given scatter of the data, the value of t will get larger, the greater the difference between the means.

So, at what probability should you reject the NH? Normally, the threshold is arbitrarily set at 5 per cent – you quite often see descriptions like 'the sample means were significantly different ($P < 0.05$)'. At this 'significance level' there is still up to a 5 per cent chance of the t value arising by chance, so about one in twenty times, on average, the conclusion will be wrong. If P turns out to be lower, then this kind of error is much less likely.

Tabulated probability levels are generally given for 5, 1 and 0.1 per cent significance levels (see Table 66.2). Note that this table is designed for 'two-tailed' tests, i.e. where the treatment or sampling strategy could have resulted in either an increase or a decrease in the measured values. These are the most likely situations you will deal with in biology.

Examine Table 66.2 and note the following:

- The larger the size of the samples (i.e. the greater the 'degrees of freedom'), the smaller t needs to be to exceed the critical value at a given significance level.
- The lower the probability, the greater t needs to be to exceed the critical value.

The mechanics of the test

A calculator that can work out means and standard deviations is helpful.

1. **Work out the sample means $\bar{Y}_1$ and $\bar{Y}_2$ and calculate the difference between them.**
2. **Work out the sample standard deviations s_1 and s_2.** (NB if your calculator offers a choice, chose the '$n-1$' option for calculating s – see p. 420).
3. **Work out the sample standard errors $SE_1 = s_1/\sqrt{n_1}$ and $SE_2 = s_2/\sqrt{n_2}$; now square each, add the squares together, then take the positive square root of this** (n_1 and n_2 are the respective sample sizes, which may, or may not, not be equal).
4. **Calculate t from the formula:**

$$t = \frac{\bar{Y}_1 - \bar{Y}_2}{\sqrt{((SE_1)^2) + ((SE_2)^2)}} \quad [66.1]$$

The value of t can be negative or positive, depending on the values of the means; this does not matter and you should compare the modulus (absolute value) of t with the values in tables.

5. **Work out the degrees of freedom $= (n_1 - 1) + (n_2 - 1)$.**
6. **Compare the t value with the appropriate critical value (see e.g. Table 66.2) and decide on the significance of your findings (see p. 427).**

Box 66.2 provides a worked example – use this to check that you understand the above procedures.

Assumptions that must be met before using the test

The most important assumptions are:

- The two samples are independent and randomly drawn (or if not, drawn in a way that does not create bias). The test assumes that the samples are quite large.
- The underlying distribution of each sample is normal. This can be tested with a special statistical test, but a rule of thumb is that a frequency distribution of the data should be (a) symmetrical about the mean and (b) nearly all of the data should be within 3 standard deviations of the mean and about two-thirds within 1 standard deviation of the mean (see p. 429).
- The two samples should have uniform variances. This again can be tested (by an F-test), but may be obvious from inspection of the two standard deviations.

Box 66.2 Worked example of a t-test

Suppose the following data were obtained in an experiment (the units are not relevant):

Control: 6.6, 5.5, 6.8, 5.8, 6.1, 5.9
Treatment: 6.3, 7.2, 6.5, 7.1, 7.5, 7.3

Using the steps outlined in Box 66.1, the following values are obtained (denoting control with subscript 1, treatment with subscript 2):

1. $\bar{Y}_1 = 6.1167$; $\bar{Y}_2 = 6.9833$: difference between means $= \bar{Y}_1 - \bar{Y}_2 = -0.8666$
2. $s_1 = 0.49565$; $s_2 = 0.47504$
3. $SE_1 = 0.49565/2.44949 = 0.202348$
   $SE_2 = 0.47504/2.44949 = 0.193934$
4. $t = \dfrac{-0.8666}{\sqrt{(0.202348^2 + 0.193934^2)}} = \dfrac{-0.8666}{0.280277} = -3.09$
5. d.f. $= (5 + 5) = 10$
6. Looking at Table 66.2, we see that the modulus of this t value exceeds the tabulated value for $P = 0.05$ at 10 degrees of freedom ($= 2.23$). We therefore reject the NH, and conclude that the means are different at the 5% level of significance. If the modulus of t had been ≤ 2.23, we would have accepted the NH. If the modulus of t had been > 3.17, we could have concluded that the means are different at the 1% level of significance.

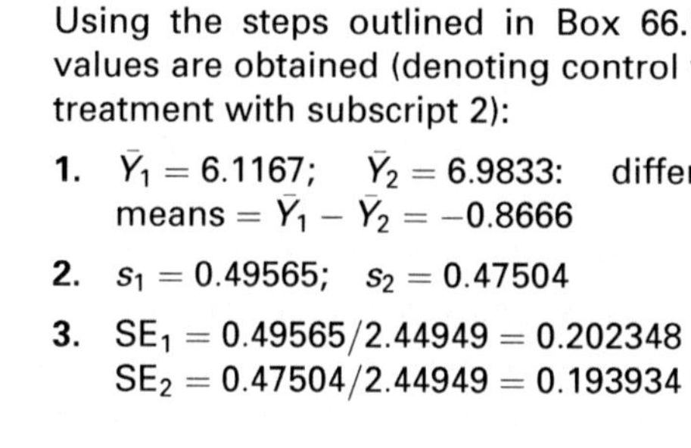

Fig. 66.6 Graphical representation of confidence limits as 'error bars' for (a) a sample mean in a plotted curve, where both upper and lower limits are shown; and (b) a sample mean in a histogram, where, by convention, only the upper value is shown. For data that are assumed to be symmetrically distributed, such representations are often used in preference to the 'box and whisker' plot shown on p. 419. Note that SE is an alternative way of representing sample imprecision/error (e.g. Fig. 62.1).

Confidence limits for statistics other than the mean – consult an advanced statistical text (e.g. Sokal and Rohlf, 1994) if you wish to indicate the reliability of estimates of e.g. population variances.

Suitable non-parametric comparisons of location for paired data (sample size ≥ 6) include Wilcoxon's signed rank test, which is used for quantitative data and assumes that the distributions have similar shape. Dixon and Mood's sign test can be used for paired data scores where one variable is recorded as 'greater than' or 'better than' the other.

Non-parametric comparisons of location for three or more samples include the Kruskal–Wallis H-test. Here, the number of samples is without limit and they can be unequal in size, but again the underlying distributions are assumed to be similar. The Friedman S-test operates with a maximum of five samples and data must conform to a randomised block design. The underlying distributions of the samples are assumed to be similar.

Comparing dispersions (e.g. variances)

If you wish to compare the variances of two sets of data that are normally distributed, use the F-test. For comparing more than two samples, it may be sufficient to use the $F_{\max}$-test, on the highest and lowest variances. The Scheffé–Box (log-anova) test is recommended for testing the significance of differences between several variances. Non-parametric tests exist but are not widely available: you may need to transform the data and use a test based on the normal distribution.

Determining whether frequency observations fit theoretical expectation

The χ^2 test (Box 53.2) is useful for tests of 'goodness of fit', e.g. comparing expected and observed progeny frequencies in genetic experiments or comparing observed frequency distributions with some theoretical function. One limitation is that simple formulae for calculating χ^2 assume that no expected number is less than 5. The G-test ($2I$-test) is used in similar circumstances.

Comparing proportion data

When comparing proportions between two small groups (e.g. whether 3/10 is significantly different from 5/10), you can use probability tables such as those of Finney et al. (1963) or calculate probabilities from formulae; however, this can be tedious for large sample sizes. Certain proportions can be transformed so that their distribution becomes normal.

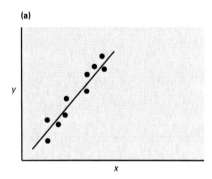
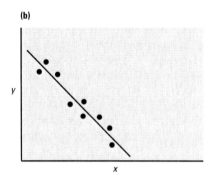
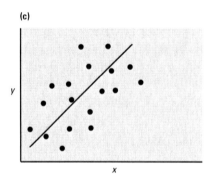
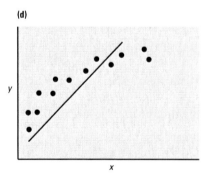

Fig. 66.7 Examples of correlation. The linear regression line is shown. In (a) and (b), the correlation between x and y is good: for (a) there is a positive correlation and the correlation coefficient, r, would be close to 1; for (b) there is a negative correlation and the correlation coefficient would be close to -1. In (c) there is a weak positive correlation and r would be close to 0. In (d) the correlation coefficient may be quite large, but the choice of linear regression is clearly inappropriate.

Placing confidence limits on an estimate of a population parameter

On many occasions, a sample statistic is used to provide an estimate of a population parameter, and it is often useful to indicate the reliability of such an estimate. This can be done by putting confidence limits on the sample statistic, i.e. by specifying an interval around the statistic within which you are confident that the true value (the population parameter) is likely to fall, at a specified level of probability. The most common application is to place confidence limits on the mean of a sample taken from a population of normally distributed data values. In practice, you determine a confidence factor for a particular level of probability that is added to and subtracted from the sample mean ($\bar{Y}$) to give the upper confidence limit and lower confidence limit respectively. These are calculated as:

$$\bar{Y} + (t_{P[n-1]} \times \text{SE}) \text{ for the upper limit and}$$
$$\bar{Y} - (t_{P[n-1]} \times \text{SE}) \text{ for the lower limit} \quad [66.2]$$

where $t_{P[n-1]}$ is the tabulated critical value of Student's t statistic for a two-tailed test with $n - 1$ degrees of freedom at a specified probability level (P) and SE is the standard error of the sample mean (p. 420). The 95 per cent confidence limits (i.e. $P = 0.05$) tells you that, on average, 95 times out of 100 the interval between the upper and lower limits will contain the true (population) value. Confidence limits are often shown as 'error bars' for individual sample means plotted in graphical form. Figure 66.6 illustrates how this is applied to plotted curves and histograms (note that this can be carried out for data series within a Microsoft Excel graph (chart) using the *Format data series* and *Y error bars* commands).

Correlation and regression

These methods are used when testing the relationship between data values for two variables. Correlation is used to measure the extent to which changes in the two sets of data values occur together in a linear manner. If one variable can be assumed to be dependent upon the other (i.e. a change in X causes a particular change in Y), then regression techniques can be used to provide a mathematical description of the underlying relationship between the variables, e.g. to find a line of best fit for a data series. If there is no a priori reason to assume dependency, then correlation methods alone are appropriate.

A correlation coefficient measures the strength of the linear relationship between two variables, but does not describe the relationship. The coefficient is expressed as a number between -1 and $+1$: a positive coefficient indicates a direct relationship, where the two variables change in the same direction, while a negative coefficient indicates an inverse relationship, where one variable decreases as the other increases (Fig. 66.7). The nearer the coefficient is to -1 or $+1$, the stronger the linear relationship between the variables, i.e. the less 'scatter' there would be about a straight line of best fit (note that this does *not* imply that one variable is dependent upon the other). A coefficient of 0 implies that the two variables show no linear association and therefore the closer the correlation coefficient is to zero, the weaker the linear relationship. The importance of graphing data is shown by the case illustrated in Fig. 66.7(d).

Pearson's product moment correlation coefficient (r) is the most commonly used statistic for testing correlations. The test is valid only if both variables are normally distributed. Statistical tests can be used to decide whether the correlation is significant (e.g. using a one-sample t-test to see whether r is significantly different from zero, based on the equation:

$$t = r \div \sqrt{[(1 - r^2) \div (n - 2)]} \text{ at } n - 2 \text{ degrees of freedom,} \qquad [66.3]$$

where n is the number of paired observations. If one or both variables are not normally distributed, then you should calculate an alternative non-parametric coeffcient, e.g. Spearman's coefficient of rank correlation (r_s) or Kendall's coefficient of rank correlation (τ). These require the two sets of data to be ranked separately, and the calculation can be complex if there are tied (equal) ranks. Spearman's coefficient is said to be better if there is any uncertainty about the reliability of closely ranked data values.

If underlying theory or empirical graphical analysis indicate a linear relationship between a dependent and an independent variable, then linear regression can be used to estimate the mathematical equation that links the two variables. Model I linear regression is the standard approach, and is available within general-purpose software programs such as Microsoft Excel (Box 66.3), and on some scientific calculators. It is suitable for experiments where a dependent variable Y varies with an *error-free* independent variable X in accordance with the relationship $Y = a + bX + e_Y$, where e_Y represents the residual (error) variability in the Y variable. For example, this relationship might apply in a laboratory procedure where you have carefully controlled the independent variable and the X values can be assumed to have zero error (e.g. in a calibration curve, see Chapter 50, or in a time course experiment where measurements are made at exact time points). The regression analysis gives estimates for a and b (equivalent to the slope and intercept of the line of best fit, p. 411): computer-based programs usually provide additional features, e.g. residual values for Y (e_Y), estimated errors for a and b, predicted values of Y along with graphical plots of the line of best fit (the trend line) and the residual values. In order for the model to be valid, the residual (error) values should be normally distributed around the trend line and their variance should be uniform (homogeneous), i.e. there should be a similar scatter of data points around the trend line along the x-axis (independent variable).

If the relationship is not linear, try a transformation (see p. 385). For example, this is commonly done in analysis of enzyme kinetics (see Fig. 52.4). However, you should be aware that the transformation of data to give a straight line can lead to errors when carrying out linear regression analysis: take care to ensure that (a) the assumptions listed in the previous paragraph are valid for the transformed data set and (b) the data points are evenly distributed throughout the range of the independent variable. If these criteria cannot be met, non-linear regression may be a better approach, but for this you will require a suitable computer program, e.g. GraphPad Prism.

The strength of the relationship between Y and X in Model I linear regression is best estimated by the coefficient of determination (r^2 or R^2), which is equivalent to the square of the Pearson correlation coefficient. The coefficient of determination varies between 0 and +1 and provides a measure of the goodness of fit of the Y data to the regression line: the closer the value is to 1, the better the fit. In effect, r^2 represents the fraction of the variance in Y that can be accounted for by the regression equation. Conversely, if you subtract this value from 1, you will obtain the residual

---

Using more advanced types of regression – these include:

- Model II linear regression, which applies to situations where a dependent variable Y varies with an independent variable X, and where both variables may have error terms associated with them.
- Multiple regression, which applies when there is a relationship between a dependent variable and two or more independent variables.
- Non-linear regression, which extends the principles of linear regression to a wide range of functions. Technically, this method is more appropriate than transforming data to allow linear regression.

Advanced statistics books should be consulted for details of these methods, which may be offered by some statistical computer programs.

Example If a regression analysis gives a value for r^2 of 0.75 (i.e. $r = 0.84$), then 75% of the variance in Y can be explained by the trend line, with $1 - r^2 = 0.25$ (25%) remaining as unexplained (residual) variation.

Box 66.3 Using a spreadsheet (Microsoft Excel) to calculate hypothesis-testing statistics

Presented below are three examples of the use of Microsoft Excel to investigate hypotheses about specific data sets. In each case, there is a brief description of the problem; a table showing the data analysed; an outline of the Microsoft Excel commands used to carry out the analysis and an annotated table of results from the spreadsheet.

Example 1: A *t*-test

As part of a project, a student applied a chemical treatment to a series of flasks containing fungal cultures with nutrient solution. An otherwise similar set of control flasks received no chemical treatment. After three weeks' growth, she measured the wet mass of the filtered cultures:

Wet mass of samples (g)

Replicate	1	2	3	4	5	6	7	8	Mean	Variance
Treated with ZH52	2.342	2.256	2.521	2.523	2.943	2.481	2.601	2.449	2.515	0.042
Control	2.658	2.791	2.731	2.402	3.041	2.668	2.823	2.509	2.703	0.038

The student proposed the null hypothesis that there was no difference between the two means and tested this using a *t*-test, as she had evidence from other studies that the fungal masses of replicate flasks were normally distributed. She also established, by calculation, that the assumption that the populations had homogeneous variances was likely to be valid. Using the *Tools > Data Analysis > t-Test: Two-Sample Assuming Equal Variance* option, with *Hypothesized Mean Difference* = 0 and *Alpha* $(=P) = 0.05$, and adjusting the number of significant figures displayed, the following table was obtained:

t-test: Two-sample assuming equal variances

	Variable 1	Variable 2
Mean	2.515	2.703
Variance	0.042	0.038
Observations	8	8
Pooled variance	0.040	
Hypothesized mean difference	0	
d.f.	14	
t Stat	−1.881	
P(T <= t) one-tail	0.040	
t Critical one-tail	1.761	
P(T <= t) two-tail	0.081	
t Critical two-tail	2.145	

The value of *t* obtained was −1.881 (row 7, '*t* stat') and the probability of obtaining this value for a two-tailed test (row 10) was 0.081 (or 8.1%), so the student was able to accept the null hypothesis and conclude that ZH52 had no significant effect on fungal growth in these circumstances.

Example 2: An ANOVA test

A biochemist made six replicate measurements of four different batches (A–D) of alcohol dehydrogenanse, obtaining the following data:

Alcohol dehydrogenase activity (U l^{-1})

Batch/Replicate	1	2	3	4	5	6	Mean	Variance
A	0.562	0.541	0.576	0.545	0.542	0.551	0.552833	0.000189
B	0.531	0.557	0.537	0.521	0.559	0.538	0.540500	0.000221
C	0.572	0.568	0.551	0.549	0.564	0.559	0.560500	0.000085
D	0.532	0.548	0.541	0.538	0.547	0.536	0.540333	0.000039

The biochemist wanted to know whether the observed differences were statistically significant, so he carried out an ANOVA test, assuming the samples were normally distributed and the variances in the three populations were homogeneous. Using the *Tools > Data Analysis > Anova: Single Factor* option, with *Alpha* $(=P) = 0.05$, and adjusting the number of significant figures displayed, the following table was obtained:

ANOVA: Single Factor

SUMMARY

Groups	Count	Sum	Average	Variance
A	6	3.317	0.552833	0.000189
B	6	3.243	0.5405	0.000221
C	6	3.363	0.5605	8.51E-05
D	6	3.242	0.540333	3.95E-05

ANOVA

Source of Variation	SS	d.f.	MS	F	P-value	F crit
Between groups	0.001761	3	0.000587	4.397856	0.015669	3.098391
Within groups	0.002669	20	0.000133			
Total	0.00443	23				

The *F* value calculated was 4.397856. This comfortably exceeds the stated critical value (F_{crit}) of 3.098391, and the probability of obtaining this result by chance (*P*-value) was calculated as 0.015669 (1.57% to three significant figures); hence the biochemist was able to reject the null hypothesis and conclude that there was a significant difference in average enzyme activity between the four batches, since $P < 0.05$. Such a finding might lead on to an investigation into why there was batch variation, e.g. had they been stored differently?

(continued)

Box 66.3 (continued)

Example 3: Testing the significance of a correlation

A researcher wanted to know whether his observations of earthworm casts on the surface of closely mown grass were related to how wet the soil was. He took weekly measurements of precipitation using a rain gauge and counted the mean number of casts per m^2 taking mean results from nine quadrats per weekly observation:

Observation	Precipitation in previous week (mm)	Mean density of casts (m^{-2})
1	11	4.4
2	1	3.1
3	0	2.3
4	5	4.6
5	8	4.5
6	2	3.3
7	4	3.5
8	15	6.4

The researcher used the Microsoft Excel function PEARSON(array1, array2) to obtain a value of +0.927 857 674 for Pearson's product moment correlation coefficient r, specifying the precipitation data as array1 and the cast density as array2. He then used a spreadsheet to calculate the t statistic (p. 435) for this r value, using eqn. [66.3]. The value of t was 7.037, with six degrees of freedom. The critical value from tables (e.g. Table 66.2) at $P = 0.001$ is 5.96, so he concluded that there was a very highly significant positive correlation between his two sets of observations. The investigator moved on from this observation and next investigated the effect of artificial hosepipe rainfall in a sheltered grass plot, to test whether there was a causal relationship involved.

(error) component, i.e. the fraction of the variance in Y that cannot be explained by the line of best fit. Multiplying the values by 100 allows you to express these fractions in percentage terms.

Using computers to calculate hypothesis-testing statistics

As with the calculation of descriptive statistics (p. 423), specialist statistical packages such as SPSS and MINITAB can be used to simplify the calculation of hypothesis-testing statistics. The correct use of the software and interpretation of the output requires an understanding of relevant terminology and of the fundamental principles governing the test, which is probably best obtained by working through one or more examples by hand before using these tools (e.g. Box 53.2, Box 66.2). Spreadsheets offer increasingly sophisticated statistical analysis functions, three examples of which are provided in Box 66.3.

Text references and sources for further study

Finney, D.J., Latscha, R., Bennett, B.M. and Hsu, P. (1963) *Tables for Testing Significance in a 2 × 2 Table*. Cambridge University Press, Cambridge.

Heath, D. (1995) *An Introduction to Experimental Design and Statistics for Biology*. UCL Press, London.

Schmuller, J. (2005) *Statistical Analysis with Excel for Dummies*. Wiley, Hoboken, New Jersey.

Sokal, R.R. and Rohlf, F.J. (1994) *Biometry*, 3rd edn. W.H. Freeman, San Francisco.

Wardlaw, A.C. (2000) *Practical Statistics for Experimental Biologists*, 2nd edn. Wiley, New York.

Study exercises

66.1 Calculate 95 per cent confidence limits. What are the 95 per cent confidence limits of a sample with a mean = 24.7, standard deviation = 6.8 and number of data values = 16? (Express your answer to three significant figures.)

66.2 Use the Poisson distribution. In a sample of 15 snails, a researcher finds the following number of parasite larvae per snail: 0, 0, 0, 0, 1, 1, 1, 3, 4, 5, 5, 7, 7, 9, 9. Using the rule of thumb on p. 429, decide whether the parasites are 'clumped' or 'repulsed' in distribution on their host. What might this mean in biological terms?

66.3 Practise using a *t*-test. A biology student examines the effect of adding a plant hormone to pea plants. She dissolves an appropriate amount of the compound in ethanol and applies 25 µl of this to the uppermost stipules of the treated plants. With the controls, she applies the same amount of pure ethanol. After three days, she measures the distance between the 2nd and 3rd internodes of the plants and obtains the results shown at the top of the next column. Carry out a *t*-test on the data and draw appropriate conclusions.

Internode distance in cm

| Control | 7.5 | 8.1 | 7.6 | 6.2 | 7.5 | 7.8 | 8.9 |
| Treatment | 5.6 | 7.5 | 8.2 | 6.7 | 3.5 | 6.5 | 5.9 |

66.4 Interpret the output from Excel linear regression analysis. The following output represents a regression analysis for an experiment measuring the uptake of an amino acid by a cell suspension (in $pmol\,cell^{-1}$) against time (in minutes). Based on this output, what is the form and strength of the underlying linear relationship? (Express the coefficients to three significant figures.)

Summary output from Microsoft Excel spreadsheet linear regression analysis

Regression statistics

Multiple R	0.985335951
R square	0.970886937
Adjusted R square	0.963608672
Standard error	2.133876419
Observations	6

ANOVA

	d.f.	SS	MS	F	Significance F
Regression	1	607.4062857	607.4063	133.3954	0.000320975
Residual	4	18.21371429	4.553429		
Total	5	625.62			

	Coefficients	Standard error	t Stat	P-value	Lower 95%	Upper 95%	Lower 95.0%	Upper 95.0%
Intercept	1.171428571	1.544386367	0.758507	0.490383	−3.11648428	5.459341423	−3.11648428	5.459341423
X variable 1	2.945714286	0.255047014	11.54969	0.000321	2.237588784	3.653839787	2.237588784	3.653839787

Chemistry

21 Procedures in volumetric analysis

Volumetric analysis, also known as titrimetric analysis, is a quantitative technique used to determine the amount of a particular substance in a solution of unknown composition.

This requires:

- A standard solution, which is a solution of a compound of accurately known concentration, that reacts with the substance to be analysed.
- The test solution, containing an unknown concentration of the substance to be analysed.
- Some means of detecting the end-point of the reaction between the standard and test solutions, e.g. a chemical indicator or, in the case of potentiometric titrations, a pH electrode (see Chapter 34). Some reactions exhibit a colour change at the end-point without the addition of an indicator.

Definition

A stoichiometric titration is one with a known reaction path, for which a chemical reaction can be written, and having no alternative or side reactions.

The volume of standard solution that reacts with the substance in the test solution is accurately measured. This volume, together with a knowledge of concentration of the standard solution and the stoichiometric relationship between the reactants, is used to calculate the amount of substance present in the test solution. Specific examples of the different types of calculations involved are shown in Chapters 22 to 25.

Classification of reactions in volumetric analysis

There are four main types of reaction

1. Acid–base or neutralization reactions, where free bases are reacted with a standard acid (or vice versa). These reactions involve the combination of hydrogen and hydroxide ions to form water.
2. Complex formation reactions, in which the reactants are combined to form a soluble ion or compound. The most important reagent for formation of such complexes is ethylenediamine tetra-acetic acid, EDTA (as the disodium salt).
3. Precipitation reactions, involving the combination of reactants to form a precipitate.
4. Oxidation–reduction reactions, i.e. reactions involving a gain (reduction) or loss (oxidation) of electrons. The standard solutions used here are either oxidizing agents (e.g. potassium permanganate) or reducing agents (e.g. iron (II) compounds).

What can be measured by titration?

- The concentration of an unknown substance e.g. $0.900 \, mol \, L^{-1}$.
- Percentage purity, e.g. 56%.
- Water of crystallization, e.g. $(NH_4)_2SO_4 \cdot nH_2O$.
- Percentage of a metal in a salt, e.g. 12% Fe in a salt.
- Water hardness, e.g. determination of the concentration of calcium and magnesium ions.

Examples of the types of calculations used in volumetric analysis are shown in Box 21.1.

Box 21.1 Types of calculations used in volumetric analysis – titrations

In titrations you react a solution of a known concentration with a solution of an unknown concentration.

If you know the mole ratio of the two reacting chemicals in solution, you can calculate the amount (the number of moles and thus the number of grams) of the solute in the solution of unknown concentration.

Let's look at the reaction between NaOH and HCl:

$$HCl + NaOH = NaCl + H_2O$$

Since the equation is balanced we know that 1 mol (36.5 g) of HCl will react with 1 mol (40 g) of NaOH. We know that a 1.0 M solution of HCl contains 1 mol of HCl in 1000 mL (1 litre) of water. Then:

1000 mL of 1.0 M HCl solution   is equivalent to 1.0 mol of NaOH
                                                  is equivalent to 1000 mL of 1.0 M NaOH solution
                                                  is equivalent to 40 g of NaOH
                                                  is equivalent to 23 g of Na^+ ions
                                                  is equivalent to 17 g of OH^- ions

Similarly for the reaction between potassium hydroxide and sulphuric acid:

$$H_2SO_4 + 2KOH = K_2SO_4 + 2H_2O$$

Since *1 mol* of H_2SO_4 reacts with *2 mol* of KOH, then:

1000 mL of 1.0 M H_2SO_4 solution   is equivalent to 2.0 mol of KOH
                                                  is equivalent to 2×1000 mL of 1.0 M KOH solution
                                                  is equivalent to 2×56 g of KOH
                                                  is equivalent to 2×39 g of K^+ ions
                                                  is equivalent to 2×17 g of OH^- ions

To work out the results of titrations you *must* always:

- Work out the balanced equation to find out the ratio of moles reacting.
- Decide what you are trying to calculate.

Example: 25.00 mL of sodium hydroxide solution were titrated by 24.00 mL of 0.1 M HCl solution. Calculate the concentration of the sodium hydroxide solution.

- $HCl + NaOH = NaCl + H_2O$
- Concentration of NaOH, i.e. moles of NaOH in 1000 mL, since concentration is mol L^{-1}.

Now:

1000 mL of 1.0 M HCl solution is equivalent to 1.0 mol of NaOH

but the concentration of HCl is only 0.1 M:

1000 mL of *0.1 M* HCl solution is equivalent to 0.1×1.0 mol of NaOH

but only 24.00 mL of HCl solution were used:

1.0 mL of 0.1 M HCl solution is equivalent to $\dfrac{1.0 \times 1.0 \times 1.0}{1000}$ mol of NaOH

and

24 mL of 0.1 M HCl solution is equivalent to $\dfrac{24 \times 1.0 \times 0.1 \times 1.0}{1000}$ mol of NaOH

$$= 2.4 \times 10^{-3} \text{ mol of NaOH}$$

but 25.00 mL of NaOH solution were used:

25.00 mL of NaOH solution contains 2.4×10^{-3} mol of NaOH

Box 21.1 (continued)

Then

1.0 mL of NaOH solution contains $\dfrac{2.4 \times 10^{-3}}{25}$ mol of NaOH

and

1000 mL of NaOH solution contains $\dfrac{1000 \times 2.4 \times 10^{-3}}{25}$ mol of NaOH

$= 0.096$ mol of NaOH

Therefore concentration of NaOH solution is *0.096 mol L^{-1}*.

Using this set of equations you can calculate directly the mass of NaOH per litre, the mass or moles of sodium ions and the mass or moles of hydroxide ions.

Note: The expression $[C_1]V_1 = [C_2]V_2$ was not used, even though it is applicable in this case.

Problems arise when $[C_1]V_1 = [C_2]V_2$ is used for reactions which are not 1 : 1, e.g.:

$$H_2SO_4 + 2KOH = K_2SO_4 + 2H_2O$$

or

$$2MnO_4^- + 16H^+ + 5C_2O_4^{2-} = 2Mn^{2+} + 10CO_2 + 8H_2O$$

or

$$IO_3^- + 5I^- + 6H^+ = 3I_2 + 3H_2O$$

Titrations

The process of adding the standard solution to the test solution is called a titration, and is carried out using a burette (see below). The point at which the reaction between the standard solution and the test substance is just complete is called the equivalence point or the theoretical (or stoichiometric) end-point. This is normally detected by a visible change, either of the standard solution itself or, more commonly, by the addition of an indicator.

Definition

A primary standard should be easy to obtain in a pure form. It should be unaffected in air during weighing, be capable of being tested for impurities and be readily soluble under the conditions used. Finally, the reaction with the standard solution should be stoichiometric and instantaneous.

Standard solutions

A standard solution can be prepared by weighing out the appropriate amount of a pure reagent and making up the solution to a particular volume, as described on p. 24. The concentration of a standard solution is expressed in terms of molarity (p. 19). A substance used in a primary standard should fulfil the following criteria:

- It should be obtainable in high purity ($>99.9\%$).
- It should remain unaltered in air during weighing (i.e. it should not be hygroscopic).
- It must not decompose when dried by heating or vacuum.
- It should be capable of being tested for impurities.
- It should be readily soluble in an appropriate solvent.
- It must react with the test substance stoichiometrically and rapidly.

Preparing a standard solution

The molarity of a solution is the concentration of the solution expressed as mol L^{-1}. If x g of a substance of molecular weight M_r is dissolved in y mL of distilled water, the moles of substance dissolved $= \dfrac{x}{M_r} = m$. Therefore,

$$\text{molarity (mol L}^{-1}) = \dfrac{m \times 1000}{y} \qquad [21.1]$$

Primary standards prepared from solid compounds should be weighed out using the 'weighing by difference' method as described in Chapter 4, and accurately made up to volume using a volumetric flask. Complete transfer of the substance from the weighing vessel to the volumetric flask is best achieved by inserting a funnel into the neck of the flask (Fig. 21.1). As much of the solid as possible should be transferred *via* the funnel. The funnel should be washed with distilled water prior to removal. Distilled water is then added to the flask, with occasional swirling to help to dissolve the solid. This is continued until the meniscus is about 1 cm below the volume mark. At this point a stopper is inserted and the flask is inverted several times to ensure the solid is completely dissolved. Finally, using a Pasteur pipette, distilled water is added up to the volume mark. The solution should be thoroughly mixed before use.

If the solid is not readily soluble in cold water it may be possible to dissolve it by stirring in warm water in a beaker. After allowing the solution to cool to room temperature, it can be transferred to the volumetric flask using a glass rod and filter funnel (Fig. 21.1) followed by several rinses of the glass rod/filter funnel. Finally, the solution is made up to the mark with distilled water.

Filling a burette

> For accurate readings – always remember to position your burette vertically.

- Clamp a clean 50.00 mL burette (p. 10) in a laboratory stand. Place a beaker on a white tile immediately below the outlet of the burette (Fig. 21.2).
- Place a small filter funnel on top of the burette and, with the tap open, carefully pour in the standard solution (or titrant) until it starts to drain into the beaker.
- Close the burette tap, and fill the burette with the standard solution until the meniscus is about 1–2 cm above the zero mark. Remove the funnel.
- Open the tap and allow the solution to drain until the meniscus falls to the zero mark. The burette is then ready for the titration.

Note that to avoid contamination the solution in the beaker should be discarded, rather than recycled.

Using a pipette

> Titrand – this is the solution of unknown composition in the conical flask. The titrant is the standard solution in the burette.

> **Safety note** Never mouth pipette.

A clean 25.00 mL pipette (p. 10) is required together with a suitable pipette filler. Various designs of pipette filler are available. The most common type is based on a rubber-bulb suction device. It is best to evaluate a range of pipette fillers, if available in the laboratory, for ease of use and performance. The pipetting procedure is as follows:

- Pour the solution of unknown composition (the titrand) into a beaker. Never place the pipette in the volumetric flask containing the solution as this can lead to contamination of the solution from the external surface of the pipette.
- Using your pipette filler, draw the titrand to just beyond the graduation mark (Fig. 21.3a). Remove the pipette filler, and invert the pipette to allow the solution to drain out. This ensures that the titrand used subsequently will be undiluted and uncontaminated by any residue or liquids in the pipette.

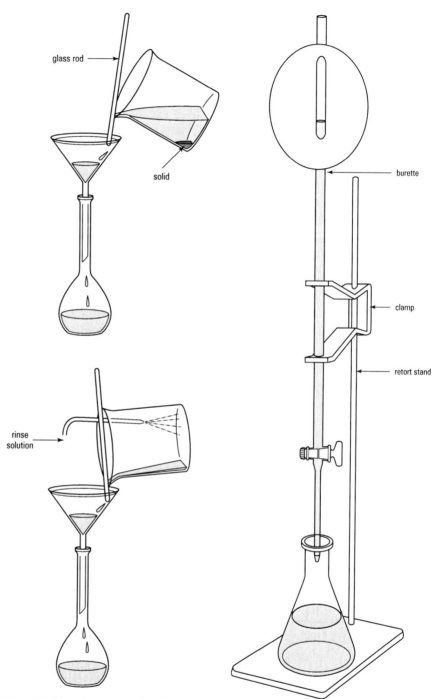

Fig. 21.1 Quantitative transfer of a solid to a volumetric flask.

Fig. 21.2 Apparatus for a titration.

- Refill the pipette until the meniscus of the titrand is above the graduation mark (Fig. 21.3a). Remove the pipette filler and block the hole at the top of the pipette with the index finger of your right hand (if right-handed; Fig. 21.3b). Carefully raise the pipette to eye level, and allow the titrand to drain out into a beaker by lifting your finger slightly from the top of the pipette. Continue until the bottom of the meniscus is on the graduation mark.

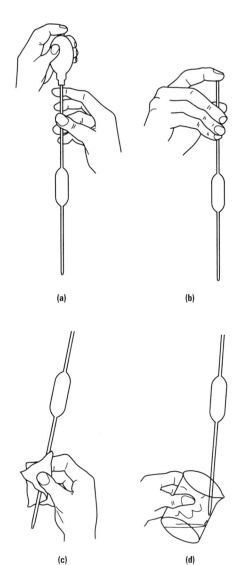

Fig. 21.3 Using a pipette.

- Wipe the outside of the pipette with a tissue (Fig. 21.3c). Be careful not to touch the point of the pipette with the tissue otherwise solution will be lost by capillary action.
- Allow the pipette's contents to drain into a conical flask.
- Finally, touch the end of the pipette on the wall of the flask (Fig. 21.3d), and rinse the inside of the neck of the flask with distilled water. This will ensure that exactly 25.00 mL of the test solution has been delivered by the pipette.

Note that it is normal for a small quantity of solution to remain in the pipette tip. This volume is taken into account when pipettes are calibrated, so do not attempt to 'blow out' this liquid into the conical flask.

Performing a titration

Add one or two drops of indicator to the titrand contained in the conical flask. For a right-handed person the process is as follows:

- Hold the conical flask containing the titrand and the indicator in your right hand, and control the tap of the burette with your left hand. The burette should be arranged so that the tap is on the opposite side of the burette to your palm. In this way, your left hand also supports the body of the burette (Fig. 21.4).
- Add the titrant by opening the tap, and simultaneously swirl the contents of the conical flask. This may take a bit of practice, so do not worry if you cannot do it straight away. The titrant can be added quickly at first, but as the end-point approaches, additions should be made drop-wise.
- The end-point is indicated when the appropriate colour change takes place. When the end-point is reached, one drop of titrant should be sufficient to cause the colour change. You should note the volume used for the titration to the nearest 0.1–0.05 mL. This is done by reading off the volume of titrant used (Fig. 21.5).
- Refill the burette to the zero mark using a funnel ready for the next titration.

> Washing the test solution – any distilled water used for washing the test solution from the walls of the conical flask has no effect on the titration or the calculation.

> Reading a burette – your eye-line should be level with the bottom of the meniscus. Then, record the volume used to one decimal place.

> Rough titration – always carry out an initial rough titration to determine the approximate volume of titrant required for the end-point. This allows you to anticipate the end-point in subsequent titrations to determine the accurate volume.

Volumetric analysis – calculating the concentration of a test substance

The calculation should be carried out in a logical order as follows:

1. Write the balanced equation for the reaction between the standard and the test substance.
2. From the stoichiometry of the reaction, determine how many moles of the test substance react with 1 mole of the standard substance. For example, in the reaction between an H_2SO_4 standard solution and an NaOH test solution:

$$H_2SO_4 + 2NaOH \rightarrow Na_2SO_4 + 2H_2O \qquad [21.2]$$

Therefore 2 moles of NaOH react with 1 mole of H_2SO_4.

3. Calculate the number of moles of standard substance used to reach the end-point of the reaction. This can be determined from knowledge of the concentration of the standard solution (mol L^{-1}) and the volume of titrant used (mL). Remember to take care with units – in this instance division by a factor of 1000 is required to convert mL to L:

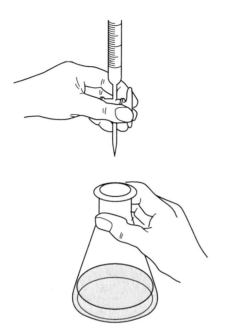

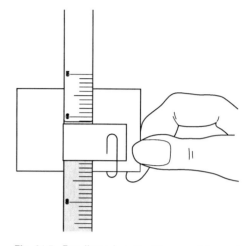

Fig. 21.5 Reading a burette. Place a white card below the level of the meniscus. This allows an accurate reading to be made.

Fig. 21.4 Performing a titration.

> For consistency of results – always perform two or three titrations (or until consistent results are obtained, i.e. titre values within 0.1–0.5 mL).

$$\text{Number of moles} = \frac{\text{concentration (mol L}^{-1}) \times \text{volume (mL)}}{1000}$$

4. The number of moles of test substance present in the titrand is then obtained from knowledge of the equivalences. In the example given above (point 2) the number of moles of test substance is twice the number of moles of standard substance. Therefore, if X moles of H_2SO_4 are used (as calculated in point 3), 2X moles of NaOH were present in the initial volume of test solution.

5. Finally, the concentration of the test solution can be calculated using the formula:

$$\frac{\text{Concentration of test}}{\text{solution (mol L}^{-1})} = \frac{1000 \times \text{amount of test substance (mol)}}{\text{initial volume of test solution (mL)}}$$

Again, the factor of 1000 is used to convert mL to L.

22 Acid–base titrations

The titration of an acid solution with a standard solution of alkali will determine the amount of alkali which is equivalent to the amount of acid present (or vice versa). The point at which this occurs is called the equivalence point or end-point. For example, the titration of hydrochloric acid with sodium hydroxide can be expressed as follows:

$$NaOH_{(aq)} + HCl_{(aq)} \rightarrow NaCl_{(aq)} + H_2O_{(l)} \qquad [22.1]$$

Safety note Never pipette by mouth.

If both the acid and alkali are strong electrolytes, the resultant solution will be neutral (pH 7). If on the other hand either the acid or alkali is a weak electrolyte the resultant solution will be slightly alkaline or acidic, respectively. In either case, detection of the end-point requires accurate measurement of pH. This can be achieved either by using an indicator dye, or by measuring the pH with a glass electrode (described in Chapter 7).

Acid–base indicators

Typical acid–base indicators are organic dyes that change colour at or near the equivalence or end-point. They have the following characteristics:

- They show pH-dependent colour changes.
- The colour change occurs within a fairly narrow pH range (approximately 2 pH units).
- The pH at which a colour change occurs varies from one indicator to another, and it is possible to select an indicator which exhibits a distinct colour change at a pH close to the equivalence or end-point.

Selected common indicators together with their pH ranges and colour changes are shown in Table 22.1. Examples for thymol blue and phenolphthalein are shown in Fig. 22.1.

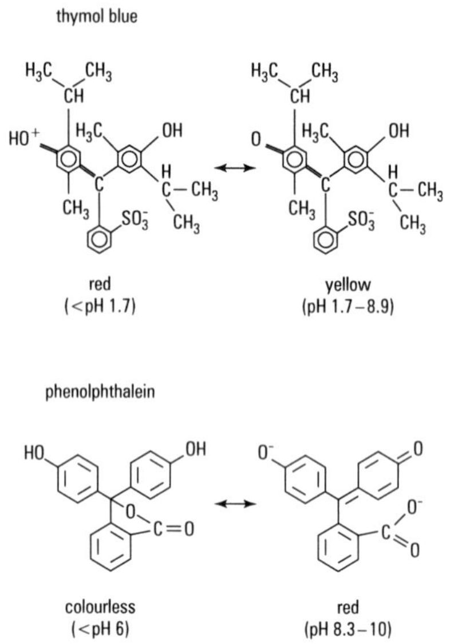

Figure 22.1 Examples of indicators used in acid-base titrations: thymol blue and phenolphthalein

Table 22.1 Colour changes and pH range of selected indicators

Indicator	pH range	Colour in acid solution	Colour in alkaline solution
Thymol blue	1.2–6.8	Red	Yellow
Methyl orange	2.8–4.0	Red	Yellow
Methyl red	4.3–6.1	Red	Yellow
Phenol red	6.8–8.2	Yellow	Red
Phenolphthalein	8.3–10.0	Colourless	Pink/red

Neutralization curves

A plot of pH against the volume of alkali added (mL) is known as a neutralization or titration curve (Fig. 22.2). The curve is generated by a 'potentiometric titration' in which pH is measured after each addition of alkali (or acid). The significant feature of the curve is the very sharp and sudden change in pH near to the equivalence point of the titration. For a strong acid and alkali this will occur at pH 7. If either the acid or base concentration is unknown, a preliminary titration is necessary to find the approximate equivalence point followed by a more accurate titration as described on p. 146. The ideal pH range for an indicator is 4.5–9.5.

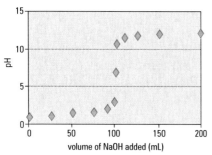

Figure 22.2 A typical neutralization curve: 0.1 M HCl with 0.1 M NaOH.

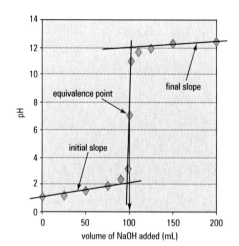

Figure 22.3 Determination of the equivalence point.

Subscripts – 'aq' is used to represent aqueous, 'l' the liquid state.

Determination of the equivalence point

From the neutralization curve (Fig. 22.2), the initial and final slopes are drawn (Fig. 22.3) and a parallel line is drawn such that the mid-point is on the curve. This is the equivalence point, producing a titration value of x mL.

Example calculations

Standardization of a sodium hydroxide solution

What is the molarity of a solution of sodium hydroxide, 25.0 mL of which requires 21.0 mL of a standard solution of hydrochloric acid of concentration 0.100 mol L^{-1} for neutralization?

Following the sequence in Box 21.1:
1. Write the equation:

$$NaOH_{(aq)} + HCl_{(aq)} \rightarrow NaCl_{(aq)} + H_2O_{(l)} \quad (22.2)$$

2. Determine the equivalences.
   Equation [22.2] shows that 1 mole of NaOH requires 1 mole of HCl for neutralization, i.e. 1 mole of NaOH is equivalent to 1 mole of HCl.
3. Calculate the number of moles of standard used.

$$\text{Number of moles of HCl} = \frac{\text{concentration (mol L}^{-1}) \times \text{volume (mL)}}{1000}$$

$$= \frac{0.100 \text{ mol L}^{-1} \times 21.0 \text{ mL}}{1000}$$

$$= 2.1 \times 10^{-3} \text{ mol}$$

4. Calculate the number of moles of NaOH in the initial volume of test solution.
   As indicated in point 2 above, 1 mole of NaOH is equivalent to 1 mole of HCl. Therefore:

   no. of moles of NaOH in initial volume = no. of moles of HCl used

   $$= 2.1 \times 10^{-3} \text{ mol}$$

5. Determine the concentration of the NaOH solution.

$$\text{Concentration of test solution (mol L}^{-1}) = \frac{1000 \times \text{amount of test substance}}{\text{initial volume of test solution (mL)}}$$

$$= \frac{1000 \times 2.1 \times 10^{-3} \text{(mol)}}{25 \text{ (mL)}}$$

$$= 0.084 \text{ mol L}^{-1}$$

Standardization of a sodium hydroxide solution using potassium hydrogen phthalate as a primary standard (p. 143)

An accurately weighed amount (5.1100 g) of potassium hydrogen phthalate (KHC$_8$H$_4$O$_4$) was dissolved in distilled water (250.00 mL). This solution (25.00 mL) required sodium hydroxide solution (23.50 mL) to reach equivalence. What is the molarity of the sodium hydroxide solution? (Note that, in this case, 25.00 mL of the standard solution was used as the titrand, whereas the test solution (NaOH) was the titrant.)

Following the sequence in Box 21.1:

1. Write the equation.

$$NaOH_{(aq)} + KHC_8H_4O_{4(aq)} \rightarrow NaKC_8H_4O_{4(aq)} + H_2O_{(l)} \qquad [22.3]$$

2. Determine the equivalences.

    Equation [22.3] shows that 1 mole of NaOH is equivalent to 1 mole of $KHC_8H_4O_4$.

3. Calculate the number of moles of standard used.

    Firstly, the concentration of the standard solution of potassium hydrogen phthalate must be calculated.

    The molecular weight of $KHC_8H_4O_4$ is $204.22\,g\,mol^{-1}$. Therefore 5.1100 g of $KHC_8H_4O_4$ is equivalent to $5.1100\,(g)/204.22\,(g\,mol^{-1}) = 0.025\,mol$.

    We can now calculate the concentration of the standard solution:

$$\text{Concentration of standard solution (mol L}^{-1}) = \frac{1000 \times \text{amount of } KHC_8H_4O_4 \text{ (mol)}}{\text{volume of standard solution (mL)}}$$

$$= \frac{1000 \times 0.025\,(\text{mol})}{250\,(\text{mL})}$$

$$= 0.100\,\text{mol L}^{-1}$$

The number of moles of standard used in the titration is as follows.

$$\text{Number of moles of } KHC_8H_4O_4 = \frac{\text{concentration (mol L}^{-1}) \times \text{volume (mL)}}{1000}$$

$$= \frac{0.100\,\text{mol L}^{-1} \times 25.0\,\text{mL}}{1000}$$

$$= 2.5 \times 10^{-3}\,\text{mol}$$

4. Calculate the number of moles of NaOH used to reach equivalence.

    As indicated in point 2 above, 1 mole of NaOH is equivalent to 1 mole of $KHC_8H_4O_4$. Therefore:

$$\text{No. of moles of NaOH used} = \text{no. of moles of } KHC_8H_4O_4 \text{ in the titrand}$$

$$= 2.5 \times 10^{-3}\,\text{mol}$$

5. Determine the concentration of the NaOH solution:

$$\text{Concentration of NaOH solution (mol L}^{-1}) = \frac{1000 \times \text{amount of NaOH (mol)}}{\text{volume of NaOH solution used (mL)}}$$

$$= \frac{1000 \times 2.5 \times 10^{-3}\,(\text{mol})}{23.50\,(\text{mL})}$$

$$= 0.106\,\text{mol L}^{-1}$$

28 Infrared spectroscopy

> **Identifying compounds** – the combination of techniques described in this and the following chapters can often provide sufficient information to identify a compound with a low probability of error.
>
> **Definitions**
>
> **Spectroscopy** – any technique involving the production and subsequent recording of a spectrum of electromagnetic radiation, usually in terms of wavelength or energy.
>
> **Spectrometry** – any technique involving the measurement of a spectrum, e.g. of electromagnetic radiation, molecular masses, etc.
>
> **Interpreting spectra** – the spectrum produced in UV–vis, IR and NMR spectroscopy is a plot of wavelength or frequency or energy (x-axis) against absorption of energy (y-axis). Convention puts high frequency (high energy, short wavelength) at the left-hand side of the spectrum.

In addition to ultraviolet–visible (UV–vis) spectroscopy (p. 164), there are three other essential techniques that you will encounter during your laboratory course. They are:

1. *Infrared (IR) spectroscopy*: this is concerned with the energy changes involved in the stretching and bending of covalent bonds in molecules.
2. *Nuclear magnetic resonance (NMR) spectroscopy*: this involves the absorption of energy by specific atomic nuclei in magnetic fields and is probably the most powerful tool available for the structural determination of molecules (Chapter 29).
3. *Mass spectrometry (MS)*: this is based on the fragmentation of compounds into smaller units. The resulting positive ions are then separated according to their mass-to-charge ratio (m/z) (Chapter 30).

As with UV–vis spectroscopy, IR and NMR spectroscopy are based on the interaction of electromagnetic radiation with molecules, whereas MS is different in that it relies on high-energy particles (electrons or ions) to break up the molecules. The relationship between the various types of spectroscopy and the electromagnetic spectrum is shown in Table 28.1.

Infrared spectroscopy

A covalent bond between two atoms can be crudely modelled as a spring connecting two masses and the frequency of vibration of the spring is defined by Hooke's law (eqn [28.1]), which relates the frequency of the vibration (v) to the strength of the spring, expressed as the force constant (k), and to the masses (m_1 and m_2) on the ends of the spring (defined as the reduced mass $\mu = (m_1 \times m_2) \div (m_1 + m_2)$).

$$v = \frac{1}{2\pi}\sqrt{\frac{k}{\mu}} \qquad [28.1]$$

In simple terms, this means that:

- the stretching vibration of a bond between two atoms will increase in frequency (energy) if on changing from a single bond to a double bond and then to a triple bond between the same two atoms (masses), i.e. the spring gets stronger. For example,

Table 28.1 The electromagnetic spectrum and types of spectroscopy

Type of radiation	Origin	Wavelength	Type of spectroscopy
γ-rays	Atomic nuclei	< 0.1 nm	γ-ray spectroscopy
X-rays	Inner shell electrons	0.01–2.0 nm	X-ray fluorescence (XRF)
Ultraviolet (UV)	Ionization	2.0–200 nm	Vacuum UV spectroscopy
UV/visible	Valency electrons	200–800 nm	UV/visible spectroscopy
Infrared	Molecular vibrations	0.8–300 μm	IR and Raman spectroscopy
Microwaves	Molecular rotations Electron spin	1 mm to 30 cm	Microwave spectroscopy Electron spin resonance (ESR)
Radio waves	Nuclear spin	0.6–10 m	Nuclear magnetic resonance (NMR)

$$v \text{ for } C\equiv C > v \text{ for } C=C > v \text{ for } C-C$$

- as the masses of the atoms on a bond increases, the frequency of the vibration decreases, i.e the effect of reducing the magnitude of μ; for example,

$$v \text{ for } C-H > v \text{ for } C-C; v \text{ for } C-H > v \text{ for } C-D; v \text{ for } O-H > v \text{ for } S-H$$

Bonds can also bend, but this movement requires less energy than stretching and thus the bending frequency of a bond is always *lower* than the corresponding stretching frequency. When IR radiation of the same frequency as the bond interacts with the bond it is absorbed and increases the amplitude of vibration of the bond. This absorption is detected by the IR spectrometer and results in a peak in the spectrum. For a vibration to be detected in the IR region the bond must undergo a change in dipole moment when the vibration occurs. Bonds with the greatest change in dipole moment during vibration show the most intense absorption, e.g. C=O and C-O.

Since bonds between specific atoms have particular frequencies of vibration, IR spectroscopy provides a means of identifying the type of bonds in a molecule, e.g. all alcohols will have an O-H stretching frequency and all compounds containing a carbonyl group will have a C=O stretching frequency. This property, which does not rely on chemical tests, is extremely useful in diagnosing the functional groups within a covalent molecule.

> **IR absorption bands** – since the frequency of vibration of a bond is a specific value you would expect to see line spectra on the chart. However, each vibration is associated with several rotational motions and bands (peaks) are seen in the spectrum.

IR spectra

A typical IR spectrum is shown in Fig. 28.1 and you should note the following points:

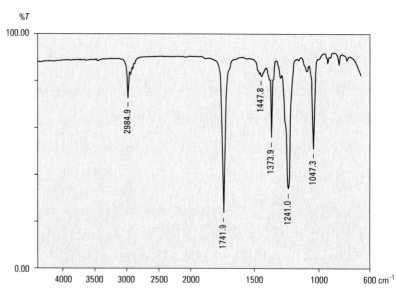

Fig. 28.1 IR spectrum of ethyl ethanoate $CH_3COOCH_2CH_3$ as a liquid film.

- The *x*-axis, the wavelength of the radiation, is given in wavenumbers ($\bar{v}$) and expressed in reciprocal centimetres (cm^{-1}). You may still see some spectra from old instruments using microns (μ, equivalent to the SI unit 'micrometres', μm, at 1×10^{-6} m) for wavelength; the conversion is given by eqn [28.2]:

$$\text{wavenumber (cm}^{-1}) = 1/\text{wavelength (cm)} = 10\,000/\text{wavelength (}\mu m\text{)}$$
[28.2]

> **The use of wavenumber** – this is an old established convention, since high wavenumber = high frequency = high energy = short wavelength. Expression of the IR range, $4000\,cm^{-1}$ to $650\,cm^{-1}$, is in 'easy' numbers and the high energy is found on the left-hand side of the spectrum. Note that IR spectroscopists often refer to wavenumbers as 'frequencies', e.g. 'the peak of the C=O stretching 'frequency' is at $1720\,cm^{-1}$'.

- The y-axis, expressing the amount of radiation absorbed by the molecule, is usually shown as % transmittance (p. 164). When no radiation is absorbed (all is transmitted through the sample) we have 100% transmittance and 0% transmittance implies all radiation is absorbed at a particular wavenumber. Since the y-axis scale goes from 0 to 100% transmittance, the absorption peaks are displayed *down* from the 100% line; this is *opposite* to most other common spectra.
- The cells holding the sample usually display imperfections and are not completely transparent to IR radiation, even when empty. Therefore the base line of the spectrum is rarely set on 100% transmittance and quantitative applications of IR spectroscopy are more complex than for UV–vis (p. 166).

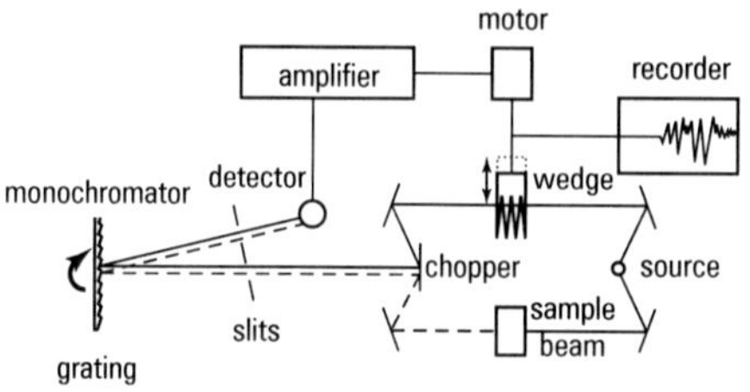

IR spectrometers

There are two general types:

1. Double-beam or dispersive instruments in which the IR radiation from a single source is split into two identical beams. One beam passes through the sample and the other is used as a reference and passes through air or the pure solvent used to dissolve the sample. The difference in intensity of the two beams is detected and recorded as a peak; the principal components of this type of instrument are shown in Fig. 28.2. The important controls on the spectrometer are:

    (a) scan speed: this is the rate at which the chart moves – slower for greater accuracy and sharp peaks;
    (b) wavelength range: the full spectrum or a part of the IR range may be selected;
    (c) 100% control: this is used to set the pen at the 100% transmittance line when no sample is present the base line. It is usual practice to set the pen at 90% transmittance at $4000\,cm^{-1}$ when the sample is present, to give peaks of the maximum deflection.

    You should remember that this is an electromechanical instrument and you should always make sure that you align the chart against the calibration marks on the chart holder. In the more advanced instruments an on-board computer stores a library of standard spectra, which can be compared with your experimental spectrum.

2. Fourier transform IR (FT–IR) spectrometer: the value of IR spectroscopy is greatly enhanced by Fourier transformation, named after the mathematician J.B. Fourier. The FT is a procedure for interconverting frequency functions and time or distance functions. The IR beam, composed of all the frequencies in the IR range, is passed through the sample and generates interference patterns, which are then transformed electronically into a normal IR spectrum. The advantages of FT–IR are:

    (a) rapid scanning speed – typically four scans can be made per minute, allowing addition of the separate scans to enhance the signal-to-noise ratio and improve the resolution of the spectrum;
    (b) simplicity of operation – the reference is scanned first, stored and then subtracted from the sample spectrum;
    (c) enhanced sensitivity: the facility of spectrum addition from multiple scans permits detection of smaller quantities of chemicals;

Fig. 28.2 Schematic diagram of a double-beam IR spectrometer.

> Using double-beam instruments – you can identify the sample beam by quickly placing your hand in the beam. If the pen records a peak, this is the sample beam, but if the pen moves up, then this is the reference beam.

> Using the 100% control – if you use this control to set the base line for the sample, you *must* turn down the 100% control when you remove the sample, otherwise the pen-drive mechanism may be damaged in trying to drive off the top of the chart.

Box 28.1 How to run an infrared spectrum of a liquid or solid film, mull or KBr disk

A. Double-beam spectrometer

1. **Ensure that the instrument is switched on** and that it has had a few minutes to warm up.
2. **Make sure that the chart is aligned with the calibration marks on the chart bed or chart drum.** Most spectrometers scan from $4000\,cm^{-1}$ to $650\,cm^{-1}$ and the pen should be at the $4000\,cm^{-1}$ mark.
3. **Adjust the 100% transmittance control to about 90%**, if necessary.
4. **Place the sample cell in the sample beam and adjust the 100% transmittance control to 90%**, or the highest value possible.
5. **Select the scan speed.** You must balance the definition required in the spectrum with the time available for the experiment. For most qualitative applications the fastest setting is satisfactory.
6. *(missing)*
7. **Press the 'scan' or 'start' button to run the spectrum.** The spectrum will be recorded and the spectrometer will automatically align itself at the end of the run. *Do not press* any other buttons while the spectrum is running or the instrument may not realign itself at the end of the run.
8. **Adjust the 100% transmittance control to about 50%**, remove the sample cell from the spectrometer and turn the 100% transmittance control to about 90%.
9. **Enter all of the following data on the spectrum:** name, date, compound and phase (liquid film, Nujol® mull, KBr disk, etc.).

B. FT–IR spectrometer

1. **Make sure that the sample compartment is empty** and close the lid.
2. **Select the number of scans**; usually four is adequate for routine work.
3. **Select 'background' on the on-screen menu**, and scan the background. *Do not press* any other buttons or icons while the spectrum is running.
4. **Place the sample cell in the beam**, close the lid, select 'sample' and scan the sample. *Do not press* any other buttons or icons while the spectrum is running.
5. **Select 'customize'**, or a similar function, and enter all the data – name, date, compound, phase (liquid film, Nujol® mull, KBr disk, etc.) – on the spectrum.
6. **Select 'print'**, to produce the spectrum from the printer.

Problems with IR spectra

These are usually caused by poor sample preparation and the more common faults are:

1. **The large peaks have tips below the bottom of the chart or the large peaks have 'squared tips' near the bottom of the chart**: the sample is too thick; remove some sample from the cell and rerun the spectrum.
2. **The spectrum is 'weak', i.e. few peaks**: the sample is too thin – add more sample or remake the KBr disk.
3. **The base line cannot be adjusted to 90% transmittance**: the NaCl plates or KBr disk are 'fogged', scratched or dirty – replace or remake the KBr disk.
4. **The pen tries to 'go off' the top of the spectrum**: obviously due to some absorption at $4000\,cm^{-1}$ when you were setting the base line. Repeat base-line set-up but at 80% transmittance and bear in mind that dirty plates, above, can be the cause.

(d) the integral computer system enables the use of libraries of spectra and simplifies spectrum manipulation, such as the subtraction of contaminant or solvent spectra.

The procedures for running IR spectra on double-beam and FT spectrometers are described in Box 28.1.

Sample handling

You can obtain IR spectra of solids, liquids and gases by use of the appropriate sample cell (sample holder). The sample holder must be completely transparent to IR radiation; consequently glass and plastic cells cannot be used. The most common sample cells you will encounter are made from sodium chloride or potassium bromide and you cannot use aqueous

---

IR spectra of aqueous solutions – special sample cells made from CaF_2 are available for aqueous solutions, but they are expensive and only used in specific applications.

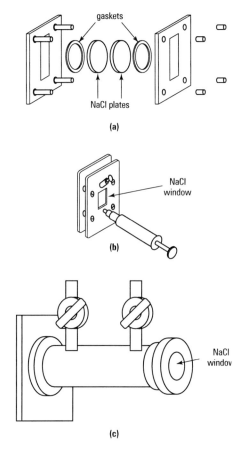

Fig. 28.3 Cells for IR spectroscopy: (a) demountable cell for liquid and solid films and mulls; (b) solution cell; (c) gas cell.

> **Storing IR sample cells and KBr powder** – cells are always stored in desiccators to prevent 'fogging' by absorption of moisture. KBr powder must be dried in the oven, cooled and kept in a desiccator.

> **Handling NaCl plates and KBr disks** – NaCl plates are delicate and easily damaged by scratching, dropping or squeezing. Hold them only by the edges and place them on filter paper or tissue when adding chemicals. KBr disks should be handled using tweezers.

solutions or very wet samples, otherwise the sample cells will dissolve. A typical range of sample cells is shown in Fig. 28.3 and for routine qualitative work you will regularly use NaCl plates and KBr disks to obtain spectra of solids and liquids. Solution cells and gas cells are utilized in more specialized applications and require specific instructions and training.

Liquid samples

The most convenient way to obtain the IR spectrum of a pure, dry liquid is to make a thin liquid film between two NaCl disks (plates). Since the film thickness is unknown, this procedure is not applicable to quantitative work.

Solid samples

If you were to place a fine powder between two NaCl plates, a usable spectrum would not be obtained because the IR radiation would be scattered by diffraction at the edges of the particles and would not pass through to the detector. There are *three* solutions to this problem:

1. *Mulls*: in which the finely ground solid is mixed with a liquid, usually Nujol® (liquid paraffin) or, less frequently, HCB (hexachloro-1,3-butadiene). This mulling liquid does not dissolve the chemical but fills the gaps round the edges of the crystals preventing diffraction and scattering of the IR radiation. Remember that these mulling liquids have their own IR spectrum, which is relatively simple, and can be subtracted either 'mentally' or by the computer. The choice of mulling liquid depends upon the region of the IR spectrum of interest: Nujol® is a simple hydrocarbon containing only C–H and C–C bonds, whereas HCB has no C–H bonds, but has C–Cl, C=C and C–C bonds. Examination of the separate spectra of your unknown compound in each of these mulling agents enables the full spectrum to be analysed.

2. *KBr disks*: here the finely ground solid compound is mixed with anhydrous KBr and squeezed under pressure. The KBr becomes fluid and forms a disk containing the solid compound dispersed evenly within it and suitable for obtaining a spectrum. The advantage of the KBr disk technique is the absence of the spectrum from the mulling liquid, but the disadvantages are the equipment required (Fig. 28.4) and the practice required to obtain suitable transparent disks, which are very delicate and rapidly absorb atmospheric moisture.

3. *Thin solid films*: here a dilute solution of the compound in a low-boiling-point solvent such as dichloromethane or ether is allowed to evaporate on a NaCl plate producing a thin transparent film. This method gives excellent results but is slightly limited by solubility factors.

When you are recording spectra of mulls, KBr disks and thin solid films air is used as the reference and they are suitable for qualitative analysis only. The procedure for the preparation of liquid and solid films and mulls is described in Box 28.2 and that for KBr disks in Box 28.3.

Interpretation of IR spectra

To identify compounds from their IR spectrum you should know at which frequencies the stretching and bending vibrations occur. A detailed analysis can be achieved using the correlation tables found in specialist textbooks. For interpretation, the spectrum is divided into three regions.

Box 28.2 How to prepare liquid and solid films and mulls

A. Preparing a liquid film

1. **Select a pair of clean NaCl plates from a desiccator**, clean them by wiping with a soft tissue soaked in dichloromethane and place them on the bench on a piece of filter paper or tissue paper to prevent scratching by the bench surface.

2. **Using a glass rod or boiling stick, place a small drop of liquid in the centre of one of the plates.** Do not use a Pasteur pipette, which may scratch the surface of the plate.

3. **Carefully, holding it by the edge, place the other plate on top and see if a thin film spreads between the plates, covering the centres.** Do not press to force the plates together. If there is not enough liquid, carefully separate the plates by lifting at the edge and add another drop of liquid. If there is too much liquid, separate the plates and wipe the liquid from one of them using a soft tissue.

B. Preparing a thin solid film

1. **Dissolve the sample (about 5 mg) in a suitable low-boiling-point solvent (about 0.25 mL)**, such as DCM or ether.

2. **Place two drops of the solution onto the centre of a NaCl plate** and allow the solvent to evaporate. Use a Pasteur pipette, but *do not* touch the surface of the plate.

3. **If the resulting thin film of solid does not cover the centre of the plate**, add a little more solution.

4. **Mount the *single* NaCl plate in the spectrometer and run the spectrum.** Note that the NaCl plate can rest on the 'V'-shaped wedge on the sample holder in the spectrometer.

C. Preparing a mull

1. **Grind a small sample of your compound (about 5 mg) using a small agate mortar and pestle for at least 2 minutes.** The powder should be as fine as possible.

2. **Add one drop of mulling agent (Nujol® or HCB) and continue grinding until a smooth paste is formed.** If the mull is too thick, add another drop of mulling agent, or if it is too thin, add a little more solid. Only experience will give you the correct consistency of the mull and the key to a good spectrum is a mull of the correct fluidity.

3. **Transfer the mull to the centre or along the diameter of an NaCl plate**, on a piece of filter paper or tissue paper to prevent scratching by the bench surface, using a small plastic spatula or a boiling stick.

4. **Carefully, holding it by the edge, place the other plate on top and very gently press to ensure that the mull spreads as a thin film between the plates.** If there is not enough liquid, carefully separate the plates by lifting at the edge and add another drop of mull. If there is too much liquid (poor spectrum), separate the plates and wipe the mull from one of them using a tissue.

C. Setting up the cell holder for liquid films and mulls

1. **Place the back-plate of the cell holder on the bench, position the rubber gasket, place the NaCl plates on the gasket and then put the second gasket on top of the plates.** These gaskets are essential to prevent fracture of the plates when you tighten the locking nuts.

2. **Carefully place the cell holder top-plate on the top gasket, drop the locking nuts into place and carefully tighten each in rotation.** These are safety nuts and if you over-tighten them or if the back- and top-plates are not parallel, they will spring loose to prevent the NaCl plates being crushed.

3. **Transfer the cell holder assembly to the spectrometer and make sure it is securely mounted in the cell compartment.**

4. **Clean the plates in the fume cupboard by wiping them with a tissue soaked in DCM**, stand them on filter paper or tissue paper to allow the solvent to evaporate and put them in the desiccator. Allow the DCM to evaporate from the tissue swab and dispose of it in the chemical waste.

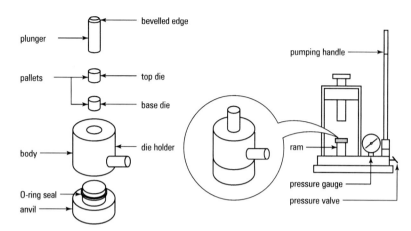

Fig. 28.4 Equipment for preparation of a KBr disk.

Region 1 *(4000–2000 cm⁻¹)*: this region contains the high frequency vibrations such as C–H, N–H and O–H stretching, together with C=C and C≡N stretching vibrations.

Region 2 *(2000–1500 cm⁻¹)*: this is known as the 'functional group region' and includes the stretching frequencies for C=C, C=O, C=N, N=O and N–H bending vibrations.

Region 3 *(1500–650 cm⁻¹)*: this region contains stretching bands for C–O, C–N, C–Hal and the C–H bending vibrations. It is known as the 'fingerprint region' because it also contains complex low-energy vibrations resulting from the overall molecular structure and these are unique to each different molecule. Fig. 28.5 shows the spectra of 1-propanol and 1-butanol, both of which show almost identical peaks for the O–H, C–H and C–O stretching frequencies and the C–H bending frequencies, but the spectra are different in the number and intensity of the peaks between 1500 and 650 cm⁻¹, resulting from the presence of the additional CH₂ in 1-butanol. Conversely, these highly specific bands in the 'fingerprint' region are useful for identification of molecules by comparison with authentic spectra via a database.

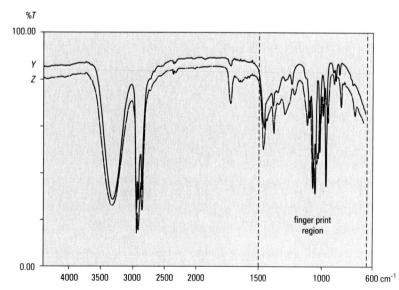

Fig. 28.5 IR spectra of 1-propanol (Y) and 1-butanol (Z).

Box 28.3 How to prepare a KBr disk

1. **Take spectroscopic-grade KBr powder from the oven** and allow it to cool in a desiccator.

2. **Grind your compound (1–2 mg) in an agate mortar for 2 minutes, then add the KBr (0.2 g) and continue grinding to a fine powder.** Put the KBr powder back into the oven.

3. **Obtain a 'disk kit'** and make sure that:
   (a) it is complete – comprising a plunger, two dies (base and top), a die holder and an anvil, as shown in Fig. 28.4;
   (b) the components are for the same device – they are not interchangeable with another disk kit – and should be numbered.

4. **Press the die holder onto the anvil ensuring a proper fit.**

5. **Lower the base die, numbered-side down, into the die holder** and make sure it slides into a depth of about 50 mm.

6. **Pour the compound/KBr powder mixture, about one-third to one-half of the amount prepared, into the die holder** and tap gently to produce an even layer on the base die.

7. **Lower the top die, numbered-side up, on top of the KBr mixture and make sure it slides down onto the powder.**

8. **Slide the plunger, with the bevelled edge at the top, into the die holder** ensuring that it is touching the top die and press down gently so that the dies slide to the bottom, ensuring that you do not then push off the anvil.

9. **Place the assembled disk kit in the hydraulic press** and tighten the top screw so that it touches the top of the plunger.

10. **Connect the anvil to a source of vacuum**, e.g. a rotary vacuum pump.

11. **Close the hydraulic release valve on the side of the press and gently pump the handle** until the pressure gauge reads between 8 and 10 tons and leave for 30 seconds.

12. **Open the hydraulic release valve gently** and, when the pressure has fallen to zero, disconnect the vacuum from the anvil.

13. **Loosen the top screw** and remove the disk kit from the press.

14. **Turn the disk kit upside down and carefully pull off the anvil.** Make sure that the plunger does not slide out by supporting it in the palm of your hand.

15. **Gently push the plunger and the base die will emerge from the die holder.** Take off the base die leaving the KBr disk exposed.

16. **Carefully slide the KBr disk into the special disk holder** using a microspatula.

17. **Run the IR spectrum immediately**, because the disk will begin to cloud over as it absorbs atmospheric moisture.

18. **Clean the disk kit components with a tissue and check that all parts are present.**

19. **If the dies or the plunger stick in the die holder, tell your instructor.**

A simple correlation chart indicating the three regions of the spectrum and their associated bond vibrations is shown in Fig. 28.6. You can obtain most diagnostic information from spectral regions 1 and 2, since these are the simplest regions containing the peaks related to specific functional groups, while region 3 is normally used for confirmation of your findings. Another important aspect of the IR spectrum is the relative intensities of the commonly found peaks and you should become familiar with peak sizes. A chart indicating the positions, general shapes and relative intensities of commonly found peaks is shown in Fig. 28.7. When you are attempting to interpret an IR spectrum you should use the approach described in Box 28.4.

If you are studying complexes formed from metals and organic ligands, the metal–ligand stretching vibration will occur below $600 \, cm^{-1}$ and special IR spectrometers are used to observe this region. However, changes in the IR spectrum of the organic ligand on complexation can be detected in the normal $4000–650 \, cm^{-1}$ range.

Infrared Spectroscopy 533

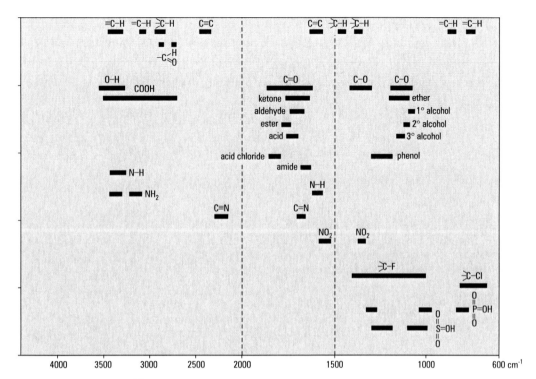

Fig. 28.6 Simplified correlation chart of functional group absorptions.

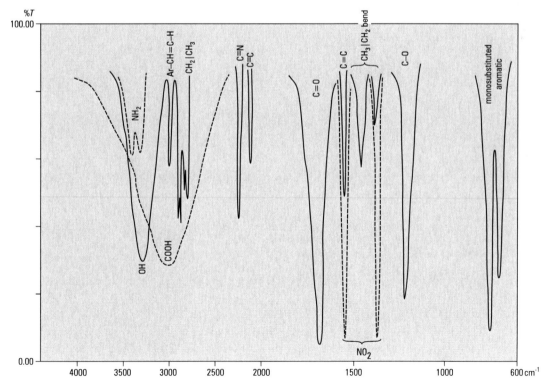

Fig. 28.7 Idealized intensities of some IR bands of common functional groups.

Box 28.4 How to interpret an IR spectrum

1. **Note the conditions under which the spectrum was obtained**, which should be written on the spectrum as 'phase'. If it is a solution or a mull, you will need to identify and 'subtract' the spectrum of the mulling agent or solvent.

2. **Consider carefully the reaction you have carried out.** You should know, from the correlation table, the functional groups and peaks in the starting materials and those expected in the product.

3. **Remember that the absence of peaks may be as useful in interpretation as the presence of peaks.**

4. *Do not attempt to identify all the peaks*, just those which are relevant to your interpretation. Go for the large peaks first.

5. **Many sharp peaks of medium to strong intensity** throughout the spectrum generally indicate an aromatic compound.

6. **Examine region 1 (4000–2000 cm^{-1}).** It is useful to draw a line on the chart at 3000 cm^{-1}: just above the line (3000–3100 cm^{-1}) you will find the stretching frequencies for C_{sp}–H and C_{sp^2}–H indicating unsaturation, whilst just below (2980–2800 cm^{-1}) you will find the C_{sp^3}–H stretching frequencies for CH_3, CH_2 and CH in saturated systems. Other bands for O–H, N–H, C≡C and C≡N are obvious.

7. **Examine region 2 (2000–1500 cm^{-1}).** Here you will find C=O stretch, usually the most intense band in the spectrum; C=C and C=N stretches, less intense and sharper; N=O stretch (from NO_2) intense and sharp and with a twin band in region 3; N–H bending vibrations – do not confuse with C=O.

8. **Examine region 3 (1500–650 cm^{-1}).** The large bands here are C–O, C–N, C–Cl, S=O, P=O, N=O (twin from region 2) stretches and C–H 'breathing' bands (900–700 cm^{-1}), which indicate the number and position of substituents on a benzene ring. Medium-intensity peaks of importance include the CH_3 and CH_2 bands at 1460 cm^{-1} and 1370 cm^{-1} from the carbon skeleton which are also found in Nujol®.

9. **Tabulate your results and make the appropriate deductions**, after consulting the detailed correlation table. Remember to correlate the spectroscopic data with the chemical data.

29 Nuclear magnetic resonance spectroscopy

Electromagnetic radiation (typically at radio frequencies of 60–600 MHz) is used to identify compounds in a process known as nuclear magnetic resonance (NMR) spectroscopy. This is possible because of differences in the magnetic states of atomic nuclei, involving very small transitions in energy levels. The atomic nuclei of the isotopes of many elements possess a magnetic moment. When these magnetic moments interact with a uniform external magnetic field, they behave like tiny compass needles and align themselves in a direction 'with' or 'against' the field. The two orientations, characteristic of nuclei with a nuclear spin quantum number $I = \frac{1}{2}$, have two different energies: the orientation aligned 'with' the field has a lower energy than that aligned 'against' the field (Fig. 29.1).

> **Nuclear spin quantum numbers and NMR** – other common nuclei with non-zero quantum numbers are ^{14}N, ^{2}D ($I = 1$) and ^{11}B and ^{35}Cl ($I = \frac{3}{2}$). Their NMR spectra are not used on a *routine* basis.

Typical magnetic nuclei of general use to chemists and biochemists are ^{1}H, ^{13}C, ^{19}F and ^{31}P, all of which have nuclear spin quantum numbers $I = \frac{1}{2}$. The energy difference between the two levels (ΔE) corresponds to a precise electromagnetic frequency (ν), according to similar quantum principles for the excitation of electrons (p. 163). When a sample containing an isotope with a magnetic nucleus is placed in a magnetic field and exposed to an appropriate radio frequency, transitions between the energy levels of magnetic nuclei will occur when the energy gap and applied frequency are in *resonance* (i.e. when they are matched exactly in energy). Differences in energy levels, and hence resonance frequencies (ν_0), depend upon the magnitude of the applied magnetic field (B_0) and the magnetogyric ratio (γ), according to the equation:

$$\nu_0 = \gamma B_0 / 2\pi \qquad [29.1]$$

For a given value of the applied field (B_0), nuclei of different elements have different values of the magnetogyric ratio (γ) and will give rise to resonance at various radio frequencies. The principal components of an NMR spectrometer are shown in Fig. 29.2.

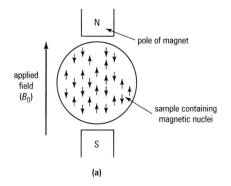

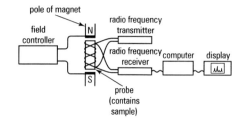

Fig. 29.2 Components of an NMR spectrometer.

Fig. 29.1 Effect of an applied magnetic field, B_0, on magnetic nuclei. (a) Nuclei in magnetic field have one of two orientations – either with the field or against the field (in the absence of an applied field, the nuclei would have random orientation). (b) Energy diagram for magnetic nuclei in applied magnetic field.

> **Example** For an external magnetic field of 2.5 T (tesla), ΔE for ^{1}H is 6.6×10^{-26} J and since $\Delta E = h\nu$, the corresponding frequency (ν) is 100 MHz; for ^{13}C in the same field, ΔE is 1.7×10^{-26} J, and ν is 25 MHz.

For magnetic nuclei in a given molecule, an NMR spectrum is generated because, in the presence of the applied field, different nuclei of the same atoms experience small, different, local magnetic fields depending on the arrangement of electrons, i.e. in the chemical bonds, in their vicinity. The effective field at the nucleus can be expressed as:

$$B = B_0(1 - \sigma) \qquad [29.2]$$

where σ (the shielding constant) expresses the contribution of the small secondary field generated by the nearby electrons. The magnitude of σ

depends on the electronic environment of a nucleus, so nuclei of the same isotope give rise to small different resonance frequencies according to the equation:

$$v_0 = \gamma B_0 (1 - \sigma)/2\pi \qquad [29.3]$$

KEY POINT **The variation of resonance frequencies with surrounding electron density is crucial to the usefulness of the NMR technique. If it did not occur, *all nuclei* of a single isotope would come into resonance at the same combination of magnetic field and radio frequency and only *one peak* would be observed in the spectrum.**

Chemical shift

The separation of resonance frequencies resulting from the different electronic environments of the nucleus of the isotope is called the *chemical shift*. It is expressed in dimensionless terms, as parts per million (ppm), against an internal standard, usually *tetramethylsilane* (TMS). By convention, the chemical shift is positive if the sample nucleus is less shielded (lower electron density in the surrounding bonds) than the nucleus in the reference and negative if it is more shielded (greater electron density in the surrounding bonds). The chemical shift scale (δ) for a nucleus is defined as:

$$\delta = [(v_{sample} - v_{reference}) \times 10^6]/(v_{reference}) \qquad [29.4]$$

This means that the chemical shift of a specific nucleus in a molecule is at the same δ value, no matter what the operating frequency of the NMR spectrometer.

An NMR spectrum is a plot of chemical shift (δ) as the *x*-axis against absorption of energy (resonance) as the *y*-axis. On the right-hand side of the spectrum at $\delta = 0$ ppm there may be a small peak, which is the reference (TMS). A typical ^{1}H–NMR spectrum is shown in Fig. 29.3.

> Measuring chemical shifts – ppm is *not* a concentration term in NMR but is used to reflect the small frequency changes that occur relative to the reference standard, measured in proportional terms.

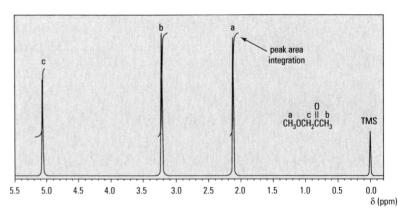

Fig. 29.3 ^{1}H–NMR spectrum of 1-methoxypropanone.

NMR spectrometers

These can operate at different radio frequencies and magnetic fields and are usually referred to in terms of radio frequency, e.g. 60 MHz, 270 MHz and 500 MHz spectrometers. Spectrometers operating above 100 MHz require expensive superconducting magnets to generate the high magnetic fields. In routine laboratory work 60 MHz and 90 MHz instruments are common but 270 MHz machines are becoming more affordable. Increasing the operating frequency of the spectrometer effectively increases the resolution of the chemical shifts of the nuclei under examination. For example, the difference in frequency between 0 and 1δ is 60 Hz in a 60 MHz spectrometer but 270 Hz for a 270 MHz instrument.

Spectrometers can be divided into two types:

1. *Continuous-wave (CW) spectrometers*, which use a permanent magnet or an electromagnet, usually operating at 60 or 90 MHz. In practice the radio frequency is held constant and small electromagnets on the faces of the main magnet (sweep coils) vary the magnetic field over the chemical shift range. The spectrometer sweeps through the spectroscopic region plotting resonances (absorption peaks) on a chart recorder (cf. dispersion IR spectrometers). CW spectrometers are usually dedicated to observation of a specific nucleus such as ^{1}H.

2. *Fourier transform (FT) spectrometers*, using superconducting magnets containing liquid nitrogen and liquid helium for cooling. Here the magnetic field is held constant and the sample is irradiated with a radio frequency pulse containing all the radio frequencies over the chemical shift region of the nucleus being examined, cf. FT–IR (p. 182). Computer control allows rapid repeat scans to accumulate spectra, presenting the data as a standard CW-type spectrum via FT processing. Simple variation of the radio frequencies permits observation of different nuclei (multinuclear NMR spectrometers). Thus an FT–NMR spectrometer can be used for obtaining ^{1}H, ^{13}C, ^{19}F, ^{15}N and ^{31}P NMR spectra.

Sample handling

The majority of NMR spectra are obtained from samples in solution and therefore the solvent should preferably not contain atoms of the nuclei being observed (except in the case of ^{13}C–NMR). The most common solvents are those in which the hydrogen atoms have been replaced by deuterium, which is not observed under the conditions under which the spectrum is obtained. CDCl$_3$ (deuteriochloroform, chloroform-d) is often the solvent of choice, but others such as dimethylsulphoxide-d^6, [(CD$_3$)$_2$SO], propanone-d^6 [(CD$_3$)$_2$CO], methanol-d^4 (CD$_3$OD) and deuterium oxide (D$_2$O) are in common use.

> Using deuterated solvents – these are expensive and should not be wasted. CDCl$_3$ is 100 times more expensive than spectroscopic-grade CHCl$_3$ and the others are at least 10–15 times more expensive than CDCl$_3$.

> Using CDCl$_3$ – when using this solvent a peak in the spectrum at $\delta = 7.26$ ppm is seen. This is due to the presence of CHCl$_3$ as an impurity in the CDCl$_3$.

As it is unlikely that you will be allowed 'hands-on' use of an NMR spectrometer, the best approach you can take to obtain a good spectrum is to ensure good sample preparation. The quality of an NMR spectrum is degraded by:

- inappropriate solvent;
- inappropriate concentration of solute;
- inappropriate solvent volume;
- solid particles in the solution;
- water in the sample (inefficient drying);
- paramagnetic compounds.

Sample preparation for NMR spectroscopy is described in Box 29.1.

Interpreting NMR spectra

As a matter of routine in your laboratory work you will be required to interpret ^{1}H–NMR spectra (also known as proton spectra). ^{13}C–NMR spectra are becoming more common, while ^{19}F and ^{31}P spectra may be obtained in specialized experiments. Therefore you should concentrate on the interpretation of ^{1}H and ^{13}C spectra in the first instance.

> **Box 29.1 How to prepare a sample for NMR spectroscopy**
>
> 1. **Make sure that your compound is free from water and solvent** (p. 39).
>
> 2. **Test the solubility of your compound in cold CH_2Cl_2.** If it is soluble you can use $CDCl_3$ as the solvent for the NMR experiment. If it is insoluble, consult your instructor for the availability of other deuterated solvents.
>
> 3. **Dissolve your compound $CDCl_3$ (about 2 mL) in a clean, dry sample tube.** Use about 10 mg of sample for CW–NMR or 5 mg of sample for FT–NMR. Check to see if the solvent contains TMS; if it does not, consult your instructor.
>
> 4. **Make a simple filter** in a new Pasteur pipette to remove insoluble material and water (Fig. 29.4). Check that your compound does not react with cotton wool and neutral alumina (alcohols and acids are strongly adsorbed on neutral alumina). If it does, replace the cotton wool with glasswool and do not use alumina. You *must* wear gloves when handling glasswool.
>
> 5. **Put the filter into a suitable clean, dry NMR tube** and, using a clean, dry Pasteur pipette, filter the solution into the NMR tube.
>
> 6. **Fill the NMR tube to the appropriate level**: between 30 and 50 mm in height is sufficient.
>
> 7. **Cap the NMR tube with the correct-size tube cap,** making sure that it is correctly fitted to prevent oscillation when the tube is spinning in the spectrometer. Make sure that the cap is fitted correctly so that it will not fall off when the tube is in the spectrometer.
>
> 8. **Wipe the outside of the tube with a clean, dry tissue** to make sure that the spectrometer will not be contaminated. Cleaning the spectrometer probe is a very difficult task.

^{1}H–NMR spectra

These normally cover the range between $\delta = 0$ and 10 ppm but the range is increased to $\delta = 15$ ppm when acidic protons are present in the molecule. The ^{1}H–NMR spectrum of a molecule gives three key pieces of information about the structure of a molecule:

1. Chemical shift (δ): the peak positions indicate the chemical (magnetic) environment of the protons, i.e. different protons in the molecule have different chemical shifts.
2. Integration: the relative size of peak area indicates how many protons have the δ value shown.
3. Coupling: the fine structure on each peak (coupling) indicates the number of protons on adjacent atoms.

These three features make ^{1}H-NMR a powerful tool in structure determination and there are two extreme approaches to it:

1. Prediction of the spectrum of the expected compound from theoretical knowledge and then comparison with the spectrum obtained. You should recognize 'patterns' (e.g. triplet and quartet for an ethyl group; a singlet of peak area six for two identical methyl groups), which were present in the starting materials, but the δ_H values may have changed in the 'new' molecule. There are computer programs, such as g-NMR®, which will simulate the NMR spectrum from a structural formula.
2. Interpretation of the spectrum from correlation tables, but this is very difficult for the inexperienced.

In practice a combination of the two approaches is used with cross-referencing and checking the proposed structure with tabulated δ_H values and reference spectra until a satisfactory answer is found.

> ^{1}H-NMR spectra – most of the spectra shown in this chapter do not extend over the normal spectral range $\delta = 0$–10 ppm. They are expanded to show the details of coupling patterns.

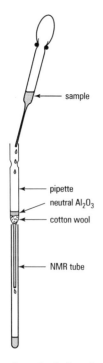

Fig. 29.4 Filtration of solutions for NMR.

KEY POINT Always make sure that your predicted structure is consistent with the spectrum.

Factors affecting chemical shift (δ_H)

The δ values of protons can be predicted to a general approximation from knowledge of the effects which produce variations in chemical shift.

1. The hybridization of the carbon atom to which the hydrogen atom is attached:

    (a) sp^3 hybridized carbon: peaks occur between $\delta = 0.9$ and 1.5 ppm in simple hydrocarbon systems. The peaks move downfield with change of structure from CH_3 to CH_2 to CH.

    (b) sp hybridized carbon: peaks occur at about $\delta = 1.5-3.5$ ppm in alkynes.

    (c) sp^2 hybridized carbon: in alkenes the resonances occur around $\delta = 4-8$ ppm and the C–H peaks of aromatic rings are found between $\delta = 6$ and 9 ppm. The large downfield shifts of these C_{sp2}–H nuclei result from deshielding of the protons by fields set up by circulation of the π-electrons in the magnetic field. The proton of the aldehyde group (CHO) is particularly deshielded by this effect and is found at $\delta = 9-10$ ppm.

2. Electron attraction or electron release by substituent atoms attached to the carbon atom. Electron attracting atoms, such as N, O, Hal attached to the carbon, attract electron density from the C–H bonds and thus deshield the proton. This results in movement of the chemical shift to higher δ values (Table 29.1). Conversely, electron-releasing groups produce additional shielding of the C–H bonds resulting in upfield shifts of δ values.

3. All the protons in benzene are identical and occur at $\delta = 7.27$ ppm. In substituted aromatic compounds, the overall electron-attracting or releasing effect of the substituent(s) alters the δ values of the remaining ring protons making them non-equivalent. The *ortho* protons are affected most.

4. For protons attached to atoms other than carbon: the chemical shifts of protons attached to oxygen increase with increasing acidity of the O–H group; thus $\delta = 1-6$ ppm for alcohols, 4–12 ppm for phenols and 10–14 ppm for carboxylic acids. Hydrogens bound to nitrogen (1° and 2° amines) are found at $\delta = 3-8$ ppm. The approximate chemical shift regions are shown in Fig. 29.5.

> Proton chemical shifts – only hydrogen atoms bonded to carbon will be considered in this simplified treatment.

> Interpreting NMR spectra: changes of δ – the terms used to indicate the movement of a particular peak with change in its chemical (magnetic) environment are: *upfield* – towards $\delta = 0$ ppm; *downfield* – towards $\delta = 10$ ppm; *shielded* – increased electron density near the proton; *deshielded* – decreased electron density near the proton.

Table 29.1 Chemical shifts of methyl protons

Compound	Chemical shift (ppm)
$(CH_3)_4Si$	0.00
CH_3R	0.90
CH_3I	2.16
CH_3Br	2.65
CH_3Cl	3.10
CH_3OR	3.30
CH_3F	4.26

Integration of peak areas

The area of each peak gives the relative number of protons and is produced directly on the spectrum (Fig. 29.6). On CW–NMR spectrometers the height of the peak area integration line must be measured using a ruler, whereas on FT–NMR machines the area is calculated and displayed as a number. You must remember that:

- The areas are *ratios*, not absolute values, and you must find a peak attributable to a specific group to obtain a reference area, e.g. a single peak at $\delta = 1.0$ ppm is likely to be a CH_3 group and thus the area displayed or measured is equal to three protons.
- You must ensure that you include integrations from all the fine-structure (coupling) peaks in the peak area.

> Integration of coupled peaks – the area under the singlet, doublet, triplet, quartet, etc., is still that of the type of hydrogen being considered. For example, if the peak for the three protons of a *methyl group* is split into a triplet by an adjacent methylene group, the area of the triplet is *three*.

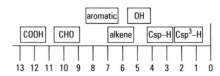

Fig. 29.5 Approximate chemical shift positions in the ^{1}H–NMR spectrum.

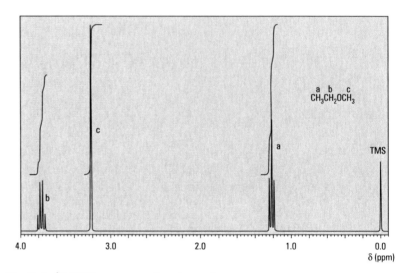

Fig. 29.6 ^{1}H–NMR spectrum of methoxyethane.

- Do not expect the peak area integrations to be exact whole numbers, e.g. an area of 2.8 is probably three protons (CH_3), 5.1 is probably five protons (e.g. a C_6H_5 group), but 1.5 is probably a CH_3 and all the peak area integrations must be doubled.

Coupling (spin–spin splitting)

Many ^{1}H–NMR signals do not consist of a single line but are usually associated with several lines (splitting patterns). Protons giving multiline signals are said to be *coupled*. This coupling arises from the magnetic influence of protons on one atom with those on an adjacent atom(s). Thus information about the nature of adjacent protons can be determined and fed into the structural elucidation problem. To a simple first approximation the following three general points are useful in the interpretation of coupling patterns:

1. Aliphatic systems: if *adjacent* carbon atoms have *different types* of protons (a and b), then the protons will couple. If a proton is coupled to n ($n = 1, 2, 3, 4, 5$, etc.) other protons on an adjacent carbon atom, the number of lines observed is $n + 1$, as shown in the examples below.

    $CH_3CH_2OCH_3$ Protons a are coupled to two protons b: $n = 2$;
      a   b   c   therefore the peak for protons a is split into three lines (a triplet).

    Protons b are coupled to three protons a: $n = 3$; therefore the peak for protons b is split into four lines (a quartet).

    Protons c have no adjacent protons and therefore are not coupled and give a single line (singlet) (Fig. 29.6).

CH₃CHBrCH₂Br   Protons a are coupled to one proton b: $n = 1$;
 a   b   c   therefore the peak for protons a is split into two lines (doublet)

Protons b are coupled to three protons a and two protons c: $n = 5$; therefore the peak for protons b is split into six lines (sextet).

Protons c are coupled to one proton b: $n = 1$; therefore the peak for protons c is split into two lines (doublet).

Protons a and protons c are not adjacent and do not couple (Fig. 29.7).

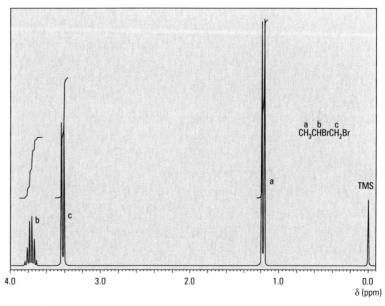

Fig. 29.7   ¹H–NMR spectrum of 1,2-dibromopropane.

```
                1 ----------- one line (singlet or s)
            1       1 --------- two lines (doublet or d)
        1       2       1 ------ three lines (triplet or t)
    1       3       3       1 --- four lines (quartet or q)
                            etc.
```

Fig. 29.8 Intensities of coupled peaks from Pascal's triangle.

The intensity of each peak in the resulting singlet, doublet, triplet, quartet, etc., is calculated from Pascal's triangle (Fig. 29.8).

The separation between the coupled lines is called the coupling constant, J, and, for aliphatic protons CH, CH_2 and CH_3, it is usually $\sim 8\,\text{Hz}$.

KEY POINT   The $(n + 1)$ rule only applies in systems where the coupling constant (J) between the protons is the same. Fortunately this is common in aliphatic systems.

2. Alkene hydrogens: hydrogen atoms on double bonds have different coupling constants depending upon the stereochemistry of the alkene. Alkene hydrogens in the Z (cis) configuration have $J = 5-14\,\text{Hz}$, whereas those in the E (trans) configuration have $J = 11-19\,\text{Hz}$ (Figs 29.9a and b).

3. Aromatic hydrogens: coupling of hydrogens, which are non-adjacent, is readily observed in aromatic compounds. Different protons *ortho* to each other couple with $J = 7-10\,\text{Hz}$, while those in a *meta* relationship have $J = 2-3\,\text{Hz}$. *Para* coupling ($J = 0-1\,\text{Hz}$) is not usually seen on the spectrum. The types of aromatic compound you are likely to meet most often are:

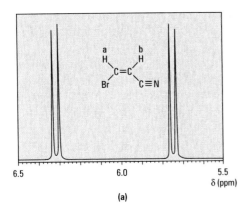

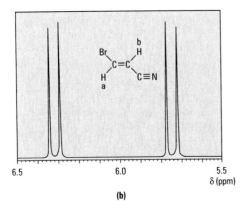

Fig. 29.9 ^{1}H–NMR spectra of: (a) (Z)-3-bromopropanonitrile; (b) (E)-3-bromopropanonitrile.

(a) Monosubstituted aromatic compounds, in which three basic patterns are found in the aromatic region of the spectrum. If the substituent exerts a weak electronic effect on the ring, the δ values of the ring protons are similar and the protons appear as a single peak of area five (Fig. 29.10a). If the group is strongly electron releasing (OH, NH$_2$, OCH$_3$, etc.), the protons appear as complex multiplets (*ortho* and *meta* coupled), below $\delta = 7.27$ ppm of relative areas two to three (Fig. 29.10b). If the group is electron attracting (e.g. NO$_2$, COOH, etc.), then the complex multiplets have $\delta > 7.27$ ppm (Fig. 29.10c);

(b) *para* disubstituted aromatic compounds, which are of two types. If the substituents are the same, then all the ring protons are identical and a singlet, of relative area four, is seen (Fig. 29.10d). If the substituents are different, then the pairs of hydrogens *ortho* to each substituent are different and *ortho*-couple to give what appears to be pair of doublets, each of relative area two (Fig. 29.10e).

(c) Increasing numbers of substituents, which decrease the number of aromatic hydrogens and the spectrum becomes simpler. Thus the common 1,2,4-trisubstituted pattern (Fig. 29.10f) is recognized easily as an *ortho*-coupled doublet, a *meta*-coupled doublet and a doublet of doublets (coupled *ortho* and *meta*).

The chemical shifts of aromatic protons can be calculated from detailed correlation tables.

^{13}C-NMR spectra

The ^{13}C nucleus has $I = \frac{1}{2}$, like ^{1}H, and the ^{13}C–NMR spectrum of a compound can be observed using a different radio frequency range (in the same magnetic field) to that for ^{1}H. The ^{13}C spectrum will give peaks for each different type of carbon atom in a molecule but the properties of the ^{13}C nucleus give some important and useful differences in the spectrum obtained:

- The natural abundance of ^{13}C is only 1.1% compared with 98.9% for ^{12}C – in any molecule no two adjacent atoms are likely to be ^{13}C and therefore coupling between ^{13}C nuclei will not be seen, giving a very simple spectrum.
- In a sample of a compound, which contains many molecules, the ^{13}C isotope is randomly distributed and all the different carbon atoms in a sample of a compound will be seen in the ^{13}C–NMR spectrum.
- The sensitivity of the ^{13}C nucleus is low and this, together with its low natural abundance, means that FT–NMR is the only practical system to produce a spectrum by accumulation of spectra by repetitions. Larger sample size in bigger NMR tubes also assist in solving the sensitivity/abundance problem.
- The chemical shift range for ^{13}C is greater ($\delta_C = 0$–250 ppm) than for ^{1}H ($\delta_H = 0$–15 ppm) giving greater spectral dispersion, i.e. the peaks for carbons with very slight differences in chemical shifts are separated and do not overlap.
- ^{13}C nuclei will couple with the ^{1}H nuclei to which they are directly bonded, e.g. CH$_3$ will appear as a quartet, CH$_2$ as a triplet, CH as a doublet, but C with no hydrogen atoms attached will appear as a singlet. This introduction of complexity in the ^{13}C–NMR spectrum is removed by broadband decoupling (see p. 198).

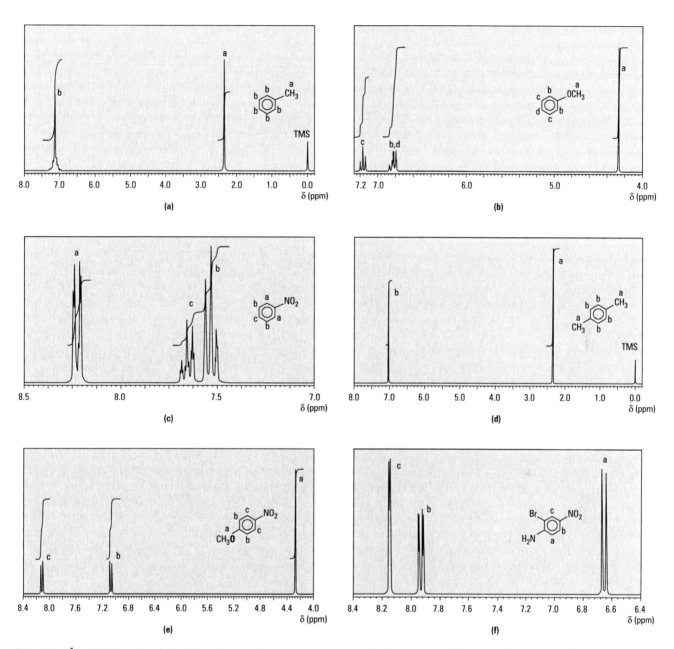

Fig. 29.10 ¹H–NMR spectra of (a) methylbenzene; (b) methoxybenzene; (c) nitrobenzene; (d) 1,4-dimethylbenzene; (e) 4-methoxynitrobenzene; (f) 4-amino-3-bromonitrobenzene (NH₂ protons not shown).

- The peak areas of the different carbon atoms are *not* related to the number of carbon atoms having the same chemical shift, as in the case for ¹H–NMR spectra.

Interpreting ^{13}C–NMR spectra

Normally you will be given two ¹³C–NMR spectra (Fig. 29.11). The upper spectrum, which is more complex (more lines) is called the *off-resonance decoupled* spectrum and shows the ¹³C–¹H coupling to enable you to determine which carbon signals are CH₃, CH₂, CH and C. Then overlapping of peaks may make the identification of different carbon atoms difficult. The

> Interpretation of ¹³C–NMR spectra – the spectrum is that of all the *carbon atoms* in the molecule. It is easy to forget that the peaks for carbon atoms carrying no hydrogen atoms are present.

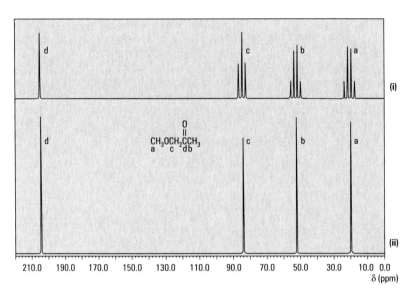

Fig. 29.11 ^{13}C–NMR spectra of 1-methoxypropanone: (i) off-resonance decoupled; (ii) broadband decoupled.

lower spectrum is a *broadband decoupled* spectrum in which the molecule is irradiated with a second radio frequency range for the protons in the molecule and effectively removes all the ^{13}C–^{1}H couplings from the spectrum. The resulting simplicity of the spectrum makes identification of the different types of carbon in the molecule relatively easy.

The chemical shifts of ^{13}C atoms (δ_C) vary in the same manner as those of protons (Fig. 29.12):

1. δ_C moves downfield as the hybridization of the carbon atom changes from sp^3 (0–50 ppm) to sp (75–105 ppm) to sp^2 (100–140 ppm);
2. for sp^3 hybridized carbon: δ_C moves further downfield with the change from CH$_3$ to CH$_2$ to CH to C;
3. for sp^2 hybridized carbon: aromatic carbons occur further downfield ($\delta_C = 115–145$ ppm) than alkene carbon atoms ($\delta_C = 100–140$ ppm);
4. bonding more electronegative atoms to carbon deshields the carbon atom and moves the peaks downfield, e.g CH$_3$–C ($\delta_C \sim 6$ ppm) and CH$_3$–O ($\delta_C \sim 55$ ppm), C=C ($\delta_C \sim 123$ ppm) and C=O ($\delta_C \sim 205$ ppm).

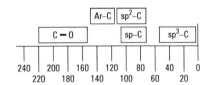

Fig. 29.12 Approximate chemical shift positions in ^{13}C–NMR.

30 Mass spectrometry

> **Understanding mass spectrometry** – since this technique does not involve the production and measurement of electromagnetic spectra and is not based on quantum principles, it should not really be referred to as a spectroscopic technique.

> **Mass-to-charge ratios** – in the overwhelming majority of simple cases the ion detected is a monopositive cation; thus $z = 1$ and the peaks seen on a low-resolution spectrometer equate to the mass of the ion.

> **Determination of exact molecular mass** – high-resolution instruments enable the molecular formula of a compound to be determined by summation of the masses of the individual isotopes of atoms, e.g. both ethane and methanal have integral mass values of 30, but the accurate values are 30.046 950 and 30.010 565 respectively.

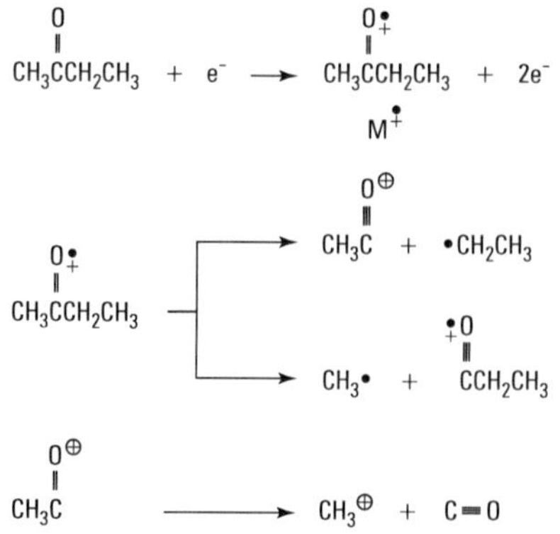

$$CH_3CCH_2CH_3 + e^- \longrightarrow CH_3CCH_2CH_3^{+\bullet} + 2e^-$$
$$M^{+\bullet}$$

$$CH_3CCH_2CH_3^{+\bullet} \longrightarrow CH_3C^{\oplus}=O + {}^\bullet CH_2CH_3$$
$$\longrightarrow CH_3^\bullet + {}^{+\bullet}O=CCH_2CH_3$$

$$CH_3C^{\oplus}=O \longrightarrow CH_3^\oplus + C\equiv O$$

Fig. 30.1 Formation and fragmentation of a molecular ion ($M^{+\bullet}$).

Mass spectrometry (MS) involves the bombardment of molecules, in the gas phase, with electrons. An electron is lost from the molecule to give a cation, the molecular ion (M^+), which then breaks down in characteristic ways to give smaller fragments, which are cations, neutral molecules and uncharged radicals (Fig. 30.1).

The mixture of molecular ion and fragments is accelerated to specific velocities using an electric field and then separated on the basis of their different masses by deflection in a magnetic or electrostatic field. Only the cations are detected and a mass spectrum is a plot of mass-to-charge ratio (m/z) on the x-axis against the number of ions (relative abundance, RA, %) on the y-axis. A schematic of the components of a mass spectrometer is shown in Fig. 30.2 and an example of a line-graph-type mass spectrum in Fig. 30.3.

There are many types of mass spectrometer, from high-resolution double-focusing instruments, which can distinguish molecular and fragment masses to six decimal places, to 'bench-top' machines with a quadrupole mass detector which can resolve masses up to about $m/z = 500$, but only in whole-number differences. Routinely you are most likely to encounter data from 'bench-top' instruments and therefore only this type of spectrum will be considered.

Sample handling

For low-resolution spectra obtainable from a 'bench-top' MS, samples should be presented in the same form and quantity as demanded for gas chromatographic analysis (p. 211). For high-resolution spectra contamination of any sort must be avoided and samples (typically less than 500 μg) should be submitted in glass sample tubes with screwcaps containing an aluminium-foil insert. MS is so sensitive that the plasticizers from plastic tubes or plastic push-on caps will be detected, as will contaminating grease from ground-glass joints and taps.

Mass spectra

The standard low-resolution mass spectrum (Fig. 30.3) is computer generated, which allows easy comparison with known spectra in a computer database for identification. The peak at the highest mass number is the molecular ion (M^+), the mass of the molecule minus an electron. The peak at RA = 100%, the base peak, is the most abundant fragment in the spectrum and the computer automatically scales the spectrum to give the most abundant ion as 100%. The mass spectrum of a compound gives the following information about its chemical structure:

- molecular ion mass, which includes information on the number of nitrogen atoms and the presence of chlorine and bromine atoms (see below), – which is not easily obtained from IR and NMR spectra;
- the most stable major fragment (base peak), which can be correlated to the structure of the molecule;
- other important fragment ions, which may give information on the structure;

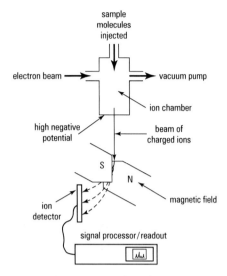

Fig. 30.2 Components of an electron-impact mass spectrometer.

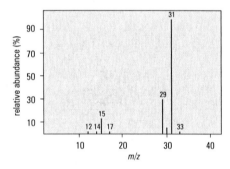

Fig. 30.3 Mass spectrum for methanol; m/z = mass-to-charge ratio.

Identification of isotope peaks – the natural abundance of ^{13}C is 1.1%. For a molecule containing n carbon atoms the probability is that $1.1 \times n\%$ of these atoms will be ^{13}C. Thus the mass spectrum of hexane (six carbons) gives a molecular ion (M$^+$) at $m/z = 86$ and a peak at $m/z = 87$ ($M + 1$) which is 6.6% the intensity of the M$^+$ peak.

- the detailed fragmentation pattern, which can be used to confirm a structure by reference to a library database, cf. the 'fingerprint' region in IR spectrometry (p. 186).

Molecular ions

The m/z value of the molecular ion is the summation of all the atomic masses in the molecule, *including the naturally occurring isotopes*. For organic molecules you will find a small peak ($M + 1$) above the apparent molecular ion mass (M$^+$) value due to the presence of ^{13}C. The importance of isotope peaks is the detection of chlorine and bromine in molecules since these two elements have large natural abundances of isotopes, e.g. $^{35}Cl:^{37}Cl = 3:1$ and $^{79}Br:^{81}Br = 1:1$. The mass spectra produced by molecules containing these atoms are very distinctive with peaks at M + 2 and even M + 4 and M + 6 depending on how many chlorine or bromine atoms are present. The identification of the number and type of halogen atoms is illustrated in Box 30.1.

Since the low-resolution mass spectrum produces integer values for m/z, the mass of M$^+$ indicates the number of nitrogen atoms in the molecule. If m/z for M$^+$ is an *odd integer*, there is an *odd number* of N atoms in the molecule and, if the value is an *even number*, then there is an *even number* of N atoms.

Base peak

The molecular ion M$^+$ fragments into cations, radicals, radical cations and neutral molecules of which only the positively charged species are detected. There are several possible fragmentations for each M$^+$ but the base peak represents the most *energetically favoured* process with the m/z value of the base peak representing the mass of the most *abundant* (and therefore most stable) positively charged species. The fragmentation of M$^+$ into the base peak follows the simplified rules outlined in Box 30.2, and for a more detailed interpretation you should consult the correlation tables to be found in the specialist texts referred to at the end of the section.

Fragmentation patterns

The mass spectrum of a molecule is unique and can be stored in a computer. A match of the spectrum with those in the computer library is made in terms of molecular weight and the 10 most abundant peaks and a selection of possibilities will be presented. At this point you need to correlate all the information obtained from the spectroscopic techniques described in Chapters 26, 28, 29 and 30 together with the chemistry of the molecule to attempt to identify the structure of the molecule.

When you attempt to interpret the mass spectrum remember that:

- Only the base peak is almost certain to be derived from the molecular ion.
- Some lesser peaks may result from alternative fragmentation pathways, but these may be useful in assigning structural features.
- MS is often used to confirm information from IR and NMR spectra; interpretation of the mass spectrum alone is very difficult, except for the simplest molecules.

Interfacing mass spectrometry

Bench-top mass spectrometers are to be found interfaced with other analytical instruments such as gas chromatographs (p. 215) and high-

Box 30.1 How to identify the number of bromine or chlorine atoms in a molecule from the molecular ion

1. Since Cl and Br have isotopes two mass numbers apart, their presence in a molecule will produce peaks at m/z values above M^+, which are two mass numbers apart, i.e. $M+2$, $M+4$, etc.

2. The expression for the number and intensities of these peaks is given by the expansion of the formula:

   $(a+b)^n$

   where a and b are the ratio of the two atom isotopes, and n is the number of atoms.

Example 1: If the molecule contains one chlorine atom then:

$(a+b)^n = (3+1)^1 = 3+1$

Thus the mass spectrum of CH_3Cl would show M^+ at $m/z = 50$ ($CH_3{}^{35}Cl$) and $M+2$ at $m/z = 52$ ($CH_3{}^{37}Cl$) and the heights of these two peaks will be in the approximate ratio 3:1 (Fig. 30.4a).

Example 2: If the molecule contains two chlorine atoms then:

$(a+b)^n = a^2 + 2ab + b^2 = (3+1)^2 = 9 + 6 + 1$

Thus the mass spectrum of CH_2Cl_2 would show M^+ at $m/z = 84$ ($CH_2{}^{35}Cl_2$), $M+2$ at $m/z = 86$ ($CH_2{}^{35}Cl^{37}Cl$) and $M+4$ at $m/z = 88$ ($CH_2{}^{37}Cl_2$) and the heights of these peaks will be the approximate ratio 9:6:1 (Fig. 30.4b).

Example 3: If the molecule contains one bromine atom then:

$(a+b)^n = (1+1)^1 = 1+1$

Thus the mass spectrum of CH_3CH_2Br would show M^+ at $m/z = 108$ ($CH_3CH_2{}^{79}Br$) and $M+2$ at $m/z = 110$ ($CH_3CH_2{}^{81}Br$) and the heights of these peaks will be in the approximate ratio 1:1 (Fig. 30.4c).

Example 4: If the molecule contains three bromine atoms then:

$(a+b)^3 = a^3 + 3a^2b + 3ab^2 + b^3 = (1+1)^3 = 1+3+3+1$

Thus the mass spectrum of $CHBr_3$ would show M^+ at $m/z = 250$ ($CH^{79}Br_3$), $M+2$ at $m/z = 252$ ($CH^{79}Br_2{}^{81}Br$), $M+4$ at $m/z = 254$ ($CH^{79}Br^{81}Br_2$) and $M+6$ at $m/z = 256$ ($CH^{81}Br_3$) and the heights of the peaks will be in the approximate ratio 1:3:3:1 (Fig. 30.4d).

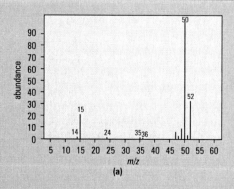

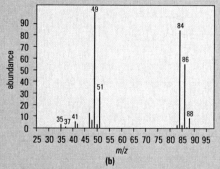

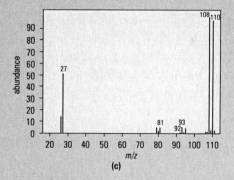

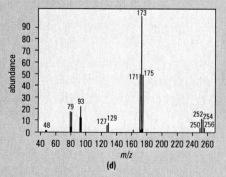

Fig. 30.4 Mass spectra of: (a) CH_3Cl; (b) CH_2Cl_2; (c) CH_3CH_2Br; (d) $CHBr_3$.

Box 30.2 Idealized fragmentation processes for the molecular ion (M⁻)

1. **α-Cleavage**: this involves breaking the 'next but one bond' to a hetero-atom (N, O, Hal, etc.) in the functional group of a molecule. The following examples illustrate the general principles:

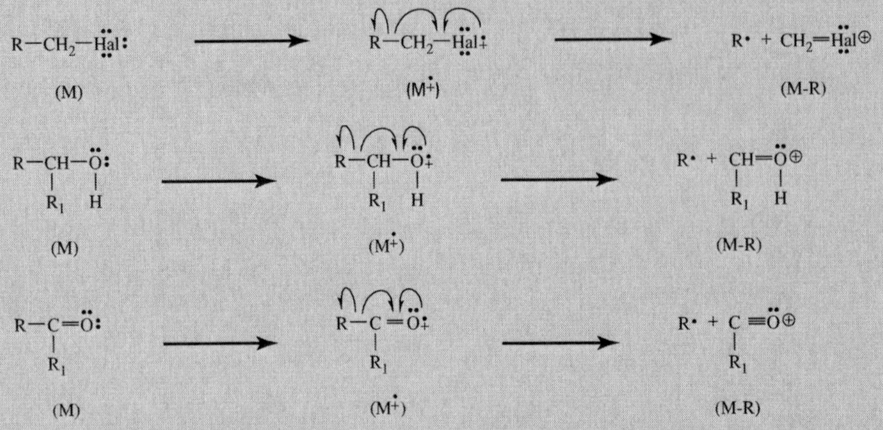

    R—CH₂—Ḧal: ⟶ R—CH₂—Ḧal⁺· ⟶ R• + CH₂=Ḧal⊕
    (M)   (M⁺·)   (M-R)

    R—CH—Ö: R—CH—Ö⁺·   R• + CH=Ö⊕
    | | ⟶ | | ⟶ |
    R₁ H   R₁ H   R₁ H
    (M)   (M⁺·)   (M-R)

    R—C=Ö: R—C=Ö⁺·   R• + C≡Ö⊕
    | ⟶ | ⟶ |
    R₁   R₁   R₁
    (M)   (M⁺·)   (M-R)

2. **σ-Bonds in alkanes**: C–C bonds break in preference to C–H bonds and the most stable carbocation will be formed as the base peak. For example, 2,2-dimethylpentane will give the stable $(CH_3)_3C^+$ cation as the base peak instead of the less stable propyl cation $CH_3CH_2CH_2^+$.

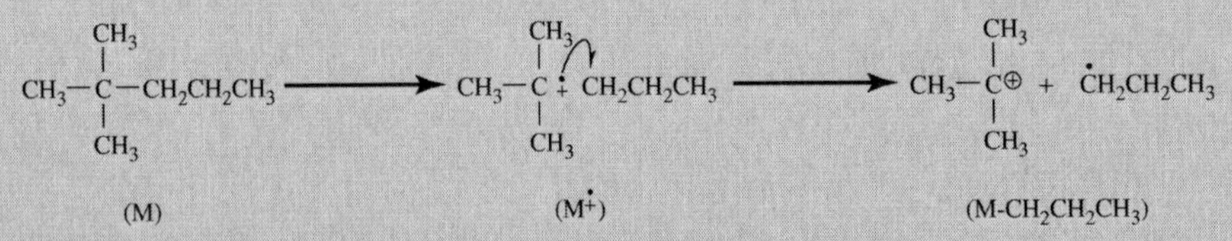

    $$CH_3-\underset{\underset{CH_3}{|}}{\overset{\overset{CH_3}{|}}{C}}-CH_2CH_2CH_3 \longrightarrow CH_3-\underset{\underset{CH_3}{|}}{\overset{\overset{CH_3}{|}}{C}}\dotplus CH_2CH_2CH_3 \longrightarrow CH_3-\underset{\underset{CH_3}{|}}{\overset{\overset{CH_3}{|}}{C}}^{\oplus} + \dot{C}H_2CH_2CH_3$$

    (M)   (M⁺·)   (M-CH₂CH₂CH₃)

3. **Aromatic compounds**: simple aromatics cleave to give a phenyl cation, $m/z = 77$, as the base peak which then loses ethyne to give $m/z = 51$. Aromatics with CH₂ next to the ring give the stable tropylium cation $m/z = 91$, and then lose ethyne to $m/z = 65$.

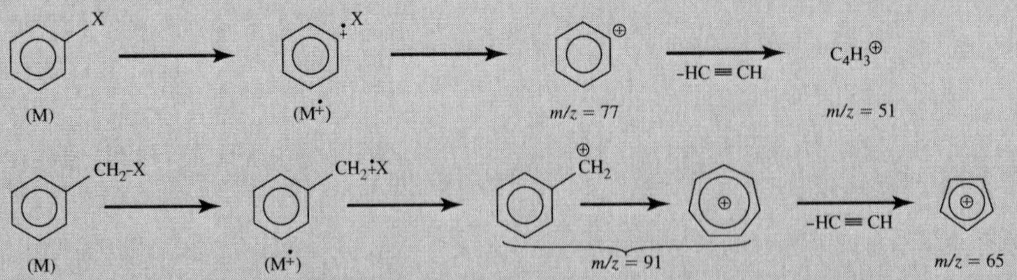

    Ph–X ⟶ [Ph–X]⁺· ⟶ Ph⊕ (m/z = 77) ⟶ (−HC≡CH) C₄H₃⊕ (m/z = 51)

    Ph–CH₂–X ⟶ [Ph–CH₂–X]⁺· ⟶ PhCH₂⊕ ⟶ tropylium (m/z = 91) ⟶ (−HC≡CH) (m/z = 65)

4. **β-Cleavage or McLafferty rearrangement**: this is applied to molecules with a carbonyl group. If there is a hydrogen atom on the carbon atom four away from the carbonyl oxygen (γ carbon atom), a rearrangement of the molecular ion occurs and a neutral alkene is lost from M⁺. This process occurs concurrently with the α-cleavage:

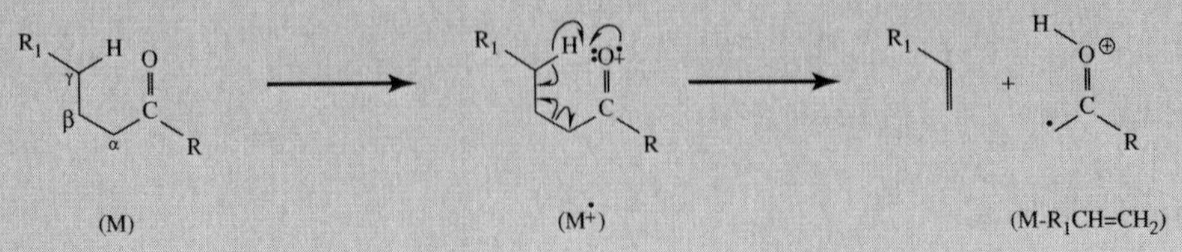

    (M)   ⟶   (M⁺·)   ⟶   (M-R₁CH=CH₂)

performance liquid chromatographs (p. 222) giving rise to the 'hyphenated techniques', GC–MS and LC–MS, respectively.

In both techniques the separated compounds from the chromatograph are sampled automatically and the mass spectrum of each compound is recorded. Comparison of the mass spectra with a library in an 'on-board' database permits identification of the components of the mixture.

Index

academic transcripts 29
accuracy 119
acid-base titrations 522–4
acids 108
actinomycetes 185, 186
   see also microbes
adsorption chromatography 308, 360–61
affinity chromatography (AC) 311–12
agar 206, 209, 234, 239
   see also cell cultures; microbial culture
agarose 324–5
agarose gel electrophoresis 433–4
agglutination tests 260
agglutinins 311
alkalis 108
allosteric enzymes 391–2
Ames test 241
amino acids
   analysis 357
   properties 354
ammonium sulphate 378–9
amperometry 341
ampholytes 108
analysis of variance (ANOVA) 505, 510
animal specimens *see* specimens

animal taxonomy *see* classification
animal tissues and organs 225–6
   see also cell cultures
ANOVA (analysis of variance) 505, 510
anthrone 366
antibiotics 228
antibody production
   monoclonal antibodies 260
   polyclonal antibodies 259
antibody structure 259
anxiety 35–6
apparatus diagrams 148, 149, 150
arithmetic mean 490, 491, 496
artefacts 121
assertiveness 9
assessment 22–4
   examinations *see* examinations
   coursework 29
   modular 29
   poster displays 47, 49
   problem-based learning 30
atmospheric conditions
   gas composition 97
   pressure 97
atomic mass 100
atomic spectroscopy 288–9
audio-visual aids 50–2
autopipettors 85

bacteria
   E.coli 453
bacterial cultures *see* cell cultures; microbialculture
bacterial genetics 422
   conjugation 426–7

DNA transfer 424–8
   genetic analysis using F' plasmids 428
   mapping with Hfr donors using interrupted conjugation 427–8
   mutants 423–4
   natural transformation 424–5
   transduction 425–6
bacterial identification 174–6, 185
   carbohydrate utilisation tests 188, 189, 219, 220
   catalase test 188, 219
   cell shape 186, 217
   classification and nomenclature 221–3
   colony characteristics 185–6, 216–17
   direct observation 185–6, 216
bacterial identification
   Gram staining 186–7, 217–8
   identification kits 189, 220
   identification tables 188, 219–20
   immunological tests 189, 221
   microscopic examination 186–7, 217
   molecular approaches 189, 220
   motility 186, 218–9
   oxidase test 187–8, 219
   typing methods 176, 190, 221
   see also microbes; microbial culture
bacteriophages 209–11
   genetics 428–9, 451
balances 87, 96
bar charts 463, 465, 466
bases 108
batch culture methods 235–6
beakers 87
bias 119
binomial distributions 501–2
bioassays 240–2
   Ames test 241
biochemical analysis 351–3
bioethics 77–8
biological population 158
biomolecules
   amino acids *see* amino acids
   carbohydrates *see* carbohydrates
   enzymes *see* enzymes
   lipids *see* lipids
   nucleic acids *see* nucleic acids
   peptides *see* peptides
   photosynthesis *see* photosynthesis
   proteins *see* proteins
   respiration *see* respiration
   transmembrane processes *see* membrane transport processes
biosensors 344, 345
Biuret method 355
blotting 330, 435–6
body plans 148, 149
body rhythm 10
bottles 87
bovine serum albumin (BSA) 356
brainstorming 13

buffers 112–14
burettes 85, 518

calculators 79
calibration 253
    amount or concentration 255
    calibration curves 253–5, 257
    linear regression plots 256
    standards 255
capillary electrophoresis 333–4, 368
    capillary gel electrophoresis (CGE) 335–6
    capillary isoelectric focusing (CIEF) 336
    capillary zone electro-phoresis (CZE) 334
    chiral capillary electro-phoresis (CCE) 335
    micellar electrokinetic chromatography (MEKC) 334
carbohydrate utilisation tests 188, 189, 219, 220
carbohydrates 365
    extraction and analysis 366
        capillary electrophoresis 368
        chemical methods 366
        chromatography 367–8
        enzymatic methods 366–7
    glycosides 366
    monosaccharides 365
    polysaccharides 368
carbon dating 275
catalase test 188, 219
cell cultures 225, 227, 234
    animal cell culture systems 227–8, 230, 231
    bioassays 240–2
        Ames test 241
    growth in liquid media 235
        batch culture 235–6
        continuous culture 236–7
    growth on solidified media 234
    measuring growth
        agar-based medium 239
        counting chamber 238
        culture-based counting methods 237, 239–40
        direct microscopic counts 237
        electronic particle counters 237
        haemocytometer 238
        plate counts of bacteria 239
    plant protoplasts 232–3
    plant tissue and cell culture systems 230, 232
    sterile technique 229
    see also microbial culture
cell diagrams 147, 148
cell disruption 245
    fractionation and isolation of organelles 248–9
    homogenising media 245–6, 247
    mechanised methods 247–8
    non-mechanical methods 246–7
    protein denaturation 247
    susceptibility 245
cell membrane transport see membrane transport processes
cell shape 186, 217
cellulose acetate 324

Central Limit Theorem 504
centrifugation 299
    balancing the rotor 303
    calculating acceleration 300
    centrifuge tubes 303
    continuous flow centrifuges 301
    density barrier 301
    density gradient 300
    differential sedimentation 300
    high-speed 301
    low-speed 301, 302
    microcentrifuges 301
    pelleting 300
    rotors 301–2
    safe practice 303–4
    ultracentrifuges 301
chain termination DNA sequencing method 438
charts 150
chemicals
    dilutions see dilutions
    Merck Index 90
    mixing 95
    preparing solutions 90–93
    safety 90, 91
    selection 90
    separating components 95
    spillages 95
    storing 95
chi^2 ($[*c]^2$) test 419–20, 507
chiral capillary electro-phoresis (CCE) 335
chlorophyll fluorescence 404–6
cholesterol 363
chromatographic immunoassays 264, 265
chromatography 305
    adsorption chromatography 308, 360–61
    affinity chromatography (AC) 311–12
    column chromatography 306
    covalent chromatography 312
    electron capture detector (ECD) 318
    flame ionisation detector (FID) 317
    fluorescence detectors 316–17
    gas chromatography (GC) 308, 318, 363, 367–8
    gas chromatography detectors 317–18
    gel permeation chromatography (GPC) 309–10
    high performance liquid chromatography (HPLC) 307, 318, 367
    hydrophobic interaction chromatography (HIC) 310–11
    immobilised metal affinity chromatography (IMAC) 312–13
    interpreting chromatograms 319
    ion exchange chromatography 309
    liquid chromatography detectors 316–17
    optimising separations 313–14

    paper chromatography 305–6, 367
        partition chromatography 308–9
        quantititative analysis 320
        recording detector output 318–19
        thermal conductivity detector (TCD) 317

thin-layer chromatography (TLC) 305–6, 361–2, 367
   UV/visible detectors 316
chromogenic enzyme substrates 384–5
chromosome mapping 418
chromosome number 415
chromosome walking 435
chromosomes 431
   bacteria *see* bacterial genetics
   linked genes 416–17
citations 72
Clark (Rank) oxygen electrodes 341–3
classification
   bacteria *see* bacterial identification
   hierarchical system 173–4
   identification guides 178–9
      advice for using keys 181
      bracketed keys 179
      comparing specimens with descriptions 182
      computerised keys 181
      flowchart keys 180
      indented keys 179
      multi-access keys 180–81
   nomenclature 174–5
   phenetic taxonomy 173
   phylogenetic taxonomy 173
   species 174
   taxa below the rank of species 175–6
   viruses 176
coefficient of dispersion 503
coefficient of variation 491, 494
colligative properties 103–4
column chromatography 306
communal records 128
communication 12
   see also writing skills
competences 4, 38
complement-based assays 266–7
computer programs *see* Excel; PowerPoint; word-processing
concentration 133, 255
   see also solution chemistry
conferences 46
confidence limits 508
confounding variables 137
conical flasks 87
conjugate pairs 108
continuous culture systems 236–7
consistency 119
correlation 508–9, 511
cosmids 451
coulometry 346
counter-current electrophoresis 262
counting 122
counting chambers 238
coursework 29
covalent chromatography 312
cross-over electrophoresis 262
crystallisation 381
curriculum vitae
   skills and personal qualities 38–9

structure and presentation 39–41
cyclodextrins 335
cystic fibrosis 420

data analysis
   exponents 484
   geometry and trigonometry 483
   linear functions and straight lines 485–6
   logarithms 485
   numerical problems 478–80, 486–7
   percentages and proportions 484
   rounding off 480, 481–2
   scientific notation 484
   statistics *see* descriptive statistics; statistical tests
data collection
   note-taking *see* note-taking
   observations 121
      artefacts 121
      counting 122
      developing observational skills 121–2
      during examinations 122
      observer effects 121
      perception 121, 122
      precision and error 121
data manipulation 457
   displaying distributions 458
   graphs *see* graphs
   organising numbers 457–8
   transforming data 458–9
data measurement
   accuracy 119
   bias 119
   consistency 119
   error 119–20
   precision 119
   SI units *see* SI units
   systematic error 119
   variables
      attributes 117
      discontinuous 117
      measurement scales 118
      qualitative 117
      quantitative 117
      ranked 117
   see also experimental design
data presentation
   graphs *see* graphs
   tables 473–6
decimal places 481–2
deoxyribonucleic acid *see* DNA
descriptive statistics 489–92
   coefficient of variation 491, 494
   computer programs 497–8
   frequency distributions 495–6
   measuring dispersion 492–4
   median 490, 491
   mean 490, 491, 496
   mode 490–2
   range 491, 492
   rank 490

semi-interquartile range 491, 492–3
standard deviation 491, 493–4
standard error of sample mean 494–5
variance 491, 493–4
diagrams 64, 79
apparatus diagrams 148, 149, 150
body plans 148, 149
cell diagrams 147, 148
drawing techniques 150–52
morphological diagrams 148, 149
tissue diagrams 147, 148
dictionaries 61
diffusion 395–7
dilutions
dilution series 93–5
harmonic dilution series 94–5
linear dilution series 94
logarithmic dilution series 94
single dilution 93
dip-stick immunoassays 264, 265
discontinuous variables 117
dispersion 492–4
dissection 191
equipment 192
accessories 194
forceps 192
mounted needles 193
scalpels 192
scissors 192
seekers 194
gross, normal and fine-scale 191–2
humane treatment 191
improving technique 194
maximum benefit 191
preparation 191
stages of an animal dissection 193
dissertations 66
distributions 458
Dixon and Mood's sign test 507
DNA (deoxyribonucleic acid) 370
assaying relative proportions of base pairs 374
bacteria *see* bacterial genetics
isolation processes 371
measuring nucleic acid content 372–3
separating nucleic acids 372
structure 370, 371
see also genetic engineering; molecular genetics
DNA fingerprinting 444–5
DNA microarrays 439
DNA sequencing 437–9
dot blotting 436
drawings 147, 149
see also diagrams
dye-binding (Bradford) method 356
dyes 312
fluorescence 445–6
see also staining

E. coli 453
edge effects 160, 166

electroanalytical techniques 338
coulometric methods 346
oxygen electrodes 341–3
chart recorder traces 344
Clark (Rank) oxygen electrodes 341–3
glucose electrodes 345–6
oxygen probes 344–5
potentiometry and ion-selective electrodes 339–40
gas-sensing electrodes 340–41
glass membrane electrodes 340
liquid and polymer membrane electrodes 341
solid-state membrane electrodes 341
voltammetric methods 341
electrochemical assays 384
electrodes 109, 110–11
oxygen electrodes
chart recorder traces 344
Clark (Rank) oxygen electrodes 341–3
glucose electrodes 345–6
measuring respiratory activity 408–9
oxygen probes 344–5
potentiometry *see* potentiometry
electrolysis 341
electrolytic cells 341
electrolytic dissociation 100
electromagnetic spectrum 278
electron capture detector (ECD) 318
electron impact ionisation (EI) 296
electron spin resonance (ESR) 295
electronic balances 87, 96
electronic particle counters 237
electrophoresis 322
agarose 324–5
basic apparatus 323
blotting 330
capillary 333–4, 368
capillary gel electro-phoresis (CGE) 335–6
capillary isoelectric focusing (CIEF) 336
capillary zone electro-phoresis (CZE) 334
chiral capillary electro-phoresis (CCE) 335
micellar electrokinetic chromatography (MEKC) 334
destaining 329
detection of enzymes 330
fixing 329
gel electrophoresis 433–5
isoelectric focusing (IEF) 332
protein purification 378
recording and quantification of results 330
separation of proteins 332–3
separating nucleic acids 372
staining 329
supporting media 323–4
cellulose acetate 324
handling 328–9
polyacrylamide gels 325–8
two-dimensional 333
ELISA (enzyme-linked immuno-sorbent assay) 263–4, 265
elution 311

employability 5–6
emulsions 360
enzymatic carbohydrate analysis 366–7
enzyme detection 330
enzyme immunoassays (EIA) 263–4, 265
enzymes 383
   activity 383
      cofactors 387
      pH 388
      substrate concentration 388
      temperature 388
   chromogenic substrates 384–5
   continuous assays 385–6
   discontinuous assays 386
   DNA manipulation 449–50
   electrochemical assays 384
   fixed time assays 386
   fluorogenic substrates 384–5
   inhibition 388
      competitive 390
      identifying the type of inhibitor 389–91
      irreversible 388–9
      non-competitive 390
      reversible 389
      uncompetitive 390–91
   kinetic assays 385–6
   kinetics 386–7
   measuring reactions 383
   protein purification 376, 378
   radioisotopic assays 384
   regulatory 391
      allosteric enzymes 391–2
      covalently modified 392
   spectrophotometric assays 384
equivalent mass 103
Erlenmeyer flasks 87
error 119–20, 121
essay questions 31–2, 33
   constructing an outline 63–4
   content 63, 64
   diagrams 64
   examples 63, 64
   handwriting 65
   lecturers' and tutors' comments 65
   organising time 63
   relevance 64
   Ten Golden Rules 64
   title 63
   see also writing skills
ethics 77–8, 143
   animal experimentation 225, 226
   dissection 191
ethidium bromide 373
examinations 24
   anxiety 35–6
   essay questions 31–2, 33
   guessing 32
   information-processing exams 34
   multiple-choice questions 32, 33
   oral exams 34–5

   past papers 26
   practical exams 34
      observations 122
   short-answer questions 32, 33
   summative exams 30–31
   see also writing skills
Excel 256, 465–7, 468, 475–6, 497–8, 509, 510–11
experimental design 136, 144
   checklist 138
   confounding variables 137
   constraints 138, 139
   multifactorial experiments 141
   nuisance variables 137
   pairing and matching subjects 141
   randomisation 139–40
   repetition 141
   replicates 138–9
experiments 68
exponential notation 484

F-test 507
fast atom bombardment-mass spectrometry (FAB-MS) 296–7
fatty acids 359
feedback 25
fieldwork 154
   checklist 156
   excursions 154–5
   mobile phones 77
   note-taking 128
   safety 156–7
   site inspection 155, 156
filtration 206
fixing specimens see specimens
flame absorption spectro-photometry 288–9
flame ionisation detector (FID) 317
flame photometry 288
flasks 87
Fick's first law of diffusion 395
flowcharts 180
fluorescence 286–7
   chlorophyll 404–6
fluorescence detectors 316–17
fluorescent dyes 445
fluorogenic enzyme substrates 384–5
Folin-Ciocalteau reagent 356
forceps 192
formative assessments 24
Fourier transformation 291, 527
frequency distributions 495–6
frequency tables 458
Friedman S-test 507
fungi 185
   see also microbes

G-test 507
galvanic cells 338
gamma-ray spectrometry 273
gas chromatography (GC) 308, 318, 363, 367–8
gas chromatography detectors 317–18

gas composition 97
gas exchange 401
    see also photosynthesis; respiration
gas-sensing glass electrodes 340–41
Gaussian distributions 503–4
Geiger-Müller tube 271
gel electrophoresis 433–5
gel permeation chromatography (GPC) 309–10
gene libraries 451–2
genetic crosses 417–19
genetic engineering 82, 448
    creating a gene library 451–2
    enzymatic manipulation of DNA 449–50
    extraction and purification of plasmid DNA 448–9
selection and detection of transformants 452–3
    transferring rDNA to a suitable host cell 452
    vectors 450
        cosmids 451
        phages 451
        plasmids 450–1
genetic equilibrium 420
genetics
    bacteria *see* bacterial genetics
    Mendelian *see* Mendelian genetics
    molecular *see* molecular genetics
    phages 428–9
glass membrane electrodes 340
glass vessels 87–8, 97
glossaries 71
glucose electrodes 345–6
glucosidase assays 367
glycerol 363
glycoproteins 357
glycosides 366
goals 8
grade descriptors 22, 23
Gram staining 186–7, 217–18
graphs 68, 79, 150, 458, 461
    drawing 461–3
    histograms 457, 463, 466–7, 469
        interpretation 469
        misleading 470–72
        paper 462–3
        plotted curve 463–6, 468
        size 462
        types 463
        very large or small numbers 462

haemocytometers 238
handwriting 65
    see also presentation
Hardy-Weinberg Principle 420
harmonic dilution series 94–5
hazardous substances 81, 82, 84, 90, 91, 100, 108, 109
    microbiology 207, 208
health and safety
    animal and plant specimens
        collecting 167
        fixing and preserving 169, 170
        anthrone 366

atomic spectroscopy 288
basic rules 82
catalase reagents 188, 219
centrifugation 302, 303–4
chemicals 90, 91
dissection equipment 192
ethidium bromide 373
fieldwork 156–7
genetic engineering and molecular genetics 82
glassware 88, 97
Gram staining 186
hydrolysis of proteins 357
legislation 81
lipid stains 361
mould cultures 185
oxidase reagents 188, 219
polyacrylamide gels 325
radioactive isotopes 276–7
risk assessment 81–2
    see also hazardous substances
solvents 360
tissue cultures 228
heat sterilisation 205–6
high performance liquid chromatography (HPLC) 307, 318, 367
highly sulphated cyclodextrins 335
Hill reaction 406
histograms 150, 457, 463, 466–7, 469
homogenisation 245
    fractionation and isolation of organelles 248–9
    mechanical methods 247–8
    media 245–6, 247
    non-mechanical methods 246–7
    protein denaturation 247
    protein purification 376
    susceptibility 245
hydraulic potential 106
hydrophobic interaction chromatography (HIC) 310–11
hydrolysis 357
hypotheses 135–6
hypothesis-testing statistics *see* statistical tests

immunoelectrophoretic assays 262
immunological tests 189, 221, 259
    agglutination tests 260
    antibody production
        monoclonal antibodies 260
        polyclonal antibodies 259
    antibody structure 259
    chromatographic immunoassays 264, 265
    complement-based assays 266–7
    counter-current electro-phoresis 262
    cross-over electrophoresis 262
    dip-stick immunoassays 264, 265
    enzyme immunoassays (EIA) 263–4, 265
    immunodiffusion assays 260–61
    immunoelectrophoretic assays 262
    immunoradiometric assay (IRMA) 263
    Laurell rocket immunoelectro-phoresis 262

precipitin tests 260
quantitative immunoelectro-phoresis 262
radioimmunoassay (RIA) 262–3
immunoradiometric assay (IRMA) 263
index cards 71
indicator dyes 109
information-processing exams 34
infra-red (IR) radiation 278
infra-red (IR) spectroscopy 291–2, 525–7
   interpretation of spectra 529, 531, 532–4
   KBr disks 529, 532
   liquid samples 529, 530
   sample handling 528–9
   solid samples 529, 530
   spectrometers 527–8
inoculating loops 207
Internet 143
interpersonal skills 12
interviews 34–5
iodine number 362
ion-exchange chromatography 309
ion-selective electrodes *see* potentiometry
ionophores 341, 397
irradiance 279
isoelectric focusing (IEF) 332
isoelectric point 375
isotope ratio mass spectroscopy (IRMS) 296

Kendall's coefficient of rank correlation 509
Kolmogorov-Smirnov test 505
Kruskal-Wallis *H*-test 507

lab partners 14–15
laboratory skills 3–5
   atmospheric conditions
      gas composition 97
      pressure 97
   calculators 79
   chemicals
      Merck Index 90
      mixing 95
      preparing 90–93
      safety 90, 91
      selection 90
      separating components 95
      spillages 95
      storing 95
   dilutions
      harmonic dilution series 94–5
      linear dilution series 94
      logarithmic dilutions series 94
      single dilution 93
   electronic balances 87, 96
   ethical and legal aspects 77–8
   health and safety
      basic rules 82
      chemicals 90, 91
      genetic engineering and molecular genetics 82
      glassware 88, 97
      legislation 81
      risk assessment 81–2
      see also hazardous substances
   liquids
      balances 87, 96
      beakers 87
      bottles and vials 87
      burettes 85, 518
      cleaning vessels 87, 88
      conical (Erlenmeyer) flasks 87
      glass or plastic vessels 87–8
      holding and storing 87
      light-sensitive chemicals 87
      measuring and dispensing 84–7
      measuring cylinders 84
      organic constituents 87
      Pasteur pipettes 84
      pipettes 85, 518–20
      pipettors 85, 86
      syringes 85, 87
      test tubes 87
      volumetric flasks 84
   mobile phones 77
   preparation 77
   presenting results 79, 80
   recording practical results 79
   solution chemistry *see* solution chemistry
   temperature
      cooling 97
      heating 96
      maintaining constant 97
      thermometers 97
   textbook use 77
   timers 98
laminar flow cabinets 207
Laurell rocket immunoelectro-phoresis 262
learning outcomes 22, 23, 24
learning styles 20–22
lectins 311
lecture notes 16–18, 24
legislation 77–8, 165, 225, 242, 275
light 133–4, 278
   absorption 282–3
light measurement
   electromagnetic spectrum 278
   irradiance 279
   light meters 280
   photometric 278
   photon flux density 279
   radiometers 279, 280
   spectral distribution 280
light meters 280
light microscopy *see* microscopy
light-sensitive chemicals 87
linear dilution series 94
linear functions 485–6
linked genes 416–17
lipids 359
   adsorption chromatography 360–61
   cholesterol content 363
   complex, compound or conjugated 359–60

gas chromatography 363
glycerol content 363
oxidation 361
quantitative assay 362–3
simple or neutral 359
solvent extraction 360
staining 361
thin layer chromatography 361–2
lipoproteins 357
liquid chromatography detectors 316–17
liquid membrane electrodes 341
liquids
balances 87, 96
beakers 87
bottles and vials 87
burettes 85, 518
cleaning vessels 87, 88
conical (Erlenmeyer) flasks 87
glass or plastic vessels 87–8
holding and storing 87
light-sensitive chemicals 87
measuring and dispensing 84–7
measuring cylinders 84
organic constituents 87
Pasteur pipettes 84
pipettes 85, 518–20
pipettors 85, 86
syringes 85, 87
test tubes 87
volumetric flasks 84
literature surveys
balancing opposing views 72
citations 72
definitions 72
glossaries 71
index cards 71
making a timetable 71
references 71, 72
scanning the literature 71
selecting a topic 71
structure and content 72
logarithmic dilution series 94
logarithms 485
Lowry (Folin-Ciocalteau) method 356
luminescence 287–8
lytic cycle 210

magnetic fleas 93, 95
magnetic resonance imaging (MRI) 295
Mann-Whitney U-test 505
mass 133
mass spectrometry (MS) 296–7, 545–6
base peak 546
fragmentation patterns 546
interface with chromatography 318, 546, 549
molecular ions 546, 547–8
mathematical transformations 458–9
mathematics
exponents 484
geometry and trigonometry 483

linear functions and straight lines 485–6
logarithms 485
numerical problems 478–80, 481, 486–7
percentages and proportions 484
rounding off 480, 481–2
scientific notation 484
statistics *see* descriptive statistics; statistical tests
mean 490, 491, 496
measurement *see* data measurement; laboratory skills; quantitative analysis
measuring cylinders 84
median 490, 491
meiosis 415
membrane transport processes 394
active transport 398–9
facilitated diffusion 396–7
measuring solute transport 394–5
passive transport 395–7
patch-clamp techniques 399
simple diffusion 395–6
see also photosynthesis; respiration
Mendelian genetics 415–17
analysis of crosses 417–19
chi^2 ([*c*]^2) test 419–20
linked genes 416–17
numerical problems 487
population genetics 419
Punnett square 417
Merck Index 90
mercury thermometers 97
micellar electrokinetic chromatography (MEKC) 334
microbes 212
identification 185
carbohydrate utilisation tests 188, 189, 219, 220
catalase test 188, 219
cell shape 186, 217
classification and nomenclature 221–3
colony characteristics 185–6, 216–17
direct observation 185–6, 216
Gram staining 186–7, 217–18
immunological tests 189, 221
kits 189, 220
microscopic examination 186–7, 217
molecular approaches 189, 220
motility 186, 218–19
oxidase test 187–8, 219
tables 188, 219–20
typing methods 176, 190, 221
sampling 212–13
microbial culture
isolation
selective and enrichment methods 214–16
separation methods 213–14
sterile technique 205
chemical agents 206
containers 207
filtration 206
hazards 207, 208
heat treatment 205–6
inoculatory loops 207

laminar flow cabinets 207
media 206–7
phages 209–11
pour plate 209, 210
radiation 206
spread plate 209
streak dilution plate 208
working area 206
microbiological strains 176
microfuges 301
microscopy 196
chemical fixation 196
decalcifying specimens 196
dehydration and clearing 197
embedding and sectioning 197–8
measuring growth in cell cultures 237
microbes 186–7, 217
mounting sections 200–201
staining 198–200
Microsoft
Word 475
see also Excel; PowerPoint
Mind Maps 16, 17, 26
mitosis 415
mobile phones 77
mode 490–2
modular assessment 29
modulus 501
molality 101
molarity 100–101
molecular genetics 82, 431–2
blotting techniques 435–6
probes 436–7
DNA fingerprinting 444–5
DNA microarrays 439
DNA sequencing 437–9
genetic engineering see genetic engineering
polymerase chain reaction (PCR) 441–5
producing DNA fragments 432–3
separation of nucleic acids using gel electro-phoresis 433–5
molecular mass 100
monoclonal antibodies 260
monosaccharides 365
morphological diagrams 148, 149
motility 186, 218–19
mould cultures 185
multifactorial experiments 141
multiple-choice questions 32, 33
mutagenicity 241
mutant bacteria 423–4
mutant phages 429

narcotisation 168, 169
normal distributions 503–4
Northern blotting 330, 436
note-taking
books and journal papers 18–19
lectures 16–18
practical work 125
communal records 128
fieldwork 127–8
primary data 125–6
primary record 126, 127
secondary record 126
project work 144
revision 26
skimming 18, 19
nuclear magnetic resonance (NMR) 293–5, 535–6
chemical shift 536, 539
coupling (spin-spin splitting) 540–42
integration of peak areas 539–40
interpreting spectra 537–8, 542–4
sample handling 537, 538
spectrometers 536–7
nucleic acids 370
assaying relative proportion of base pairs in DNA 374
extraction and purification 371
DNA isolation procedures 371
RNA isolation procedures 371–2
separating nucleic acids 372
measuring nucleic acid content 372–3
structure 370–71
see also molecular genetics
nucleosides 370
nucleotides 370
nuisance variable 137
null hypothesis 500–501
numerical problems 178–80, 481, 486–7

objectives 8, 22, 24
observations 121
artefacts 121
counting 122
developing observational skills 121–2
during examinations 122
observer effects 121
perception 121, 122
precision and error 121
Ohm's law 323
optical illusions 121, 122
oral exams 34–5, 66
organelles 248–9
organic material 87
osmolality 103
osmolarity 103
osmometry 103–4
osmotic coefficients 103
osmotic effects 100
see also membrane transport processes
osmotic properties
osmotic pressure 104–5
water activity 105
oxidase test 187–8, 219
oxidation 338
lipids 361
oxygen electrodes
chart recorder traces 344
Clark (Rank) oxygen electrodes 341–3

glucose electrodes 345–6
measuring respiratory activity 408–9
oxygen probes 344–5

paper chromatography 305–6, 367
parameters 158
parametric distributions 501
paraphrasing 18
partition chromatography 308–9
past papers 26
Pasteur pipettes 84
Pearson's product moment correlation coefficient 508–9
peer assessment 12
peptides
  assays
    Biuret method 355
    bovine serum albumin (BSA) 356
    direct measurement of UV absorbence 355
    dye-binding (Bradford) method 356
    Lowry (Folin-Ciocalteau) method 356
    Warburg-Christian method 355
  properties 355, 356–7
percentages 484
perception 121, 122
personal development planning (PDP) 5, 38
personal qualities 38
personality types 13–14
pH measurement 108–9
  buffers 112–14
  electrodes 109, 110–11
  indicator dyes 109
phagemids 451
phages 209–11
  genetics 428–9, 451
phenetic taxonomy 173
phosphorescence 287
photometric measurements 278
photon flux density 279
photosynthesis 401–2
  carbon dioxide uptake 402
  chlorophyll fluorescence 404–6
  electron transport 406–7
  oxygen production 402
  radiocarbon fixation 403
  stable carbon isotopes 403–4
photosynthetic irradiance 279
phylogenetic taxonomy 173
physical constants 132
pipettes 84, 85, 518–20
pipettors 85, 86
plagiarism 13
plant specimens *see* specimens
plant taxonomy *see* classification
plant tissues and organs 225, 226
  see also cell cultures
plasmids 448, 450–1
plastic vessels 87–8
plotted curves 463–6, 468
Poisson distributions 502–3

poly(U)-agarose 312
polyacrylamide gel 325–8
  electrophoresis 433–4
polyclonal antibodies 259
polymer membrane electrodes 341
polymerase chain reaction (PCR) 441–5
polymers 380
polysaccharides 368
population genetics 419
populations 158
poster displays 45
  assessment 47, 49
  colour 46–7
  conclusions 47
  conferences 46
  graphs and diagrams 47
  handouts 47
  introduction 47
  layout 45–6
  materials and methods 47
  PowerPoint 46, 48
  results 47
  scientific meetings 47, 48
  subtitles and headings 46
  text 46
  title 46
potentiometry
  ion-selective electrodes 339–40
    gas-sensing glass electrodes 340–41
    glass membrane electrodes 340
    liquid and polymer membrane electrodes 341
    solid-state membrane electrodes 341
pour plate 209, 210
PowerPoint
  lecture notes 17–18
  poster displays 46, 48
  presentations 51
practical exams 34
  observations 122
practical reports 66, 67–8
practical work
  note-taking 125
    communal records 128
    fieldwork 128
    primary data 125–6
    primary record 126, 127
    secondary record 126
  see also project work
precipitin tests 260
precision 119, 121
presentation 79, 80
  see also handwriting
presentations
  audience 52
  audio-visual aids 50–2
  checklist 54
  concluding remarks 53, 54
  introductory remarks 52
  main message 53
  PowerPoint 51

preliminary information 50
summary
  slides 53
  time allocation 53
preserving specimens *see* specimens
pressure 97
primary data 125–6
primary record 126, 127
problem-based learning (PBL) 30
project reports 66, 67, 145
project work
  analysing results 144
  deciding on a topic 143
  designing experiments 144
  getting started 144
  keeping notes 144
  laboratory work 144
  planning 143
  supervision 143, 144
  writing the report 66, 67, 145
  see also practical work
protein A 312
protein denaturation 247
protein purification
  ammonium sulphate precipitation 378–9
  concentration by ultrafiltration 380
  crystallisation 381
  differential solubility separation techniques 378–80
  heat denaturation 380
  homogenisation 376
  monitoring 377–8
  objectives 375
  precipitation by changing pH 379–80
  preserving enzyme activity 376
  removing salts and changing the buffer 380–81
  solubilisation 376
  solvent and polymer precipitation methods 380
  source materials 376
  strategy 376–7
proteins
  amino acid analysis 357
  assays
    Biuret method 355
    bovine serum albumin (BSA) 356
    direct measurement of UV absorbence 355
    dye-binding (Bradford) method 356
    Lowry (Folin-Ciocalteau) method 356
    Warburg-Christian method 355
  hydrolysis 357
  primary, secondary and tertiary structure 357
  properties 354–5, 356–7
  sequence analysis 357
proteomics 332
protoplasts 232–3
pulsed field gel electro-phoresis (PFGE) 434–5
Punnet square 417
pyrolysis-mass spectrometry (PY-MS) 296

qualitative variables (attributes) 117
quantitative analysis
  biochemical 351–3
    see also biomolecules
  calibration *see* calibration
  centrifugation *see* centrifugation
  chromatography *see* chromatography
  electroanalytical techniques *see* electroanalytical techniques
  electrophoresis *see* electro-phoresis
  immunological tests *see* immunological tests
  light *see* light measurement
  radioactive isotopes *see* radioactive isotopes
  spectrometry *see* spectrometry
  spectroscopy *see* spectroscopy
  validity 352
  volumetric analysis *see* volumetric analysis
quantitative immunoelectro-phoresis 262
quantitative variables 117

radiation 206, 278
radioactive isotopes 269
  biological applications 274–5
  measuring radioactivity 271
    autoradiography 273
    gamma-ray spectrometry 273
    gamma-ray spectrometry 273
    Geiger-Müller tube 271
    scintillation counters 271–3, 274
    specific activity of an experimental solution 272
  radioactive decay 269–70
  safety procedures 276–7
  working practices 275–7
radioimmunoassay (RIA) 262–3
radioisotopic assays 384
radiometry 279, 280
Raman spectroscopy 291–2
randomisation 139–40
ranked variables 117
recombinant proteins 376
reduction 338
references 71, 72, 145
regression 508–9, 511
relative atomic mass 100
relative molecular mass 100
replicates 138–9
report writing 66
  abstract 66
  conclusions 66
  correct tense 68
  graphs and tables 68
  key words 66
  oral assessments 66
  practical and project reports 66, 67–8, 145
  presenting results 68
  producing a scientific paper 69
  repeating experiments 68
  results and discussion 66
  theses and dissertations 66
research projects *see* project work
respiration 401, 407
  measuring respiratory activity 408

chloroplasts and mitochondria 409–10
    electron transport 408–9
    oxygen electrode 408–9
    proton-motive force 409–10
revision 24–7
ribonucleic acid *see* RNA
risk assessment 81–2
  see also hazardous substances
RNA (ribonucleic acid) 370
  isolation procedures 371–2
  measuring nucleic acid content 372–3
  separating nucleic acids 372
  structure 370, 371
  see also molecular genetics

safety *see* health and safety
sampling 158
  animal and plant specimens *see* specimens
  edge effects 160, 166
  in time 161
  microbes 212–13
  population factors 158
  protocol 159
    dimensions of the area 160
    locating samples 159
    number of units per sample 160–61
  strategy 158–9
  subsampling 161
Sanger DNA sequencing method 438
saponification 362
scalpels 192
Schetté-Box test 507
scientific meetings 47, 48
scientific method 135–6
  experimental design 136, 144
    checklist 138
    confounding variables 137
    constraints 138, 139
    multifactorial experiments 141
    nuisance variables 137
    pairing and matching subjects 140
    randomisation 139–40
    repetition 141
    replicates 138–9
scientific notation 484
scientific papers 69
scientific reports *see* report writing
scientific writing *see* writing skills
scintillation counters 271–3, 274
secondary record 126
short-answer questions 32, 33
silica gel 361
SI units 130–31
  amount of substance 133
  compound expressions 131
  concentration 133
  conversion factors 132
  light 133–4
  mass 133
  physical constants 132
  prefixes 131
  temperature 133
  time 133
  volume 133
skill categories 3–5
skills
  personal qualities 38
  transferability 3–5
skimmming 18, 19
slot blotting 436
solid-state membrane electrodes 341
solution chemistry 100
  colligative properties 103–4
  concentration 100
    activity 102–3
    equivalent mass 103
    molality 101
    molarity 100–101
    normality 103
    osmolality 103
    osmolarity 103
    per cent 102
    per million/billion 102
  electrolytic dissociation 100
  hydraulic potential 106
  ideal/non-ideal behaviour 100
  osmometry 103–4
  osmotic effects 100
    osmotic pressure 104–5
    water activity 105
  pH measurement 108–9
    buffers 112–14
    electrodes 109, 110–11
    indicator dyes 109
  water potential 106
solutions 90–93, 95, 255
solvents 360, 380
Southern blotting 330, 435–6
Spearman's coefficient of rank correlation 509
species classification 174
specimens
  classification *see* classification
  collecting 165
    equipment 165–7
    safety 167
  dissection *see* dissection
  fixation 168
    fixatives 169–70
    for microscopy 196
    safety 169, 170
  identification guides 178–9
    advice for using keys 181
    bracketed keys 179
    comparing specimens with descriptions 182
    computerised keys 181
    flowchart keys 180
    indented keys 179
    multi-access keys 180–81
  microscopic examination *see* microscopy
  narcotisation 168, 169

preservation 168
  dry preservation 171
  preservatives 169–70
  safety 169, 170
  wet preservation 170–71
  storage 171
spectral distribution 280
spectrofluorimetry 373
spectrometry
  electron spin resonance (ESR) 295
  infra-red 527–8
  mass spectrometry (MS) 296–7, 545–6
    base peak 546
    fragmentation patterns 546
    interface with chromatography 318, 546, 549
    molecular ions 546, 547–8
  nuclear magnetic resonance (NMR) 293–5, 536–7
spectrophotometry 282–6, 384
  measuring nucleic acid content 372–3
spectroscopy 282
  atomic 288–9
  fluorescence 286–7
  luminescence 287–8
  phosphorescence 287
  Raman spectrum 291–2
  UV/visible spectrophotometry 282–6
  see also infra-red spectroscopy; nuclear magnetic resonance
spiked samples 352
spillages 95
spoken presentations see presentations
spread plate 209
spreadsheets 3, 5, 256
  descriptive statistics 497–8
  graphs 465–7
SQ3R technique 18, 19
staining
  dyes 312
  electrophoresis 329
  for microscopy 198–200
  Gram staining 186, 217–18
  lipids 361
standard deviation 491, 493–4
standard error 494–5
statistical calculations 487
statistical tests 500
  analysis of variance (ANOVA) 510
  binomial distributions 501–2
  Central Limit Theorem 504
  chi^2 ($[*c]^2$) test 419–20, 507
  choosing a suitable test 505–11
  coefficient of dispersion 503
  comparing dispersions (variances) 507
  comparing locations (means) 505, 507
  comparing proportion data 507
  computer programs 509, 510–11
  confidence limits 508
  correlation and regression 508–9, 511
  Dixon and Mood's sign test 507
  frequency observations and theoretical expectation 507
  F-test 507
  Friedman S-test 507
  G-test 507
  Gaussian distributions 503–4
  Kendall's coefficient of rank correlation 509
  Kolmgorov-Smirnov test 505
  Kruskal-Wallis H-test 507
  Mann-Whitney U-test 505
  normal distributions 503–4
  null hypothesis 500–501
  parametric distributions 501–4
  Poisson distributions 502–3
  Scheffé-Box test 507
  Spearman's coefficient of rank correlation 509
  t-test 505, 506–7, 510
  Wilcoxon's signed rank test 507
statistics 158
  descriptive see descriptive statistics
  sampling see sampling
  variables see variables
steroisomers 335
sterile technique 205, 229
  chemical agents 206
  containers 207
  filtration 206
  heat treatment 205–6
  inoculating loop 207
  laminar flow cabinets 207
  media 206–7
  microbiological hazards 207, 208
  phages 209–11
  pour plate 209, 210
  radiation 206
  spread plate 209
  streak dilution plate 208
  working area 206
stock solutions 92–3, 255
streak dilution plate 208
Student's t-test 505, 506–7, 510
subsampling 161
subspecies 175–6
summative assessments 24
summative exams 30–31
syringes 85, 87
systematic error 119

t-test 505, 506–7, 510
tables 68
taxonomy see classification
teamwork 12
  brainstorming 13
  collaboration 13
  delegation and sharing of tasks 12
  dynamics 13–14
  effective listening and communication 12
  interpersonal skills 12
  lab partners 14–15
  peer assessment 12

temperature
    cooling 97
    heating 96
    maintaining constant 97
    SI units 133
    thermometers 97
test tubes 87
thermal conductivity detector (TCD) 317
thermometers 97
thesauri 61
theses 66
thin-layer chromatography (TLC) 305–6, 361–2, 367
thinking processes 22, 23
time 133
time management 8
    analysing current activities 8–9
    assertiveness 9
    avoiding time-wasting activities 9
    body rhythm 10
    checklists 10
    goal setting 8
    organising tasks 9–10
    quality 9
    revision 24, 25
    writing skills 56
timers 98
tissue cultures 225, 227
    animal cell culture systems 227–8, 230, 231
    plant protoplasts 232–3
    plant tissue and cell culture systems 230, 232
    sterile technique 229
    see also cell cultures
tissue diagrams 147, 148
tissue disruption see cell disruption
titration curves 332
titrations 515–18
    acid-base titrations 522–4
transcripts 29
transferable skills 3–5
transformations 458–9
transmembrane processes see membrane transport processes

UV (ultraviolet) absorbence 355
UV radiation 278
UV/visible chromatography detectors 316
UV/visible spectrophotometry 282–6

validity 352
variables
    confounding 137
    discontinuous 117
    measurement scales 118
    nuisance 137
    qualitative (attributes) 117
    quantitative 117
    ranked 117
variance 491, 493–4
VARK system 20–22
vials 87
virus taxonomy 176
    bacteriophages 209–11
viva voce see oral exams
voltammetry 341
volume 133

volumetric analysis 515
    burettes 518
    calculating concentration of test substance 520–21
    pipettes 518–20
    titrations 515–18
        acid-base 522–4
volumetric flasks 84
volumetric scales 84
vortex mixers 95

Warburg-Christian method 355
water activity 105
water potential 106
Western blotting 330
Wilcoxon's signed rank test 507
word-processing 475
writing skills
    analytical writing 60
    choice of words and phrases 59
    common errors 62
    comparative writing 60
    descriptive writing 60
    dictionaries 61
    getting started 58
    guides for written English 61
    organising information and ideas 56–8
    organising time 56
    paragraphs 59–60
    punctuation 59
    revising a text 61
    scientific style 58
    sentences 59
    technique 58
    thesauri 61
    see also essay questions

yeasts 185
    see also microbes

zwitterions 112–13